Monte Carlo Calculations in Nuclear Medicine (Second Edition)

Therapeutic applications

IPEM–IOP Series in Physics and Engineering in Medicine and Biology

About the Series

The series in Physics and Engineering in Medicine and Biology will allow the Institute of Physics and Engineering in Medicine (IPEM) to enhance its mission to 'advance physics and engineering applied to medicine and biology for the public good'.

It is focused on key areas including, but not limited to:

- clinical engineering
- diagnostic radiology
- informatics and computing
- magnetic resonance imaging
- nuclear medicine
- physiological measurement
- radiation protection
- radiotherapy
- rehabilitation engineering
- ultrasound and non-ionising radiation.

A number of IPEM–IOP titles are being published as part of the EUTEMPE Network Series for Medical Physics Experts.

Monte Carlo Calculations in Nuclear Medicine (Second Edition)

Therapeutic applications

Edited by
Habib Zaidi
Division of Nuclear Medicine, Geneva University Hospital, Geneva, Switzerland

IOP Publishing, Bristol, UK

ISBN 978-0-7503-2694-0 (ebook)
ISBN 978-0-7503-2692-6 (print)
ISBN 978-0-7503-2695-7 (myPrint)
ISBN 978-0-7503-2693-3 (mobi)

DOI 10.1088/978-0-7503-2694-0

Version: 20220601

IOP ebooks

British Library Cataloguing-in-Publication Data: A catalogue record for this book is available from the British Library.

Published by IOP Publishing, wholly owned by The Institute of Physics, London

IOP Publishing, Temple Circus, Temple Way, Bristol, BS1 6HG, UK

US Office: IOP Publishing, Inc., 190 North Independence Mall West, Suite 601, Philadelphia, PA 19106, USA

Contents

Preface

It is difficult to overstate how dramatically the radiopharmaceutical therapy (RPT) landscape has changed since the last edition of this book. The FDA has approved four RPT agents, more if constructs that are defined as devices are included. Small and large pharma is investing heavily in this area and there is a robust pipeline of both alpha-particle and beta-particle emitting RPT agents. It is also no exaggeration to say that interest in dosimetry as a potential approach to improving the efficacy and safety of these agents has skyrocketed. At last count there were 10 early stage and established vendors for dosimetry software. Efforts to standardize and validate dosimetry practice are ongoing. Clinical trials to examine the potential of dosimetry to improve clinical outcome have been conducted and more are being considered. All dosimetry methods are ultimately founded upon or validated by Monte Carlo calculations. This book provides a review of the Monte Carlo method as it is applied in the field of therapeutic nuclear medicine. Driven in part by the remarkable increase in computing power and its ready and inexpensive availability, this is a relatively new yet rapidly expanding field. As in the original edition, an effort has, been made to place the reviews provided in this book in a broader context. The effort to do this is reflected by the inclusion of chapters that do not directly address Monte Carlo techniques, but rather provide an overview of issues that are closely related to therapeutic nuclear medicine and to the potential role of Monte Carlo in this field. A brief overview of each chapter is provided below.

Chapters 1 and 8 review the fundamental theory of Monte Carlo, techniques that have been used in nuclear medicine dosimetry and Monte Carlo codes currently available for carrying out absorbed dose calculations. Unlike external radiotherapy or brachytherapy, the kinetics and biodistribution of internally administered radionuclides must be measured. Imaging is, therefore, an integral part of nuclear medicine dosimetry. As such, imaging considerations are incorporated in almost every chapter of the book. Recent advances in nuclear medical imaging techniques as applied in radiation therapy are specifically reviewed in chapter 2. The anatomical input required for Monte Carlo-based dose calculations may be provided either as idealized geometric representations of anatomy or as discrete, voxel-based representations that reflect the actual shape and composition of a human. Again, these considerations are present throughout the book, but chapters 5 and 19 focus on this specifically for the whole-body and for the bone marrow, respectively; the latter being a particularly important and challenging portion of the anatomy. The role of Monte Carlo in providing a means for validating the overall scheme of imaging-based, patient-specific absorbed dose calculations is reviewed in chapter 20. Chapters 7, 16, 17 and 19 examine specific implementations of the Monte Carlo approach in the special cases of microdosimetry, selective internal radiation therapy, boron neutron capture therapy and Auger electrons used in molecularly targeted radionuclide therapy, respectively. Microdosimetry is becoming increasingly important as high linear energy transfer (LET) (see chapter 7) emissions are being considered for therapeutic use. In alpha-emitter therapy, for example, a small number of decays can substantially influence cell survival, the Monte Carlo method, is therefore indispensable in simulating such a stochastic process and thereby understanding experimental cell survival studies and potentially translating such results into the clinic. Intravascular radiation therapy using radionuclides is a relatively new area of radionuclide therapy. The potentially efficacy of this approach is heavily dependent upon the spatial distribution of absorbed dose relative to the anatomical distribution of the cells involved in restenosis. The absorbed dose distribution is, in turn, dependent upon the radionuclide emissions, the configuration of emitters and the heterogeneous environment that is traversed in going from the source region to the target regions. These aspects are best addressed by Monte Carlo calculations.

Conventional, non-Monte Carlo-based approaches to radionuclide dosimetry are described in chapters 3 and 4. Chapter 3, in particular, provides a review of the dose schema, developed by the Medical Internal Radionuclide Dose (MIRD) Committee that is most widely implemented for radionuclide dosimetry. Chapter 4 describes the point-kernel methodology which has been used extensively in patient-specific absorbed dose calculations and which may be thought of as a pre-cursor to a full Monte Carlo implementation of imaging-based patient-specific dosimetry. Chapters 9 to 14 briefly review few software packages that have been developed for carrying out radionuclide dosimetry in therapeutic nuclear medicine. This includes the popular OLINDA/EXM2.0 package, which is commercially available, with the aim to facilitate the implementation of the MIRD schema while also allowing for realistic calculations that make it possible to incorporate simple tumour geometries within the MIRD formalism. The second category, including the 3D-RD, RAPID, VRAK, Raydose and VIDA, describes imaging-guided voxel-based dosimetry approaches to patient-specific dosimetry. Novel approaches for voxel-based dosimetry, including novel artificial intelligence-assisted techniques are discussed in chapter 15.

Chapter 6 provides a comprehensive review of the radiobiology relevant to therapeutic nuclear medicine. This chapter is essential in understanding the biological aspects critical to the design of successful radionuclide therapy for cancer. It serves as a reminder that regardless of how precisely the distribution of ionising energy density (i.e., absorbed dose) is estimated, this information will predict biologic response only if the pertinent biologic considerations are also incorporated.

Finally, we would like to thank all of the contributors for their invaluable hard work and for keeping to a very tight schedule. The topic of this book is rapidly evolving and the editors felt it important to minimise the time required to get this book into press. That this problem was so very well avoided speaks to the commitment and dedication of the contributors. We found compilation of this book to be a rewarding and educational experience and hope that the reader is left with the same experience.

George Sgouros and Habib Zaidi

Acknowledgement

The Editor would like to thank all of the contributors for their invaluable hard work and for keeping to a very tight schedule.

Editor's biography

Habib Zaidi

Professor Habib Zaidi is Chief physicist and head of the PET Instrumentation and Neuroimaging Laboratory at Geneva University Hospital and faculty member at the medical school of Geneva University. He is also a Professor of Medical Physics at the University of Groningen (Netherlands), Adjunct Professor of Medical Physics and Molecular Imaging at the University of Southern Denmark, Adjunct Professor of Medical Physics at Shahid Beheshti University and visiting Professor at Tehran University of Medical Sciences. He is actively involved in developing imaging solutions for cutting-edge interdisciplinary biomedical research and clinical diagnosis in addition to lecturing undergraduate and postgraduate courses on medical physics and medical imaging. His research is supported by the EEC, Swiss National Foundation, private foundations and industry (Total 8.8 M US$) and centres on hybrid imaging instrumentation (PET/CT and PET/MRI), deep learning for various imaging applications, modelling medical imaging systems using the Monte Carlo method, development of computational anatomical models and radiation dosimetry, image reconstruction, quantification and kinetic modelling techniques in emission tomography as well as statistical image analysis, and more recently on novel design of dedicated PET and PET/MRI scanners. He was guest editor for 12 special issues of peer-reviewed journals dedicated to *Medical Image Segmentation*, *PET Instrumentation and Novel Quantitative Techniques*, *Computational Anthropomorphic Anatomical Models*, *Respiratory and Cardiac Gating in PET Imaging*, *Evolving medical imaging techniques*, *Trends in PET quantification (2 parts)*, *PET/MRI Instrumentation and Quantitative Procedures and Clinical Applications*, *Nuclear Medicine Physics and Instrumentation*, and *Artificial Intelligence* and serves as founding Editor-in-Chief (scientific) of the *British Journal of Radiology (BJR)|Open*, Deputy Editor for *Medical Physics*, and member of the editorial board of the *Journal of Nuclear Cardiology*, *International Journal of Imaging Systems and Technology*, *Clinical and Translational Imaging*, *American Journal of Nuclear Medicine and Molecular Imaging*, *Brain Imaging Methods (Frontiers in Neuroscience and Neurology)*, *Cancer Translational Medicine* and the *IAEA AMPLE Platform in Medical Physics*. He has been elevated to the grade of fellow of the IEEE, AIMBE and the AAPM and was elected liaison representative of the *International Organization for Medical Physics (IOMP)* to the World Health Organization (WHO) and member of the International Medical Physics Certification Board (IMPCB) and the Imaging Physics Committee of the *American Association of Physicists in Medicine (AAPM)* in addition to being affiliated to several international medical physics and nuclear medicine organisations. He is developer of physics web-based instructional modules for the RSNA and Editor of IPEM's Nuclear Medicine web-based instructional modules. He is involved in the evaluation of research proposals for European and International granting organisations and participates in the organisation of international symposia and conferences. His academic accomplishments in the area of quantitative PET imaging have been well recognized by his peers and by the medical imaging community at large since he is a recipient of many awards and distinctions among which the prestigious 2003 *Bruce Hasegawa Young Investigator Medical Imaging Science Award* given by the *Nuclear Medical and Imaging Sciences Technical Committee of the IEEE*, the 2004 *Mark Tetalman Memorial Award* given by the *Society of Nuclear Medicine*, the 2007 *Young Scientist Prize in Biological Physics* given by the *International Union of Pure and Applied Physics (IUPAP)*, the prestigious (100 000$) 2010 *kuwait Prize of Applied sciences* (known as the *Middle Eastern Nobel Prize*) given by the *Kuwait Foundation for the Advancement of Sciences (KFAS)* for '*outstanding accomplishments in Biomedical technology*', the 2013 *John S. Laughlin Young Scientist Award* given by the *AAPM*, the 2013 *Vikram Sarabhai Oration Award* given by the *Society of Nuclear Medicine, India (SNMI)*, the 2015 *Sir Godfrey Hounsfield Award* given by the *British Institute of Radiology (BIR)*, the 2017 *IBA-Europhysics Prize* given by the European Physical Society (EPS) and the 2019 *Khwarizmi International Award* given by the Iranian Research Organization for Science and Technology (IROST). Professor Zaidi has been an invited speaker of over 160 keynote lectures and talks at an International level, has authored over 740+ publications (he is the senior or first author in a majority of these publications), including 350 peer-reviewed journal articles in prominent journals (ISI-h index=52|67 Web of Science™|Google scholar, >16 600+ citations), 340 conference proceedings and 41 book chapters and is the editor of four textbooks on *Therapeutic Applications of Monte Carlo Calculations in Nuclear Medicine*, *Quantitative Analysis in Nuclear Medicine Imaging*, *Molecular Imaging of Small Animals and Computational anatomical animal models.*

List of contributors

Pedro Andreo
Department of Medical Radiation Physics and Nuclear Medicine, Karolinska University Hospital, Stockholm, Sweden

Manuel Bardiès
Institut de Recherche en Cancérologie de Montpellier, UMR U1194 INSERM/ ICM/Université de Montpellier, Montpellier, France

Bryan Bednarz
Department of Medical Physics, University of Wisconsin—Madison, Madison, WI, USA

Salvatore Berenato
Department of Medical Physics, Velindre Cancer Centre, Cardiff, UK

Abigail Besemer
Department of Radiation Oncology, University of Nebraska Medical Center, Omaha, NE, USA

Wesley Bolch
University of Florida, Gainsville, FL, USA

Yuni K Dewaraja
Department of Radiology, University of Michigan, Ann Arbor, MI, USA

Arnaud Dieudonné
Nuclear Medicine Department, APHP.Nord, Beaujon, Universite Paris Cité, Paris, France
Equipe QuantIF, LITIS, Centre Henri Becquerel, Rouen, France

Nadia Falzone
GenesisCare, Theranostics, Alexandria, NSW, 2015, Australia

Pablo Minguez Gabiña
Department of Medical Physics and Radiation Protection, Gurutzeta-Cruces University Hospital/Biocruces Health Research Institute, Barakaldo, Spain

Joseph Grudzinski
Department of Radiology, University of Wisconsin—Madison, Madison, WI, USA

Michel Hesse
Cliniques Universitaires Saint-Luc, Brussels, Belgium

Jacob Hesterman
Invicro, Needham, MA, USA

Robert F Hobbs
Johns Hopkins University, Baltimore, MD, USA

John L Humm
Department of Medical Physics, Memorial Sloan Kettering Cancer Center, New York, USA

Price Jackson
Peter MacCallum Cancer Centre, Melbourne, Victoria, Australia

Amin I Kassis
Harvard Medical School, Boston, MA, USA

Susan Kost
Cleveland Clinic, Cleveland, OH, USA

Michael Ljungberg
Medical Radiation Physics, Lund, Lund University, Lund, Sweden

Ian Marsh
Department of Radiology, University of Wisconsin—Madison, Madison, WI, USA

Leyla Moghaddasi
Ballarat Regional Integrated Cancer Centre, Ballarat, Victoria, Australia

Graeme O'Keefe
Austin Hospital, Heidelberg, Victoria, Australia

Keryn Gresco
Invicro, Needham, MA, USA

Nick Patterson
Cardiff University School of Engineering, Cardiff, UK

John C Roeske
Department of Radiation Oncology, Stritch School of Medicine, Cardinal Bernardin Cancer Center, Loyola University Chicago, Maywood, IL, USA

Manuel Sanchez-Garcia
SERGAS, Hospital Clínico Universitario de Santiago, Santiago de Compostela, Spain

George Sgouros
Johns Hopkins University, Baltimore, MD, USA

Emiliano Spezi
Cardiff University School of Engineering, Cardiff, UK

Michael G Stabin
NV5/Dade Moeller, Richland, WA, USA

Lidia Strigari
IRCCS Azienda Ospedaliero-Universitaria di Bologna, Bologna, Italy

Stephan Walrand
Cliniques Universitaires Saint-Luc, Brussels, Belgium

Tianwu Xie
Institute of Radiation Medicine, Fudan University, Shanghai, China

Pat Zanzonico
Memorial Sloan Kettering Cancer Center, New York, USA

Habib Zaidi
Division of Nuclear Medicine and Molecular Imaging, Geneva University Hospital, Geneva, Switzerland

Chapter 1

Monte Carlo techniques in nuclear medicine dosimetry

Habib Zaidi and Pedro Andreo

1.1 Introduction

The use of the Monte Carlo method to simulate radiation transport has for decades been considered as the most accurate means of predicting absorbed dose distributions and other quantities of interest in the radiation treatment of cancer patients using either external or radionuclide radiotherapy. The same trend has occurred for the estimation of the absorbed dose in diagnostic procedures using radionuclides. There is broad consensus in accepting that the earliest Monte Carlo calculations in medical radiation physics were made in the field of nuclear medicine, where the technique was used for dosimetry modelling and computations. Formalism and data based on Monte Carlo calculations, developed by the Medical Internal Radiation Dose (MIRD) committee of the Society of Nuclear Medicine, have been published in a series of supplements to the *Journal of Nuclear Medicine,* the first one being released in 1968 [1]. Some of these pamphlets have made extensive use of Monte Carlo calculations to derive specific absorbed dose fractions for photon and electron sources uniformly distributed in organs of mathematical phantoms. Interest in Monte Carlo-based dose calculations with α- and β-emitters was revived decades ago with the application of radiolabelled monoclonal antibodies to radioimmunotherapy (RIT), leading to a generalized use of molecular-targeted radionuclide therapy [2] sometimes using the same molecule/target agent for imaging and therapy. As a consequence, questions were raised about the need and potential of Monte Carlo techniques, as well as on their accuracy and validation in a clinical environment, and on the wide availability of the technique to the nuclear medicine community. Developments in recent years have answered these and other questions on the grounds of the broad implementation of Monte Carlo simulations for clinical calculations and research.

doi:10.1088/978-0-7503-2694-0ch1

The main purpose of this chapter is to introduce the framework for the application of the Monte Carlo method to a wide range of problems in nuclear medicine dosimetry. Emphasis is given mainly to applications where photon, electron and positron transport in matter are simulated, but also certain applications with other particle types will be mentioned. An introductory review is presented of the different Monte Carlo codes and some of their applications in imaging and therapeutic nuclear medicine, with emphasis on internal dose calculations in radionuclide therapy; this will be extended to specific nuclear medicine applications in the different chapters of the book. Interested readers are encouraged to consult comprehensive reviews [3–6] and books [7–10] published on the subject. Some computational aspects of the Monte Carlo method, mainly related to the generation of random numbers, sampling and variance reduction techniques are given in [3, 6, 11].

1.2 Monte Carlo techniques in medical radiation physics

Monte Carlo techniques have become one of the most commonly used numerical tools in the different areas of medical radiation physics following the development and subsequent implementation of powerful computing systems in clinical sites. They have been extensively applied to simulate all kinds of processes involving radiation interactions and to calculate physical quantities and parameters that are difficult, or even impossible, to measure or to calculate analytically or using deterministic techniques. With the rapid development of computer technology, notably including low-cost PC-based clusters and sometimes GPUs, Monte Carlo calculations have become of practical use.

The application of the Monte Carlo method in medical radiation physics expands over almost all areas in the field and numerous comprehensive reviews have been published. In addition to those for the entire field [3, 12], reviews on the specialties of radiation protection and health physics [13], nuclear medicine [4, 6] and internal dosimetry [5, 14], radiotherapy physics [15] and dosimetry [11, 16], and diagnostic radiology [11, 17] deserve being cited. The use of Monte Carlo techniques in external beam radiotherapy and diagnostic radiology falls outside the scope of this review, whereas applications in radiation protection and nuclear medicine imaging and radionuclide dosimetry are presented in the following sections. It should be emphasized, however, that simulations in radiotherapy dosimetry have contributed substantially to the development and further improvements of the Monte Carlo method, particularly with regard to the simulation of electron transport in the presence of boundaries and inhomogeneities.

A number of Monte Carlo computer codes and packages have been and still are being used in the field of nuclear medicine imaging and internal dosimetry with many of them available as open source codes in the public domain [18]. Overall, the available computer codes can be divided in two major classes, namely general purpose Monte Carlo codes developed for high-energy physics research or general medical physics applications and dedicated software packages developed specifically for nuclear medicine imaging and/or internal dosimetry applications. These are highly sophisticated tools requiring advanced computer skills and familiarity with radiation transport theory. Each category has its own advantages and drawbacks; the incentives for the selection of a particular code being mostly dictated by the user's requirements, availability of code and documentation, technical support and maintenance, experience and skills in computer programming and available computational resources.

Two major groups of Monte Carlo systems and codes are available, which can be classified into 'general-purpose' and 'specific':

(a) The first group corresponds to comprehensive systems, in general publicly available, developed in research centres and universities worldwide. They have been designed to simulate a broad energy range of electromagnetic cascades in any material and include advanced geometry packages (table 1.1) [18]. Although currently of limited use, the pioneer ETRAN code [19] deserves recognition for being one of the first codes based on accurate photon–electron transport; its limited geometry possibilities were superseded by the ITS system [20], which allowed complex geometries. Today's most widely used Monte Carlo systems for photon–electron–positron transport are EGSnrc [21] and PENELOPE [22], both including rather similar advanced physics interaction models and variance-reduction techniques. PENELOPE adds the simulation of radioactive sources [23] on the basis of evaluated nuclear data from the international nuclide database [24]. To model intricate geometrical structures the EGSnrc system includes a general-purpose C++ geometry library package [25], whereas in PENELOPE bodies are limited by quadric surfaces. Other general-purpose Monte Carlo systems add the simulation of different types of particles, such as neutrons, protons and other heavy-charged particles; the widely spread MCNP [26], MCNPX [27] and GEANT [28] pertain to this group. Most of these systems and codes have been used for specific nuclear medicine calculations. For the simulation of SPECT and PET detectors, important applications have been developed with the widely used GATE/GEANT [29, 30] and with PENELOPE [31]. Quantitative SPECT imaging and calculation of dose-rate volume histograms have received considerable attention [32]. One investigation [33] on the accuracy of tumour absorbed dose estimates for ^{111}In, ^{177}Lu, ^{131}I and ^{90}Y using the now obsolete EGS4 [34] concluded, however, that a full Monte Carlo simulation is justified when the particle ranges are much longer than the spatial resolution of the image from which the absorbed dose calculations are made. With regard to dosimetry, the pioneer analytical calculations of photon isotropic dose kernels by Berger [1] have been superseded by several Monte Carlo calculations, including those made with EGS4 and with EGSnrc [11]. Electron kernels, also pioneered by Berger [35] for mono-energetic sources and β-emitters using Monte Carlo calculations, were the subject of subsequent EGS4 calculations, [36] and others, finding minor differences with Berger kernels. A comparison of GATE with EGSnrc and MCNP for electron dose kernels [37] showed that GATE and EGSnrc agreed to within better than 3% but the agreement was less good with MCNP. GATE has

Table 1.1. Key features of general-purpose public domain Monte Carlo codes used in modelling nuclear medicine imaging systems.

Monte Carlo code	Brief description	Representative references
EGS4 [54]/EGSnrc [55]	Coupled photons/electrons transport in any material through user specified geometries. Simulation of nuclear imaging systems not specifically included and requires an extensive amount of user programming in Mortran.	[56–78]
ITS including TIGER CYLTRAN and ACCEPT [79]	Coupled photons/electrons transport in any material through slabs, cylinders or combinatorial. Simulation of nuclear imaging systems not specifically included and requires an extensive amount of user programming in Fortran.	
MCNP [80]/MCNPX [81]	Coupled neutrons/photons/electrons transport in any material through user-generalized geometry. Simulation of nuclear imaging systems not specifically included and requires an extensive amount of user manipulation of input date files to model complex geometries.	[67, 72, 82–114]
GEANT [115–117]	Coupled photons/electrons transport in any material through combinatorial geometry. Simulation of nuclear imaging systems not specifically included and requires an extensive amount of user programming in C/C++.	[72, 114, 118–139]
PENELOPE [140, 141]	Coupled photons/clcctrons transport in any material through combinatorial geometry. Simulation of nuclear imaging systems not specifically included and requires an extensive amount of user programming in C/C++.	[93, 142–146] [143, 144, 146]
FLUKA [147, 148]	Coupled photons/electrons transport in any material through combinatorial geometry. Simulation of nuclear imaging systems not specifically included and requires an extensive amount of user programming in C/C++.	[102, 131, 149–154]

also been used to evaluate the performance of SPECT for determining the position and distribution of ^{225}Ac in targeted α-therapy [38], whereas MCNPX was used to compare α- and β-targeted therapy of bone metastases [39]. Cellular dosimetry has been the subject of detailed Monte Carlo calculations; MCNP was used to evaluate S-values for ^{111}In [40] and ^{64}Cu [41] in subcellular distributions and estimate the absorbed dose to the nucleus in Auger electron therapy targeting peptide receptors while results obtained with GEANT were reported for ^{99m}Tc, ^{123}I, and ^{111}In in [42]. Patient-specific dosimetry and related radiobiology have been reviewed in [43].

(b) The second group refers to 'in-house' Monte Carlo codes specifically designed for nuclear medicine applications, where the interaction physics, materials and geometry are usually tailored to the intended application (table 1.2) [18]. The code SIMIND [44], whose base is the description of a standard SPECT camera, has been widely spread. It has been used for the optimization of the energy window in quantitative ^{111}In imaging [45], showing good agreement between measurements and simulations, both quantitatively and qualitatively. In combination with GATE it has been used to study the effect of energy and source location on a gamma camera

Table 1.2. Key features of dedicated Monte Carlo codes used to simulate nuclear medical imaging systems.

MC code	General description
SIMSET [155, 156]	Photon transport in any material through voxel-based phantoms. Simulation of SPECT and PET imaging systems included. User modules written in C could be linked.
SIMIND [157, 158]	Photon transport in any material through shape- or voxel-based phantoms. Simulation of SPECT imaging systems included. User modules written in Fortran could be linked.
Unnamed [159, 160]	Photon transport in any material through shape-based phantoms. Simulation of SPECT imaging systems included. User modules written in Fortran/C could be linked.
MCMATV [161, 162]	Photon transport in any material through voxel-based phantoms. Simulation of SPECT imaging systems included. User modules written in Fortran could be linked.
PETSIM [163, 164]	Photon transport in any material through shape-based phantoms. Simulation of PET imaging systems included. User modules written in Fortran could be linked.
PET-EGS [59]	Photon transport in any material through shape-based phantoms. Simulation of cylindrical PET imaging systems included. User modules written in Fortran could be linked.
EIDOLON [165, 166]	Photon transport in any material through shape- or voxel-based phantoms. Simulation of 3D PET imaging systems included. User modules written in C/Objective-C could be linked.
PET-SORTEO [167]	Photons transport in any material through shape- or voxel-based phantoms. Simulation of 3D PET and transmission imaging systems included.
GATE [123, 136]	Coupled photons/electrons transport in any material through shape- or voxel-based phantoms. Simulation of nuclear imaging systems/internal dosimetry specifically included and requires user programming in C/C++.
GAMOS [168]	Coupled photons/electrons transport in any material through shape- or voxel-based phantoms. Simulation of nuclear imaging systems/internal dosimetry specifically included and requires user programming in C/C++.

intrinsic and extrinsic spatial resolution [46]. SIMIND has also been used to generate response functions and kernels for quantitative ^{90}Y bremsstrahlung SPECT [47]. For PET, the codes SIMSET [48] and PET-EGS [49] were used to validate and provide a unified description for PET simulations [50], finding that only PET-EGS provided accurate modelling of count rates while SIMSET was more efficient from a computational view. Other image-based dosimetry codes are based on the use of standard geometrical phantoms, like DOSE3D [51], or patient-specific voxel phantom data like 3D-ID [52] and SIMDOS [53]. Additional codes for internal radionuclide dosimetry are described in detail in chapters 9–14 of this book.

The potential efficacy of Neutron Capture Therapy (NCT), predominantly with the use of boron (BNCT), led scientists to develop this experimental technique and a comprehensive review was published by the IAEA [169]. In BNCT, a tumour-seeking compound containing ^{10}B, a non-radioactive isotope, is introduced into the brain and given time to accumulate in the tumour. The patient is then exposed to a beam of neutrons, usually in a nuclear reactor facility, which are captured by the ^{10}B. Capturing neutrons breaks the boron nuclei, resulting in the emission of alpha radiation and lithium nuclei. Both α-particles and Li have high LET but short ranges, and as a result they potentially can destroy the malignant cells in which the ^{10}B is imbedded without damaging the adjacent healthy cells. BNCT methods use an epithermal neutron beam in conjunction with tumour-targeting boron carrier compounds for the irradiation of glioblastomas and metastatic melanomas. Hence, BNCT can be considered a 'hybrid' procedure that involves the dosimetry of the external neutron beam and that of the internal boron compounds distribution; only the latter is close to the nuclear medicine field. During the years following the initial enthusiasm for BNCT, efforts have been focused on the search for suitable compounds that yield a high concentration ratio between the tumour and the surrounding healthy tissue. In addition to the external beam dosimetry of the incident neutrons, patient dose calculations are carried out with specific Monte Carlo-based treatment planning systems developed by several institutions worldwide [170], which often use PET imaging to determine the boron concentration in the tumour and surrounding tissues. SPECT prompt-gamma detection of the 478 keV photons created in most capture reactions, used to visualize and quantify the dose distribution during treatment, has received considerable attention [171].

A remark on Monte Carlo-based patient-specific absorbed dose calculations is at hand. An important constraint for these calculations is that the detailed mass density and composition of the different sub-volumes (voxels) in a patient or anatomical phantom geometry are required. The mass density of geometry voxels can be obtained with reasonable accuracy from a CT-number calibration, N_{CT}, where values for reference materials of known density are used to derive the mass or electron density of patient voxel N_{CT}'s. Note that a given voxel might include different tissue types, but still be related to the calibration-derived density of the voxel. Due to the dominance of photon Compton interactions, electron density is the preferred quantity to be used for scaling distances and thickness. A non-trivial problem is assigning body-tissue types and their composition to CT-densities, for which look-up tables, usually based on compositions given in ICRU or ICRP publications, are commonly used for each tissue type as a function of the voxel density. Major drawbacks of this method are the differences in patient-to-patient body tissue compositions and the assumption that the compositions given by ICRU or ICRP are 'standard'. These two aspects conflict with a critical statement given by ICRU Report 44 [172]: *'It is imperative that body-tissue compositions are not given the standing of physical constants and their reported variability is always taken into account'*. The rationale for such clarification is that tabulated compositions were derived from small human samples and they may as well vary with gender and ethnics. An unexplored method for proper tissue characterization could be MR spectroscopy. Additionally, mass stopping powers for charged particles require the mean excitation energy of the target material, a key physical parameter that even if the material composition is known still needs an approximate method for its estimation (the Bragg-additivity rule) [173]. That is the main reason why body-tissue mass stopping powers have an uncertainty significantly larger than for most elements or reference materials, being up to the order of 10%–15% (see ICRU Report 37 [174]). Further details on these constraints can be found in [175]. The controversy on the type of tissue to be assigned to tumor cells embedded in a specific medium or organ should be added to these constraints, as for example a tumor in bone tissue can be more water-like than bone-like.

1.3 Applications of Monte Carlo techniques in quantitative imaging for radiation dosimetry

Advances and recent innovations in nuclear medical imaging instrumentation including single-photon emission computed tomography (SPECT) and positron emission tomography (PET) systems, as well as hybrid or concurent functional and structural imaging systems are ideal for Monte Carlo modelling techniques because of the stochastic nature of radiation emission, transport and detection processes. Factors, which have contributed to the increased popularity and wider use of Monte Carlo techniques, include improved and more accurate models of radiation transport processes, the practicality of its application taking advantage of acceleration schemes and the improved speed of modern scientific computing platforms. This section summarizes potential areas of application of Monte Carlo techniques in quantitative imaging for radiation dosimetry. An overview of existing simulation programs was presented in the previous section, whereas examples of useful features of such sophisticated tools in connection with extracting advanced quantitative imaging features for personalized radiation dosimetry calculations will be discussed here.

Recent developments in nuclear medicine instrumentation and multiple-processor parallel processing systems have created new opportunities for Monte Carlo simulation in diagnostic and therapeutic nuclear medicine imaging. One of the aims of the medical physicist involved in nuclear medical imaging research is to optimize the design of imaging systems and to improve the quality and quantitative accuracy of emission images. Several degrading factors affect the quality and the accuracy of nuclear medicine images, including the geometry of the imaging system, properties of the detectors, collimator (for SPECT) and gantry design, attenuation and scatter compensation, partial volume and image reconstruction algorithms [177]. Integrating improvements in these topics with available tracers and sensitive and specific molecular imaging probes under development/approval, will provide major advantages to the general nuclear medicine clinician and research investigator (figure 1.1). Mathematical modelling is necessary for the assessment of the impact of various parameters in nuclear medical imaging systems since no analytical solution

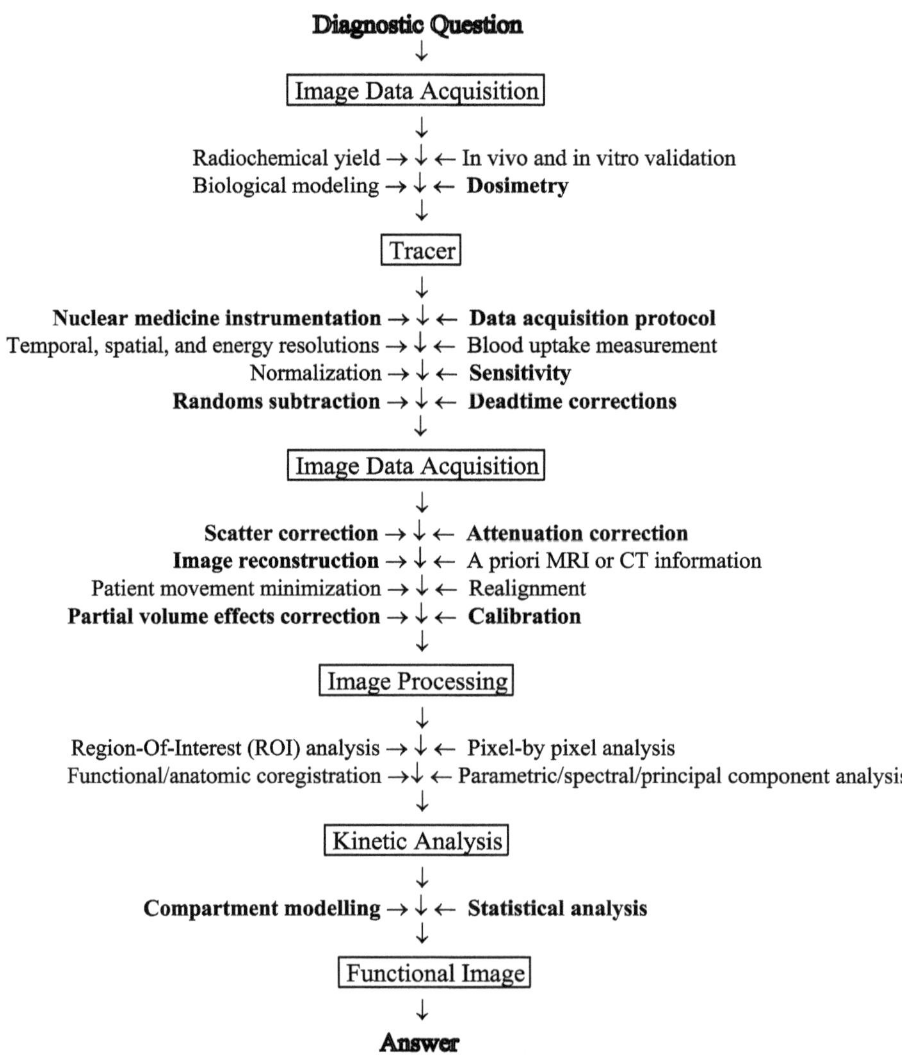

Figure 1.1. Scientific and technical strategy for recording quantitatively accurate functional/molecular images. In bold, the parts where Monte Carlo simulation plays an important role (Adapted with permission from [176] Wiley. Copyright (1999) American Association of Physicists in Medicine).

enables to solve Boltzmann's radiation transport equation, describing the interaction of photons with non-uniformly attenuating body structures and complex detector geometries and gantries.

There has been an enormous increase and interest in the use of Monte Carlo techniques in all aspects of nuclear imaging, including planar imaging [178], SPECT [158, 160, 179] and PET [136, 166, 180–183]. However, due to computer limitations, the method has not yet fully lived up to its potential. With the advent of high-speed supercomputers, the field has received increased attention, particularly with parallel algorithms, which have much higher execution rates.

Figure 1.2 illustrates the conceptual design of Monte Carlo or statistical simulation as applied to a cylindrical PET scanner. Assuming that the behaviour of the imaging system can be described by probability density functions (*pdf's*), then the Monte Carlo simulation can proceed by sampling from these *pdf's*, which necessitates a fast and effective way to generate random numbers uniformly distributed. Photon emissions are generated within the phantom and are transported by sampling from *pdf's* through the scattering medium and detection system until they are absorbed or escape the volume of interest without hitting the crystal. The outcomes of these random samplings, or trials, must be accumulated or tallied in an appropriate manner to produce the desired result, but the essential characteristic of Monte Carlo is the use of random sampling techniques to arrive at a solution of the physical problem.

The Monte Carlo method is a widely used research tool in different areas of nuclear imaging, including detector modelling and systems design, image correction and reconstruction techniques, internal dosimetry and pharmacokinetic modelling [176]. The method has proven to be very useful for solving complex problems that cannot be modelled by computer codes using deterministic methods or when experimental measurements may be impractical. Monte Carlo simulation of detector responses and determination of their efficiencies is one of the areas, which has

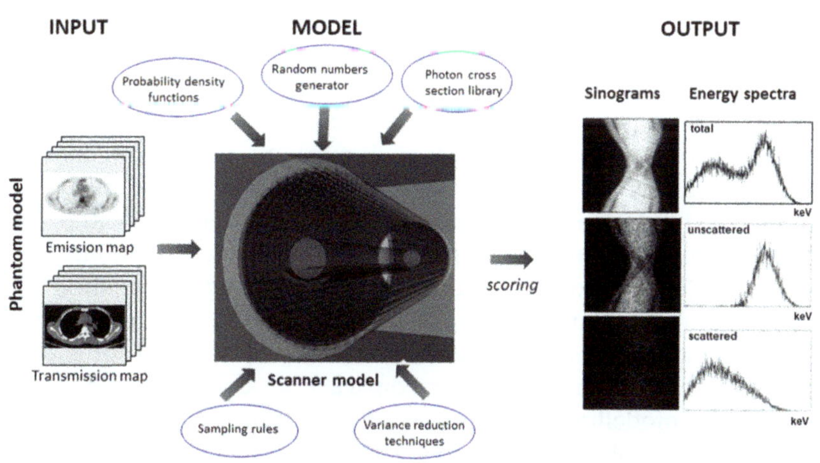

Figure 1.2. Conceptual design and main components of a Monte Carlo software package dedicated for simulation of cylindrical multi-ring PET scanners.

received considerable attention. Many detector modelling applications have been developed in the PET field including the pioneering work of Derenzo [184] who simulated arrays of detectors of different materials and sizes to study the effect of the inter-crystal septa and later on to optimize the optical coupling between BGO crystals and PMTs by taking into account the reflection and scattering along the detection system [185]. The search for an appropriate crystal for this imaging modality has been conducted in a comparative study of several crystals including BGO, CsF and NaI(Tl) [186], BaF_2 used in time-of-flight PET [187], liquid Xenon [188], LuYAP [189] and more recently GLuGAG and LFS-3 [190].

Optimizing the design of SPECT and PET systems using the Monte Carlo method has received considerable attention and a large number of applications were the result of such investigations. During the last few decades, simulation of scintillation camera imaging using both deterministic and Monte Carlo methods have been developed to assess qualitatively and quantitatively the image formation process and interpretation [191] and to assist in the development of collimators [192–194]. Several researchers have used Monte Carlo simulation methods to study potential novel designs of dedicated small-animal PET scanners [195, 196]. An important conclusion drawn from these studies is that unlike human imaging where both sensitivity and spatial resolution limitations significantly affect the quantitative imaging performance of a tomograph, the imaging performance of dedicated animal tomographs is almost solely based upon its spatial resolution limitations.

For instance, the *Eidolon* Monte Carlo simulator [166], developed using modern software development tools, is an efficient tool that can be used to generate data sets in a controllable manner in order to assess different reconstruction algorithms and scatter correction methods [197]. As the 'best' algorithm can only be selected with respect to a certain task, different 'basic' performance measures can be used. Image degrading effects are illustrated using simulated projections of the digitized 3D Hoffman brain phantom [198]. A slice of this phantom is shown in figure 1.3(a). The ratio between the activity in white, grey matter and ventricles has been chosen as 1:4:0, respectively. The projections of this phantom at different levels of fidelity are generated. The strengths of the image degrading factors are characteristic of an [^{18}F]-FDG-brain study. Figures 1.3(f)–(i) show the effects of different aspects of image degradation on filtered backprojection reconstructions. The loss of resolution caused by detector blurring (FWHM = 4 mm) on projection data and FBP reconstructions is shown in figures 1.3(c) and (g), respectively, while in figures 1.3(d) and (h), effects of detector blurring, attenuation and scatter are included in the simulation and no corrections performed on the simulated data sets. Finally, in figures 1.3(e) and (i), effects of detector blurring, attenuation and scatter are included and appropriate corrections for attenuation and scatter applied.

Monte Carlo simulations have been shown to be very useful for the validation and comparative evaluation of image reconstruction techniques. Smith *et al* [199] used Monte Carlo modelling to study photon detection kernels, which characterize the probabilities that photons emitted by radioisotopes in different parts of the source region will be detected at particular projection pixels of the projection images for the case of parallel-hole collimators. The authors also proposed a reconstruction

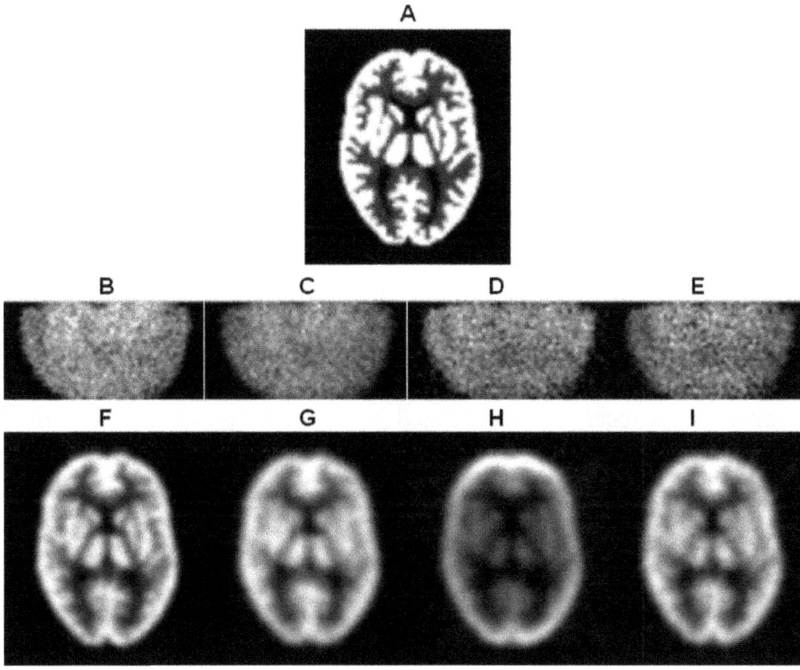

Figure 1.3. (a) Transaxial slice of the digital 3D Hoffman brain phantom. Monte Carlo simulation of the 3D Hoffman brain phantom in different imaging situations showing 2D projections (middle row) and their filtered backprojection reconstructions (bottom row). Cases (b) and (f), are a 2D projection and reconstructed image neglecting attenuation, scatter and detector blurring. In (c) and (g), effects of detector blurring are included, while in (d) and (h), effects of detector blurring, attenuation and scatter are included in the simulation and no corrections performed on the simulated data sets. Finally, in (e) and (i), effects of detector blurring, attenuation and scatter are included and appropriate corrections for attenuation and scatter applied.

method using the 3D kernels in which projection measurements in three adjacent planes are used simultaneously to estimate the source activity of the centre plane. The search for unified reconstruction algorithms led to the development of inverse Monte Carlo (IMC) reconstruction techniques. The principal merits of IMC are that, like direct Monte Carlo, the method can be applied to complex and multivariable problems, and variance reduction procedures can be applied [200]. Floyd *et al* [201] used IMC to perform tomographic reconstruction for SPECT with simultaneous compensation for attenuation, scatter, and distance-dependent collimator resolution. This concept was further investigated later on by few research groups but didn't reach the maturity required for clinical adoption [78, 202, 203].

For internal radiation dose estimates, the biodistribution of a trace ^{131}I-labeled monoclonal antibody is generally used to predict the biodistribution of a high-dose administration for therapy. Imaging therapeutic doses would further confirm the hypothesis that the biodistribution is similar, however, current generation scintillation cameras are unable to handle accurately the corresponding high counting rate. Monte Carlo calculations have been used in the development of a method for imaging therapeutic doses of ^{131}I by using thick Pb sheets placed on the front surface of a high-energy parallel-hole collimator [204]. The accuracy of ^{131}I tumour quantification after RIT has been further investigated with an ultra-high-energy (UHE) collimator designed for imaging 511-keV photons [205]. It has been shown that the difference in tumour size, relative to the size of a calibration sphere, has the biggest effect on accuracy, and recovery coefficients are needed to improve quantification of small tumours.

There has been renewed interest in pinhole collimation for high-resolution imaging of small organs and structures (e.g. the thyroid) and for regionally administered monoclonal antibody imaging since it provides an improved spatial resolution and an increase in sensitivity as the distance between the source and the pinhole aperture decreases [206]. Huili *et al* [207] simulated point response functions for pinhole apertures with various aperture span angle, hole size, and materials. The point responses have been parameterized using radially circular symmetric two-dimensional exponential functions, which can be incorporated into image reconstruction algorithms that compensate for the penetration effect. The effect of pinhole aperture design parameters on angle-dependent sensitivity for high-resolution pinhole imaging has been also investigated using Monte Carlo modelling [208]. Simulated ^{131}I SPECT studies for uniform cylinders showed that activity concentrations have been underestimated toward the outside of the cylinders when a $\sin^3\theta$ rather than the correct $\sin^x\theta$ sensitivity correction has been applied in image reconstruction, where x is a parameter describing the sensitivity and θ is the angle of the incident ray with the surface of the detector crystal.

Another promising application of Monte Carlo calculations is to achieve accurate quantitative imaging through appropriate modeling and correction for Compton scattering and photon attenuation in nuclear medicine imaging since the user has the ability to separate the detected photons into their components: primary events, scatter events, contribution of down-scatter events, etc. Monte Carlo modelling thus allows a detailed investigation of the spatial and energy distribution of Compton scatter, which would be difficult to perform using present experimental techniques, even with very good energy resolution detectors [209, 210]. A Monte Carlo study of the acceptance of scattered events in a depth-encoding large-aperture camera made of position-encoding blocks modified to resolve the depth of interaction through a variation in the photopeak pulse-height has been performed by Moison *et al* [211]. Monte Carlo simulations have also been extensively used to evaluate and compare scatter correction schemes in both SPECT [212, 213] and PET [197, 214] imaging.

Figure 1.4 shows the spectral energy distribution obtained by Monte Carlo simulation of a 20 cm diameter water-filled cylindrical phantom containing a uniform distribution of ^{18}F as positron-emitting source [215]. The scattered events in the energy pulse-height distribution have been separated into scattered and unscattered events and where the latter were separated again according to the order of Compton scattering.

The simulation of transmission scanning allowed studying the effect of down-scatter from emission (^{99m}Tc) into transmission (^{153}Gd) energy window in SPECT [216] and to investigate detected scattered photons in single-photon transmission scanning using ^{137}Cs single-photons sources for PET [138]. A comparative

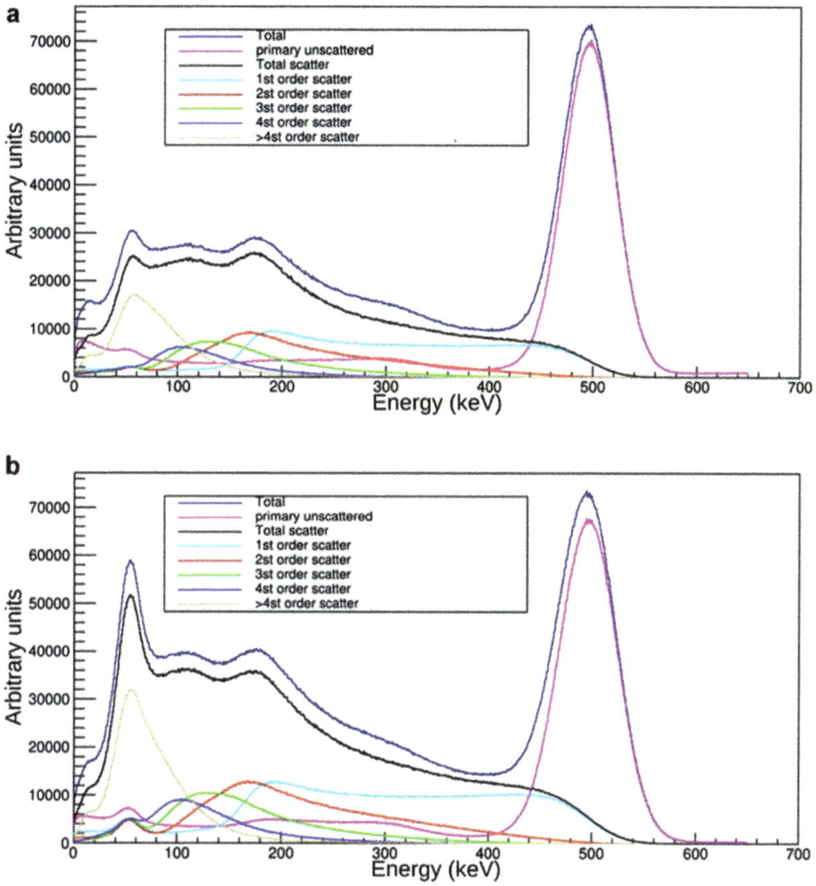

Figure 1.4. Plots of the energy spectra calculated by the Monte Carlo method in (a) the mCT and (b) large axial field-of-view (LAFOV) PET scanners, showing the contributions to each from unscattered and scattered photons. The simulated phantom is a cylinder with a diameter of 20 cm and lengths of 70 cm and 200 cm for the mCT and LAFOV scanners, respectively, uniformly filled with an ^{18}F solution. The spectrum of unscattered photons is shown in magenta, while the spectra due to scattered photons (solid black line) is separated into the contributions from different orders of photon scattering (shown in a range of colours). The distributions of scattered and unscattered photons are added to give the total spectra (solid blue line). The mCT and LAFOV scanner energy resolutions are 11.7% and 13%, respectively. Reprinted with permission from [215], copyright (2021), with permission from Elsevier.

evaluation of pre- and post-processing attenuation correction methods in lung SPECT has been also conducted using Monte Carlo simulations of a digital thorax phantom containing a homogeneous activity distribution in the lung [217]. Likewise, a number of studies evaluated the impact of Compton scattering in current cone-beam and potential future generation cone-beam CT scanners on the accuracy of CT-based attenuation correction in hybrid PET/CT scanners. For instance, the MCNP Monte Carlo code was used to generate data corresponding to the geometry of hypothetical cone-beam CT mimicking the geometry of a prototype flat panel detector-based CT scanner adjusted to the dimensions of large area panel PET detectors [218]. A comparison of the simulated attenuation profiles for primary and total projection data of a cylindrical water phantom in both fan- and cone-beam CT geometries is shown in figure 1.5(a) [219]. The high magnitude of the scattered component results in noticeable underestimation of attenuation profiles for the cone-beam geometry. The central profile of the corresponding reconstructed CT images is shown in figure 1.5(b). The characteristic cupping artefact visible on the reconstructed CT images for the cone-beam geometry depicts the impact of scattered radiation. It should be emphasized that the effect of beam hardening was removed from the profiles prior to image reconstruction [220]. The central profile of the corresponding CTAC-based μmap of the cylindrical water phantom at 511 keV is shown in figure 1.5(c). The central profiles of attenuation correction factors computed using calculated attenuation correction considering the theoretical linear attenuation coefficient of 511 keV photons in water (0.096 cm^{-1}) and the generated attenuation maps using CT-based attenuation correction are compared in figure 1.5(d) [219].

The imaging properties of therapeutic radionuclides and optimized imaging protocols have been evaluated using Monte Carlo simulation studies. Different strategies are being developed to improve image quality and quantitative accuracy in

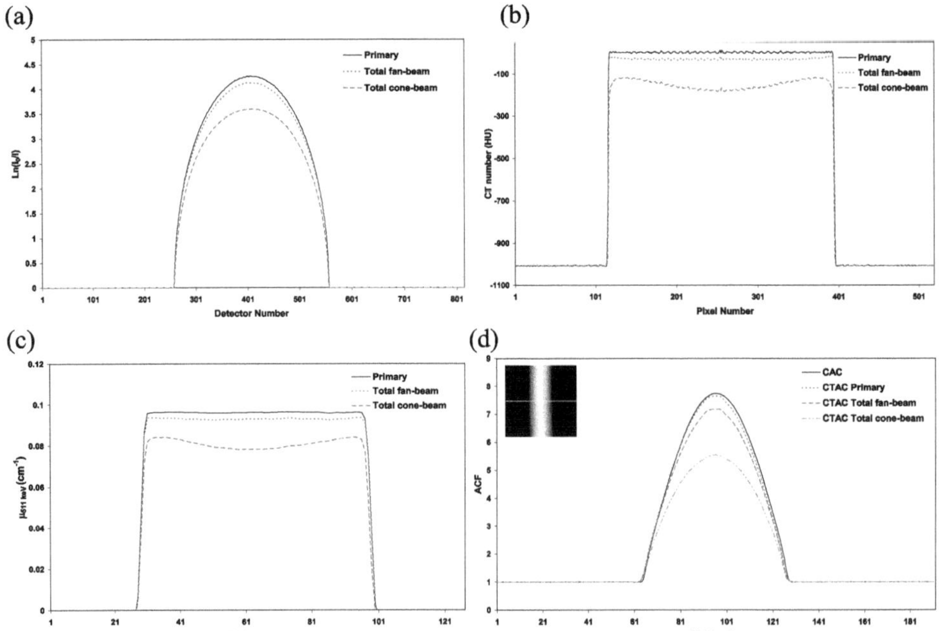

Figure 1.5. Monte Carlo simulations of the cylindrical water phantom (200 mm diameter) in both fan- and cone-beam CT geometries showing: (a) attenuation profiles of primary and total photons; (b) central profiles through reconstructed CT images; (c) Central profiles through attenuation maps at 511 keV generated using the bi-linear transformation method; and (d) central profiles through ACF sinograms calculated from attenuation maps shown in (c). Reprinted from [219], copyright (2006) with permission of Springer.

tumour SPECT imaging, including collimator-detector response compensation and high-energy scatter correction techniques. In an elegant study by Dewaraja *et al* [221], Monte Carlo simulations have been used to evaluate how object shape influences 'spill-out' and 'spill-in', which are major sources of quantification errors associated with the poor spatial resolution of ^{131}I SPECT and to characterize energy and spatial distributions of scatter and penetration [222]. Similar studies were performed for other radionuclides, including β-emitters, such as ^{90}Y [223, 224] and ^{177}Lu [225, 226] and α-emitters, e.g. ^{211}At/^{225}Ac [227] and ^{209}At [228].

1.4 Monte Carlo techniques in nuclear medicine dosimetry

1.4.1 Calculation of absorbed fractions

The origins of the *absorbed fraction* concept adopted by the MIRD committee for internal dose calculations trace back to the earlier work by Ellet *et al* [229] in the sixties. Many of the pamphlets published by the MIRD committee made extensive use of Monte Carlo calculations to derive specific absorbed fractions for photon sources uniformly distributed in organs of mathematical phantoms [230]. This has been extended later on to electrons and beta particles [231, 232].

Cristy [233] demonstrated that the reciprocity theorem, which states that for any pair of regions in a uniform isotropic or uniform scatterless model, the specific absorbed fraction is independent of which region is designated source and which is designated target, may also be valid for heterogeneous phantoms for certain conditions. Other approaches using the MIRD formalism have also been proposed [234]. Poston *et al* [235] calculated photon specific absorbed fractions for both the Cristy and Eckerman gastrointestinal tract and their revised model and reported differences between electron absorbed fraction values with and without electron tracking. Previously calculated absorbed fractions for unit density spheres in an infinite unit density medium for photon and electron emitters have been recently re-evaluated using both the EGS4 and MCNP4B Monte Carlo codes [236]. Calculation of absorbed fractions for positron emitters relevant to neurologic studies has also been reported [237].

The application of the Monte Carlo method to internal radiation dosimetry is further emphasized in two MIRD Pamphlets. In the MIRD Pamphlet No. 15 [238], the EGS4 Monte Carlo radiation transport code has been used to revise substantially the dosimetric model of the adult head and brain originally published in MIRD Pamphlet No. 5. Pamphlet No. 17 [239] crystallizes the utility of the MIRD formalism for the calculation of the non-uniform distribution of radiation-absorbed dose in different organs through the use of radionuclide-specific S values defined at the voxel level. Skeletal *S values* and absorbed fractions estimates to both marrow and endosteum in trabecular and cortical bone have been further improved through the use of electron transport models [240].

Mathematical and computerized anthropomorphic phantoms are continuously being improved [241, 242]. Current developments are aimed at computational models that are flexible while providing accurate modelling of patient populations (figure 1.6) [243]. An important contribution came from Bouchet and Bolch [244] making available a series of five dosimetric head and brain models developed to

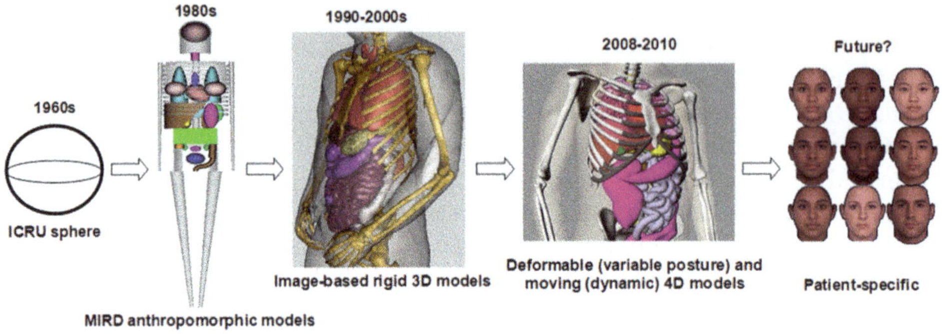

Figure 1.6. Evolution of computational models of the human anatomy from the crude ICRU spherical model to more realistic and complex (person-specific) computational models. Copyright IEEE. Reprinted with permission from [243].

allow more precise dosimetry in paediatric neuroimaging procedures. A new rectal model [245] and dynamic urinary bladder model [246] have also been proposed more recently. To develop more patient-specific dosimetry, new mathematical models for adults of different height have been developed using anthropometric data [247]. The use of dynamic anthropomorphic phantoms in Monte Carlo simulations is becoming possible due to the increasing availability of computing power. This includes the development of appropriate primitives that allow accurate modelling of anatomical variations and patient motion, like superquadrics [248] and non-uniform rational B-spline surfaces [249]. An image-based whole-body model, called VIP-Man, has been developed using high-resolution transversal colour photographic images obtained from the National Library of Medicine's Visible Human Project [250]. More recent developments are covered in Chapter 5 of this volume. The EGS4 code has been also used to estimate specific absorbed fractions from internal electron emitters with energies from 100 keV to 4 MeV [251].

1.4.2 Derivation of dose-point kernels

In most cases, Monte Carlo calculations are used to simulate the random distribution of sources or targets, whereas the actual dosimetric calculation is performed using the so-called dose-point kernels. Such kernels, usually spherical and calculated for mono-energetic photon or electron sources describe the pattern of energy deposited at various radial distances from photon [252–254] and electron or β [255, 256] point sources. Dose-point kernels can be calculated using analytical or Monte Carlo methods. Hybrid approaches (analytical calculations using Monte Carlo data) have also been considered to decrease the computation time [257]. Three Monte Carlo systems have mainly been used for this purpose, namely, ETRAN [255], the ACCEPT code of the ITS system [258] used by Sgouros *et al* [259], and EGS4 [256, 260]. Limitations and constraints of some of these codes have been reported in the literature whose impact on the calculated kernels is difficult to evaluate. ETRAN, for instance, had an incorrect sampling of the electron energy-loss straggling that has been corrected for in the ITS3 system (based on ETRAN);

EGS4 did not include the accurate electron physics and transport algorithms which have been incorporated into the recent EGSnrc system [261]. The EGS4 system has also been used to characterize the spatial and energy distribution of bremsstrahlung radiation from beta point sources in water of relevance to RIT [262]. This study provided the initial data required for modelling and analysing the scatter, attenuation, and image formation processes in quantitative imaging of bremsstrahlung for RIT dosimetry. Furhang [253] generated photon point dose kernels and absorbed fractions in water for the full photon emission spectrum of radionuclides of interest in nuclear medicine, by simulating the transport of particles using Monte Carlo techniques. The kernels have been then fitted to a mathematical expression.

A unified approach for photon and β-particle dosimetry has been proposed by Leichner [258] by fitting Berger's tables for photons and electrons to generate an empirical function that is valid for both photons and β-particles. Therefore, both point-kernel and Monte Carlo techniques can be effectively employed to calculate absorbed dose to tissue from radionuclides that emit photons or electrons. The latter are computationally much more intensive; however, point-kernel methods are restricted to homogeneous tissue regions that can be mathematically described by analytical geometries, whereas Monte Carlo methods have the advantage of being able to accommodate heterogeneous tissue regions with complex geometric shapes. This limitation has been addressed in more recent works [263–265]. More recent developments are covered in chapter 4 of this volume.

1.4.3 Pharmacokinetic modelling

Pharmacokinetic modelling is a useful component for estimating the cumulated activity in various source organs in the body. A few applications of Monte Carlo techniques have been reported in the field. In particular, Casciari [266] developed a compartmental model of [F-18]-fluoromisonidazole transport and metabolism to compute the volume average kappa in tissue regions from [F-18]-fluoromisonidazole PET time-activity data and characterized it using Monte Carlo simulations and PET time-activity data. This model has been able to accurately determine kappa for a variety of computer generated time-activity curves including those for hypothetical heterogeneous tissue regions and poorly perfused tissue regions. Compartmental models allow also the *in vivo* analysis of radioligand binding to receptor sites in the human brain. Benzodiazepine receptor binding has been studied using a three-compartmental model [267]. The validity of the results of the coefficient of variation of each parameter has been verified with statistical results provided by Monte Carlo simulation. Burger and Buck [268] examined the possibility of using a mathematical metabolite correction, which might obviate the need for actual metabolite measurements. Mathematical metabolite correction has been implemented by estimating the input curve together with kinetic tissue parameters. The general feasibility of the approach has been evaluated in a Monte Carlo simulation using a two-tissue compartment model. A simplified approach involving linear-regression straight-line parameter fitting of dynamic scan data has been developed for both specific and non-specific models [269]. Monte Carlo simulations have been used to evaluate parameter standard deviations, due to data noise, and much smaller noise-induced biases. The authors reported good agreement between regression and traditional methods.

Welch [270] investigated and quantified the effect of typical SPECT system resolution and photon counting statistics on the bias and precision of dynamic cardiac SPECT parameters. Simulation of dynamic SPECT projection data has been performed using a realistic human torso phantom assuming both perfect system resolution and a system resolution typical of a clinical SPECT system. The results showed that the rate constant characterizing the washing of activity into the myocardium is more sensitive to the position of the region of interest than the washout rate constant and that the main effect of increased photon noise in the projection data is to decrease the precision of the estimated parameters.

Computer simulations demonstrate that estimation of the kinetic parameters directly from the projections is more accurate than the estimation from the reconstructed images [271]. A strategy for joint estimation of physiological parameters and myocardial boundaries has been proposed and evaluated by simulated myocardial perfusion studies based on a simplified heart model [272]. A method allowing the estimation of kinetic parameters directly from SPECT cone-beam projections has also been proposed and validated with a simulated chest phantom [273]. The results showed that myocardial uptake and washout parameters estimated by conventional analysis of noiseless simulated cone-beam data had biases ranging between 3%–26% and 0%–28%, respectively, while uncertainties of parameter estimates with this method ranged between 0.2%–9% for the uptake parameters and between 0.3%–6% for the washout parameters.

1.5 Monte Carlo techniques in radiation protection

There is a considerable debate between scientists working in medical radiation physics and in the field of radiation safety on the term 'radiation protection' as applied to radiation medicine, and in particular to patients undergoing diagnostic medical procedures or radiation treatment. The determination of the dose delivered to different organs, for example, pertains to the standard clinical procedure or quality assurance process performed in each medical specialty, be it a diagnostic nuclear medicine or radiology technique, or a radiotherapy treatment. For a hospital physicist, radiation dosimetry is not radiation protection but clinical practice. The radiation safety officer calls the task radiation protection. As the authors of this chapter belong to the first group, it has been a natural decision not to consider dosimetry applications within this section, but as a clinical speciality in nuclear medicine (see section 1.4). Other applications of the Monte Carlo method linked to radiation protection aspects are briefly described below. It should be emphasized, however, that considering the large uncertainties allowed in radiation protection, and the considerable labour and time required to perform Monte Carlo simulations, many of the problems analysed by the existing applications could have been solved using rather simple analytical procedures. The use of Monte Carlo techniques in radiation protection should only be justified in cases of complex situations, either due to the presence of complicated mixed radiation sources or geometries.

1.5.1 Shielding calculations

Monte Carlo techniques have been used to determine detector calibration factors and to predict gamma-ray spectra for well-defined measurements. The techniques have sometimes been used to predict shielding requirements and critical structural specifications that must conform radiation protection rules. Different Monte Carlo systems have been used to evaluate shielding including EGS4 [274] and MCNP for diagnostic x-ray rooms [275], and ITS to calculate door thicknesses for megavoltage radiotherapy [276]. More sophisticated applications include quantifying the effect of lead shielding on dose distributions during radiotherapy treatment planning with electron beams [277] and to design the moderator-collimator-shielding system of a neutron beam optimized for radiography that utilizes a ^{252}Cf source [278]. Monte Carlo methods have also been used to estimate the neutron and photon dose rates in a variety of locations in the vicinity of an accelerator-based neutron capture therapy facility, as well as the shielding configuration required when the device is run at maximum current [279]. Likewise, they were used for shielding calculations for cylclotrons [103] and PET/CT facilities to consider the effects of self-absorption on the photon energy spectrum [89].

1.5.2 Characterization of detectors and radiation monitoring instruments

Monte Carlo simulations have been shown to be useful in the evaluation of the design of collimated detectors used to measure ^{125}I or ^{131}I uptake in the thyroid gland [280]. Two detector sizes have been simulated for each radioisotope, and activity has been placed both in the gland and in the remainder of the body in varying amounts to assess the efficacy of the collimation. This study has been followed by a detailed report where the authors assessed the uncertainty introduced into an activity estimate of radioiodine in the thyroid when the size and shape of the gland differs from that of the calibration phantom [281]. The depth-dose distribution in TLD dosimeters irradiated with beta rays from a ^{90}Sr-^{90}Y source has also been investigated using the Monte Carlo method [282]. A pilot study has been conducted at the Lawrence Livermore National Laboratory to demonstrate the applicability of Monte Carlo techniques to calibrate *in vivo* measurement systems and investigate the effects of the source geometry and the detector size on the measured efficiency [282]. The development of a calibration phantom for *in vivo* measurements of actinides using the MCNP Monte Carlo system to determine the point source distribution that closely approximates a homogeneous bone contamination was described in detail [283]. The use of advanced computational models for virtual calibration of devices use for in vivo monitoring of contamination has also been more recently reported [84].

1.5.3 Radiation dose calculations to staff and population

A few studies have reported the use of the Monte Carlo method to calculate the dose received by medical and technical staff in a clinical environment. A recent publication estimated the dose due to radiation scattered from the roof of an adjacent radiotherapy facility where only the primary x-ray radiation had been accounted for in the shielding calculations [284]. Another group has calculated coefficients for converting dosimeter readings to equivalent dose to the foetus using the equivalent dose to the uterus and Monte Carlo modelling to simulate the equivalent dose to the foetus during the first two months of pregnancy [285]. Pattison [286] used gamma-ray and beta-particle dose rates modelled using distributed point sources and empirically-based Monte Carlo approaches, respectively, to determine the dose received by the skin of the fingers of clinical and laboratory staff during injections of ^{153}Sm. Conversion coefficients per air-kerma free-in-air for the personal dose equivalent, $H_p(10)$ have also been calculated using Monte Carlo methods for various dosimeter positions in the trunk of a voxel model of an adult male versus the direction of incidence of broad parallel photon beams with energies between 10 keV and 10 MeV [287]. Fluence-to-effective dose equivalent conversion factors as a function of radiation angles and sex for mono-energetic photon beams of 0.08, 0.3, and 1.0 MeV have also been reported elsewhere [288]. Monitoring of occupational radiation doses to staff from interventional radiology procedures [289–291] and nuclear medicine procedures [292] have also been reported.

Application of the Monte Carlo method to determine release criteria for patients administered therapeutic amounts of radioactivity has been the subject of an interesting debate between supporters [293] and detractors [294] of the utility of the technique to solve a problem that can be easily resolved otherwise. Johnson and Barnhart [294] suggested that accurate results could have been derived by using 'pen-and-paper' instead of performing an expensive Monte Carlo simulation. Siegel *et al* [295], on the other hand, argued that it would be difficult to show that the approximations involved in the calculation by hand are accurate enough without having the Monte Carlo or other 'gold standard' results (such as a phantom study) for comparison. It should be recognized that holding a Geiger-Muller counter at 1 m from the umbilicus of a patient who has been administered a known amount of ^{131}I allows checking if the model predicts the readings or not. The results obtained by Johnson and Barnhart [294] indicated that their analytical model accurately predicted the reading without having to resort in a Monte Carlo simulation. Moreover, the limits of integration of their most accurate model (model #3) could be changed to accommodate an individual with different torso dimensions, and obtain an answer in a fraction of the time that it would take to adapt the target geometry and run a Monte Carlo simulation.

1.6 Current and future applications of Monte Carlo calculations in nuclear medicine dosimetry

Comprehensive 'general-purpose' Monte Carlo codes appear to be well consolidated and except for minor updates, no substantial changes should be expected. Their dissemination and use by the nuclear medicine community is, on the other hand, expected to increase due to the availability of faster and more powerful computer technology at clinical sites. Due to this proliferation and to the extended use of new nuclear medicine techniques, additional specific codes optimized for particular techniques are expected to be developed. These will have to be benchmarked against the 'general-purpose' systems, but will make it easier to implement and perform 'ad-hoc' clinical calculations.

The trend towards Monte Carlo patient-specific image-based internal dosimetry, and related cell dosimetry and radiobiology, is expected to grow considerably. Most likely, this will motivate a patient dose escalation that requires optimization of the dose delivery consistent with the so-called AHARA principle (As High As Reasonably Achievable) [296] to minimize complications and morbidity in therapeutic applications. Constraints like tissue characterization and the uncertainty they pose to calculated absorbed doses must, however, be kept in mind so that the advantages of the Monte Carlo method are not over-emphasized. Techniques aiming at improving tissue characterization must be on the agenda horizon.

1.6.1 Patient-specific dosimetry and treatment planning

Currently, the preferred strategy with various forms of targeted molecular radiotherapy is to use personalized patient dosimetry and this approach may become routinely employed clinically. The dose distribution pattern is often calculated by generalizing a point source dose distribution [259, 297], but direct calculation by Monte Carlo techniques is also frequently reported because it allows media of inhomogeneous density to be considered [298–305].

The development of a 3D treatment planner based on SPECT/PET imaging is an area of considerable research interest and several dose calculation algorithms have been developed. The essential steps required in developing a 3D treatment planning program for voxel-based dosimetry calculations were extensively discussed in the literature [306–309]. Projection data acquired from an emission tomographic imaging system are processed to reconstruct transverse section images which yields a count density map of source regions in the body. This count density is converted to an activity map using the sensitivity derived from a calibration phantom. In the final step, this activity distribution is converted to a dose rate or dose map either by convolving the activity distribution with dose-point kernels or by direct Monte Carlo calculations. To elaborate a treatment plan for an individual patient, prospective dose estimates can be made by using a tracer activity of radiolabelled antibody to obtain biodistribution information prior to administration of a larger therapeutic activity. The clinical implementability of treatment planning algorithms will depend to a significant extent on the time required to generate absorbed dose estimates for a particular patient. For more details, the reader is referred to chapters 9–15, which fully cover the development steps and clinical experience gained using these tools.

1.6.2 On-line PET monitoring of radiotherapy beams

On-line monitoring of the positron-emitting activity created in patient tissues undergoing radiotherapy treatment has been a goal pursued by several investigators since the 1990s. Researchers at the Research Center of Rossendorf in Dresden [310–312] and at the University of Michigan [313] have made important contributions to this technique in radiotherapy treatments with photons, protons and light and heavy ions. Whereas the clinical application of on-line PET monitoring in photon radiotherapy has so far been limited by the reduced activity produced in a patient using currently available clinical accelerators, its use in heavy-particle radiotherapy has become a useful technique to visualize β^+ activity distributions that help to determine the effective range of heavy particles in the patient, as well as to evaluate blood flow in some organs [314].

The rationale for the preferred use of light- ($Z \leqslant 10$) and heavy-ions over conventional radiation beams is twofold. First, the dose distribution increases along the penetration depth of the beam ending with a sharp maximum at the end of the particle range (the Bragg peak), at the time that the very low scattering properties produce a very narrow penumbra in the beam. Second, there is an intense ionization pattern along the particle path, and notably at the end of its range, which results in localized bursts of energy deposition at microscopic level yielding increased cell killing and thus a radiobiological effect largely superior to that of conventional radiotherapy beams, even in conditions of hypoxia [315–317]. The resolution achieved is of 2–3 mm, which poses strict demands in target localization. Because of their lower production costs, the use of protons in radiotherapy has become well established and today the number of clinical studies published on proton therapy exceeds 800 in the over 90 particle therapy facilities with the number of studies expanding every year. the number of patients treated with this modality worldwide is close to 170 000 with over 75 000 treated in the United States. From the two properties mentioned above, the physical dose distribution of protons, enabling accurate dose conformation delivery, poses a clear advantage over conventional radiotherapy beams; however their radiobiological effect is only about 8%–10% higher than that of photons or electrons. The radiobiological superiority of ions heavier than protons, with a RBE augmentation of the order of 3–4 times higher than that of conventional beams, led to an increased interest worldwide that resulted in the construction of the Japanese HIMAC clinical facility in Chiba, near Tokyo. HIMAC started the treatment of patients mainly with carbon ions in 1994 [318], and more than 1000 patients have been treated in this facility. It had become clear that whereas heavy-ions like the neon beams used in the early seventies in Berkeley have the largest radiobiological effects, these also appear in regions close to the beam entrance and in the plateau region, where usually normal tissue is situated. In addition, due to the large penetration of the fragmentation products released by the incident ions, the tail of the dose distribution beyond the Bragg peak may be too high for sparing normal tissue beyond the primary ion range. These two aspects suggest that the ideal ions for radiotherapy are the light-ions, and for this reason carbon has dominated the clinical applications at HIMAC. The GSI heavy-ion physics research facility in Darmstadt, Germany, initiated clinical treatments also with carbon ions in 1997 [319]. GSI uses advanced 3-D beam scanning techniques and PET imaging to monitor the delivery of dose by visualizing the positrons emitted by ^{10}C and ^{11}C ions produced in the nuclear reactions of the incident carbon ions with tissue, together with radiobiologically optimized treatment planning. Plans for building additional radiotherapy light-ion facilities have followed in Europe, and feasibility studies have been conducted for such facilities in Germany, Austria, Italy, France, Sweden, and Switzerland among others [320].

Several groups have reported the applicability of PET to *in vivo* dosimetry for proton radiotherapy using the Monte Carlo method [321]. Different Monte Carlo codes have been used to investigate this challenging field including Berger's PTRAN [322], used by Del Guerra *et al* [323], FLUKA [151], and specially designed software [313]. Parodi and Enghardt [151], in particular, made Monte Carlo simulations of the number and spatial distribution of the positron emitters created by therapeutic protons in PMMA. Since proton therapy requires a particle fluence up to 20 times higher than that required in carbon therapy for the same physical dose level, the authors concluded that the total activity produced within the irradiated volume should be twice as intense as that actually used at the carbon ion facility at GSI. Moreover, the cdered ratio between the fluence of protons and carbon ions could increase considerably further due to the higher relative biological effectiveness of carbon ions, leading to a lower physical dose than for protons at the same level of biological dose. The field is progressing at a rapid and exciting pace and novel techniques taking advantage of Monte Carlo calculations are being introduced [102, 315, 324–326]. For instance, the potential of Monte Carlo simulations in proton computed tomography (pCT), a promising tomographic imaging modality permitting direct reconstruction of proton relative stopping power required for proton therapy dose calculations was comprehensively reviewed recently [327].

References

[1] Berger M J 1968 Energy deposition in water by photons from point isotropic sources. MIRD Pamphlet No. 2 *J. Nucl. Med.* **6** 15–25

[2] Baum R P 2014 *Therapeutic Nuclear Medicine* (Berlin: Springer)

[3] Andreo P 1981 Monte Carlo techniques in medical radiation physics *Phys. Med. Biol.* **36** 861–920

[4] Buvat I and Castiglioni I 2002 Monte Carlo simulations in SPET and PET *Q. J. Nucl. Med.* **46** 48–61

[5] Stabin M G 2006 Nuclear medicine dosimetry *Phys. Med. Biol.* **51** R187–202

[6] Zaidi H 1999 Relevance of accurate Monte Carlo modeling in nuclear medicine imaging *Med. Phys.* **26** 574–608

[7] Andreo P, Burns D, Nahum A, Seuntjens J and Attix F 2017 *Fundamentals of Ionizing Radiation Dosimetry* (Weinheim: Wiley-VCH) 1000

[8] Jenkins T M, Nelson W R and Rindi A 1998 *Monte Carlo Transport of Electrons and Photons* (New York: Plenum)

[9] Ljungberg M, Strand S-E and King M A (ed) 2013 Monte Carlo calculations in nuclear medicine *Applications in Diagnostic Imaging* 2nd edn (Boca Raton, FL: CRC Press)

[10] Zaidi H and Sgouros G (ed) 2002 *Therapeutic Applications of Monte Carlo Calculations in Nuclear Medicine* (Bristol: Institute of Physics Publishing)

[11] Andreo P, Burns D T, Nahum A E, Seuntjens J and Attix F H 2017 *Fundamentals of Ionizing Radiation Dosimetry* (Weinheim: Wiley-VCH)

[12] Raeside D E 1976 Monte Carlo principles and applications *Phys. Med. Biol.* **21** 181–97

[13] Turner J E, Wright H A and Hamm R N 1985 A Monte Carlo primer for health physicists *Health Phys.* **48** 717–33

[14] Strigari L, Konijnenberg M, Chiesa C, Bardies M, Du Y and Gleisner K S *et al* 2014 The evidence base for the use of internal dosimetry in the clinical practice of molecular radiotherapy *Eur. J. Nucl. Med. Mol. Imaging* **41** 1976–88

[15] Mackie T R 1990 Applications of the Monte Carlo method in radiotherapy ed K R Kase, B E Bjärngard and F H Attix *The Dosimetry of Ionizing Radiation* vol 3 (New York: Academic) pp 541–620

[16] Rogers D W O 2006 Fifty years of Monte Carlo simulations for medical physics *Phys. Med. Biol.* **51** R287–301

[17] Boone J M, Buonocore M H and Cooper V N 2000 Monte Carlo validation in diagnostic radiological imaging *Med. Phys.* **27** 1294–304

[18] Zaidi H 2006 Monte Carlo modeling in nuclear medicine imaging ed H Zaidi *Quantitative Analysis in Nuclear Medicine Imaging. Biosciences* (New York: Springer) pp 358–90

[19] Seltzer S M 1988 An overview of ETRAN Monte Carlo methods ed T M Jenkins, W R Nelson and A Rindi *Monte Carlo Transport of Electrons and Photons* (New York: Plenum) pp 153–81

[20] Halbleib J 1988 Structure and operation of the ITS code system ed T M Jenkins, W R Nelson and A Rindi *Monte Carlo Transport of Electrons and Photons* (New York: Plenum) pp 249–62

[21] Kawrakow I, Mainegra-Hing E, Rogers D W O, Tessier F and Walters B R B 2019 *The EGSnrc Code System: Monte Carlo Simulation of Electron and Photon Transport* (Ottawa: National Research Council Canada)

[22] Salvat F 2018 *PENELOPE-2018: A Code System for Monte Carlo Simulation of Electron and Photon Transport* (Issy-les-Moulineaux: OECD Nuclear Energy Agency)

[23] García-Toraño E, Peyres V and Salvat F 2019 PenNuc: Monte Carlo simulation of the decay of radionuclides *Comp. Phys. Comm.* **245** 106849

[24] Bé M M, Coursol N, Duchemin B, Lagoutine F, Legrand J and Debertin K *et al* 2011 *Table of Radionuclides—Introduction. Saclay* (Gif-sur-Yvette Cedex: Laboratoire National Henri Becquerel)

[25] Kawrakow I 2005 egspp*: the EGSnrc C++ class library* (Ottawa: National Research Council Canada)

[26] Hughes G 2014 Recent developments in low-energy electron/photon transport for MCNP6 *Prog. Nucl. Sci. Technol.* **4** 454–8

[27] Pelowitz D B, Durkee J W, Elson J S, Fensin M L and Hendricks J S 2011 *MCNPX 2.7.0 Extensions* (Los Alamos, NM: Los Alamos National Laboratory)

[28] Agostinelli S, Allison J, Amako K, Apostolakis J and Araujo H 2003 GEANT4—a simulation toolkit *Nucl. Instr. Meth. A.* **506** 250–303

[29] Descourt P, Carlier T, Du Y, Song T Y, Buvat I and Frey E C *et al* 2010 Implementation of angular response function modelling in SPECT simulations with GATE *Phys. Med. Biol.* **55** N253–66

[30] Jan S, Santin G, Strul D, Staelens S, Assié K and Autret D *et al* 2004 GATE: a simulation toolkit for PET and SPECT *Phys. Med. Biol.* **49** 4543–61

[31] España S, Herraiz J L, Vicente E, Vaquero J J, Desco M and Udias J M 2009 PeneloPET, a Monte Carlo PET simulation tool based on PENELOPE: features and validation *Phys. Med. Biol.* **54** 1723–42

[32] Zaidi H 2006 *Quantitative Analysis in Nuclear Medicine Imaging* (New York: Springer)
[33] Ljunberg M and Sjögreen-Gleisner K 2011 The accuracy of absorbed dose estimates in tumours determined by quantitative SPECT: a Monte Carlo study *Acta Oncol.* **50** 981–9
[34] Nelson W R, Hirayama H and Rogers D W O 1985 *The EGS4 Code System* (Standford, CA: Stanford Linear Accelerator Center)
[35] Berger M J 1971 Distribution of absorbed dose around point sources of electrons and beta particles in water and other media MIRD Pamphlet No. 7 *J. Nucl. Med.* **9** 5–23
[36] Simpkin D J and Mackie T R 1990 EGS4 Monte Carlo determination of the beta dose kernel in water *Med. Phys.* **17** 179–86
[37] Maigne L, Perrot Y, Schaart D R, Donnarieix D and Breton V 2011 Comparison of GATE/GEANT4 with EGSnrc and MCNP for electron dose calculations at energies between 15 keV and 20 MeV *Phys. Med. Biol.* **56** 811–27
[38] Lee T, Kim M, Lee W, Kim B, Lim I and Song K *et al* 2019 Performance evaluation of a Compton SPECT imager for determining the position and distribution of ^{225}Ac in targeted alpha therapy: a Monte Carlo simulation based phantom study *Appl. Rad. Isotop.* **154** 108893
[39] Sadremomtaz A and Masoumi M 2019 Comparison between targeted radionuclide therapy of bone metastases based on b-emitting and a-emitting radionuclides *J. Med. Imaging Rad. Sci.* **50** 272–9
[40] Cai Z, Pignol J-P, Chan C and Reilly R M 2010 Cellular dosimetry of ^{111}In using Monte Carlo n-particle computer code: comparison with analytic methods and correlation with in vitro cytotoxicity *J. Nucl. Med.* **51** 462–70
[41] Cai Z, Kwon Y L and Reilly R M 2017 Monte Carlo n-particle (MCNP) modeling of the cellular dosimetry of 64Cu: comparison with MIRDcell S values and implications for studies of its cytotoxic effects *J. Nucl. Med.* **58** 339–45
[42] Freudenberg R and Kotzerke J 2010 Cellular dosimetry using the Geant4 Monte Carlo toolkit *J. Nucl. Med.* **51** 1488–9
[43] Sgouros G and Hobbs R F 2014 Patient-specific dosimetry, radiobiology, and the previously-treated patient ed R P Baum *Therapeutic Nuclear Medicine* (Berlin: Springer) pp 737–45
[44] Ljunberg M and Strand S E 1989 A Monte Carlo program for the simulation of scintillation camera characteristics *Comput. Methods Programs Biomed.* **29** 257–72
[45] Holstensson M, Hindorf C, Ljungberg M, Partridge M and Flux G D 2011 Optimization of energy-window settings for scatter correction in quantitative ^{111}In imaging: comparison of measurements and Monte Carlo simulations *Cancer Biother. Radiophar.* **22** 136–42
[46] Holstensson M, Partridge M, Buckley S E and Flux G D 2010 The effect of energy and source location on gamma camera intrinsic and extrinsic spatial resolution: an experimental and Monte Carlo study *Phys. Med. Biol.* **55** 1735–51
[47] Rong X, Du Y, Ljunberg M, Rault E, Vandenberghe S and Frey E C 2012 Development and evaluation of an improved quantitative ^{90}Y bremsstrahlung SPECT method *Med. Phys.* **39** 2346–58
[48] Harrison R L, Vannoy S D, Haynor D R, Gillipsie S B, Kaplan M S and Lewellen T K 1993 Preliminary experience with the photon history generator module of a public domain simulation system for emission tomography *Conf. Rec. IEEE Nucl. Sci. Symp. Med. Imaging Conf.* **2** 1154–8

[49] Castiglioni I, Cremonesi O, Gilardi M C, Bettinardi V, Rizzo G and Savi A *et al* 1999 Scatter correction techniques in 3D PET: a Monte Carlo evaluation *IEEE Trans. Nucl. Sci.* **46** 2053–8

[50] Buvat I, Castiglioni I, Feuardent J and Gilardi M C 2005 Unified description and validation of Monte Carlo simulators in PET *Phys. Med. Biol.* **50** 329–46

[51] Clairand I, Ricard M, Gouriou J, Di Paola M and Aubert B 1999 DOSE3D: EGS4 Monte Carlo code-based software for internal radionuclide dosimetry *J. Nucl. Med.* **40** 1517–23

[52] Kolbert K S, Sgouros G, Scott A M, Bronstein J E, Malane R A and Zhang J *et al* 1997 Implementation and evaluation of patient-specific three-dimensional internal dosimetry *J. Nucl. Med.* **38** 301–8

[53] Dewaraja Y K, Wilderman S J, Ljungberg M, Koral K F, Kenneth Z and Kaminiski M S 2005 Accurate dosimetry in ^{131}I radionuclide therapy using patient-specific, 3-dimensional methods for SPECT reconstruction and absorbed dose calculation *J. Nucl. Med.* **46** 840–9

[54] Nelson W R, Hirayama H and Rogers D W O 1985 *The EGS4 Code System* (Stanford, CA: Stanford Linear Accelerator Center) Report No.: SLAC-256

[55] Kawrakow I, Mainegra-Hing E and Rogers D W O 2003 *The EGSnrc Code System: Monte Carlo Simulation of Electron and Photon Transport* (Ottawa: Ionizing Radiation Standards, National Research Council of Canada) NRCC Report PIRS-701

[56] Adam L E, Karp J S and Brix G 1999 Investigation of scattered radiation in 3D whole-body positron emission tomography using Monte Carlo simulations *Phys. Med. Biol.* **44** 2879–95

[57] Adam L E, Karp J S and Freifelder R 2000 Energy-based scatter correction for 3-D PET scanners using NaI(T1) detectors *IEEE Trans. Med. Imaging* **19** 513–21

[58] Bollini D, Del Guerra A, Di Domenico G, Galli G, Gambaccini M and Zavattini G 1997 Sub-millimeter planar imaging with positron emitters: EGS4 code simulation and experimental results *IEEE Trans. Nucl. Sci.* **44** 1499–502

[59] Castiglioni I, Cremonesi O, Gilardi M-C, Bettinardi V, Rizzo G and Savi A *et al* 1999 Scatter correction techniques in 3D PET: a Monte Carlo evaluation *IEEE Trans. Nucl. Sci.* **46** 2053–8

[60] Castiglioni I, Cremonesi O, Gilardi M-C, Savi A, Bettinardi V and Rizzo G *et al* 2002 A Monte Carlo model of noise components in 3D PET *IEEE Trans. Nucl. Sci.* **49** 2297–303

[61] Cheng J C K, Boellaard R and Laforest R 2015 Evaluation of the effect of magnetic field on PET spatial resolution and contrast recovery using clinical PET scanners and EGSnrc simulations *IEEE Trans. Nucl. Sci.* **62** 101–10

[62] Conti M, Casey M E, Eriksson L and Eriksson M 2002 Benchmarking a Monte Carlo simulation code on a prototype LSO scanner *IEEE Trans. Nucl. Sci.* **49** 644–8

[63] Del Guerra A and Nelson W R 1988 Positron emission tomography applications of EGS ed T Jenkins M NWRaRA *Monte Carlo Transport of Electrons and Photons* (New York: Plenum) pp 469–84

[64] Denis-Bacelar A M, Romanchikova M, Chittenden S, Saran F H, Mandeville H and Du Y *et al* 2013 Patient-specific dosimetry for intracavitary 32P-chromic phosphate colloid therapy of cystic brain tumours *Eur. J. Nucl. Med. Mol. Imaging* **40** 1532–41

[65] Di Domenico G, Zavattini G, Cesca N, Auricchio N, Andritschke R and Schopper F *et al* 2007 SiliPET: an ultra-high resolution design of a small animal PET scanner based on stacks of double-sided silicon strip detector *Nucl. Instr. Meth. A.* **571** 22–5

[66] Fabbri C, Sarti G, Cremonesi M, Ferrari M, Di Dia A and Agostini M *et al* 2009 Quantitative analysis of 90Y Bremsstrahlung SPECT-CT images for application to 3D patient-specific dosimetry *Cancer Biother. Radiopharm.* **24** 145–54
[67] Hadid L, Desbrée A, Schlattl H, Franck D, Blanchardon E and Zankl M 2010 Application of the ICRP/ICRU reference computational phantoms to internal dosimetry: calculation of specific absorbed fractions of energy for photons and electrons *Phys. Med. Biol.* **55** 3631–41
[68] Hough M, Johnson P, Rajon D, Jokisch D, Lee C and Bolch W 2011 An image-based skeletal dosimetry model for the ICRP reference adult male-internal electron sources *Phys. Med. Biol.* **56** 2309–46
[69] Motta A, Righi S, Guerra A D, Belcari N, Vaiano A and Domenico G D *et al* 2004 A full Monte Carlo simulation of the YAP-PEM prototype for breast tumor detection *Nucl. Instr. Meth. A.* **527** 201–5
[70] Narita Y, Eberl S, Iida H, Hutton B F, Braun M and Nakamura T *et al* 1996 Monte Carlo and experimental evaluation of accuracy and noise properties of two scatter correction methods for SPECT *Phys. Med. Biol.* **41** 2481–96
[71] Narita Y, Iida H, Eberl S and Nakamura T 1997 Monte Carlo evaluation of accuracy and noise properties of two scatter correction methods for 201Tl cardiac SPECT *IEEE Trans. Nucl. Sci.* **44** 2465–72
[72] Pacilio M, Lanconelli N, Lo M S, Betti M, Montani L and Torres A L *et al* 2009 Differences among Monte Carlo codes in the calculations of voxel S values for radionuclide targeted therapy and analysis of their impact on absorbed dose evaluations *Med. Phys.* **36** 1543–52
[73] Padilla L, Lee C, Milner R, Shahlaee A and Bolch W E 2008 Canine anatomic phantom for preclinical dosimetry in internal emitter therapy *J. Nucl. Med.* **49** 446–52
[74] Pafundi D, Rajon D, Jokisch D, Lee C and Bolch W 2011 An image-based skeletal dosimetry model for the ICRP reference newborn--internal electron sources *Phys. Med. Biol.* **55** 1785–814
[75] Palmer M R, Zhu X and Parker J A 2005 Modeling and simulation of positron range effects for high resolution PET imaging *IEEE Trans. Nucl. Sci.* **52** 1391–5
[76] Petoussi-Henss N, Schlattl H, Zankl M, Endo A and Saito K 2012 Organ doses from environmental exposures calculated using voxel phantoms of adults and children *Phys. Med. Biol.* **57** 5679–713
[77] Zankl M, Schlattl H, Petoussi-Henss N and Hoeschen C 2012 Electron specific absorbed fractions for the adult male and female ICRP/ICRU reference computational phantoms *Phys. Med. Biol.* **57** 4501–26
[78] Zhang L, Staelens S, Van Holen R, De Beenhouwer J, Verhaeghe J and Kawrakow I *et al* 2010 Fast and memory-efficient Monte Carlo-based image reconstruction for whole-body PET *Med. Phys.* **37** 3667–76
[79] Halbleib J A, Kensek R P, Valdez G D, Seltzer S M and Berger M J 1992 ITS: the integrated TIGER Series of electron/photon transport codes—version 3.0 *IEEE Trans. Nucl. Sci.* **39** 1025–30
[80] Briesmeister J F 2000 *MCNP—A General Monte Carlo N-Particle Transport Code. Version 4C* (Los Alamos, NM: Los Alamos National Laboratory) Report No.: LA-13709-M
[81] Hendricks J S, McKinney G W, Waters L S, Roberts T L, Egdorf H W and Finch J P *et al* 2004 *MCNPX, VERSION 2.5.e* (Los Alamos, NM: Los Alamos National Laboratory, NM) Report No.: LA-UR-04–0569

[82] Asgari A, Ashoor M, Sohrabpour M, Shokrani P and Rezaei A 2015 Evaluation of various energy windows at different radionuclides for scatter and attenuation correction in nuclear medicine *Ann. Nucl. Med.* **29** 375–83

[83] Barquero R, Garcia H P, Incio M G, Minguez P, Cardenas A and Martinez D *et al* 2017 131I activity quantification of gamma camera planar images *Phys. Med. Biol.* **62** 909–26

[84] Chen Y, Qiu R, Li C, Wu Z and Li J 2016 Construction of Chinese adult male phantom library and its application in the virtual calibration of in vivo measurement *Phys. Med. Biol.* **61** 2124–44

[85] Dieudonne A, Hobbs R F, Bolch W E, Sgouros G and Gardin I 2010 Fine-resolution voxel S values for constructing absorbed dose distributions at variable voxel size *J. Nucl. Med.* **51** 1600–7

[86] Divoli A, Chiavassa S, Ferrer L, Barbet J, Flux G D and Bardies M 2009 Effect of patient morphology on dosimetric calculations for internal irradiation as assessed by comparisons of Monte Carlo versus conventional methodologies *J. Nucl. Med.* **50** 316–23

[87] Du Y, Frey E C, Wang W T, Tocharoenchai C, Baird W H and Tsui B M W 2002 Combination of MCNP and SimSET for Monte Carlo simulation of SPECT with medium- and high-energy photons *IEEE Trans. Nucl. Sci.* **49** 668–74

[88] Elbast M, Saudo A, Franck D, Petitot F and Desbree A 2012 Microdosimetry of alpha particles for simple and 3D voxelised geometries using MCNPX and Geant4 Monte Carlo codes *Radiat. Prot. Dosim.* **150** 342–9

[89] Elschot M, de Wit T C and de Jong H W A M 2010 The influence of self-absorption on PET and PET/CT shielding requirements *Med. Phys.* **37** 2999–3007

[90] Fernandez M, Hanscheid H, Mauxion T, Bardies M, Kletting P and Glatting G *et al* 2013 A fast method for rescaling voxel S values for arbitrary voxel sizes in targeted radionuclide therapy from a single Monte Carlo calculation *Med. Phys.* **40** 082502–8

[91] Jacquemin M, Broggio D, Franck D and Desbree A 2019 Development of a dosimetric model for in vitro labelled cells with beta (+) emitters in PET tracking studies *Phys. Med. Biol.* **64** 155015

[92] Konijnenberg M, Melis M, Valkema R, Krenning E and de Jong M 2007 Radiation dose distribution in human kidneys by octreotides in peptide receptor radionuclide therapy *J. Nucl. Med.* **48** 134–42

[93] Lanconelli N, Pacilio M, Lo Meo S, Botta F, Di Dia A, Torres and Aroche L A *et al* 2012 A free database of radionuclide voxel S values for the dosimetry of nonuniform activity distributions *Phys. Med. Biol.* **57** 517–33

[94] Larsson E, Ljungberg M, Strand S E and Jonsson B A 2011 Monte Carlo calculations of absorbed doses in tumours using a modified MOBY mouse phantom for pre-clinical dosimetry studies *Acta Oncol.* **50** 973–80

[95] Lehmann J, Hartmann Siantar C, Wessol D E, Wemple C A, Nigg D and Cogliati J *et al* 2005 Monte Carlo treatment planning for molecular targeted radiotherapy within the MINERVA system *Phys. Med. Biol.* **50** 947–58

[96] Mauxion T, Barbet J, Suhard J, Pouget J-P, Poirot M and Bardies M 2013 Improved realism of hybrid mouse models may not be sufficient to generate reference dosimetric data *Med. Phys.* **40** 052501–11

[97] Miller W H, Hartmann-Siantar C, Fisher D, Descalle M A, Daly T and Lehmann J *et al* 2005 Evaluation of beta-absorbed fractions in a mouse model for ^{90}Y, ^{188}Re, ^{166}Ho, ^{149}Pm, ^{64}Cu, and ^{177}Lu radionuclides *Cancer Biother. Radiopharm.* **20** 436–49

[98] Mok G S, Du Y, Wang Y, Frey E C and Tsui B M 2010 Development and validation of a Monte Carlo simulation tool for multi-pinhole SPECT *Mol. Imaging Biol.* **12** 295–304
[99] Olguin E, President B, Ghaly M, Frey E, Sgouros G and Bolch W E 2020 Specific absorbed fractions and radionuclide S-values for tumors of varying size and composition *Phys. Med. Biol.* **65** 235015
[100] Petitguillaume A, Bernardini M, Hadid L, de Labriolle-Vaylet C, Franck D and Desbree A 2014 Three-dimensional personalized Monte Carlo dosimetry in ^{90}Y resin microspheres therapy of hepatic metastases: nontumoral liver and lungs radiation protection considerations and treatment planning optimization *J. Nucl. Med.* **55** 405–13
[101] Sanchez-Garcia M, Gardin I, Lebtahi R and Dieudonne A 2015 Implementation and validation of collapsed cone superposition for radiopharmaceutical dosimetry of photon emitters *Phys. Med. Biol.* **60** 7861–76
[102] Seravalli E, Robert C, Bauer J, Stichelbaut F, Kurz C and Smeets J *et al* 2012 Monte Carlo calculations of positron emitter yields in proton radiotherapy *Phys. Med. Biol.* **57** 1659–73
[103] Sheu R D, Chen C C, Sheu R J, Kao C H and Jiang S H 2005 The refined shielding design for the cyclotron room of the Buddhist Tzu Chi General Hospital *Radiat. Prot. Dosimetry.* **115** 216–21
[104] Song H, He B, Prideaux A, Du Y, Frey E and Kasecamp W *et al* 2006 Lung dosimetry for radioiodine treatment planning in the case of diffuse lung metastases *J. Nucl. Med.* **47** 1985–94
[105] Song T Y, Choi Y, Chung Y H, Jung J H, Choe Y S and Lee K-H *et al* 2003 Optimization of pinhole collimator for small animal SPECT using Monte Carlo simulation *IEEE Trans. Nucl. Sci.* **50** 327–32
[106] Stabin M G and Yoriyaz H 2002 Photon specific absorbed fractions calculated in the trunk of an adult male voxel-based phantom *Health Phys.* **82** 21–44
[107] Wayson M, Lee C, Sgouros G, Treves S T, Frey E and Bolch W E 2012 Internal photon and electron dosimetry of the newborn patient-a hybrid computational phantom study *Phys. Med. Biol.* **57** 1433–57
[108] Xie T, Liu Q and Zaidi H 2012 Evaluation of S-values and dose distributions for (90)Y, (131)I, (166)Ho, and (188)Re in seven lobes of the rat liver *Med. Phys.* **39** 1462–72
[109] Xie T, Park J S, Zhuo W and Zaidi H 2020 Development of a nonhuman primate computational phantom for radiation dosimetry *Med. Phys.* **47** 736–44
[110] Xie T and Zaidi H 2014 Effect of respiratory motion on internal radiation dosimetry *Med. Phys.* **41** 112506–10
[111] Xie T, Zanotti-Fregonara P, Edet-Sanson A and Zaidi H 2018 Patient-specific computational model and dosimetry calculations for a patient pregnant with twins undergoing a PET/CT examination *J. Nucl. Med.* **59** 1451–8
[112] Yanch J C and Dobrzeniecki A B 1993 Monte Carlo simulation in SPECT: complete 3-D modeling of source, collimator and tomographic data acquisition *IEEE Trans. Nucl. Sci.* **40** 198–203
[113] Yanch J C, Dobrzeniecki A B, Ramanathan C and Behrman R 1992 Physically realistic Monte Carlo simulation of source, collimator and tomographic data acquisition for emission computed tomography *Phys. Med. Biol.* **37** 853–70
[114] Yoriyaz H, Moralles M, Siqueira Pde T, Guimaraes Cda C, Cintra F B and dos Santos A 2009 Physical models, cross sections, and numerical approximations used in MCNP and

GEANT4 Monte Carlo codes for photon and electron absorbed fraction calculation *Med. Phys.* **36** 5198–213

[115] Agostinelli S, Allison J, Amako K, Apostolakis J, Araujo H and Arce P *et al* 2003 GEANT4-a simulation toolkit *Nucl. Instrum. Meth. A.* **506** 250–303

[116] Allison J, Amako K, Apostolakis J, Araujo H, Arce Dubois P and Asai M *et al* 2006 Geant4 developments and applications *IEEE Trans. Nucl. Sci.* **53** 270–8

[117] Allison J, Amako K, Apostolakis J, Arce P, Asai M and Aso T *et al* 2016 Recent developments in Geant4 *Nucl. Instr. Meth.* A **835** 186–225

[118] Besemer A E, Titz B, Grudzinski J J, Weichert J P, Kuo J S and Robins H I *et al* 2017 Impact of PET and MRI threshold-based tumor volume segmentation on patient-specific targeted radionuclide therapy dosimetry using CLR1404 *Phys. Med. Biol.* **62** 6008–25

[119] Besemer A E, Yang Y M, Grudzinski J J, Hall L T and Bednarz B P 2018 Development and validation of RAPID: a patient-specific Monte Carlo three-dimensional internal dosimetry platform *Cancer Biother. Radiopharm.* **33** 155–65

[120] Branco S, Jan S and Almeida P 2007 Monte Carlo simulations in small animal PET imaging *Nucl. Instrum. Meth.* A **580** 1127–30

[121] Gottschlag H, Kösters T, Vernekohl D, Reygers K, Schäfers K and Wübbeling F *et al* 2013 Towards quantitative image reconstruction using Monte-Carlo simulations in multi-wire proportional chamber-based small animal PET. Nuclear Science *IEEE Trans.* **60** 3343–54

[122] Handler W B, Gilbert K M, Peng H and Chronik B A 2006 Simulation of scattering and attenuation of 511 keV photons in a combined PET/field-cycled MRI system *Phys. Med. Biol.* **51** 2479–91

[123] Jan S, Benoit D, Becheva E, Carlier T, Cassol F and Descourt P *et al* 2011 GATE V6: a major enhancement of the GATE simulation platform enabling modelling of CT and radiotherapy *Phys. Med. Biol.* **56** 881–901

[124] Jan S, Santin G, Strul D, Staelens S, Assie K and Autret D *et al* 2004 GATE: a simulation toolkit for PET and SPECT *Phys. Med. Biol.* **49** 4543–61

[125] Kost S D, Dewaraja Y K, Abramson R G and Stabin M G 2015 VIDA: a voxel-based dosimetry method for targeted radionuclide therapy using Geant4 *Cancer Biother. Radiopharm.* **30** 16–26

[126] Kraus R, Delso G and Ziegler S I 2012 Simulation study of tissue-specific positron range correction for the new Biograph mMR whole-body PET/MR system *IEEE Trans. Nucl. Sci.* **59** 1900–9

[127] McNamara A L, Toghyani M, Gillam J E, Wu K and Kuncic Z 2014 Towards optimal imaging with PET: an in silico feasibility study *Phys. Med. Biol.* **59** 7587–600

[128] Michel C, Bol A, Spinks T, Townsend D W, Bailey D and Grootoonk S *et al* 1991 Assessment of response function in two PET scanners with and without interplane septa *IEEE Trans. Med. Imaging* **10** 240–8

[129] Moghaddasi L, Bezak E and Harriss-Phillips W 2016 Monte-Carlo model development for evaluation of current clinical target volume definition for heterogeneous and hypoxic glioblastoma *Phys. Med. Biol.* **61** 3407–26

[130] Prasad R and Zaidi H 2014 Scatter characterization and correction for simultaneous multiple small-animal PET imaging *Mol. Imaging Biol.* **16** 199–209

[131] Robert C, Dedes G, Battistoni G, Bohlen T T, Buvat I and Cerutti F *et al* 2013 Distributions of secondary particles in proton and carbon-ion therapy: a comparison between GATE/Geant4 and FLUKA Monte Carlo codes *Phys. Med. Biol.* **58** 2879–99

[132] Rodríguez-Villafuerte M, Yang Y and Cherry S R 2014 A Monte Carlo investigation of the spatial resolution performance of a small-animal PET scanner designed for mouse brain imaging studies *Phys. Med.* **30** 76–85

[133] Salim R and Taherparvar P 2020 Cellular S values in spindle-shaped cells: a dosimetry study on more realistic cell geometries using Geant4-DNA Monte Carlo simulation toolkit *Ann. Nucl. Med.* **34** 742–56

[134] Sanaat A, Arabi H, Ay M and Zaidi H 2020 Novel preclinical PET geometrical concept using a monolithic scintillator crystal offering concurrent enhancement in spatial resolution and detection sensitivity: a simulation study *Phys. Med. Biol.* **65** 045013

[135] Sanaat A, Zafarghandi M S and Ay M R 2019 Design and performance evaluation of high resolution small animal PET scanner based on monolithic crystal: a simulation study *J. Instrum.* **14** P01005-P

[136] Sarrut D, Bala M, Bardies M, Bert J, Chauvin M and Chatzipapas K *et al* 2021 Advanced Monte Carlo simulations of emission tomography imaging systems with GATE *Phys. Med. Biol.* **66** 10TR03

[137] Surti S, Zou W, Daube-Witherspoon M E, McDonough J and Karp J S 2011 Design study of an in situ PET scanner for use in proton beam therapy *Phys. Med. Biol.* **56** 2667–85

[138] Wegmann K, Adam L-E, Livieratos L, Zaers J and Brix L B D G 1999 Investigation of the scatter contribution in single photon transmission measurements by means of Monte Carlo simulations *IEEE Trans. Nucl. Sci.* **46** 1184–90

[139] Zeraatkar N, Ay M R, Kamali-Asl A R and Zaidi H 2011 Accurate Monte Carlo modeling and performance assessment of the X-PET subsystem of the FLEX Triumph preclinical PET/CT scanner *Med. Phys.* **38** 1217–25

[140] Salvat F, Fernandez-Varea J, Costa E and Sempau J 2001 PENELOPE—A code system for Monte Carlo simulation of electron and photon transport. *Workshop Proceedings, Issy-les-Moulineaux, France, 5–7 November 2001*

[141] Sempau J, Acosta E, Baro J, Fernandez-Varea J M and Salvat F 1997 An algorithm for Monte Carlo simulation of the coupled electron-photon transport *Nucl. Instr. Meth. B.* **132** 377–90

[142] Abushab K M, Herraiz J L, Vicente E, Cal-González J, Espana S and Vaquero J J *et al* 2016 Evaluation of PeneloPET simulations of Biograph PET/CT scanners *IEEE Trans. Nucl. Sci.* **63** 1367–74

[143] Cot A, Sempau J, Pareto D, Bullich S, Pavia J and Calvino F *et al* 2002 Evaluation of the geometric, scatter, and septal penetration components in fan-beam collimators using Monte Carlo simulation *IEEE Trans. Nucl. Sci.* **49** 12–6

[144] Cot A, Sempau J, Pareto D, Bullich S, Pavia J and Calvino F *et al* 2004 Study of the point spread function (PSF) for 123I SPECT imaging using Monte Carlo simulation *Phys. Med. Biol.* **49** 3125–36

[145] Espana S, Herraiz J L, Vicente E, Vaquero J J, Desco M and Udias J M 2009 PeneloPET, a Monte Carlo PET simulation tool based on PENELOPE: features and validation *Phys. Med. Biol.* **54** 1723–42

[146] Sanchez-Crespo A, Andreo P and Larsson S A 2004 Positron flight in human tissues and its influence on PET image spatial resolution *Eur. J. Nucl. Med. Mol. Imaging* **31** 44–51

[147] Aiginger H, Andersen V, Ballarini F, Battistoni G, Campanella M and Carboni M *et al* 2005 The FLUKA code: new developments and application to 1 GeV/n iron beams *Adv. Space Res.* **35** 214–22

[148] Fasso A, Ferrari A and Sala P R 2001 Electron-photon transport in FLUKA: status ed A Kling, F Barao, M Nakagawa, L Tavora and P Vaz *Proceedings of the Monte Carlo 2000 Conference* (Berlin: Springer) pp 159–64

[149] Cambraia Lopes P, Bauer J, Salomon A, Rinaldi I, Tabacchini V and Tessonnier T *et al* 2016 First in situ TOF-PET study using digital photon counters for proton range verification *Phys. Med. Biol.* **61** 6203–30

[150] Fourkal E, Veltchev I, Lin M, Koren S, Meyer J and Doss M *et al* 2013 3D inpatient dose reconstruction from the PET-CT imaging of 90Y microspheres for metastatic cancer to the liver: feasibility study *Med. Phys.* **40** 081702–10

[151] Parodi K and Enghardt W 2000 Potential application of PET in quality assurance of proton therapy *Phys. Med. Biol.* **45** N151–6

[152] Parodi K, Enghardt W, Haberer T and In-beam P E T 2002 measurements of beta+ radioactivity induced by proton beams *Phys. Med. Biol.* **47** 21–36

[153] Parodi K, Paganetti H, Cascio E, Flanz J B, Bonab A A and Alpert N M *et al* 2007 PET/CT imaging for treatment verification after proton therapy: a study with plastic phantoms and metallic implants *Med. Phys.* **34** 419–35

[154] Sommerer F, Cerutti F, Parodi K, Ferrari A, Enghardt W, Aiginger H and In-beam P E T 2009 Monitoring of mono-energetic ^{16}O and ^{12}C beams: experiments and FLUKA simulations for homogeneous targets *Phys. Med. Biol.* **54** 3979–96

[155] Harrison R L, Vannoy S D, Haynor D R, Gillispie S B, Kaplan M S and Lewellen T K (ed) Preliminary experience with the photon history generator module for a public-domain simulation system for emission tomography *1993 Records of IEEE Nuclear Science Symposium and Medical Imaging Conference* 1993 (*San Francisco*) (New York: IEEE)

[156] Lewellen T, Harrison R L and Vannoy S 1998 The SIMSET program ed M Ljungberg, S-E Strand and M A King *Monte Carlo Calculations in Nuclear Medicine: Applications in Diagnostic Imaging. Series in Medical Physics and Biomedical Engineering* (Bristol: Institute of Physics Publishing) pp 77–92

[157] Ljungberg M 1998 The SIMIND Monte Carlo program ed M Ljungberg, S-E Strand and M A King *Monte Carlo Calculations in Nuclear Medicine: Applications in Diagnostic Imaging. Series in Medical Physics and Biomedical Engineering* (Bristol: Institute of Physics Publishing) pp 145–63

[158] Ljungberg M and Strand S-E 1989 A Monte Carlo program for the simulation of scintillation camera characteristics *Comput. Methods Programs Biomed.* **29** 257–72

[159] de Vries D and Moore S 1998 Monte Carlo simulation of photon transport in gamma camera collimators ed M Ljungberg, S-E Strand and M A King *Monte Carlo Calculations in Nuclear Medicine: Applications in Diagnostic Imaging. Series in Medical Physics and Biomedical Engineering* (Bristol: Institute of Physics Publishing) pp 125–44

[160] De Vries D J, Moore S C, Zimmerman R E, Mueller S P, Friedland B and Lanza R C 1990 Development and validation of a Monte Carlo simulation of photon transport in an Anger camera *EEE Trans. Med. Imaging.* **9** 430–8

[161] Smith M F 1998 Vectorized Monte Carlo code for modelling photon transport in nuclear medicine ed M Ljungberg, S-E Strand and M A King *Monte Carlo Calculations in Nuclear Medicine: Applications in Diagnostic Imaging. Series in Medical Physics and Biomedical Engineering* (Bristol: Institute of Physics Publishing) pp 93–109

[162] Smith M F, Floyd C E and Jaszczak R J 1993 A vectorized Monte Carlo code for modeling photon transport in SPECT *Med. Phys.* **20** 1121–7

[163] Thompson C and Picard Y 1998 PETSIM: Monte Carlo simulation of positron imaging systems ed M Ljungberg, S-E Strand and M A King *Monte Carlo Calculations in Nuclear Medicine: Applications in Diagnostic Imaging. Series in Medical Physics and Biomedical Engineering* (Bristol: Institute of Physics Publishing) pp 233–48
[164] Thompson C J, Cantu J-M and Picard Y 1992 PETSIM: Monte Carlo program simulation of all sensitivity and resolution parameters of cylindrical positron imaging systems *Phys. Med. Biol.* **37** 731–49
[165] Zaidi H, Labbé C and Morel C 1998 Implementation of an environment for Monte Carlo simulation of fully 3D positron tomography on a high-performance parallel platform *Parallel Comput.* **24** 1523–36
[166] Zaidi H, Scheurer A H and Morel C 1999 An object-oriented Monte Carlo simulator for 3D cylindrical positron tomographs *Comput. Methods Programs Biomed.* **58** 133–45
[167] Reilhac A, Lartizien C, Costes N, Sans S, Comtat C and Gunn R N *et al* 2004 PET-SORTEO: a Monte Carlo-based simulator with high count rate capabilities *IEEE Trans. Nucl. Sci.* **51** 46–52
[168] Cañadas M, Arce P and Mendes P R 2010 Validation of a small-animal PET simulation using GAMOS: a GEANT4-based framework *Phys. Med. Biol.* **56** 273–88
[169] Levin V, Andreo P and Dodd B 2001 *Current Status of Neutron Capture Therapy* (Vienna: International Atomic Energy Agency)
[170] Savolainen S, Kortesniemi M, Timonen M, Reijonen V, Kuusela L and Uusi-Simola J *et al* 2013 Boron neutron capture therapy (BNCT) in Finland: technological and physical prospects after 20 years of experiences *Phys. Med.* **29** 233–48
[171] Minsky D M, Valda A A, Kreiner A J, Green S, Wojnecki C and Ghani Z 2011 First tomographic image of neutron capture rate in a BNCT facility *Appl. Radiat. Isot.* **69** 1858–61
[172] ICRU 1989 *Tissue Substitutes in Radiation Dosimetry and Measurement* (Bethesda, MD: International Commission on Radiation Units and Measurements)
[173] Bar E, Andreo P, Lalonde A, Royle G and Bouchard H 2018 Optimized I-values for use with the Bragg additivity rule and their impact on proton stopping power and range uncertainty *Phys. Med. Biol.* **63** 165007
[174] ICRU 1984 *Stopping Powers for Electrons and Positrons* (Bethesda, MD: International Commission on Radiation Units and Measurements)
[175] Andreo P 2015 Dose to 'water-like' media or dose to tissue in MV photons radiotherapy treatment planning: still a matter of debate *Phys. Med. Biol.* **60** 309–37 Corrigendum in 2619
[176] Zaidi H 1999 Relevance of accurate Monte Carlo modeling in nuclear medical imaging *Med. Phys.* **26** 574–608
[177] Zaidi H 1996 Quantitative SPECT: recent developments in detector response, attenuation and scatter correction techniques *Phys. Med.* **XII** 101–17
[178] Webb S, Binnie D M, Flower M A and Ott R J 1992 Monte Carlo modelling of the performance of a rotating slit-collimator for improved planar gamma-camera imaging *Phys. Med. Biol.* **37** 1095–108
[179] Holstensson M, Smedby O, Poludniowski G, Sanchez-Crespo A, Savitcheva I and Oberg M *et al* 2019 Comparison of acquisition protocols for ventilation/perfusion SPECT-a Monte Carlo study *Phys. Med. Biol.* **64** 235018
[180] Dahlbom M, Rosenquist G, Eriksson L and Bohm C 1989 A study of the possibility of using multi-slice PET systems for 3D imaging *IEEE Trans. Nucl. Sci.* **36** 1066–71

[181] Keller N A and Lupton J R 1983 PET detector ring aperture function calculations using Monte Carlo techniques *IEEE Trans. Nucl. Sci.* **30** 676–80

[182] Lupton L R and Keller N A 1983 Performance study of single-slice positron emission tomography scanners by Monte Carlo techniques *IEEE Trans. Med. Imaging* **2** 154–68

[183] Paredes-Pacheco J, Lopez-Gonzalez F J, Silva-Rodriguez J, Efthimiou N, Ninerola-Baizan A and Ruibal A *et al* 2021 SimPET—An open online platform for the Monte Carlo simulation of realistic brain PET data. Validation for (18) F-FDG scans *Med. Phys.* **48** 2482–93

[184] Derenzo S E 1981 Monte Carlo calculations of the detection efficiency of NaI(Tl), BGO, CsF, Ge and plastic detectors for 511 keV photons *IEEE Trans. Nucl. Sci.* **28** 131–6

[185] Derenzo S E and Riles J K 1982 Monte Carlo calculations of the optical coupling between bismuth germanate crystals and photomultiplier tubes *IEEE Trans. Nucl. Sci.* **29** 191–5

[186] Bottigli U, Guzzardi R, Mey M, Bellazzini R, Giannetti P and Giorgi M A *et al* 1985 Monte Carlo simulation and experimental tests on BGO, CsF and NaI(Tl) crystals for positron emission tomography *J. Nucl. Med. Allied Sci.* **29** 221–7

[187] Bice A N, Lewellen T K, Miyaoka R S, Harrison R L, Haynor D R and Pollard K R *et al* 1990 Monte Carlo simulation of BaF2 detectors used in time-of-flight positron emission tomography *IEEE Trans. Nucl. Sci.* **37** 696–701

[188] Lopes M I, Chepel V, Carvalho J C, Ferreira Marques R and Policarpo A J P 1995 Performance analysis based on a Monte Carlo simulation of a liquid Xenon PET detector *IEEE Trans. Nucl. Sci.* **42** 2298–302

[189] Eriksson L, Townsend D, Eriksson M, Melcher C, Schmand M and Bendriem B *et al* 2004 Experience with scintillators for PET: towards the fifth generation of PET scanners *Nucl. Instr. Meth.* A **525** 242–8

[190] Ghabrial A, Franklin D and Zaidi H 2018 A Monte Carlo simulation study of the impact of novel scintillation crystals on performance characteristics of PET scanners *Phys. Med.* **50** 37–45

[191] Bradshaw J, Burnham C and Correia J 1985 Application of Monte Carlo methods to the design of SPECT detector systems *IEEE Trans. Nucl. Sci.* **32** 753–7

[192] Boutaghane N, Bouzid B and Zaidi H 2019 Conceptual design of a large pixelated CZT detector with four-hole collimator matched pixel detector for SPECT imaging: a Monte Carlo simulation study *J. Instrum.* **14** P02026-P

[193] Kimiaei S, Ljungberg M and Larsson S A 1997 Evaluation of optimally designed planar-concave collimators in single-photon emission tomography *Eur. J. Nucl. Med.* **24** 1398–404

[194] Peterson M, Strand S E and Ljungberg M 2015 Using Rose's metal alloy as a pinhole collimator material in preclinical small-animal imaging: a Monte Carlo evaluation *Med. Phys.* **42** 1698–709

[195] Moses W W, Virador P R G, Derenzo S E, Huesman R H and Budinger T F 1997 Design of a high-resolution, high-sensitivity PET camera for human brains and small animals *IEEE Trans. Nucl. Sci.* **41** 1487–91

[196] Pavlopoulos S and Tzanakos G 1996 Design and performance evaluation of a high-resolution small animal positron tomograph *IEEE Trans. Nucl. Sci.* **43** 3249–55

[197] Zaidi H 2000 Comparative evaluation of scatter correction techniques in 3D positron emission tomography *Eur. J. Nucl. Med.* **27** 1813–26

[198] Hoffman E J, Cutler P D, Digby W M and Mazziotta J C 1990 3-D phantom to simulate cerebral blood flow and metabolic images for PET *IEEE Trans. Nucl. Sci.* **37** 616–20

[199] Smith M F, Floyd C E, Jaszczak R J and Coleman R E 1992 Three-dimensional photon detection kernels and their application to SPECT reconstruction *Phys. Med. Biol.* **37** 605–22
[200] Dunn W L 1981 Inverse Monte Carlo analysis *J. Comput. Phys.* **41** 154–66
[201] Floyd C E, Jaszczak R J, Greer K L and Coleman R E 1986 Inverse Monte Carlo as a unified reconstruction algorithm for ECT *J. Nucl. Med.* **27** 1577–85
[202] Beekman F J, de Jong H W A M and van Geloven S 2002 Efficient fully 3-D iterative SPECT reconstruction with Monte Carlo-based scatter compensation *IEEE Trans. Med. Imaging* **21** 867–77
[203] Lazaro D, El Bitar Z, Breton V, Hill D and Buvat I 2005 Fully 3D Monte Carlo reconstruction in SPECT: a feasibility study *Phys. Med. Biol.* **50** 3739–54
[204] Pollard K R, Bice A N, Eary J F, Durack L D and Lewellen T K 1992 A method for imaging therapeutic doses of iodine-131 with a clinical gamma camera *J. Nucl. Med.* **33** 771–6
[205] Dewaraja Y K, Ljungberg M and Koral K F 2000 Accuracy of 131I tumor quantification in radioimmunotherapy using SPECT imaging with an ultra-high-energy collimator: Monte Carlo study *J. Nucl. Med.* **41** 1760–7
[206] Zaidi H 1996 Assessment of thyroid volume with pinhole emission computed tomography *Phys. Med.* **XII2** 97–100
[207] Huili W, Jaszczak R J and Coleman R E 1996 Monte Carlo modeling of penetration effect for iodine-131 pinhole imaging *IEEE Trans. Nucl. Sci.* **43** 3272–7
[208] Smith M F and Jaszczak R J 1997 The effect of gamma ray penetration on angle-dependent sensitivity for pinhole collimation in nuclear medicine *Med. Phys.* **24** 1701–9
[209] Dresser M M and Knoll G F 1973 Results of scattering in radioisotope imaging *IEEE Trans. Nucl. Sci.* **20** 266–72
[210] Zaidi H 2001 Scatter modelling and correction strategies in fully 3-D PET *Nucl. Med. Commun.* **22** 1181–4
[211] Moisan C, Tupper P, Rogers J G and de Jong J K 1996 A Monte Carlo study of the acceptance to scattered events in a depth encoding PET camera *IEEE Trans. Nucl. Sci.* **43** 1974–80
[212] El Fakhri G N, Buvat I, Pelegrini M, Benali H, Almeida P and Bendriem B *et al* 1999 Respective roles of scatter, attenuation, depth-dependent collimator response and finite spatial resolution in cardiac single-photon emission tomography quantitation: a Monte Carlo study *Eur. J. Nucl. Med.* **26** 437–46
[213] Hutton B F, Buvat I and Beekman F J 2011 Review and current status of SPECT scatter correction *Phys. Med. Biol.* **56** R85–112
[214] Zaidi H and Koral K F 2004 Scatter modelling and compensation in emission tomography *Eur. J. Nucl. Med. Mol. Imaging* **31** 761–82
[215] Ghabrial A, Franklin D R and Zaidi H 2021 A Monte Carlo simulation study of scatter fraction and the impact of patient BMI on scatter in long axial field-of-view PET scanners *Z. Med. Phys.* **31** 305–315
[216] Ljungberg M, Strand S-E, Rajeevan N and King M A 1994 Monte Carlo simulation of transmission studies using a planar source with a parallel collimator and a line source with a fan-beam collimator *IEEE Trans. Nucl. Sci.* **41** 1577–84
[217] Gustafsson A, Bake B, Jacobsson L, Johansson A, Ljungberg M and Moonen M 1998 Evaluation of attenuation corrections using Monte Carlo simulated lung SPECT *Phys. Med. Biol.* **43** 2325–36

[218] Conti M, Bendriem B, Casey M, Eriksson L, Jakoby B and Jones W F *et al* 2006 Performance of a high sensitivity PET scanner based on LSO panel detectors *IEEE Trans. Nucl. Sci.* **53** 1136–42

[219] Ay M and Zaidi H 2006 Assessment of errors caused by x-ray scatter and use of contrast medium when using CT-based attenuation correction in PET *Eur. J. Nucl. Med. Mol. Imaging* **33** 1301–13

[220] Kanamori H, Nakamori N, Inoue K and Takenaka E 1985 Effect of scattered x-ray on CT images *Phys. Med. Biol.* **30** 239–49

[221] Dewaraja Y K, Ljungberg M and Koral K F 2001 Monte Carlo evaluation of object shape effects in iodine-131 SPET tumor activity quantification *Eur. J. Nucl. Med.* **28** 900–6

[222] Dewaraja Y K, Ljungberg M and Koral K F 2000 Characterization of scatter and penetration using Monte Carlo simulation in 131I imaging *J. Nucl. Med.* **41** 123–30

[223] Chun S Y, Nguyen M P, Phan T Q, Kim H, Fessler J A and Dewaraja Y K 2020 Algorithms and analyses for joint spectral image reconstruction in Y-90 Bremsstrahlung SPECT *IEEE Trans. Med. Imaging* **39** 1369–79

[224] Takahashi A, Himuro K, Yamashita Y, Komiya I, Baba S and Sasaki M 2015 Monte Carlo simulation of PET and SPECT imaging of 90Y *Med. Phys.* **42** 1926–35

[225] Kayal G, Chauvin M, Vergara-Gil A, Clayton N, Ferrer L and Moalosi T *et al* 2021 Generation of clinical 177Lu SPECT/CT images based on Monte Carlo simulation with GATE *Phys Med.* **85** 24–31

[226] Sekikawa Y, Funada K, Akamatsu G, Himuro K, Takahashi A and Baba S *et al* 2021 Monte Carlo simulation of the acquisition conditions for 177Lu molecular imaging of hepatic tumors *Ann. Nucl. Med.* **35** 823–33

[227] Tranel J, Feng F Y, James S S and Hope T A 2021 Effect of microdistribution of alpha and beta-emitters in targeted radionuclide therapies on delivered absorbed dose in a GATE model of bone marrow *Phys. Med. Biol.* **66** 035016

[228] Crawford J R, Robertson A K H, Yang H, Rodríguez-Rodríguez C, Esquinas P L and Kunz P *et al* 2018 Evaluation of209At as a theranostic isotope for209At-radiopharmaceutical development using high-energy SPECT *Phys. Med. Biol.* **63** 045025

[229] Ellett W H and Humes R M 1971 Absorbed fractions for small volumes containing photon-emitting radioactivity *J. Nucl. Med.* **Suppl 5** 25–32

[230] Snyder W, Ford M R and Warner G 1978 *Estimates of Specific Absorbed Fractions for Photon Sources Uniformly Distributed in Various Organs of a Heterogeneous Phantom. Pamphlet No. 5, Revised* (New York: Society of Nuclear Medicine)

[231] Akabani G, Poston J W and Bolch W E 1991 Estimates of beta absorbed fractions in small tissue volumes for selected radionuclides *J. Nucl. Med.* **32** 835–9

[232] Siegel J A and Stabin M G 1994 Absorbed fractions for electrons and beta particles in spheres of various sizes *J. Nucl. Med.* **35** 152–6

[233] Cristy M 1983 Applying the reciprocal dose principle to heterogeneous phantoms: practical experience from Monte Carlo studies *Phys. Med. Biol.* **28** 1289–303

[234] Cristy M and Eckerman K F 1987 Specific absorbed fractions of energy at various ages from internal photon sources *I Methods, II One Year Old, III Five Year Old, IV Ten Year Old, V Fifteen Year Old Male and Adult Demale, VI New-born and VII Adult Male* (Oak Ridge, TN: Oak Ridge National Laboratory) Report No.: ORNL/TM 8381/V1-V7

[235] Poston J W, Kodimer K A and Bolch W E 1996 A revised model for the calculation of absorbed energy in the gastrointestinal tract *Health Phys.* **71** 307–14

[236] Stabin M G and Konijnenberg M W 2000 Re-evaluation of absorbed fractions for photons and electrons in spheres of various sizes *J. Nucl. Med.* **41** 149–60
[237] Bice A N, Links J M, Wong D F and Wagner H N 1985 Absorbed fractions for dose calculations of neuroreceptor PET studies *Eur. J. Nucl. Med.* **11** 127–31
[238] Bouchet L G, Bolch W E, Weber D A, Atkins H L and Poston J W 1999 MIRD Pamphlet No. 15: radionuclide S values in a revised dosimetric model of the adult head and brain. Medical Internal Radiation Dose *J. Nucl. Med.* **40** 62S–101S
[239] Bolch W E, Bouchet L G, Robertson J S, Wessels B W, Siegel J A and Howell R W *et al* 1999 MIRD pamphlet No. 17: the dosimetry of nonuniform activity distributions—radionuclide S values at the voxel level. Medical Internal Radiation Dose Committee *J. Nucl. Med.* **40** 11S–36S
[240] Bouchet L G, Bolch W E, Howell R W and Rao D V 2000 S values for radionuclides localized within the skeleton *J. Nucl. Med.* **41** 189–212
[241] Xu X G 2014 An exponential growth of computational phantom research in radiation protection, imaging, and radiotherapy: a review of the fifty-year history *Phys. Med. Biol.* **59** R233–302
[242] Zaidi H and Xu X G 2007 Computational anthropomorphic models of the human anatomy: the path to realistic Monte Carlo modeling in medical imaging *Annu. Rev. Biomed. Eng.* **9** 471–500
[243] Zaidi H and Tsui B M W 2009 Review of computational anthropomorphic anatomical and physiological models *Proc. IEEE* **97** 1938–53
[244] Bouchet L G and Bolch W E 1999 Five pediatric head and brain mathematical models for use in internal dosimetry *J. Nucl. Med.* **40** 1327–36
[245] Mardirossian G, Tagesson M, Blanco P, Bouchet L G, Stabin M and Yoriyaz H *et al* 1999 A new rectal model for dosimetry applications *J. Nucl. Med.* **40** 1524–31
[246] Thomas S R, Stabin M G, Chen C T and Samaratunga R C 1999 MIRD Pamphlet No. 14 revised: a dynamic urinary bladder model for radiation dose calculations. Task Group of the MIRD Committee, Society of Nuclear Medicine *J. Nucl. Med.* **40** 102S–23S
[247] Clairand I, Bouchet L G, Ricard M, Durigon M, Di-Paola M and Aubert B 2000 Improvement of internal dose calculations using mathematical models of different adult heights *Phys. Med. Biol.* **45** 2771–85
[248] Peter J, Gilland D R, Jaszczak R J and Coleman R E 1999 Four-dimensional superquadric-based cardiac phantom for Monte Carlo simulation of radiological imaging systems *IEEE Trans. Nucl. Sci.* **46** 2211–7
[249] Segars W P, Lalush D S and Tsui B M W 2001 Modeling respiratory mechanics in the MCAT and spline-based MCAT phantoms *IEEE Trans. Nucl. Sci.* **48** 89–97
[250] Xu X G, Chao T C and Bozkurt A 2000 VIP-Man: an image-based whole-body adult male model constructed from color photographs of the Visible Human Project for multi-particle Monte Carlo calculations *Health Phys.* **78** 476–86
[251] Chao T C and Xu X G 2001 Specific absorbed fractions from the image-based VIP-Man body model and EGS4-VLSI Monte Carlo code: internal electron emitters *Phys. Med. Biol.* **46** 901–27
[252] Berger M J 1968 MIRD pamphlet 2: energy deposition in water by photons from point isotropic sources *J. Nucl. Med.* **9** 15–25
[253] Furhang E E, Sgouros G and Chui C S 1996 Radionuclide photon dose kernels for internal emitter dosimetry *Med. Phys.* **23** 759–64

[254] Mackie T R, Bielajew A F, Rogers D W and Battista J J 1988 Generation of photon energy deposition kernels using the EGS Monte Carlo code *Phys. Med. Biol.* **33** 1–20
[255] Berger M J 1971 MIRD pamphlet 7: distribution of absorbed doses around point sources of electrons and beta particles in water and other media *J. Nucl. Med.* **12** 5–23
[256] Simpkin D J and Mackie T R 1990 EGS4 Monte Carlo determination of the beta dose kernel in water *Med. Phys.* **17** 179–86
[257] Bardies M and Myers M J 1996 Computational methods in radionuclide dosimetry *Phys. Med. Biol.* **41** 1941–55
[258] Leichner P K 1994 A unified approach to photon and beta particle dosimetry *J. Nucl. Med.* **35** 1721–9
[259] Sgouros G, Chiu S, Pentlow K S, Brewster L J, Kalaigian H and Baldwin B *et al* 1993 Three-dimensional dosimetry for radioimmunotherapy treatment planning *J. Nucl. Med.* **34** 1595–601
[260] Johnson J C, Langhorst S M, Loyalka S K, Volkert W A and Ketring A R 1992 Calculation of radiation dose at a bone-to-marrow interface using Monte Carlo modeling techniques (EGS4) *J. Nucl. Med.* **33** 623–8
[261] Kawrakow I 2000 Accurate condensed history Monte Carlo simulation of electron transport. II. Application to ion chamber response simulations *Med. Phys.* **27** 499–513
[262] Simpkin D J, Cullom S J and Mackie T R 1992 The spatial and energy dependence of bremsstrahlung production about beta point sources in H_2O *Med. Phys.* **19** 105–14
[263] Dieudonne A, Hobbs R F, Lebtahi R, Maurel F, Baechler S and Wahl R L *et al* 2013 Study of the impact of tissue density heterogeneities on 3-dimensional abdominal dosimetry: comparison between dose kernel convolution and direct Monte Carlo methods *J. Nucl. Med.* **54** 236–43
[264] Khazaee Moghadam M, Kamali Asl A, Geramifar P and Zaidi H 2016 Evaluating the application of tissue-specific dose point kernels instead of water dose point kernels in internal dosimetry: a Monte Carlo study *Cancer Biother. Radiopharm.* **31** 367–79
[265] Papadimitroulas P, Loudos G, Nikiforidis G C and Kagadis G C 2012 A dose point kernel database using GATE Monte Carlo simulation toolkit for nuclear medicine applications: comparison with other Monte Carlo codes *Med. Phys.* **39** 5238–47
[266] Casciari J J, Graham M M and Rasey J S 1995 A modeling approach for quantifying tumor hypoxia with [F-18]fluoromisonidazole PET time-activity data *Med. Phys.* **22** 1127–39
[267] Millet P, Delforge J, Pappata S, Syrota A and Cinotti L 1996 Error analysis on parameter estimates in the ligand-receptor model: application to parameter imaging using PET data *Phys. Med. Biol.* **41** 2739–56
[268] Burger C and Buck A 1996 Tracer kinetic modelling of receptor data with mathematical metabolite correction *Eur. J. Nucl. Med.* **23** 539–45
[269] Thie J A, Smith G T and Hubner K F 1997 Linear least squares compartmental-model-independent parameter identification in PET *IEEE Trans. Med. Imaging* **16** 11–6
[270] Welch A, Smith A M and Gullberg G T 1995 An investigation of the effect of finite system resolution and photon noise on the bias and precision of dynamic cardiac SPECT parameters *Med. Phys.* **22** 1829–36
[271] Zeng G L, Gullberg G T and Huesman R H 1995 Using linear time-invariant system theory to estimate kinetic parameters directly from projection measurements *IEEE Trans. Nucl. Sci.* **42** 2339–46
[272] Ping-Chun C, Rogers W L, Clinthorne N H, Fessler J A and Hero A O 1994 Model-based estimation for dynamic cardiac studies using ECT *IEEE Trans. Med. Imaging* **13** 217–26

[273] Huesman R H, Reutter B W, Zeng G L and Gullberg G T 1998 Kinetic parameter estimation from SPECT cone-beam projection measurements *Phys. Med. Biol.* **43** 973–82

[274] Simpkin D J 1989 Shielding requirements for constant-potential diagnostic x-ray beams determined by a Monte Carlo calculation *Health Phys.* **56** 151–64

[275] Metzger R, Richardson R and Van_Riper K A 1993 A Monte Carlo model for retrospective analysis of shield design in a diagnostic x-ray room *Health Phys.* **65** 164–71

[276] Biggs P J 1991 Calculation of shielding door thicknesses for radiation therapy facilities using the ITS Monte Carlo program *Health Phys.* **61** 465–72

[277] Verhaegen F, Buffa F M and Deehan C 2001 Quantifying effects of lead shielding in electron beams: a Monte Carlo study *Phys. Med. Biol.* **46** 757–69

[278] da Silva A X and Crispim V R 2001 Moderator-collimator-shielding design for neutron radiography systems using 252Cf *Appl. Radiat. Isot.* **54** 217–25

[279] Howard W B and Yanch J C 1995 Shielding design and dose assessment for accelerator based neutron capture therapy *Health Phys.* **68** 723–30

[280] Kramer G H, Chamberlain M J and Yiu S 1997 A study of thyroid radioiodine monitoring by Monte Carlo simulations: implications for equipment design *Phys. Med. Biol.* **42** 2175–82

[281] Kramer G H and Crowley P 2000 The assessment of the effect of thyroid size and shape on the activity estimate using Monte Carlo simulation *Health Phys.* **78** 727–38

[282] Rodriguez-Villafuerte M, Gamboa deBuen I and Brandan M E 1997 Monte Carlo simulation of depth-dose distributions in TLD-100 under 90Sr-90Y irradiation *Health Phys.* **72** 574–8

[283] Gualdrini G, Battisti P, Biagini R, De_Felice P, Fazio A and Ferrari P 2000 Development and characterisation of a head calibration phantom for in vivo measurements of actinides *Appl. Radiat. Isot.* **53** 387–93

[284] Zavgorodni S F 2001 A method for calculating the dose to a multi-storey building due to radiation scattered from the roof of an adjacent radiotherapy facility *Med. Phys.* **28** 1926–30

[285] Osei E K and Kotre C J 2001 Equivalent dose to the fetus from occupational exposure of pregnant staff in diagnostic radiology *Br. J. Radiol.* **74** 629–37

[286] Pattison J E 1999 Finger doses received during 153Sm injections *Health Phys.* **77** 530–5

[287] Zankl M 1999 Personal dose equivalent for photons and its variation with dosimeter position *Health Phys.* **76** 162–70

[288] Xu X G, Reece W D and Poston J W 1995 A study of the angular dependence problem in effective dose equivalent assessment *Health Phys.* **68** 214–24

[289] Almén A, Andersson M, O'Connor U, Abdelrahman M, Camp A and García V *et al* 2021 Personal dosimetry using Monte-Carlo simulations for occupational dose monitoring in interventional radiology: the results of a proof of concept in a clinical setting *Radiat. Prot. Dosimetry* **195** 391–98

[290] Bhar M, Mora S, Kadri O, Zein S, Manai K and Incerti S 2021 Monte Carlo study of patient and medical staff radiation exposures during interventional cardiology *Phys. Med.* **82** 200–10

[291] Siiskonen T, Tapiovaara M, Kosunen A, Lehtinen M and Vartiainen E 2007 Monte Carlo simulations of occupational radiation doses in interventional radiology *Br. J. Radiol.* **80** 460–8

[292] Villoing D, Cuthbert T A, Kitahara C M and Lee C 2020 NCINM: organ dose calculator for patients undergoing nuclear medicine procedures *Biomed. Phys. Eng. Express* **6** 055010

[293] Sparks R B, Siegel J A and Wahl R L 1998 The need for better methods to determine release criteria for patients administered radioactive material *Health Phys.* **75** 385–8

[294] Johnson T K and Barnhart T 1999 An alternative to Monte Carlo in determining release criteria for patients administered radioactive material *Health Phys.* **77** 724–6

[295] Siegel J A, Sparks R B and Wahl R L 1999 Response to Johnson and Barnhart *Health Phys.* **77** 726–7

[296] Andreo P 2007 Radiation dosimetry: where do we stand and where do we go from here? ed S I Kim and T S Suh *World Congress on Medical Physics and Biomedical Engineering 2006. IFMBE Proceedings* vol 1 (Berlin: Springer) p 7

[297] Akabani G, Hawkins W G, Eckblade M B and Leichner P K 1997 Patient-specific dosimetry using quantitative SPECT imaging and three-dimensional discrete Fourier transform convolution *J. Nucl. Med.* **38** 308–14

[298] Dewaraja Y K, Frey E C, Sgouros G, Brill A B, Roberson P and Zanzonico P B *et al* 2012 MIRD Pamphlet No. 23: quantitative SPECT for patient-specific 3-dimensional dosimetry in internal radionuclide therapy *J. Nucl. Med.* **53** 1310–25

[299] Furhang E E, Chui C S, Kolbert K S, Larson S M and Sgouros G 1997 Implementation of a Monte Carlo dosimetry method for patient-specific internal emitter therapy *Med. Phys.* **24** 1163–72

[300] Hobbs R F, Wahl R L, Lodge M A, Javadi M S, Cho S Y and Chien D T *et al* 2009 124I PET-based 3D-RD dosimetry for a pediatric thyroid cancer patient: real-time treatment planning and methodologic comparison *J. Nucl. Med.* **50** 1844–7

[301] Ljungberg M, Celler A, Konijnenberg M W, Eckerman K F, Dewaraja Y K and Sjogreen Gleisner K 2015 MIRD Pamphlet No. 26: joint EANM/MIRD guidelines for quantitative 177Lu SPECT applied for dosimetry of radiopharmaceutical therapy *J. Nucl. Med.* **57** 151–62

[302] Ljungberg M, Sjögreen and Gleisner K 2018 3-D image-based dosimetry in radionuclide therapy *IEEE Trans. Rad. Plasma Med. Sci.* **2** 527–40

[303] Ljungberg M, Sjogreen K, Liu X, Frey E, Dewaraja Y and Strand S E 2002 A 3-dimensional absorbed dose calculation method based on quantitative SPECT for radionuclide therapy: evaluation for (131)I using Monte Carlo simulation *J. Nucl. Med.* **43** 1101–9

[304] Tagesson M, Ljungberg M and Strand S E 1996 A Monte-Carlo program converting activity distributions to absorbed dose distributions in a radionuclide treatment planning system *Acta Oncol.* **35** 367–72

[305] Yoriyaz H, Stabin M G and dos Santos A 2001 Monte Carlo MCNP-4B-based absorbed dose distribution estimates for patient-specific dosimetry *J. Nucl. Med.* **42** 662–9

[306] Roncali E, Capala J, Benedict S H, Akabani G, Bednarz B and Bhadrasain V *et al* 2021 Overview of the first NRG-NCI workshop on dosimetry of systemic radiopharmaceutical therapy (RPT) *J. Nucl. Med.* **62** 1133–39

[307] Sgouros G, Bodei L, McDevitt M R and Nedrow J R 2020 Radiopharmaceutical therapy in cancer: clinical advances and challenges *Nat. Rev. Drug Discovery* **19** 589–608

[308] St James S, Bednarz B, Benedict S, Buchsbaum J C, Dewaraja Y and Frey E *et al* 2021 Current status of radiopharmaceutical therapy *Int. J. Radiat. Oncol. Biol. Phys.* **109** 891–901

[309] Xiao Y, Roncali E, Hobbs R, James S S, Bednarz B and Benedict S *et al* 2021 Toward individualized voxel-level dosimetry for radiopharmaceutical therapy *Int. J. Radiat. Oncol. Biol. Phys.* **109** 902–4

[310] Enghardt W, Debus J, Haberer T, Hasch B G, Hinz R and Jakel O *et al* 1999 The application of PET to quality assurance of heavy-ion tumor therapy *Strahlenther. Onkol.* **175** 33–6

[311] Enghardt W, Fromm W D, Geissel H, Keller H, Kraft G and Magel A *et al* 1992 The spatial distribution of positron-emitting nuclei generated by relativistic light ion beams in organic matter *Phys. Med. Biol.* **37** 2127–31

[312] Enghardt W, Fromm W D, Manfrass P and Schardt D 1992 Limited-angle 3D reconstruction of PET images for dose localization in light ion tumour therapy *Phys. Med. Biol.* **37** 791–8

[313] Litzenberg D W, Becchetti F D and Roberts D A 1997 On-line PET monitoring of radiotherapy beams: image reconstruction and Monte Carlo simulations of detector geometries *IEEE Trans. Nucl. Sci.* **44** 1646–57

[314] Paganetti H, Beltran C, Both S, Dong L, Flanz J and Furutani K *et al* 2021 Roadmap: proton therapy physics and biology *Phys. Med. Biol.* **66** 05RM1

[315] Faddegon B, Ramos-Méndez J, Schuemann J, McNamara A, Shin J and Perl J *et al* 2020 The TOPAS tool for particle simulation, a Monte Carlo simulation tool for physics, biology and clinical research *Phys. Med.* **72** 114–21

[316] Kurz C, Bauer J, Unholtz D, Richter D, Herfarth K and Debus J *et al* 2016 Initial clinical evaluation of PET-based ion beam therapy monitoring under consideration of organ motion *Med. Phys.* **43** 975–82

[317] Parodi K 2020 Latest developments in in-vivo imaging for proton therapy *Br. J. Radiol.* **93** 20190787

[318] Kanai T, Endo M, Minohara S, Miyahara N, Koyama_ito H and Tomura H *et al* 1999 Biophysical characteristics of HIMAC clinical irradiation system for heavy-ion radiation therapy *Int. J. Radiat. Oncol. Biol. Phys.* **44** 201–10

[319] Kraft G 2000 Tumor therapy with heavy charged particles *Prog. Part. Nucl. Phys.* **45** S473–544

[320] Rosso D 2000 A boost for hadron therapy *Phys. World.*

[321] Parodi K 2015 Vision 20/20: positron emission tomography in radiation therapy planning, delivery, and monitoring *Med. Phys.* **42** 7153–68

[322] Berger M J 1993 *Proton Monte Carlo Transport Program PTRAN* (Bethesda, MD: NIST) *Report No.: Report NISTIR* 5113

[323] Del Guerra A, Di Domenico G and Mukhopadhayay D 1997 PET dosimetry in proton radiotherapy: a Monte Carlo Study *Appl. Radiat. Isot.* **18** 1617–24

[324] Knopf A, Parodi K, Paganetti H, Cascio E, Bonab A and Bortfeld T 2008 Quantitative assessment of the physical potential of proton beam range verification with PET/CT *Phys. Med. Biol.* **53** 4137–51

[325] Pinto M, Kröniger K, Bauer J, Nilsson R, Traneus E and Parodi K 2020 A filtering approach for PET and PG predictions in a proton treatment planning system *Phys. Med. Biol.* **65** 095014

[326] Sommerer F, Parodi K, Ferrari A, Poljanc K, Enghardt W and Aiginger H 2006 Investigating the accuracy of the FLUKA code for transport of therapeutic ion beams in matter *Phys. Med. Biol.* **51** 4385–98

[327] Dedes G, Dickmann J, Giacometti V, Rit S, Krah N and Meyer S *et al* 2022 The role of Monte Carlo simulation in understanding the performance of proton computed tomography *Z. Med. Phys.* **32** 23–38

Chapter 2

Advances in imaging for radiation therapy

Stephan Walrand and Michel Hesse

2.1 SPECT

2.1.1 Large CZT SPECT

The replacement of the NaI(Tl) crystal and photomultiplier tubes (PMTs) by solid-state cadmium–zinc–telluride (CZT) detectors in commercial gamma cameras, started more than two decades ago [1]. With the CZT technology the gamma photon is directly converted into electron–hole pairs that are collected by electrodes producing a measurable electric signal. At first limited to small fields of view (FOV) because of fabrication difficulties and costs, two large FOV CZT SPECT cameras are now commercially available. The discovery NM/CT 870 CZT (GE Healthcare, Haifa, Israel), whose gantry design is the same as for conventional Anger SPECT cameras, (see www.gehealthcare.com) and the Veriton-CT (Spectrum Dynamics, Haifa, Israel), whose design is revolutionary with 12 CZT detector columns equally distributed on a ring, with independent radial and swivel motions, greatly improving the system sensibility and spatial resolution (see www.spectrum-dynamics.com).

CZT detectors benefit from several advantages compared to conventional NaI (Tl)-PMT detectors [2]. Their pixelated nature improves the intrinsic spatial resolution (currently at 2.46 mm), and the count rate capability. CZT detectors also benefit from a better energy resolution (less than 6%) improving image contrast by reducing the impact of scattered photons, and allowing a better separation of energy peaks in multiple isotopes acquisitions. On the other hand, the limited thickness of current CZT detectors rendered them inappropriate for imaging high energy isotopes: with a CZT thickness of 7.25 mm, the GE Discovery 870 system can image isotopes up to 250 keV (with recently developed medium energy collimators), while the SD Veriton, with its 6 mm-thick CZT and fixed low-energy collimators, is limited to gamma energies below 200 keV. The compactness of CZT detectors also reduces the camera's footprint, and their smaller detection dead space enables them to come closer to the patient which improves further the image quality.

doi:10.1088/978-0-7503-2694-0ch2

Because of their limited detection efficiency of high energy gamma photons, CZT SPECT cameras are not adapted for imaging all therapeutic radioisotopes, like for example ^{131}I or alpha emitters that often emit (very) high energy gammas. However they can be, and have been, used for radioisotopes emitting at least one gamma in their energy range, of course with reduced sensibility and additional corrections to take care of higher energy gamma penetration and scatter [3]. For example the uptake of ^{177}Lu, used in combination with DOTATOC or DOTATATE for neuroendocrine tumours [4], or with PSMA for prostate cancers [5], can be imaged and quantified on CZT SPECT cameras using the lowest energy peak at 113 keV [6, 7].

While imaging the distribution of the injected therapeutic agent in the patient tissues is important to evaluate the absorbed doses, molecular radiotherapies require also treatment planning to determine the activity to be injected. This is usually done by performing an injection of a surrogate that can be imaged with a PET/CT or SPECT/CT. For example, neuroendocrine radiotherapies with ^{90}Y-DOTATOC can be planned from ^{111}In-DOTATOC [8]. Similarly to ^{177}Lu, ^{111}In emits one gamma above 200 keV, and one below, at 171 keV, that can be imaged with CZT cameras [9]. As indicated before, while imaging with CZT SPECT is quite straightforward, quantifying the uptake may prove challenging due to the corrections needed to remove the events from the ^{111}In 245 keV gammas. These corrections will depend on the physical properties and design of the CZT SPECT camera, but also on the specific ^{111}In distribution inside the patient.

Another form of treatment planning that could benefit from the improved image quality of CZT is that for ^{90}Y liver radioembolization. Before the treatment a simulation of the biodistribution of ^{90}Y microspheres in the hepatic arterial tree is performed by injecting ^{99m}Tc-MAA particles following the same injection procedure as for therapy [10]. The distribution of ^{99m}Tc-MAA is then imaged by SPECT/CT in order to estimate the lung shunt, to rule out extrahepatic uptake and eventually for dosimetric planning [11, 12]. The CZT camera is in that case perfectly adapted as the imaging target is the ^{99m}Tc and its 140 keV gamma.

2.1.2 ^{90}Y bremsstrahlung SPECT

The first bremsstrahlung imaging was performed in 1967 using a rectilinear scanner after a ^{90}Y microspheres liver radioembolization [13]. During four decades bremsstrahlung x-rays imaging was the only modality used to image ^{90}Y radiotherapy, but it was superseded in recent years by the introduction of ^{90}Y TOF-PET imaging [14, 15]. Current Monte Carlo (MC) based bremsstrahlung SPECT reconstruction quality approaches that of ^{90}Y of TOF-PET imaging [16, 17].

Even if bremsstrahlung SPECT acquisitions are less expensive that PET ones, bremsstrahlung SPECT suffers from two drawbacks. It requires accurate modelling of the camera-collimator system which makes it hardly marketable. No hardware improvement is made in SPECT regarding the bremsstrahlung imaging, while ^{90}Y TOF-PET imaging continuously improves together with the Coincidence Time Resolution (CTR) (see section 2.2.4) which now reaches 210 ps [18].

However, SPECT of a ^{90}Y hot–cold spheres phantom using a ½ inch thick NaI Anger camera equipped with a medium energy pinhole collimator provided imaging

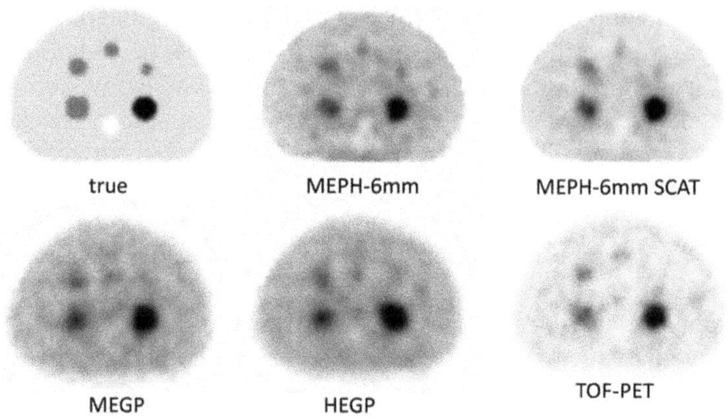

Figure 2.1. Hot and cold sphere phantoms. The figure shows transverse slices passing through the spheres' centre for the different acquisitions with reconstructions of four iterations × eight subsets. The true activity distribution is represented with the same voxel size as the reconstructions. Reprinted from [19] with permission of the authors, copyright (2011).

quality outweighing that of ^{90}Y TOF-PET [19]. Figure 2.1 shows that with scatter corrected reconstruction, 6 mm aperture MEPH SPECT allowed the visualization of all the spheres. This results from the fact that the thick pinhole housing avoids most of the high energy x-ray septa penetration contaminations present when using an MEGP, or even HEGP collimator. Reconstruction of phantom modelling liver post ^{90}Y radioembolization proved 3 min acquisitions sufficient to get an accurate activity distribution assessment. That should allow real time monitoring of spheres delivery in the catheterization room.

MC simulations with Gate showed that a significant improvement can further be obtained by using a 3 cm thick pinhole housing and replacing the NaI crystal by 25 mm-thick pixelated BGO-PMT blocks [20]. The high stopping power of BGO removed the high energy x-ray back-scattered component originating from the camera back compartment. Figure 2.2 shows that in this setup total scattering contamination is significantly smaller than the primary x-rays up to 350 keV.

In a way similar to TOF-PET (see section 2.2.1.2), in this setup the angular and energy information are redundant: for each angle the acquired data can be expressed as a Laplace transform of the line of response (LOR) activity (see equation (6) in [20]). This transform could be inverted to directly get the activity distribution from a single angle acquisition. However Laplace inversion is an ill-posed problem, such as TOF deconvolution, but this additional information will likely improve the condition number for the SPECT system such as TOF does for PET (see section 2.2.1.3).

2.2 TOF-PET

2.2.1 Motivation

Although still far from providing an accurate localization of the annihilation (CTR $\approx$ 3 cm), TOF information provides several benefits improving the reconstructed image quality.

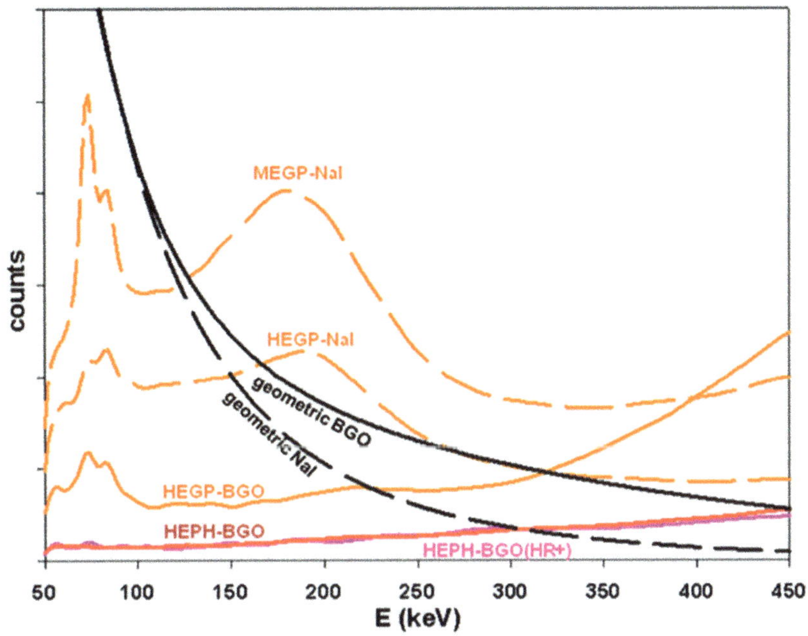

Figure 2.2. Comparison of the total scatter components (coloured curves) with the geometric x-rays (black curves) obtained by MC simulations for a ^{90}Y point source surrounded by 1 cm of perpex and set 10 cm in front of the collimator. Reprinted from [20] with permission of the authors, copyright (2014) Walrand, Hesse, Wojcik, Lhommel and Jamar.

2.2.1.1 Signal-to-noise ratio (SNR)

Randoms have a uniform TOF distribution, as a result, a large part of them are located by the TOF information outside the patient and thus, do not contribute to the noise, as shown and explained in [21].

2.2.1.2 Information redundancy

Angular and TOF information are redundant. Indeed, after division by the LOR attenuation, each LOR time data set is the convolution of the LOR activity with the time Gaussian resolution. Thus, the activity distribution could be obtained by the deconvolution of the LOR time sets corresponding to a single angle. This redundancy allows simultaneously reconstructing the emission distribution and the attenuation map together [22–24].

It was also noted and shown using simulations, that TOF-PET reconstruction was more robust in the presence of inconsistent data [25, 26], especially for emission-attenuation mismatch, which is always present due to the patient respiration. This property was mathematically proved in the case of a uniform additional background and thus inconsistent with the patient attenuation [27], which is also always present due to imperfections and noise in randoms corrections.

2.2.1.3 Condition number improvement

PET is an ill-posed problem [28] for which there exist non-null solutions for null acquired data. Obviously, providing additional information is always a good idea in such a problem, which is to some extent the purpose of regularization by priors or penalty incorporation. In this regard, TOF acts as a penalty, but which is really satisfied by the activity distribution. As TOF information is a convolution, thus itself an ill-posed problem, it is not sufficient to fully regularize the problem. However, it has been shown that the condition number of the problem which measures the ill-posedness, i.e. the largest to the smallest eigenvalue ratio, improves together with the CTR [29].

2.2.1.4 Ultimate goal

When the CTR reaches 25 ps, the TOF will allow one to directly localize the annihilation event in a 4 mm pixel, avoiding any reconstruction process. As a result the images will only be affected by an uncorrelated Poisson noise which facilitates image interpretation by a human observer [30]. Lastly, reconstructed low activity regions of interest will no longer be impacted by high taking up neighbour regions. However, reaching such CTR value will likely require development of a new detection concept [31].

2.2.2 CTR issues in crystal-photodetector blocks

A cascade of different physical effects impacts the scintillation photons' detection time. Photons are emitted randomly with a probability slowly decreasing with time (half decay time τ_d of about some tens of ns). Due to the narrow solid angle of the photodetector pixel in thick pixelated crystal, the photons have a high probability of emission towards a lateral crystal surface. Thus, a random number of reflections is needed to reach the photodetector surface, resulting in a dispersion d of the travel time. The different physical processes involved in the photodetector also introduce a random delay of FWHM J of about some tens of ps between the photon entry and the output pulse.

Derenzo *et al* performed extensive MC simulation of light transport in different crystal-photodetector setups and found a sophisticated formula providing an accurate estimation of the total CTR [32]. For $LaBr_3$ and also Lu based crystals where $\tau_d > \tau_r + d + \mathrm{J}$ (τ_r is the light rise time), this formula approximatively reduces to:

$$\mathrm{CTR} \approx 2.20\sqrt{\frac{\tau_d\,(\tau_r + \mathrm{d} + 0.95\,\mathrm{J})}{N}} \tag{2.1}$$

where N is the number of photons detected. This contribution results from the fact that in any random process measurement, fluctuations decrease when the number of measurements increases, even if the goal is just to keep the first measured time.

N is the product of the crystal scintillation yield Y per 511 keV, of the optical transport efficiency F and of the photodetector efficiency PDE, i.e.:

$$N = Y \times F \times \mathrm{PDE} \tag{2.2}$$

The main photon killing actor is the reflective material covering the crystal. Indeed the best often used reflective material, Teflon®, has for a 0.3 mm-thickness an absolute reflective factor of 0.944 [33], which means that after five reflections already 25% of the photons are lost. Reducing the number of reflection improves the CTR by reducing d and increasing N at the same time.

Obviously, an easy way to be free of this issue is to use thin crystal. Although this solution is not usable regarding the quadratic dependence of PET efficiency versus that of the crystal, this solution is often used to assess the intrinsic timing performance of the crystal-photodetector block.

In thick crystal, depending on its depth of interaction (DOI), the scintillation photons will have a shorter or longer path to travel reaching the photodetector. Between two interactions, gamma rays travel through any material at the speed of light c in vacuum, while, due to the high refractive index of scintillation crystals (>1.8), scintillation photons travel the crystal with about half that speed. Thus, according to the DOI in the two opposing 20 mm-thick L[Y]SO crystal blocks, the TOF difference between the two coincidence photons can be delayed up to 106 ps.

Figure 2.3 shows the typical photons dispersion in a $3 \times 3 \times 20$ mm^3 LSO crystal covered by Teflon® obtained by optical MC transport. Note the very small number of photons reaching the photodetector surface.

2.2.3 State-of-the-art PET instrumentation

The first commercial TOF-PET was proposed early in 1988 by Scanditronix and already reached an impressive TOF resolution of 540 ps [34]. However, due to the low system sensitivity, resulting from the low stopping power of the BaF_2 crystal used, the system was never able to transcend the imaging quality of conventional 25 mm-thick BGO systems.

The interest in TOF-PET development re-emerged at the end of the nineties with the introduction of Lu based crystal having good timing performance and a stopping power almost as good as that of BGO. The first commercial TOF-PET success was

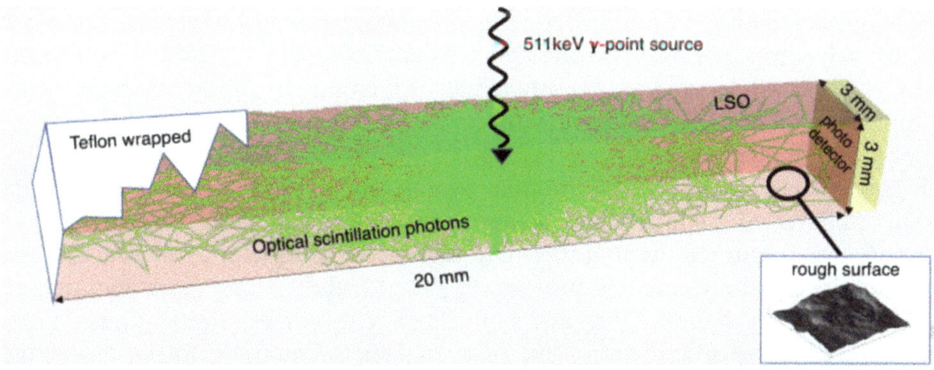

Figure 2.3. Example of photo-dispersion in LSO crystal obtained by the optical MC transport LUTDavis model [Roncali and Cherry 2013]. Figure available at: https://github.com/OpenGATE/GateContrib/tree/master/imaging/LUTDavisModel courtesy of Dr E Roncali.

launched in 2006 by Philips. Using a 22 mm-thick LYSO-PMT bloc detector, the CTR was 573 ps [35]. The next breakthrough to a CTR of 310 ps was again achieved by Philips in 2014 using 19 mm-thick LYSO and replacing the PMTs by SiPMs allowing a digital counting of the scintillation photons [36]. The current best CTR of 210 ps is obtained by the Siemens TOF-PET biograph vision based on 20 mm-thick LSO crystal coupled to a new generation SiPM [18]. Detailed information is not provided about the crystal doping and wrapping, but crystal pixels were individually selected (Siemens), which makes sense regarding the light output variations observed in similar LSO samples.

TOF resolution improvement in commercial PET was thus mainly obtained by using more efficient photodetectors, and also by reducing the crystal thickness which decreases the system sensitivity. Development of new crystal-photodetector block design in order to achieve better TOF resolution without crystal thickness reduction is the subject of numerous investigations [37].

2.2.4 ^{90}Y therapy PET imaging

Within less than a decade, TOF-PET became the standard modality for the therapy quality and dosimetry assessment post ^{90}Y liver radioembolization [14, 38]. Bremsstrahlung SPECT with MC based reconstruction methods provides ^{90}Y image quality approaching that obtained using TOF-PET system [16]. However, they have the drawback of requiring an accurate camera-collimator modelling. They are thus hardly marketable regarding the continuous launching of new camera models.

Even more than in ^{18}F-fluoro-deoxy-glucose (FDG) imaging, the CTR has a significant impact on ^{90}Y image quality, especially using Lu based crystal systems, which exhibit disturbing artefactual extrahepatic counts without TOF information [39]. This effect results from the fact that the randoms, mainly originating from Lu in low counts rate studies, have a uniform distribution inconsistent with the patient attenuation. It was shown mathematically that when using TOF information the residual randoms correction errors are not amplified in the reconstructed image by the attenuation correction [27].

The improvement of the CTR for the liver imaging is also of prime importance [40]. MC dynamic microsphere transport simulations [41, 42], *ex vivo* autoradiography [43] and MRI imaging [44] prove microspheres distribution heterogeneities of centimetric scale (figures 2.4(a)–(c)), much better modelled by a hot rods phantom than by a homogeneous hot spheres one.

Figure 2.4 shows the reconstruction of a 4 mm-thick slice inside the hot bars region of a Jaszczak Deluxe phantom representative of such heterogeneity scale. The phantom was acquired with a ^{90}Y chloride activity resulting in 100 Gy mean absorbed dose in the sector regions (figures 2.4(g) and (h)), and again acquired after decaying to 50 Gy (figures 2.4(d) and (e)). These doses correspond to clinical normal liver recommended dose in resin and glass spheres radioembolization, respectively. The acquisition time on a Philips Gemini TF (CTR = 550 ps) and on a Philips Vereos (CTR = 320 ps) was 20 min per bed position as usually performed in clinical post liver radioembolization (see [45] for the detailed acquisition setup). For

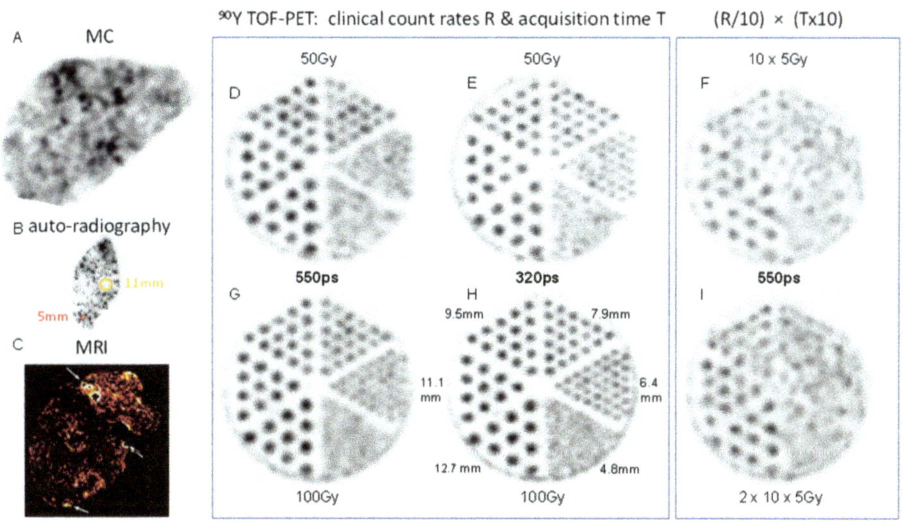

Figure 2.4. (a) MC sphere simulation (see [41], copyright SNMMI). (b) Autoradiography of a normal liver (NL) tissue post resin spheres liver radioembolization (reprinted from [43], copyright (2014), Högberg *et al*; licensee Springer. With permission of Springer). (c) MRI imaging post ^{166}Ho loaded spheres liver radioembolization (reprinted from [44] with the permission of Dr M Lam, copyright (2012), with permission fo Springer). (d–i) 4 mm-thick ^{90}Y reconstructed slice inside the hot sectors of a Jaszczak Deluxe phantom acquired by the authors (see text). All images were scaled together.

comparison, the phantom was again acquired after decaying to 5 Gy, a typical dose sometimes used in reported ^{90}Y phantom studies. The acquisition time was accordingly increased to get the same total number of counts (figures 2.4(f) and (i)).

The reconstructions, using the parameters recommended by the manufacturer for FDG studies, show a significant image improvement for the 320 ps resolution TOF-PET versus the 550 ps one: five sectors were completely visualized. This explains the success of ^{90}Y TOF-PET based Equivalent Uniform Dose (EUD) in predicting the tumour dose-response in liver radioembolization [46–48].

In contrast, although compensated by a tenfold longer acquisition time, due to the constant Lu randoms, the reconstructed slices using a tenfold lower activity resulted in a poor image quality. Use of the generalized TOF-SNR expression [21] combined with the randoms constant rate originating from the Lu based crystal clearly explains this issue [40]. This proves the necessity to performed phantom evaluations using specific activities similar to those obtained in clinical liver radioembolization [45]. Partial activity filling of large NEMA phantoms can be easily implemented in order to avoid PET saturation issue [49].

Figure 2.5 shows that for low specific ^{90}Y activities, such as obtained for kidneys in PRRT, old BGO PET performances outweighed those of a modern LYSO TOF-PET [50]. For a 45-min acquisition and an actual cortex absorbed dose of 13.8 Gy, the dose estimation was 13.4 ± 0.4 Gy for the BGO PET and 11.3 ± 0.5 Gy for the LYSO TOF-PET.

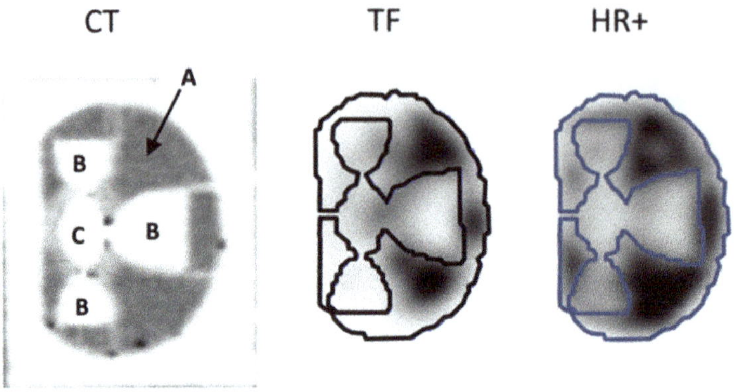

Figure 2.5. Coronal slices of kidney based on the MIRD fillable kidney phantom model, obtained by averaging reconstructions of eight 45-min acquisitions for Gemini TF (TF) and Exact HR+ (HR+) systems. (a) Renal cortex. (b) Medullar pyramids. (c) Renal pelvis. There is a lack of reconstructed activity on kidney (left side facing spine) for Gemini TF reconstruction (reprinted from [50] with permission of SNMMI).

2.2.5 CTR improvements: current investigations

Many tracks are under evaluation in order to improve TOF-PET, an exhaustive list of which can be found in [37]. We will focus our attention on three main ones, where MC can be of prime interest in order to speed up the development.

2.2.5.1 Light transport optimization

There are two parameters that control the light reflection on the crystal boundaries: the crystal surface polishing and the reflective material used.

When the reflector surface is perfectly smooth the light is reflected symmetrically to the surface normal (specular reflection). When the surface is rough, or if the light is reflected inside the reflector such as for polytetrafluoroethylene (PTFE) (Teflon®), the reflection is diffuse and covers all the directions. The Lambertian is a special case of diffuse reflection in which it appears equally bright from all reflection angles. Common reflector materials were extensively analysed in [33, 51]. Pure specular reflectors are VM2000, aluminum. Lu mirror, glossy PTFE, mat PTFE and titanium dioxide have a Lambertian reflection plus an increasing specular component for incident angle larger than 5°, 10°, 40° and 50°, respectively.

A photon may undergo a lot of reflections to reach the photodetector, it is of prime importance to have a reflection coefficient as close as possible to 1. As no reflector has a lossless reflection [33], it was proposed to leave an air gap between the crystal and the reflector layer in order to profit from total lossless reflection above the critical angle (34° for L[Y]SO). MC simulation showed that this setup improves light collection and CTR [52]. Note that to improve simulations accuracy, the MC code should include the reflection delay which can reach about 0.1 ns near the critical angle [53].

The dual Lambertian-specular behaviour is helpful for the crystal lateral surface: a photon travelling towards the photodetector has a high probability to continue after a lateral reflection (see figure 2.6(a)), and a photon emit towards the lateral

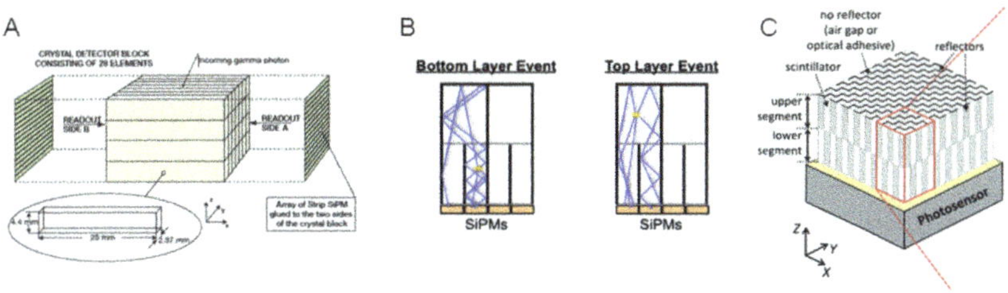

Figure 2.6. DOI assessment methods. (a) Two lateral SiPMs array (reprinted from [59], copyright IOP Publishing. CC BY 3.0). (b) Pixel light sharing (reprinted from [60], copyright IOP publishing. All rights reserved). (c) mismatched pixelated crystal layers (reprinted from [62], copyright IOP publishing. All rights reserved).

surface at low incident angle has at least a small probability to be reflected towards the photodetector (see figure 2.6(b)), rather than being an inefficient ping-pong play between the two opposite surfaces.

An additional possibility to adjust the diffuse-specular ratio of the reflection is to tune the crystal surface roughness. Experiments showed that all well-polished crystal surfaces give higher detected photon number than all roughened surfaces [54]. This feature property was again recently confirmed [55], but interestingly this study showed that hybrid surface with only roughing one crystal lateral surface in the small DOI region gives the best light detection and CTR.

2.2.5.2 DOI assessment in pixelated block

The first concept of DOI assessment was based on a detector block made of two different crystal layers, the layers in which the photoelectric effect occurs was identified based on the scintillation light pulse shape. Amazingly, it was one of the first proposed application of the LSO crystal being an additional first layer of a BGO PET [56]. However, this method had the major drawback of allowing only a two-DOI regions discrimination.

With the apparition of a position sensitive avalanche photodiode, it was proposed to set such a device on the crystal surface opposite to the PMT [57]. The ratio between the light detected in the front and in the back detector continuously varies with the DOI. This dual-end solution is still nowadays investigated for TOF-PET using L[Y]SO sandwiched in between two SiPM arrays [58], and is used in a DOI 4.4 MeV Compton camera developed for proton beam therapy imaging (see section 2.6.3). An elegant solution to avoid disturbing Compton scattering in the front SiPM is to rotate the detector bloc (figure 2.6(a)): in this setup the light position on the SiPM gives the DOI and the Y position, the ratio between the two lateral SiPMs counts provides the x position [59]. However, these two options have the drawback to multiply by two the gantry electronics and especially the costly SiPMs.

A low cost solution is to allow light sharing between two or more crystal pixels in the low DOI region (figure 2.6(b)), the light distribution between the coupled pixels depending on the DOI [60]. Recent MC simulations show that an improvement of about 100 ps can be achieved using this method [61]. Lastly, two [62] or up to four

[63] layers of pixelated crystal, partially isolated by reflectors, can be stacked with a reflector position mismatch between the different layers (figure 2.6(c)): photoelectric effect occurring in different layers produces different light distributions on the SiPM.

Many DOI assessment methods are under investigation, each one having its own benefits and drawbacks, and it is still unclear to see which one will be implemented in commercial TOF-PET.

2.2.5.3 Faster photons

A detailed analysis of the different existing options in order to get faster gamma-to-light conversion can be found in [64]. Crystal scintillation is a very complex process including several steps [65]. Very schematically, after thermalization of electron–hole pairs resulting from the photoelectron energy deposition, these pairs have to migrate to luminescent centres in order to produce an optical photon. This migration makes a significant contribution in the scintillation pulse shape.

One possibility is to use faster transient processes occurring in the crystal at the beginning of the gamma–light conversion process. The reader can find a detailed and exhaustive analysis of these transient processes in [65]. Amongst these transient processes, Cherenkov light emission appears promising. The Cherenkov emission already starts with the motion of the photoelectron, or of the recoil electron, are emitted within ≈10 ps and is thus one of the fastest transient emission. For crystal with a refractive index $n = 1.6$ the Cherenkov threshold is 144 keV, and 63 keV for $n = 2.2$.

Cherenkov photons are unfortunately emitted in small number, e.g. ≈ 17 for a 511 keV interaction in BGO with a emission spectrum intensity decreasing from 310 nm to 700 nm by a factor 6 while the SiPM PDE maximum is around 500 nm. Despite these unfavourable features, a promising CTR of 277 ps has been observed using a 2 × 2 × 20 mm BGO coupled to a fast SiPM [66]. As BGO has no background emission, the performances of a BGO TOF-PET with such CTR should already outweigh those of the state of the art 210 ps LSO TOF-PET, especially in low count rate studies, such as post ^{90}Y therapy imaging.

A promising concurrent to BGO is the cheap gamma ray semi-conductor detector TlBr. Like BGO, TlBr is transparent and has no background radiation but has a better stopping power than BGO and a higher refractive index ($n = 2.6$) which increases the Cherenkov emission. TlBr has another major benefit: by using a multistrip anode set transversally to the gamma incoming direction it is possible to get a DOI estimation in order to correct the Cherenkov emission detection time (figure 2.7(a)). A preliminary experimental setup gave a CTR = 330 ps for a 4 × 4 × 5 mm [67].

Note that, similarly to the scintillation photons, the light transport in the crystal has to be optimized in order to fully benefit of the fast emission property of the Cherenkov photons.

2.2.6 Prospective: wave optics approach

Crystal-photodetector coupling and the dual Lambertian-specular behavior could perhaps still be improved by taking benefit from the wave–particle duality of individual photon. Let's recall that, as described by quantum mechanics, experiments prove

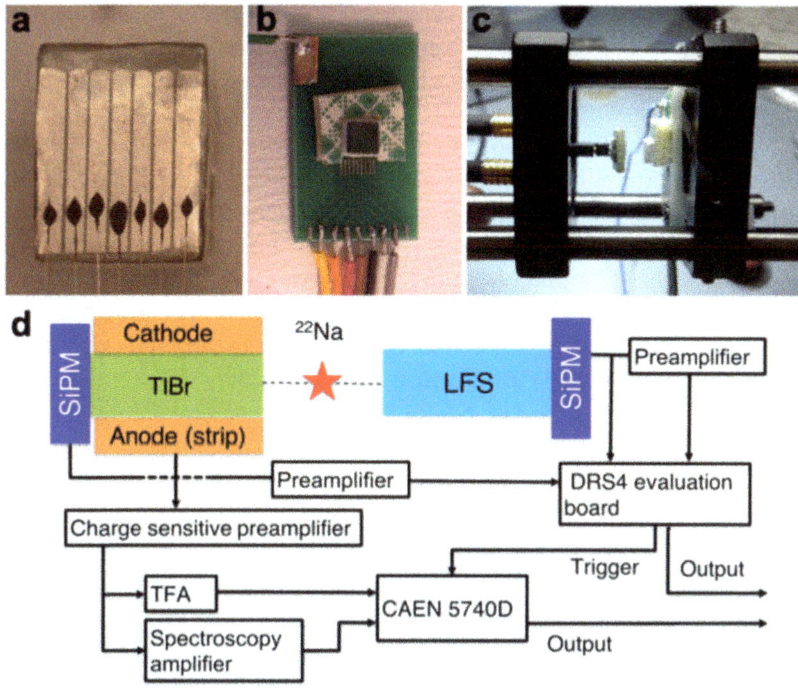

Figure 2.7. (a) Segmented electrode of the CCI TlBr detector used in the measurements. (b) CCI TlBr detector mounted on a custom PCB board. (c) Mechanism used to couple the SiPM to a CCI TlBr detector. (d) Schematic of the readout setup (reprinted from [67], copyright IOP Publishing, all rights reserved).

individual particles reproduce wave interference and diffraction patterns, even when they travel one by one.

2.2.6.1 Crystal-photodetector coupling

In order to avoid disturbing reflections at the crystal-photodetector interface, air should be avoided between the two components. It is the major purpose of filling this interface by optical grease. However, as a consequence of the wave nature of light, whatever its refractive index, an interface medium with a thickness much smaller than the wavelength has no effect on the light propagation. So if the two component surfaces are very much in plane form, good transmission is also obtained with dry assembling.

When the interface thickness is about 1/4e wavelength in its medium, i.e. about $100/n$ nm where n is the refractive index, the two successive reflections on the two interface boundaries are in phase opposition, and thus partially cancel out, improving the light transmission. The cancellation and light transmission are maximum when the interface refractive index is the geometric mean of the crystal and photodetector window ones, which is $n = 1.58$ for L[Y]SO-glass interface of respective indexes $n = 1.8$ and $n = 1.4$. Many modern crystal-photodetector blocks are thus now assembled with the Meltmount glue ($nr = 1.58$). However, accurately tuning the glue thickness to 0.06 μm appears quite challenging.

2.2.6.2 Light transport optimization

Periodic blazed grating surface has the interesting property of preferably diffracting the light into a specific direction (figures 2.8(c)–(e)) when the grating shape is adjusted to the photon wavelength [68].

Figure 2.9 shows the efficiency of the different diffraction orders n for a triangular blazed gratings as a function of the angle of incidence θ_i in the longitudinal plane (grating amplitude $h = 0.55\lambda$, period $p = 3\lambda$ and silver material). Negative angle of incidence corresponds to the direction towards the photodetector. By leaving an air gap between the crystal boundary and the blazed grating, only photons with an angle of incidence between −33° and +33° will reach the blazed grating, all the other ones will undergo a total lossless specular reflection on the crystal–air interface. The quantified diffraction angles θ_d are given by the relation:

$$-\sin\theta_d = \sin\theta_i - \frac{n\lambda}{p} = \sin\theta_i' \tag{2.3}$$

where θ_i' is the novel angle of incidence that the photon will have if it reaches another lateral crystal boundary. Order $n = 0$ is similar to the conventional specular reflection. Equation (2.3) and figure 2.9 show that the order $n = 1$ redirects about 87% of the photons coming around the normal incidence, i.e. $\theta_i \approx 0$, towards the photodetector with a novel incidence angle $\theta_i' \approx -20°$, the following grating diffraction gives $\theta_i' \approx -42°$ with a 75% probability which results in a total lossless reflection cascade towards the photodetector. By comparison, Lambertian reflectors redirect less than 50% of such photons towards the photodetector while the other half is redirected in the wrong direction, with this random half distribution again occurring for the following reflections [51].

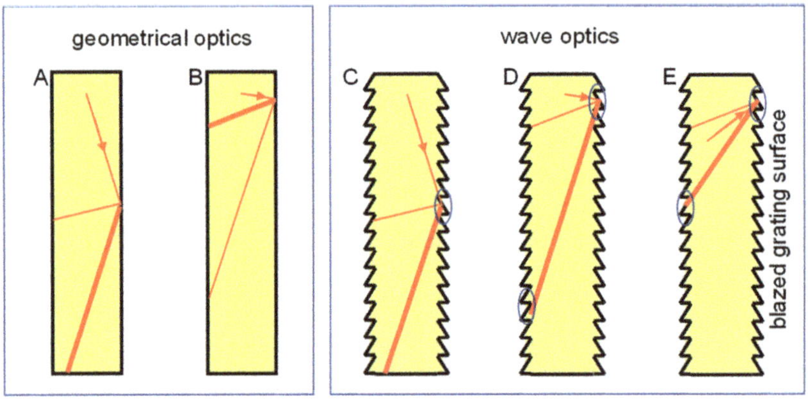

Figure 2.8. (a and b) Reflecting property of plane, or random rough surface, presenting a Lambertian + specular reflection such as Teflon® (only two reflection possibilities amongst the continuous possible range are drawn). (c–e) Diffracting property of a periodic blazed grating surface (the actual grating shape is about the photon wavelength, i.e. 400 nm, and is thus highly magnified in the figure. The black ellipse emphasizes that, due to its wavelength, the photon 'feels' the pattern periodicity). The thickness of the reflected or diffracted ray corresponds to the reflection probability (note that the reflection angles are quantified for the blazed grating device).

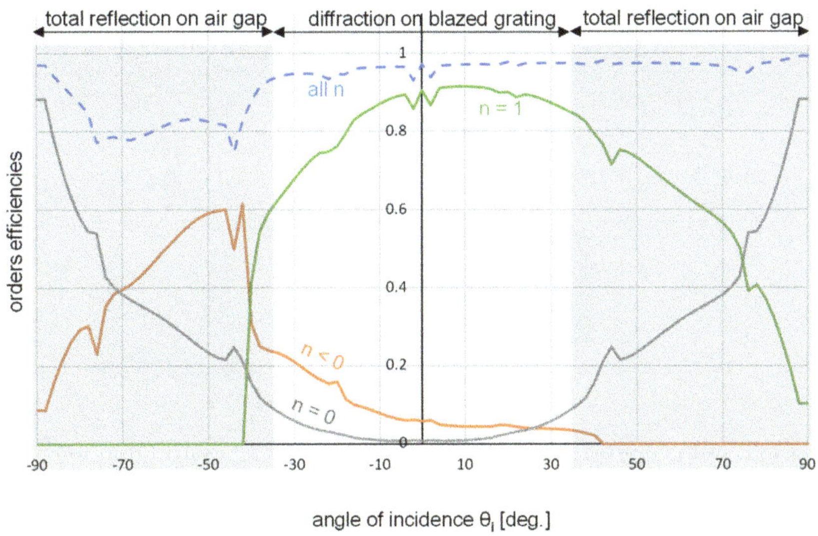

Figure 2.9. Total efficiency, i.e. first Stokes vector component, of diffraction orders for a triangular blazed grating as a function of the angle of incidence for unpolarized light. (computed with SCATMECH developed by T A Germer and available at https://pages.nist.gov/SCATMECH/docs/index.htm). Order $n = 0$ is similar to the specular reflection, while positive orders tend to redirect the photons towards the photodetector.

A description of available grating techniques can be found in [69] (optical firms commercialize this service).

MC codes for photon transport in crystal including full wave optics should be developed in order to evaluate the potential improvement of oriented blazed grating versus conventional reflectors. The complexity of such codes will be increased by the fact that blazed grating diffraction is dependent on the photon polarization and needs thus to be described using Mueller matrix and Stokes vector. Open software, such as SCATMECH using the rigorous coupled-wave analysis (RCWA), enables their computation for any grating shape (figure 2.9).

More general grating shapes will also have to be investigated in the prospect of dumping the small but disturbing negative orders component that redirect the photons away from the photodetector (orange curve in figure 2.9).

2.3 Compton camera

2.3.1 Motivation

Electronic collimation makes PET sensitivity and spatial resolution better than SPECT by a factor of about 100 and 3, respectively. Early in 1974 the use of two detectors in series was proposed to perform an electronic collimation of single photon [70] and by the way to perform 3D acquisition such as in PET. The first detector is optimized to favour Compton scattering while the second is a classical γ camera that records the scattered gamma ray. The Compton relation allows deriving information about the incoming direction and energy of the primary gamma ray.

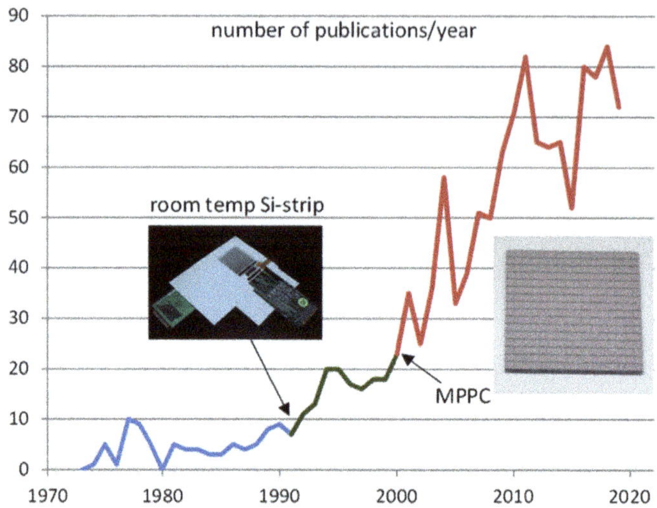

Figure 2.10. Number of scientific publications per year in scholar.google.com containing Compton AND (camera OR telescope OR imager OR imaging) in the title.

However, use of thin crystal coupled to PMTs for the scattering layer resulted in very poor imaging due to disturbing scattering in the PMT material. Interest in developing Compton camera really started in 1990 (figure 2.10) after the introduction of crossed Si-strips detector [71] directly recording the scattering position and the recoil electron energy, the electronics of which being located beside the detection area (see section 2.4.2). A second boost in development was triggered by the introduction of thin multi-pixel photo counter (MPPC) highly transparent to gamma rays [72] allowing one to use any thin scintillation crystal as scattering layer.

Up to 2005 Compton cameras were mainly developed and used for astronomy applications, in balloon or in satellite, where their capability to acquire in one go a large part of the sky was of prime importance. Afterwards they were investigated for medical applications, the first promising imaging occurring only recently for mice [73] and very recently in humans [74]. This last imaging was performed directly using one of the numerous mobile Compton cameras developed after 2011 for the Fukushima plan survey using helicopters or drones.

Being in the development phase for clinical applications, MC simulations of Compton camera are crucial and several codes have been developed to take into account specific features involved in Compton camera, such as a GUI-based software [75], or a Gate module CCMod [76].

Compton camera can be made so compact, that the laparoscopic Compton camera is even under development for laparoscopic tumours resection [77].

2.3.2 Principle

Figure 2.11 shows four principles of an electronically collimated single-photon camera.

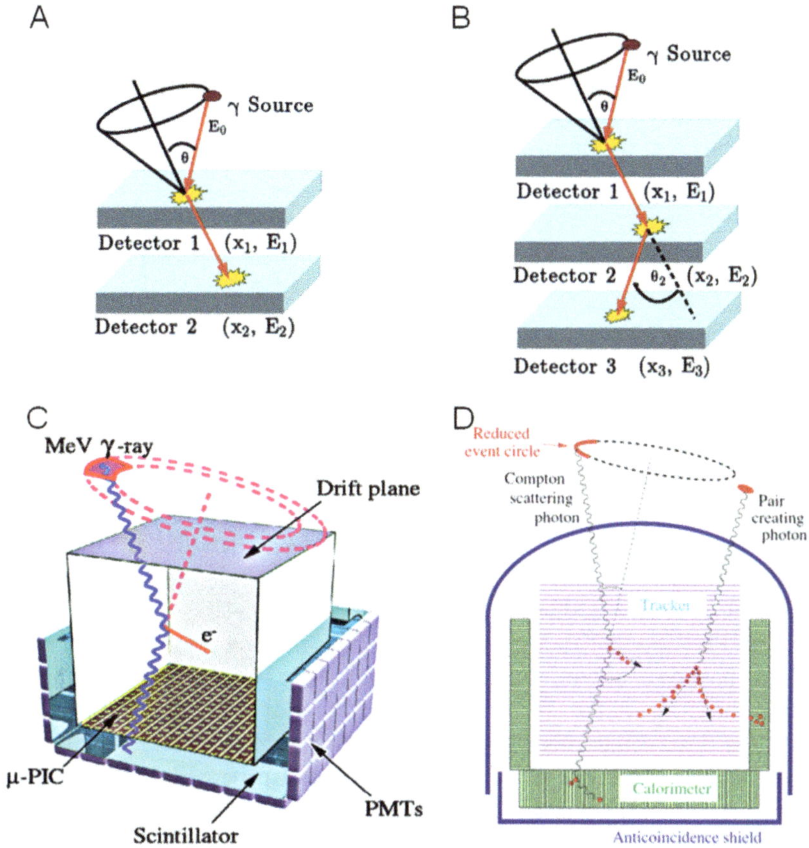

Figure 2.11. Four principles of electronically collimated single-photon camera. (a and b) Multilayers Compton camera working in two or three interactions events acquisition, respectively (reprinted from [78], copyright IOP publishing, all rights reserved). (c) Modern gaseous e-tracking Compton camera (reprinted from [79], copyright (2005) with permission of Elsevier). (d) Si-strip e-tracking and pair-creation camera (adapted from [80], copyright (2006) with permission of Springer Nature).

The historic Compton camera (figure 2.11(a)) proposed by [70] was based on two interactions events detections (note that the camera can have more than two scattering layers in order to increase the detection probability). When the scattered gamma rays are fully absorbed by the calorimeter layer, usually made of a thick crystal, the primary gamma ray emission is known to be on the cone with direction axis $\overrightarrow{x_1x_2}$ and revolution angle θ given by the Compton and energy conservation relations:

$$\cos\theta = 1 - \frac{m_e c^2}{E_0}\frac{E_1}{E_0 - E_1} \tag{2.4}$$

$$E_0 = E_1 + E_2 \tag{2.5}$$

Even when using mono-energetic gamma emitter, the second relation is needed to allow rejection of gamma rays scattered inside the patient body.

For high energy, the probability that the scattered gamma ray will only depose a fraction of its energy by Compton scattering and escape the calorimeter increases. Special spectral reconstruction techniques have been developed in order to compensate this issue [81]. An exact way to solve the issue is to limit the detection of three interactions events (figure 2.11(b)), in this case the energy E_0 is given by the relation [82]:

$$E_0 = E_1 + \frac{E_2}{2}\left(1 + \sqrt{1 + \frac{m_e c^2}{E_2} \frac{4}{1 - \cos\theta_2}}\right) \tag{2.6}$$

The drawback of this solution is a significant reduction of the camera sensitivity.

An alternative, is to apply on the two interactions events camera (figure 2.11(a)) the concept of active shielding [83]. The principle is to shield the back and lateral sides of the camera by a thick BGO rather than by lead or tungsten. Even for high energy, the probability that a gamma ray hits or crosses the BGO without any interaction is low. Thus gamma rays escaping the calorimeter layer after a partial interaction can be rejected by anti-coincidence with the BGO scintillation. Obviously, gamma rays that are back-scattered through the camera aperture are not rejected by this active shielding, i.e. around 40% [84].

Similarly to PET, but with a higher impact, a Compton camera using thick calorimeter suffers from parallax errors. Different systems are used to assess the DOI, e.g. sandwiching the calorimeter crystal by two MPPC [84] or by using calorimeter crystal pixelated in the three directions with additional lateral photodetector [85].

However, even if high sensitivity improvement can be obtained using multilayer cameras (figures 2.11(a) and (b)), the situation is, however, not so favourable as that in PET, where the electronic collimation determines an LOR. With a multilayers camera the electronic collimation gives a surface of response (figures 2.11(a) and (b)), which makes the tomographic process still more ill-posed than that of SPECT or PET. This issue is overcome by also tracking the recoil electron which results in an LOR. The electron tracking can be performed using a gaseous time projection chamber (TPC) (figure 2.11(c)) [86] or using several tens of silicon strip detectors (figure 2.11(d)) [80]. For gamma rays above 1.022 MeV, this last setup can also take advantage of pair-creation events by tracking the positron and the electron.

2.3.3 Reconstruction

Even if the incoming direction is not exactly known, a non-e-tracking Compton camera records gamma ray on all the LORs crossing the active target and the scattering layer surface. Even for a single camera position, this constitutes already a tomography acquisition, described by the conical radon transform. This transform has an analytical inversion [87] and an overview of different reconstruction algorithms can be found in [88]. Similarly to PET, list-mode MLEM based iterative algorithms providing higher imaging quality have been developed [89]. Specifically, this algorithm takes into account that for large cone revolution angle θ an event can come from outside the imaging space.

Contrary to Anger camera, the energy resolution not only affects the rejection capability of intra-patient scattered gamma rays, but also alters the angular accuracy of the cone revolution angle θ assessment as shown by equation (2.4), i.e. the conic surface is replaced by a thin conic ring. See [90] for a detailed discussion about how different factors govern Compton camera performance, such as: interactions position uncertainty, energy resolution, Doppler broadening effect, backward scattering detection. Ideally, all these effects should be included in the reconstruction algorithm [91].

Obviously an e-tracking Compton camera keeps the property of recording gamma rays on all the LORs crossing the target and the e-tracking chamber, but with the additional benefit to directly determine the LOR. This greatly facilitates the reconstruction process and dramatically reduce the ill-posedness of the tomographic system [92].

2.3.4 First clinical radionuclide imaging

Compton camera have several benefits. In most applications where the target of interest is the major radioactivity source, lateral shielding is not needed. Joined with the absence of collimator, Compton camera can be very light, i.e. a few kilograms. Furthermore as the camera solid FOV angle is almost 2π, the detector size can be much smaller than the transverse target section. These features allow realization of handheld camera. Figure 2.12 shows, to our knowledge, the first proof of concept in clinical use of a Compton camera [74], furthermore with an 22×22 cm^2 handheld camera initially developed for environmental survey of Fukushima plant by helicopter.

The planar image was acquired on the collimator lead fluorescence peak (80 keV) which benefits from the higher sensitivity of the crystal versus directly acquiring in the higher energy gamma rays energy windows. The scattered and calorimeter layers of the handheld Compton camera were made of Ce:GaGG scintillator coupled with several MPPCs allowing DOI assessment in the calorimeter. The camera was located 30 cm under the patient and the energy acquisition window was 318–378 keV. The acquisition shows compatible intestinal activity with the conventional planar view regarding intestines mobility and attenuation difference. The image is

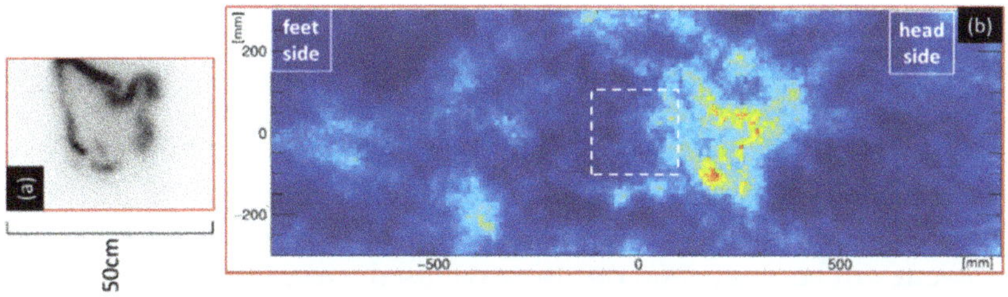

Figure 2.12. Image of a patient injected with 2.9 MBq of ^{223}Ra (reprinted from [74], copyright (2019) with permission of Elsevier). (a) 30 min planar image acquired with a MGEP Anger camera. (b) 2D whole-body reconstruction on the 30 cm far plane from a 10 min acquisition using the handheld Compton camera. (The left image was rotated to respect orientation and a dashed square was added by the authors to visualize the size and position of the camera).

still noisier, but this could easily be improve by optimizing the distance between the scatter layer and the calorimeter to human acquisition geometry. Also, utilization of several cameras located at different longitudinal and angular positions will improve the sensitivity and also will limit the reconstruction noise by improving the tomographic sampling completeness.

2.3.5 Perspectives: new radionuclides imaging

A Compton camera will not only improve radionuclide imaging but will also open access of radionuclides that cannot be imaged using a conventional gamma camera. Indeed, the gamma rays optimal energy range for a Compton camera is 500 keV to 4 MeV, in which the Compton scattering effect is dominant. The Compton camera can also be used outside this range, but with a sensitivity reduction due to loss of events by photoelectric effect in the scattering layer (or medium) below 500 keV, or by pair-creation above 4 MeV. Note that the electron and pair-creation tracking camera (figure 2.11(d)) could in theory be used to any upper energy limit.

Amongst high energy radionuclides of interest, and depending that a routine production can be developed, we could list: pure e-capture ^{95}Tc (20 h, 0.8 MeV) and ^{96}Tc(103 h, 0.8 MeV) allowing study of slow pharmacokinetics process, and beta emitter ^{59}Fe (44 d, 1.1 and 1.3 MeV) allowing long follow up of iron metabolism.

Quantitative *in vivo* imaging by Compton camera of the complex ^{225}Ac cascade decay should allow a dosimetry after a first therapy cycle in order to plan the following ones. Indeed the low activity injected, typically 6 MBq, makes difficult whole-body imaging with a conventional camera. Furthermore, the ^{225}Ac cascade decay has the alpha emitter ^{213}Bi having 440 and 800 keV gamma rays emissions. Due to its half-life of 46 min, ^{213}Bi likely follows a different pharmacokinetics than that of ^{225}Ac, making its pharmacokinetics assessment of interest.

2.3.6 Perspectives: multi-coincidence Compton camera

(This section is reprinted by permission from Springer Nature Customer Service Centre GmbH: Perspectives in Nuclear Medicine Tomography: A Physicist's Point of View by S Walrand and F Jamar [*93*]*: section* 5.2.4: 2013)

Using several Compton cameras positioned along a polygon setup around the patient, all the 360° directions are covered removing the need to rotate the camera. In addition, contrary to multi-head SPECT using mechanical collimators, the system can record all the γ-rays originating from a single decay.

In 1987, Liang *et al* [94] pointed out that, using these Compton cameras in coincidence, the decay must have occurred on the intersection of several cones (figure 2.13). With an isotope emitting two γ-rays in cascade such as ^{111}In or ^{67}Ga, the decay event will be located on one or two closed curve lines being the intersection of the two cones (figure 2.13(a)). Using an isotope emitting three γ-rays in cascade such as ^{130}I or ^{94}Tc, the intersection of the three cones will reduce to a finite number of points, in the most favourable case only one point (figure 2.13(b)). Of course, increasing the coincidence multiplicity decreases the system efficiency: keeping only triple coincidences events efficiency is expected to fall to that of conventional SPECT

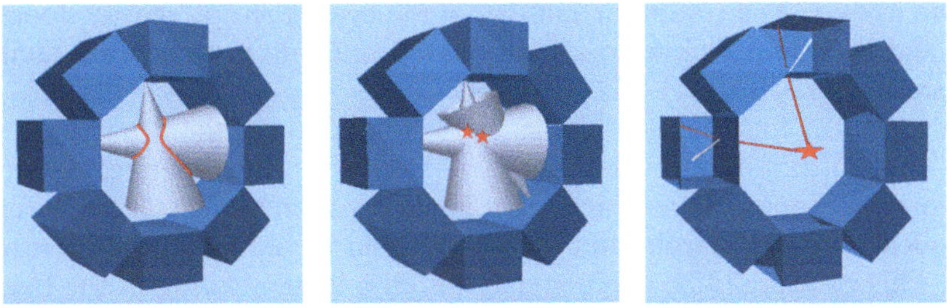

Figure 2.13. Improved accurate position information can be obtained using an octagon of eight Compton cameras working in coincidence. (a and b) Conventional Compton camera with isotopes emitting at least two, or at least three γ-rays, respectively. (c) Recoil electron tracking Compton camera with isotope emitting at least two γ-rays (top and left detectors are shown open to illustrate the recoil photon and the recoil electron tracking).

($\approx 10^{-4}$) [95]. However, each of such recorded events will be much more informative that in no coincidence mode, and by only considering triple-coincidence-intersections reducing to one point will make the reconstruction process superfluous (the benefits of this feature have already been discussed in section 2.1).

Using several recoil electron tracking Compton cameras like the one proposed by Kabuki all the two γ-rays coincidences will directly provide the decay position (figure 2.13(c)). To this end the coincidence triggering has to be performed using the first detector, i.e. the Compton camera that is a fast silicon detector. A recent prototype of such a Compton camera working at atmospheric pressure showed a single-photon efficiency of 1.5 10^{-4} for energies above 150 keV and for a $10 \times 10 \times 10$ cm^3 detection volume [96], equivalent to an efficiency of about 0.4% for $30 \times 30 \times 30$ cm^3 detection volume. Simulations have shown that at higher pressure values this detector volume could provide an efficiency of about 2% [97]. Surrounding the patient by such detectors, such as in figure 2.13, the single-photon efficiency should be around 10%. When working in two γ-rays coincidences the sensitivity should be similar to that of 3D PET, i.e. ≈ 1%, but again without the need of using a reconstruction process. Substances labelled with ^{94}Tc(4.9 h), ^{95}Tc(20 h), ^{96}Tc(103 h), ^{111}In(67 h), ^{198}Tl(7 h), ^{199}Tl(7 h), ^{67}Ga(78 h) will be usable and, as a consequence, also most of the radiopharmaceuticals developed for SPECT. For isotopes emitting n γ-rays in cascade per decay, the efficiency of detecting at least two photons in coincidences is $n \times \text{eff} \times (n\text{-}1) \times \text{eff}$, where eff is the detector efficiency. Thus, for ^{96}Tc, that emits at least three γ-rays of energy around 800 keV, the efficiency will thus further be improved by a factor three versus a two-photons emitter. Obviously, such sensitivity estimations are based on preliminary simulations and have still to be confirmed by real prototypes.

2.4 Proton beam therapy imaging

2.4.1 Motivation

An ion beam has the unique feature to allow shaping the absorbed dose within a part of the beam path. This feature allows preserving tissues located behind the tumour and ahead as well. As a result, the number of ion beam therapy centres doubled each decade, and is currently close to a hundred [98], mainly proton beam facilities for cost reasons.

However, many uncertainties challenge the prediction of the Bragg peak position [99]: patients positioning, translation to CT Hounsfield into chemical composition, and tumour positions change which is a major limitation in the soft tissue region. As a consequence, safety margin and dose limitation have to be implemented in the therapy. This issue could be overcome by an on-line monitoring of the Bragg peak position, permitting the patient repositioning or the beam energy tuning.

During ion therapy, positron emitters are produced in very low abundance from nuclear fragmentation, mainly ^{15}O (122 s), ^{11}C (20 min) and ^{13}N (10 min). However, the use of PET to monitor the Bragg peak meets several issues. Monte Carlo simulations show that for proton therapy there is a mismatch between the Bragg peak position and the positron emitters production (figure 2.14(a)). Even if a deep learning machine can localize the Bragg peak [100], the isotopes half-lives hamper a second PET imaging after a quick patient repositioning or ion energy tuning.

Imaging of prompt gamma rays emitted by the 4.4 MeV $^{12}C^{2+}$ excited state solves these two issues. As the cross-section of $^{12}C^{2+}$ production has a low energy threshold and sharp peak (figure 2.14(b)), the 4.4 MeV gamma rays production matches the Bragg peak position (figure 2.14(a)). Furthermore, their emissions instantaneously stop when switching the proton beam off ($^{12}C2^{+}$ lifetime≈60 fs [101]), allowing an easy novel monitoring after therapy setup tuning. We will focus on two techniques to image the absorbed dose in proton beam therapy via gamma rays acquisition (the

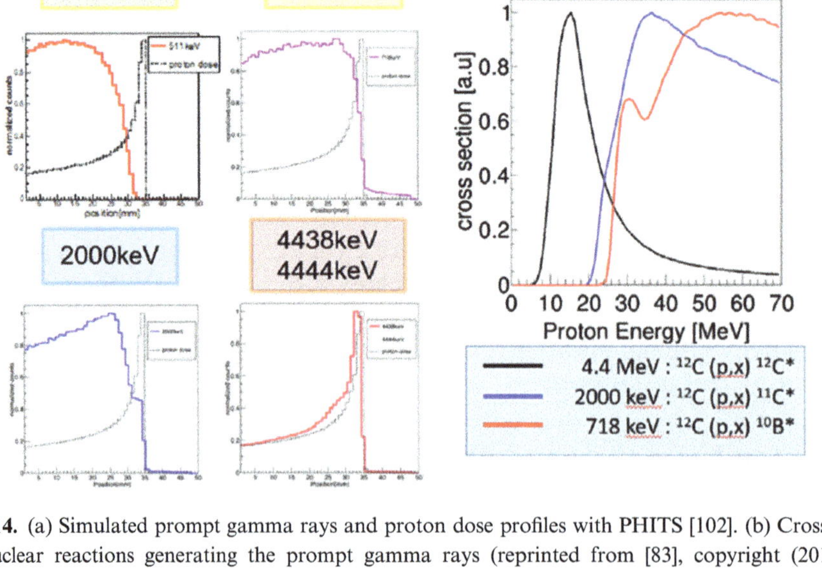

Figure 2.14. (a) Simulated prompt gamma rays and proton dose profiles with PHITS [102]. (b) Cross-section of the nuclear reactions generating the prompt gamma rays (reprinted from [83], copyright (2019) with permission of Elsevier).

reader can find in [102] an exhaustive list with a short description of the different techniques).

Beside prompt gamma rays imaging, a novel technique based on acoustic wave production by the protons interaction is in development keeping the temporal and spatial properties of the 4.4 MeV gamma rays imaging.

2.4.2 Compton camera

Figure 2.15(a) shows current development in 4.4 MeV imaging using a Compton camera [84]: the scatterer is made of two layers of $0.5 \times 0.5 \times 3$ mm^3 Ce:GACG voxels and the absorber made of 10 layers of $2 \times 2 \times 2$ mm^3 Ce:GACG voxels. Each scattering layer is coupled to an MPPC array while the absorber one is sandwiched between two MPPCs allowing DOI assessment. Only events giving the same position on these two MPPC are used, which eliminates 90% of multi-scattered events for which the DOI position cannot be assessed. BGO is set on the back and lateral sides of the camera as an active shielding for incoming background, but also as escape detector allowing the rejection of 60% of the partial absorber interactions (see section 2.3.2).

The 3.5 *h* acquisition for a 70 MeV 2.5 pA beam targeting in eight PMMA slab phantoms clearly shows the wrong position of the positron emitters (figure 2.15(b)), while the Bragg peak position is rightly localized using the 4.4 MeV prompt gamma rays (figure 2.15(c)) as was expected from MC simulations (figure 2.15(a)). The elongation of the 4.4 MeV gamma rays emission distribution beyond the phantom plate resulted from the worse sampling in very decentred regions versus the detector surface. Preferably, the detector should be centred on the expected Bragg peak position. The next step should be developing a more sensitive system by increasing the detection area, currently 3×3 cm^2, and by setting several cameras around the target, and capable operating at higher counts rate.

The imaging of 4.4 MeV gamma rays could be highly improved by the use of a hodoscope [104]. Such a device will be placed in the proton beam and should give the individual proton transverse position and time. As a result the reconstruction

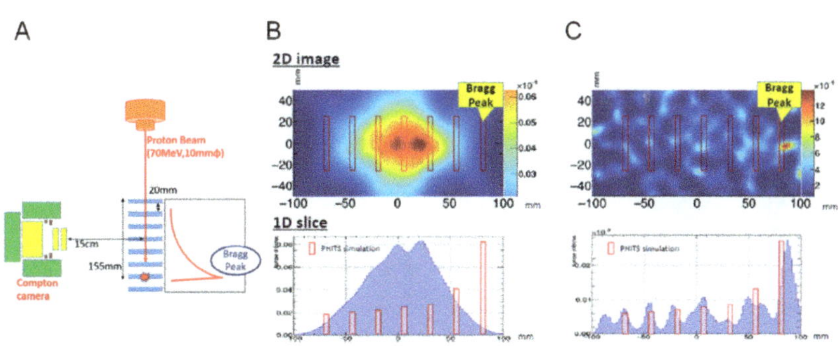

Figure 2.15. (a): Acquisition setup. (b and c) Reconstructed slice and profiles of 511 keV and 4.4 MeV gamma ray production distribution, respectively. The red rectangles in the slice indicate the phantom positions. The red bars in the profiles show the absorbed dose simulated with PHITS (reprinted from [84], copyright (2019) with permission of Elsevier).

process is no longer needed since the gamma ray emission will be located at the intersection of the Compton camera cone response surface with the line joining the ion beam aperture to the measured proton transverse position. A time coincidence window between the hodoscope and the Compton camera will further reduce background contamination. As a result the counts statistics needed to obtain an accurate imaging could be drastically reduced.

An interesting analysis on Compton camera challenges in prompt gamma rays imaging can be found in [105], together with a detailed references list.

2.4.3 Acoustics waves imaging

Localization of the Bragg peak using proton-induced acoustic signal was originally proposed in 1957 [106]. For the time being, experimental proton beam acoustic wave acquisitions and reconstructions have only been performed in a water tank, the acoustic sensors being set directly in the water [107, 108].

Figure 2.16 shows recent simulation proton absorbed dose assessment by TOF reconstruction of the acoustics signal [109], this paper also contained a detailed reference list on the topics. A CT based patient phantom was used. The absorbed dose from 2×10^7 115 MeV protons beam was simulated by GATE while the acoustic waves were simulated using the instantaneous thermoelastic expansion model [110]. The database [111] was used for the acoustic waves propagation in the different tissues. Ideal sensors were used in the simulations.

The 32 sensors reconstructed absorbed dose (figure 2.16(f)) is accurately in line with the simulated absorbed dose. However, acoustic sensors imperfections and skin-detector interface wave reflections were not modelled in the simulation. Nevertheless, this technique looks promising, as it will be cheap, compact and not impacted by particles and photons background.

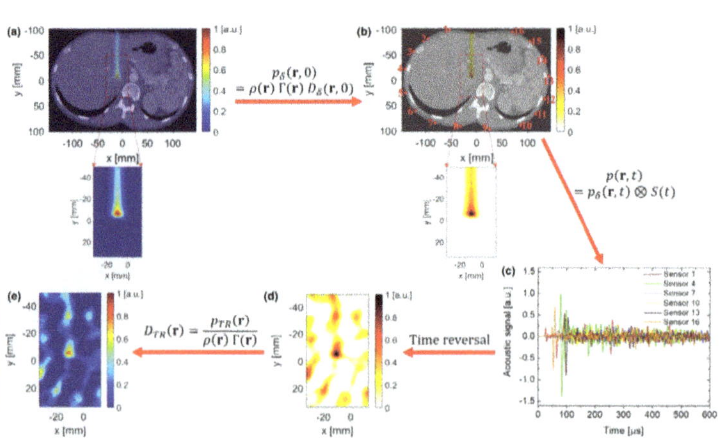

Figure 2.16. Schematic image reconstruction. (a) The original dose distributions. (b) The corresponding original pressure distributions and the 16 sensor positions. (c) 6 out of 16 acoustic time series signals. (d) The reconstructed pressure distribution. (e and f) The corresponding reconstructed dose distribution using 16 and 32 sensors (not shown), respectively (reprinted from [109] with permission of Wiley & Sons, copyright (2019) American Association of Physicists in Medicine).

References

[1] Butler J F *et al* 1997 CdZnTe solid-state gamma camera *IEEE Nucl. Sci. Conf.* R. 1 281–5

[2] Amrami R, Shani G, Hefetz Y, Blevis I and Pansky A 2000 A comparison between the performance of a pixellated CdZnTe based gamma camera and Anger NaI(TI) scintillator gamma camera *Proc. 22nd Annual Int. Conf.* (Washington, DC: IEEE) 352–55

[3] Wacholz C, Hruska C and Veriton O C M 2020 Multi-CZT detector SPECT/CT system acceptance testing *J. Nucl. Med.* **61** 3003

[4] Strosberg J *et al* 2017 Phase 3 trial of ^{177}Lu-dotatate for midgut neuroendocrine tumors *New Engl. J. Med.* **376** 125–35

[5] Kratochwil C *et al* 2019 EANM procedure guidelines for radionuclide therapy with ^{177}Lu-labelled PSMA-ligands (^{177}Lu-PSMA-RLT) *Eur. J. Nucl. Med. Mol. Imag.* **46** 2536–44

[6] Kennedy J, Lugassi R, Reizberg I and Keidar Z 2017 Quantitation of Lu177 using general purpose whole-body solid state CZT and standard SPECT/CT cameras *J. Nucl. Med.* **58** 398

[7] Kennedy J A, Lugassi R, Gill R and Keidar Z 2020 Digital solid-state SPECT/CT quantitation of absolute ^{177}Lu-radiotracer concentration: *in vivo*/*in vitro* validation *J. Nucl. Med.* **61** 1381–7

[8] de Jong M *et al* 1997 Yttrium-90 and indium-111 labelling, receptor binding and biodistribution of [DOTA 0, d-Phe 1, Tyr 3] octreotide, a promising somatostatin analogue for radionuclide therapy. *Eur. J. Nucl. Med.* **24** 368–71

[9] Caobelli F, Wollenweber T, Bavendiek U, Kühn C, Schütze C, Geworski L, Thackeray J T, Bauersachs J, Haverich A and Bengel F M 2017 Simultaneous dual-isotope solid-state detector SPECT for improved tracking of white blood cells in suspected endocarditis *Eur. Heart J.* **38** 436–43

[10] Padia S A *et al* 2017 Radioembolization of hepatic malignancies: background, quality improvement guidelines, and future directions *JVIR.* **28** 1

[11] Garin E, Rolland Y, Laffont S and Edeline J 2016 Clinical impact of ^{99m}Tc-MAA SPECT/CT-based dosimetry in the radioembolization of liver malignancies with 90 Y-loaded microspheres. *Eur. J. Nucl. Med. Mol. I.* **43** 559–75

[12] Ho S, Lau W Y, Leung T W, Chan M, Ngar Y K, Johnson P J and Li A K 1996 Partition model for estimating radiation doses from yttrium-90 microspheres in treating hepatic tumours *Eur. J. Nucl. Med.* **23** 947–52

[13] Simon N and Feitelberg S 1967 Scanning bremsstrahlung of yttrium-90 microspheres injected intra-arterially *Radiology* **88** 719–24

[14] Lhommel R, Goffette P, Van den Eynde M, Jamar F, Pauwels S, Bilbao J I and Walrand S 2009 Yttrium-90 TOF PET scan demonstrates high-resolution biodistribution after liver SIRT *Eur. J. Nucl. Med. Mol. Imag.* **36** 1696

[15] Wright C L, Zhang J, Tweedle M F, Knopp M V and Hall N C 2015 Theranostic imaging of yttrium-90 *BioMed Res. Int.* **28** 2015

[16] Dewaraja Y K, Chun S Y, Srinivasa R N, Kaza R K, Cuneo K C, Majdalany B S, Novelli P M, Ljungberg M and Fessler J A 2017 Improved quantitative ^{90}Y bremsstrahlung SPECT/CT reconstruction with Monte Carlo scatter modeling *Med. Phys.* **44** 6364–76

[17] Xiang H, Lim H, Fessler J A and Dewaraja Y K 2020 A deep neural network for fast and accurate scatter estimation in quantitative SPECT/CT under challenging scatter conditions. *Eur. J. Nucl. Med. Mol. Imag.* **47** 2956

[18] Reddin J S, Scheuermann J S, Bharkhada D, Smith A M, Casey M E, Conti M and Karp J S 2018 Performance evaluation of the SiPM-based siemens biograph vision PET/CT system. *IEEE Nucl. Sci. Symp. Med. Imag. Conf. (NSS/MIC)* p 1
[19] Walrand S, Hesse M, Demonceau G, Pauwels S and Jamar F 2011 Yttrium-90-labeled microsphere tracking during liver selective internal radiotherapy by bremsstrahlung pinhole SPECT: feasibility study and evaluation in an abdominal phantom *EJNMMI Res* **1** 32
[20] Walrand S, Hesse M, Wojcik R, Lhommel R and Jamar F 2014 Optimal design of Anger camera for bremsstrahlung imaging: Monte Carlo evaluation *Front. Oncol.* **4** 149
[21] Conti M 2006 Effect of randoms on signal-to-noise ratio in TOF PET *IEEE Trans. Nucl. Sci.* **53** 1188–193
[22] Rezaei A, Defrise M and Nuyts J 2014 ML-reconstruction for TOF-PET with simultaneous estimation of the attenuation factors *IEEE Trans. Nucl. Imag.* **33** 1563–72
[23] Mehranian A and Zaidi H 2015 Joint estimation of activity and attenuation in whole-body TOF PET/MRI using constrained Gaussian mixture models *IEEE Trans. Med. Imag.* **34** 1808–21
[24] Nuyts J, Rezaei A and Defrise M 2018 The validation problem of joint emission/ transmission reconstruction from TOF-PET projections. *IEEE Trans. Rad. Plas. Med. Sci.* **2** 273–78
[25] Conti M 2010 Why is TOF PET reconstruction a more robust method in the presence of inconsistent data? *Phys. Med. Biol.* **56** 155
[26] Conti M 2011 Focus on time-of-flight PET: the benefits of improved time resolution *Eur. J. Nucl. Med. Mol. Imag.* **38** 1147–157
[27] Walrand S, Hesse M, Jamar F and Lhommel R 2018 The origin and reduction of spurious extrahepatic counts observed in 90Y non-TOF PET imaging post radioembolization *Phys. Med. Biol.* **63** 075016
[28] Bertero M and Boccacci P 1998 *Introduction to Inverse Problems in Imaging* (Boca Raton, FL: CRC Press)
[29] Llacer J 1982 Tomographic image reconstruction by eigenvector decomposition: its limitations and areas of applicability *IEEE Trans. Nucl. Imag.* **1** 34–42
[30] Burgess A E 2011 Visual perception studies and observer models in medical imaging *Sem. Nucl. Med.* **41** 419–36
[31] Schaart D R, Ziegler S and Zaidi H 2020 Achieving 10 ps coincidence time resolution in TOF-PET is an impossible dream *Med. Phys.* **47** 2721–24
[32] Derenzo S E, Choong W S and Moses W W 2014 Fundamental limits of scintillation detector timing precision *Phys. Med. Biol.* **59** 3261
[33] Janecek M 2012 Reflectivity spectra for commonly used reflectors *IEEE Trans. Nucl. Sci.* **59** 490–97
[34] Lewellen T K, Bice A N, Harrison R L, Pencke M D and Link J M 1988 Performance measurements of the SP3000/UW time-of-flight positron emission tomograph *IEEE Trans. Nucl. Sci.* **35** 665–69
[35] Muzic R F and Kolthammer J A 2006 PET performance of the GEMINI TF: a time-of-flight PET/CT scanner *IEEE Nucl. Sci. Conf. R.* **3** 1940–44
[36] Miller M, Griesmer J, Jordan D, Laurence T, Muzic R, Narayanan M, Natarajamani D, Su K H and Wang S 2014 Initial characterization of a prototype digital photon counting PET system *J. Nucl. Med.* **55** 658

[37] Berg E and Cherry S R 2018 Innovations in instrumentation for positron emission tomography *Sem. Nucl. Med.* **48** 311–31
[38] Pasciak A S, Bradley Y and McKinney J M (ed) 2016 *Handbook of Radioembolization: Physics, Biology, Nuclear Medicine, and Imaging* (Boca Raton, FL: CRC Press)
[39] Conti M and Bendriem B 2019 The new opportunities for high time resolution clinical TOF PET *Clin. Trans. Imag.* **7** 139–47
[40] Walrand S and Hesse M 2019 PET/CT specificities in ^{90}Y imaging post radioembolization *PET Clin.* **14** 469–76
[41] Walrand S, Hesse M, Chiesa C, Lhommel R and Jamar F 2014 The low hepatic toxicity per Gray of ^{90}Y glass microspheres is linked to their transport in the arterial tree favoring a nonuniform trapping as observed in posttherapy PET imaging *J. Nucl. Med.* **55** 135–40
[42] Crookston N R, Fung G S and Frey E C 2018 Development of a customizable hepatic arterial tree and particle transport model for use in treatment planning *IEEE Trans. Radiat. Plasma Med. Sci.* **3** 31–7
[43] Hogberg J, Rizell M and Hultborn R *et al* 2014 Heterogeneity of microsphere distribution in resected liver and tumour tissue following selective intrahepatic radiotherapy *EJNMMI Res.* **4** 48
[44] van de Maat G H, Seevinck P R and Elschot M *et al* 2013 MRIbased biodistribution assessment of holmium-166 poly(L-lactic acid) microspheres after radioembolisation *Eur. Radiol.* **23** 827–35
[45] Hesse M and Walrand S 2019 Significant artefactual noise in ^{90}Y TOF-PET imaging of low specific activity phantoms arises despite increased acquisition time *EJNMMI Phys.* **6** 1–4
[46] d'Abadie P, Hesse M, Jamar F, Lhommel R and Walrand S 2018 90 Y TOF-PET based EUD reunifies patient survival prediction in resin and glass microspheres radioembolization of HCC tumours *Phys. Med. Biol.* **63** 245010
[47] Chiesa C, Bardiès M and Zaidi H 2019 Voxel-based dosimetry is superior to mean absorbed dose approach for establishing dose-effect relationship in targeted radionuclide therapy *Med. Phys.* **46** 5403–6
[48] Dewaraja Y K, Devasia T, Kaza R K, Mikell J K, Owen D, Roberson P L and Schipper M J 2020 Prediction of tumor control in ^{90}Y radioembolization by logit models with PET/CT-based dose metrics *J. Nucl. Med.* **61** 104–11
[49] Van Elmbt L, Vandenberghe S, Walrand S, Pauwels S and Jamar F 2011 Comparison of yttrium-90 quantitative imaging by TOF and non-TOF PET in a phantom of liver selective internal radiotherapy *Phys. Med. Biol.* **56** 6759
[50] Walrand S, Jamar F, van Elmbt L, Lhommel R, Bekonde E B and Pauwels S 2010 4-Step renal dosimetry dependent on cortex geometry applied to ^{90}Y peptide receptor radiotherapy: evaluation using a fillable kidney phantom imaged by ^{90}Y PET *J. Nucl. Med.* **51** 1969–73
[51] Janecek M and Moses W W 2008 Optical reflectance measurements for commonly used reflectors *IEEE Trans. Nucl. Sci.* **55** 2432–37
[52] Harmon E S, Thompson M O, Schmidtlein C R, Turner J N and Krol A 2019 Towards 50 ps TOF-PET for brain imaging *Med. Imaging* **10953** 03
[53] Chauvat D, Bonnet C, Dunseath K, Emile O, Le and Floch A 2005 Timing the total reflection of light *Phys. Lett.* A **336** 271–73
[54] Huber J S, Moses W W, Andreaco M S, Loope M, Melcher C L and Nutt R 1999 Geometry and surface treatment dependence of the light collection from LSO crystals *Nucl. Instrum. Meth. A.* **437** 374–80

[55] Berg E, Roncali E and Cherry S R 2015 Optimizing light transport in scintillation crystals for time-of-flight PET: an experimental and optical Monte Carlo simulation study *Biomed. Opt. Express* **6** 2220–30
[56] Schmand M, Eriksson L, Casey M E, Andreaco M S, Melcher C, Wienhard K, Flugge G and Nutt R 1998 Performance results of a new DOI detector block for a high resolution PET-LSO research tomograph HRRT *IEEE Trans. Nucl. Sci.* **45** 3000–6
[57] Burr K C, Ivan A, Castleberry D E, Thompson R A, LeBlanc J W, Shah K S and Farrell R 2004 Depth-of-interaction-encoding high-resolution PET detector with sub-nanosecond timing *IEEE Symp. Conf. R. Nucl. Sci.* **6** 3880–84
[58] Li M and Abbaszadeh S 2019 Depth-of-interaction study of a dual-readout detector based on TOFPET2 application-specific integrated circuit *Phys. Med. Biol.* **64** 175008
[59] Doroud K, Liu Z and Williams M C 2019 A new geometry of scintillating crystals with strip SiPMs: a PET detector with precise position and time determination *J. Instr.* **14** P01016
[60] Bieniosek M F, Cates J W and Levin C S 2016 A multiplexed TOF and DOI capable PET detector using a binary position sensitive network *Phys. Med. Biol.* **61** 7639
[61] Walrand S, Hesse M and Jamar F 2020 Achieving sub-100 ps TOF resolution in thick-LSO PET while reducing the system cost: a Monte Carlo study *Phys. Med. Biol.* **65** 205009
[62] Son J W, Lee M S and Lee J S 2016 A depth-of-interaction PET detector using a stair-shaped reflector arrangement and a single-ended scintillation light readout *Phys. Med. Biol.* **62** 465
[63] Furumiya T, Tsuda T, Tonami H, Satoh M, Nakazawa M, Ohi J and Kitamura K 2014 Development of a SiPM based MR-compatible DOI-TOF-PET detector *IEEE Nucl. Sci. Symp. Med. Imag. Conf.(NSS/MIC)* **8** 1–3
[64] Lecoq P *et al* 2020 Roadmap toward the 10 ps time-of-flight PET challenge *Phys. Med. Biol.* **65** 21
[65] Korzhik M, Tamulaitis G and Vasil'ev A N 2020 *Physics of Fast Processes in Scintillators* (Berlin: Springer Nature)
[66] Gundacker S, Turtos R M, Kratochwil N, Pots R H, Paganoni M, Lecoq P and Auffray E 2020 Experimental time resolution limits of modern SiPMs and TOF-PET detectors exploring different scintillators and Cherenkov emission *Phys. Med. Biol.* **65** 025001
[67] Ariño-Estrada G, Mitchell G S, Kim H, Du J, Kwon S I, Cirignano L J, Shah K S and Cherry S R 2019 First Cerenkov charge-induction (CCI) TlBr detector for TOF-PET and proton range verification *Phys. Med. Biol.* **64** 175001
[68] Harvey J E and Pfisterer R N 2019 Understanding diffraction grating behavior: including conical diffraction and Rayleigh anomalies from transmission gratings *Opt. Eng.* **58** 087105
[69] Oliva M 2014 High efficiency blazed gratings in resonance domain *Doctoral Dissertation* https://db-thueringen.de/receive/dbt_mods_00027044
[70] Todd R W, Nightingale J M and Everett D B 1974 A proposed γ camera *Nature* **251** 132–34
[71] Kamae T, Hanada N and Enomoto R 1988 Prototype design of multiple Compton gamma-ray camera *IEEE Trans. Nucl. Sci.* **35** 352–55
[72] Buzhan P *et al* 2002 The advanced study of silicon photomultiplier *Adv. Tech. Part. Phys.* 717–28
[73] Kishimoto A *et al* 2017 First demonstration of multi-color 3-D *in vivo* imaging using ultra-compact Compton camera *Sci. Rep.* **7** 1–7
[74] Fujieda K *et al* 2020 First demonstration of portable Compton camera to visualize 223-Ra concentration for radionuclide therapy *Nucl. Instrum. Meth. A.* **958** 162802

[75] Lee S H, Seo H, Park J H, Park S H, Lee J S, Lee J H, Lee C S and Kim C H 2009 CIS: a GUI-based software system for Monte Carlo simulation of Compton camera *Nucl. Technol.* **168** 55–60
[76] Etxebeste A, Dauvergne D, Fontana M, Létang J M, Llosá G, Munoz E, Oliver J F, Testa É and Sarrut D 2020 CCMod: a GATE module for Compton camera imaging simulation *Phys. Med. Biol.* **65** 055004
[77] Nakamura Y, Shimazoe K, Takahashi H, Yoshimura S, Seto Y, Kato S, Takahashi M and Momose T 2016 Development of a novel handheld intra-operative laparoscopic Compton camera for ^{18}F-Fluoro-2-deoxy-2-D-glucose-guided surgery *Phys. Med. Biol.* **61** 5837
[78] Muñoz E, Barrio J, Etxebeste A, Ortega P G, Lacasta C, Oliver J F, Solaz C and Llosá G 2017 Performance evaluation of MACACO: a multilayer Compton camera *Phys. Med. Biol.* **62** 7321
[79] Takada A *et al* 2005 Development of an advanced Compton camera with gaseous TPC and scintillator *Nucl. Instrum. Meth. A.* **546** 258–62
[80] Andritschke R, Zoglauer A, Kanbach G, Bloser P F and Schopper F 2005 The Compton and pair creation telescope MEGA *Exp. Astron.* **20** 395–403
[81] Muñoz E, Barrientos L, Bernabéu J, Borja-Lloret M, Llosá G, Ros A, Roser J and Oliver J F 2020 A spectral reconstruction algorithm for two-plane Compton cameras *Phys. Med. Biol.* **65** 025011
[82] Dogan N and Wehe D 1993 Optimization and angular resolution calculations for a multiple Compton scatter camera *Proc. IEEE Nucl. Sci. Symp. Med. Imag. Conf. R* 269–73
[83] Vallerga J V, Ricker G R, Schnepple W S and Ortale C 1982 A bismuth germanate-shielded mercuric iodide x-ray detector for space applications *IEEE Trans. Nucl. Sci.* **29** 151–54
[84] Mochizuki S, Kataoka J, Koide A, Fujieda K, Maruhashi T, Kurihara T, Sueoka K, Tagawa L, Yoneyama M and Inaniwa T 2019 High-precision Compton imaging of 4.4 MeV prompt gamma-ray toward an on-line monitor for proton therapy *Nucl. Instrum. Meth. A.* **936** 43–5
[85] Koide A *et al* 2018 Precision imaging of 4.4 MeV gamma rays using a 3-D position sensitive Compton camera *Sci. Rep.* **8** 1–9
[86] Tanimori T, Nishi Y, Ochi A and Nishi Y 1999 Imaging gaseous detector based on micro-processing technology *Nucl. Instrum. Meth. A.* **436** 188–95
[87] Nguyen M K and Truong T T 2002 Exact inversion of a compound conical Radon transform and a novel nuclear imaging principle *C.R. Math.* **335** 213–17
[88] Smith B 2005 Reconstruction methods and completeness conditions for two Compton data models *J. Opt. Soc. Am.* A **22** 445–59
[89] Zoglauer 2005 First light for the next generation of Compton and pair telescopes *PhD Thesis* Technische Universitat Munchen
[90] Kishimoto A 2017 Study of ultra-compact Compton camera based on 3-D position sensitive scintillator for environmental and medical gamma-ray imaging *Doctoral Dissertation* Waseda University
[91] Wilderman S J, Fessler J A, Clinthorne N H, LeBlanc J W and Rogers W L 2001 Improved modeling of system response in list mode EM reconstruction of Compton scatter camera images *IEEE Trans. Nucl. Sci.* **48** 111–16
[92] Inagaki M, Ogawa K and Tanimori T 2019 Image reconstruction with an electron tracking Compton camera *IEEE Nucl. Sci. Symp. Med. Imag. Conf.(NSS/MIC)* 1–3

[93] Walrand S and Jamar F 2013 Perspectives in nuclear medicine tomography: a physicist's point of view *Imaging in Nuclear Medicine* (Berlin, Heidelberg: Springer) pp 69–96
[94] Liang Z, Hart H and Schoenfeld A 1987 Triple gamma coincidence tomographic imaging without image processing *IEEE Conf. Eng. Med. Biol.* **9** 825–26
[95] Andreyev A, Sitek A and Celler A 2010 Spatial resolution of the multiple coincidences Compton camera *Nuclear Science Symp. Conf. Record (NSS/MIC)* 2099–103
[96] Takada A, Kubo H and Nishimur H *et al* Observation of diffuse cosmic and atmospheric gamma rays at balloon altitudes with an electron-tracking Compton camera http://arxiv.org/PS_cache/arxiv/pdf/1103/1103.3436v2.pdf
[97] Kabuki S, Hattori K and Kohara R *et al* 2007 Development of electron tracking Compton camera using micro pixel gas chamber for medical imaging *Nucl. Inst. Meth. Phys. Res. Sect.* A **580** 1031–35
[98] PTCOG web page https://ptcog.ch/index.php/facilities-in-operation
[99] Paganetti H 2012 Range uncertainties in proton therapy and the role of Monte Carlo simulations *Phys. Med. Biol.* **57** R99
[100] Hu Z, Li G, Zhang X, Ye K, Lu J and Peng H 2020 A machine learning framework with anatomical prior for online dose verification using positron emitters and PET in proton therapy *Phys. Med. Biol.* **65** 185003
[101] Stocki T J, Measday D F, Gete E, Saliba M A, Moftah B A and Gorringe T P 2002 Gamma rays from muon capture in ^{14}N *Nucl. Phys.* A **697** 55–74
[102] Wrońska A 2020 Prompt gamma imaging in proton therapy-status, challenges and developments *J. Phys. Conf. Ser.* **1561** 012021
[103] Sato T *et al* 2018 Features of particle and heavy ion transport code system (PHITS) version 3.02 *J. Nucl. Sci. Tech.* **55** 684–90
[104] Gallin-Martel M L *et al* 2018 A large area diamond-based beam tagging hodoscope for ion therapy monitoring *EPJ Web. Conf.* **170** 09005
[105] Hueso-González F, Fiedler F, Golnik C, Kormoll T, Pausch G, Petzoldt J, Römer K E and Enghardt W 2016 Compton camera and prompt gamma ray timing: two methods for *in vivo* range assessment in proton therapy *Front. Oncol.* **6** 80
[106] Askaryan G A 1957 Hydrodynamic radiation from the tracks of ionizing particles in stable liquids *Sov. J. At. Energy.* **3** 921–23
[107] Jones K C, Vander S F, Sehgal C M and Avery S 2016 Acoustic time-of-flight for proton range verification in water *Med. Phys.* **43** 5213–224
[108] Nie W, Jones K C, Petro S, Kassaee A, Sehgal C M and Avery S 2018 Proton range verification in homogeneous materials through acoustic measurements *Phys. Med. Biol.* **63** 36–49
[109] Yu Y, Li Z, Zhang D, Xing L and Peng H 2019 Simulation studies of time reversal-based protoacoustic reconstruction for range and dose verification in proton therapy *Med. Phys.* **46** 3649–62
[110] Ahmad M, Xiang L, Yousefi S and Xing L 2015 Theoretical detection threshold of the proton-acoustic range verification technique *Med. Phys.* **42** 5735–744
[111] Hasgall P, Di Gennaro F and Baumgartner C *et al* 2018 Database for thermal and electromagnetic parameters of biological tissues. Vers 4.0. itis.swiss/database

IOP Publishing

Monte Carlo Calculations in Nuclear Medicine (Second Edition)
Therapeutic applications
Habib Zaidi

Chapter 3

Basic concepts of internal radiation dosimetry

Pat Zanzonico and Wesley Bolch

3.1 Introduction

Radiation dosimetry deals with the determination of the amount, as well as the spatial and temporal distributions, of energy deposited in matter by ionizing radiation. Internal radionuclide dosimetry deals specifically with radiation energy deposition in tissue for radionuclides within the body. Internal dosimetry has been applied to occupational exposures in radiation protection, environmental exposures in radiation epidemiology, and diagnostic and therapeutic exposures in nuclear medicine. With the increasing therapeutic application of internal radionuclides in medicine, larger administered activities are used and higher normal-tissue doses, with associated radiation injury, may result. Treatment planning for radionuclide therapy thus requires, in addition to reliable dose-response relationships for target tissues and dose-toxicity relationships for normal tissues, increasingly more accurate and precise dose estimates for target tissue and at-risk normal tissues. Radiation dosimetry in nuclear medicine thus continues to evolve from (a) the estimation of population-averaged and organ-averaged doses for risk estimation for stochastic effects to (b) the calculation of individualized patient organ doses that reflect the heterogeneity of the dose distribution within both tumors and non-targeted normal organs, and their associated probabilities of tumor control and tissue complications (deterministic effects), respectively [1].

The calculation of radiation absorbed dose to an organ of the body from an internally distributed radionuclide requires knowledge of a number of factors and rests on a number of assumptions. Fundamentally, one needs to know the following:

- the amount of radioactivity administered—the administered activity A_0;
- the rate of radioactive decay of the administered radionuclide—the physical half-life (or decay constant);
- the types of radiations emitted by the decaying radionuclide and their respective frequencies and energies of emission—together providing the energy emitted per decay;

doi:10.1088/978-0-7503-2694-0ch3

- the fraction of the administered activity which localizes in each tissue or organ (the 'source regions') with the respective uptakes or, more completely, the time–activity functions;
- the length of time the radioactive material resides in each tissue or organ—the effective half-time—as derived from the time–activity function;
- the total number of decays (nuclear transitions) which occur in each tissue or organ (or source region)—the time-integrated activity;
- the fraction of radiation energy that is absorbed in the tissue or organ itself as well as in other tissues and organs (or 'target regions')—the absorbed fractions; and
- the mass of each tissue or organ ('target regions').

As reviewed in this chapter, the standard computational methods developed for internal radiation dosimetry, such as the MIRD schema, have historically been based on 'average' kinetics and anatomic (i.e., anthropomorphic) models. Strategies for individualized patient dosimetry, more appropriate for radionuclide therapy, are presented in other chapters of this textbook and are only briefly discussed in the current chapter.

3.2 Radiation quantities and units

3.2.1 Stochastic versus deterministic quantities

Dosimetric quantities may be characterized as 'stochastic' or 'deterministic' [1–4]. A quantity subject to statistical fluctuations is termed 'stochastic,' while the mean of a large number of determinations of the quantity is termed 'deterministic.' Each stochastic quantity, then, has a corresponding deterministic quantity. The field of 'microdosimetry' deals with the number, size, and spatial and temporal distributions of individual energy-deposition events, particularly in microscopic structures of the order of molecules, macromolecules, and supramolecular complexes in size, characterizing such events in terms of inherently stochastic quantities [3, 4].

3.2.2 Definitions of dosimetric quantities

A physically and biologically meaningful specification of radiation 'dose' is a problematic concept and has been specified in a variety of ways. The following definitions are extracted from references [2, 5, 6].

'Exposure X,' is defined as follows:

$$X = \frac{Q}{m} \tag{3.1}$$

where Q = the absolute value of the total charge of ions of one sign produced in air when all the electrons (negatrons and positrons) liberated by photons (x- and γ-rays) in air are completely stopped in air,

and m = the mass of air in which the electrons (negatrons and positrons) were liberated by photons (x-rays and γ-rays).

Exposure is essentially the ionization equivalent of air kerma (defined below) and, as in the case of kerma, it may often be convenient to refer to a value of exposure at a point inside a material different from air. In such a case, the exposure specified will be that which would be determined for a small quantity of air placed at the point of interest. Exposure is a deterministic quantity and thus corresponds to a sufficiently large irradiated volume and/or a sufficiently large amount of radiation to yield insignificant statistical fluctuations in its measured value. The traditional unit of exposure is the roentgen (R) which equates to 1 electrostatic unit (esu) per cm^3 of air at STP (standard pressure and temperature). The SI unit of exposure is 1 C kg^{-1} of air (regardless of temperature or pressure) such that 1 C kg^{-1} = 3876 R.

The 'absorbed dose D,' is probably the most widely used quantity to characterize 'dose' and is defined as follows:

$$D = \frac{\overline{E}}{m} \tag{3.2}$$

where E = the mean energy imparted by ionizing radiation to matter
and m = the mass of matter to which the energy is imparted.

Absorbed dose is the deterministic correlate of the stochastic quantity, 'specific energy, z':

$$z = \frac{\varepsilon}{m} \tag{3.3}$$

where ε = the energy imparted to matter by a single energy-deposition event
and m = the mass of the matter.

The traditional unit of absorbed dose is the rad which equates to 100 ergs of energy per gram of tissue. The SI unit of absorbed dose is the gray (Gy) which equates to 1 J kg^{-1}; thus, 1 Gy = 100 rad.

It is conceptually straightforward to relate absorbed dose to exposure [7]. The average energy required to produce an ion pair in air is nearly constant for all electron energies, corresponding to a value of the mean energy per ion pair (W) of 33.7 eV/ion pair (IP) [8]. We may relate the quantities exposure and absorbed dose in either traditional or SI units as follows:

$$(1\ \mathrm{R})\left(\frac{\frac{\mathrm{C}}{\mathrm{kg}}}{3876\ \mathrm{R}}\right)\left(\frac{\mathrm{IP}}{1.602\times10^{-19}\ \mathrm{C}}\right)\left(\frac{33.7\ \mathrm{eV}}{\mathrm{IP}}\right)\left(\frac{1.602\times10^{-12}\ \mathrm{erg}}{\mathrm{eV}}\right)$$

$$\left(\frac{\mathrm{kg}}{1000\ \mathrm{g}}\right)\left(\frac{\mathrm{rad}}{100\ \mathrm{erg/g}}\right) = 0.869\ \mathrm{rad}$$

$$\left(1\ \frac{\mathrm{C}}{\mathrm{kg}}\right)\left(\frac{\mathrm{IP}}{1.602\times10^{-19}\ \mathrm{C}}\right)\left(\frac{33.7\ \mathrm{eV}}{\mathrm{IP}}\right)\left(\frac{1.602\times10^{-19}\ \mathrm{J}}{\mathrm{eV}}\right)\left(\frac{\mathrm{Gy}}{\mathrm{J/kg}}\right) = 33.7\ \mathrm{G}y \tag{3.4}$$

Thus, the relationship between exposure to air (in units of R or C kg^{-1}) to the absorbed dose to air (in units of rad or Gy) would be given as:

$$D_{\text{air}}\ (\text{rad}) = 0.869\ X_{\text{air}}\ (R)$$

$$D_{\text{air}}\ (\text{Gy}) = 33.7\ X_{\text{air}}\ (\text{C kg}^{-1}) \tag{3.5}$$

Under conditions of charge-particle equilibrium, the photon absorbed dose to any non-air medium (soft tissue) may be computed by scaling the corresponding absorbed dose to air by the ratio of the mass-energy absorption coefficients (MEAC) (μ_{en}/ρ) such that equation (3.5) may be written as [7]:

$$D_{\text{tissue}}\ (\text{rad}) = 0.869 \left[\frac{(\mu_{\text{en}}/\rho)_{\text{tissue}}}{(\mu_{\text{en}}/\rho)_{\text{air}}} \right] \text{X}_{\text{air}}\ (R) \ = \ \left(f_{\text{tissue}}^{\text{traditional}}\right) X_{\text{air}}\ (R)$$

$$D_{\text{tissue}}\ (\text{Gy}) = 33.7 \left[\frac{(\mu_{\text{en}}/\rho)_{\text{tissue}}}{(\mu_{\text{en}}/\rho)_{\text{air}}} \right] X_{\text{air}}\ (\text{C kg}^{-1}) = \left(f_{\text{tissue}}^{\text{SI}}\right) X_{\text{air}}\ (\text{C kg}^{-1}) \tag{3.6}$$

As the MEAC ratios (tissue-to-air) vary between 1.00 and 1.12 over a wide range of photon energies (0.01 to 10 MeV) in air, water, and soft tissue [7, 8], values of $\left(f_{\text{tissue}}^{\text{traditional}}\right)$ correspondingly vary in value between 0.87 and 0.97. Thus, fortuitously, the absorbed dose (in rad) may practically be considered to be equal to the air exposure value (in R). The value of $\left(f_{\text{tissue}}^{\text{SI}}\right)$ varies between 33.7 and 37.7.

'Kerma K,' is defined as follows:

$$K = \frac{\overline{E}_{\text{tr}}}{m} \tag{3.7}$$

where $\overline{E}_{\text{tr}}$ = the sum of the initial kinetic energies of all the charged ionizing particles liberated by uncharged ionizing particles (including photons) in matter
and m = the mass of matter in which the charged ionizing particles were liberated.

In the case in which the matter is air, kerma is often referred to as 'air kerma' or 'free air kerma.'

The quality as well as the quantity of radiation are important determinants of the frequency and/or severity of radiogenic effects. The 'quality' of a radiation is related to the characteristics of the microscopic spatial distribution of energy-deposition events, determined by the mass, charge, and energy (i.e., velocity) of the charged particles composing the radiation field or, in the case of x-rays, γ-rays, and neutrons, the secondary charged particles they produce. Sparsely ionizing radiations such as x-rays and γ-rays and intermediate-to-high-energy electrons and β-particles are characterized as 'low-quality' radiations, while densely ionizing radiations such as very-low-energy electrons (e.g., Auger electrons), protons, neutrons, and α-particles are typically characterized as 'high-quality' radiations. Importantly, for the same absorbed dose, the frequency and/or severity of radiogenic biological effects are generally less for sparsely ionizing, low-quality radiations than for densely ionizing, high-quality radiations.

The quality of radiation is quantitatively characterized by the 'restricted linear energy transfer, L_Δ or LET_Δ, of a material for charged particles such that:

$$L_\Delta = \left(\frac{dE_\Delta}{dl}\right) \tag{3.8}$$

where dE_Δ = the energy lost by a charged particle in traversing a distance in matter due to collisions in which the energy loss is less than the energy cut-off Δ,
and dl = the distance traversed in matter.

The restricted linear energy transfer (L_Δ) requires the specification of a cut-off energy (Δ), necessitated by primary energy-deposition events resulting in relatively high-energy, relatively long-range secondary electrons (i.e., delta (δ) rays) whose energy-deposition events may be considered as separate from those along the track of the primary radiation. In radiation protection, this feature is generally disregarded by specification of the 'unrestricted linear energy transfer, $L\infty$ or LET∞' (also known simply as the 'linear energy transfer, L or LET'), where the energy cut-off (Δ) is set equal to infinity (∞).

LET is the deterministic correlate of the stochastic quantity, 'lineal energy, y':

$$y = \frac{\varepsilon}{\bar{l}} \tag{3.9}$$

where ε = the energy imparted to matter by a single energy-deposition event,
and l = the mean chord length of the volume of matter in which that energy is imparted.

As previously noted, for the same absorbed dose, the frequency and/or severity of radiogenic biological effects are generally less for low-LET radiations than for high-LET radiations. The influence of radiation quality on the frequency and/or severity of biological effects is quantified by the 'relative biological effectiveness, or RBE$_x$' of radiation 'x':

$$\mathrm{RBE}_x = \left(\frac{D_{\mathrm{ref}}}{D_x}\right)_{\text{same biological effect}} \tag{3.10}$$

where D_{ref} = the absorbed dose of reference radiation (typically a widely available sparsely ionizing radiation such as cobalt-60 γ-rays) required to produce a specific, quantitatively expressed frequency and/or severity of a specific biological effect,
and D_x = the absorbed dose of radiation x required to produce the same specific, quantitatively expressed frequency and/or severity of the same specific biological effect
with all pertinent parameters, except the radiation itself, maintained as identically as possible. Because the relative RBE represents a ratio of absorbed doses, it is a dimensionless quantity.

Two additional dosimetric quantities, both derived from the absorbed dose, are the 'equivalent dose' and the 'effective dose'. The organ or tissue-equivalent dose is a quantity that accounts for the biological effectiveness of different radiations in inducing stochastic effects (i.e., cancer and germ-cell mutation). It is used for calculation of the effective dose, which is widely used as the basis for formulating radiation protection standards, that is, the maximum permissible doses for occupationally exposed individuals and for members of the general public. Equivalent dose $H(r_T)$ to an organ or tissue r_T is:

$$H(r_T) = \sum_R w_R\, D_R(r_T) \tag{3.11}$$

where w_R = the radiation weighting factor for radiation type R,
and $D_R(r_T)$ = the absorbed dose from radiation type R to organ or tissue r_T.

The unit for equivalent dose is the J kg^{-1}, with the special named unit of sievert (Sv) [6]. Current values of w_R recommended by the International Commission on Radiological Protection (ICRP) are 1.0 for photons, electrons, positrons, and beta particles and 20 for alpha particles [9]. Some radionuclides used in nuclear medicine (such as ^{99m}Tc, ^{123}I, ^{125}I, and ^{201}Tl) emit Auger electrons. The radiation-weighting factors assigned to these low-energy, very-short-range electrons may be higher than 1.0 when the radionuclide is incorporated into the DNA of the cell nucleus. The ICRP does not give specific recommendations on the value of w_R for Auger electrons, but recommends that the value be determined on a case-by-case basis. Further guidance, however, has been provided by the American Association of Physicists in Medicine (AAPM) [10, 11]. The AAPM recognized that the RBE (for deterministic effects) for Auger electrons emitted by DNA-incorporated radionuclides is similar to that seen for high-LET alpha particles and thus has recommended a radiation weighting factor of 20 for such Auger-electron emitters.

Equivalent dose is, in turn, used to compute the effective dose. The effective dose is a doubly-weighted quantity derived from the mean organ absorbed dose that accounts for both the radiation weighting factor w_R and the relative likelihood of radiation-induced detriment such as cancer for a given organ equivalent dose, thus accounting for organ-specific radiation sensitivity. It is widely used as the basis for formulating radiation protection standards, that is, the maximum permissible doses for occupationally exposed individuals and for members of the general public. Both equivalent dose and effective dose apply only to stochastic biological effects and do not represent any measure of the short-term deterministic biological effects associated with the high organ doses that may be encountered in radiopharmaceutical therapy. The calculated values of effective dose do not apply to specific individuals. Numerically, effective dose is the sum of all the weighted equivalent doses for all irradiated organs and tissues of the body of a reference anatomic model, whether the body is irradiated uniformly or non-uniformly. It is calculated as:

$$E = \sum_T w_T \frac{H^M(r_T) + H^F(r_T)}{2} \tag{3.12}$$

where $H^M(r_T)$, $H^F(r_T)$ = the dose equivalents to target tissue r_T in the reference male ('M') and reference female ('F') individual, respectively,
and w_T = tissue weighting factor for tissue T, defined as the fraction of total stochastic risk (or detriment) associated with tissue T.

Table 3.1. ICRP Publication 103 tissue weighting factors [9].

Tissue	wT	∑wT
Bone-marrow, breast, colon, lung, stomach, remainder tissues[a]	0.12	0.72
Gonads	0.08	0.08
Urinary bladder, oesophagus, liver, thyroid	0.04	0.16
Endosteum, brain, salivary glands, skin	0.01	0.04

[a]Remainder tissues: adrenals, extrathoracic (ET) regions of the respiratory tract, gall bladder, heart, kidneys, lymphatic nodes, muscle, oral mucosa, pancreas, prostate (male), small intestine, spleen, thymus, uterus/cervix (female).

Values of tissue weighting factors recommended in ICRP Publication 103 [9] are given in table 3.1. At the present time, values of the tissue weighted factor are provided as a single set to include averaging of the relative detriment across a range of patient ages and both sexes. These values of w_T may be applied to sex-averaged equivalent organ doses for not only reference adults, but also various pediatric reference anatomic models (see section 3.5).

3.3 The MIRD schema

One of the most widely used approaches for internal dose calculations in medicine was developed, and continues to be refined, by the Medical Internal Radiation Dose Committee of the Society of Nuclear Medicine in a methodology generally referred to as the 'MIRD schema' or 'MIRD formalism' [12–20]. The MIRD schema, including notation, terminology, mathematical methodology, and reference data, has been disseminated in the form of the collected MIRD Pamphlets and associated publications. The ICRP has developed a similar methodology and similar reference data [21], and the MIRD schema and the ICRP schema have now been largely harmonized (see MIRD Pamphlet No. 21 and ICRP Publication 130).

As it is applied to diagnostic radiopharmaceuticals, historically, its most common application, the MIRD schema implicitly assumes that activity and time-integrated activity are uniformly distributed within source regions and that radiation energy is uniformly deposited within target regions. Importantly, however, the scale of source and target regions can vary from the organ to sub-organ (i.e., image voxel) to cellular and sub-cellular dimensions [14, 15, 22–24]. Dosimetry for diagnostic radiopharmaceuticals is generally based on (a) average time–activity data in animal models translated to humans and/or in a small cohort of human subjects and (b) age- and sex-specific 'average' models of human anatomy. In light of the relatively low doses and radiogenic risks associated with diagnostic nuclear medicine procedures, any inaccuracies with applying dose estimates derived using such average data to individual patients are insignificant. Importantly, however, this may not be the case in applying such normal-organ dose estimates to individual patients for planning of radiopharmaceutical therapy. The anatomic models used in the MIRD schema do not incorporate tumors as either source or target regions, although the self-dose to tumors, modeled as tissue spheres, can be calculated using the MIRD schema [25].

The basic calculation in the MIRD schema, yielding the mean absorbed dose to target region r_T from the activity (i.e., the time-integrated activity) in source region r_S, is as follows [12, 13]:

$$\overline{D}(r_T \leftarrow r_S) = \widetilde{A}(r_S) \sum_i \frac{E_i \, Y_i \, \phi(r_T \leftarrow r_S)_i}{M_T} \tag{3.13a}$$

$$= \widetilde{A}(r_S) \sum_i \Delta_i \, \Phi(r_T \leftarrow r_S)_i \tag{3.13b}$$

$$= \widetilde{A}(r_S) \, S(r_T \leftarrow r_S) \tag{3.13c}$$

where $\widetilde{A}(r_S)$ = the time-integrated in source region r_S, that is, the total number of decays in source region r_S,

M_T = the mass of target region r_T,

Δ_i = the average energy emitted per decay in the form of radiation i (formerly known as the equilibrium dose constant for radiation i) [19, 20],

$\phi(r_T \leftarrow r_S)_i$ = the absorbed fraction in target region r_T for radiation i emitted in source region r_S, that is, the fraction of energy of radiation i emitted in source region r_S that is absorbed in target region r_T,

$\Phi(r_T \leftarrow r_S)_i$ = the specific absorbed fraction in target region r_T for radiation i emitted in source region r_S, that is, the fraction of energy of radiation i emitted in source region r_S that is absorbed per unit mass in target region r_T,

$$\Phi(r_T \leftarrow r_S)_i = \frac{\phi(r_T \leftarrow r_S)_i}{M_T} \tag{3.14}$$

and $S(r_T \leftarrow r_S)$ = the radionuclide-specific S value for target region r_T and source region r_S, that is, the absorbed dose to target region r_T per unit time-integrated activity in source region r_S,

$$S(r_T \leftarrow r_S)_i = \sum_i \Delta_i \, \Phi(r_T \leftarrow r_S)_i \tag{3.15}$$

An important simplification of absorbed dose calculations was introduced by combining the energy emitted per decay Δ_i, the source region-to-target region absorbed fraction $\phi(r_T \leftarrow r_S)_i$, and the target region mass M_T into a single quantity, the S value, as defined by equation (3.15) [15]. As a result, given the time-integrated activity $\widetilde{A}(r_S)$ in a given source region r_S, one can use the tabulated S values to yield, by a single multiplication, the absorbed dose contribution $\overline{D}(r_T \leftarrow r_S)$ to a target region r_T. Besides computationally simplifying the determination of absorbed dose, the S value conceptually clarified this task by isolating all non-time-dependent and, largely, non-biology-dependent dosimetric factors into a single parameter.

The total mean absorbed dose $\overline{D}(r_T)$ to target region r_T is then calculated by summing of the absorbed dose contributions from all source regions r_S:

$$\bar{D}(r_T) = \sum_{r_S} \tilde{A}(r_S) \sum_i \frac{E_i \; Y_i \; \phi(r_T \leftarrow r_S)_i}{M_T} \tag{3.16a}$$

$$= \sum_{r_S} \tilde{A}(r_S) \sum_i \Delta_i \; \Phi(r_T \leftarrow r_S)_i \tag{3.16b}$$

$$= \sum_{r_S} \tilde{A}(r_S) \; S(r_T \leftarrow r_S) \tag{3.16c}$$

3.4 Time-integrated activities

An essential component of the MIRD schema and of internal dosimetry in general is the determination of the time-integrated activity in source region r_S, $\tilde{A}(r_S)$, that is, the total number of decays in source region r_S (equations (3.13a)–(3.13c) and equations (3.16a)–(3.16c)). There are several commonly used approaches to the determination of the time-integrated activity: (a) curve-fitting and analytical integration, (b) numerical integration, and (c) compartmental modeling. All generally use a series of discrete source-region activity measurements.

3.4.1 Curve-fitting and analytical integration

The time-integrated activity, $\tilde{A}(r_S)$ in a given source region r_S is calculated by integration from the time of administration of the radiopharmaceutical ($t = 0$) to the time of its complete elimination (t = ∞) of the fitted time–activity function, $A(r_S, t)$:

$$\tilde{A}(r_S) = \int_0^{\infty} A(r_S, t)\mathrm{d}t. \tag{3.17}$$

The time–activity function $A(r_S, t)$ includes both the physical decay of the radionuclide and the biological clearance from the source region r_S (see below). More generally, the time-integrated activity may be determined for some finite period of time τ following administration of the radiopharmaceutical rather than for the time for its complete elimination ($t = \infty$). In this case, equation (3.17) becomes:

$$\tilde{A}(r_S, \tau) = \int_0^{\tau} A(r_S, t)\mathrm{d}t. \tag{3.18}$$

Despite the complexity of the multiple underlying biological processes, the time–activity data of many radiopharmaceuticals in a tissue or organ can often be accurately represented by one or the sum of multiple exponential terms. A model for $A(r_S, t)$ that includes three exponential terms would be given as:

$$A(r_S, t) = A_0\left[f_{S_1}\mathrm{e}^{-\lambda_{S_1}t} + f_{S_2}\mathrm{e}^{-\lambda_{S_2}t} + f_{S_3}\mathrm{e}^{-\lambda_{S_3}t}\right] = A_0 \sum\nolimits_{j=1}^{3} f_{S_j}\mathrm{e}^{-\lambda_{S_j}t} \tag{3.19}$$

where A_0 = the administered activity of the radiopharmaceutical (at $t = 0$)

f_{S_j}= the fraction of the administered activity that is assigned to the *j*th exponential component of the time–activity function for source region r_S extrapolated to time $t = 0$,

and λ_{S_j}= the clearance constant of the *j*th exponential component of the time–activity function in source region r_S, that is, the fraction per unit time of activity eliminated for the *j*th exponential component of the time–activity function in source region r_S.

Although other functions may in some instances model time–activity data more accurately, a sum of exponential terms—as in equation (3.19)—is a suitable mathematical representation of the time–activity data for dosimetric analyses of most radiopharmaceuticals. Values of f_{S_j} are positive under the assumption of instantaneous uptake by the source region from blood and thus each exponential term models radiopharmaceutical clearance (or washout) from the source region. A negative value of f_{S_1} for the first term can, however, be assigned to allow for modeling of an initial uptake phase (rise from zero to some maximum fraction of injected activity) of the radiopharmaceutical from circulating blood post-injection at $t = 0$.

Because the time–activity data, $A(r_S, t)$, represents the actual radiopharmaceutical activity in source region r_S at time t post-administration (rather than the activity decay-corrected to the time of administration, that is, time $t = 0$), the clearance constant, λ_{S_j}, is actually the 'effective (*e*)' clearance constant as it includes the effects of *both* biological clearance and physical (i.e., radioactive) decay:

$$\lambda_{S_j} = \lambda_{e_j}^{S} = \lambda_p + \lambda_{b_j}^{S} \tag{3.20}$$

where $\lambda_{e_j}^{S}$ = the effective (*e*) clearance constant of the radiopharmaceutical in the *j*th compartment of source region r_S

λ_p = the physical (*p*) decay constant of the radionuclide of the radiopharmaceutical

and $\lambda_{b_j}^{S}$ = the biological (*b*) clearance constant of the *j*th exponential component of the time–activity function in source region r_S, that is, the fraction per unit time of activity biologically eliminated for the *j*th exponential component of the time–activity function in source region r_S.

For *each* exponential component (i.e., each exponential term in equation (3.19)), the physical and biological half-life (or half-time) is inversely related to its corresponding physical decay or biological clearance constant, respectively:

$$T_p = \frac{\ln 2}{\lambda_p} \approx \frac{0.693}{\lambda_p} \quad \text{and} \quad T_b = \frac{\ln 2}{\lambda_b} \approx \frac{0.693}{\lambda_b} \tag{3.21}$$

Furthermore, the expression for the biological half-time T_b can be applied for each of the *j*th compartments of the time–activity function, as well as for different source regions r_S in the radiopharmaceutical patient.

In combining equations (3.20) and (3.21), we may define the effective half-time, T_e, in terms of its constituent physical and biological half-lives (which can also be extended to different source regions and different time–activity compartments):

$$T_e = \frac{1}{\frac{1}{T_p} + \frac{1}{T_b}} = \frac{T_p\, T_b}{T_p + T_b} \tag{3.22}$$

where T_p = the physical half-life of the radionuclide
and T_b = the biological half-time of the jth exponential component of the time–activity function for source region r_S (identifiers j and s are not shown).

Integration of the time–activity function $A(r_S, t)$ (equation (3.19)) to infinite time yields the time-integrated activity $\tilde{A}(r_S)$ in source region r_S:

$$\tilde{A}(r_S) = \sum_j \frac{A_0 f_{S_j}}{\lambda_{e_j}^S} = 1.443\, A_0 \sum_j f_{S_j}\, T_{e_j}^S \tag{3.23}$$

The time-integrated activity, $\tilde{A}(r_S)$, is often normalized to the administered activity A_0 (e.g., activity of the radiopharmaceutical injected in units of MBq or mCi). The ratio of $\tilde{A}(r_S)$ by A_0 is units of time (seconds, minutes, or hours typically) and is defined as the time-integrated activity coefficient $\tilde{a}(r_S)$ (or TIAC).

$$\tilde{a}(r_S) = \frac{\tilde{A}(r_S)}{A_0} = \sum_j \frac{f_{S_j}}{\lambda_{e_j}^S} = 1.443 \sum_j f_{S_j}\, T_{e_j}^S \tag{3.24}$$

By convention of both the ICRP and MIRD schemata, lower-case letters note normalized quantities (as in this case the normalization parameter is A_0). Prior to MIRD Pamphlet No. 21, the TIAC was termed the 'residence time' and given the symbol τ. Tau is now reserved for the symbol representing the upper limit of the time integrations of the function $A(r_S, t)$.

In a similar way, we can go back to equation (3.16c) (the basic expression in the MIRD schema for computing the mean absorbed dose), and also normalize to the administered activity A_0, thus defining an absorbed dose coefficient $\bar{d}(r_T)$ to target region r_T:

$$\bar{d}(r_T) = \frac{\bar{D}(r_T)}{A_0} = \sum_{r_S} \tilde{a}(r_S)\, S(r_T \leftarrow r_S) \tag{3.25}$$

In an analogous fashion, we can normalize the effective dose to the patient for a given radiopharmaceutical by the administered activity to get the effective dose coefficient:

$$e = \frac{E}{A_0} = \frac{1}{A_0} \sum_T w_T \frac{H^M(r_T) + H^F(r_T)}{2} \tag{3.26}$$

The availability of organ dose and effective dose coefficients for a variety of radiopharmaceuticals, organs/tissues of interest, and reference computational models of the patient conveniently allow for assessing these quantities (organ dose and effective dose) in patients by simply scaling the dose coefficients by the administered activity. Thus, $\bar{D}(r_T) = A_0 \bar{d}(r_T)$ and $E = A_0 e$.

3.4.2 Numerical integration

A common method of numerical, or 'manual,' integration involves connecting consecutive data points by straight lines (figure 3.1) and integrating by the so-called trapezoidal rule. The function *f*(*t*), shown as a dashed line in figure 3.1, may be approximated by a straight-line (solid line in figure 3.1) between time points *a* and *b*. In the trapezoidal approach to integration, the area under the curve is: $\int_a^b f(t)\,\mathrm{d}x \approx (b - a)[f(a) + f(b)]/2$. The trapezoidal rule extends only to the final measured datum. Beyond this final data point (i.e., to infinity), the integral is unknown by the trapezoidal rule and must be estimated by other means. For example, one can conservatively assume elimination of the radiopharmaceutical by physical decay only [26]. An important advantage of numerical integration is its generality: it is adaptable to non-monotonic time–activity curves and no simplifying assumptions are introduced regarding the analytic form (e.g., a sum of exponential terms) of the time–activity curve. On the other hand, a disadvantage of the assumption of radiopharmaceutical elimination by physical decay only following the last measurement is that it results in a systematic overestimation, perhaps substantial, of the area under the time–activity curve, that is, the time-integrated activity, by ignoring any long-term components of radiopharmaceutical tissue clearance.

3.4.3 Compartmental modeling

The third approach to estimation of time-integrated activities. A compartmental model (figure 3.2) is a mathematical representation of the exchange of a material (such as a radiopharmaceutical or other tracer) among compartments in a biological system (such organs or even organ sub-regions in the human body) and between the biological system and its external environment (such as ingestion and excretion). The definition, number, and inter-relationships of compartments in a model must be

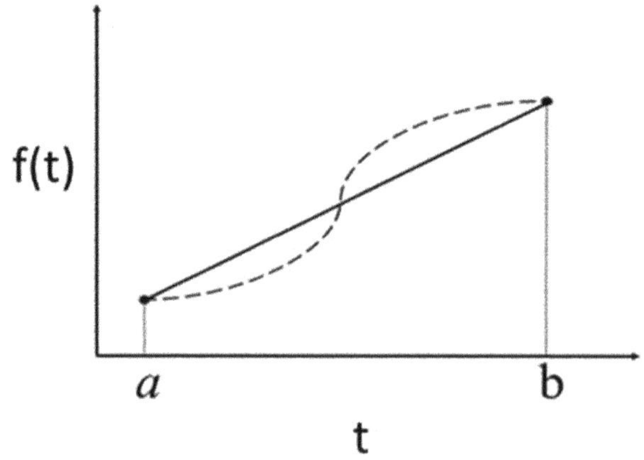

Figure 3.1. The function dashed line) may be approximately represented by a straight-line (solid line) between points *a* and *b*.

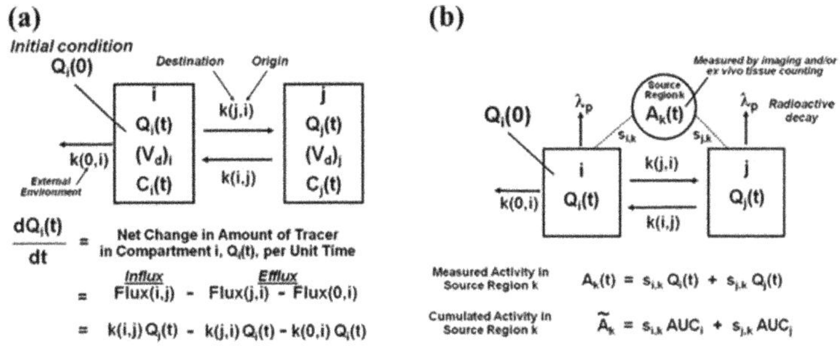

Figure 3.2. (a) Representation and nomenclature of a compartmental model. Each compartment i is represented by a box (or a circle), where $Q_i(t)$ is the amount of tracer in compartment i at time t post-administration, $(V_d)_i$ is the volume of distribution of the tracer in compartment i, and $C_i(t) = Q_i(t)/(V_d)_i$ the concentration of tracer in compartment i at time t post-administration. The external environment (into which the tracer is excreted or otherwise eliminated) is often labeled 0. The initial condition, $Q_i(0)$, is the amount of tracer in compartment i at time $t = 0$; for an intravenous bolus injection of a radiopharmaceutical, compartment $i = 1$ corresponds to blood (or plasma) and $Q_i(0)$ corresponds to the administered activity. The parameter $k(j,i)$ is the exchange rate from compartment i to compartment j, that is, the fraction of the amount of tracer transferred from compartment i to compartment j per unit time. The convention for identifying exchange rates is to place the number of the destination compartment first and that of the origin compartment second. The product $k(j,i)\ Q_i(t)$ is the flux from compartment i to compartment j, that is, the (absolute) amount of tracer transferred from compartment i to compartment j per unit time. (b) Adaptation of the generic two-compartment model in (a) to radiopharmaceutical data. The effect of radioactive decay may be explicitly incorporated into the model by appending to each compartment an elimination exchange rate equal to the physical decay constant λ_p of the radionuclide in the radiopharmaceutical. As indicated by the first equation in (b), the activity $A_k(t)$ in a source region k (such as a tumor or organ) is a weighted sum of the activities in the compartments i and j, where the weighting factors $s_{i,k}$ and $s_{j,k}$ are the fractions of $Q_i(t)$ and $Q_j(t)$, respectively, in source region k; $A_k(t)$ are measured by non-invasive imaging or by *ex vivo* counting of tissues.

consistent with established biological properties and characteristics. In general terms, a compartment is an ensemble of pharmacokinetically identical entities (such as molecules of the radiopharmaceutical) that may be localized to an identifiable organ, physiologic volume (such as the extracellular fluid space), or other anatomic entity. However, a compartment is a functional, rather than a structural concept, and thus may *not* actually correspond to any anatomic entity. Implicit in the identification of a compartment is that the tracer is instantaneously and uniformly dispersed within the compartment.

In figure 3.2, conservation of mass results in a differential equation for the time-rate of change of the amount of tracer in each compartment i. The resulting system of linked differential equations represents the mathematical formulation of a model. For compartmental models of many biological phenomena, the exchange rates $k(i,j)$ are constant, or time-invariant, that is, their values are independent of tracer concentrations in the various compartments. In such cases, the exchange rates are also known as 'rate constants,' and the system of differential equations is linear. For saturable-binding phenomena such as antibody-antigen or receptor-ligand interactions, the exchange rates vary as the concentrations of the reactants vary.

This results in time-varying exchange rates and a system of non-linear differential equations. Once the topography of a compartmental model has been specified, based on some a priori knowledge of the pertinent biology, solving the model amounts to deriving the set of values of the exchange rates, $k(i,j)$, which yield the closest agreement between the measured and the model-derived source region r_S time–activity data. Except for relatively simple models (with at most three compartments), analytic solution of the corresponding system of differential equations is onerous, if not intractable; in general, therefore, models are solved iteratively (i.e., numerically) using an appropriate computer program.

An advantage of compartmental modeling is its ability to yield time–activity data and time-integrated activities in source regions that cannot be measured directly in the clinical setting. These include microscopic source regions such as the cell surface (e.g., with radiolabeled antibodies bound to antigens) or the cell nucleus (e.g., with radioligands bound to nuclear receptors). Such microscopic distributions of time-integrated activities are likely important for radiation dosimetry of ultra-short-range radiations such as Auger electrons and α particles.

3.5 Radionuclide *S*-values

3.5.1 Reference phantoms *S*-values

As discussed, radionuclide *S*-values, $S(r_T \leftarrow r_S)$, define the absorbed dose to a target region r_T per nuclear decay of the radiopharmaceutical in a source region r_S. For patient-specific dosimetry, radionuclide *S* values for all relevant source and target tissues can be computed by Monte Carlo radiation transport or other computational means using whole-body CT or MR image sets. Only then would one have truly 'patient-specific' *S* values for the dose computation, the derivation of which is often impractical in the present clinical environment. While such an approach is being pursued at certain research centers, radiopharmaceutical dosimetry is generally based on pre-computed tables of radionuclide *S* values for source/target organ pairs in a limited series of anatomic computerized models (or phantoms) of the 'average' patient. These 'average' patients are typically taken to be 'reference' persons defined, for example, by the ICRP (1–3), and including males and females at roughly 50th percentile heights and weights for a range of fixed and defined ages (e.g., newborn, 1-year-old, 5-year-old, 10-year-old, 15-year-old, and adult). Figure 3.3, for example, displays the Oak Ridge National Laboratory (ORNL) series of age-dependent stylized phantoms [27]. These phantoms are hermaphrodites (having both sex organs at each reference age), and include models of organ boundaries and the outer body contours based upon mathematical surfaces of regularly shaped three-dimensional objects such as spheres, ellipsoids, cones, and tapered cylinders. One limitation with these forms of anatomical models is that the regular shapes of the internal organs result in unrealistically large inter-organ tissue spaces as compared to those present in actual human anatomy. Photon cross-dose (when $r_T \neq r_S$) is therefore under-estimated, especially at lower photon energies. Furthermore, values of $\phi(r_T \leftarrow r_S)$ for electrons and other particulate radiations have traditionally been set to unity, $\phi(r_T \leftarrow r_S) = 1$ when $r_T = r_S$ and to zero, $\phi(r_T \leftarrow r_S) = 0$ when $r_T \neq r_S$. For small

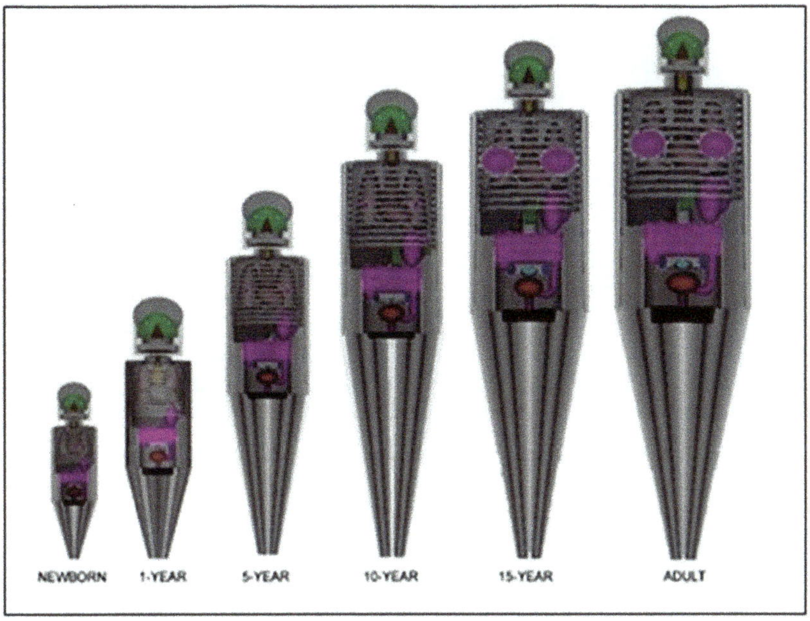

Figure 3.3. Family of stylized reference phantoms from the Oak Ridge National Laboratory. Each phantom above is a hermaphrodite with both male and female sex organs present. Note that in these stylized phantom the external anatomy and internal organs are represented by regular three-dimensional shapes such as spheres and ellipsoids. [27].

organs and higher-energy electrons and beta particles, that assumption tends to over-estimate organ self-dose (i.e., when $r_T = r_S$).

The ICRP has now adopted the series of age- and sex-dependent voxel-based reference phantoms shown in figure 3.4 [28, 29]. These phantoms are based on CT images of human cadavers and thus the internal organ anatomy is significantly more realistic than that given by the ORNL stylized phantom series. In ICRP Publication 133 [30], new energy-specific values of specific absorbed fractions, $\Phi(r_T \leftarrow r_S)$, are given for internally emitted photons, electrons, and alpha particles incorporating some 79 different source tissue regions r_S and some 41 different target tissue regions r_T in the adult reference phantoms. A similar publication by the ICRP will appear in late 2022 giving specific absorbed fractions for the pediatric reference phantoms of the ICRP.

3.5.2 Patient *S*-values

Differences in body size and internal organ anatomy between those of a reference phantom and those of an individual patient can potentially introduce large errors in estimates of mean organ doses to the patient, mainly due to errors in estimation of the self-dose (which typically accounts for the largest contribution to the organ dose). To a first approximation, self-dose *S* values scale inversely with organ mass; therefore, an individual's self-dose *S* values may be obtained from the respective reference-phantom *S* values by scaling by the ratio of the phantom-to-individual organ mass, where the individual's organ masses may be extracted from CT or MR imaging.

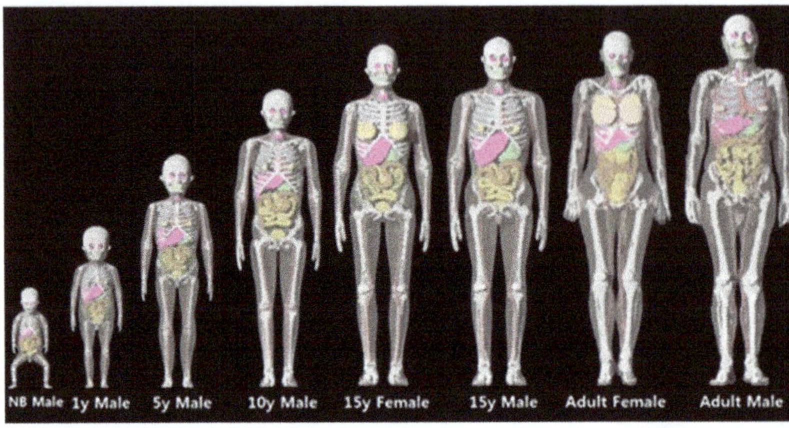

Figure 3.4. Family of voxel-based reference phantoms from the ICRP. A total of 12 reference phantoms are in the ICRP reference series, with only the male versions shown above at ages below 15 years. In these voxel-based phantoms, the external anatomy and internal organs are more anatomically realistic than in the earlier-generation stylized phantoms [28, 29].

Re-scaling of specific absorbed fraction $\Phi(r_T \leftarrow r_S, E_o, m_{\text{ref}})$ for radiation i for a reference (ref) phantom organ of mass m_{ref} to the specific absorbed fraction $\Phi(r_T \leftarrow r_S, E_o, m_{\text{non-ref}})$ for a non-reference (non-ref) (i.e., patient) organ of mass $m_{\text{non-ref}}$ has been based on the inverse ratio of the reference and non-reference organ masses to a power scaling factor $R(E_o, m_{\text{ref}}, m_{\text{non-ref}})$:

$$\Phi(r_T \leftarrow r_S, E_o, \mathrm{m}_{\text{non-ref}}) = \Phi(r_T \leftarrow r_S, E_o, m_{\text{ref}})\left(\frac{m_{\text{non-ref}}}{m_{\text{ref}}}\right)^{R(E_o,\, m_{\text{ref}},\, m_{\text{non-ref}})} \tag{3.27}$$

In MIRD Pamphlet No 11 [15], the scaling ratio $R(E_o, m_{\text{ref}}, m_{\text{non-ref}})$ was set at a constant (i.e., energy-independent) value of -1 for electrons and of $-2/3 \approx -0.67$ for photons. A more robust approach to patient-specific adjustments to reference phantom self-dose S-values was proposed by Wayson and Bolch [31] in which photon and electron power scaling factors were computed as a function of energy for ranges of reference and non-reference organ masses, as shown for photons and electrons in figures 3.5(a) and (b), respectively. The MIRD Pamphlet No 11 constant of -0.67 is shown to be reasonably accurate across the photon energy range of ~200 keV to 1 MeV, but with values of R as low as -0.996 for 1-keV photons and as high as -0.357 for 4-keV photons (figure 3.5(a)). The MIRD Pamphlet No 11 constant of -1.0 for electron self-dose scaling of the specific absorbed fraction is appropriate for low-to-intermediate electron energies, but it then increases above that value at energies above several hundred keV, especially for smaller organ masses. In another study, Wayson and Bolch [32] explored patient-specific rescaling of both self-dose and cross-dose photon and electron specific absorbed fractions given various external-body morphometry metrics for the patient and the reference phantom (to include sitting height and waist circumference).

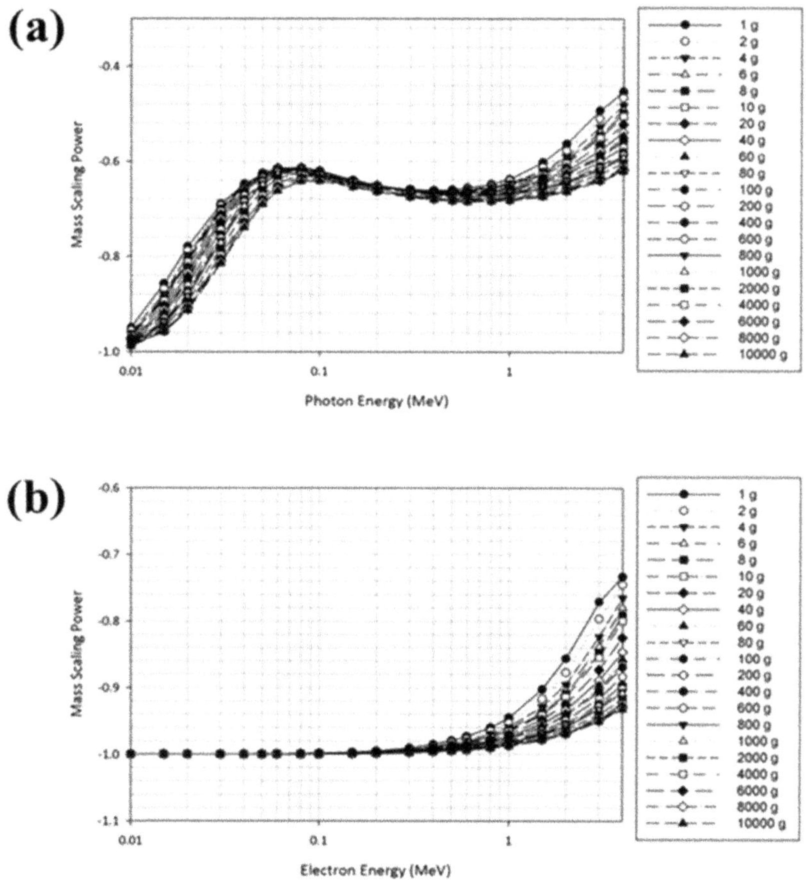

Figure 3.5. Mass-ratio scaling powers $R(E_o, m_{\text{ref}}, m_{\text{non-ref}})$ for (a) photon and (b) electron organ self-dose as a function of both reference organ mass and initial radiation energy. Results are averaged over non-reference organ sizes. Reproduced from [31], copyright IOP Publishing. All rights reserved.

3.6 Radiopharmaceutical therapy: adiopharmaceutical therapy: non-uniform dose distributions

With the ongoing emergence of radiopharmaceutical therapy [33, 34], spatial non-uniformity of the resulting radiation dose has become increasingly important at both the macroscopic and microscopic tissue levels. A single average-dose value for a tissue or tumor may not adequately represent the dose distribution within the tissue or tumor volume and may, therefore, poorly predict biological outcome. In external-beam radiotherapy, tumor dose is relatively uniform and normal-organ doses notably non-uniform, with significant doses delivered to only limited portions of organs at risk immediately adjacent to the prescribed tumor volume. In contrast, in nuclear medicine, the distribution of the radiopharmaceutical within tumors and organs is dependent on a number of factors (e.g., molecular weight and tissue penetration characteristics of the radiopharmaceutical, tissue perfusion, vascular patency, hypoxia) that will impact the activity and absorbed dose distributions and typically make them non-uniform. From a response standpoint, a tumor with a tumoricidal average absorbed dose may proliferate if a sub-volume of the tumor receives a sub-tumoricidal absorbed dose. O'Donoghue [35], for example, modeled the impact of dose non-uniformity [36, 37] on radiocurability of tumors and demonstrated that tumor response is poorer (i.e., tumor cell survival is greater) as the dose becomes more non-uniform. A substantial fraction of tumor cells will thus receive sub-lethal doses and the lesion may therefore recur even if the average tumor dose is sufficiently high to otherwise expect a significant therapeutic response.

Traditionally, the MIRD schema has been applied to non-uniform activity distributions only to the extent that radioactivity localizes in different source organs, and, within such source regions the activity has been assumed to be uniformly distributed. However, there is nothing intrinsic in the MIRD schema that precludes its extension to smaller regions, down to cell and sub-cellular source regions. There are at least three approaches to the calculation of non-uniform dose distributions: dose point-kernel convolution [22, 38], voxel S values [22], and Monte Carlo simulation [39]. As practical alternatives to more computationally intensive Monte Carlo-based approaches, the dose point-kernel and voxel S-value approaches are perhaps the most widely used methods for 3D internal dosimetry. A photon dose point kernel, $\Phi(x)$, is the radial distance (x)-dependent absorbed dose about an isotropic point source in an infinite homogeneous medium (typically, a soft tissue-equivalent medium such as water) [22, 38]:

$$\Phi(x) = \left[\frac{\mu_{\text{en}}}{\rho} \bullet \frac{1}{4\pi\, x^2} \bullet e^{-\mu x}\right] B_{\text{en}}(\mu x) \tag{3.28}$$

where μ_{en} = the photon linear energy absorption coefficient,
ρ = the mass density of the absorbing medium,
μ = the photon linear attenuation coefficient,
and $B_{\text{en}}(\mu x)$ = the energy absorption buildup factor, which takes into account the contribution of scattered photons to the energy deposited at distance μx mean-free-paths from the source point [38].

Dose calculations are then performed by the mathematical convolution of the dose point kernel with the image-derived activity distribution in the source volume.

In another approach to non-uniform dosimetry under the MIRD schema, one may define the voxel S value as the mean absorbed dose to a target voxel per decay in a source voxel, both of which are contained in an infinite homogeneous tissue medium [22]. The absorbed dose to voxel k, $D(\text{voxel}_k)$, within a tumor or organ thus equals the voxel dose contributions, $D(\text{voxel}_k \leftarrow \text{voxel}_h)$, summed over the N voxels comprising the tumor or organ:

$$D(\text{voxel}_T) = \sum_{S=1}^{N} D(\text{voxel}_T \leftarrow \text{voxel}_S) \tag{3.29a}$$

$$= \sum_{S=1}^{N} \tilde{A}(\text{voxel}_S)\, S(\text{voxel}_T \leftarrow \text{voxel}_S) \tag{3.29b}$$

where $\tilde{A}(\text{voxel}_S)$ = the time-integrated activity in source voxel S.

The voxel S-value approach thus suffers from the same limitations as the dose point-kernel convolution methods in that tissue inhomogeneities are generally not accommodated. This approach is further limited by the spatial resolution and resulting voxel dimensions of the imaging modalities (i.e., SPECT and PET) used to derive activity distribution data. Tabulations of voxel S values are available [22, 40–42].

With the availability of high-speed desktop computers and of compatible simulation codes, the use of Monte Carlo analysis in the clinical setting has become practical [40, 43–47]. Monte Carlo-based dosimetry can more accurately account for tissue variations in mass density and atomic number as well as edge effects, which may be important at the periphery of the body and at soft tissue-lung and soft tissue-bone interfaces [46, 48].

Once a dose distribution has been calculated by any of the foregoing methods, a corresponding 'dose-volume histogram (DVH)' can be constructed [35, 49, 50]. Basically, a DVH is a graph of the fraction or percentage of the tumor or organ volume receiving a specified dose versus the dose (differential form) or the fraction or percentage of the tumor or organ volume receiving less than a specified dose versus the dose (integral, or cumulative, form). It graphically presents the minimum, mean, and maximum doses and the dispersion about the mean dose. The greater this dispersion, the more non-uniform is the dose distribution. DVHs, routinely encountered in external-beam radiation therapy, may be particularly useful in radiopharmaceutical therapy, where the activity and therefore the dose distributions within tumors may be markedly non-uniform.

Due to the differential delivery to and uptake by cells of administered radiopharmaceuticals, the resulting radiation exposures among different cells and cell populations may be highly non-uniform. Further, the biological effects of short-range particulate radiations such as alpha particles and Auger electrons may be highly dependent on the location of the radioactive decays on and within the cell (e.g., membrane versus cytoplasm versus nucleus). Organ-level and even sub-organ-level absorbed dose estimates may therefore not reliably predict the radiation response of distinct cells and cell populations to radiopharmaceuticals emitting such short-range radiations. Accordingly, to assist in characterizing the biological response and potentially developing and optimizing treatment plans for therapeutic radiopharmaceuticals, the MIRD Committee has compiled radionuclide S values for cell-level source and target regions for isolated cells and cell clusters over a range of cellular and nuclear dimensions [51]. Using these cellular S values, a web-based applet, MIRDcell [24], has been developed which automates radionuclide cell-level dosimetry and radiation-response modeling; it can be accessed at http://mirdcell.njms.rutgers.edu.

3.7 Concluding remarks

This chapter has reviewed the basic concepts associated with the computational approaches to internal radiation dosimetry, that is, the determination of the amount and the spatial and temporal distributions of energy deposited by a radionuclide within the body. Internal radiation doses must be calculated, rather than measured.

These calculational methods have evolved over many decades from relatively simple approaches to those with a high level of sophistication. Even so, Dr Robert Loevinger, one of the originators of the MIRD schema, cautioned that, '...there is in principle no way of attaching a numerical uncertainty to the profound mismatch between the patient and the model (the totality of all assumptions that enter into the dose calculation). The extent to which the model represents in some meaningful way a patient, or a class of patients, is always open to question, and it is the responsibility of the clinician to make that judgment' [13]. In contrast to *diagnostic* radiopharmaceuticals, *therapeutic* radiopharmaceuticals engender much smaller risk–benefit ratios and therefore markedly reduced tolerances for any such inaccuracies in dose estimation. To the extent, then, that specific patients deviate kinetically and anatomically from the respective kinetic and anatomic averages, tissue dose estimates will be inaccurate and individualized (i.e., patient-specific) dosimetry therefore required.

References

[1] Zanzonico P B 2000 Internal radionuclide radiation dosimetry: a review of basic concepts and recent developments *J. Nucl. Med.* **41** 297–308

[2] ICRU 1980 Radiation Quantities and Units. Int. Commission on Radiation Units and Measurements (ICRU) Report No 33 (Bethesda, MD)

[3] ICRU 1983 Microdosimetry (Bethesda, MD: Int. Commission on Radiation Units and Measurements)

[4] Rossi H 1968 Microscopic energy distributions in irradiated manner *Radiation Dosimetry* ed F Attix, W Roesch and E Tochlin (New York: Academic) pp 43–92

[5] ICRU 1998 *Fundamental Quantities and Units for Ionizing Radiation. Int. Commission on Radiation Units and Measurements (ICRU) Report 60* (Bethesda, MD: Int. Commission on Radiation Units and Measurements (ICRU))

[6] ICRU 2011 Fundamental quantities and units for ionizing radiation. ICRU Report No 85 *J. ICRU* **11** 1–20

[7] Johns H and Cunningham J 1974 *The Physics of Radiology* (Springfield, IL: Charles C Thomas)

[8] Attix F 1986 *Introduction to Radiological Physics and Radiation Dosimetry* (New York: Wiley)

[9] ICRP 2007 Recommendations of the Int. Commission on Radiological Protection. Int. Commission on Radiological Protection (ICRP) Publication 103 *Ann. ICRP* **2007** 1–332

[10] Howell R W 1992 Radiation spectra for Auger-electron emitting radionuclides: Report No. 2 of AAPM Nuclear Medicine Task Group No. 6 *Med. Phys.* **19** 1371–83

[11] Humm J L, Howell R W and Rao D V 1994 Dosimetry of Auger-electron-emitting radionuclides: Report No. 3 of AAPM Nuclear Medicine Task Group No. 6 *Med. Phys.* **21** 1901–15

[12] Bolch W E, Eckerman K F, Sgouros G and Thomas S R 2009 MIRD pamphlet No. 21: a generalized schema for radiopharmaceutical dosimetry—standardization of nomenclature *J. Nucl. Med.* **50** 477–84

[13] Loevinger R, Budinger T, Watson E, Atkins H, Blau M, Lathrop K, Poston J, Robertson J, Thomas S and Weber D 1991 *MIRD Primer for Absorbed Dose Calculations (Revised edition)* (New York: Society of Nuclear Medicine) p 128

[14] Snyder W, Ford M, Warner G and Jr H F 1969 Estimates of absorbed fractions for monoenergetic photon sources uniformyly distributed in various organs of a heterogeneous phantom. MIRD pamphlet No 5 *J. Nucl. Med.* **10** 5–52
[15] Snyder W, Ford M, Warner G and Watson S 1975 Sabsorbed dose per unit cumulated activity for selected radionuclides and organs *Medical Internal Radiation Dose (MIRD) Pamphlet No 11* (New York: Society of Nuclear Medicine)
[16] Howell R W *et al* 1999 The MIRD perspective 1999. Medical Internal Radiation Dose Committee *J. Nucl. Med.* **40** 3S–10S
[17] Loevinger R 1989 The MIRD perspective *Dosimetry of Administered Radionuclides* ed S Adelstein, A Kassis and R Burt (Washington, DC: American College of Nuclear Physicians/Department of Energy) pp 29–43
[18] Watson E and Foreward J 1999 *Nucl Med* **40** 1S–2S
[19] Eckerman K F and Endo A 1989 *MIRD: Radionuclide Data and Decay Schemes* 2nd edn (Reston, VA: Society of Nuclear Medicine) p 447
[20] Weber D, Eckerman K, Dillman L and Ryman J 1989 *MIRD: Radionuclide Data and Decay Schemes* (New York: Society of Nuclear Medicine) p 447
[21] ICRP 1988 *Radiation Dose to Patients from Radiopharmaceuticals. Int. Commission on Radiological Protection (ICRP) Publication 53* (New York: Pergamon)
[22] Bolch W E *et al* 1999 MIRD pamphlet No. 17: the dosimetry of nonuniform activity distributions--radionuclide S values at the voxel level. Medical Internal Radiation Dose Committee *J. Nucl. Med.* **40** 11S–36S
[23] Howell R W 1994 The MIRD schema: from organ to cellular dimensions *J. Nucl. Med.* **35** 531–3
[24] Vaziri B, Wu H, Dhawan A P, Du P, Howell R W and Committee S M 2014 MIRD pamphlet No. 25: MIRDcell V2.0 software tool for dosimetric analysis of biologic response of multicellular populations *J. Nucl. Med.* **55** 1557–64
[25] Olguin E, President B, Ghaly M, Frey E, Sgouros G and Bolch W E 2020 Specific absorbed fractions and radionuclide S-values for tumors of varying size and composition *Phys. Med. Biol.* **65** 235015
[26] Benua R, Cicale N and Sonenberg M 1962 The relation of radioiodine dosimetry to results and complications in the treatment of metastatic thyroid cancer *Am. J. Roentgenol.* **87** 171–82
[27] Cristy M and Eckerman K 1987 Specific absorbed fractions of energy at various ages from internal photon sources (I–VII) *Oak Ridge National Laboratory Report ORNL/TM-8381/V1–7* (Springfield, VA: National Technical Information Service, Dept of Commerce)
[28] ICRP 2002 Basic anatomical and physiological data for use in radiological protection—reference values. ICRP Publication 89 *Ann. ICRP* **32** 1–277
[29] ICRP 2009 Adult reference computational phantoms. ICRP Publication 110 *Ann. ICRP* **39** 1–165
[30] ICRP 2016 ICRP Publication 133: The ICRP computational framework for internal dose assessment for reference adults: specific absorbed fractions *Ann. ICRP* **45** 1–74
[31] Wayson M B and Bolch W E 2018 Individualized adjustments to reference phantom internal organ dosimetry—scaling factors given knowledge of patient internal anatomy *Phys. Med. Biol.* **63** 085006
[32] Wayson M B and Bolch W E 2018 Individualized adjustments to reference phantom internal organ dosimetry—scaling factors given knowledge of patient external anatomy *Phys. Med. Biol.* **63** 085007

[33] Sgouros G 2019 Radiophramaceutical therapy *Health Phys.* **116** 175–78
[34] Divgi C 2018 The current state of radiopharmaceutical therapy *J. Nucl. Med.* **59** 1706–7
[35] O'Donoghue J A 1999 Implications of nonuniform tumor doses for radioimmunotherapy *J. Nucl. Med.* **40** 1337–41
[36] Humm J L, Macklis R M, Bump K, Cobb L M and Chin L M 1993 Internal dosimetry using data derived from autoradiographs *J. Nucl. Med.* **34** 1811–7
[37] Humm J L, Macklis R M, Lu X Q, Yang Y, Bump K, Beresford B and Chin L M 1995 The spatial accuracy of cellular dose estimates obtained from 3D reconstructed serial tissue autoradiographs *Phys. Med. Biol.* **40** 163–80
[38] Berger M J 1968 Energy deposition in water by photons from point isotropic sources *J. Nucl. Med.* Suppl **1** 17–25
[39] Zaidi H and Andreo P T 2002 The Monte Carlo method: theory and computational issues *Therapeutic Applications of Monte Carlo Calculations in Nuclear Medicine* ed H Zaidi and G Sgouros (Boca Raton, FL: CRC Press)
[40] Dieudonne A, Hobbs R F, Bolch W E, Sgouros G and Gardin I 2010 Fine-resolution voxel S values for constructing absorbed dose distributions at variable voxel size *J. Nucl. Med.* **51** 1600–7
[41] Lanconelli N, Pacilio M, Lo Meo S, Botta F, Di Dia A, Aroche L A T, Perez M A C and Cremonesi M 2012 A free database of radionuclide voxel S values for the dosimetry of nonuniform activity distributions *Phys. Med. Biol.* **57** 517–33
[42] Pacilio M, Botta F, Aroche L T, Amato E, Diaz N C, Lanconelli N, Perez M C, Basile C and Cremonesi M 2012 Impact of voxel S-values calculation method on 3D dosimetry for radionuclide therapy: application to SIRT and IART treatments *Eur. J. Nucl. Med. Mol. Imaging* **39** S320–320
[43] Furhang E E, Chui C S, Kolbert K S, Larson S M and Sgouros G 1997 Implementation of a Monte Carlo dosimetry method for patient-specific internal emitter therapy *Med. Phys.* **24** 1163–72
[44] Furhang E E, Chui C S and Sgouros G 1996 A Monte Carlo approach to patient-specific dosimetry *Med. Phys.* **23** 1523–9
[45] Tagesson M, Ljungberg M and Strand S E 1996 A Monte-Carlo program converting activity distributions to absorbed dose distributions in a radionuclide treatment planning system *Acta Oncol.* **35** 367–72
[46] Lee M S, Kim J H, Paeng J C, Kang K W, Jeong J M, Lee D S and Lee J S 2018 Whole-body voxel-based personalized dosimetry: the multiple voxel S-value approach for heterogeneous media with nonuniform activity distributions *J. Nucl. Med.* **59** 1133–39
[47] Papadimitroulas P, Loudos G, Nikiforidis G C and Kagadis G C 2012 A dose point kernel database using GATE Monte Carlo simulation toolkit for nuclear medicine applications: comparison with other Monte Carlo codes *Med. Phys.* **39** 5238–47
[48] Dieudonne A, Hobbs R F, Lebtahi R, Maurel F, Baechler S, Wahl R L, Boubaker A, Le Guludec D, Sgouros G and Gardin I 2013 Study of the impact of tissue density heterogeneities on 3-dimensional abdominal dosimetry: comparison between dose kernel convolution and direct Monte Carlo methods *J. Nucl. Med.* **54** 236–43
[49] Utsunomiya D, Nakaura T, Honda T, Shiraishi S, Tomiguchi S, Kawanaka K, Morishita S, Awai K, Ogawa H and Yamashita Y 2005 Object-specific attenuation correction at SPECT/CT in thorax: optimization of respiratory protocol for image registration *Radiology* **237** 662–9

[50] Zanzonico P 2006 PET-based biological imaging for radiation therapy treatment planning *Crit. Rev. Eukaryot. Gene Expr* **16** 61–101

[51] Goddu S *et al* 1997 *MIRD Cellular S Factors: Self-Absorbed Dose per Unit Cumulated Activity for Selected Radionuclides and Monoenergetic Electrons and Alpha Particle Emitters Incorporated into Different Cell Compartments* (Reston, VA: Society of Nuclear Medicine) p 183

Chapter 4

Dose point-kernels for radionuclide dosimetry

Manuel Bardiès and Lidia Strigari

4.1 Introduction

Dose point-kernels (DPKs) are widely used for radionuclide dosimetry. More than six decades ago, Loevinger in [1] justified the use of 'Point Source Dose Functions' (ancestors of DPK) for beta emissions: 'The problem of determining the absorbed dose around a localized β-particle source in tissue can be separated in two steps. First, the distribution of (absorbed) dose around a (hypothetical) point source of β-particles in tissue is determined. Second, the (absorbed) dose distribution in and around a localized β-particle source is calculated by a suitable summation of the elementary point source distribution.'

The superposition principle is implicit in this approach: If any source volume can be broken down into a set of point sources, the determination of the absorbed dose distribution around point sources allows source volumes to be treated by 'suitable summation'.

For Berger in [2], again in the case of beta emissions, the specific absorbed fraction $\Phi(x)$ used to calculate the absorbed dose rate at some distance x from an isotopic point source can also be regarded as a 'point kernel' because of its use in applications of the superposition principle. Thus,

$$R(x) = A\, n\, k\, \overline{E}\; \Phi(x) \tag{4.1}$$

where

$R(x)$ is the mean absorbed dose rate (in cGy s^{-1}),
x is the distance to the source (in cm),
A is the activity (in Bq),
n is the number of beta particles emitted per decay,
$k = 1.6\ 10^{-8}$ g cGy MeV^{-1},
E is the mean energy from beta particles emitted by the source (in MeV),
$\Phi(x)$ is the point kernel (in g^{-1}).

doi:10.1088/978-0-7503-2694-0ch4

The relation is apparent here between the DPK and the specific absorbed fraction, as defined by the MIRD Committee [3].

MIRD pamphlet No. 7 [4] expanded the formalism of the calculation of absorbed doses delivered at some distance from point sources of electrons or beta particles in water and other media. The definition of the variables used differentiated the specific absorbed fraction (for sources of electrons or beta particles) from 'energy-dissipation distribution' as defined by Spencer [5] and, in particular, allowed the introduction of 'scaled absorbed-dose distribution'.

Thus, for a point source of electrons of energy E_0, the absorbed dose rate $R(x, E_0)$ is given by:

$$R(x,\ E_0) = A\ n\ k\ E_0\ \Phi(x,\ E_0) \tag{4.2}$$

This formulation (very close to the preceding one) defines $\Phi(x, E_0)$ as the 'point isotropic specific absorbed fraction' expressed in g^{-1}. Thus, the quantity $4\pi\rho\, x^2\, \Phi(x, E_0)\, dx$ is the fraction of emitted energy absorbed in a spherical layer of radius x and thickness dx in a medium of density ρ (g cm^{-3}). If it is assumed that all the energy emitted is finally absorbed by the medium, then:

$$4\pi\rho \int_0^\infty x^2\ \Phi(x,\ E_0)dx = 1 \tag{4.3}$$

which constitutes the normalization requirement for $\Phi(x, E_0)$.

Dimensionless 'energy-dissipation distribution' $J(u, E_0)$, as defined according to [6], gives the energy dissipated by the primary electrons (which is not strictly equivalent to the energy absorbed in the medium) in the spherical layers of radius u around a point source:

$$J(u,\ E_0) = 4\pi\rho x^2\ \frac{E_0}{L(E_0)}\ \Phi(x,\ E_0) \tag{4.4}$$

where $L(E_0)$ is the mean energy loss of an electron per unit path length (in MeV cm^{-1}) at source energy E_0, and $u = x/r(E_0)$ is the distance to the source expressed as a fraction of the initial electron range $r(E_0)$, in cm, according to the continuous-slowing-down-approximation (csda).

The distance $r(E_0)$ is often noted as r_0 in the literature. Normalization of the distance (u) and introduction of the factor $E_0/L(E_0)$ allow the dependence of $J(u, E_0)$ on E_0 to be reduced.

The scaled absorbed dose distribution, $F(\xi,\ E_0)$, is defined by:

$$F(\xi,\ E_0) = 4\pi\rho x^2\ \Phi(x,\ E_0)\frac{dx}{d\xi} = 4\pi\rho x^2 x_{90}\ \Phi(x,\ E_0) \tag{4.5}$$

where $\xi = x/x_{90}$ is the ratio of the distance from the source x to the 90-percentile distance.

$F(\xi,\ E_0)$ is dimensionless and relatively independent of the medium crossed (even if x_{90} itself depends on the medium crossed). No explicit definition of the DPK is provided in the article, but the concept of scaled absorbed dose distribution was frequently applied in subsequent works and presented as the DPK. A minor modification often encountered relates to normalization, which can be performed relative to x_{90} or r_0, depending on the author. In this chapter, $F(\xi, E_0)$ will be used for DPK, although the terminology and the normalization factor actually used in the articles cited may differ slightly.

This equation can be generalized to beta particle sources, which introduces:

$$F_\beta(\xi) = 4\pi\rho x^2\, \Phi_\beta(x)\frac{\mathrm{d}x}{\mathrm{d}\xi} = 4\pi\rho x^2 x_{90}\, \Phi_\beta(x) \tag{4.6}$$

The modification of the variables is self-explanatory.

In principle, the resolution of any dosimetric problem consists in considering the contribution of all emissions from a source volume to an entire target volume. This implies the integration in six dimensions of an *absorbed dose deposition function* at some distance from a point source [5]. This function can be defined as the DPK. It is readily apparent that the value of this breakdown relates to the possibility of using a single function applicable to the entire area of interest. This is feasible for an isotropic homogeneous medium, in which case the determination of absorbed dose deposition at some distance from a point source allows any dosimetric problem to be solved by 'suitable summation.' Thus, DPK approaches are relevant mainly in the case of a homogeneous medium, which can be considered as a restriction in terms of application domains.

Another limitation relates to the calculation mode for DPK. In general, the *mean* absorbed dose (a non-stochastic value) deposited at some distance from a point source is considered, which places implicitly the calculation in the domain of *macrodosimetry, regardless of the scale at which the computation is made (i.e. clinical, preclinical/tissue, or even cellular).*

This chapter is divided into four parts: the first concerns how DPKs are obtained (the methods used and the limitations of these methods); the second considers the impact of the medium; the third describes the various data sets available in the literature (photons or electrons); and the fourth considers the domains for the use of DPK in radionuclide dosimetry.

4.2 Methods used to generate dose point-kernels

The methods used to generate DPKs have kept pace with changes in the experimental data available and benefited from progress in the field of numerical calculation. This section describes the different methods used to generate point-kernels and their changes over time.

4.2.1 Photons

According to MIRD formalism, photon DPK in units of [cGy/decay] at a distance r from a monoenergetic gamma source of energy E (MeV) in an unbounded homogeneous medium can be expressed as:

$$K(r) = 1.6\ 10^{-8}\ E\ \frac{\mu_{\mathrm{en}}}{4\pi r^2 \rho} \mathrm{e}^{-\mu r} B_{\mathrm{en}}(\mu r) \tag{4.7}$$

where μ and μ_{en} are, respectively, the linear attenuation and energy absorption coefficients of the photon in the medium. The quantity B_{en} is the energy absorption build-up factor, defined as the ratio between the absorbed dose rate due to both primary and scattered photons and that due to primary photons alone.

Using the moments method of Spencer and Fano, Berger [6] published B_{en} tables for photon energies from 15 keV to 3 MeV for distances μr from 0.05 to 20 mean free paths in water (1 mean free path = $1/\mu$). At individual photon energies within the energy range, Berger also expressed the build-up factor as a ten-term polynomial of distance. Coefficients of the polynomial have no obvious simple energy dependence. It has been shown by Brownell *et al* [7] that Berger's method gives results comparable to those obtained using a Monte Carlo code for photon transport in a large soft-tissue volume.

Berger's tabulated B_{en} values were fitted to analytic functions other than polynomials by Kwok *et al* [8] and Leichner [9]. Such fittings of data minimize computer storage space. The parameters of the fitting function have a simple relationship with photon energy for Kwok *et al*, while a common expression applies to the absorbed fractions for photons and beta particles for Leichner.

The DPK for any specific gamma-emitting nuclide in a homogeneous medium can be obtained by summing the monoenergetic photon DPK K_i over the photon energy spectrum (n_i, E_i), where n_i is the probability of emitting gamma energy E_i per decay and μ_i and $\mu_{\mathrm{en},i}$ are, respectively, the linear attenuation and energy absorption coefficients of the photon with energy E_i in the medium.

$$K_i(r) = 1.6\ 10^{-8}\ n_i E_i\ \frac{\mu_{\mathrm{en},\,i}}{4\pi r^2 \rho} e^{-\mu_i r} B_{\mathrm{en}}(\mu_i r) \tag{4.8}$$

Several groups of researchers used Monte Carlo codes to derive photon DPK in water, namely ETRAN [10], EGS4 [11, 12] and MCNP [13, 14]. The agreement between the codes themselves and with Berger's calculation is within 3%–5% [14–16].

4.2.2 Electrons

At the end of the 1950s, Spencer [17] resolved electron transport equations numerically (according to the csda), while taking multiple diffusion into account according to the moments method. These calculations, as validated from experimental measurements in different media and for different energies [18], considered energy distribution as a function of the distance for isotropic point sources of monoenergetic electrons (20 keV to 10 MeV) in different homogeneous media. As a result of adoption of the csda, statistical variations in energy loss straggling were not considered. Consequently, there was a tendency to underestimate an absorbed dose deposited at some distance from the emission point.

Cross *et al* [19] and [20] conducted important studies for measurement and calculation of DPK (planar source or point source) based on the moments method of

Spencer. Important progress was made by Berger [21], who used the Monte Carlo method to simulate electron transport. Berger was able to take both multiple scattering and energy loss straggling into account. Energy loss from delta rays and bremsstrahlung production were also incorporated.

Although the point-kernels presented in MIRD pamphlet No. 7 [4] are derived from the work of Spencer, those discussed later [22] were calculated using the ETRAN Monte Carlo code [10].

The results obtained with ETRAN versions prior to 1986 may differ from those obtained with EGS4, due to incorrect sampling of energy loss straggling in ETRAN, however, there are no important differences now in the results obtained with these two codes [10]. As an example, in a comparative study between various Monte Carlo codes (PENELOPE, GEANT4, MCNPX, and ETRAN), relative differences between electron DPK in water for monoenergetic electrons were lower than 2%, 2.5%, and 7% for ETRAN, GEANT4, and MCNPX, respectively, when compared to PENELOPE [23].

The DPK proposed for electrons rarely descend below 10–20 keV, which for most simulation codes corresponds to the limit for validity of the multiple diffusion theory.

Of note, an overestimation in the total Auger electron (AE) energy output by MIRD-RADTABS has been reported leading to higher predicted energy deposition by AE emitting radionuclides, especially in the immediate vicinity of the decaying radionuclides. This should be considered when MIRD RADTABS data are used to simulate biological damage at nanoscale dimensions [24].

From a practical point of view, 10 keV electrons have a x_{90} range of around 2 μm in water. This means that most applications will fall in the domain of cellular/infra-cellular dosimetry. Generating DPK for electrons with an energy lower than 10–20 keV is feasible, and it is possible to simulate step-by-step interactions. Various track structure codes have been proposed for this purpose [25]. DPK may be used to verify the relevance of radiation transport algorithms [26, 27]. However, their use is generally limited to energy depositions very near the emission point (e.g. for Auger emitters) and their application to radiobiology. However, the concept of mean absorbed dose is put in question when statistical fluctuations around the mean absorbed energy become pronounced [28]. This is the domain of microdosimetry and DPK, adapted only to macrodosimetry, should not be used in that context.

4.2.3 Beta emitters

Until 1955, the only experimental data available concerned a dozen radionuclides in air. It has been shown that these data could be represented by the formula [1, 7]:

$$\Phi(x) = \frac{k'}{(\mu x)^2}\left\{c\left[1 - \frac{\mu x}{c}\exp^{\left(1-\frac{\mu x}{c}\right)}\right] + \mu x \exp^{(1-\mu x)}\right\} \quad \mu x < c$$

$$\Phi(x) = \frac{k'}{\mu x}\exp^{(1-\mu x)} \quad \mu x \geqslant c \tag{4.9}$$

where k' is a normalization constant, μ the apparent coefficient of absorption for the medium, and c a dimensionless parameter. The coefficients μ and c were given for soft tissues and beta emitters:

$$\mu = 18.6\,(E_0 - 0.036)^{-1.37}\left(2 - \frac{\overline{E}}{\overline{E}^*}\right)\,\mathrm{cm^2\,g^{-1}} \tag{4.10}$$

$$c = \begin{cases} 2 & 0.17 \leqslant E_0 \leqslant 0.5\ \mathrm{MeV} \\ 1.5 & 0.5 \leqslant E_0 \leqslant 1.5\ \mathrm{MeV} \\ 1 & 1.5 \leqslant E_0 \leqslant 3\ \mathrm{MeV} \end{cases} \tag{4.11}$$

where E is the mean energy of the beta spectrum, E_0 the maximal energy, and $\overline{E}^*$ the mean energy of a spectrum representing an allowed shape of maximum energy E_0.

For beta emitters, the contributions from the β spectra must be taken into account by sampling emitted energy over the emission spectra and consider the contribution from every energy bin. Two major types of approaches are possible, depending on the problem involved:

- The problem to be resolved may consist in comparing absorbed doses (or absorbed fractions) delivered to the same target by *different radionuclides*. In this case, it seems preferable to calculate, once and for all, tables of absorbed fractions for monoenergetic electrons and then integrate these results into the emission spectra of the radionuclides considered.
- Conversely, when various geometric configurations need to be studied for the *same radionuclide*, it may be preferable to compute directly the DPK of the radionuclide considered.

In both cases, the application of data for monoenergetic electrons to all electron emissions of a radionuclide should be performed with due care. First of all, it is essential to consider the beta-emission spectra and not mean energy [29]. Some spectra can be neglected if their emission percentage is too low to contribute significantly to the delivered absorbed dose, provided that this choice is indicated and clearly documented. Monoenergetic emissions (Auger or conversion electrons) should be taken into account if they contribute significantly to the delivered absorbed dose. In addition, it must be noted that generating DPK for electrons below a given energy (range) may not be useful: depending on spatial sampling of the geometry, the (low) energy of these electrons can be considered as locally absorbed.

For example, the electron emission values given in table 4.1 for ^{131}I have been published by the Laboratoire Henri Becquerel (formerly *French Primary Laboratory of Ionizing Radiations*) [30]. A comparison of these values with those cited by the MIRD Committee [31] (in boldface in the table) indicates that the difference represents less than 1% of the energy emitted by transition.

Regardless of the method and code employed, current results are concordant enough to allow DPK to be considered as reliable data usable for more complex geometries in the input of calculation programs, at least for energies and spatial

Table 4.1. Electron emission data for ^{131}I.

	Energy/particle E_i	Particles/transition 100 . n_i	Energy/transition $\Delta_i = n_i E_i$	
	keV	%	g.cGy/μCi.h	fJ/Bq.s
e_{AL}	3.9 ± 1.5	5.7 ± 0.5	0.0004	0.031
e_{AK}	29.02 ± 5.51	0.7 ± 0.1	0.0004	0.027
ec_{1K}	**45.622 ± 0.016**	**3.55 ± 0.08**	**0.0035**	**0.259**
ec_{1L}	**75.065 ± 0,335**	**0.47 ± 0.01**	**0.0007**	**0.056**
ec_{1MN}	79.65 ± 0.45	0.105 ± 0.002	0.0002	0.013
ec_{3K}	129.27 ± 0.03	0.48 ± 0.09	0.0013	0.099
ec_{4K}	142.649 ± 0.016	0.050 ± 0.002	0.0002	0.011
ec_{3L}	158.48–159.15	0.23 ± 0.04	0.0008	0.058
ec_{3MN}	162.9–163.8	0.065 ± 0.010	0.0002	0.017
ec_{4L}	171.16–172.43	0.0110 ± 0.0004	0.0000	0.003
ec_{7K}	**249.74 ± 0.03**	**0.254 ± 0.009**	**0.0013**	**0.102**
ec_{7L}	278.85–279.52	0.045 ± 0.002	0.0003	0.020
ec_{7MN}	283.3–284.2	0.0112 ± 0.0005	0.0000	0.005
ec_{14K}	**329.92 ± 0.02**	**1.55 ± 0.05**	**0.0109**	**0.819**
ec_{14L}	**359.03–359.70**	**0.25 ± 0.01**	**0.0019**	**0.144**
ec_{17K}	602.417 ± 0.016	0.0285 ± 0.0003	0.0004	0.028
$\beta_1^-{}_{(max)}$	**247.9 ± 0.6**	**2.11 ± 0.03**	**0.0031**	**0.235**
$\beta_2^-{}_{(max)}$	**309.9 ± 0.6**	**0.63 ± 0.01**	**0.0012**	**0.088**
$\beta_3^-{}_{(max)}$	**333.8 ± 0.6**	**7.21 ± 0.07**	**0.0148**	**1.116**
$\beta_4^-{}_{(max)}$	**606.3 ± 0.6**	**89.9 ± 0.6**	**0.368**	**27.59**
$\beta_5^-{}_{(max)}$	629.7 ± 0.6	0.072 ± 0.007	0.0003	0.026
$\beta_6^-{}_{(max)}$	**806.9 ± 0.6**	**~0.1**	**0.0006**	**0.045**

sampling relevant for radionuclide dosimetry. As a matter of fact, generating DPK can be an elegant way to verify that several codes (or different versions of the same code) give the same results.

4.2.4 Voxel *S* values

Bolch *et al* [5] extended the DPK concept to cubic geometry by presenting voxel *S* value (VSV) tables. The objective was to use these VSVs in cases in which (cumulated) activity is determined from tomoscintigraphic images or quantitative autoradiographies, i.e. from numerical images sampled in three dimensions (voxels). For each target voxel *k*, the mean absorbed dose is obtained by summing the contributions of each surrounding source voxel *h*:

$$\bar{D}_{(k)} = \sum\nolimits_{h=0}^{n} \tilde{A}_h \bullet S_{(k \leftarrow h)} \tag{4.12}$$

The determination of $S_{(k \leftarrow h)}$ was performed by Monte Carlo calculation using the EGS4 code in a homogeneous medium (soft tissue). The calculation was performed for different radionuclides (^{32}P, ^{89}Sr, ^{90}Y, ^{99m}Tc, and ^{131}I) for cubic voxels with 3- and 6-mm sides (i.e. for PET and SPECT, respectively) and for ^{131}I for cubic voxels with 0.1 mm sides (for autoradiography).

There are different ways to obtain VSV. One is to compute directly (most often using Monte Carlo simulation) how the energy emitted in a source voxel is absorbed in a 3D array. The geometry (homogeneous source distribution in the source voxel, isotropic emission, energy scoring in the source and surrounding voxels) is usually easy to define with current Monte Carlo codes [32]. However, a potential issue is that any change in activity spatial sampling requires the generation of new voxel *S* values.

A workaround is to generate VSV with a small spatial sampling, in order to be able to resample for arbitrary voxel sizes. This was proposed by Dieudonné *et al* for ^{18}F, ^{90}Y, ^{99m}Tc, ^{111}In, ^{123}I, ^{131}I, ^{177}Lu, ^{186}Re, and ^{201}Tl [33]. Fast rescaling of VSV was also proposed by Fernandez *et al* [34].

An alternative is to consider finely sampled DPK, and generate VSV based on the DPK. This has the advantage to require only one DPK per radionuclide [35].

Once VSVs are calculated, dosimetry can be performed by using convolution techniques [36].

To increase the speed of the process, the entire activity in the source voxel h can be considered to be concentrated at the centre of the voxel [37]. The absorbed dose rate is then calculated as the contribution to the centre of the target voxel k from the centre of the source voxel (centre-to-centre approximation).

A workflow for identify the possible strategy for using the Local energy deposition (LED), kernel convolution or full Monte Carlo calculation is shown in figure 4.1.

4.3 Impact of the medium

Calculations performed using point-kernels implicitly assume that the absorbed dose is deposited within a uniform, isotropic medium (usually water). The effect of accounting for differences in density and atomic number variations and the potential errors associated with neglecting such variations have been studied for photon dosimetry using Monte Carlo calculations [12].

The results are abstracted in table 4.2. The first column of values (Φ) presents calculations performed assuming a homogeneous water phantom with air beyond the outer contour of the reference man phantom. The second column (Φ_ρ) presents calculations accounting for density (ρ) variations, and the third ($\Phi_{\rho,Z}$) for both density and atomic number (Z) variations.

The table shows that density and atomic number variations may be neglected for geometries in which the source, target and separating medium are similar. The pancreas, spleen and separating medium, for example, have similar densities and atomic numbers. Correspondingly, the magnitude of the error in neglecting ρ and Z variations is small. Conversely, vertebral bone separates the left and right kidneys, so that failure to correct for ρ and Z could lead to absorbed dose overestimations of more than one order of magnitude (depending on the photon energy spectrum).

The effect of accounting for differences in density and atomic number variations on photon DPK was also addressed by [38] and [14]. Both of these groups

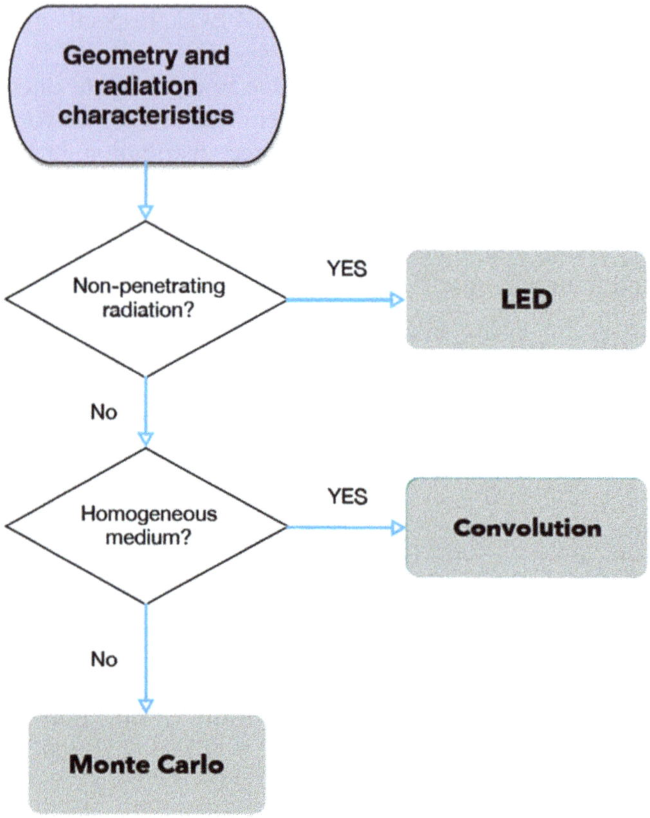

Figure 4.1. Workflow to identify the need of using the Local energy deposition (LED), kernel convolution or full Monte Carlo calculation. Adapted from [79].

Table 4.2. Variation of photon absorbed fractions for selected radionuclides and organs when making different assumptions regarding organs' density and atomic number. Reproduced from [12] John Wiley & Sons. Copyright (1996) American Association of Physicists in Medicine.

Radionuclide	Source	Target	Φ	Φ_ρ	$\Phi_{\rho,Z}$
^{125}I	Pancreas	Spleen	$1.8\ 10^{-4}$	$1.7\ 10^{-4}$	$1.7\ 10^{-4}$
	L Kidney	Liver	$5.9\ 10^{-7}$	$4.9\ 10^{-7}$	$3.3\ 10^{-7}$
	L Kidney	R Kidney	$7.0\ 10^{-6}$	$4.0\ 10^{-6}$	$4.4\ 10^{-7}$
^{131}I	Pancreas	Spleen	$7.2\ 10^{-5}$	$6.7\ 10^{-5}$	$6.7\ 10^{-5}$
	L Kidney	Liver	$5.0\ 10^{-6}$	$4.8\ 10^{-6}$	$4.8\ 10^{-6}$
	L Kidney	R Kidney	$1.4\ 10^{-5}$	$1.2\ 10^{-5}$	$1.2\ 10^{-5}$

introduced a correction factor F to allow for a change in the effective path length between a photon point source and the point where the absorbed dose is evaluated. If the path passes through different materials j of varying thickness t_j and a linear attenuation coefficient μ_j, and μ_w is the linear attenuation coefficient of water.

F is multiplied by the DPK for water to become a modified DPK. This approach also assumes that an equivalent thickness of water, producing the same primary photon attenuation as the actual path through a heterogeneous geometry, predicts scatter build-up correctly.

$$F = e^{-\sum_j (\mu_j - \mu_w) t_j} \tag{4.13}$$

The effect of tissue inhomogeneity is much more complicated on electron/beta DPK than on photon DPK, because of rapid variation of the electron scattering cross-section with the atomic number of the scattering medium and electron energy. Kwok *et al* [8, 39] and Nunes *et al* [40, 41] used both Monte Carlo calculations and experiments to assess the effect of bone and air on absorbed dose distribution of point sources of low-energy electrons and beta emitters near tissue boundaries. When a point source of ^{147}Pm (maximum beta energy 225 keV) or ^{204}Tl (maximum beta energy 763 keV) was positioned at a planar interface of cortical bone (CB) and red marrow (RM), the absorbed dose in RM within 6 μm from the interface was increased by 9±1% (standard error SE) for both sources, as compared to that in the same region inside a homogeneous volume of RM. The absorbed dose in a spherical shell of RM 20 μm thick next to a spherical interface of CB/RM received from a monoenergetic point source of electrons at the center of a sphere of RM with a radius of 500 μm was increased by up to 22±1% (SE) compared to that for the same scoring region in the homogeneous case. The maximum absorbed dose increase occurred at 400 keV electron energy. The dependence of absorbed dose enhancement on the curvature of the interface was also demonstrated. Conversely, when a ^{32}P point source was positioned at a planar interface of air and soft tissue, the absorbed dose in soft tissue 5.5 μm from the boundary was reduced by 26.44 ± 0.02% (SE) from that in the homogeneous situation. There was no simple modification of electron/beta DPK due to tissue inhomogeneity.

More recently, a new method for whole-body voxel-based personalized dosimetry in heterogeneous media with non-uniform activity distributions has be proposed and referred as the multiple absorbed dose point kernel or VSV approach. Instead of using only a single VSV, as found in water, the method uses multiple numbers (N) of VSVs to cover media of various density ranges, as found in the whole. In the phantom and patient studies, the multiple VSV approach showed significant decreases in voxel-level errors, especially for the lung and bone regions yielding results comparable to those of the direct Monte Carlo approach [42].

4.4 Examples of absorbed dose point-kernels available for radionuclide dosimetry

The tables in MIRD pamphlet No. 7 [4] provide $F(\xi, E_0)$ values for nine energies in water ranging from 25 keV to 4 MeV. These DPKs were generated from Spencer's energy-dissipation distributions. Insofar as the data show little variation as a function of energy, it is possible to interpolate these values, if necessary. The MIRD pamphlet No. 7 also give $F_\beta(\xi)$ values for 75 radionuclides in water.

The values used for β spectra were taken essentially from a compilation by [43]. These data can be extrapolated to other media than water.

Berger [22] published an article initially intended as a 'preliminary report,' but which became one of the most frequently cited references in the DPK field. These 'improved point-kernels' were obtained by Monte Carlo simulation using the ETRAN code [21]. Angular deflections and energy loss straggling due to multiple Coulomb scattering (by atoms and orbital electrons) were taken into account, and energy transport by secondary bremsstrahlung was included. The $F(\xi, E_0)$ tables were given for 36 energies ranging from 0.5 keV to 10 MeV, which in practice allowed scanning of all of the necessary energies in the domain of beta-emitting radionuclides. These data have been widely used in dosimetric models or in DPK calculations for radionuclides. Subsequently, it appeared that an incorrect sampling of energy-loss straggling in ETRAN affected the validity of these data [10, 23]. The corrected values now available extend from 20 keV to 20 MeV [10].

The tables for 'beta-ray dose distributions' in water published by [20, 44] were calculated using the ACCEPT code (derived from ETRAN). For monoenergetic emissions, comparison with values obtained with the EGS4 code [45] for six energies ranging from 50 keV to 3.0 MeV indicated that differences between the two data sets were low (a few percent). $F(\xi, E_0)$ values were given for nine energies from 25 keV to 4 MeV.

Leichner [9] proposed a unified approach to photon and beta particle dosimetry based on a fit of Berger tables for photons [6] and electrons [4]. The empirical function proposed is also valid for photons and beta particles.

Other point-kernels generated by the EGS4 Monte Carlo code for 14 radionuclides in an aqueous medium (^{123}I, ^{124}I, ^{125}I, ^{131}I, ^{111}In, ^{64}Cu, ^{67}Cu, ^{67}Ga, ^{68}Ga, ^{186}Re, ^{188}Re, ^{153}Sm, ^{117m}Sn, and ^{99m}Tc) have also been parameterized to analytic functions for easy computation on personal computers [21]. The results are expressed by providing for each radionuclide the coefficients of a series of polynomials and exponentials approximating the calculated values. Once again, the equations proposed are easy to include in more complex calculation programs.

The monoenergetic DPK values published by [22] were used by [46] to calculate DPK for six radionuclides of interest in nuclear medicine (^{32}P, ^{67}Cu, ^{90}Y, ^{131}I, ^{186}Re and ^{188}Re). Only the beta emissions of each radionuclide were considered. The results are presented in table form as well as in analytic form (via a fit of the data by a sum of log-normal and exponential functions).

The EGS4 Monte Carlo code [47] has also been used to generate point kernels for monoenergetic electrons (50 keV to 3 MeV) and for eight radionuclides (^{32}P, ^{67}Cu, ^{90}Y, ^{105}Rh, ^{131}I, ^{153}Sm, ^{186}Re, and ^{188}Re) [15]. The calculation was performed for beta spectra and monoenergetic emissions.

The tables for beta-ray absorbed dose distributions in water published by [20, 44] were calculated using the ACCEPT code (derived from ETRAN). The integration of monoenergetic DPK for 147 radionuclides was performed for beta emissions with a relative intensity greater than 1% and conversion or Auger electrons with energy above 30 keV and intensity greater than 1%. The results are in agreement with those obtained experimentally or by simulation using other codes (ETRAN or EGS4).

As experimental values measured in water were not available, validation was performed by extrapolation of values measured in other media (air, gas, plastic or metal scintillators). The results are shown as the product of the absorbed dose rate at distance r from a point source multiplied by the squared distance, i.e. r^2. $J'(r)$, in nGy h^{-1} Bq^{-1} cm^2, which in fact amounts to the product obtained by multiplying r^2 by an absorbed dose per decay (similar to an S-factor according to MIRD nomenclature) [48].

Cross [49] proposed empirical expressions for calculation of absorbed doses delivered at some distance from point sources The principle consists in adding an 'end-of-range' term to an expression similar to that proposed by Loevinger. The adjustment was made for 60 radionuclides using data derived from the ACCEPT code. The error involved is of the order of a few percent in most cases.

Databases of DPK or VSV are now available and can be used for absorbed dose calculations [50–53].

The use of approximated relations for data (experimental or calculated), which were employed in the past to reduce calculation time, seems less justified today. Insofar as any data fit introduces an element of uncertainty into the initial scheme, it seems more logical to use raw data directly.

The calculation of raw data is generally performed from Monte Carlo simulation codes using homogenous medium. A review of current studies indicates that the various codes proposed give similar results and are sufficiently concordant with experimental values for DPK, so that they can be considered as reliable and used as input data in more elaborate calculation models.

4.5 Use of dose point-kernels in absorbed dose calculations

It is rather difficult to present the many approaches for the use of DPKs in radionuclide dosimetry presented in the literature. The range of applications is rather broad (calculation of absorbed doses, absorbed fractions or S-factors), the scales considered extend from the cell to the organ, and the methods used are numerous (numerical integration or convolution; hybrid methods associating Monte Carlo simulation and the convolution of radioactive sources by DPK). The classification adopted here is based on the scale of the problem, from cellular to organ dimensions as summarized in figure 4.2 and described in the following paragraphs.

4.5.1 Absorbed fractions on a cellular scale

DPKs are frequently used for calculations on a cellular scale, for which the medium can often be considered as homogeneous (generally equivalent to water) and isotropic.

Models have been developed to evaluate the dosimetric properties of different beta emitters in the context of targeted radiotherapy. For example, in the case of homogeneous uptake of electron emitters (energy E_0) throughout the cell (radius R_s), Berger [54] has shown that the self-absorbed fraction of energy can be calculated by:

$$\phi(R_S \leftarrow R_S) = \frac{1}{r_0}\int_0^{2R_S}\left[1-1.5\frac{x}{2R_S} + 0.5\left(\frac{x}{2R_S}\right)^3\right]F(\xi,\, E_0)dx \tag{4.14}$$

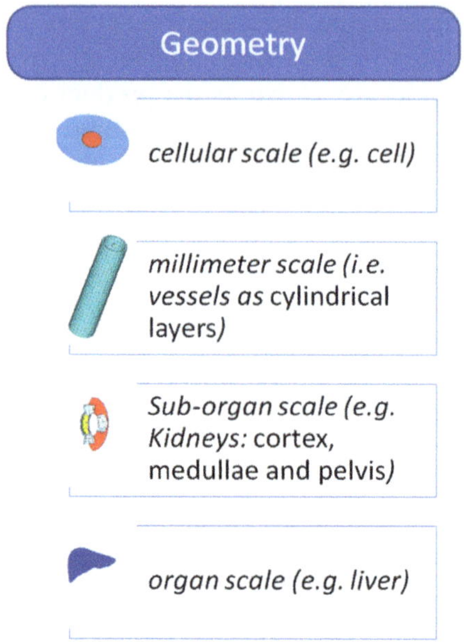

Figure 4.2. The classification of different dosimetric approaches using DPK/VSV, based on the scale of the problem, from cellular to organ dimensions.

Likewise, for uptake on the cell surface (e.g. in radioimmunotherapy), the self-absorbed fraction [46] is given by:

$$\phi(R_S \leftarrow R_S) = \frac{1}{2r_0}\int_0^{\infty}\left[1 - \frac{x}{2R_S}\right]F(\xi,\, E_0)dx \tag{4.15}$$

where $F(\xi, E_0)$ is the scaled point kernel (considered above) and r_0 the maximal electron range according to the csda.

Various publications have proposed absorbed fractions on a cellular scale. All describe the geometry of the problem analytically (even though the integration or convolution used to approach the result is generally performed numerically). They often apply a geometrical reduction factor Ψ (ranging from 0 to 1) describing the efficacy of irradiation at some distance from the source point in terms of the fraction of the sphere centred on the emission point intersecting target volume. Berger [54] noted that 'One can interpret $\Psi(x)$ very simply as the fraction of the spherical shell of radius x around the point of observation that lies within the source region'.

The frequent use of DPK on a cellular scale can be explained at least in part by the simplification introduced through spherical symmetries, both for the geometrical model and energy deposition at some distance from a point source.

4.5.2 Dosimetry on a millimetre scale

When the absorbed dose is delivered on a millimetre scale, e.g. in the case of tumour mass during targeted therapy trials in animals, the use of DPK can be envisaged (generally for beta emitters), provided that the medium is always homogeneous.

In this case, various possibilities can be considered, depending on how activity distribution is determined.

If the activity is described analytically, the preceding case applies, i.e. appropriate integration of DPK [55]. Thus, it is possible to convolve activity distribution by DPK [8]. If the activity is described discretely, e.g. in the case of serial autoradiographs of tumour sections, it is possible to convolute each volume element (voxel) by the DPK to obtain the absorbed dose throughout the volume [56, 57].

A mouse model for calculating cross-organ beta absorbed doses for yttrium-90-labeled immunoconjugates was proposed by Hui *et al* [58] This approach used the DPK convolution proposed in MIRD pamphlet No. 7 [4] to calculate self-absorbed fractions for organs with a density close to that of water, and Monte Carlo simulation [47] for organs with a different density (lungs, bone and marrow).

It is possible to simulate the geometry of the problem (size, relative position of source regions and targets) by a Monte Carlo approach [59] and then use DPK to calculate the absorbed doses delivered at some distance from the emission points. This approach has been used to design a 'mouse bone marrow dosimetry model' [60] or in the case of dosimetry of follicular lymphomas [58].

Tumour blood vessels have been modelled [61] as a group of micro-cylindrical layers comprising endothelial cells (one-cell thick, 10 μm diameter), viable tumour cells (25 cell thick, 250 μm radius), and necrotic tumour region (>250 μm from any blood vessel). Sources of radioactivity were assumed to be uniformly distributed either in endothelial cells or in concentric cylindrical shells within the viable tumour-cell region. The number of radioactive decays required to deposit cytocidal absorbed doses (>100 Gy) in the vascular endothelial cells (endothelial cell mean absorbed dose) or, alternatively, at the tumour edge of adjacent viable tumour cells was determined for beta (^{32}P, ^{33}P, ^{67}Cu, ^{90}Y, ^{131}I, and ^{188}Re) and alpha (^{211}At and ^{213}Bi) particle emitters. The EGSnrc Monte Carlo simulation code was used for direct beta particle dosimetry, but a DPK method was chosen for alpha particle dosimetry.

4.5.3 Organ scale dosimetry

Absorbed doses delivered by gamma emissions are generally considered on an organ scale, in which case beta emissions are regarded as non-penetrating. The calculation of *S*-factors for the best-known anthropomorphic phantoms is based in part on DPK [62–70]. In this case, activity is considered to be uniformly distributed in organs modelled by simplified geometric forms.

DPKs have also been proposed for customized dosimetry in the case of therapeutic applications of radionuclides [71–73]. Activity is determined from quantitative tomoscintigraphy. Thus, three-dimensional discrete distribution is considered, and activity is regarded as uniform in each volume element or voxel. The calculation is performed by DPK convolution of activity distribution.

In the case of discrete convolutions, the use of Fast Fourier Transform (FFT) [72] or Fast Hartley Transform (FHT) [74] allows a significant reduction of calculation time, which could facilitate the use of these methods in a clinical context. However, it should be recalled that these methods apply to a homogeneous medium.

Absorbed dose point kernels have been used directly to derive the absorbed dose in homogeneous medium [75].

More recently, the hepatic arterial tree of the patient was segmented from the cone-beam CT (CBCT) to predict the microsphere transport using multiscale CFD modelling. To calculate the absorbed dose distribution, the predicted microsphere distribution between the liver segments using flow analysis was convolved with ^{90}Y DPK. The absorbed dose distribution was found to be consistent with the ^{90}Y PET images, supporting the feasibility of this approach [76].

Several examples of clinical dosimetry software (academic or commercial) implement DPK (or VSV) convolution in order to speed up the absorbed dose calculation step [37]. These include Mr Voxel, MIM SurePlan™ MRT, Hybrid Viewer Dosimetry, PLANET Dose, QDOSE, Simplicity, Q-Suite™, Nukdose, Hermes, etc.

A question arises which dosimetry method is optimal given the limitations in reconstruction- and quantification procedures.

Recent approaches using multiple VSV in heterogeneous media have recently been proposed to overcome the limit usually associated with DPK convolution [77]. Recent studies use VSV as input data for clinical dosimetry based on deep learning approaches [78].

4.6 Conclusion

Absorbed DPKs constitute one of the tools available for performing dosimetric calculations in nuclear medicine [59]. In general, absorbed doses delivered by beta emitters are considered on a microscopic scale and those by gamma emitters on an organ scale. Thus, DPK dosimetric methods can be applied on a microscopic or macroscopic scale.

DPKs constitute the most suitable approach when the medium is homogeneous and isotropic but provide only a stop-gap solution when it is heterogeneous and an estimation of delivered absorbed doses is required within a reasonable time. In fact, the relevant alternative to using DPK in a heterogeneous medium consists in simulating particle transport and energy depositions integrally by Monte Carlo methods. This alternative is being used increasingly in nuclear medicine. However, the methods applied still require abilities in physics and computer science and involve a computing power that limits their extension. In the near future, it is likely that simulation methods will satisfy current needs for customized dosimetry on a per patient basis in targeted radiotherapy and other applications. Alternatively, hybrid approaches will be able to consider DPK in heterogeneous media.

In principle, the field of DPK applications considers cases in which the mean absorbed dose at some distance from a point source is valid, representing a stochastic and macrodosimetric magnitude. When absorbed dose calculations are made on a cellular scale, it may be necessary to verify that the conditions of application of macrodosimetry are met. Otherwise, microdosimetric methods should be implemented, as reported in [28].

DPKs have been used for more than 60 years in the field of radionuclide dosimetry. They represent the energy transfer function at some distance from a point source and can be used in many dosimetric approaches, by applying the superposition principle. Values available in the literature both for gamma and electron/beta emissions can be considered as reliable. The large number of approaches proposed, whether for simple or complex geometries described analytically or discretely, are indicative of the value and versatility of this type of data in medical physics.

References

[1] Loevinger R, Japha E M and Brownell G L 1956 *Discrete Radioisotope Sources in Radiation Dosimetry* (New York: Academic)

[2] Meredith W J 1970 Medical radionuclides: radiation dose and effects *Int. J. Radiat. Biol. Related Stud. Phys. Chem. Med.* **18** 399

[3] Loevinger R and Berman M 1968 MIRD Pamphlet No. 1: a schema for absorbed-dose calculation for biologically distributed radionuclides *J. Nucl. Med.* **9** 97–14

[4] Berger M J 1971 MIRD pamphlet No. 7: distribution of absorbed doses around point sources of electrons and beta particles in water and other media *J. Nucl. Med.* **12** 5–23

[5] Bolch W E B L *et al* 1999 MIRD pamphlet No. 17: the dosimetry of nonuniform activity distributions—radionuclide *S* values at the voxel level. Medical Internal Radiation Dose Committee *J. Nucl. Med.* **40** 11S–36S

[6] Berger M J 1968 Energy deposition in water by photons from point isotropic sources MIRD Pamphlet No. 2 *J. Nucl. Med.* **1** 15–25

[7] Brownell G F, Ellett W H and Reddy A R 1968 Absorbed fractions for photon dosimetry *J. Nucl. Med.* **Suppl.** 29–39

[8] Kwok C S, Prestwich W V and Wilson B C 1985 Calculation of radiation doses for nonuniformity distributed beta and gamma radionuclides in soft tissue *Med. Phys.* **12** 405–12

[9] Leichner P K 1994 A unified approach to photon and beta particle dosimetry *J. Nucl. Med.* **35** 1721–9

[10] Seltzer S M 1988 An overview of ETRAN Monte Carlo methods *Monte Carlo Transport of Electrons and PhotonsEttore Majorana International Science Series* vol 38 (Boston, MA: Springer) pp 153–81

[11] Luxton G 1994 Comparison of radiation dosimetry in water and in solid phantom materials for I-125 and Pd-103 brachytherapy sources: EGS4 Monte Carlo Study *Med. Phys.* **21** 631–41

[12] Furhang E E, Sgouros G and Chui C S 1996 Radionuclide photon dose kernels for internal emitter dosimetry *Med. Phys.* **23** 759–64

[13] Tsoulfanidis N and Shrestha B 1997 Photon dose equivalent rate from a cylindrical source using a point kernel technique *Health Phys.* **72** 931–5

[14] Janicki C, Duggan D M and Rahdert D A 2001 A dose-point-kernel model for a low energy gamma-emitting stent in a heterogeneous medium *Med. Phys.* **28** 1397–405

[15] Lubberink M, Lundqvist H, Westlin J E, Tolmachev V, Schneider H, Lövqvist A and Sundin A *et al* 1999 Positron emission tomography and radioimmunotargeting—aspects of quantification and dosimetry *Acta Oncol.* **38** 343-9

[16] Luxton G and Jozsef G 1999 Radial dose distribution, dose to water and dose rate constant for monoenergetic photon point sources from 10 keV to 2 MeV:EGS4 Monte Carlo model calculation *Med. Phys.* **26** 2531–8

[17] Spencer L V 1955 Theory of electron penetration *Phys. Rev.* **98** 1597

[18] Spencer L V C 1959 Energy Dissipation by Fast Electrons: US Department of Commerce, National Bureau of Standards

[19] Cross W G and Williams G 1967 The distribution of absorbed energy from a point beta source *Can. J. Phys.* **45** 2021–40

[20] Cross W, Freedman N and Wong P 1992 Tables of beta-ray dose distributions in water *Technical report* AECL-10521 (Chalk River, ON: Chalk River Nuclear Labs, Atomic Energy of Canada Ltd)

[21] Berger M J 1973 Monte Carlo calculation of the penetration of diffusion of fast charged particles *Methods Comput. Phys.* 135–215

[22] Berger M J 1973 *Improved Point Kernels for Electron and Beta-ray Dosimetry* (NBSIR 73-107 (Gaithersburg, MD: National Bureau of Standards))

[23] Rogers D W O and Bielajew A F 1988 *A Comparison of EGS4 and ETRAN in Monte carlo Transport of Electrons and Photons* (New York: Plenum Press) pp 323–44

[24] Falzone N, Lee B Q, Fernandez-Varea J M, Kartsonaki C, Stuchbery A E and Kibedi T *et al* 2017 Absorbed dose evaluation of Auger electron-emitting radionuclides: impact of input decay spectra on dose point kernels and S-values *Phys. Med. Biol.* **62** 2239–53

[25] Paretzke H, Goodhead D, Kaplan I and Terrissol M 1995 Track structure quantities. Atomic and molecular data for radiotherapy and radiation research. International Atomic Energy Agency. Contract No. IAEA-TECDOC-799 (International Atomic Energy Agency)

[26] Champion C, Incerti S, Perrot Y, Delorme R, Bordage M C and Bardies M *et al* 2014 Dose point kernels in liquid water: an intra-comparison between GEANT4-DNA and a variety of Monte Carlo codes *Appl. Radiat. Isot.* **83** 137–41

[27] Bordage M C, Bordes J, Edel S, Terrissol M, Franceries X and Bardies M *et al* 2016 Implementation of new physics models for low energy electrons in liquid water in Geant4-DNA *Phys Med.* **32** 1833–40

[28] ICRU 1983 Microdosimetry *ICRU Report 36* (Bethesda, MD: International Commission on Radiation Units and Measurements)

[29] Akabani G, Poston J W S R and Fau-Bolch W E 1991 Estimates of beta absorbed fractions in small tissue volumes for selected radionuclides *J. Nucl. Med.* **32** 835–9

[30] Lagoutine F, Legrand J and Coursol N 1983 Table de radionucléides: Commissariat à l'Energie Atomique

[31] Weber D A, Eckerman K F and Dillman L T 1989 *MIRD—Radionuclide Data and Decay Schemes* (New York: Society of Nuclear Medicine)

[32] Strigari L, Menghi E, D'Andrea M and Benassi M 2006 Monte Carlo dose voxel kernel calculations of beta-emitting and Auger-emitting radionuclides for internal dosimetry: a comparison between EGSnrcMP and EGS4 *Med. Phys.* **33** 3383–9

[33] Dieudonné A, Hobbs R F, Bolch W E, Sgouros G and Gardin I 2010 Fine-resolution voxel *S* values for constructing absorbed dose distributions at variable voxel size *J. Nucl. Med.* **51** 1600–7

[34] Fernández M, Hänscheid H, Mauxion T, Bardiès M, Kletting P and Glatting G *et al* 2013 A fast method for rescaling voxel S values for arbitrary voxel sizes in targeted radionuclide therapy from a single Monte Carlo calculation *Med. Phys.* **40** 082502

[35] Franquiz J M, Chigurupati S and Kandagatla K 2003 Beta voxel *S* values for internal emitter dosimetry *Med. Phys.* **30** 1030–2

[36] Strigari L, D'Andrea M, Maini C L, Sciuto R and Benassi M 2006 Biological optimization of heterogeneous dose distributions in systemic radiotherapy *Med. Phys.* **33** 1857–66

[37] Pérez P and Valente M 2019 DOSIS: an integrated computational tool for patient-specific dosimetry in nuclear medicine by Monte Carlo and dose point kernel approaches *Appl. Radiat. Isot.* **150** 135–40

[38] Monroe J I, Dempsey J F, Dorton J A, Mutic S, Markman J and Williamson J F *et al* 2001 Experimental validation of dose calculation algorithms for the GliaSite RTS, a novel ^{125}I liquid-filled balloon brachytherapy applicator *Med. Phys.* **28** 73–85
[39] Kwok C S, Bialobzyski P J and Yu S K 1990 Effect of tissue inhomogeneity on dose distribution of continuous activity of low-energy electrons in bone marrow cavities with different topologies *Med. Phys.* **17** 786–93
[40] Nunes J, Fau P W and Kwok C S 1993 An evaluation of the EGS4 and CYLTRAN Monte Carlo codes with regard to boundary beta-ray dosimetry by comparison with experimental beta-ray dose backscatter factors *Med. Phys.* **20** 1243–50
[41] Nunes J, Fau P W and Kwok C S 1993 Experimental determination of 32P dose backscatter factors at and near soft-tissue boundaries *Med. Phys.* **20** 223–31
[42] Lee M S, Kim J H, Paeng J C, Kang K W, Jeong J M and Lee D S *et al* 2018 Whole-body voxel-based personalized dosimetry: the multiple voxel s-value approach for heterogeneous media with nonuniform activity distributions *J. Nucl. Med.* **59** 1133–9
[43] Martin M J and Blichert-Toft P 1970 Radioactive atoms: Auger-electron, α-, β, γ-, and x-ray data *At. Data Nucl. Data Tables* **8** 1–198
[44] Cross W G, Freedman N O and Wong P Y 1992 Beta-ray dose distributions from point sources in an infinite water medium *Health Phys.* **63** 160–71
[45] Simpkin D J and Mackie T R 1990 EGS4 Monte Carlo determination of the beta dose kernel in water *Med. Phys.* **17** 179–86
[46] Prestwich W V, Nunes J and Kwok C S 1989 Beta dose point kernels for radionuclides of potential use in radioimmunotherapy *J. Nucl. Med.* **30** 1036–46
[47] Nelson W R, Hirayama H and Rogers D W 1985 *EGS4 code system* (Menlo Park, CA: Stanford Linear Accelerator Center)
[48] Loevinger R, Budinger T and Watson E 1991 *MIRD Primer for Absorbed Dose Calculations.* Revised (New York: The Society of Nuclear Medicine)
[49] Cross W G 1997 Empirical expressions for beta ray point source dose distributions *Radiat. Prot. Dosim.* **69** 85–96
[50] Papadimitroulas P, Loudos G, Nikiforidis G C and Kagadis G C 2012 A dose point kernel database using GATE Monte Carlo simulation toolkit for nuclear medicine applications: comparison with other Monte Carlo codes *Med. Phys.* **39** 5238–47
[51] Lanconelli N, Pacilio M, Lo Meo S, Botta F, Di Dia A and Aroche A T *et al* 2012 A free database of radionuclide voxel S values for the dosimetry of nonuniform activity distributions *Phys. Med. Biol.* **57** 517–33
[52] Pacilio M, Lanconelli N, Lo M S, Betti M, Montani L and Torres A L *et al* 2009 Differences among Monte Carlo codes in the calculations of voxel S values for radionuclide targeted therapy and analysis of their impact on absorbed dose evaluations *Med. Phys.* **36** 1543–52
[53] Graves S A, Flynn R T and Hyer D E 2019 Dose point kernels for 2,174 radionuclides *Med. Phys.* **46** 5284–93
[54] Berger M 1970 Beta ray dosimetry calculations with the use of point kernels *Medical Radionuclides: Radiation Dose and Effects* (Tennessee: USAEC) pp 63–86
[55] Langmuir V K and Sutherland R M 1988 Dosimetry models for radioimmunotherapy *Med. Phys.* **15** 867–73
[56] Roberson P L, Buchsbaum D J, Heidorn D B and Ten Haken R K 1992 Three-dimensional tumor dosimetry for radioimmunotherapy using serial autoradiography *Int. J. Radiat. Oncol. Biol. Phys.* **24** P329–34

[57] Roberson P L, Heidorn D B, Buchsbaum D J and Buchsbaum D J 1994 Three-dimensional reconstruction of monoclonal antibody uptake in tumor and calculation of beta dose-rate nonuniformity *Cancer* **73** 912–8
[58] Hui T E, Fisher D R, Press O W, Eary J F, Weinstein J N, Badger C C and Bernstein I D *et al* 1992 Localized beta dosimetry of ^{131}I-labeled antibodies in follicular lymphoma *Med. Phys.* **19** 97–104
[59] Strigari L, Sciuto R, D'Andrea M, Pasqualoni R, Benassi M and Maini C L 2007 Radiopharmaceutical therapy of bone metastases with $^{89}SrCl_2$, ^{186}Re-HEDP and ^{153}Sm-EDTMP: a dosimetric study using Monte Carlo simulation *Eur. J. Nucl. Med. Mol. Imaging* **34** 1031–8
[60] Muthuswamy M S, Roberson P L and Buchsbaum D J 1998 A mouse bone marrow dosimetry model *J. Nucl. Med.* **39** 1243–7
[61] Zhu X, Palmer M R, Makrigiorgos G M and Kassis A I 2010 Solid-tumor radionuclide therapy dosimetry: new paradigms in view of tumor microenvironment and angiogenesis *Med. Phys.* **37** 2974–84
[62] Snyder W S 1975 *Committee SonmMIRD. 'S,' Absorbed Dose Per Unit Cumulated Activity for Selected Radionuclides and Organs* (New York: Society of Nuclear Medicine)
[63] Ford M R, Snyder W S, Dillman L T and Watson S B 1977 Maximum permissible concentration (MPC) values for spontaneously fissioning radionuclides *Health Phys.* **33** 35–43
[64] Cristy M and Eckerman K F 1987 *Specific Absorbed Fractions of Energy at Various Ages from Internal Photon Sources: 1, Methods* (TN: Oak Ridge National Lab). Report No.: ORNL/TM-8381/V1; Other: ON: DE87013345 United States 10.2172/6233735 Other: ON: DE87013345 NTIS, PC A06/MF A01; 1. ORNL English
[65] Cristy M and Eckerman K F 1987 *Specific Absorbed Fractions of Energy at Various Ages from Internal Photon Sources: 2, One-Year-Old* (TN: Oak Ridge National Lab.). Report No.: ORNL/TM-8381/V2; Other: ON: DE87013346 United States 10.2172/6203023 Other: ON: DE87013346 NTIS, PC E03/MF A01; 1. ORNL English
[66] Cristy M and Eckerman K F 1987 *Specific Absorbed Fractions of Energy at Various Ages From Internal Photon Sources: 3, Five-Year-Old* (TN: Oak Ridge National Lab.). Report No.: ORNL/TM-8381/V3; Other: ON: DE87013347; TRN: 87-033042 United States 10.2172/6263443 Other: ON: DE87013347; TRN: 87-033042 NTIS, PC E03/MF A01; 1. ORNL English
[67] Cristy M and Eckerman K F 1987 *Specific Absorbed Fractions of Energy at Various Ages from Internal Photon Sources: 4, Ten-Year-Old* (TN: Oak Ridge National Lab.). Report No.: ORNL/TM-8381/V4; Other: ON: DE87013348; TRN: 87-033043 United States 10.2172/6233673 Other: ON: DE87013348; TRN: 87-033043 NTIS, PC E03/MF/A01; 1. ORNL English
[68] Cristy M and Eckerman K F 1987 *Specific Absorbed Fractions of Energy at Various Ages from Internal Photon Sources: 5, Fifteen-Year-Old Male and Adult Female* (TN: Oak Ridge National Lab.). Report No.: ORNL/TM-8381/V5; Other: ON: DE87013349; TRN: 87-033044 United States 10.2172/6263426 Other: ON: DE87013349; TRN: 87-033044 NTIS, PC E03/MF A01; 1. ORNL English
[69] Cristy M and Eckerman K F 1987 *Specific Absorbed Fractions of Energy at Various Ages from Internal Photon Sources: 6, Newborn* (TN: Oak Ridge National Lab.). Report No.: ORNL/TM-8381/V6; Other: ON: DE87013350 United States 10.2172/6202949 Other: ON: DE87013350 NTIS, PC E03/MF A01; 1. ORNL English

[70] Cristy M and Eckerman K F 1987 *Specific Absorbed Fractions of Energy at Various Ages From Internal Photon Sources: 7, Adult Male* (TN: Oak Ridge National Lab.). Report No.: ORNL/TM-8381/V7; Other: ON: DE87013351 United States 10.2172/6233638 Other: ON: DE87013351 NTIS, PC E03/MF A01; 1. ORNL English

[71] Giap H B, Macey D J, Bayouth J E and Boyer A L 1995 Validation of a dose-point kernel convolution technique for internal dosimetry *Phys. Med. Biol.* **40** 365–81

[72] Akabani G, Hawkins W G, Eckblade M B and Leichner P K 1997 Patient-specific dosimetry using quantitative SPECT imaging and three-dimensional discrete Fourier transform convolution *J. Nucl. Med.* **38** 308–14

[73] Lampinen J S, Pohjonen H K, Savolainen S E and Savolainen S E 1998 Calculating internal dose by convolution from SPECT/MR fusion images *Ann. Nucl. Med.* **12** 1–5

[74] Erdi A K, Yorke E D, Loew M H, Erdi Y E, Sarfaraz M and Wessels B W 1998 Use of the fast Hartley transform for three-dimensional dose calculation in radionuclide therapy *Med. Phys.* **25** 2226–33

[75] Liu A, Williams L E, Wong J Y and Raubitschek A A 1998 Monte Carlo-assisted voxel source kernel method (MAVSK) for internal beta dosimetry *Nucl. Med. Biol.* **25** 423–33

[76] Roncali E, Taebi A, Foster C and Vu C T 2020 Personalized dosimetry for liver cancer Y-90 radioembolization using computational fluid dynamics and Monte Carlo simulation *Ann. Biomed. Eng.* **48** 1499–510

[77] Sanchez-Garcia M, Gardin I, Lebtahi R and Dieudonné A 2014 A new approach for dose calculation in targeted radionuclide therapy (TRT) based on collapsed cone superposition: validation with (90)Y *Phys. Med. Biol.* **59** 4769–84

[78] Lee M S, Hwang D, Kim J H and Lee J S 2019 Deep-dose: a voxel dose estimation method using deep convolutional neural network for personalized internal dosimetry *Sci. Rep.* **9** 10308

[79] Bardiès M and Vergara Gil A 2022 Absorbed dose calculations in Dosimetry for Radiopharmaceutical Therapy IAEA at press

IOP Publishing

Monte Carlo Calculations in Nuclear Medicine (Second Edition)
Therapeutic applications
Habib Zaidi

Chapter 5

Computational models of human anatomy

Wesley Bolch

5.1 Introduction

Assessment of the absorbed dose to organs and tissues within the nuclear medicine patient—both in the context of diagnostic imaging and of cancer therapy—cannot be achieved experimentally and thus must be made via computation. As outlined in other chapters of this book, computational dosimetry in therapy nuclear medicine can be accomplished by Monte Carlo radiation transport simulation, by dose-point-kernel convolution, or by application of the dosimetry schema of the Medical Internal Radiation Dose (MIRD) Committee of the Society of Nuclear Medicine and Molecular Imaging (SNMMI). In the latter, the radiation-absorbed dose to a target organ (or region) r_T can be computed as $D(r_T) = \sum_S \widetilde{A}(r_S)S(r_T \leftarrow r_S)$, where $\widetilde{A}(r_S)$ is the time-integrated activity (total number of decays of the radiopharmaceutical) within one or more source organs (or regions) r_S and $S(r_T \leftarrow r_S)$ is the radionuclide S-value (defined as the radiation-absorbed dose to r_T per radionuclide decay in r_S). The former is obtained via planar or 3D (SPECT or PET) imaging of the patient, potentially in combination with anatomic CT, and more recently, MR imaging. The latter is generally computed via radiation transport simulations (Monte Carlo modeling or dose-point-kernel convolution) in a 3D anatomic representation of the patient. This anatomic patient model is what we refer to as a *computational human phantom* (or CHP). The purpose of this chapter is to review the history of CHPs in general, with some reference to their past, present, and even future applications in radiopharmaceutical therapy. As will be seen, there has been significant expansion of CHPs since their early introduction in the early 1960s. The reader is encouraged to explore comprehensive reviews of CHPs to include a textbook on the topic by Xu and Eckerman [1], and comprehensive review articles by Xu [2] and by Kainz *et al* [3]. As noted in the Xu review [2], there has been an exponential increase in the construction of whole-body CHPs over the past three decades, and their applications to dosimetry of radiation protection, medical imaging, and radiotherapy of cancer patients. Due to this large number of CHPs developed to date, this chapter presents only a select sampling of CHPs available in the medical research and clinical community.

5.2 Phantom format types and morphometric categories

5.2.1 Phantom format types

There are two general attributes of computational human phantoms: (1) format types and (2) morphometric categories [4]. Format types describe the computational and/or mathematical techniques applied to describe both the virtual patient's outer body contour as well as their internal organ structure. To date, CHPs have been produced along three format types: stylized (or mathematical), voxel (or tomographic), and mesh (polygon-mesh, tetrahedral-mesh, or non-uniform B-spline surfaces—NURBS). The earliest—or first generation of CHPs—were introduced in the early 1960s and termed stylized phantoms where organ/body surfaces are described via 3D geometric expressions (spheres, ellipsoids, toroids, truncated cones, etc). Beginning in the late 1980s, and coupled with the clinical maturation of CT and MR imaging, computational human phantoms appeared in their second generation of format type in which all body regions and internal organs are represented as contiguous collections of cuboidal volume elements (voxels) each with a unique organ/tissue identifier for elemental composition and mass density. In many respects, CT-based patient models used in external beam radiotherapy are voxel phantoms of the cancer patient to be treated. Finally, and starting in the early 2000s, mesh-based phantoms emerged as the thirrd generation of CHPs where body contours and organ boundaries are described either by an array of control points (defining NURBS surfaces) or by arrays of triangles (defining a polygon-mesh surface). Another variation of the mesh-type phantom is one in which the organ and body region volumes—initially defined by their polygon-mesh surfaces—are filled in 3D by tetrahedrons (defining a tetrahedral-mesh phantom) [5]. There was a period of time—between the early 2000s and the early 2010s—when mesh-type phantoms were available but radiation transport codes could not accommodate the CHP in this phantom format, and thus these mesh-type phantoms had to be converted into a corresponding voxel-type phantom for radiation transport simulations. This process, however, allowed the user full control over the final voxel size, which is not the case for a CT/MR-based voxel phantom where the voxel size is pre-determined by the original patient image resolution. Fortunately, nearly all Monte Carlo-based radiation transport codes can now efficiently handle computational human phantoms directly in their mesh-type formats, thus making obsolete the need for the voxelization process [6].

5.2.2 Phantom morphometric categories

For each of the three phantom format types, one can design a computational human phantom along one of four different morphometric categories. These are: (1) reference phantoms; (2) patient-dependent phantoms; (3) patient-sculpted phantoms; and finally, (4) patient-specific phantoms. *Reference phantoms* were defined originally for radiation protection purposes by the International Commission on

Radiological Protection (ICRP). ICRP reference phantoms include males and females of only a few select ages—0-year (newborn), 1-year, 5-year, 10-year, 15-year, and adult (nominally 35-years of age). Furthermore, body sizes of reference phantoms are defined by standing heights, total body masses, and internal organ masses fixed at roughly their 50th percentile values across select populations (mostly North Americans and Western Europeans) as given in ICRP Publication 23 [7] and later ICRP Publication 89 [8].

With the development of an increasing number of voxel phantoms, and more recently with the scalability of mesh phantoms, these limitations on 50th percentile phantom heights/weights could be lifted, thus permitting the development of *patient-dependent phantoms* of different combinations of non-50th percentile heights and weights [9]. By definition, patient-dependent phantoms are members of phantom libraries, from which pre-computed databases of organ dose coefficients can be generated across various medical imaging modalities (CT, fluoroscopy, nuclear medicine, and radiography). Use of these phantoms and/or dose coefficient libraries in patient dosimetry are made through a morphometric matching of the individual patient to select members of the phantom library [10]. The simplest, and in many cases most accurate, matching process is by closest pairing of patient-to-phantom by standing height and total body weight. Pre-computed dose coefficient libraries are limited to those of reference and patient-dependent phantoms.

The third morphometric category is the *patient-sculpted phantom*. Here, a member of a phantom library is first selected to represent the medical patient, but a further step is then imposed—3D sculpting of the outer body contour to closely, or even exactly, match that of the real patient. In so doing, internal organs of the patient-dependent phantom are either uniformly enlarged or reduced as different body circumferences of the real patient are matched. In this process, the final phantom appears on the exterior to fully represent the patient, however, no guarantee is made that internal organ structures are equivalently matched. Residual differences in organ size, shape, and internal position remain, with resulting baseline uncertainties in organ doses from medical imaging or therapy procedures (~10% in the case of CT imaging [10]).

Of course, a full match of patient-to-phantom appears in the *patient-specific phantom*—one in which both the outer body shape, but also all internal organ anatomy of the phantom, exactly matches those of the real patient. The patient-specific phantom, which can only be developed from 3D imaging of the patient (CT or MR) followed by detailed organ and body surface contouring, is thus the true *virtual twin* to the actual medical patient. In modern-day radiopharmaceutical treatment planning—in which SPECT/CT and PET/CT images are used to both define radiopharmaceutical time-activity curves and patient organ anatomy—organ dosimetry is indeed based upon patient-specific phantoms. Limitations exist, however, in that: (1) the CT or MR image of the patient might only be covering a portion of the total body (thus complicating assessment of radiation dose to organs out of the imaging field as well as to distributed organs such as the active bone marrow); and (2) automated image segmentation of the full complement of internal organs might not be realized. These limitations, however, might eventually be overcome by way of: (1) use of phantoms constructed as a fusion of pre-existing computational phantom anatomy and inserted partial-body patient-specific CT/MR-based anatomy [11]; and (2) AI-driven optimized algorithms for rapid and accurate anatomy image segmentation [12]. Nevertheless, short of whole-body imaging and fully automated organ segmentation, radiopharmaceutical therapy treatment planning will continue to make use of either patient-dependent phantom libraries, or patient-sculpting of select members of these virtual patient libraries.

5.3 Historical developments

5.3.1 Early approaches to dose assessment

The first scientists to be involved in estimating the dose due to internally distributed radionuclides were very familiar with the techniques used for dosimetry of external radiation sources and radium implants [13]. These techniques were already relatively well developed. However, unlike external radiation exposure, the radiation-absorbed dose due to an internally deposited radionuclide usually had to be calculated rather than measured. Procedures to be followed in these dose calculations probably began with the formulations of Marinelli and his colleagues in the 1940s [14, 15]. A widely used elaboration of the Marinelli formulation was due to Loevinger and his colleagues that was published in the 1950s [16, 17].

Early approaches to such calculations were very simple and took into account the activity and the 'effective half-life' of the radionuclide in the organ, the energy deposited in the organ, and an approximate organ mass. Typically, it was assumed that the radionuclide was uniformly distributed in the homogeneous organ of interest. Usually, the calculations were made of the absorbed dose rate from beta-emitting and gamma-emitting radionuclides separately. In the beta-radiation calculations, it was assumed that all the beta-radiation energy was absorbed in the organ containing the radionuclide. For gamma-emitters, a 'geometry factor' was applied to account for the radiation energy escaping the organ. Dose calculations were completed for simple shapes such as right circular cylinders as well as models for specific organs such as the thyroid gland and the total body [16, 17].

5.3.2 Need for mathematical models in internal assessment

While these relatively simple approaches to dosimetry provided reasonable estimates of the doses due to internally deposited radionuclides, researchers and practitioners longed for better representations of individual organs as well as the human body and techniques that could provide more accurate dose estimates. There was a constant desire to model the organs of the body, and the entire body itself, in a more realistic fashion. This desire led to a progression of developments that ultimately produced the very realistic and complex mathematical models (called phantoms) in use today. However, this progression took some time, starting with very simple shapes and very simple calculational approaches in the late 1950s and early 1960s.

The International Commission on Radiological Protection (ICRP) in its recommendations on internal exposure assumed that organs of the body could be represented by spheres of varying radii [18]. Each organ was represented by a sphere and the effective radius of the sphere was specified. In these simple calculations, the radionuclide source was assumed to be at the center of the sphere and the 'effective absorbed energy' was calculated for each organ. Corrections were made for photon energy lost from the sphere. In addition, the ICRP specified a radius for the total body and assumed that it could be represented by a sphere with a radius of 30 cm. It is interesting to note that the same 30 cm radius as used for the organs designated the 'muscle,' the gastrointestinal tract, and the small intestine.

Increases in the size and speed of digital computers allowed improvements in the models and the approaches taken to radiation transport calculations. In addition, as the number of radionuclides in use in nuclear medicine increased and more was understood about the distribution of these materials in the organs, there was a need for more detailed descriptions of the organs in which specific regions or structures within the organ were delineated.

5.3.3 Simple anatomic models of Brownell, Ellett, and Reddy

Brownell *et al* published perhaps the most extensive, early compilation of data for photon dosimetry in MIRD Pamphlet No. 3 [19]. This pamphlet, which summarized much of their earlier research, provided estimates obtained via Monte Carlo calculations of the photon-absorbed fractions of energy for centrally located point sources in three phantoms with simple shapes. Phantom shapes included elliptical cylinders, right circular cylinders, and spheres. All phantoms were assumed to consist of a homogeneous tissue composed of carbon, hydrogen, oxygen, and nitrogen, with trace amounts of chlorine and sodium.

For phantoms shaped as elliptical cylinders, Brownell and his colleagues considered thirteen monoenergetic photon sources as well as the radionuclide radium. Photon energies ranged from 0.02 to 2.75 MeV. There were nineteen elliptical cylinders ranging in mass from 2 to 200 kg (although one wonders why such a large mass was of interest). For right circular cylinders, only six energies were considered (ranging from 0.04 to 1.46 MeV) but the same nineteen masses were included in the calculations. For spherical shapes, photon energies ranged from 0.02 to 2.75 MeV and phantoms with the same nineteen masses were considered.

In addition, these authors calculated absorbed fractions of energy for uniform distributions of activity in some simple phantoms. Results were presented for ellipsoids, small spheres, thick ellipsoids, and flat ellipsoids. Again, monoenergetic photon sources in the range 0.02–2.75 MeV and phantoms of various masses were considered. For ellipsoids, the masses ranged from 2 to 200 kg, as in previous calculations. For small spheres, thick ellipsoids, and flat ellipsoids only masses in the range of 0.3–6 kg were considered.

5.3.4 Early models developed by Snyder

In the late 1950s and 1960s, many research groups were developing simple anatomic phantoms and Monte Carlo techniques to be used in dose assessment. One major effort was led by W S Snyder at the Oak Ridge National Laboratory (ORNL). Initially, the ORNL research group focused on the use of a right circular cylinder with dimensions suitable for simulating the trunk of an adult human (30 cm diameter with a 60 cm height). This phantom was subdivided into a large number of regions and energy deposition was scored in each of these regions. Both external and internal sources of radiation were considered and absorbed dose to a specific organ was derived from consideration of the dose to the subregions in which the organ was assumed to reside.

5.3.5 The Snyder–Fisher phantom

In 1966, Fisher and Snyder, at the Oak Ridge National Laboratory (ORNL), reported on the development of an adult phantom for use in radiation-absorbed dose calculations [20]. The adult phantom was assumed to be standing erect with the arms at the sides of the body. Three specific regions were defined: the head and neck, the trunk including the arms, and the legs. The head and neck were represented by a 14 cm × 20 cm elliptical cylinder with a height of 24 cm. The trunk and arms were modeled as a larger elliptical cylinder, 20 cm × 40 cm with a height of 70 cm. The legs below the buttocks were modeled as a truncated elliptical cone with a height 80 cm. Regions of little dosimetric concern were not included, e.g., the hands, feet, ears, nose, etc. The composition of the phantom was assumed to be tissue distributed homogeneously throughout. No attempt was made to model the lungs or skeleton or to define the locations of specific organs in the phantom. Approximately 120 subregions were defined in the phantom that were used to assign approximate values of the absorbed doses to organs located within specific regions. In some cases, absorbed dose estimates for large organs required the evaluation of the doses deposited in several of these regions.

Also included in this research was the development of phantoms for humans of other ages. These phantoms represented humans with ages 0- (newborn), 1-, 5-, 10-, and 15-years of age. These phantoms will be discussed in more detail in section 5.3.8.

In 1967, Fisher and Snyder reported on the development of an adult phantom with 22 internal organs and more than 100 subregions [21]. This phantom represented the next step in the development of anthropomorphic phantoms for use in dose calculations. However, this phantom also was homogeneous and the skeleton and lung regions were not defined—in terms of the assignment of different densities and elemental compositions for these regions. This point was very misleading in that many of the early drawings indicated such regions. However, a careful reading of the research reports from the ORNL group will confirm the homogeneous composition. Estimates of the absorbed dose to the skeleton were obtained by summing all the energy deposited in the entire skeleton and dividing by the skeletal volume. No special treatments were applied to obtain absorbed dose estimates for the lungs. Calculations using the adult phantom agreed well with those of Ellett *et al* [22, 23] but were about 50% lower than those obtained using the ICRP spherical organ methods [18].

Even though the original phantom was designed for use with internally-deposited radionuclides, Snyder saw many other applications. In addition, in 1967, he used the phantom to study the distribution of dose in the body from external, point sources of gamma rays [24]. He studied four photon energies (0.07, 0.15, 0.5 and 1.0 MeV) and four different source locations at distances of one and two meters from the center of the phantom.

A heterogeneous phantom was the next logical step in the development of anthropomorphic phantoms. Snyder and his colleagues [25] published an extensive compilation of calculations of the absorbed fractions of energy for monoenergetic photons sources uniformly distributed in selected organs in this phantom. This phantom was composed of three regions: skeleton, lungs, and the remainder (soft tissue). The densities of these regions were about 1.5, 0.3 and 1.0 g cm^{-3}, respectively. The organ masses were selected to follow as closely as possible the data of Reference Man [7]. This phantom ultimately became known in the nuclear medicine community as the 'MIRD Phantom'.

5.3.6 The MIRD-5 Phantom

The MIRD Phantom was developed by Snyder's research group at ORNL and, even though Snyder chose to call the phantom a 'standard man approximation,' it was based on information being compiled for ICRP Publication 23 [7] on Reference Man. In reality, there was a parallel effort at the time to provide improved estimates of absorbed dose and dose equivalent for the ICRP. These estimates were later published as the ICRP Publication 30 series of dose coefficients [26].

The heterogeneous phantom contained three principal idealizations:

1. It had simple geometrical shapes that approximated the organ and body shapes and dimensions. Twenty-two internal organs were included in the design but other unimportant structures (the nose, hands, feet, etc) were ignored. It was assumed that each organ or tissue was homogeneous in terms of composition and density. However, different compositions and densities were used for the skeletal region, the lungs, and the remainder of the body (soft tissue).
2. The monoenergetic radiation source was assumed to be uniformly distributed in one of the organs (i.e., the source organ).
3. The Monte Carlo method was applied in the calculations to obtain a set of photon histories upon which estimates of the energy deposited in the organs and tissues of the phantom could be based. Only photon transport was considered and energy transferred to electrons by photon interactions was assumed to be deposited locally (i.e., the kerma approximation).

Fifteen source organs and 25 target organs or regions were included in the set of calculations. Twelve photon energies were considered ranging from 0.01 to 4.0 MeV. A limited number of histories were followed in each calculation (25,000 to 50,000 histories) and, for low energy photons, the absorbed fractions of energy for many organs were considered unreliable and were not included in the tabulations.

In 1978, Snyder *et al* published the results of an extensive set of calculations of specific absorbed fractions of energy using an improved heterogeneous phantom [27]. As with previous calculations, these results were for monoenergetic photon sources uniformly distributed in organs of the heterogeneous phantom. However, significant improvements had been made in the phantom. These improvements included:

1. The head section was represented by a right elliptical cylinder topped by half an ellipsoid.
2. The leg regions consisted of the frustums of two circular cones.
3. The male genitalia were moved to a position outside the trunk and on the front of the revised model of the legs.
4. Detailed descriptions of the scapulae and the clavicles were included.
5. The stomach and upper and lower large intestine were modeled as organs with walls and contents.
6. A source and target region called the 'skin' was added to the phantom.

Twenty source organs and 20 target organs or regions were included in the set of calculations. Twelve photon energies were considered ranging from 0.01 to 4.0 MeV. The number of histories followed was increased to 60,000 in hopes of increasing the reliability of the results. However, as before, calculations for low-energy photons were very unreliable and other methods were used to provide an estimate of the specific absorbed fractions in these cases.

Over the years, a number of changes (improvements) to the heterogeneous phantom have been made (see section 5.4). However, the fundamental use of the phantom for internal dose calculations has been essentially unchanged since its inception.

5.3.7 Photon and electron transport

Initially, Monte Carlo radiation transport codes used in internal dose assessment were capable only of transporting photons. Perhaps one of the most widely used computer codes was the ALGAM code developed at ORNL [28]. This code was used with the original Snyder–Fisher phantom and continued to be the basic code used for calculations by the ORNL group as the complexity of the phantom increased. In these calculations, it was assumed that electrons, created by photon interactions, deposited their energy at the point of creation. That is, the photon transport code was not capable of transporting electrons.

Computer codes, developed over the intervening years, now have the capability of transporting both photons and electrons, including the production of bremsstrahlung x-rays, and considering other types of photon and electron interactions [29–33]. Subsequently, investigations using a coupled electron/photon transport code indicated that for most situations, this apparent deficiency was not significant, especially in the larger organs of adults [34, 35].

5.3.8 Similitude and pediatric phantoms

Development of the adult human phantom by Snyder and his colleagues (see section 5.3.5) was paralleled by the development of phantoms representing humans of other ages [20]. These phantoms represented children with ages of 0-year (newborn), 1-year, 5-years, 10-years, and 15-years of age. These early designs were assumed to have outer dimensions that represented the average height, surface area, and body mass of a child of the particular age. All 'pediatric phantoms' were obtained by applying a set of simple transformations to the axes of the Cartesian coordinate system in which the adult phantom was located. These phantoms became known as the 'similitude phantoms' because of their resemblance to the adult. This approach has its limitations because children generally are not just 'little adults.' However, these phantoms were the first developed to answer a real need in the nuclear medicine community [36].

Improvements in the pediatric models was closely linked with the development of the heterogeneous adult phantom. Even though these new phantoms were heterogeneous, the pediatric phantoms were obtained through the same transformation method. That is, these phantoms also were similitude phantoms. The outside dimensions were obtained by applying a series of transformations to the coordinate system and no consideration was given to the actual organ sizes or shapes of the 'transformed' organs. Although the masses of these 'transformed organs' had to be known for the calculation of absorbed dose, these masses were never published [37].

The limitations associated with transforming the major axes of the adult phantom should be clear. Children are not simply small adults and their organs are not necessarily 'smaller adult organs.' Some organs are larger in a child than in the adult and get smaller as the individual approaches adulthood, e.g., the thymus. In addition, it was important that the actual size, shape, location, and mass of each organ be known for accurate dose calculations. For these reasons, and others, a significant effort was undertaken at ORNL during the mid-1970s to develop individual pediatric phantoms based upon a careful review of the existing literature for each particular age. This effort produced the next generation of mathematical phantoms that, although they appeared to be modeled after the adult, were designed independently.

Three 'individual phantoms' were designed by Hwang *et al* [38, 39]. These were the newborn, the one-year-old, and the five-year-old models. A separate effort was undertaken by Jones *et al* [40] for the 15-year-old and Deus and Poston [41] undertook the design of a ten-year-old after the other four designs were complete. The development of the ten-year-old was significantly different from those for the other four ages. In fact, this design was intended to point the way to the next generation of more realistic phantoms. Even though the design was completed and used for a limited number of dose calculations, it was not popular because of the very complex geometry and other approaches to the development of phantoms were followed [42, 43].

5.3.9 Development of MIRDOSE codes

Internal dose calculations using a stylized model generally are performed by combining tabulated dose conversion coefficients (e.g., particle and energy dependent specific absorbed fractions or radionuclide S values) with radionuclide activity information in source organs or regions [44]. Because this process can be tedious to perform and involves repetitive tasks, it lends itself to treatment with computer programs. The MIRDOSE computer program was developed to perform internal dose calculations using specific stylized anatomic models. The software was originally presented in 1984 by Watson and Stabin [45] and was continuously supported and updated since then, with version 2 available in 1987 [46], and version 3 in 1994 [47]. Using MIRDOSE, dose calculations are performed by searching through internal databases of radionuclide decay data and specific absorbed fractions of energy (specific for a stylized model) to generate radionuclide S values of interest. These S values are combined with user-entered source organ residence times to calculate the dose in the different organs and regions of the body. Libraries used in version 2 of the MIRDOSE computer program were based on 59 radionuclides derived from references [48–50] and on the pediatric phantom series of Cristy and Eckerman [42]. In version 3, libraries were updated to use 224 radionuclides [48] for 10 mathematical phantoms (the six ORNL phantom series [42] and the four pregnant female phantom series at different gestation times [51]) with 28 source and 28 target regions. In addition, special models for the gastrointestinal tract [26] and the bladder [52] were implemented in both MIRDOSE 2 and 3. To replace the ICRP Publication 30 bone model [26] used in MIRDOSE 2 for skeletal dosimetry, the new model of the skeletal system developed by Eckerman [53] was implemented in MIRDOSE3, thus allowing for the calculation of doses in fifteen different bone sites. To support tumor dosimetry, the ability to calculate absorbed doses to small, unit density spheres was added to MIRDOSE3 using absorbed fractions of energy for both photon and electron sources [19, 54, 55].

5.4 Stylized anatomic models

5.4.1 The ORNL phantom series

The Oak Ridge National Laboratory (ORNL) in 1980 developed a new series of stylized models of various ages [42]. The series included an adult male, a newborn and individuals of ages 1, 5, 10 and 15 years developed from anthropological data (legs, trunk and head) and from age specific organ masses published in ICRP Publication 23 [7]. Figure 5.1 shows frontal views of the entire series, while figure 5.2 shows representative transverse cross-sectional view of the adult and newborn models. Although some of the organ shapes and centroids were still obtained using the similitude rule from the Snyder–Fisher adult model, these phantoms represented a great improvement for pediatric dosimetry over the similitude pediatric phantoms. These phantoms also presented new regions and improvements such as a new stylized breast tissue region for all ages, the inclusion of the new model of the heart developed by Coffey [56, 57], and a new model of the thyroid. While the ORNL pediatric model series was initially published in 1981 [42], these models were not readily utilized until 1987 with the publication of ORNL/TM-8381 [43]. In this report, the only major change in the phantom series was that the 15-year-old model was assumed also to represent the average adult female. For this purpose, the breast, uterus and ovaries were modified according to published reference average values [7]. The phantoms were used with the ETRAN Monte Carlo photon transport code [58, 59] to calculate specific absorbed fractions of energy in all five pediatric phantoms, as well as in the adult male, for 12 photon energies (0.01 to 4.0 MeV).

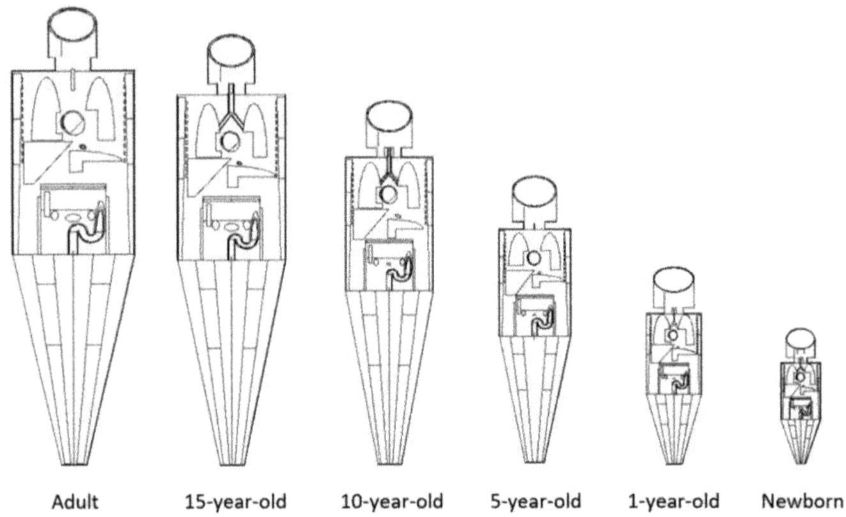

Figure 5.1. Front views of the series of ORNL stylized hermaphrodite computational phantoms.

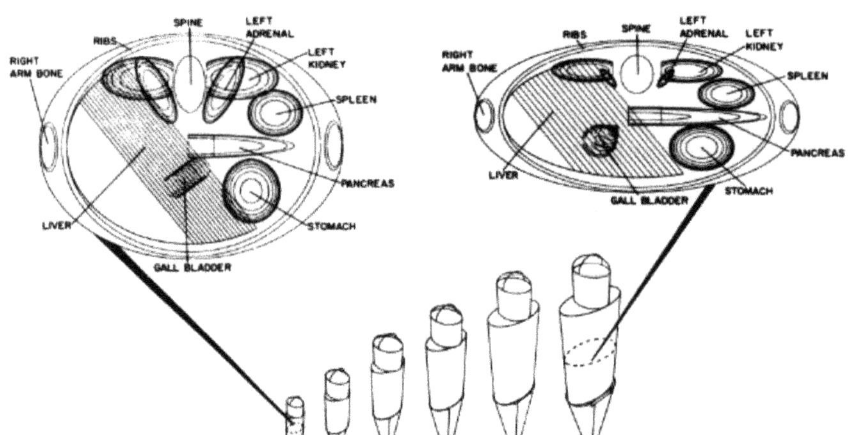

Figure 5.2. Transaxial views of the ORNL newborn and adult male phantoms. Reprinted with permission from [43]. Copyright ORNL.

Electron transport was not considered in these simulations and electron energy was assumed to be locally deposited.

5.4.2 The MIRD stylized models

Since the publication of the stylized dosimetric model of Snyder *et al* in MIRD Pamphlet 5 Revised [27], the MIRD Committee has refined several internal organs to support the development of radioimaging tracers and therapeutic nuclear medicine. Modifications to the MIRD stylized model have been published as MIRD Pamphlets, which include equations of the new geometries, tabulated absorbed fractions of energy for monoenergetic photons and electrons, and tabulated radionuclide *S* values.

In 1999, the MIRD Committee adopted six new age-specific models of the head and brain [60] representing average reference brain and head for a newborn, one-year-old, five-year-old, ten-year-old, fifteen-year-old (also representing the average adult female), and adult male. These phantoms were intended to provide better dosimetric tools in response to the increased number of neuroimaging radio-pharmaceuticals [61, 62]. Due to the regional uptake of these new agents within the brain, accurate absorbed dose calculations required the use of a detailed model of the subregions of the brain not available with previous models. Similar to previous stylized models, simplistic geometrical shapes (intersection of ellipsoids, elliptical cylinders, tori and planes) were used to represent the different regions of the head and brain, with volumes derived from published reference masses [7, 63] and shapes from analysis of MRI images. Twenty-one subregions were modeled within the head and neck, including five regions representing bony structures (simulated as a homogenized mixture of bone and red marrow with a density of 1.4 g cm^{-3}) and sixteen tissue regions (density of 1.04 g cm^{-3}). Within the brain, eight subregions were delineated: the caudate nuclei, cerebellum, cerebral cortex, lateral ventricles, lentiform nuclei (a composite region of both the putamen and the globus pallidus), thalami, third ventricle, and white matter. Other regions considered within the head included the cranial cerebrospinal fluid (CSF), cranium, eyes, mandible, spinal cord, spinal cerebrospinal fluid, spinal skeleton, teeth, thyroid, upper face region, and skin.

In both the ORNL phantoms of Cristy and Eckerman [42, 43] and in the MIRD adult model developed by Snyder [27], the kidneys were represented as two symmetric ellipsoids cut by a plane with no differentiation of their internal structure. However, because of their unique physiology, the kidneys are seen to concentrate radioactivity nonuniformly [64]. Snyder and Ford [65] designed a dosimetric model of the human kidney to study the administration of the mercurial diuretic chlormerodrin (neohydrin) labelled with ^{197}Hg and ^{203}Hg. Besides the kidneys, other internal organs of interest included the ovaries and the bladder. The kidneys were assumed to be of equal size (with a total volume of 305.8 cm^3) and to be composed of three separate regions, the cortex, the medulla and the papillary region. Dose estimates were provided for the cortex, medulla, ovaries and bladder per millicurie-hour of residence of ^{197}Hg and ^{203}Hg in the cortex, in the medulla, in the bladder and in the total body (blood).

MacAfee published in 1970 a multi-region kidney model of an adult [66] representing the renal cortex and medulla as two concentric elliptical shells, and the renal pelvis as a wedge-shaped hollow structure at the center of each kidney. In 1975, the MIRD committee used this model in MIRD Dose Estimate Report No. 6 to calculate the dose to the renal cortex and renal medulla from ^{197}Hg- and ^{203}Hg-labeled-Clormerodrin [67].

Patel described a multi-compartment kidney model in 1988 [68]. This model, which was similar to the model used by Snyder and Ford [65], consisted of three regions (the cortex, medulla and papillae) that served as both the source and target regions. The geometry of the kidney was the same as in the original Snyder–Fisher phantom, i.e., the kidney was assumed to be an ellipsoid cut by a plane parallel to the *z*-axis of the phantom. This model was incorporated into the Snyder–Fisher heterogeneous phantom in a modified version of the ALGAM transport code [28] and absorbed fractions and specific absorbed fractions of energy were calculated for 12 monoenergetic photon sources in the range 0.01 to 4.0 MeV. These results were used to obtain *S* values for selected radionuclides for the three regions of the kidney as the sources. The radionuclides considered were ^{32}P, ^{51}Cr, ^{57}Co, ^{67}Ga, ^{99}mTc, ^{111}In, ^{123}I, ^{131}I, ^{127}Xe, ^{133}Xe, and ^{201}Tl.

A revised kidney model was adopted by the MIRD Committee in 2003 and published as MIRD Pamphlet No. 19 [69]. Following the increased use of radiopharmaceuticals in therapeutic nuclear medicine and recent associated kidney toxicity [70], the MIRD Committee developed six advanced stylised models of the kidney (newborn, one-year-old, five-year-old, ten-year-old, fifteen-year-old and adult male). The outer dimensions of these models conformed to those used in the ORNL single-region kidney models while twelve interior structures were defined for each kidney: five medullary pyramids, five papillae, the renal pelvis, and the renal cortex. Although the number of medullary pyramids in these models was less than that seen in the real anatomy (6–18 pyramids), it represented a compromise between the mathematical simplicity needed for Monte Carlo transport calculations, and the need for an improved anatomic representation over the concentric ellipsoid-shell model of McAfee [66]. Each region was derived from dimensions and volumes given in ICRP Publication 23 [7] for both the newborn and the adult, and assumed constant volume ratios between the different kidney subregions for the other ages. In these models each medullary pyramid was modeled by half-ellipsoids (two vertical and three horizontal) with the papillae at its tip, the pelvis by a portion of an ellipsoid within the whole kidney, and the cortex was the remainder of the kidney. The MIRD 19 model was subsequently used in 2008 as the basis for exploring further predictive models of renal toxicity based on the dosimetric quantity BED—Biologically Effective Dose [71].

In both MIRD Pamphlet Nos. 15 and 19, the EGS4 Monte Carlo transport code [29, 72] was used for photon and electron transport. In these two Pamphlets, absorbed fractions of energy were tabulated for selected source and target combinations (twelve energies were simulated between 10 keV and 4 MeV). Following the MIRD method of internal dose calculation [73], mean absorbed doses to the target regions per unit cumulated activity in the source region (*S* values) were tabulated for specific radionuclides.

5.4.3 Stylized models of the lower abdomen

The development of nuclear medicine imaging and therapy over the past several decades has resulted in the need for more accurate dosimetry in regions that either were not represented or misrepresented in the MIRD and ORNL stylized phantoms. The lower abdomen is a particular region of the anatomy that is difficult to model via mathematical surface equations due to the intricate geometry of its organs.

Many assumptions were made in modeling these organs leading to either the organ not being represented, or being over-simplified. Development of new radio-agents with specific uptake in the prostate or in the wall of the gastrointestinal tract led to a need to modify the dosimetric model of the lower abdomen.

In 1994, Stabin developed a mathematical model of the prostate gland and included it in the ORNL stylized model [74]. This new organ was modeled as a single sphere located just below the urinary bladder, with a volume consistent with Reference Man of ICRP Publication 23 [7]. This model was used to calculate absorbed fractions of energy and *S* values for selected radionuclides.

In 1999, Mardirossian *et al* recognized that the relative spatial position of the urinary bladder, rectum and prostate were incorrectly modeled in the ORNL phantom series. They developed a new model of the lower part of the colon [75], and separated the rectum from the lower large intestine. This new model included an explicitly defined rectum, an anatomically corrected sigmoid and descending colons, and a prostatic urethra and seminal duct. These modifications were implemented in the ORNL phantom series, after changing the position of the bladder and prostate gland to properly model the relative positions of these organs. These models were developed not only for the adult male model, but also for the other phantoms in the ORNL series using physiological and anatomical descriptions published in ICRP Publication 23 [7].

Because the intestinal tract and curvatures cannot be modeled with simple geometrical shapes that can be easily coupled to a Monte Carlo transport code, all models of this region have relied on thickening the wall region to preserve anatomical wall and content mass. The critical cells for these organs have been identified as the cells within the mucosal layer. In 1996, Poston *et al* developed a revision of the GI tract of the MIRD adult phantom to better represent these sensitive targets [34, 35]. The actual wall of the GI tract was divided in its thickness into four regions of varying radiosensitivities; these layers were very difficult to model because the thickness of each layer varied from one section to another along the different regions of the GI tract. Poston *et al* developed two methods to model this wall. The first method divided the tissue wall into ten small, concentric layers (100 μm thick for the adult) and the dose to each layer was recorded separately. Then, the determination of the mucosal layer thickness for each section would give directly the dose to the mucosa. However, since it is not possible to determine directly the mucosal layer thickness for a specific patient, the subdivision into 10 regions has not been used for medical dose calculation. In a second method, Poston *et al* measured the average thickness of the mucosal layer along the GI tract from anatomic slides of cross-sections of a human GI tract. Different mucosal thicknesses for the stomach, the small intestine and the large intestine were obtained. This layer was included in the GI wall of the adult mathematical phantom and coupled to the EGS4 Monte Carlo transport code [29].

Stubbs *et al* [76] presented calculations of the radiation-absorbed dose to the walls of hollow organs. These authors studied all four sections of the gastrointestinal tract but only for four radionuclides important to nuclear medicine. They presented *S* values (in units of Gy/Bq-s) for ^{90}Y, ^{99m}Tc, ^{123}I and ^{131}I. They concluded, as did

Poston *et al*, that the ICRP 'one-half' assumption was overly conservative for non-penetrating radiation.

5.4.4 Updates to the ORNL and MIRD stylized human phantoms

Other modifications and additions to the ORNL and MIRD stylized models include a peritoneal cavity [77], a revised model of the nasal cavity and major airway [78], and a new model of the long bones [79]. The peritoneal cavity model was developed in 1989 by Watson *et al* [77]. They modified the MIRD phantom to include a region representing the peritoneal cavity to support the dosimetry associated with several therapeutic and diagnostic techniques involving injection of radioactive material into the peritoneal cavity. Similarly, in 1997, Deloar *et al* developed a model of the nasal cavity and major airway to support the dosimetry associated with ^{15}O-labeled gases PET imaging agents [78]. In 2000, Clairand *et al* modified the model of the long bones of the ORNL phantom series to properly differentiate the cortical bone, trabecular bone, and medullary cavity [79]. This effort was undertaken to support bone dosimetry for photon sources and to allow more accurate marrow dose calculations for children. Because in children, the active marrow is not only found in the trabecular bone but also in the medullary cavities, the stylized models of the long bones of the legs and arms (truncated circular cones) of the ORNL phantom series did not allow for accurate bone dosimetry calculations.

In 2006, Han *et al* published extensive revisions to the adult and pediatric series of ORNL stylized phantoms [80]. Their revisions were made to: (1) incorporate recent developments in stylized models of the head, brain, kidneys, rectosigmoid colon, and extra-pulmonary airways; (2) incorporate new models of the salivary glands and the mucosa layer of the urinary bladder, alimentary tract organs, and respiratory airways; (3) adopt reference values of elemental tissue compositions and mass densities from ICRP Publication 89 and ICRU Report 46; (4) provide for explicit treatment of left and right organs within organ pairs; (5) provide for a systematic tabulation of electron absorbed fractions as a function of energy and subject age for all internal organs; and (6) provide for methods of deriving patient-specific values of the specific absorbed fraction for both electrons and photons through interpolation/extrapolation of their phantom-derived values. At the time of its publication, it was argued that while voxel computational phantoms (see section 5.4) provided improved anatomic realism given the CT or MR image sets used in their construction, there did not yet exist a comprehensive series of reference pediatric voxel phantoms, nor the ability to simulate very fine anatomic structures as can be modeled via mathematical approximation. The authors noted that these revised stylized pediatric phantoms would fill this data need in medical dosimetry.

5.4.5 Use of stylized models in therapeutic nuclear medicine

Because stylized models are based on an average individual (Reference Man), associated dose calculations can only represent the average dose to a patient population. Although this average dose can sometimes be applied to specific individuals, large variations in patient weight, size, and organ mass are expected, and can significantly change the dose received by an organ. Consequently, Reference Man S values have to be scaled to match a specific individual. In MIRD Pamphlet 11, Snyder *et al* discussed methods of scaling radionuclide S values from the 70-kg Reference Man to individual patients [81]. For target organs r_T which are sufficiently distant from the source organ r_S, values of the specific absorbed fraction $\Phi(r_T \leftarrow r_S)$ for photon irradiation are reasonably independent of organ mass, and thus no scaling from Reference Man to the patient is required. In these cases, increases (or decreases) in target organ mass m_T are compensated (scaled) by proportional increases (or decreases) in the absorbed fraction $\phi(r_T \leftarrow r_S)$. As the target organ approaches the source organ, however, this compensation no longer exists for photons. When r_T and r_S coincide (the target organ becomes the source organ), the dose from photon emitters above 100 keV becomes proportional to $(m_s)^{-2/3}$. For electron sources, in which full-energy absorption is typically assumed, the dose from electrons scales as the inverse first power of the source organ mass $(m_s)^{-1}$.

A more robust approach to patient-specific adjustments to reference phantom self-dose S-values was proposed by Wayson and Bolch [82] in which photon and electron power scaling factors were computed as a function of energy for ranges of reference and non-reference organ masses. The MIRD Pamphlet No. 11 constant of −2/3 is shown to be reasonably accurate across the photon energy range of ~200 keV to 1 MeV, but with values as low as −0.996 for 1-keV photons and as high as −0.357 for 4-keV photons. The MIRD Pamphlet No. 11 constant of −1.0 for electron self-dose scaling of the specific absorbed fraction was found to be appropriate for low-to-intermediate electron energies, but it then increases above that value at energies above several hundred keV, especially for smaller organ masses. In another study, Wayson and Bolch [83] explored patient-specific rescaling of both self-dose and cross-dose photon and electron specific absorbed fractions given various external-body morphometry metrics for the patient and the reference phantom (to include sitting height and waist circumference).

To better match a stylized model to a specific patient, models of different sizes also can be used as proposed by Clairand *et al* [84]. Using the MIRD stylized model, they developed six new models of different heights: three male models of heights 160, 170 and 180 cm and three female models of heights 150, 160 and 170 cm. Using a statistical analysis of anthropometric data gathered from 684 forensic autopsy cases for the years 1987–1991 (all subjects selected were Caucasian adults who died of external causes and showed no pathological changes), they derived organ masses as a function of height [85]. Only organs weighed during autopsy were taken into account in this study (heart, kidneys, liver, lungs, pancreas, spleen, testes, and thyroid), and other organs were obtained from the ORNL models.

To support the dosimetry in therapeutic nuclear medicine, simple tumor models also have been introduced in stylized phantoms [86–88]. These approaches allow for the determination of the photon dose from a tumor to adjacent organs, but also provide the photon dose from body organs to the tumor, which for some radionuclide can be non-negligible [89]. Clairand *et al* developed a computer program (DOSE3D) to simulate spherical tumors inside the ORNL adult phantom [86]. With this program, the user can select one or more tumors of different sizes, and the EGS4

Monte Carlo transport code is used for the particle transport. Similarly, Johnson *et al* developed the MABDOSE software to simulate spherical tumors inside the ORNL phantom series [87]. In this code, the stylized phantom series is first voxelized at a user-selectable resolution (10, 5, 2, or 1 mm), and the ETRAN Monte Carlo photon transport code is then used for the particle transport.

5.5 Voxel-type anatomic models

As discussed above, stylized mathematical models of human anatomy provide a relatively efficient geometry for use with Monte Carlo radiation transport codes. Nevertheless, these models are only approximations to the true anatomical features of individuals for which dose estimates are required. An alternative class of anatomic models is based upon 3D imaging techniques such as magnetic resonance (MR) imaging or computed tomography (CT). These voxel (or tomographic) models of human anatomy represent large arrays of image voxels that are individually assigned both a tissue type (e.g., soft tissue, bone, air, etc) and an organ identity (heart wall, femur shaft, tracheal airway, etc). Thus, image segmentation is needed to process the original image into a format acceptable for radiation transport using codes.

5.5.1 Methods of construction

Two general anatomic sources exist for the construction of tomographic models of human anatomy: (1) cadavers and (2) live subjects, typically medical patients. Each data source has its distinct advantages and disadvantages. Cadaver imaging generally offers a substantially improved opportunity for full anatomic coverage including the extremities. With CT image acquisition, higher resolution scans can be performed, as radiation dose considerations and patient motion are not of concern. For these same reasons, multiple scans on the same cadaver can be performed using different technique factors (kVp, mA, filtration, etc). Problems associated with cadaver imaging include tissue shrinkage, body fluid pooling, air introduction to the gastrointestinal tract, collapse of the lungs, and general organ settling. Perhaps the greatest disadvantage, however, is that cadaver imaging will most likely not involve the use of CT tissue contrast agents needed for soft-tissue image segmentation.

Computed tomography is perhaps the imaging modality of choice for construction of full-body tomographic computational models. Skeletal tissues are more readily defined under CT imaging, whereas image distortions of skeletal tissues are problematic under MR imaging. With live patient imaging, CT contrast agents provide acceptable definitions of soft tissue borders needed for image segmentation, and thus MR does not offer a distinct advantage over CT for this data source. In cadaver imaging, the lack of contrast agents hinders border definitions for soft-tissue organs. To compensate, MR imaging of the cadaver might be considered. However, the frozen state of the cadaver tissues can alter both spin–lattice and spin–spin relaxation times (T1 and T2, respectively), complicating and distorting MR images of the subject. Thus, CT imaging is recommended for both live and deceased subjects upon whom tomographic models are to be constructed. With modern-day multi-detector, multi-slice CT imaging, scan times are minimized, offering improved opportunities for live-subject imaging as potential data sources. CT-MR image fusion is another possibility for input data to image segmentation. However, patient motion artifacts within MR images, and their generally lower image resolutions, currently restrict opportunities for CT-MR co-registration to the head region where patient motion is more easily minimized, as is the case for radiosurgery treatment planning of brain tumors.

Following image acquisition, the next step in model construction is image segmentation. In cases where the CT image provides strong tissue contrast, automated pixel-growing methods of image segmentation can be applied to rapidly delineate organ boundaries. In these methods, a central pixel is tagged and the algorithm groups all neighboring pixels within a defined gray-level interval (e.g., skeletal regions, air lumen of nasal passages, certain soft-tissue organs under CT contrast). For those organs with poor tissue contrast, organ boundaries must be segmented manually. Final review of the tomographic model by medical personnel trained in radiographic anatomy is highly recommended. Standardized software packages, such as IDL (Research Systems, Boulder, CO) and 3D Slicer (https://www.slicer.org/) can offer all necessary tools for construction of voxel-based computational models.

5.5.2 Review of representative voxel phantoms

Several research groups have been involved in the development of voxel computational models over the past three decades. Most efforts have been directed toward adult models, but a few important studies have resulted in pediatric tomographic models for radiation dosimetry studies.

In 1994, Zubal and colleagues published their work in manually segmenting 129 CT transverse slices of a living male adult [90]. Their goal was to create a tomographic computational model for use in simulating nuclear medicine imaging data [91]. A total of 36 tissue regions were identified within the 3D voxel array. The original CT images were obtained in two imaging sessions: one of 78 slices from the neck to mid-thigh with 10-mm slice thicknesses (pixel size of 1 mm), and one of 51 slices from the head to the neck region with 5-mm slices (pixel size of 0.5 mm). Image processing techniques were applied to achieve a final isotropic cubic voxel resolution of 4 mm on a side. Similar techniques also have been applied in the creation of a high-resolution, MRI-based voxel head model [92]. Stabin *et al* subsequently coupled the Zubal total-body model to the MCNP radiation transport code [93]. Yoriyaz *et al* later reported estimates of cross-organ photon dose based on MCNP calculations within the Zubal model [94, 95]. Tagesson *et al* used the segmented Zubal head model to estimate absorbed fractions and radionuclide S values for a variety of radiopharmaceuticals used in neuroimaging [96].

In 1997, Jones reported research performed at the National Radiation Protection Board (NRPB) on a voxel-based computational model of the adult (NORMAN) [97]. This model was used to estimate organ doses from external photon sources over a range of energies and irradiation geometries. When comparing the results to those obtained using a MIRD-type stylized model, differences in organ doses were found to range from a few percent to over 100% at photon energies between 10 and 100 keV; differences were less than 20% at photon energies exceeding 100 keV. A subsequent study found that estimates of organ doses from internal photon sources differed widely as assessed using two different models, and differences in cross-organ doses generally increased with decreasing photon energy [98].

The National Library of Medicine's Visible Human Project also provides an excellent data source from which one may construct tomographic computational models for use in diagnostic x-ray simulations for adult patients. This data set consists of MRI, CT, and anatomical images. The CT data consist of axial CT scans of the entire body taken at 1-mm intervals at a resolution of 512 × 512 pixels. Dr George Xu and his students reported on their efforts to merge this data set with both the MCNP and EGS4 radiation transport codes [99]. Computational results were subsequently reported for internal electron sources [100], external photon sources [101], neutrons [102, 103], and electrons [104].

Some of the very first pediatric voxel computational phantoms were developed at the GSF Laboratory in Munich, Germany in the late 1980s. These models were constructed for specific use in pediatric dosimetry studies [105, 106]. Two of these are the BABY and CHILD models developed at by Zankl and colleagues. BABY is constructed from a whole-body CT exam of an 8-week-old female 24 h after death. CHILD was constructed from a CT exam of a live 7-year-old female who was undergoing subsequent whole-body irradiation prior to bone marrow transplantation. Image segmentation was performed which assigned voxels to 54 regions in the baby and 64 regions in the child. The models have been used to determine pediatric organ doses in a variety of medical exposure scenarios including nuclear medicine [107], projection radiography [108–111], and computed tomography [112, 113]. Another early pediatric model, ADELAIDE, was based upon 44 consecutive CT scans of a 14-year-old female [114]. This particular model is restricted to the torso only, and has been used with the EGS4 transport code to assess organ doses received during CT exams.

The first true newborn voxel phantom was developed at the University of Florida (UF) and published in 2002 [115]. This model was used to explore organ dosimetry in both fluoroscopy and CT imaging with organ doses compared to those given by the ORNL stylized newborn phantom [116–119]. Additional pediatric voxel phantoms were developed at UF to include the UF Series A and Series B pediatric voxel phantoms. In 2005, the UF Series A voxel phantoms were published to include five phantoms of different ages—9-month male, 4-year female, 8-year female, 11-year male, and a 14-year male [120]. They were constructed from CT imaging data of live patients using an IDL-based image segmentation tool. Lungs, bones, and adipose tissue were automatically segmented through window leveling of the CT numbers. Additional organs were segmented either semi-automatically or manually with the aid of both anatomic knowledge and available image-processing techniques. The phantoms were created from fused images taken from head and chest-abdomen-pelvis CT exams of the same individuals or of two different individuals of the same sex and similar age. As a result, they were partial body phantoms without arms and legs (which were outside the original scan range). In 2006, the UF Series B pediatric voxel phantoms were published to include: (1) inclusion of arms and legs via rescaling of the extremities of a health Korean adult model; and (2) resizing of internal organs and body height/weight as guided by age/sex-dependent interpolation of ICRP Publication 89 reference values. The UF Series B phantoms were later used as the basis for the creation of the UF series of hybrid phantoms (see section 5.6) and later the ICRP Publication 143 reference voxel phantoms [121].

5.5.3 ICRP adult and pediatric voxel-based reference computational phantoms (VCRPs)

Nuclear medicine dose coefficients for various radiopharmaceuticals have been computed and tabulated by the ICRP within its Task Group 36 based upon: (1) development of reference models of the radiopharmaceutical biokinetics (mostly in the form of sums of exponential terms); and (2) radionuclide *S*-values derived from the source-target organ-specific absorbed fractions (SAFs) coming from radiation transport simulations in the ORNL stylized phantom series [122]. While not fully considered ICRP-endorsed computational phantoms of the ICRP reference persons, the ICRP in effect had adopted the ORNL phantoms for that purpose. Even the most recent compilation of nuclear medicine dose coefficients (organ dose per administered activity) in ICRP Publication 128 [123] continue to use photon SAFs from the original 1987 transport results in ORNL TM-8381 [43] and to assume (for most source/target organs) that all beta particle and electron energy is localized deposited in the source organs.

In the late 1990s, the ICRP Main Commission made the decision to move forward with the development—within ICRP Committee 2—of its own reference computational phantom in the voxel-type format. This work was primarily conducted by Maria Zankl and colleagues at the GSF laboratory in Munich, Germany, as part of a task group of ICRP Committee 2. The source images of what would eventually become the ICRP Publication 110 voxel-type reference computational phantoms (VRCPs) of the ICRP Publication 89-defined adult male and adult female were the following [124]. For the adult male model (called Golem), whole-body clinical CT images of a 38-year-old male of height 176 cm and body mass of slightly under 70 kg was selected. This individual had been treated by whole-body radiation therapy for leukemia, yet did not have obvious signs of anatomical abnormalities that would have precluded their use in model development. The CT image set was of the person lying supine with the arms parallel to the body torso. The image data set consisted of 220 CT image slices of an in-plane pixel matrix of 256 by 256. The original voxel size was 8 mm along the *z*-axis (cranial-to-caudal direction), with a resulting voxel volume of 34.6 mm^3. In a similar manner, the female reference voxel phantom (called Laura) was constructed from a CT image set of a 43-year-old individual of standing height 167 cm and total body mass of 59 kg. This data set consisted of 174 CT image slices of 5-mm in voxel height (head and trunk) and 43 CT image slices of 20-mm voxel height (legs), with both having an in-plane image matrix of 256 by 256 pixels. Voxel volumes were 17.6 mm^3.

Segmentation of the image sets were performed by both in-house and commercial software. Following a series of voxel-based adjustments to the Golem and Laura voxels, the official ICRP-adopted voxel-based reference computational phantoms were finalized to include 141 different internal tissue structures, to include a skeletal tissue compartmentalized to trabecular spongiosa, cortical bone, and for the long bone shafts—medullary marrow cavities. Figure 5.3 below shows frontal views of both the ICRP adult male and adult female reference voxel phantoms, along with internal views of the male model.

These adult reference voxel phantoms from ICRP have been used in a series of subsequent publications including the reporting of dose coefficients for external radiation fields found in occupational settings [125], in space activities within low-earth orbit [126], in commercial aviation [127], and in regions of environmental radionuclide contamination [128]. For internal exposures, the ICRP 110 voxel phantoms were used to compute a comprehensive series of energy-dependent SAFs for internally emitted photons, electrons, alpha particles, and neutrons (the latter to accommodate internal intake of radionuclide which decay in part through spontaneous fission). This work is summarized in ICRP Publication 133 [129], and was used to assemble a wide series of dose coefficients for internally inhaled or ingested radionuclides. These internal dose coefficients have appeared in a series of ICRP publications under the title OIR—Occupational Intakes of Radionuclides [130–133]. ICRP Publication 133 included a major departure from the prior SAF datasets used by ICRP from ORNL TM-8381 in that, for the first time, an explicit accounting for electron transport—to include both collisional losses and bremsstrahlung x-ray production—was made thus avoiding the simplistic assumption of full-energy absorption of beta particles and electrons in the source organ. Of this writing,

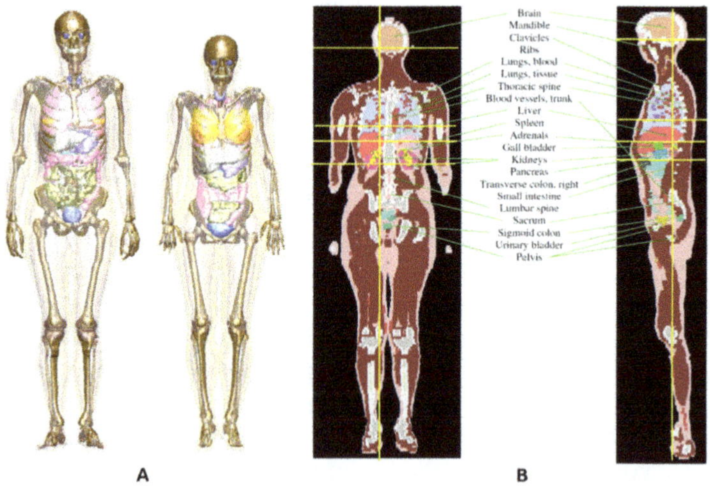

Figure 5.3. (a) Frontal views of the ICRP Publication 110 adult voxel reference computational phantoms (VRCPs) with the adult male on the left and the adult female on the right. (b) Selected coronal and sagittal views of the adult male phantom. Reprinted with permission from [125]. Copyright 2012 ICRP. Published by Elsevier Ltd.

ICRP Task Group 36 is undertaking a major revision to the nuclear medicine dose coefficients of its Publication 128 to include both: (1) updated SAFs from ICRP Publication 133; and (2) a revision of all biokinetic models of radiopharmaceuticals from simple sums of exponential terms to full compartmental models to include explicit consideration of circulating blood as a separate and independent source region for dose estimation.

In 2019, Akhavanaliaf *et al* [134] reported on the development of an extensive library of adult voxel-type computational phantoms—479 in total—which encompassed the breadth of heights and weights in an adult population of patients across the age range of 20 to 80 years. Anthropometric parameters were extracted from the National Health and Nutrition Examination Survey (NHANES) database to include standing height, total body weight, and body mass index. A computer code was written to adjust standing height and percentage of fat free mass of the ICRP Publication 110 reference voxel phantoms via 3D scaling. The waist circumference and total body mass were further adjusted. The diversity of organ masses due to anthropometric differences were shown to deviate from reference or mean values from 3% to up to 21%. Figure 5.4 below shows four members from this expanded voxel phantom library in relation to the 'anchor' phantoms from ICRP Publication 110.

In 2010, Lee *et al* [135] published the UF series of pediatric reference phantoms, derived from the prior UF Series B voxel phantoms, but adjusted to fully match ICRP heights, weights, and internal organ masses given in ICRP Publication 89 for the reference male and female newborn, 1-year-old 5-year-old, 10-year-old, and 15-year-old. In 2020, voxelized versions of these mesh-based reference phantoms were published by the ICRP in its Publication 143. Figure 5.5 shows frontal views of the ICRP pediatric voxel-based reference computational phantoms. Pediatric reference SAFs for photons, electrons, alpha particles, and internally emitted neutrons were being prepared for release by the ICRP in late 2021.

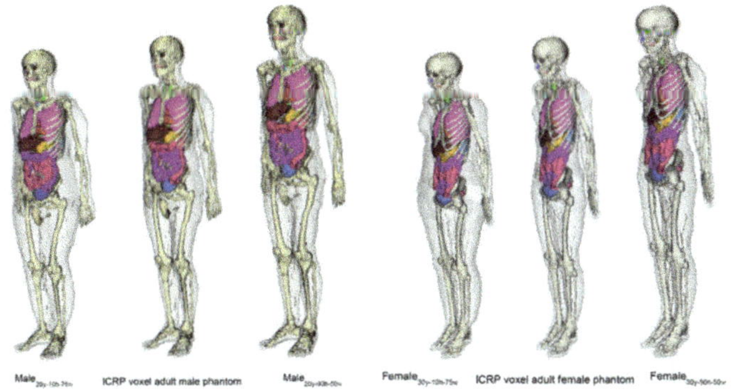

Figure 5.4. Rotated views of the adult male (second from left) and adult female (second from right) ICRP Publication 110 reference voxel phantoms along two scaled versions of each side. For both the adult male and adult female phantoms, the two additional phantoms are at 10th height/75th weight percentile and at 90th height/50th weight percentile, respectively, for age 20 years (male) and age 30 years (female). Copyright (2019) IEEE. Reprinted with permission from [134].

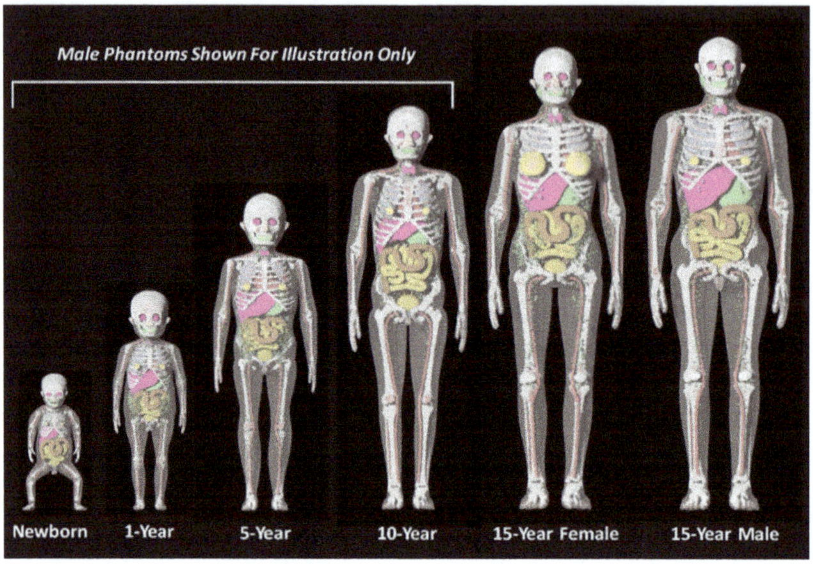

Figure 5.5. Frontal views of the ICRP Publication 143 pediatric voxel reference computational phantoms (VCRPs). Reprinted with permission from [121].

5.5.4 Efforts to develop country-specific or population-specific reference phantoms

It was clearly stated in both ICRP Publication 23 [7] and later in ICRP Publication 89 [8] that anatomic parameters defining the ICRP reference person (newborn to adult) were based upon a review of body morphometry data (standing height and total body mass) and of organ autopsy data (organ masses) which were heavily weighted by the broader availability of literature on persons in North American and Western European countries. As such, the ICRP reference persons would be clearly 'oversized' in comparison to the average person in many Asian nations. To address this issue, the International Atomic Energy Agency (IAEA) funded a Coordinated Research Program during the period 1988 to 1993 to define the Reference Asian Person. The results of this program were published in two volumes in 1998 as IAEA-TECDOC-1005 with Volume 1 giving a data summary and conclusions, and Volume 2 giving specific reports across nine different nations in Asia and the Indian Subcontinent [136]. Later, in the mid-to-late 2000s, various countries launched efforts to develop full-scale voxel phantoms of their country-specific reference adults. Some examples include: (1) the CNMAN phantom of the Chinese Visible Human Project [137]; (2) the HDRK-Man from the Korean Visible Human Project [138]; and (3) the TARO and HANAKO MR-based voxel phantoms from Japan [139].

5.5.5 Comparison to stylized mathematical models

The obvious difference between voxel-type and stylized anatomic models is the improved geometric descriptions of organ position, organ shape, and body contour offered by the former. In estimating internal organ absorbed dose, however, only the absorbed fraction of energy and the target organ mass are of consequence. When the voxel computational model results in estimates of the absorbed fractions and organ masses that are essentially equivalent to those in the stylized model, any improvement in anatomic description is of no consequence. When differences in dosimetry are noted, they must be considered in regard to the development of each model. As noted above, stylized models are generally constructed based upon estimates of average organ size and position as documented in ICRP Publication 89 [8]. With a voxel model, organ positions and locations can be very different in as much as the individual does not represent the average characteristics of reference subject populations. The voxel model is highly patient-specific to the individual that was imaged. When applied to other patients, however, dosimetry data gathered from a single voxel model could be less desirable than that obtained from a reference-stylized model, particularly when the current patient differs substantially from the voxel model subject. Ideally, an entire series of whole-body voxel computational models would be needed. Patient-specific estimates of internal organ doses could be made through selecting one of an array of voxel models that best describes the current patient of interest.

Early attempts in the development voxel models were severely hindered by data storage and computer memory restrictions during radiation transport. At present, advances in computer hardware and software have all but eliminated these issues, except in extreme cases. Nevertheless, certain limitations of voxel models over stylized models continue to exist that are of particular note. Voxel dimensions within voxel models are initially restricted to those used in acquiring the medical image. Image slice interpolations algorithms subsequently may be applied to reduce voxel dimensions, although residual errors may persist as a result of partial-volume averaging within the original image data set. Small anatomic structures such as organ epithelial layers and GI mucosa may be missed altogether within the segmented regions of the voxel model. For these structures, the simplicity of geometric representation within the stylized mathematical models is a distinct advantage, although this feature—modeling fine tissue structures—has now been solved with the advent of mesh-type models (see section 5.6). Peter *et al* reviewed problems related to discretization errors within voxel computational models used in simulation studies of radiographic imaging [107]. Rajon *et al* similarly reviewed the dosimetric impact of stair-stepped artifacts in voxel-based models of trabecular spongiosa via microCT imaging of bone [140].

One organ of particular importance to radionuclide therapy is the skeletal system. This anatomic region is particularly difficult to construct using stylized geometrical descriptions, and thus voxel computational models offer a distinct advantage in providing macroscopic details of bone site contours, shapes, and locations. Nevertheless, voxel dimensions are much larger than microscopic tissue sites of interest (marrow cavities, bone trabeculae, endosteal tissues, etc). Consequently, microimaging technologies applied to *ex vivo* samples are necessary to produce usable images for charged-particle transport within trabecular regions of the skeleton [108]. Final assessment of marrow and endosteal tissue doses from photon sources external to the skeleton can be made via coupling of the photon transport within the whole-body tomographic computational model, while transport of photon-produced secondary electrons will invoke microscopic voxelized geometries of trabecular bone regions. Eckerman *et al* used this dual-transport technique for skeletal dosimetry utilizing: (1) chord-based models of electron transport in trabecular bone; and (2) the ORNL series of stylized models for photon transport [38]. This overall dual-transport methodology has been extended to the use of macroscopic and microscopic voxel models of the skeletal system [141]. These concepts are discussed further in chapter 19.

5.5.6 Use of voxel models in therapeutic nuclear medicine

In many forms of radiation cancer therapy, to include external beam radiotherapy and brachytherapy, radiation dosimetry is an integral component to treatment planning, where the objective is to maximize the tumor absorbed dose while avoiding or minimizing normal tissue toxicities. In therapeutic nuclear medicine, tumor dosimetry may be problematic owing to: (1) lack of imaging data to define the tumor, especially for disseminated and diffuse disease; (2) the dynamic nature of radiopharmaceutical uptake, retention, and washout; (3) non-uniformities in the spatial distribution of the agent at the cellular level; and (4) time-dependent dose rates. Consequently, there is a paucity of data on tumor dose-response relationships, upon which values of prescribed tumor dose may be assessed. Exceptions do occur, such as in treatment of solid tumors and malignant and benign thyroid disease, but even here, there are no standardized clinically accepted protocols for radionuclide treatment planning based upon delivery of a dose prescription to the tumor.

Resultantly, current dosimetry practice in therapeutic nuclear medicine is to assess the absorbed dose to radiosensitive tissues, and based upon a general understanding of toxicity thresholds (taken primarily from previous experience in external beam radiotherapy), adjust administered activities to the cancer patient to maximize uptake and dose to the tumor while avoiding normal organ toxicities. The dose to the tumor is neither quantified nor prescribed. In contrast to diagnostic nuclear medicine dosimetry, more sophisticated methods of dosimetry are generally required in therapeutic nuclear medicine in order to provide predictive indices of tissue toxicities. In the context of therapy, patient variability in terms of both agent pharmacokinetics and body morphometry must be explicitly considered on an individual patient basis to assure optimized treatment.

Prior to the era of hybrid nuclear medicine imaging, organ dosimetry needed for prediction of normal organ toxicity was performed via the MIRD schema and pre-computed radionuclide S values derived originally from stylized anatomic models, but later by individual voxel models created from image segmentation of CT or MR image sets. Obviously, the accuracy of these S values for the individual patient could be questioned owing to anatomical differences between organ size, shape and location in the cancer patient and those within the voxel phantom from which S values were derived. With the introduction of hybrid imaging—SPECT/CT and PET/CT—the possibility for construction of patient-specific voxel phantoms using the CT portion of the hybrid image became possible. As described in the various chapters of this text, a number of different research and commercial software codes are based upon the CT-derived patient-specific voxel phantom resulting from image segmentation of the CT image. Tissue heterogeneity is accounted for via variations in voxel-specific CT numbers of the image. The issue still remains that: (1) many of these CT images are partial body, thus complicating the assignment of tissue dose to out-of-scan and distributed organs; and (2) the boundaries of voxelized organs of interest must be segmented either manually, automatically, or through combination of both techniques. These two approaches to nuclear medicine organ dosimetry—the use of pre-computed S values from whole-body stylized/voxel phantoms versus the direct computation of organ dose via Monte Carlo radiation transport or dose-point-kernel convolution in a CT-devised patient-specific voxel phantom—are clearly justified in the context of diagnostic versus therapeutic uses of radiopharmaceuticals.

5.6 Mesh-type anatomic models

5.6.1 Methods of construction

As noted previously, mesh-type phantoms have been available since the early 2000s. They allow for the scalability and deformability of the modeled anatomy similar to that provided by stylized phantoms, yet they retain the anatomical realism of voxel phantoms. In the coupling of mesh phantoms to radiation transport codes, however, a final step of voxelization had to be performed as the particle tracking algorithms employed at that time did not recognize NURBS or polygon-mesh surfaces. Mesh phantom voxelization thus entailed filling these surfaces with an array of voxels of user-defined dimensions. A second advantage of mesh phantom voxelization was a resolution of potential surface overlaps and intersections introduced during phantom construction, rescaling, and/or deformation. The voxelization processes, by definition, eliminated these tissue incongruencies. Within the past few years, however, significant advances have been made in particle tracking algorithms so as to now enable the direct use of meshed geometries during MC radiation transport simulation. These developments were initially introduced into the MCNP code in 2009 [142], into the GEANT4 code in 2013 [143], and into the PHITS code in 2015 [6, 32].

There are in general three methods by which a mesh-type phantom can be created. The first is identical to that for the construction of a voxel-type computational phantom—segmentation of medical images of either live subjects or cadavers. Here, body contour and internal organ boundaries are delineated using segmentation tools within either a commercial program (such as 3D-Doctor—https://www.3d-doctor.com/) or open-source code (such as 3D-Slicer—https://www.slicer.org/). Next, these contour surfaces may be represented as either NURBS or polygon-mesh surfaces using either a commercial program (such as Rhinoceros 3D—https://www.rhino3d.com/) or open-source code (such as Blender—https://www.blender.org/).

The second method is the conversion of a previously constructed voxel-type phantom into a mesh-based format. A significant example of this process is described in ICRP Publication 145 in which members of ICRP Task Group 103 developed mesh-based twins of the reference voxel phantoms from ICRP Publication 110 (see below in section 5.6.6). This process was very laborious and time-consuming. Consequently, researchers have sought more robust and automated means of converting voxel phantoms to a polygon-mesh structure. One such approach was outline by Brown *et al* in which a boundary detection algorithm was implemented in conjunction with polygon detection to form high-quality mesh data suitable for radiation transport simulations [144].

The third method of creating a mesh-type phantom is simply to adopt an existing publicly or commercially available model and then alter that model via combinations of uniform and non-uniform scaling. An example of this approach is the construction of the RPI-AM (adult male) and RPI-AF (adult female) reference phantoms created at Rensselaer Polytechnic Institute (RPI) and reported by Zhang *et al* [145]. These mesh-type phantoms were created from a set of 140 organ-specific surface mesh models provided by the Anatomium 3D P1 V2.0 model library (500 models were available for this purpose). Organ masses were carefully adjusted to agree within 0.5% error to reference organ mass given in ICRP Publication 89 [8].

Below, we explore a small sample of the many mesh-type phantoms that have been developed since the early 2000s to the present day. These are given for illustration purposes only—the reader is encouraged to refer to more comprehensive reviews by Xu [2] and by Kainz *et al* [3].

5.6.2 The NCAT and XCAT phantom series

The very first mesh-type anatomic model applied to medical imaging was the NCAT phantom, developed and conceived by Paul Segars within his PhD dissertation at the University of North Carolina in 2001 [146]. NCAT—which stands for NURBS-based Cardiac-Torso phantom—was an extension from the previous MCAT (stylized or mathematically based) phantom used to model nuclear medicine images from a known torso anatomy under the influence of both cardiac and respiratory motion. Over the years—at both Johns Hopkins University and now at Duke University—this phantom has been extended to include anatomy beyond the torso and to include a library of patient-specific NURBS-based models called the XCAT (eXtended CArdiac-Torso) phantoms [147]. The XCAT phantoms have been developed for a variety of multimodality imaging research applications to include PET, SPECT, and CT imaging, as well as fluoroscopic and tomosynthesis imaging. Example images from the XCAT phantom library is shown in figure 5.6, along with synthetic CT images of the phantom anatomy.

5.6.3 The University of Florida phantom series

In 2010, the University of Florida series of mesh-type phantoms was presented by Lee *et al* [135]. This phantom series—shown in figure 5.7—was constructed as a combination of NURBS surfaces with more complex organ structures (such as the spine) left in polygon-mesh format. In a collaboration with the National Cancer Institute (NCI), the UF series of reference phantoms was rescaled to present a population of anatomic models across a broad range of heights and weights, as

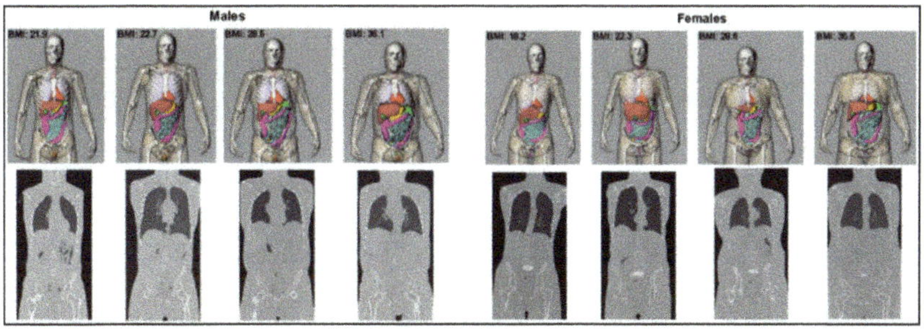

Figure 5.6. Four adult male (left) and four adult female (right) XCAT phantoms with CT simulation results shown at the bottom. Reproduced from [147] John Wiley & Sons. Copyright (2013) American Association of Physicists in Medicine

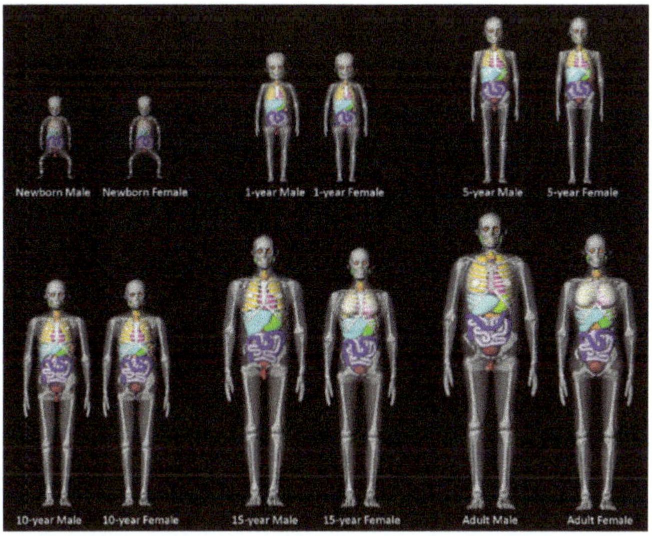

Figure 5.7. The University of Florida (UF) series of reference phantoms. Reproduced from [135]. Copyright IOP Publishing Ltd. All rights reserved.

shown in figure 5.8. The resulting UF/NCI phantom library as reported by Geyer *et al* consists of four sub-libraries—100 adult males, 93 adult females, 85 pediatric males, and 73 pediatric females. The UF/NCI phantom library has been utilized in a variety of dosimetry studies to include CT imaging [148], fluoroscopic imaging [149], and diagnostic nuclear medicine [150].

5.6.4 IT'IS virtual family phantom series

In a collaboration between the IT'IS Foundation, the U.S. Food and Drug Administration, and other groups, a series of mesh-based anatomic models within the Virtual Family and Virtual Population were developed to include male and female adults and children, an obese male, an elderly male, a pregnant female, and a newborn child [152]. These models were all developed from MR data of healthy

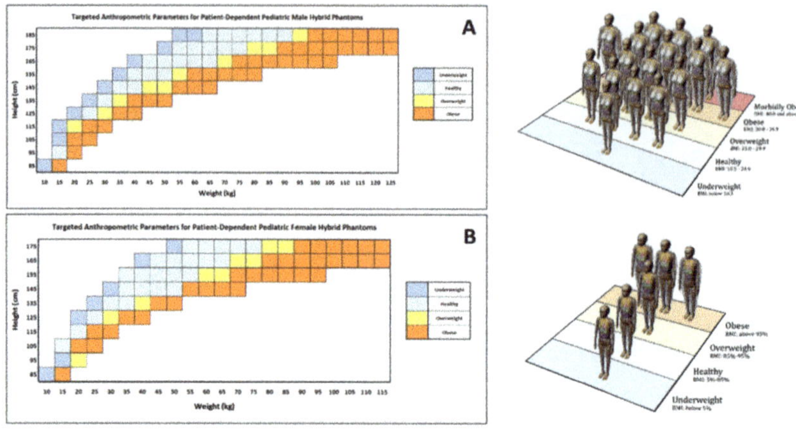

Figure 5.8. The UF/NCI series of hybrid computational phantoms. Distribution of phantom heights and weights for (a) pediatric males and (b) pediatric females. Top-right: representative members of the adult female phantoms. Bottom-right: representative members of the pediatric male phantoms. Reproduced from [151].

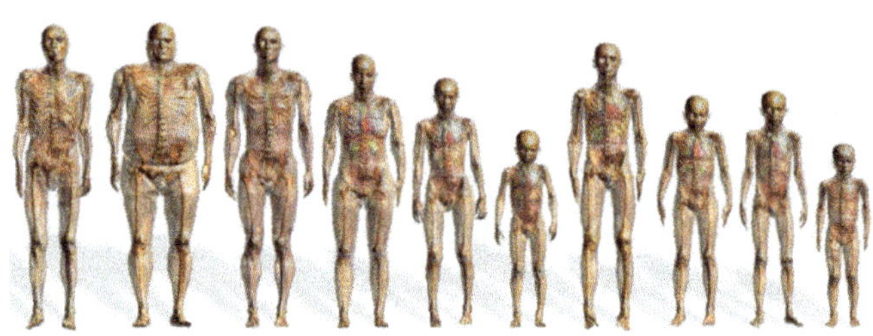

Figure 5.9. The complete Virtual Population representing humans (males and females) of different age groups and body shapes. Reproduced from [152].

volunteers. The latest version of the Virtual Population series is based on higher resolution segmentation (0.5 mm voxels). Members of the series are able to be morphed and repositioned based upon realistic biomechanical models to simulate extremity articulation and tissue deformation, expansion, and shrinkage. Representative models are shown in figure 5.9.

5.6.5 RPI phantom series

Under the direction of Dr George Xu, researchers at RPI extended their work with image data set of the Visible Human Project to convert their VIP-Man voxel model to a mesh-based format with the ability to model both respiratory and cardiac motion [99]. Following the development of the RPI-AM and RPI-AF polygon-mesh reference phantoms [145], Ding *et al* extended their anatomic modeling developed to produce a set of ten phantoms (five males and five females) covering body mass indices from 23.5 kg m^{-2} (normal body weight) to 46.4 kg m^{-2} (morbidly obese body weight). These phantoms—shown below in figure 5.10—were used to explore variations in organ dosimetry under CT x-ray imaging [153].

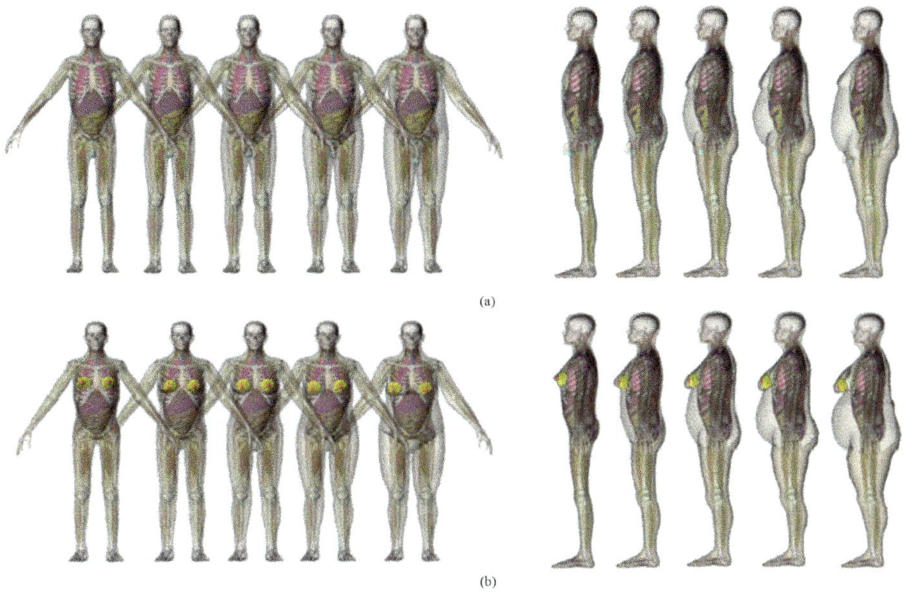

Figure 5.10. Anterior and lateral views of the RPI BMI-adjustable (a) male and (b) female phantoms. The phantoms have the same height (1.76 m in the male and 1.63 m in the female) but differ in weight. From left to right the weight classifications are normal-weight, overweight, obese level-I, obese level-II and morbidly obese. Reproduced from [153]. Copyright IOP Publishing Ltd. All rights reserved.

5.6.6 The ICRP publication 145 mesh-type adult reference phantoms

In May of 2016, the ICRP established its Task Group 103 with the intended goal of extending its adult reference computational phantoms from a voxel-based format to a mesh-based format. This work was completed in 2020—driven to a significant extent by Dr Chan Kim and his students at Hanyang University in the Republic of Korea—resulting in the release of ICRP Publication 145 [154]. These mesh-type reference computational phantoms (MRCPs) of the ICRP reference adult male and adult female were to be mesh-based twins of their original voxel-based reference computational phantoms (VRCPs) given in Publication 110 [124]—see figure 5.11. Two key developments were noted by the Task Group in the course of the voxel-to-mesh conversion process. First, the newer technology would improve the anatomical integrity of the anatomic model. As the Publication 110 VCRPs were created from direct segmentation of contiguous CT image slices of fairly course slice thickness (8 mm for the male and 5 mm for the female), stair-step artifacts were apparent in the original voxel-based format to include: (1) gaps in the voxelized skin model with external exposure of skeletal muscle and breast tissue; and (2) gaps in the voxelized cortical bone model with internal exposure of the underlying regions of trabecular spongiosa. Second, the ICRP had previously relied on supplemental stylized models of very thin target tissue layers along the respiratory and alimentary tracts, which with the mesh-based technology could now be accommodated within the whole-body phantom. Third, it became apparent that all previous phantoms developed to represent the ICRP reference series

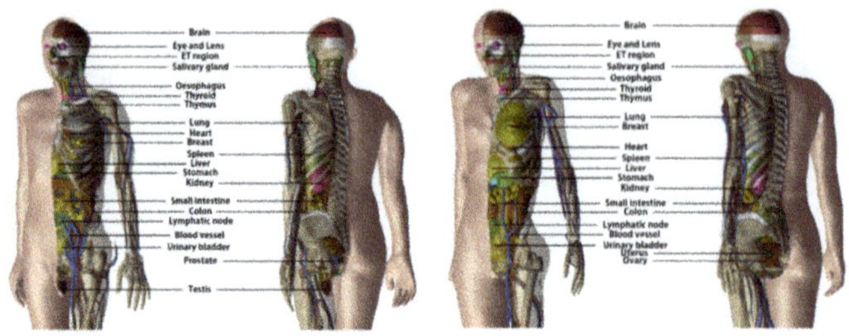

Figure 5.11. Mesh-type ICRP adult male (left) and adult female (right) computational reference phantoms (MCRP). Reprinted with permission from [154].

had intentionally targeted internal organ volumes that, when coupled with reference tissue mass densities, would agree with reference organ masses given in ICRP Publication 89 [8]. Unfortunately for prior phantom modelers, it only recently became apparent that Publication 89 reference organ masses were to represent organ parenchyma only, and thus these reference masses did not account for organ blood content. Consequently, the organ volumes with the VCRPs of ICRP Publication 110 were too small, and self-dose SAFs given in Publication 133 [129] had to be mass adjusted to account for the added organ blood masses. Thus, in the conversion from voxel to mesh-based formats under Task Group 103, all internal organ volumes could now be properly adjusted to reflect their true *in vivo* volumes.

5.6.7 HUREL phantom series

In 2020, Choi *et al* at the Hanyang University Radiation Engineering Laboratory (HUREL) reported on the development of 108 adult male and 104 adult female mesh-based anatomic models covering most body sizes representative of Caucasian and Asian populations [155]. These models were developed using an in-house program for automatic phantom adjustment using the ICRP Publication 145 MRCPs as anchor (or source) phantoms. Ten secondary anthropometric parameters with respect to standing height and body were derived from various databases and applied to the phantom library construction. Representative images of the HUREL library of mesh-type phantoms are shown in figure 5.12 below.

5.6.8 The ICRP mesh-based pediatric phantoms

With the successful conversion and expansion of the ICRP VRCPs from ICRP Publication 110 to a high-quality polygon-mesh format as given in ICRP Publication 145, Task Group 103 is presently extending this voxel-to-mesh conversion work to the ICRP pediatric phantom series. Work in the mesh-based skeletal models of ICRP pediatric phantoms has been reported by Choi *et al* [156]. The current status of the full series of mesh-type pediatric reference phantoms was presented at the 2020 Meeting of the International Radiation Protection Association (IRPA) and is described in Choi *et al* [157]. Oblique views of the mesh-type ICRP pediatric phantoms in their provision state are shown below in figure 5.13. It is anticipated that they will be published by the ICRP in an upcoming report released in 2022.

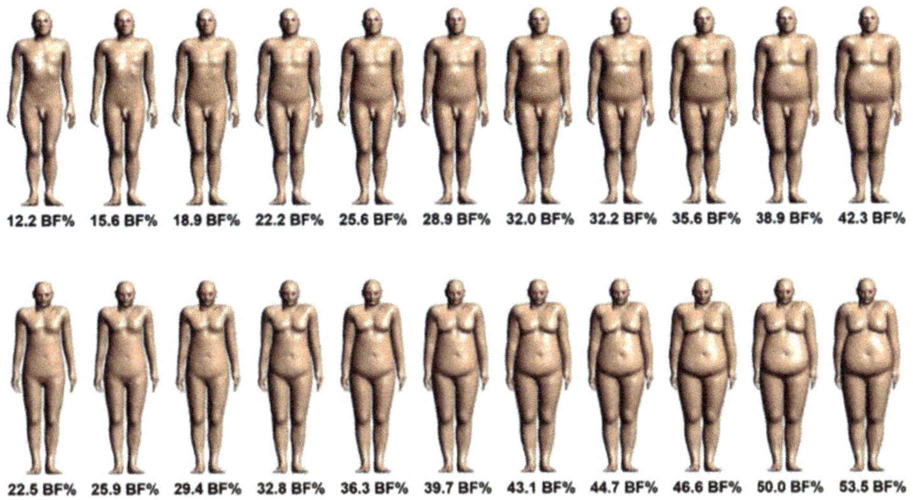

Figure 5.12. HUREL mesh-type phantom library for varying body-fat percentages (BF%s). Male adult phantoms (top row) and female adult phantoms (bottom row). Reproduced from [155]. Copyright IOP Publishing Ltd. All rights reserved.

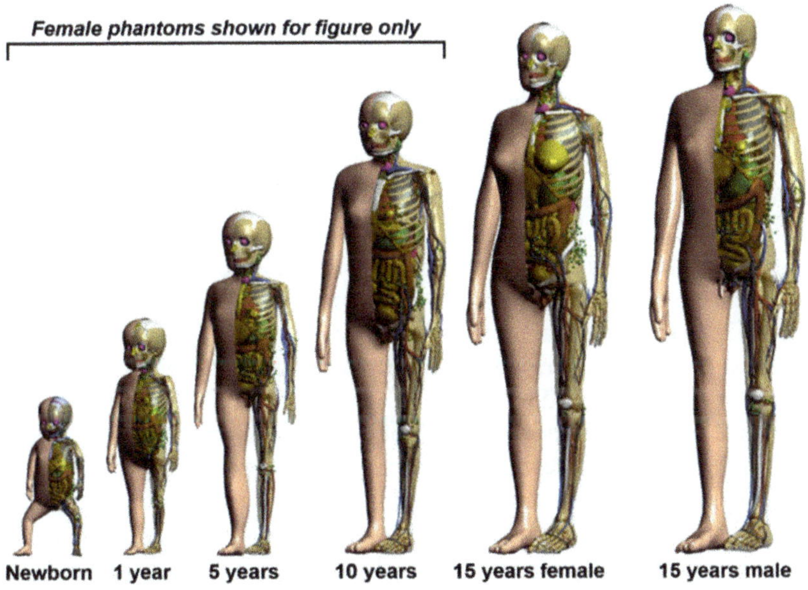

Figure 5.13. Provisional mesh-type ICRP pediatric reference phantoms prepared by ICRP Task Group 103. Reproduced from [157]. © IOP Publishing Ltd. All rights reserved.

5.6.9 Use of mesh-based models in therapeutic nuclear medicine

For assessment of radiation-absorbed dose to normal organs and tissues of concern for radiogenic toxicity, the morphometric category of phantom most appropriate to the task is that of the *patient-specific* phantom. For organ dosimetry in diagnostic medical imaging, including nuclear medicine, radiation doses are relatively small, are generally well below thresholds for organ toxicity, and thus the level of dose accuracy is not demandingly high. Consequently, use of *patient-dependent* phantom libraries (morphometrically matched to the patient) or *patient-sculpted* phantoms (with residual uncertainties in internal organ structure) are sufficient for image optimization and management of stochastic risk which themselves have large uncertainties. In radiopharmaceutical therapy, however, the requirement for dose accuracy is high as critical decisions on administered activity (and possibly dosing sequence) are dependent upon reliable models of dose-response.

As noted previously, state-of-the-art dosimetry software for radiopharmaceutical therapy are widely based upon voxel-type patient-specific models originating from CT or MR images of the therapy patient. As in other forms of radiation therapy—to include external beam and brachytherapy—voxel-based CT-derived patient models allow one to score absorbed dose in individual voxels so as to assemble dose-volume histograms across both the treated tumor and the normal organs-at-risk. Providing voxel-based organ dosimetry in radiopharmaceutical therapy allows for the combination of DVHs across different therapy modalities.

Given the utility of voxel-based patient-specific anatomic models, what are possible reasons to move toward mesh-based patient-specific anatomic models? Two factors come into this decision. The first is that for a CT-based patient-specific model, the resulting phantom is perhaps never of the entire body thus limiting dose assessment to organs within the scan range, and limiting the computation of absorbed dose to distributed organs such as the active bone marrow. The second is the need to delineate which array of CT voxels belong to different organs (organ segmentation). In a mesh-type phantom, all tissues and organs are pre-defined and available for tissue dosimetry, which is in contrast to the need for organ boundary segmentation for CT-based models.

One possible solution is the use of intensity-based deformable image registration between pre-existing whole-body mesh phantoms (reference or patient-dependent) and patient CT images. In a recent study by Carter *et al*, a linear attenuation coefficient phantom was generated by resampling the PSRK-Man Korean reference phantom onto a voxel grid and defining organ regions with corresponding Hounsfield unit (HU) reference values [158]. The resulting HU phantom was co-registered with a patient CT scan using Plastimatch B-spline deformable registration (see figure 5.14). In parallel, major organs were manually contoured to generate a 'ground truth' patient-specific phantom for comparisons. Monte Carlo derived *S*-values, which support nuclear medicine dosimetry, were calculated using both approaches and compared. The study found that deformable registration techniques can be used to create a personalized mesh-type phantom to better support nuclear medicine organ dosimetry. The method was also shown to be easier and faster than manual segmentation—as needed to create a patient-specific voxel phantom. While the study focused on the diagnostic imaging agent 18F-FDG, the application of this approach to radiopharmaceutical therapy is obvious and promising. Clearly, an additional step not discussed would be the insertion of modeled tumors, as the starting mesh-type phantom would have been tumor free by design (figure 5.14).

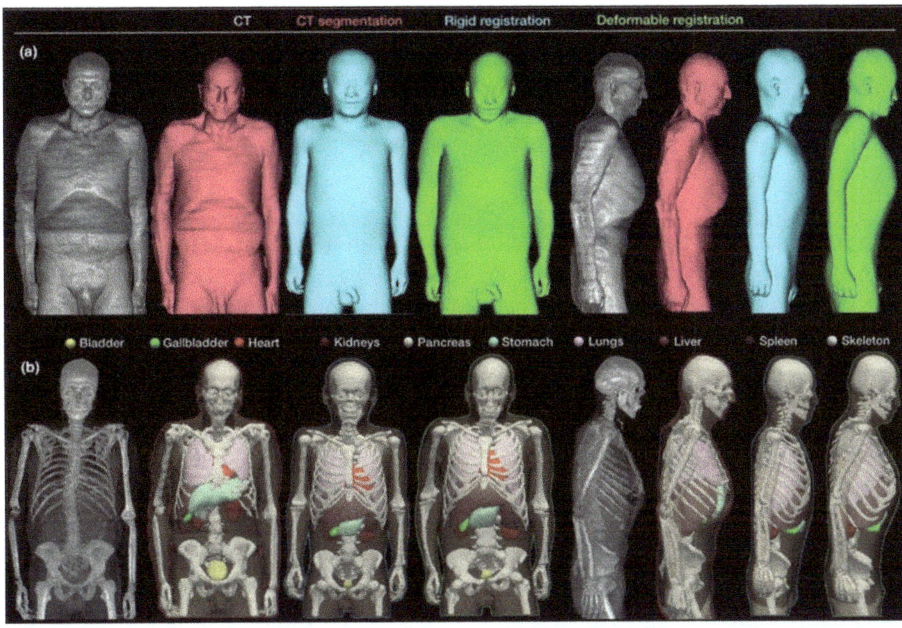

Figure 5.14. Phantoms derived from PET/CT segmentation, and morphing of a reference phantom via deformable image registration. (a) Volume-rendered CT and exterior surfaces of segmented phantom (red), rigidly co-registered reference phantom (blue), and deformably co-registered reference phantom (green). Note the improved match of exterior body contour evident by visual assessment, of the deformed phantom relative to the original (rigidly co-registered) phantom. (b) Volume-rendered CT windowed for bone contrast; interior organs of each phantom. Reproduced from [158]. Copyright IOP Publishing Ltd. All rights reserved.

5.7 Conclusions

To some extent, mathematical formulations of organs and tissues of the body have been used in the dosimetry of internally distributed radionuclides for nearly 80 years. The concepts are not new ones. However, over this time, the sophistication of these 'computational phantoms' has increased significantly. This evolution began with the specification of a single organ mass, followed by the use of simple shapes to simulate organs or the entire body of an adult human. The ability to model the entire body of a 'Reference Person' and to specify the location, shape, volume, and mass of organs in the body was the next logical step. These developments were followed, almost immediately with the development of mathematical models for humans of other ages as well as attempts to model specific regions in important organs such as the kidneys and brain. Aanatomic models, based upon 3D imaging techniques such as magnetic resonance imaging or computed tomography, have been widely developed and used in internal dosimetry. These techniques lend themselves to the design and specification of anatomic models representing the individual patient and, in addition, allow the definition of structures and/or regions in organs so that the distribution of the radionuclide or radiopharmaceutical can be better represented and the radiation-absorbed dose can be more accurately assessed. While for radiopharmaceutical therapy, CT-derived patient-specific voxel phantoms are the norm, the rapid adoption and maturation of mesh-based phantoms—stemming originally from the fields of computer animation and modeling—might be considered as a viable and time-saving alternative to both tumor and normal organ dose assessment in the growing and promising field of radiopharmaceutical therapy.

References

[1] Xu X G and Eckerman K F 2009 Handbook of anatomical models for radiation dosimetry *Series in Medical Physics and Biomedical Engineering.* (London: Taylor and Francis)

[2] Xu X G 2014 An exponential growth of computational phantom research in radiation protection, imaging, and radiotherapy: a review of the fifty-year history *Phys. Med. Biol.* **59** R233–302

[3] Kainz W *et al* 2019 Advances in computational human phantoms and their applications in biomedical engineering – a topical review *IEEE Trans. Radiat. Plasma Med. Sci.* **3** 1–23

[4] Bolch W, Lee C, Wayson M and Johnson P 2010 Hybrid computational phantoms for medical dose reconstruction *Radiat. Environ. Biophys.* **49** 155–68

[5] Han H, Yeom Y S, Choi C, moon S, Shin B, Ha S and Kim C H 2020 POLY2TET: a computer program for conversion of computational human phantoms from polygonal mesh to tetrahedral mesh *J. Radiol. Prot.* **40** 962–79

[6] Furuta T, Sato T, Han M C, Yeom Y S, Kim C H, Brown J L and Bolch W E 2017 Implementation of tetrahedral-mesh geometry in Monte Carlo radiation transport code PHITS *Phys. Med. Biol.* **62** 4798–810

[7] ICRP 1975 ICRP publication 23: report on the task group on reference man *Int. Commission on Radiological Protection* (Oxford: Pergamon)

[8] ICRP, ICRP Publication 89 2002 Basic anatomical and physiological data for use in radiological protection – reference values *Ann. ICRP.* **32** 1–277

[9] Johnson P, Whalen S, Wayson M, Juneja B, Lee C and Bolch W 2009 Hybrid patient-dependent phantoms covering statistical distributions of body morphometry in the US adult and pediatric population *Proc. IEEE.* **97** 2060–75

[10] Stepusin E J, Long D J, Marshall E L and Bolch W E 2017 Assessment of different patient-to-phantom matching criteria applied to a Monte Carlo-based CT organ dose library *Med. Phys.* **44** 5498–508

[11] Kuzmin G A, Mille M M, Jung J W, Lee C, Pelletier C, Akabani G and Lee C 2018 A novel method to extend a partial-body CT for the reconstruction of dose to organs beyond the scan range *Radiat. Res.* **189** 618–26

[12] Xie T and Zaidi H 2019 Estimation of the radiation dose in pregnancy: an automated patient-specific model using convolutional neural networks *Eur. Radiol.* **29** 6805–15

[13] NCRP 1985 *NCRP Report No. 85: The Experimental Basis for Absorbed-Dose Calculations in Medical Uses of Radionuclides.* (Bethesda, MD: National Council on Radiation Protection and Measurements)

[14] Marinelli L D 1942 Dosage determination with radioactive isotopes *Am. J. Roetengol. Rad. Therapy.* **47** 210–16

[15] Marinelli L D, Quimby. J H and Hine G J 1948 , Dosage dsetermination with radioactive isotopes II. Practical considerations in therapy and protection *Am. J. Roetengol. Rad. Therapy.* **59** 260–81

[16] Loevinger R, Hiolt J G and Hine G J 1965 Internally administered radionuclides *Radiation Dosimetry* ed G J Hine and G L Brownell (New York: Academic) 801–73

[17] Loevinger R, Japha E M and Brownell G L 1965 Discrete radioisotope sources *Radiation Dosimetry* ed G J Hine and G L Brownell (New York: Academic) pp 693–799
[18] ICRP 1960 ICRP Publication 2: Recommendations of the ICRP, Report of Committee 2 on Permissible dose for Int. Radiation *Interational Commission on Radiological Protection, Oxford, UK*
[19] Brownell G L, Ellet W H and Reddy A R 1968 MIRD Pamphlet No. 3: absorbed fractions for photon dosimetry *J. Nucl. Med.* **9** 27–39
[20] Fisher H and Snyder W 1966 *Variation of Dose Delivered by 137Cs as a Function of Body Size from Infancy to Adulthood* (Oak Ridge, TN: Oak Ridge National Laboratory)
[21] Fisher H and Snyder W 1967 *Distribution of Dose in the Body from a Source of Gamma Rays Distributed Uniformly in an Organ* (Oak Ridge, TN: Oak Ridge National Laboratory)
[22] Ellett W H, Callahan A B and Brownell G L 1964 Gamma-ray dosimetry of internal emitters. Monte Carlo calculations of absorbed dose from point sources *Br. J. Radiol.* **37** 45–52
[23] Ellett W H, Callahan A B and Brownell G L 1965 Gamma-ray dosimetry of internal emitters II: Monte Carlo calculations of absorbed dsoe form uniform sources *Brit. J. Radiol.* **38** 541–44
[24] Snyder W S 1967 *The Variation of Dose in Man From Exposure to a Point Source of Gamma Rays* (Oak Ridge, TN: Oak Ridge National Laboratory)
[25] Snyder W S, Ford M R, Warner G G and Fisher H L 1969 *MIRD Pamphlet No. 5: Estimates of Absorbed Fractions for Monoenergetic Photon Sources Uniformly Distributed in Various Organs of a Heterogeneous Phantom* (New York: Society of Nuclear Medicine)
[26] ICRP, ICRP Publication 30 1979 Part 1: limits for intakes of radionuclides by workers *Ann. ICRP.* **3** 1–116
[27] Snyder W S, Ford M R and Warner G G 1978 *MIRD Pamphlet No. 5, Revised: Estimates of Specific Absorbed Fractions for Photon Sources Uniformly Distributed in Various Organs of a Heterogeneous Phantom* (New York: Society of Nuclear Medicine)
[28] Warner G G and Craig A N J 1968 ALGAM: A Computer Program for Estimating Internal Dose in a Man Phantom. *ORNL TM-2550* (Oak Ridge, TN: Oak Ridge National Laboratory)
[29] Nelson W R, Hirayama R H and Roger D W O 1985 The EGS4 code system (Stanford, CA: Stanford Linear Accelerator Center)
[30] Briesmeister J F 1997 MCNP – A General Monte Carlo N-Particle Transport Code. (Los Alamos, NM: Los Alamos National Laboratory)
[31] Pelowitz D B *MCNPX User's Manual Version 2.7.0.* (Los Alamos, NM: Los Alamos National Laboratory: LANL)
[32] Sato T *et al* 2018 Features of particle and heavy ion transport code system (PHITS) version 3.02 *J. Nucl. Sci. Tech.* **55** 684–90
[33] Kaddouch S and El Khayati N 2017 Geant4/GATE Monte Carlo code for internal dosimetry using voxelized phantom *Moscow Univ. Phys. Bull.* **72** 658–62
[34] Poston J J W, Kodimer K A, Bolch W E and Poston S J W 1996 A revised model for the calculation of absorbed energy in the gastrointestinal tract *Health Phys.* **71** 307–14
[35] Poston J J W, Kodimer K A, Bolch W E and Poston S J W 1996 Calculation of absorbed energy in the gastrointestinal tract *Health Phys.* **71** 300–6
[36] Kereiakes J G, Seltzer R A, Blackburn B and Saenger E L 1965 Radionuclide doses to infants and children: a plea for a standard child *Health Phys.* **11** 999–1004

[37] Poston J W 1990 The development of early pediatric models and their applications to radiation absorbed dose calculations *Dosimetry of Administered Radionuclides* ed S J Adelstein and A I Kassis (Washington, D.C.: American College of Nuclear Physicians) 105–21
[38] Hwang J M L, Shoup R L, Warner G G and Poston J W 1976 *Mathematical Description of a One- and Five-Year-Old Child for Use in Dosimetry Calculations. ORNL TM-5293* (Oak Ridge, TN: Oak Ridge National Laboratory)
[39] Hwang J M L, Shoup R L and Poston J W 1976 *Mathematical Descriptions of a Newborn Human for Use in Dosimetry Calculations. ORNL TM-5453* (Oak Ridge, TN: Oak Ridge National Laboratory)
[40] Jones R M, Poston J W, Hwang J L, Jones T D and Warner G G 1976 *The Development and Use of a Fifteen-Year-Old Equivalent Mathematical Phantom for Internal Dose Calculations. ORNL/TM-5278* (Oak Ridge, TN: Oak Ridge National Laboratory)
[41] Deus S F and Poston J W 1976 The development of a mathematical phantom representing a 10-year-old for use in internal dosimetry calculations *Symp. on Radiopharmaceutical Dosimetry* ed E E Watson (Oak Ridge, TN: Food and Drug Administration)
[42] Cristy M 1980 *Mathematical Phantoms Representing Children of Various Ages for Use in Estimates of Internal Dose* (Oak Ridge, TN: Oak Ridge National Laboratory)
[43] Cristy M and Eckerman K F 1987 *Specific Absorbed Fractions of Energy at Various Ages from Internal Photon Sources* (Oak Ridge, TN: Oak Ridge National Laboratory)
[44] Bolch W E, Eckerman K F, Sgouros G and Thomas S R 2009 MIRD Pamphlet No. 21: a generalized schema for radiopharmaceutical dosimetry: standardization of nomenclature *J. Nucl. Med.* **50** 477–84
[45] Watson E E and Stabin M G 1984 Basic alternative software package for internal radiation dose calculations *17th Midyear Symp. of the Health Physics Society*; R L Kathern, D P Higby and M A McKinney (Richland, WA: Health Physics Society)
[46] Watson E E, Stabin M and Bolch W E MIRDOSE2. 1987 (Oak Ridge, TN: Oak Ridge Associated Universities)
[47] Stabin M G 1996 MIRDOSE: personal computer software for internal dose assessment in nuclear medicine *J. Nucl. Med.* **37** 538–46
[48] Weber D A, Eckerman K F, Dillman L T and Ryman J C 1989 *MIRD: Radionuclide Data and Decay Schemes* (New York: Society of Nuclear Medicine)
[49] ICRP, ICRP Publication 38: 1983 Radionuclide Transformations – Energy and Intensity of Emissions *Ann. ICRP.* **11–13** 1–1250
[50] Kocher D C 1981 Decay Data Tables *DOE TIC-11026* (Oak Ridge, TN: Oak Ridge National Laboratory)
[51] Stabin M, Watson E, Cristy M, Ryman J, Eckerman K, Davis J, Marshall D and Gehlen M 1995 *Mathematical Models and Specific Absorbed Fractions of Photon Energy in the Nonpregnant Adult Female and at the End of Each Trimester of Pregnancy. ORNL/TM-12907.* (Oak Ridge, TN: Oak Ridge National Laboratory)
[52] Cloutier R J, Smith S A, Watson E E, Snyder W S and Warner G G 1973 Dose to the fetus from radionuclides in the bladder *Health Phys.* **25** 147–61
[53] Eckerman K F 1985 Aspects of the dosimetry of radionuclides within the skeleton with particular emphasis on the active marrow *Proc. of the Fourth Int. Radiopharmaceutical Dosimetry Symp* (Oak Ridge, TN: Oak Ridge Associated Universities)

[54] Ellet W H and Humes R M 1971 *MIRD Pamphlet No. 8: Absorbed Fractions for Small Volumes Containing Photon-Emitting Radioactivity. MIRD Pamphlet No. 8.* (New York: Society of Nuclear Medicine)
[55] Siegel J A and Stabin M G 1994 Absorbed fractions for electrons and beta particles in spheres of various sizes *J. Nucl. Med.* **35** 152–6
[56] Coffey J L, Cristy M and Warner G G 1981 MIRD Pamphlet No. 13: specific absorbed fractions for photon sources uniformly distributed in the heart chambers and heart wall of a heterogeneous phantom *J. Nucl. Med.* **22** 65–71
[57] Coffey J L and Watson E E 1981 S-values for selected radionuclides and organs with the heart wall and heart contents as source organs *Third Int. Radiopharmaceutical Dosimetry Symp.* ed E E Watson *et al* (Rockville, MD: US Department of Health & Human Services) 563–94
[58] Seltzer S M 1988 An overview of ETRAN Monte Carlo methods *Monte Carlo Transport of Electrons and Photons* ed T M Jenkins, W R Nelson and A Rindi (New York: Plenum Press) pp 153–81
[59] Seltzer S M 1991 Electron-photon Monte Carlo calculations: the ETRAN code *Appl. Radiat. Isot.* **42** 917–41
[60] Bouchet L G, Bolch W E, Weber D A, Atkins H L and Poston J W Sr. MIRD Pamphlet No. 15: Radionuclide S values in a revised dosimetric model of the adult head and brain. Medical Internal Radiation Dose *J. Nucl. Med.* 1999**40** 62S–101S
[61] London E D 1993 *Imaging Drug Action in the Brain.* (Boca Raton: CRC Press)
[62] van Heertum R L and Tikofsky R S 1995 *Cerebral SPECT Imaging* 2nd edn (New York: Raven Press)
[63] Blinkov S M and Glezer I I 1968 *The Human Brain in Figures and Tables* (New York: Plenum Press)
[64] Willis K W, Martinez D A, Hedley-Whyte E T, Davis M A, Judy P F and Treves S 1977 Renal localization of 99mTc-stannous glucophetonate and ^{99m}Tc-stannous dimercaptosuccinate in the rat by frozen section autoradiography. The efficiency and resolution of technetium-99m *Radiat. Res.* **69** 475–88
[65] Snyder W S and Ford M R 1967 *A Dosimetric Study for the Administration of Neohydrin Labeled with Hg-203 and Hg-197 ORNL TM-4168* (Oak Ridge, TN: Oak Ridge National Laboratory)
[66] McAfee J G 1969 Problems in evaluating the radiation dose for radionuclides excreted by the kidneys *Medical Radionuclides: Radiation Dose and Effects. Oak Ridge, TN* (U.S. Atomic Energy Commission)
[67] Blau M, McAfee J G, Roher R H, Snyder W S and Smith E M 1975 MIRD dose estimate report No. 6: Hg-197 and Hg-203 labeled chlormerodrin *J. Nucl. Med.* **16** 1214–17
[68] Patel J S 1988 A revised model of the kidneys for medical internal radiation dose *Nuclear Engineering.* (College Station, TX: Texas A&M Universtiy)
[69] Bouchet L, Bolch W, Blanco P, Wessels B, Siegel J, Rajon D, Clairand I and Sgouros G 2003 MIRD Pamphlet No. 19: absorbed fractions and radionuclide S values for six age-dependent multi-region models of the kidney *J. Nucl. Med.* **44** 1113–47
[70] Boerman O C, Oyen W J and Corstens F H 2001 Between the Scylla and Charybdis of peptide radionuclide therapy: hitting the tumor and saving the kidney *Eur. J. Nucl. Med.* **28** 1447–49

[71] Wessels B W *et al* 2008 MIRD pamphlet No. 20: the effect of model assumptions on kidney dosimetry and response: implications for radionuclide therapy *J. Nucl. Med.* **49** 1884–99

[72] Ford R L and Nelson W R 1978 *The EGS Cose System: Computer Programs for Monte Carlo Simualtion of Electromagnetic Cascade Shower. SLAC Report 210* (Stanford, CA: Stanford Linear Accelerator Center)

[73] Loevinger R, Budinger T F and Watson E E 1991 *MIRD Primer for Absorbed Dose Calculations* 2nd edn (New York: The Society of Nuclear Medicine) 128

[74] Stabin M G 1994 A model of the prostate gland for use in internal dosimetry *J. Nucl. Med.* **35** 516–20

[75] Mardirossian G, Tagesson M, Blanco P, Bouchet L G, Stabin M, Yoriyaz H, Baza S, Ljungberg M, Strand S E and Brill A B 1999 A new rectal model for dosimetry applications *J. Nucl. Med.* **40** 1524–31

[76] Stubbs J, Evans J and Stabin M 1998 Radiation absorbed doses to the walls of hollow organs *J. Nucl. Med.* **39** 1989–95

[77] Watson E E, Stabin M G, Davis J L and Eckerman K F 1989 A model of the peritoneal cavity for use in internal dosimetry *J. Nucl. Med.* **30** 2002–11

[78] Deloar H M, Watabe H, Nakamura T, Narita Y, Yamadera A, Fujiwara T and Itoh M Internal dose estimation including the nasal cavity and major airway for continuous inhalation of C15O2, 15O2 and C15O using the thermoluminescent dosimeter method. *J. Nucl. Med.* 1997**38** 1603–13

[79] Clariand I and Bouchet and B.W. E. L G 2000 A new macroscopic model of the long bones for skeletal dosimetry – abstract *J Nucl Med.* **41** 1062

[80] Han E Y, Bolch W E and Eckerman K F 2006 Revisions to the ORNL series of adult and pediatric computational phantoms for use with the MIRD schema *Health Phys.* **90** 337–56

[81] Snyder W S, Ford M R, Warner G G and Watson S B 1975 *MIRD Pamphlet No. 11: S, Absorbed Dose Per Unit Cumulated Activity for Selected Radionuclides and Organs* (New York: Society of Nuclear Medicine)

[82] Wayson M B and Bolch W E 2018 Individualized adjustments to reference phantom internal organ dosimetry-scaling factors given knowledge of patient internal anatomy *Phys. Med. Biol.* **63** 085006

[83] Wayson M B and Bolch W E 2018 Individualized adjustments to reference phantom internal organ dosimetry – scaling factors give knowledge of patient external anatomy *Phys. Med. Biol.* **63** 085007

[84] Clairand *et al* 2000 Improvement of internal dose calculations using mathematical models of different adult heights *Phys. Med. Biol.* **45** 2771–85

[85] de la Grandmaison G L, Clairand I and Durigon M 2001 Organ weight in 684 adult autopsies: new tables for a Caucasoid population *Forensic Sci. Int.* **119** 149–54

[86] Clairand *et al* 1999 DOSE3D: EGS4 Monte Carlo code-based software for internal radionuclide dosimetry *J. Nucl. Med.* **40** 1517–23

[87] Johnson T K 1988 MABDOSE: a generalized program for internal radionuclide dosimetry *Comput. Meth. Progr. Biomed.* **27** 159–67

[88] Johnson T K, McClure D L, McCourt S L, Andl G J and Berman B D 1998 MABDOSE: a computer program for the calculation of dose *Proc. of the Sixth Int. Radiopharmaceutical Dosimetry Symp.* (Oak Ridge, TN: ORAU)

[89] Johnson T K and Colby S B 1993 Photon contribution to tumor dose from considerations of 131I radiolabeled antibody uptake in liver, spleen, and whole body *Med. Phys.* **20** 1667–74

[90] Zubal I G, Harrell C R, Smith E O, Rattner Z, Gindi G and Hoffer P B 1994 Computerized three-dimensional segmented human anatomy *Med. Phys.* **21** 299–302
[91] Zubal I G and Harell C H 1992 Voxel based Monte Carlo calculations of nuclear medicine images and applied variance reduction techniques *Image Vision Comput.* **10** 342–48
[92] Zubal I G, Harrell C R, Smith E O, Smith A L and Krischlunas P 1996 High resolution, MRI-based, segmented, computerized head phantom *Proc. of the Sixth Int. Radiopharmaceutical Dosimetry Symp.* (Oak Ridge, TN: ORAU)
[93] Stabin M G, Yoriyaz H, Brill A B and Dawant M 1999 Monte Carlo calculation of dose conversion factors for a new generation of dosimetry phantoms *J. Nucl. Med.* **40** 310–11
[94] Yoriyaz H, dos Santos A, Stabin M G and Cabezas R 2000 Absorbed fractions in a voxel-based phantom calculated with the MCNP-4B code *Med. Phys.* **27** 1555–62
[95] Stabin M G and Yoriyaz H 2002 Photon specific absorbed fractions calculated in the trunk of an adult male voxel-based phantom *Health Phys.* **82** 21–44
[96] Tagesson M, Zubal I G, Ljungberg M and Strand S E 1996 S values for subregions in the brain *Proc. of the Sixth Int. Radiopharmaceutical Dosimetry Symp.* (Oak Ridge, TN: ORAU)
[97] Jones D G 1997 A realistic anthropomorphic phantom for calculating organ doses arising from external photon irradiation *Radiat. Prot. Dosim.* **72** 21–9
[98] Jones D G 1998 A realistic anthropomorphic phantom for calculating specific absorbed fractions of energy deposited from internal gamma emitters *Radiat. Prot. Dosim.* **79** 411–14
[99] Xu X G, Chao T C and Bozkurt A 2000 VIP-Man: an image-based whole-body adult male model constructed from color photographs of the Visible Human Project for multi-particle Monte Carlo calculations *Health Phys.* **78** 476–86
[100] Chao T C and Xu X G 2001 Specific absorbed fractions from the image-based VIP-Man body model and EGS4-VLSI Monte Carlo code: internal electron emitters *Phys. Med. Biol.* **46** 901–27
[101] Chao T C, Bozkurt A and Xu X G 2001 Conversion coefficients based on the VIP-Man anatomical model and EGS4 *Health Phys.* **81** 163–83
[102] Bozkurt A, Chao T C and Xu X G 2000 Fluence-to-dose conversion coefficients from monoenergetic neutrons below 20 MeV based on the VIP-man anatomical model *Phys. Med. Biol.* **45** 3059–79
[103] Bozkurt A, Chao T C and Xu X G 2001 Fluence-to-dose conversion coefficients based on the VIP-Man anatomical model and MCNPX code for monoenergetic neutrons above 20 MeV *Health Phys.* **81** 184–202
[104] Chao T C, Bozkurt A and Xu X G 2001 Organ dose conversion coefficients for 0.1–10 MeV electrons calculated for the VIP-Man tomographic model *Health Phys.* **81** 203–14
[105] Zankl M, Veit R, Williams G, Schneider K, Fendel H, Petoussi N and Drexler G 1988 The construction of computer tomographic phantoms and their application in radiology and radiation protection *Radiat. Env. Biophys.* **27** 153–64
[106] Zankl M, Veit R, Petoussi N, Mannweiler E, Wittmann A and Drexler G 1994 Realistic computerized human phantoms *Adv. Space Res.* **14** 423–31
[107] Petoussi-Henss N and Zankl M 1998 Voxel anthropomorphic models as a tool for internal dosimetry *Radiat. Prot. Dosim.* **79** 415–18
[108] Petoussi-Henss N, Panzer W, Zankl M and Drexler G 1995 Dose-area product and body doses *Radiat. Prot. Dosim.* **57** 363–66

[109] Veit R and Zankl M 1992 Influence of patient size on organ doses in diagnostic radiology *Radiat. Prot. Dosim.* **43** 241–43
[110] Veit R and Zankl M 1993 Variation of organ doses in pediatric radiology due to patient diameter calculated with phantoms of varying voxel size *Radiat. Prot. Dosim.* **49** 353–56
[111] Zankl M 1993 Computational models employed for dose assessment in diagnostic radiology *Radiat. Prot. Dosim.* **49** 339–44
[112] Kalender W A, Schmidt B, Zankl M and Schmidt M 1999 A PC program for estimating organ dose and effective dose values in computed tomography *Eur Radiol.* **9** 555–62
[113] Zankl M, Panzer W, Petoussi-Henb and Drexler G 1995 Organ doses for children from computer tomographic examinations *Radiat. Prot. Dosim.* **57** 393–96
[114] Caon M, Bibbo G and Pattison J 1999 An EGS4-ready tomographic computational model of a 14-year-old female torso for calculating organ doses from CT examinations *Phys. Med. Biol.* **44** 2213–25
[115] Nipper J C, Williams J L and Bolch W E 2002 Creation of two tomographic voxel models of paediatric patients in the first year of life *Phys. Med. Biol.* **47** 3143–64
[116] Staton R J and Bolch W E 2004 Implementation of a newborn tomography computational model for dose assessment in pediatric radiology [Abstract] *Med. Phys.* **31** 1757
[117] Staton R J, Pazik F D, Nipper J C, Williams J L and Bolch W E 2003 A comparison of newborn stylized and tomographic models for dose assessment in pediatric radiology *Phys. Med. Biol.* **48** 805–20
[118] Staton R J, Williams J L, Arreola M M, Hintenlang D E and Bolch W E 2007 Organ and effective doses in infants undergoing upper gastrointestinal (UGI) fluoroscopic examination *Med. Phys.* **34** 703–10
[119] Pazik F D, Staton R J, Hintenlang D E, Arreola M M, Williams J L and Bolch W E 2007 Organ and effective doses in newborns and infants undergoing voiding cystourethrograms (VCUG): a comparison of stylized and tomographic phantoms *Med. Phys.* **34** 294–306
[120] Lee C, Williams J, Lee C and Bolch W 2005 The UF series of tomographic computational phantoms of pediatric patients *Med. Phys.* **32** 3537 48
[121] ICRP, ICRP Publication 143 2020 Paediatric reference computational phantoms *Ann. ICRP.* **49** 1–299
[122] Bolch W E, Petoussi-Henss N, Paquet F and Harrison J 2016 ICRP dose coefficients: computational development and current status *Ann. ICRP.* **45** 156–77
[123] ICRP, ICRP Publication 128 2015 Radiation Dose to Patients from Radiopharmaceuticals – A Compendium of Current Information Related to Frequently Used Substances *Ann. ICRP.* **44** 1–321
[124] ICRP, ICRP Publication 110 2009 Adult reference computational phantoms *Ann. ICRP.* **39** 1–165
[125] ICRP, ICRP Publication 116 2010 Conversion coefficients for radiological protection quantities for external radiation exposures *Ann. ICRP.* **40** 1–257
[126] ICRP, ICRP publication 123 2013 Assessment of radiation exposure of astronauts in space *Ann. ICRP.* **42** 1–339
[127] ICRP, ICRP Publication 132 2016 Radiological protection from cosmic radiation in aviation *Ann. ICRP.* **45** 1–48
[128] ICRP, ICRP Publication 144 2020 Age-dependent dose coefficients for external exposures to environmental sources *Ann. ICRP.* **49** 1–147

[129] ICRP, ICRP Publication 133 2016 The ICRP computational framework for internal dose assessment for reference adults: specific absorbed fractions *Ann. ICRP.* **45** 1–74
[130] ICRP, ICRP Publication 130 2015 Occupational Intakes of Radionuclides *Part 1. Ann ICRP.* **44** 1–188
[131] ICRP, ICRP Publication 134 2016 Occupational Intakes of Radionuclides *Part 2. Ann ICRP.* **45** 1–352
[132] ICRP, ICRP Publication 137 2017 Occupational intakes of radionuclides *Part 3. Ann ICRP.* **46** 1–486
[133] ICRP, ICRP Publication 141 2019 Occupational intakes of radionuclides *Part 4. Ann ICRP.* **48** 1–497
[134] Akhavanaliaf A, Xie T and Zaidi H 2019 Development of a library of adult computational phantoms based on anthropometric indexes *IEEE Trans. Med. Imaging.* **3** 65–75
[135] Lee C, Lodwick D, Hurtado J, Pafundi D, Williams J L and Bolch W E 2010 The UF family of reference hybrid phantoms for computational radiation dosimetry *Phys. Med. Biol.* **55** 339–63
[136] IAEA 1998 *Compilation of anatomical, physiological, and metabolic characteristics for Reference Asian Man.* (Vienna, Austria: Int. Atomic Energy Agency)
[137] Zhang B, Ma J, Liu L and Cheng J 2007 CNMAN: a Chinese adult male voxel phantom constructed from color photographs of a visible anatomical data set *Radiat. Prot. Dosim.* **124** 130–6
[138] Kim C H, Choi S H, Jeong J H, Lee C and Chung M S 2008 HDRK-Man: a whole-body voxel model based on high-resolution color slice images of a Korean adult male cadaver *Phys. Med. Biol.* **53** 4093–106
[139] Lee C, Nagaoka T and Lee J-K 2006 Implementation of Japanese male and female tomographic phantoms to multi-particle Monte Carlo models for ionizing radiation dosimetry *J. Nucl. Sci. Tech.* **43** 937–45
[140] Rajon D A, Jokisch D W, Patton P W, Shah A P and Bolch W E 2000 Voxel size effects in three-dimensional nuclear magnetic resonance microscopy performed for trabecular bone dosimetry *Med. Phys.* **27** 2624–35
[141] Shah A P, Bolch W E, Rajon D A, Patton P W and Jokisch D W 2005 A paired-image radiation transport model for skeletal dosimetry *J. Nucl. Med.* **46** 344–53
[142] Werner C J *et al* 2018 *MCNP Version 6.2 Release Notes LA-UR-18–20808* (Los Alamos, NM: Los Alamos National Laboratory)
[143] Collaboration G 2017 *Introduction of GEANT4 Release 10.4* (Geneva, Switzerland: CERN) (https://geant4.web.cern.ch/support/user_documentation)
[144] Brown J L, Furuta T and Bolch W E 2019 A robust algorithm for voxel-to-mesh phantom conversion *Brain and Human Body Modeling: Computational Human Modeling at EMBC 2018* ed S Markarov, G Noetscher and M Horner (Zurich: Springer Nature) ch 16
[145] Zhang J, Na Y H, Caracappa P F and Xu X G 2009 RPI-AM and RPI-AF, a pair of mesh-based, size-adjustable adult male and female computational phantoms using ICRP-89 parameters and their calculations for organ doses from monoenergetic photon beams *Phys. Med. Biol.* **54** 5885–908
[146] Segars W P 2001 Development and application of the new dynamic NURBS-based cardiac-torso (NCAT) phantom *Biomedical Engineering.* (Chapel Hill, NC: University of North Carolina)

[147] Segars W P *et al* 2013 Population of anatomically variable 4D XCAT adult phantoms for imaging research and optimization *Med. Phys.* **40** 043701

[148] Stepusin E J 2016 Pre-Computed Monte Carlo Dosimetry and Reporting for Computed Tomography (Department of Biomedical Engineering, Unversity of Florida)

[149] Marshall E, Rajderkar D, Brown J, Stepusin E, Borrego D and Bolch W 2019 A scalable database of organ doses for common diagnostic fluoroscopy examinations of children: procedures of current practice at the University of Florida *Phys. Med. Biol.* **64** 134023

[150] Sgouros G, Frey E C, Bolch W E, Wayson M B, Abadia A F and Treves S T 2011 An approach for balancing diagnostic image quality with cancer risk: application to pediatric diagnostic imaging of ^{99m}Tc-dimercaptosuccinic acid *J. Nucl. Med.* **52** 1923–9

[151] Geyer A M, O'Reilly S, Lee C, Long D J and Bolch W E 2014 The UF/NCI family of hybrid computational phantoms representing the current US population of male and female children, adolescents, and adults—application to CT dosimetry *Phys. Med. Biol.* **59** 5225

[152] Gosselin M C *et al* 2014 Development of a new generation of high-resolution anatomical models for medical device evaluation: the Virtual Population 3.0 *Phys. Med. Biol.* **59** 5287–303

[153] Ding A, Mille M M, Liu T, Caracappa P F and Xu X G 2012 Extension of RPI-adult male and female computational phantoms to obese patients and a Monte Carlo study of the effect on CT imaging dose *Phys. Med. Biol.* **57** 2441–59

[154] ICRP, ICRP Publication 145 2020 Adult mesh-type reference computational phantoms *Ann. ICRP.* **49** 1–203

[155] Choi C, Yeom Y S, Lee H, Han H, Shin B, Nguyen T T and Kim C H 2020 Body-size-dependent phantom library constructed from ICRP mesh-type reference computational phantoms *Phys. Med. Biol.* **65** 125014

[156] Choi C *et al* 2021 Development of skeletal systems for ICRP pediatric mesh-type reference computational phantoms *J. Radiol. Prot.* **41** 139

[157] Choi C, Shin B, Yeom Y S, Nguyen T T, Han H, Ha S, Chung B S, Bolch W E and Kim C H 2021 Development of pediatric mesh-type reference computational phantom series of int. commission on radiological protection *J. Radiol. Prot.* **41** S160

[158] Carter L M, Camilo Ocampo Ramos J, Bolch W E, Lewis J S and Kesner A L 2021 Technical note: patient-morphed mesh-type phantoms to support personalized nuclear medicine dosimetry – a proof of concept study *Med. Phys.* **48** 2018–26

IOP Publishing

Monte Carlo Calculations in Nuclear Medicine (Second Edition)
Therapeutic applications
Habib Zaidi

Chapter 6

Radiobiology aspects and radionuclide selection criteria in cancer therapy

Amin I Kassis

6.1 Introduction

The deposition of energy by ionizing radiation is a random process. Following energy absorption by cells, certain molecular modifications are induced that may lead to cell death and other radiologic effects. Despite the fact that this process is stochastic in nature, the death of a few cells, in general, within a tissue or an organ will not have a significant effect on its function. As the dose increases, more cells will die and eventually organ/tissue function will be impaired.

6.2 Radiobiologic effects

6.2.1 Molecular lesions

Damage to the genome of cells is the basis for most radiation effects which display themselves generally as single- and double-strand breaks within DNA (SSB and DSB, respectively), base damage, and cross-links between DNA strands and nuclear proteins [1]. Most cells types, however, are equipped with a host of enzymes that are efficient at repairing damage to DNA. To aid in their function, such cells are also outfitted with machinery that holds up the mitotic cycle until repairs can be made. Repair of DNA damage from low-linear-energy-transfer (low-LET) radiations (photons and energetic electrons) is very efficient and sensitive to dose rate [2]. On the other hand, the repair of damage from high-LET densely ionizing radiation (alpha particles and Auger effects in DNA) is more complex.

Another manifestation of damage to DNA is the appearance of chromosomal and chromatidal aberrations. Chromosome-type aberrations are induced in G_1-phase chromosomes or in unreplicated regions of S-phase chromosomes. Chromatid-type aberrations are induced in replicated regions of S-phase

doi:10.1088/978-0-7503-2694-0ch6

chromosomes and/or in G_2-phase chromosomes. The shape of the dose–effect relationship observed for chromosomal aberrations from low-LET radiation follows the general formula $\alpha D + \beta D^2$. With increasing LET, the relationship goes from linear–quadratic to linear. Lowering the dose rate also changes the low-LET aberration curve from the quadratic type to one increasingly more linear.

6.2.2 Cellular responses

6.2.2.1 Clonal survival

Several *in vitro* assays have been described to measure the ability of cells to proliferate. In practice, these assays measure the capacity of cells to successfully reproduce and, thus, to form a colony. Using a colony-forming assay, it is possible to determine the decrease in survival, expressed as a surviving fraction, as a function of a graded radiation dose.

Radiation survival curves are log–linear plots of surviving fraction (log) versus dose (linear). The shape of the survival curve constructed through such a set of survival points varies and will depend on certain biologic, physical, and chemical factors. In general, two types of dose–survival curves have been described (figure 6.1). For the exponential survival curve, the slope is always constant and can be expressed by the following equation:

$$S/S_0 = e^{\alpha D} \tag{6.1}$$

where S/S_0 is the surviving fraction of the irradiated cells, D is the dose delivered, and $\alpha = 1/D_0$, the dose needed to reduce survival to 0.37 (for this type of curve, D_0 is

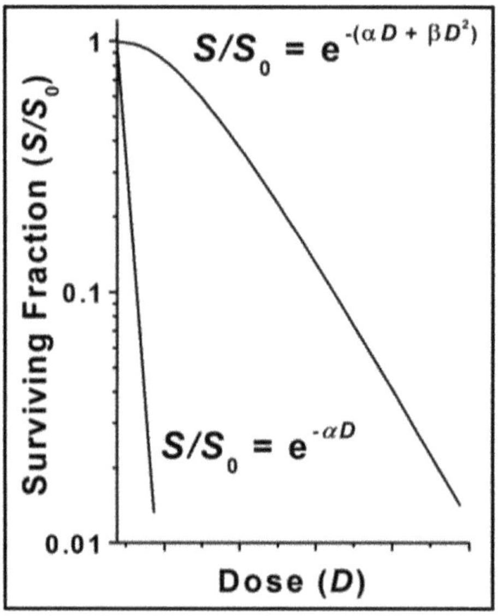

Figure 6.1. Dose–response survival curves for mammalian cells *in vitro*.

also known as D_{37}). Such exponential survival curves are observed when mammalian cells are exposed to 'densely ionizing' radiation (e.g., alpha-particle emissions, DNA-incorporated Auger electron emitters).

In the second type of dose–response relationship, expressed by the sigmoidal survival curve (figure 6.1), the efficiency of cell kill is not constant: at low doses, a slow decrease in survival is observed and the curve has a shoulder; at higher doses, an exponential decrease in survival is seen. This type of survival curve occurs routinely when mammalian cells are exposed to sparsely ionizing radiation (e.g., x-rays, beta particles, extranuclear Auger electrons). Such curves can be fitted using a second order polynomial equation:

$$S/S_0 = \alpha D + \beta D^2 \tag{6.2}$$

where α and β are the fit parameters and solving for D when $S/S_0 = 0.37$ yields the D_{37} value. Alternatively, and preferably, the curve can be fitted using a linear–quadratic equation:

$$S/S_0 = e^{-(\alpha D+\beta D^2)} \tag{6.3}$$

where α equals the rate of cell kill by a single-hit mechanism, D is the dose delivered, β equals the rate of cell kill by a double-hit mechanism, and solving for D when $S/S_0 = 0.37$ yields the D_{37} value. Rewriting equation (6.3) provides a method for calculating the rate constants α and β by graphic linearization [3]:

$$\frac{\ln(S/S_0)}{D} = -\alpha - \beta D \tag{6.4}$$

When $\ln(S/S_0)/D$ is plotted against dose D, a linear relationship is obtained in which the y intercept with zero dose is the linear inactivation constant α and the slope of the line equals the quadratic inactivation constant β. Note that the reciprocals of α and $\sqrt{\beta}$ equal, respectively, $D_{0\alpha}$ and $D_{0\beta}$, the latter representing the D_0 value of the exponentially decreasing portion of the curve. Such plots of cell inactivation can assist in making independent components readily apparent, and they have significantly improved the fit of data for mammalian cell survival curves and the resolution of problems associated with the conceptual difficulty of zero slope with zero dose. Note that if radiation is protracted or the dose rate is low, as often occurs with radionuclides, the α term predominates.

6.2.2.2 Division delay

Dividing cells pass through four phases that are based on two observable events, mitosis (M) and DNA synthesis (S). Two gap phases that occur before and after the S phase are known as G_1 and G_2, respectively. The delay in the progression of dividing cells through their cell cycle following irradiation is a well-known phenomenon. Usually it is reversible and dose-dependent, occurs only at specific points in the cell cycle, and is similar for both surviving and non-surviving cells. Cells undergoing mitosis (M phase) continue through division basically undisturbed; those in the G_1 phase of the cell cycle have very little delay, those in the S phase a moderate delay, and those in the G_2 phase maximum delay. The net result is that many cells accumulate at the G_2/M boundary, and the ratio of mitotic cells to nonmitotic cells (i.e., the mitotic index) is altered. The length of the delay and the decrease in mitotic index are both functions of dose.

6.2.2.3 Redistribution

The radiosensitivity of cells is a function of their position within the cell cycle. Cells in late S phase are most resistant (the survival curve following γ–irradiation has a wide shoulder), while those in the G_2 and M phases are most radiosensitive (the survival curves have a steep slope and no shoulder). Consequently, following irradiation, the cells in the most sensitive phase will be killed preferentially. This redistribution of cells will lead to partial cell-cycle synchrony and a change in the overall radiosensitivity of the cell cohort. One would expect a more radioresistant cell population, but in reality the population rapidly becomes desynchronised and the net effect is sensitization of the surviving population.

6.2.2.4 Repair

Mammalian cells are generally capable of repairing some of the damage induced by radiation. This phenomenon is dose-rate dependent. As the dose rate decreases, the capacity of cells to repair radiation damage increases and this is manifested by a widening of the shoulder of the survival curve accompanied by an increase in D_0 (e.g., x-ray irradiation). In essence, two types of repair have been described. The first, sublethal-damage (SLD) repair, occurs when a dose of radiation is split into two fractions and sufficient time is allowed (0.5–2 h) for any/all of the radiation-induced damage to be repaired. Naturally, if no repair is allowed to occur, for example, by the immediate application of a second dose of radiation, the cells will die. Sublethal damage and its repair have been shown to be important factors in the sparing of normal tissues during fractionated radiotherapy. The second, potentially lethal damage (PLD) repair, is observed only when mammalian cells are grown under suboptimal conditions following a single dose of radiation. Under such circumstances, an increase (rather than a decrease) in survival is observed. This phenomenon is believed to be a result of the delayed entry of the irradiated cells into mitosis, thereby allowing the cells to repair the PLD. Both SLD and PLD as well as their repair have been reported only for x-ray and gamma irradiation (i.e., low-LET-type survival curves that have considerable shoulders) and are practically non-existent for neutron and alpha-particle irradiation (i.e., high-LET radiation).

6.2.2.5 Oxygen effect

Oxygen radiosensitizes mammalian cells to the damaging effects of radiation. Consequently, hypoxic cells can be as much as threefold more radioresistant than well-oxygenated cells. It is thought that following irradiation, oxygen enhances free radical formation and/or blocks the reversal of certain reversible chemical alterations that have occurred. Here again, it is important to note that the oxygen effect is greatest for photons and high-energy beta particles (low-LET-type survival curves) and is practically absent for alpha particles and neutrons (high-LET-type survival curves).

6.2.3 Tissue responses

6.2.3.1 Heterogeneity of response

Following irradiation, the response of an organ/tissue/tumour will depend to a large degree on its inherent radiosensitivity. In the clinical situation, the term maximum tolerated dose (MTD) is used to indicate the highest dose that a normal organ can withstand. Once the MTD has been exceeded, the patient will exhibit the particular signs and symptoms associated with dysfunction of the organ being irradiated.

The MTD to irradiation of assorted tissues and organs is variable. The tolerance doses for the gonads and bone marrow are low (100–200 cGy); those for the intestine, kidneys, and heart are moderate (2000–4500 cGy); and those for mature bone and cartilage, the bladder, and the central nervous system are high (5000–7000 cGy). These doses are for external beam therapy. With radionuclides, where the dose rate is much lower, the MTD may be somewhat higher. Consequently, knowledge of the MTD for the various organs and/or tissues being irradiated is important and can help to predict with some certainty the risks associated with a radiotherapeutic dose.

6.2.3.2 Protraction and fractionation effects

Radiotherapists have long realized that dividing the radiation dose into daily or weekly fractions is more effective in eradicating tumours than giving a single dose, while reducing the undesirable effects in normal tissues. The size and number of the fractions, the treatment time, and the total dose given depend mainly on the radiosensitivity of the tumour being eliminated and the tolerance of the surrounding normal tissues and organs. Since the response of a tumour to irradiation is determined in part by the depopulation of its dividing cells, it is expected that tumour cells react to dose fractionation in a fashion similar to that of acutely responding rather than late responding normal tissues, i.e., have high α/β ratios (higher ratios indicate that a lower proportion of the damage will be repaired). In fact, Thames and Hendry [4] and Williams *et al* [5] have reported significant differences between the α/β ratios for tumours and normal tissues. These authors have observed that while growing tumours exhibit a predominant α/β ratio of 10 to 20, normal tissues show a central tendency around 3 to 5. To the extent that such animal data are relevant to human tumours, it is not surprising that dose fractionation favours tumour control and minimizes normal tissue damage.

6.2.3.3 Bystander effects

Many studies have demonstrated that a radiobiologic phenomenon termed ‘bystander effect’ can be observed in mammalian cells grown and irradiated *in vitro*. Bystander damage describes biologic effects, originating from irradiated cells, in cells not directly affected by radiation-induced ionizations. Investigators have reported that cells display lower survival and a higher rate of genetic changes than would be predicted from direct-ionization-only models [6–12] when a small fraction

Table 6.1. Physical characteristics of therapeutic radionuclides.

Decay mode	Particles	$E_{(min)}$–$E_{(max)}$	Range	LET[a](keV μm^{-1})
α^{++}	He nuclei	5–9 MeV[b]	40–100 μm	~50–100
β^{-}	Energetic electrons	0.05–2.3 MeV[b]	0.05–12 mm	~0.2
EC/IC[d]	Non-energetic electrons	eV–keV[c]	2–500 nm	~4–26

[a] Along most of the track.
[b] Monoenergetic.
[c] Average (>1% intensity), continuous distribution of energy.
[d] EC = electron capture; IC = internal conversion.

of a cell population growing *in vitro* has been traversed by α-particles. Evidence from these reports challenges the past half-century's central tenet that radiation conveys damage to DNA either through direct ionization or indirectly via, for example, hydroxyl radicals produced in water molecules in the immediate vicinity of DNA.

6.2.4 Radiation quality

Beta particle, alpha particle, and Auger electron emitters are three types of radionuclide that can be used for tumour therapy (table 6.1). Particles from these radioactive emissions produce tracks along which energy is transferred to and deposited in biologic matter. The intensity of the energy transfer varies and depends on the energy, charge, and mass of the traversing particle. The term linear energy transfer (LET) describes the transfer of energy (e.g., keV) along the track traversed (e.g., micrometer) by the particle. The LET of beta particle emitters, such as iodine-131 and yttrium-90, is low and, depending on their energy, the particles traverse several millimetres. On the other hand, the LET of alpha particles emitted from astatine-211 and bismuth-213 and of Auger electrons emitted by iodine-125 is high. Actually, for Auger emitters, the electrons produce ionizations that are clustered around the point of decay (i.e., they are not along a linear track). While the tracks of alpha particles are several cell diameters in length, those of Auger cascade electrons are localized at the site of decay in a sphere with radius of several nanometers. The tracks at low LET are sparsely ionizing and many (thousands) are needed to produce a detectable biologic effect (e.g., cell death). At high LET the distance between ionizations becomes shorter and the tracks are dense so that these particles are much more efficient at producing lethal effects. For example, several alpha-particle traversals (fewer than five) through a cell nucleus or approximately 50 to 100 DNA-incorporated iodine-125 decays are sufficient to sterilize a cell.

6.3 Targeting principles in radionuclide therapy

6.3.1 The radionuclides

6.3.1.1 Alpha particle emitters

An alpha particle that is emitted by any of the radionuclides suitable/available for therapy (table 6.2) is identical to a helium nucleus with energies ranging from 4 to 9 MeV. The corresponding tissue ranges are 40 to 100 μm. Alpha particles travel

Table 6.2. Physical characteristics of alpha particle emitters.

Radionuclide	Half-life	$_{A}{}^{a}$(MeV)	$_{A}{}^{b}$(μm)
^{211}At	7.2 h	6.79	60
^{212}Bi	60.6 min	7.80	75
^{213}Bi	45.7 min	8.32	84
^{223}Ra	11.43 d	5.64	45
^{225}Ac	10 d	6.83	61

[a] Mean energy ($_A$) of alpha particles emitted per disintegration [15].
[b] Mean range ($_A$) of alpha particles calculated using second order polynomial regression fit [16]:

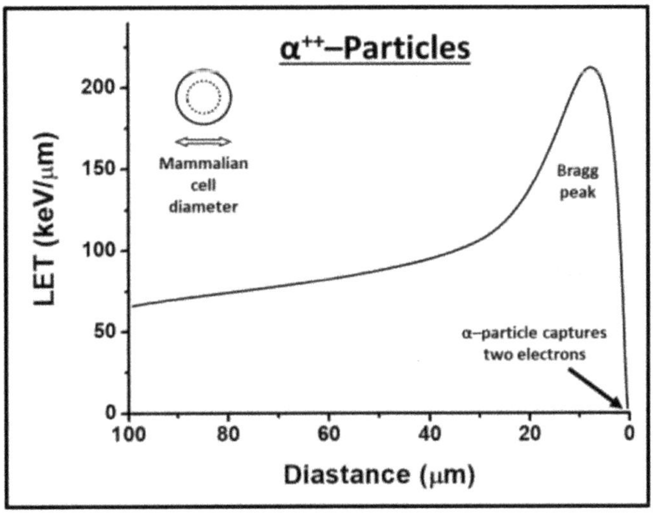

Figure 6.2. Ionization density along path of α-particle as function of traversed distance.

in straight lines, initially depositing their energy, in soft tissue, at approximately 80–100 keV μm^{-1} and increasing their rate of energy deposition (to ~300 keV μm^{-1}) toward the end of their tracks (Bragg peak, figure 6.2). Consequently, in the case of cell irradiation, the therapeutic efficacy of alpha-particle emitters depends on: (i) distance of the decaying atom from the targeted mammalian cell *nucleus*—vis-à-vis the probability of a nuclear traversal (figure 6.3); and (ii) role of heavy ion recoil of the daughter atom, in particular when the alpha-particle emitter is covalently bound to nuclear DNA [13]. Of equal importance are the contribution(s) from bystander effects and the magnitude of cross-dose (from radioactive sources associated with one cell to an adjacent/nearby cell—see below) as this will vary considerably depending on the size of the labelled cell cluster and the fraction of cells labelled [14] (figure 6.3).

$R = 3.87E + 0.75E^2 - 0.45$, where R is the range (μm) in unit density matter and E is the alpha-particle energy (MeV).

The typical energy deposition from an alpha particle traversing the diameter of a cell nucleus is about 600 keV; this translates to an absorbed dose of about 0.14 Gy to

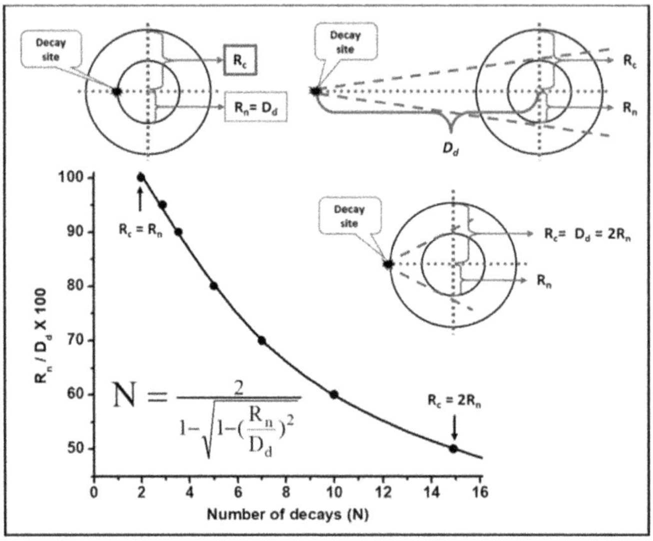

Figure 6.3. Number of radioactive atoms required to ensure traversal of cell nucleus by 1 energetic particle as function of distance of decaying atom from centre of cell (D_d). Nuclear radius (R_n) to D_d (%) is plotted as function of number of decays (*N*). Note, (a) nuclear localization of radioactive atom only will lead to 1 traversal per decaying atom; (b) when decaying atom is on nuclear membrane, >2 radioactive atoms are needed for one nuclear traversal; and (c) when decaying atoms is localized on cell membrane and diameter of cell is 2× that of nucleus, >15 radioactive atoms are necessary to ensure one nuclear traversal. Rc: cell radius.

the cell nucleus per hit. These numbers depend, among other parameters, on the size of the cell nucleus. Whereas the stochastics of energy distribution are unimportant for absorbed doses greater than 1 Gy for beta particles because each cell experiences thousands of individual particle tracks, the *average* specific energy deposited per unit mass (i.e., the absorbed dose) is not a suitable parameter for cells traversed by a few alpha particles. Several authors have thus described microdosimetric approaches for calculating the specific energy deposited in *individual cells* and predicting the response of cells to alpha-particle irradiation [17, 18]; others (e.g., [19]) have related the specific energy distributions to the fraction of cell survivors.

The magnitude of cross-dose (from radioactive sources associated with one cell to an adjacent cell) is an important factor when evaluating alpha particles for therapy. This will vary considerably depending on the size of the labelled cell cluster and the fraction of cells labelled [14].

Finally, when the alpha particle emitter is covalently bound to the nuclear DNA of tumour cells (i.e., in the case of cell self-irradiation), the contribution of heavy ion recoil of the daughter atom must also be considered in assessing the radiobiologic effects [13].

6.3.1.2 Energetic beta particle emitters

Current radionuclide therapy in humans is based almost exclusively on beta-particle-emitting isotopes. Typically, the electrons that are emitted from the nuclei of the decaying atoms (1 electron/decay) have various energies up to a maximum and thus have a distribution of ranges (table 6.3). As each particle emitted traverses matter, it follows a contorted path, loses its kinetic energy, and eventually comes to a stop. Since the LET of these light, charged particles is very low (0.2 keV μm^{-1}), *except for the few nanometers at the end of their range just before they stop* (figure 6.4), they are sparsely ionizing and quite inefficient at damaging DNA and killing cells. Consequently, their use as therapeutic agents predicates the presence of high radionuclide concentrations within the targeted tissue and the traversal of several thousand electrons per mammalian cell nucleus.

Table 6.3. Physical characteristics of beta particle emitters.

Radionuclide	Half-life	$\bar{\beta}^{-}_{(min)}{}^{a}$(keV)	$\bar{\beta}^{-}_{(min)}{}^{b}$(mm)	$E_{\beta^{-}(max)}{}^{a}$(keV)	$R_{\beta^{-}(max)}{}^{b}$(mm)
^{33}P	25.4 d	77	0.09	249	0.63
^{169}Er	9.4 d	99	0.14	350	1.1
^{177}Lu	6.7 d	133	0.23	497	1.8
^{67}Cu	61.9 h	141	0.26	575	2.1
^{131}I	8.0 d	182	0.39	610	2.3
^{153}Sm	46.8 h	224	0.54	805	3.3
^{109}Pd	13.5 h	361	1.1	1028	4.5
^{186}Re	3.8 d	349	1.1	1077	4.8
^{89}Sr	50.5 d	583	2.2	1491	7.0
^{32}P	14.3 d	695	2.8	1710	8.2
^{188}Re	17.0 h	764	3.1	2120	10.4
^{90}Y	64.1 h	935	4.0	2284	11.3

[a] Mean minimum energy of least energetic β^--particle emitted per disintegration (>1% intensity);
[b] Range in water of β^--particle ($E = 0.02$–100 keV) calculated using Cole's equation [20]: $R = 0.043(E + 0.367)^{1.77} – 0.007$, and for energetic ($E$ = 0.1–0.5 MeV) β^--particle calculated using second order fits [21]: $R = 5.3E + 0.0034E^2 – 0.93$;
[b] Maximum energy emitted/particle.

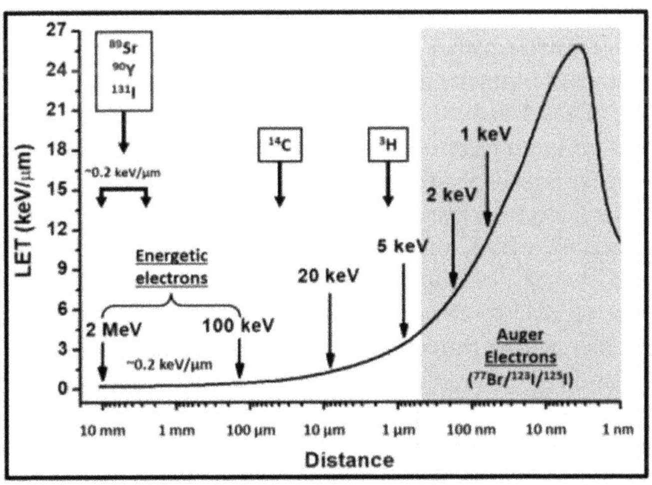

Figure 6.4. Rate of energy loss of electrons as a function of traversed signal.

An important ramification of the long range (mm) of each emitted electron is the production of cross-fire, a circumstance that negates the need to target every cell within the tumour. For microscopic disease, however, long-range and some medium-range emitters may deposit a significant fraction of the energy of their particles outside the tumour [22]. This is particularly of concern in the selection of a radiopharmaceutical for the palliation of bone pain from metastatic osseous lesions [23]. Long-range emitters, (e.g., yttrium-90) run the risk of significantly irradiating bone marrow as well as bony lesions, whereas short-range emitters (e.g., phosphorus-33) are calculated to result in a significantly lower dose to bone marrow relative to bone/bony lesions.

The matter of inhomogeneity in the distribution of a radionuclide and its consequent dose has also been addressed. O'Donoghue [24] has derived a method for calculating an equivalent uniform biologically effective dose based on the absorbed-dose distribution represented by the biologically effective dose–volume histogram. For larger tumours he postulates that a combination of radionuclide and external beam therapy may result in optimal dose distribution.

Many of the beta-particle-emitting radionuclides used for therapy also release gamma photons that generally do not add significantly to the dose delivered to the target tissue. However, these photons may contribute considerably to the whole-body dose. For example, 3.7 GBq of iodine-131 distributed throughout the whole body would deposit approximately 60 cGy per day. Since the bone marrow is usually the dose-limiting organ (200–300 cGy), the success or failure of therapy will depend on not exceeding its MTD.

6.3.1.3 Low-energy Auger electron emitters

During the decay of certain radioactive atoms, a vacancy is formed (most commonly in the K shell) as a consequence of electron capture (EC), with the prompt emission of a neutrino from the atomic nucleus and/or internal conversion (IC). Such vacancies are rapidly filled by electrons dropping in from higher shells. As a result of nuclear rearrangements, some Auger electron emitters also emit a gamma photon that may itself be converted to a nuclear vacancy resulting in a second shower of Auger electrons (e.g., iodine-125). Similarly, other metastable nuclei (e.g., technetium-99m, bromine-80m) emit a gamma photon, which may or may not be converted into an electron vacancy (only 3% of the 140 keV gamma photons of technetium-99m are converted). This process, creation of an electron vacancy within a shell and its filling up, leads to a cascade of atomic electron transitions that move the vacancy towards the outermost shell. Each inner-shell electron transition results in the emission of a characteristic x-ray photon or an Auger, Coster–Kronig, or super Coster–Kronig monoenergetic electron (collectively called Auger electrons). Typically, an atom undergoing EC and/or IC emits: (1) multiple (e.g., 5-to-50) electrons with energies ranging from a few eV to approximately 80 keV whose range in water is from a fraction of a nanometer up to a few micrometers; and (2) in some cases (e.g., iodine-125, platinum-195m), as many as 20 to 33 Auger electrons are emitted per decay on average (table 6.4). It is important to note that these low energy, light, negatively charged particles travel in contorted paths. However, unlike energetic electron emitters (e.g., ^{131}I, ^{90}Y), whose linear energy transfer (LET) is low (~0.2 keV μm^{-1}) along most of their rather long linear path

Table 6.4. Physical characteristics of auger-electron emitters.

Radionuclide	Half-life	Average electron yield per decay		
		<0.5 μm range (%)	<100 nm range (%)	<2 nm range (%)
^{125}I	60.5 d	20 (98%)	18 (86%)	8 (39%)
^{123}I	13.3 h	11 (98%)	10 (89%)	5 (40%)
^{77}Br	57 h	7 (100%)	6 (95%)	3 (51%)
^{111}In	3 d	15 (98%)	14 (91%)	8 (53%)
^{195m}Pt	4 d	33 (92%)	33 (79%)	7 (19%)
	LET[a]:	4–26	9–26	<18

[a] Fit of data by Cole (*4*).

(mm-to-cm in tissue), consequent to the ionizations occurring sparingly, the LET of Auger electrons is 10–100 fold higher (~4–~25 keV μm^{-1}) especially at very low energies (figure 6.4) with the ionizations clustered within several cubic nanometers around the point of decay [25]. The ejection of these orbital electrons also leaves the decaying atoms transiently with a high positive charge and leads to the deposition of highly localized energy around the decay site and charge transfer [26–28].

The short range of Auger electrons requires their close proximity to radiosensitive targets for radiotherapeutic effectiveness. This is essentially a consequence of the precipitous drop in energy density as a function of distance in nanometers (e.g., figure 1 in Kassis *et al* [29] and figure 4 in Kassis *et al* [27]). These Auger electron emitters need to be located in the cell nucleus, close to or incorporated into DNA.

6.3.2 Half-life

For a radiopharmaceutical with an infinite residence time in a tumour or tissue, a radionuclide with a long physical half-life will deliver more decays than one with a short half-life if both radiopharmaceuticals have the same initial radioactivity. Moreover, there can be a striking difference in the time-dependent dose rate delivered by the two. If the number of radionuclide atoms per unit of tissue mass is n and the energy emitted (and absorbed) per decay is E, then the absorbed-dose rate is proportional to nE/T where T is the half-life. The ratio E/T is an important indicator of the intrinsic radiotherapeutic potency of the radionuclide [30]. In general, for biologic reasons (e.g., repair of SLD), higher dose rates delivered over shorter treatment times are more effective than lower dose rates delivered over longer periods. Thus, a radionuclide with a shorter half-life will tend to be more biologically effective than one with similar emission energy but a longer half-life. In addressing this complicated and multifaceted phenomenon, O'Donoghue and Wheldon [31] have noted that the relative effectiveness of iodine-125 and iodine-123, when the same number of atoms of each is bound to a tumour, depends on the tumour doubling time and the rate at which the radiopharmaceutical dissociates from the target. When both are very long, the longer-lived iodine-125 is theoretically more effective; otherwise the shorter-lived iodine-123 is preferred.

6.3.3 Radionuclide residence time

The biological half-life of the isotope (i.e. removal of the radioactive compound from the target area) must also be considered. As a first approximation to this effect, we [32] have shown that if the residence time of radioactivity in the tumour is T, then the percentage (P) of atoms decayed at the target can be estimated as:

$$P = 100\left(1 - \frac{N_T}{N_0}\right) = 100\left(1 - e^{\left(-0.693\left(\frac{T}{T_{1/2}}\right)\right)}\right) \tag{6.5}$$

where N_0 is the total number of atoms initially binding to the target, N_T is the remaining (un-decayed) atoms at time T, and $T_{1/2}$ is the physical half-life of the radionuclide. Therefore, for a certain percentage of atoms to decay at the target (e.g., 10%, 25%, 50%, 75% and 99%), the expected residence time of radioactivity can be calculated as:

$$T = (T_{1/2})\left(\frac{\ln(1 - P/100)}{-0.693}\right) \tag{6.6}$$

Our results (figure 6.5) show that for radioisotopes with a short half-life, such as ^{213}Bi, a residence time of less than 10 h would be sufficient for most atoms to decay at the target; for isotopes with a long half-life, such as ^{33}P, a significant proportion of radioactive atoms might leave the target prior to their decay. These findings reiterate

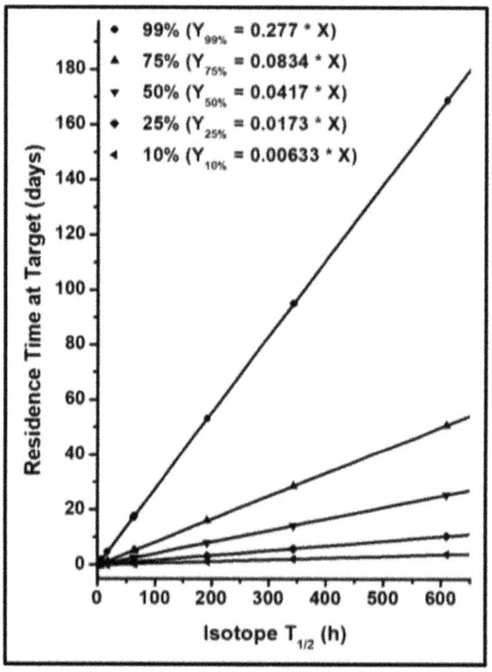

Figure 6.5. Required residence time for isotope to have 99%, 75%, 50%, 25% and 10% of atoms decayed at target. In equations, X is isotope half-life in hours, Y is required residence time in target in days.

the importance of matching the physical half-life of an isotope with that of the carrier molecule [33].

6.3.4 Choice of vector or ligand

The selection of a suitable carrier molecule rests on many factors. These include: (i) the biologic specificity and in vivo stability; (ii) the biologic mechanism(s) that bind it to target cells and the affinity of the carrier for these sites; (iii) the stability of the complex thus formed; (iv) the chemical properties of the carrier molecule which must permit the conjugation of a therapeutic radionuclide without degradation of the intrinsic characteristics of the molecule. Finally, the physical half-life of the radionuclide must be at least equal to, and preferably much longer than, the biologic half-life of the carrier molecule.

One of the simplest targeting agents is radioiodine, administered as iodide for treating functional cancer of the thyroid and, potentially, breast cancers that display the iodine transporter. In the palliation of metastatic bone pain, a simple cation, radiostrontium, has been used. Its targeting relies on the accretion of alkaline-earth cation congeners of calcium on bony surfaces. The oxides of radiophosphorus as orthophosphates and the somewhat more complex phosphor-nates labelled with radiorhenium and radiosamarium have also been proposed.

Short peptides labelled with yttrium-90 have been used to target neuroendocrine tumours [34]. Meta-[^{131}I]iodobenzylguanidine (M^{131}IBG), an iodinated neurotransmitter, is taken up by neural crest tumours [35]. Radiocolloid labelled with the alpha particle emitter astatine-211 has been successfully employed in treating ovarian ascites tumours in mice [36, 37].

Radioiodinated pyrimidine deoxyribose nucleosides (e.g., radio-iododeoxyuridine [IUdR]) have been used for the experimental treatment of ovarian, brain, and spinal cord tumours in animals [38–44]. Because, the carrier is incorporated into the DNA of dividing cells, it is most effective when labelled with Auger-electron-emitting radionuclides.

Successful therapy of lymphomas and leukemias has been achieved with intact antibodies and antibody fragments labelled with alpha- and beta-particle-emitting radionuclides [45]. Specificity in this case relies on tumour-associated antigens located on the malignant cell surface.

6.3.5 Properties of targets

Critical tumour characteristics include: (i) their accessibility; (ii) the number of carrier-molecule binding sites; and (iii) the distribution of binding sites among the targeted and non-targeted cells and their relationship to the cell cycle. The microscopic environment of the target, including tumour vascularity, vascular permeability, oxygenation, microscopic organization and architecture, is also important [46, 47].

To optimally target a radiopharmaceutical, the route of administration (e.g., intravenous, intralymphatic, intraperitoneal, intracerebral, intravesical, intrathecal, and intrasynovial) must also be considered. Some pathways may provide a mechanical means for maximizing tumour-to-nontumour ratios.

The specific activity (fraction of targeting molecules that are tagged with radioactive atoms) of the radiopharmaceutical should also be taken into account, especially when receptors can be saturated easily and, therefore, weaker binding nonspecifically to non-targeted cells can compete with the target. It should be noted that certain treatments are sometimes assisted by a mass effect, e.g., tumour uptake of radiolabeled antibodies can be enhanced by the addition of unlabelled immunoglobulin.

Tumour environment is yet another factor to be considered. A tumour is usually surrounded by—or adjacent to—normal tissue with normal vasculature. Therefore, a viable region in a tumour will depend on the normal vasculature or neovasculature or both for its supply of oxygen and nutrients. Figure 6.6 is a schematic showing the types of regions that are likely to be present within solid tumours [32]. Assuming that only tumour neovasculature is targeted, tumour cells within the neovasculature-dependent viable region (V_T in figure 6.6) can be eliminated by either anti-angiogenic agents or radionuclides. For regions that also depend on normal vasculature (V_{NT}), nonradioactive anti-angiogenic agents would not be effective; but if the neovasculature is targeted with radionuclides, such tumour cells could be killed so long as the range of the particulate emission(s) equals or exceeds the V_{NT} cell region and the energy deposited is cytocidal. However, for viable tumour tissue regions that depend solely on normal vasculature for survival (V_N region), anti-angiogenic agents (radioactive/nonradioactive) will never be curative. Consequently, if the goal is the deposition of a tumouricidal dose in every tumour cell, tumour-targeting radiopharmaceuticals are the only option. In practice, other factors should also be taken into account, such as the cross-irradiation dose contribution when long-range beta emitters are used, and the hypoxic conditions at tumour margins.

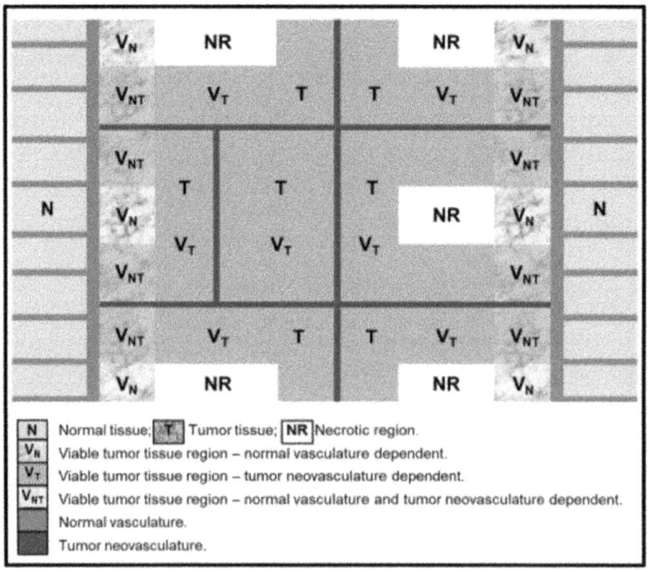

Figure 6.6. Schematic of tumour blood supply.

Cross-irradiation occurs if two neovasculatures are sufficiently close that radionuclides binding to one blood vessel could deposit considerable energy at the viable tumour region of the other vessel, leading to underestimation of radiation dose at viable tumour edges. For the viable tumour edge, which is at the end of the oxygen or nutrition diffusion range, the tumour cells might be in a hypoxic environment and relatively resistant to radiation, thereby requiring a higher radiation dose [48]. In these approximations, homogeneous radioactivity distributions were assumed for different diffusion ranges. In reality the bio-distribution is often complex and depends on the targeting mechanism of specific molecules. For tumour-targeting radiopharmaceuticals administered intravenously, the radial distribution around a vessel is more likely to be higher toward the lumen of the vessel, decreasing with the distance. While this will not affect the conclusions made with long-range radionuclides (i.e., energetic beta emitters), for shorter-range beta and alpha emitters a larger administered dose of radiopharmaceutical would be necessary for effective tumour-edge control with a tumouricidal endpoint.

6.4 Experimental therapeutics

6.4.1 Alpha particle emitters

The potential application of alpha-particle-emitting radionuclides as targeted therapeutic agents has been of interest for more than 25 years. If selectively accumulated in the targeted tissues (e.g., tumours), the decay of such radionuclides with their high LET and short range of a few (~5–10) cell diameters should result in highly localized energy deposition in the targeted tumour cells and minimal irradiation of surrounding normal host tissues. As a consequence of the LET of these particles (80 keV μm^{-1}), one to four traversals of alpha particles through a mammalian cell nucleus will kill the cell [13, 18, 49, 50]. In comparison, the LET of negatrons emitted by the decay of beta emitters is very much smaller (0.2 keV μm^{-1}) and, thus, approximately 20,000 beta particles must traverse a cell nucleus for its sterilization.

The investigation of the therapeutic potential of alpha particle emitters has focused on three radionuclides: astatine-211 (^{211}At), bismuth-212 (^{212}Bi), and bismuth-213 (^{213}Bi) (table 6.2). *In vitro* studies [18, 49] have demonstrated that the decrease in survival of mammalian cells after exposure to uniformly distributed alpha particles from these radionuclides is monoexponential (figure 6.7) but that, as predicted theoretically [51] and shown experimentally [13], these curves develop a tail when the dose is nonuniform (figure 6.7, insert). Such studies have shown that: (i) these alpha particle emitters are highly toxic [18, 49, 52–57]; (ii) cells in monolayer require more alpha-particle traversals than spherical cells [18, 49]; and (iii) only a few (~1–4) alpha-particle traversals through a mammalian cell nucleus are necessary to sterilize a cell [18, 49, 57].

At the molecular level, the traversal of alpha particles through a mammalian cell nucleus leads to the efficient production of chromosomal aberrations and DSB. For example, the incubation of mammalian cells with ^{211}At-astatide causes a significant increase in chomosomal aberrations [49], but these decline with the passage of time.

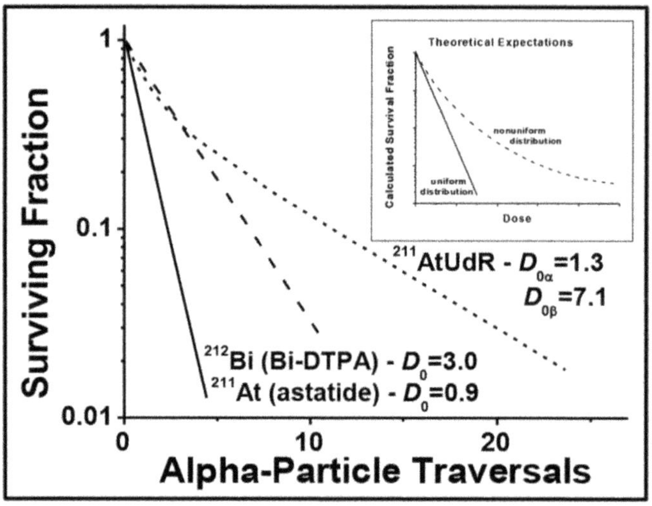

Figure 6.7. Survival of mammalian cells exposed in suspension to ^{211}At-astatide, ^{212}Bi-DTPA, or ^{211}AtUdR when 50% of cells are radiolabeled. Insert: theoretical expectations.

Interestingly, Walicka *et al* [13] have shown that more than 10 DSB are produced per decay of DNA-incorporated astatine-211, a value much higher than that obtained following the decay of the DNA-incorporated Auger electron emitter iodine-125 [58, 59].

Investigators have also assessed the therapeutic potential of alpha particle emitters in tumour-bearing animals [36, 37, 54, 60–68]. Bloomer *et al* [36, 37] have reported a dose-related prolongation in median survival when mice bearing an intraperitoneal murine ovarian tumour are treated with ^{211}At-tellurium colloid administered directly into the peritoneal cavity. While this alpha-particle-emitting radiocolloid is curative without serious morbidity, colloids of beta-particle-emitting radionuclides (phosphorus-32, dysprosium-165, yttrium-90) are not (figure 6.8). In another set of *in vivo* studies examining the therapeutic efficacy of a ^{212}Bi-labelled monoclonal antibody, the radionuclide is most effective when used with a carrier having target specificity [62]. In addition, McDevitt and co-workers [67] demonstrated that ^{225}Ac-labelled internalizing antibodies are therapeutically effective in mice bearing solid prostate carcinoma or disseminated lymphoma. The targeting ligand is radiolabelled with ^{225}Ac ($T_{1/2}$ = 10 d), a radionuclide that decays to stable bismuth-209 via a cascade of six daughters (francium-221, astatine-217, bismuth-213, polonium-213, thallium-209, lead-209) with the emission of five α- and three β-particles. While the proposed approach seems promising, the use of ^{225}Ac-labelled radiopharmaceuticals in humans is very likely to generate high normal-tissue toxicity (as a consequence of the α-particle recoil-energy-induced dissociation of the five chemically different radioactive atoms from the targeting carrier) post decay of the ^{225}Ac-labelled molecules that have not been internalized by the targeted tumour cells.

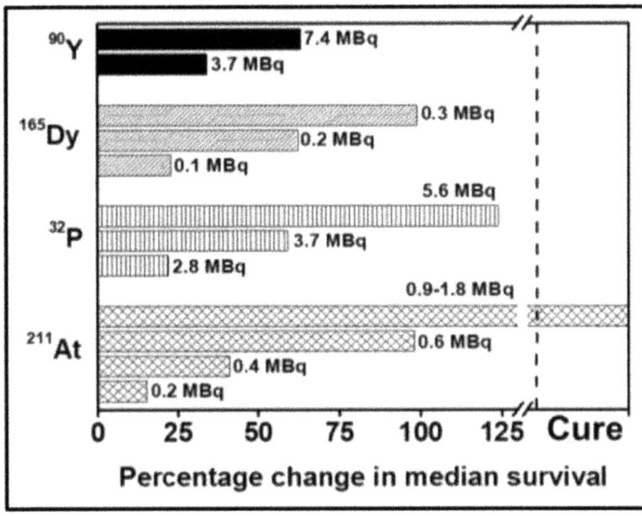

Figure 6.8. Percentage change in median survival of tumour-bearing mice treated with α- and β-emitting. From [37], reprinted with permission from AAAS.

6.4.2 Beta particle emitters

Investigators have assessed the *in vitro* toxicity of beta-particle-emitting radionuclides for many years. Early work, carried out by radiobiologists, concentrated on tritium and carbon-14, whereas later studies examined the therapeutic potential of more energetic beta particle emitters (table 6.3). In these studies (figure 6.9), the survival curves have a distinct shoulder [69–71]. Invariably, the D_0 calculated is several thousand decays. The decay of such radionuclides has been shown to lead to various molecular alterations (e.g., SSB, DSB, chromosomal aberrations), but only when the cells are exposed to very large numbers of decays (10,000–150,000) [70, 72–74].

Despite the rather low *in vitro* toxicity of beta-particle-emitting radionuclides, they continue to be pursued for targeted therapy. This is in part due to their availability and also to the long range of the emitted electrons, which can lead to the irradiation of all the cells that are within the maximum range and path of the particle (i.e., cross-fire). As mentioned above, the main advantage of cross-fire is that it negates the necessity of the radiotherapeutic agent's being present within each of the targeted cells, i.e., it permits a certain degree of heterogeneity. However, three factors will mainly determine whether the dose delivered to the targeted tissue will be therapeutically effective: (i) it is essential that the radiotherapeutic agent concentrate within foci throughout the targeted tissue; (ii) the distances between these hot foci must be equal to or less than twice the maximum range of the emitted energetic beta particles; and (iii) the concentration of the radiotherapeutic agent within each hot focus must be sufficiently high to produce a cumulative cross-fire dose to the surrounding targeted cells of approximately 10,000 cGy. Since the dose is inversely proportional to the square of the distance, it is important to note that the

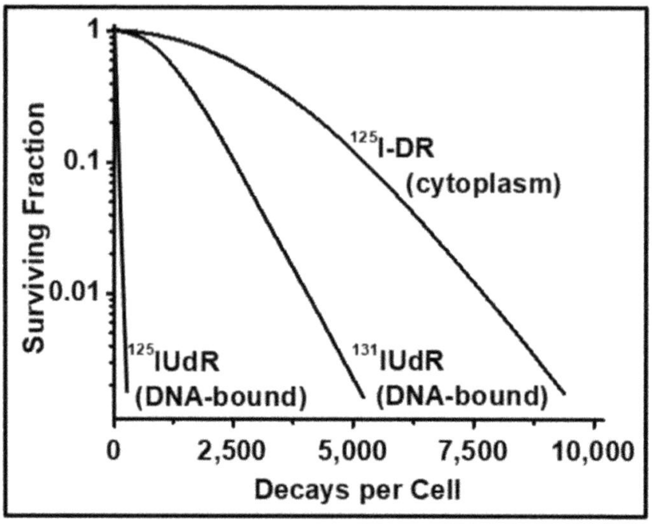

Figure 6.9. Survival of cells following exposure to decays from ^{125}IUdR or ^{131}IUdR (DNA-incorporated) or ^{125}I-DR (within cytoplasm).

concentration of the therapeutic agent needed to deposit such cytocidal doses will decrease precipitously when the distance between the hot foci decreases.

Experimentally, these predictions have been substantiated in various animal-tumour therapy studies. In many of these studies, radionuclides have been targeted after conjugation to monoclonal antibodies. For example, investigators have assessed the therapeutic efficacy of ^{131}I-labelled monoclonal antibodies in rodents bearing subcutaneous tumours. Although a substantial proportion of cells within a tumour mass show reduced or no expression of the targeted antigen and, therefore, are not targeted by the radioiodinated antibody, ^{131}I-labelled antibodies that localize in high concentrations in tumours are therapeutically efficacious and can lead to total tumour eradication in some instances [75–78]. Therefore, even when ^{131}I is not-so-uniformly distributed within a tumour, the decay of this radionuclide can lead to tumour sterilization as long as it is present in sufficiently high concentrations. Similar results have been reported with other beta-particle-emitting isotopes, in particular yttrium-90 [71, 79, 80] and copper-67 [81, 82].

Several 'two- and three-step' approaches have also been proposed to target radioactivity to a tumour [83–87]. In general, an antibody that is not internalized by the targeted cell is injected prior to the administration of a small radiolabelled molecule that has a strong affinity to the antibody. The most extensively studied approach utilizes the high avidity of avidin (Av) or streptavidin (SAv) for biotin, a vitamin found in low concentration in blood and tissues. Because the noncovalently bound Av/SAv–biotin complex has an extremely low dissociation constant (k_d) of about 10^{-15} M, investigators have used these two molecules as binding pairs to bridge molecules that have no affinities for each other and to target various radionuclides. In one study [87] examining the efficacy of this approach, 100% of tumour-bearing mice were cured of their disease when treated with a ^{90}Y-labelled biotin derivative.

6.4.3 Auger electron emitters

The toxicity to mammalian cells of radionuclides that decay by EC and/or IC has, for the most part, been established with the Auger-electron-emitting radionuclide iodine-125. Because of its predominant (93%) IC decay following EC, iodine-125 is a prolific emitter of Auger electrons (mean of ~20 per decaying atom [88–91]). Dosimetric calculations have predicted that the electrons most frequently produced by iodine-125 dissipate their energy in the immediate vicinity of the decaying atom and deposit 10^6 to 10^9 rad/decay within a two-nanometer sphere around the decay site [27, 89, 92]. Thus, the biologic effects are expected to depend critically on the proximity of the radionuclide to DNA.

The radiotoxicity of iodine-125 has been studied *in vitro* under a number of conditions including: (i) when the radioelement is incorporated into the DNA duplex as iododeoxyuridine (IUdR) [27, 70, 93–97]; (ii) when it is intercalated between the stacked bases of DNA as 3-acetamido-5-[^{125}I]iodoproflavine (A^{125}IP) [98, 99]; (iii) when it is bound to the minor and major grooves of DNA as an iodinated Hoechst dye [100]; (iv) as a radiolabelled triplex-forming oligonucleotide [101]; (v) when it is bound to transcription elements as iodinated tamoxifen and estrogen [102, 103]; and (vi) when it is located in the cytoplasm of cultured cells as iodorhodamine (^{125}I-DR) [28].

The decay of iodine-125 incorporated into cellular DNA as IUdR leads to an exponential decrease in clonal survival (figure 6.9). Similar responses are seen with the Auger electron emitters bromine-77 [29] and iodine-123 [104]. When plotted as a function of the total number of decays, the three curves have different slopes. If the dose to the cell nucleus is calculated, a single cell-survival curve is obtained (figure 6.10), suggesting that the radiation dose to the cell nucleus is a parameter that describes these observed biologic effects adequately. However, it has become apparent that the dose to the cell nucleus is not necessarily indicative of toxicity. For

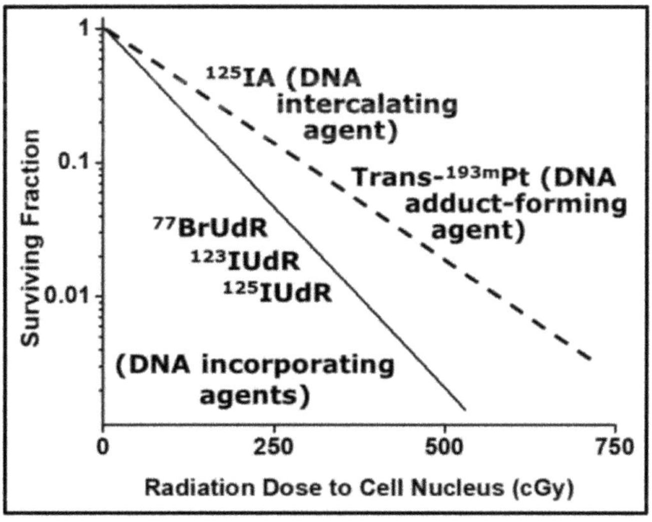

Figure 6.10. Survival fraction of cells plotted as a function of radiation dose to cell nucleus.

example, when the radiotoxicity of the DNA-intercalator ^{125}I-acridine [99] and that of the DNA-adduct-forming ^{195m}Pt-labelled *trans*-platinum [105] were assessed in mammalian cells and survival plotted as a function of radiation dose to the cell nucleus, the data points could all be fitted on a single monoexponential curve whose slope however was lower than that for the DNA-incorporated ^{125}I, ^{123}I, and ^{77}Br (figure 6.10). Other studies have also shown that when cells are incubated with the DNA-minor-groove-binder ^{125}I-Hoechst 33 342 [100] or treated with a triplex-forming ^{125}I-labelled synthetic oligonucleotide [101], the survival curves are linear–quadratic. In the latter case, the decays needed to decrease survival to the same degree was >400-fold greater in comparison to DNA-incorporated ^{125}I. These findings reiterate that nuclear localization (i) does not necessarily lead to monoexponential decrease in survival only, and (ii) is not always highly toxic to mammalian cells.

From these experiments, it has been concluded that the position of the Auger electron emitter with regard to DNA is a major determinant of radiotoxicity and that under certain circumstances decays result in a high-LET-like response. The amount of energy deposited in the nucleus is also important, and decays taking place in the cytoplasm are much less effective.

As mentioned above, internal emitters that undergo radioactive decay by electron capture and/or internal conversion result in the emission of a surge of low-energy (<1 keV) Auger electrons. Since many of these electrons traverse a very short distance (a few nm), the density of the hydroxyl radicals (•OH) generated will be very high around the decaying atom and will decrease drastically as a function of length of their tortuous path. Accordingly, when the radionuclide atoms are uniformly distributed in medium, their decay will result in the formation of randomly dispersed 'hot spots' (volumes densely traversed by electrons and occupied by •OH and other radicals) and 'cold spots' (volumes sparsely traversed by electrons and deficient in •OH and other radicals). Dosimetric calculations have also supported these expectations. For example, the decay of ^{125}I has been shown to lead to the deposition of a very high dose (~10^9 cGy/decaying atom) in the immediate vicinity (~2 nm^3) of the decay site and that there is a sharp and significant drop in the energy deposited (from ~10^9 cGy to ~10^6 cGy) as a function of increasing distance (few nm) from the decaying ^{125}I atom [25–29, 89, 91, 106, 107]. Consistent with these dosimetric expectations, early studies generally demonstrated that when ^{125}I atoms are positioned in close proximity to the naked DNA molecules (e.g., synthetic oligonucleotides, bacterial DNA, supercoiled plasmid DNA), either by the incorporation of an ^{125}I-labelled nucleoside analog into DNA or the binding of an ^{125}I-labelled DNA groove binder or intercalator, its decay leads to the efficient (~0.5–1 DSB/^{125}I decay) induction of DSB that are mainly caused by direct ionizations [108–114]. These findings are surprising since the •OH-radical scavenger DMSO had been shown to substantially (~70%) reduce the DSB yield in mammalian cell DNA post ^{125}I decay [58]. These latter studies suggest that the spatial organization of chromatin in mammalian cells supports conditions suitable for the induction of DSBs by the attack of free radical clusters at sites in DNA that is somewhat distant from the location of the ^{125}1 decay. Since the mean diffusion distance of •OH in water is ~6 nm [115], we proposed [58] that chromatin structure (highly packed DNA) provides conditions for the formation of >1 DSB per ^{125}I decay by indirect mechanism(s) and that a cluster of •OH produced by decay of ^{125}I may thus attack a DNA site present 100s of nucleotides away from the decaying atom but factually placed in close proximity to the decaying atom (for example DNA strands juxtaposed within a nucleosome or those present in two successive turns of a 30-nm DNA solenoid).

In an attempt to further understand the factors controlling (and the mechanisms underlying) DNA DSB, we addressed the role of DNA topology in the production of radiation-induced DSB [116]. In these studies, DSB yields following the decay of ^{125}I were quantified and compared in plasmid DNA molecules with three differing topologies: SC, relaxed circular (nicked, N) and linear (L) DNA. Surprisingly, and contrary to the expectations based on experimental results obtained in mammalian cells [58], our studies [116] showed that the number of DSB produced by the decay of ^{125}I in SC DNA (~0.5 DSB/^{125}I decay) was ~1/3 at seen in N and L DNA (~1.6–1.7 DSB/^{125}I decay). Additionally, the mechanism of DSB formation was also found to be affected by these differing topologies (figure 6.11). For example, whereas 100% of the DSB in SC DNA is a consequence of direct ionizations, ~50% of the DSBs formed in circular and linear plasmid DNA are inhibited by the •OH-radical scavenger DMSO and are therefore caused by indirect effects. These studies lead to the following two unexpected conclusions: (1) energy deposition in plasmid DNA (naked) is not indicative of DSB yield: it depends on DNA topology (table 6.5); and

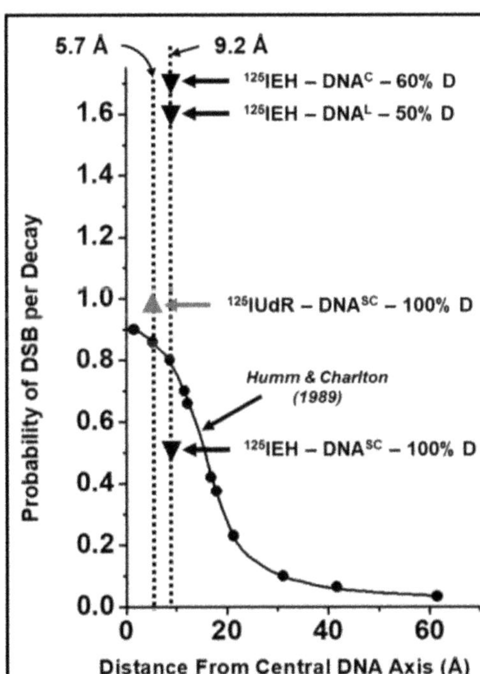

Figure 6.11. Effect of DNA topology on DSB yield in supercoiled (SC), circular (C), and linear (L) naked plasmid DNA.

Table 6.5. Topology of different forms of plasmid DNA.

Property	SC	N	L
Shape	Circular	Circular	Linear
Curvature	+++++	++	±
Compaction	Yes	Relaxed	Relaxed
Torsional energy	Yes	No	No
Twist	Under-winded	Normal(10.4 bp/turn)	Normal(10.4 bp/turn)

SC: Supercoiled; N: Nicked; L: Linear.

(2) DSB yield in DNA with similar topologies is (SC) or is not (L) proportional to the energy deposited in DNA.

Another judgment made by those of us examining the radiotoxicity of low-energy electron emitters was the perceived need to radiotarget every tumour cell (a direct consequence of the short range of the emitted electrons and, therefore, the absence of cross-fire irradiation of neighbouring cells). This too has proven to be inaccurate, as the decay of such isotopes has been shown to lead to a bystander effect, i.e. a dose-independent inhibition/retardation of mammalian cell growth in nonradiotargeted cells by signals(s) produced in Auger-electron-labelled cells. For example, *in vivo* studies have shown that the growth of subcutaneously implanted human tumour cells was influenced by the presence of ^{125}I- or ^{123}I-labelled cells that had been previously mixed with unlabelled tumour cells. Despite the fact that the electron spectra of both radionuclides are identical, the injection of a mixture of unlabelled and ^{125}I-labelled cells in mice *inhibited* the growth of unlabelled, unirradiated cells [117], whereas the mixture with ^{123}I-labelled cells *enhanced* the growth of unlabelled cells [118]. Interestingly, the percentage decrease (^{125}I-induced) and increase (^{123}I-induced) in tumour growth were quite comparable for both isotopes (figure 6.12). Finally, it should be noted that similar inhibitory (*iBE*) and stimulatory (*sBE*) bystander effects were observed when the radiolabelled cells were incubated *in vitro* with unlabelled cells [117–121].

The bystander effect induced by radioactive decay has impacted our views on risk assessment and therapeutic efficacy following the administration of radiopharmaceuticals to patients. Traditionally, dose estimations are carried out by averaging the radiation dose to cells within a tissue or tumour mass from radioactive atoms present on or within the cells (self-dose) and that from radionuclides present in/on other cells or in the extracellular fluids (cross-dose). Such absorbed dose estimates have played an important role in determining the amount of radioactivity to be administered to patients in diagnostic/therapeutic procedures. Since the decay of Auger electron emitting radionuclides can lead to either an *iBE* or an *sBE*, i.e., the actual radiobiological response will be greater/less than that predicted by dosimetric estimates alone, it is essential to assess the bystander effect of all therapeutic radiopharmaceuticals and select those exhibiting an *iBE*.

The very high *in vitro* radiotoxicity observed with DNA-incorporated Auger-electron emitters has been exploited in experimental radionuclide therapy. While many radiopharmaceuticals have been synthesized and their radiotoxic/therapeutic effects examined *in vitro* and *in vivo*, the thymidine analog 5-iodo-2′-deoxyuridine

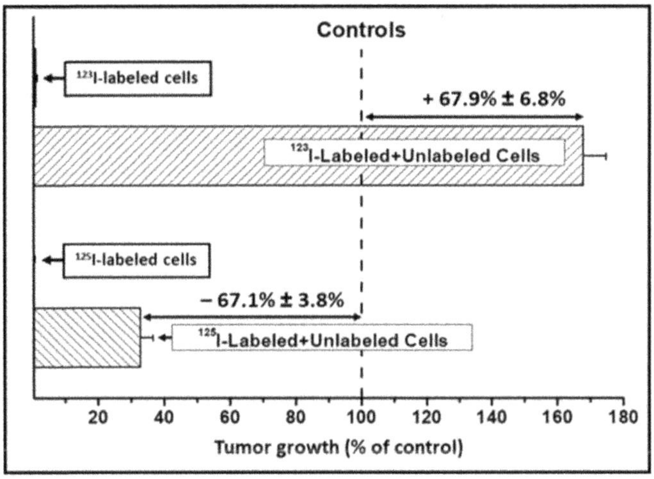

Figure 6.12. Inhibitory (^{125}IUdR) and stimulatory (^{123}IUdR) bystander effects consequent to decay of either Auger electron emitter within human tumours grown in mice.

(^{123}IUdR/^{125}IUdR) has been used in most of these studies and the results have been very promising [38, 39, 42, 44]. For example, the intraperitoneal (i.p.) injection of ^{125}IUdR [38] or ^{123}IUdR [39] into mice with i.p. ovarian cancer has led to a 4–5 log reduction in tumour cell survival. Similarly, the intrathecal (i.t.) administration of therapeutic doses of ^{125}IUdR into rats with i.t. tumours significantly delay the onset of paralysis, especially when the radiopharmaceutical was co-administered with methotrexate (MTX), an antimetabolite that enhances IUdR uptake by DNA-synthesizing cells (figure 6.13). In the latter cases, this was exemplified by a 5–6-log tumour cell kill and the curing of ~30% of the tumour-bearing rats [42, 44]. Consequent to these promising results, MTX and ^{125}IUdR were administered i.t. to a patient with pancreatic cancer metastatic to the CNS who had failed to respond to conventional therapy [122]. A dramatic drop in spinal fluid CA19.9 level was observed after the single treatment with the radiolabelled agent that was accompanied by clinical improvement. The findings observed suggest that this approach may be an effective treatment for neoplastic meningitis.

6.5 The ideal herapeutic radiopharmaceutical

Under ideal conditions, a therapeutic radiopharmaceutical would be characterized by: (i) very high specificity for tumour cells; (ii) high affinity to tumour cells; (iii) stability in blood *in vivo* (not catabolized/metabolized; if somewhat unstable, metabolism/catabolism half-life should be much greater than physical half-life ($T_{1/2P}$) of the radionuclide); (iv) biologic half-life ($T_{1/2B}$) comparable with $T_{1/2P}$ of the radionuclide; (v) rapid targeting of tumour cells; (vi) significant retention by tumour cells for period 2-to-3-fold longer than the $T_{1/2P}$ of the radionuclide; (vii) lack of catabolism by tumour cells; (viii) minimal taken up and retention by normal tissues/cells; (ix) rapid excretion from the body; (x) pharmacokinetics not altered by

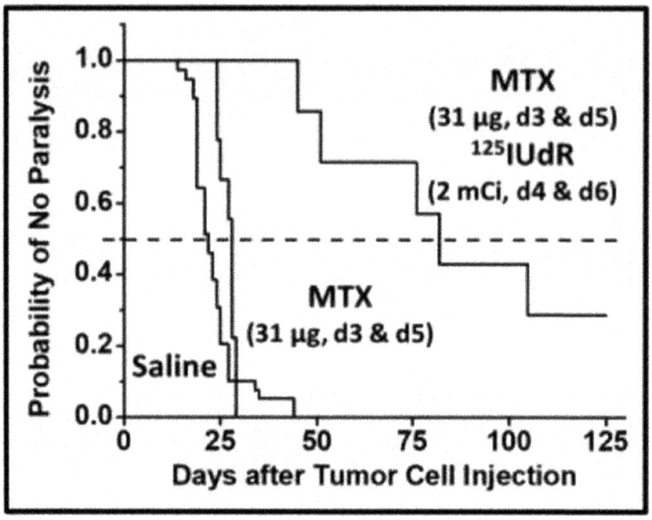

Figure 6.13. Induction of hind leg paralysis in rats bearing intrathecal tumors after treatment with methotrexate (MTX) or MTX and ^{125}IUdR.

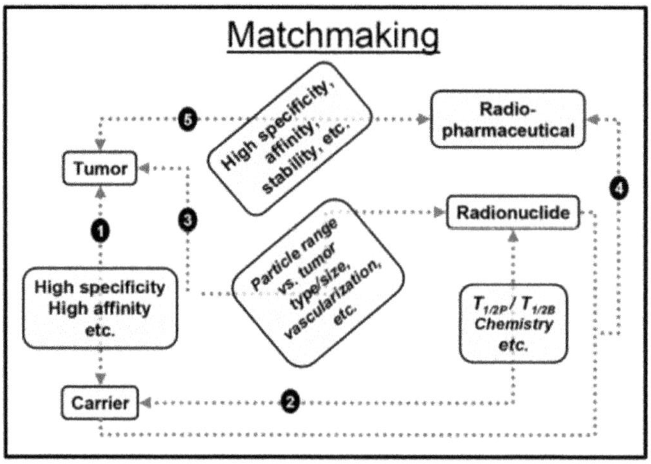

Figure 6.14. Match-making of radionuclide, radiopharmaceutical, tumour, and targeting ligand.

repeated injection; (xi) efficient cell killing, i.e., inefficient at causing other radiation-induced biologic effects (e.g. mutations, transformations); (xii) less heterogeneity within tumour cell clusters than the range of emitted particles; (xiii) deposition of a therapeutic dose to *all* tumour cells; and (xiv) delivery of a toxic dose to tumour cells *prior to* deposition of the maximum tolerated dose in normal tissues.

Finally, the successful development of a therapeutic radiopharmaceutical necessitates the multistep matching of: (i) the disease and the carrier molecule; (ii) the disease and the radionuclide; (iii) the carrier and the radionuclide; and (iv) the radiopharmaceutical and the disease (figure 6.14). During these successive steps, special attention must be paid to insure the compatibility of the individual components. Once the match-making of the radiopharmaceutical has been completed, it is ready for being assessed further *in vivo*, first in tumour-bearing mice and then in human trials.

References

[1] Ward J F 1986 Mechanisms of DNA repair and their potential modification for radiotherapy *Int. J. Radiat. Oncol. Biol. Phys.* **12** 1027–32

[2] Goodhead D T 1985 Saturable repair models of radiation action in mammalian cells *Radiat. Res.* **104** S-58–67

[3] Chapman J D 1980 Biophysical models of mammalian cell inactivation by radiation *Radiation biology in cancer research.* ed R E Meyn and H R Withers 1980 (New York: Raven Press) pp 21–32

[4] Thames H D and Hendry J H 1987 *Fractionation in radiotherapy.* (London: Taylor and Francis)

[5] Williams M V, Denekamp J and Fowler J F 1985 A review of α/β ratios for experimental tumors: implications for clinical studies of altered fractionation *Int. J. Radiat. Oncol. Biol. Phys.* **11** 87–96

[6] Nagasawa H and Little J B 1992 Induction of sister chromatid exchanges by extremely low doses of alpha-particles *Cancer Res.* **52** 6394–96

[7] Azzam E I, de Toledo S M, Gooding T and Little J B 1998 Intercellular communication is involved in the bystander regulation of gene expression in human cells exposed to very low fluences of alpha particles *Radiat. Res.* **150** 497–504

[8] Bishayee A, Rao D V and Howell R W 1976-88 Evidence for pronounced bystander effects caused by nonuniform distributions of radioactivity using a novel three-dimensional tissue culture model *Radiat. Res.* **152** 88–97

[9] Nagasawa H and Little J B 1999 Unexpected sensitivity to the induction of mutations by very low doses of alpha-particle radiation: evidence for a bystander effect *Radiat. Res.* **152** 552–7

[10] Zhou H, Randers-Pehrson G, Waldren C A, Vannais D, Hall E J and Hei T K 2000 Induction of a bystander mutagenic effect of alpha particles in mammalian cells *Proc. Natl Acad. Sci. USA* **97** 2099–104

[11] Azzam E I, de Toledo S M and Little J B 2001 Direct evidence for the participation of gap junction-mediated intercellular communication in the transmission of damage signals from a-particle irradiated to nonirradiated cells *Proc. Natl Acad. Sci. USA* **98** 473–78

[12] Brenner D J, Little J B and Sachs R K 2001 The bystander effect in radiation oncogenesis: II. A quantitative model *Radiat. Res.* **155** 402–08

[13] Walicka M A, Vaidyanathan G, Zalutsky M R, Adelstein S J and Kassis A I 1998 Survival and DNA damage in Chinese hamster V79 cells exposed to alpha particles emitted by DNA-incorporated astatine-211 *Radiat. Res.* **150** 263–68

[14] Goddu S M, Howell R W and Rao D V 1994 Cellular dosimetry: absorbed fractions for monoenergetic electron and alpha particle sources and S-values for radionuclides uniformly distributed in different cell compartments *J. Nucl. Med.* **35** 303–16

[15] Kocher D C 1981 Radioactive decay data tables: a handbook of decay data for application to radiation dosimetry and radiological assessments *Springfield, Virginia: National Technical Information Center, US Department of Energy* **1981** 221

[16] ICRU 1993 Stopping powers and ranges for protons and alpha particles *Report 49* (Bethesda, MD: Int. Commission on Radiation Units and Measurements)
[17] Charlton D E and Sephton R 1991 A relationship between microdosimetric spectra and cell survival for high-LET irradiation *Int. J. Radiat. Biol.* **59** 447–57
[18] Charlton D E, Kassis A I and Adelstein S J 1994 A comparison of experimental and calculated survival curves for V79 cells grown as monolayers or in suspension exposed to alpha irradiation from ^{212}Bi distributed in the growth medium *Radiat. Prot. Dosimetry* **52** 311–15
[19] Roeske J C and Stinchcomb T G 1999 The use of microdosimetric moments in evaluating cell survival for therapeutic alpha-particle emitters *Radiat. Res.* **151** 31–8
[20] Cole A 1969 Absorption of 20-eV to 50,000-eV electron beams in air and plastic *Radiat. Res.* **38** 7–33
[21] ICRU 1984 Stopping powers for electrons and positrons *Report 37* (Bethesda, MD: Int. Commission on Radiation Units and Measurements)
[22] Wheldon T E 1994 Targeting radiation to tumours *Int. J. Radiat. Biol.* **65** 109–16
[23] Bouchet L G, Bolch W E, Goddu S M, Howell R W and Rao D V 2000 Considerations in the selection of radiopharmaceuticals for palliation of bone pain from metastatic osseous lesions *J. Nucl. Med.* **41** 682–87
[24] O'Donoghue J A 1999 Implications of nonuniform tumor doses for radioimmunotherapy *J. Nucl. Med.* **40** 1337–341
[25] Kassis A I 2004 The amazing world of Auger electrons *Int. J. Radiat. Biol.* **80** 789–803
[26] Kassis A I, Adelstein S J, Haydock C and Sastry K S R 1980 Radiotoxicity of ^{75}Se and ^{35}S: theory and application to a cellular model *Radiat. Res.* **84** 407–25
[27] Kassis A I, Sastry K S R and Adelstein S J 1987 Kinetics of uptake, retention, and radiotoxicity of ^{125}I UdR in mammalian cells: implications of localized energy deposition by Auger processes *Radiat. Res.* **109** 78–89
[28] Kassis A I, Fayad F, Kinsey B M, Sastry K S R, Taube R A and Adelstein S J 1987 Radiotoxicity of ^{125}I in mammalian cells *Radiat. Res.* **111** 305–18
[29] Kassis A I, Adelstein S J, Haydock C, Sastry K S R, McElvany K D and Welch M J 1982 Lethality of Auger electrons from the decay of bromine-77 in the DNA of mammalian cells *Radiat. Res.* **90** 362–73
[30] O'Donoghue J A 1994 The impact of tumor cell proliferation in radioimmunotherapy *Cancer* **73** 974–80
[31] O'Donoghue J A and Wheldon T E 1996 Targeted radiotherapy using Auger electron emitters *Phys. Med. Biol.* **41** 1973–92
[32] Zhu X, Palmer M R, Makrigiorgos G M and Kassis A I 2010 Solid-tumor radionuclide therapy dosimetry: new paradigms in view of tumor microenvironment and angiogenesis *Med. Phys.* **37** 2974–84
[33] Kassis A I 2005 Radiotargeting agents for cancer therapy *Expert Opin. Drug Deliv.* **2** 981–91
[34] Smith M C *et al* 2000 OctreoTher: ongoing early clinical development of a somatostatin-receptor-targeted radionuclide antineoplastic therapy *Digestion* **62** 69–72
[35] Loh K-C, Fitzgerald P A, Matthay K K, Yeo P P B and Price D C 1997 The treatment of malignant pheochromocytoma with iodine-131 metaiodobenzylguanidine (^{131}I-MIBG): a comprehensive review of 116 reported patients *J. Endocrinol. Invest.* **20** 648–58
[36] Bloomer W D, McLaughlin W H, Neirinckx R D, Adelstein S J, Gordon P R, Ruth T J and Wolf A P 1981 Astatine-211–tellurium radiocolloid cures experimental malignant ascites *Science* **212** 340–1

[37] Bloomer W D, McLaughlin W H, Lambrecht R M, Atcher R W, Mirzadeh S, Madara J L, Milius R A, Zalutsky M R, Adelstein S J and Wolf A P 1984 ^{211}At radiocolloid therapy: further observations and comparison with radiocolloids of ^{32}P, ^{165}Dy, and ^{90}Y *Int. J. Radiat. Oncol. Biol. Phys.* **10** 341–8

[38] Bloomer W D and Adelstein S J 1977 5–^{125}I-iododeoxyuridine as prototype for radionuclide therapy with Auger emitters *Nature* **265** 620–1

[39] Baranowska-Kortylewicz J, Makrigiorgos G M, Van Den Abbeele A D, Berman R M, Adelstein S J and Kassis A I 1991 5-[^{123}I]iodo-2′-deoxyuridine in the radiotherapy of an early ascites tumor model Int. *J. Radiat. Oncol. Biol. Phys.* **21** 1541–51

[40] Sahu S K, Wen P Y C, Foulon C F, Nagel J S, Black P M, Adelstein S J and Kassis A I 1997 Intrathecal 5-[^{125}I]iodo-2′-deoxyuridine in a rat model of leptomeningeal metastases *J. Nucl. Med.* **38** 386–90

[41] Kassis A I *et al* 1998 5-[^{251}]iodo-2′-deoxyuridine in the radiotherapy of brain tumors in rats *J. Nucl. Med.* **39** 1148–54

[42] Kassis A I, Dahman B A and Adelstein S J 2000 In vivo therapy of neoplastic meningitis with methotrexate and 5-[^{125}I]iodo-2′-deoxyuridine *Acta Oncol.* **39** 731–7

[43] Kassis A I, Dahman B A and Adelstein S J 2000 Therapy of neoplastic meningitis with methotrexate and 5-[^{125}I]iodo-2′-deoxyuridine *J. Nucl. Med.* **41** 275P

[44] Kassis A I, Kirichian A M, Wang K, Semnani E S and Adelstein S J 2004 Therapeutic potential of 5-[^{125}I]iodo-2′-deoxyuridine and methotrexate in the treatment of advanced neoplastic meningitis *Int. J. Radiat. Biol.* **80** 941–6

[45] Foon K A 2000 Monoclonal antibody therapies for lymphomas *Cancer J.* **6** 273–78

[46] Jain R K 1990 Tumor physiology and antibody delivery *Front. Radiat. Ther. Oncol.* **24** 32–46

[47] Netti P A, Hamberg L M, Babich J W, Kierstead D, Graham W, Hunter G J, Wolf G L, Fischman A, Boucher Y and Jain R K 1999 Enhancement of fluid filtration across tumor vessels: implication for delivery of macromolecules *Proc. Natl. Acad. Sci. USA* **96** 3137–42

[48] Hall E J 2000 Radiobiology for the Radiologist *5th edn* (Philadelphia, PA: Lippincott Williams and Wilkins)

[49] Kassis A I, Harris C R, Adelstein S J, Ruth T J, Lambrecht R and Wolf A P 1986 The *in vitro* radiobiology of astatine-211 decay *Radiat. Res.* **105** 27–36

[50] Raju M R, Eisen Y, Carpenter S and Inkret W C 1991 Radiobiology of a particles III. Cell inactivation by a-particle traversals of the cell nucleus *Radiat. Res.* **128** 204–9

[51] Humm J L, Chin L M, Cobb L and Begent R 1990 Microdosimetry in radioimmunotherapy *Radiat. Prot. Dosim.* **31** 433–6

[52] Kurtzman S H *et al* 1988 ^{212}Bismuth linked to an antipancreatic carcinoma antibody: model for alpha-particle-emitter radioimmunotherapy *J. Natl. Cancer Inst.* **80** 449–52

[53] Larsen R H, Bruland Ø S, Hoff P, Alstad J, Lindmo T and Rofstad E K 1994 Inactivation of human osteosarcoma cells *in vitro* by ^{211}At-TP-3 monoclonal antibody: comparison with astatine-211-labeled bovine serum albumin, free astatine-211 and external-beam X rays *Radiat. Res.* **139** 178–84

[54] Larsen R H and Bruland Ø S 1998 Intratumour injection of immunoglobulins labelled with the a-particle emitter ^{211}At: analyses of tumour retention, microdistribution and growth delay *Br. J. Cancer* **77** 1115–22

[55] Strickland D K, Vaidyanathan G and Zalutsky M R 1994 Cytotoxicity of a-particle-emitting *m*-[^{211}At]astatobenzylguanidine on human neuroblastoma cells *Cancer Res.* **54** 5414–9

[56] Palm S, Bäck T, Claesson I, Delle U, Hultborn R, Jacobsson L, Köpf I and Lindegren S 1998 Effects of the alpha-particle emitter At-211 and low-dose-rate gamma-radiation on the human cell line Colo-205 as studied with a growth assay *Anticancer Res.* **18** 1671–6

[57] Nikula T K *et al* 1999 Alpha-emitting bismuth cyclohexylbenzyl DTPA constructs of recombinant humanized anti-CD33 antibodies: pharmacokinetics, bioactivity, toxicity and chemistry *J. Nucl. Med.* **40** 166–76

[58] Walicka M A, Adelstein S J and Kassis A I 1998 Indirect mechanisms contribute to biological effects produced by decay of DNA-incorporated iodine-125 in mammalian cells *in vitro*: double-strand breaks *Radiat. Res.* **149** 134–41

[59] Kassis A I, Walicka M A and Adelstein S J 2000 Double-strand break yield following ^{125}I decay: effects of DNA conformation *Acta Oncol.* **39** 721–6

[60] Kozak R W, Atcher R W, Gansow O A, Friedman A M, Hines J J and Waldmann T A 1986 Bismuth-212-labeled anti-Tac monoclonal antibody: a-particle-emitting radionuclides as modalities for radioimmunotherapy *Proc. Natl. Acad. Sci. USA* **83** 474–8

[61] Harrison A and Royle L 1987 Efficacy of astatine-211-labeled monoclonal antibody in treatment of murine T-cell lymphoma *Natl. Cancer Inst. Monogr.* **3** 157–8

[62] Macklis R M, Kinsey B M, Kassis A I, Ferrara J L, Atcher R W, Hines J J, Coleman C N, Adelstein S J and Burakoff S J 1988 Radioimmunotherapy with alpha-particle-emitting immunoconjugates *Science* **240** 1024–6

[63] Link E M and Carpenter R N 1990 ^{211}At-methylene blue for targeted radiotherapy of human melanoma xenografts: treatment of micrometastases *Cancer Res.* **50** 2963–7

[64] Zalutsky M R, McLendon R E, Garg P K, Archer G E, Schuster J M and Bigner D D 1994 Radioimmunotherapy of neoplastic meningitis in rats using an a-particle-emitting immunoconjugate *Cancer Res.* **54** 4719–25

[65] Kennel S J and Mirzadeh S 1998 Vascular targeted radioimmunotherapy with ^{213}Bi—an a-particle emitter *Nucl. Med. Biol.* **25** 241–6

[66] Adams G P, Shaller C C, Chappell L L, Wu C, Horak E M, Simmons H H, Litwin S, Marks J D, Weiner L M and Brechbiel M W 2000 Delivery of the a-emitting radioisotope bismuth-213 to solid tumors via a single-chain Fv and diabody molecules *Nucl. Med. Biol.* **27** 339–46

[67] McDevitt M R *et al* 2001 Tumor therapy with targeted atomic nanogenerators *Science* **294** 1537–40

[68] Kratochwil C, Bruchertseifer F, Giesel F L, Weis M, Verburg F A, Mottaghy F, Kopka K, Apostolidis C, Haberkorn U and Morgenstern A 2016 225 Ac-PSMA-617 for PSMA-targeted α-radiation therapy of metastatic castration-resistant prostate cancer *J. Nucl. Med.* **57** 1941–44

[69] Burki H J, Koch C and Wolff S 1977 Molecular suicide studies of ^{125}I and ^{3}H disintegration in the DNA of Chinese hamster cells *Curr. Top. Radiat. Res. Q.* **12** 408–25

[70] Chan P C, Lisco E, Lisco H and Adelstein S J 1976 The radiotoxicity of iodine-125 in mammalian cells. II. A comparative study on cell survival and cytogenetic responses to ^{125}IUdR, ^{131}IUdR, and ^{3}HTdR *Radiat. Res.* **67** 332–43

[71] Govindan S V, Goldenberg D M, Elsamra S E, Griffiths G L, Ong G L, Brechbiel M W, Burton J, Sgouros G and Mattes M J 2000 Radionuclides linked to a CD74 antibody as therapeutic agents for B-cell lymphoma: comparison of Auger electron emitters with b-particle emitters *J. Nucl. Med.* **41** 2089–97

[72] Cleaver J E and Burki H J 1974 Biological damage from intranuclear carbon-14 decays: DNA single-strand breaks and repair in mammalian cells *Int. J. Radiat. Biol.* **26** 399–403

[73] Gutiérrez S, Carbonell E, Galofré P, Creus A and Marcos R 1998 The alkaline single-cell gel electrophoresis (SCGE) assay applied to the analysis of radiation-induced DNA damage in thyroid cancer patients treated with ^{131}I *Mutat. Res.* **413** 111–9

[74] Hengstler J G *et al* 2000 Induction of DNA single-strand breaks by ^{131}I and ^{99m}Tc in human mononuclear blood cells *in vitro* and extrapolation to the *in vivo* situation *Radiat. Res.* **153** 512–20

[75] Esteban J M, Schlom J, Mornex F and Colcher D 1987 Radioimmunotherapy of athymic mice bearing human colon carcinomas with monoclonal antibody B72.3: histological and autoradiographic study of effects on tumors and normal organs *Eur. J. Cancer Clin. Oncol.* **23** 643–55

[76] Riklund K E, Edbom G, Makiya R, Johansson B, Gerdes U, Hietala S-O, Ekelund L, Stigbrand T and Stendahl U 1991 Radioimmunoscintigraphy of gynecologic tumors with ^{131}I-labeled anti-PLAP monoclonal antibodies *Acta Radiol.* **32** 375–80

[77] Buchsbaum D, Khazaeli M B, Liu T, Bright S, Richardson K, Jones M and Meredith R 1995 Fractionated radioimmunotherapy of human colon carcinoma xenografts with ^{131}I-labeled monoclonal antibody CC49 *Cancer Res.* **55** 5881s–887s

[78] Sato N, Saga T, Sakahara H, Yao Z, Nakamoto Y, Zhang M, Kuroki M, Matsuoka Y, Iida Y and Konishi J 1999 Intratumoral distribution of radiolabeled antibody and radioimmunotherapy in experimental liver metastases model of nude mouse *J. Nucl. Med.* **40** 685–92

[79] Otte A, Mueller-Brand J, Dellas S, Nitzsche E U, Herrmann R and Maecke H R 1998 Yttrium-90-labelled somatostatin-analogue for cancer treatment *Lancet* **351** 417–8

[80] Chinn P C, Leonard J E, Rosenberg J, Hanna N and Anderson D R 1999 Preclinical evaluation of ^{90}Y-labeled anti-CD20 monoclonal antibody for treatment of non-Hodgkin's lymphoma *Int. J. Oncol.* **15** 1017–25

[81] DeNardo G L, Kukis D L, DeNardo S J, Shen S, Mausner L F, O'Donnell R T, Lamborn K R, Meyers F J, Srivastava S C and Miers L A 1997 Enhancement of ^{67}Cu-2IT-BAT-Lym-1 therapy in mice with human Burkitt's lymphoma (Raji) using interleukin-2 *Cancer* **80** 2576–82

[82] DeNardo G L, Kukis D L, Shen S, Mausner L F, Meares C F, Srivastava S C, Miers L A and DeNardo S J 1997 Efficacy and toxicity of ^{67}Cu-2IT-BAT-Lym-1 radioimmunoconjugate in mice implanted with human Burkitt's lymphoma (Raji) *Clin. Cancer Res.* **3** 71–9

[83] Hnatowich D J, Virzi F and Rusckowski M 1987 Investigations of avidin and biotin for imaging applications *J. Nucl. Med.* **28** 1294–302

[84] Goodwin D A, Meares C F, McCall M J, McTigue M and Chaovapong W 1988 Pretargeted immunoscintigraphy of murine tumors with indium-111-labeled bifunctional haptens *J. Nucl. Med.* **29** 226–34

[85] Le Doussal J-M, Gruaz-Guyon A, Martin M, Gautherot E, Delaage M and Barbet J 1990 Targeting of indium 111-labeled bivalent hapten to human melanoma mediated by bispecific monoclonal antibody conjugates: imaging of tumors hosted in nude mice *Cancer Res.* **50** 3445–52

[86] Kassis A I, Jones P L, Matalka K Z and Adelstein S J 1996 Antibody-dependent signal amplification in tumor xenografts after pretreatment with biotinylated monoclonal antibody and avidin or streptavidin *J. Nucl. Med.* **37** 343–52

[87] Axworthy D B, Reno J M, Hylarides M D, Mallett R W, Theodore L J, Gustavson L M, Su F-M, Hobson L J, Beaumier P L and Fritzberg A R 2000 Cure of human carcinoma xenografts by a single dose of pretargeted yttrium-90 with negligible toxicity *Proc. Natl Acad. Sci. USA* **97** 1802–7

[88] Charlton D E and Booz J 1981 A Monte Carlo treatment of the decay of ^{125}I *Radiat. Res.* **87** 10–23

[89] Sastry K S R and Rao D V 1984 Dosimetry of low energy electrons ed D V Rao, R Chandra and M C Graham *Physics of Nuclear Medicine: Recent Advances.* (Woodbury, NY: American Institute of Physics) 1984 169–208

[90] Charlton D E, Pomplun E and Booz J 1987 Some consequences of the Auger effect: fluorescence yield, charge potential, and energy imparted *Radiat. Res.* **111** 553–64

[91] Pomplun E, Booz J and Charlton D E 1987 A Monte Carlo simulation of Auger cascades *Radiat. Res.* **111** 533–52

[92] Kassis A I, Sastry K S R and Adelstein S J 1985 Intracellular localisation of Auger electron emitters: biophysical dosimetry *Radiat. Prot. Dosim.* **13** 233–6

[93] Hofer K G and Hughes W L 1971 Radiotoxicity of intranuclear tritium, ^{125}iodine and ^{131}iodine *Radiat. Res.* **47** 94–109

[94] Porteous D D 1971 The toxicity of ^{125}IUdR in cultured mouse BP8 tumour cells *Br. J. Cancer* **25** 594–7

[95] Burki H J, Roots R, Feinendegen L E and Bond V P 1973 Inactivation of mammalian cells after disintegrations of ^{3}H or ^{125}I in cell DNA at −196 °C *Int. J. Radiat. Biol.* **24** 363–75

[96] Bradley E W, Chan P C and Adelstein S J 1975 The radiotoxicity of iodine-125 in mammalian cells I. Effects on the survival curve of radioiodine incorporated into DNA *Radiat. Res.* **64** 555–63

[97] Hofer K G, Harris C R and Smith J M 1975 Radiotoxicity of intracellular ^{67}Ga, ^{125}I and ^{3}H: nuclear versus cytoplasmic radiation effects in murine L1210 leukaemia *Int. J. Radiat. Biol.* **28** 225–41

[98] Kassis A I, Howell R W, Sastry K S R and Adelstein S J 1988 Positional effects of Auger decays in mammalian cells in culture *DNA Damage by Auger emitters* ed K F Baverstock and D E Charlton 1988 (London: Taylor and Francis) pp 1–13

[99] Kassis A I, Fayad F, Kinsey B M, Sastry K S R and Adelstein S J 1989 Radiotoxicity of an ^{125}I-labeled DNA intercalator in mammalian cells *Radiat. Res.* **118** 283–94

[100] Walicka M A, Ding Y, Roy A M, Harapanhalli R S, Adelstein S J and Kassis A I 1999 Cytotoxicity of [^{125}I]iodoHoechst 33342: contribution of scavengeable effects *Int. J. Radiat. Biol.* **75** 1579–87

[101] Sedelnikova O A, Panyutin I G, Thierry A R and Neumann R D 1998 Radiotoxicity of iodine-125-labeled oligodeoxyribonucleotides in mammalian cells *J. Nucl. Med.* **39** 1412–8

[102] Bloomer W D, McLaughlin W H, Milius R A, Weichselbaum R R and Adelstein S J 1983 Estrogen receptor-mediated cytotoxicity using iodine-125 *J. Cell. Biochem.* **21** 39–45

[103] Yasui L S, Hughes A and DeSombre E R 2001 Cytotoxicity of ^{125}I-estrogen decay in non-oestrogen receptor-expressing human breast cancer cells, MDA-231 and oestrogen receptor-expressing MCF-7 cells *Int. J. Radiat. Biol.* **77** 955–62

[104] Makrigiorgos G M, Kassis A I, Baranowska-Kortylewicz J, McElvany K D, Welch M J, Sastry K S R and Adelstein S J 1989 Radiotoxicity of 5-[^{123}I]iodo-2′-deoxyuridine in V79 cells: a comparison with 5-[^{125}I]iodo-2′-deoxyuridine *Radiat. Res.* **118** 532–44

[105] Howell R W, Kassis A I, Adelstein S J, Rao D V, Wright H A, Hamm R N, Turner J E and Sastry K S R 1994 Radiotoxicity of platinum-195m-labeled *trans*-platinum (II) in mammalian cells *Radiat. Res.* **140** 55–62
[106] Kassis A I, Adelstein S J, Haydock C and Sastry K S R 1983 Thallium-201: an experimental and a theoretical radiobiological approach to dosimetry *J. Nucl. Med.* **24** 1164–75
[107] Charlton D E and Humm J L 1988 A method of calculating initial DNA strand breakage following the decay of incorporated ^{125}I *Int. J. Radiat. Biol.* **53** 353–65
[108] Martin R F and Haseltine W A 1981 Range of radiochemical damage to DNA with decay of iodine-125 *Science* **213** 896–8
[109] Kandaiya S, Lobachevsky P N, D'Cunha G and Martin R F 1996 DNA strand breakage by ^{125}I-decay in a synthetic oligodeoxynucleotide: fragment distribution and evaluation of DMSO protection effect *Acta Oncol.* **35** 803–8
[110] Lobachevsky P N and Martin R F 1996 DNA strand breakage by ^{125}I-decay in a synthetic oligodeoxynucleotide: quantitative analysis of fragment distribution *Acta Oncol.* **35** 809–15
[111] Panyutin I G and Neumann R D 1997 Radioprobing of DNA: distribution of DNA breaks produced by decay of ^{125}I incorporated into a triplex-forming oligonucleotide correlates with geometry of the triplex *Nucleic Acids Res.* **25** 883–7
[112] Kassis A I, Harapanhalli R S and Adelstein S J 1999 Strand breaks in plasmid DNA after positional changes of Auger electron-emitting iodine-125: direct compared to indirect effects *Radiat. Res.* **152** 530–8
[113] Balagurumoorthy P, Chen K, Bash R C, Adelstein S J and Kassis A I 2006 Mechanisms underlying production of double-strand breaks in plasmid DNA after decay of ^{125}I-Hoechst *Radiat. Res.* **166** 333–44
[114] Lobachevsky P N, White J, Leung M, Skene C and Martin R F 2008 Plasmid breakage by ^{125}I-labelled DNA ligands: effect of DNA-iodine atom distance on breakage efficiency *Int. J. Radiat. Biol.* **84** 991–1000
[115] Roots R and Okada S 1975 Estimation of life times and diffusion distances of radicals involved in X-ray-induced DNA strand breaks or killing of mammalian cells *Radiat. Res.* **64** 306–20
[116] Balagurumoorthy P, Chen K, Adelstein S J and Kassis A I 2008 Auger electron-induced double-strand breaks depend on DNA topology *Radiat. Res.* **170** 70–82
[117] Xue L Y, Butler N J, Makrigiorgos G M, Adelstein S J and Kassis A I 2002 Bystander effect produced by radiolabeled tumor cells *in vivo Proc. Natl. Acad. Sci. U.S.A.* **99** 13765–70
[118] Kishikawa H, Wang K, Adelstein S J and Kassis A I 2006 Inhibitory and stimulatory bystander effects are differentially induced by iodine-125 and iodine-123 *Radiat. Res.* **165** 688–94
[119] Howell R W and Bishayee A 2002 Bystander effects caused by nonuniform distributions of DNA-incorporated ^{125}I *Micron* **33** 127–32
[120] Boyd M, Ross S C, Dorrens J, Fullerton N E, Tan K W, Zalutsky M R and Mairs R J 2006 Radiation-induced biologic bystander effect elicited *in vitro* by targeted radiopharmaceuticals labeled with alpha-, beta-, and Auger electron-emitting radionuclides *J. Nucl. Med.* **47** 1007–15
[121] Boyd M, Sorensen A, McCluskey A G and Mairs R J 2008 Radiation quality-dependent bystander effects elicited by targeted radionuclides *J. Pharm. Pharmacol.* **60** 951–58
[122] Rebischung C *et al* 2008 First human treatment of resistant neoplastic meningitis by intrathecal administration of MTX Plus 125IUdR *Int. J. Radiat. Biol.* **84** 1123–29

Chapter 7

Microdosimetry of targeted radionuclides

Pablo Mínguez Gabiña, John C Roeske and John L Humm

7.1 Introduction

Microdosimetry is the study of the stochastic process of energy deposition from radiation within small volumes. Small in the context of microdosimetry typically refers to tissue equivalent volumes commensurate with the size of the biological targets of relevance. This has commonly meant spherical volumes of between 0.3 and 20 μm—the realm in which energy deposition can be experimentally measured using a proportional counter. Such chambers are constructed of tissue equivalent plastic and operated with a tissue equivalent gas. Since these devices, used for the purpose of measuring microdosimetric spectra, were developed by Rossi [1] they are frequently referred to as 'Rossi chambers'. One of the greatest applications of the Rossi chambers is to characterize unknown mixed radiation beams found at reactor facilities. Since Rossi counters measure radiation deposits within the chamber event by event, accumulation of the magnitude of events using a multi-channel analyser results in a spectrum of the energy deposition. Because the magnitude and spectrum of energy deposition differs for radiations of different linear energy transfer (LET), microdosimetry has developed methods of curve stripping to separate the contribution to the absorbed dose from photons, electrons, neutrons (recoil protons) and heavy ions. This has been an important contribution to radiation protection, where the relative biological effectiveness (RBE) of different radiations depends upon the LET of the radiation. In fact, microdosimetry goes further than measuring the LET of the radiation field. By measuring the stochastics of individual radiation interactions within small, cell-equivalent volumes, it serves as a methodology of determining the entire spectrum of radiation deposits and LETs experienced by a cell or population of cells.

The energy deposition spectra measured by Rossi counters are presented in two formats: specific energy z and lineal energy y. Specific energy is defined $z = \varepsilon/m$, where ε is the magnitude of the energy deposit within a mass m. It is the stochastic analogue of absorbed dose D. Lineal energy is defined $y = \varepsilon/l$, where l is the mean chord length of the simulated cavity. Lineal energy is the stochastic analogue of LET.

doi:10.1088/978-0-7503-2694-0ch7

To study energy deposition within volumes <0.3 μm in diameter, track structure simulation codes are required. These codes use Monte Carlo methods to model individual particle histories as they traverse matter. Coupled with models of the biological target, these track structure codes have been used to make predictions of DNA strand break and base damage [2–8]. With the development of increasingly sophisticated radionuclide targeting techniques, an understanding of the microdosimetry for these systems will assist in the interpretation of biological effects.

Radiolabelled antibodies and other nuclear medicine radionuclide targeting compounds are used either at diagnostic or therapeutic activities. Diagnostic activities administered to patients result typically in whole-body absorbed doses on the order of 1 cGy. While at first glance one would expect that microdosimetric methods are required to characterize the stochastic nature of energy deposition, a closer evaluation reveals that this may not be necessary. For low LET radiations such as β, x- and γ-rays emitted by radionuclides, an absorbed dose of 1 cGy corresponds to an average of approximately 50 individual track traversals of a cell (depending upon the assumed cell diameter). The relative deviation in the number of hits (assuming Poisson statistics) is 14% (1/sqrt 50). Applying the criteria of Kellerer and Chmelevsky [9], who state that microdosimetry should be applied when the relative deviation of the specific energy spectrum is greater than 20%, the need for microdosimetry at these absorbed doses is borderline for diagnostic tracers. However, the application of Poisson statistics is warranted only when the radionuclide distribution is uniform. This is rarely the case for any radionuclide administered. As a consequence, an extended application of microdosimetry, which deals not only with the stochastics of the energy deposition process (classical microdosimetry), but also with the unique biodistribution of the radionuclide may be warranted in the analysis of energy deposition at all administered tracer activities. This concept is exemplified by the study of Makrigiorgos *et al* [10] who performed measurements of the microdistribution of a routine diagnostic tracer ^{99m}Tc-Microlite from liver tissue samples and showed that, relative to other cells, the colloidal agent was concentrated 200- to 1000-fold higher in liver macrophages relative to the average activity per gram of liver.

There are two types of radionuclides, for which classical microdosimetry has the greatest relevance, even at therapeutically administered activities: alpha and Auger electron emitters. Since alpha and Auger electron therapies using antibody and other carrier molecules have become a clinical reality over the past few years, the remainder of this chapter will focus on a discussion of each.

7.2 Alpha emitters

Over the last several years, researchers have taken an interest in the use of alpha-particle emitters as a potential form of therapy for micrometastatic disease [11–32]. The radiobiological advantages of alpha particles are their high LET and independence from dose rate, oxygen effects and phase cell cycle [33]. When combined with a suitable carrier (e.g., antibody), the short range of alpha particles in tissue (40–90 μm) results in a highly localized deposition of energy near the point of emission (presumably in the tumour) and limits the volume of normal tissues irradiated.

The dosimetry of alpha-particle emitters is challenging due to the stochastic nature of energy deposited in small, subcellular targets as shown schematically in figure 7.1. In this diagram, some cell nuclei receive multiple alpha particle hits, while others receive no hits whatsoever. (In this context, we define a hit as any alpha particle that intersects a cell nucleus, regardless of the energy deposited.) Additionally, an alpha particle may traverse a variety of path lengths in passing through a cell nucleus. For example, an alpha particle that intersects a spherical nucleus through the exact centre will have a path length equal to twice the radius, whereas an alpha particle that intersects nearly tangent to the cell nucleus will have little or no path length. Since the energy deposited by an alpha particle is proportional to its path length, each alpha particle in the above scenario produces one hit, but the amount of energy deposited is significantly different. Since the distribution of hits to the cell nucleus is often characterized using Poisson statistics, this problem is often referred to as a compound Poisson process—a Poisson distribution in the number of events where the magnitude of each event varies. Other effects that influence the dosimetry of alpha particles include the variation in energy loss (LET) along an individual particle's path, the size and shape of the target, and the source configuration (within the cell versus outside the cell, etc). Thus, it is generally accepted that the analysis of the cellular response to alpha-particle irradiation requires microdosimetric methods. The importance of microdosimetry in therapeutic applications (e.g., radioimmunotherapy (RIT)) is that while two different microdosimetric distributions may have the same average specific energy (absorbed dose), they can result in significantly different levels of cell kill [34, 35].

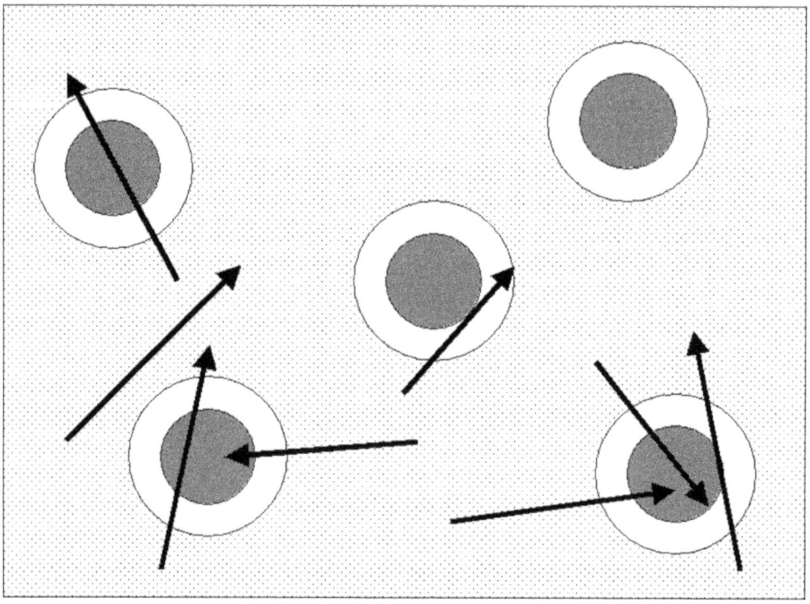

Figure 7.1. Schematic diagram illustrating the stochastic nature of energy deposited in individual cell nuclei (shaded circles) by alpha particles (lines with arrows). Some cells will receive no hits, while others will receive a number of hits based on the Poisson distribution.

7.2.1 Monte Carlo simulation of energy deposition by alpha-particle emitters

Microdosimetric spectra may be calculated using either analytical or Monte Carlo methods [36, 37]. Analytical methods use convolutions (via Fourier transforms) of the single-event spectrum to calculate multi-event distributions. The single-event spectrum represents the pattern of specific energy depositions for exactly one alpha particle hit. While analytical codes are computationally efficient, they are often limited to simple source–target geometries as the single-event spectrum must be known for each source–target configuration. Monte Carlo codes offer greater flexibility than analytical methods, and can simulate a wide variety of geometries and source configurations. While Monte Carlo codes may be computationally intensive, the increasing availability of fast desktop computers with graphics processing units (GPU) has made the calculation time less of a concern.

Alpha-particle Monte Carlo codes require the developer to specify both the source and target geometry, as well as provide a characterization of the alpha particle energy loss. Idealizations are often made to simplify the coding and reduce calculation time. A typical algorithm initially selects an alpha-particle emitter and determines if the source decays are based on its effective half-life. Next, the energy of the alpha-particle emission is determined by random sampling of the branching ratios. Based on isotropic emission, the directional cosines of the alpha-particle track are randomly assigned. In nearly all Monte Carlo codes, alpha particles are assumed to travel in straight lines. This approximation is valid for alpha particles having energies less than 10 MeV [38]. In addition, the range of delta rays and the width of the alpha-particle track (~100 nm) are often ignored since the targets that are often studied (i.e., cell nucleus) are much larger than these dimensions [39]. Using this straight-line approximation, the parametric equations [40, 41] can be solved to determine if the alpha particle intersects the cell nucleus:

$$x = x' + t\cos(\theta_1) \tag{7.1}$$

$$y = y' + t\cos(\theta_2) \tag{7.2}$$

$$z = z' + t\cos(\theta_3) \tag{7.3}$$

In equations (7.1)–(7.3), x', y', z' are the coordinates of the alpha-particle emission, and x, y, z are the intercepts of the alpha particle with the nucleus. The angles θ_1, θ_2 and θ_3 are the angles of the alpha-particle path relative to the x-, y-, and z-axes, respectively. If the nucleus is spherical (or some other simple geometry), equations (7.1)–(7.3) can be substituted into the target's analytical form to determine the distance traversed by the particle upon entering and leaving the nucleus. For example, a spherical nucleus is given by:

$$(x - x_o)^2 + (y - y_o)^2 + (z - z_o)^2 = r_n^2, \tag{7.4}$$

where x_o, y_o, and z_o are the centre coordinates of a nucleus with radius r_n. Substituting equations (7.1)–(7.3) into equation (7.4) and solving for t yields two solutions—t_1 corresponding to the distance from the source to where the alpha particle enters the nucleus, and t_2 corresponding to the distance between the source and where the alpha particle exits the nucleus. Note that when the alpha particle emission occurs within the nucleus (starter), t_1 is chosen to be zero, while t_2 is the intercept with the nuclear surface. If the alpha particle terminates its track within the nucleus (stopper) the value of t_2 is set to the range of the particle. A hit occurs when the solutions for t_1 and t_2 are both real, positive numbers.

Next, the energy deposited within the cell nucleus is calculated. This energy is given by:

$$\varepsilon_1 = \int_{R-t_2}^{R-t_1} \frac{\mathrm{d}E}{\mathrm{d}x}\mathrm{d}x. \tag{7.5}$$

The energy deposited by a single alpha-particle transit is transformed into specific energy using the relation:

$$z_1 = \varepsilon_1/m. \tag{7.6}$$

In equation (7.5), ($\mathrm{d}E/\mathrm{d}x$) is the stopping power, and the limits of integration are over the residual range of the particle upon entering ($R-t_1$) and leaving the target ($R-t_2$), respectively. The stopping power, which characterizes the rate of energy loss per unit path length, folds in the individual processes involved in alpha particle energy loss [42]. Stopping power data for a variety of media can be obtained from the literature [43–47]. Inherent in the stopping power formulation is the continuous slowing down approximation (CSDA). As the name implies, this approximation assumes that alpha particles lose energy continuously as they traverse matter. Thus, the determination of the specific energy deposited relies heavily on the choice of stopping powers used. In a comparison of the impact of stopping power formulation, Stinchcomb and Roeske [36] calculated specific energy spectra for the same source–target geometry as Humm [35]. The former case used Janni's stopping power [48] for protons (appropriately scaled for alpha particles), while in the latter case Walsh's stopping power [49] was utilized. In general, a disparity of 1%–7% was observed in the average specific energies using these two different characterizations of the stopping power. However, in practical situations, it is expected that the uncertainty in the activity distribution will be much larger than that of the stopping powers used in the Monte Carlo calculation.

The process of determining the specific energy deposited for an individual interaction can be repeated to determine the multi-event specific energy spectrum. In this case, all sources within the maximum range of the alpha-particle emission are evaluated to determine if they intersect the nucleus and the subsequent specific energy deposited is calculated and summed. This scenario represents one possible outcome. Next, the process is repeated, and most likely, a different total specific energy will be calculated. By repeating this process, the entire outcome of specific energies deposited can be determined. A histogram of these individual results is the multi-event distribution. An alternative is to determine the single-event specific energy spectrum. This distribution is calculated by examining only those

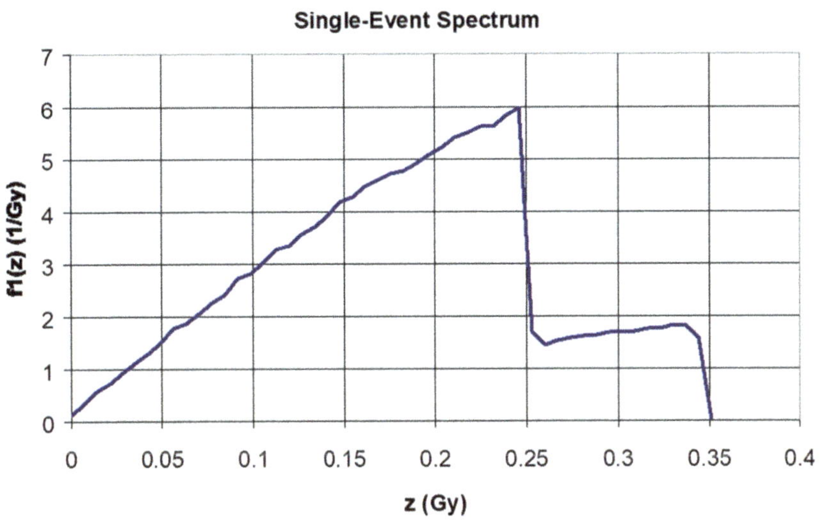

Figure 7.2. Single-event specific energy distribution for ^{212}Bi sources distributed on a cell surface. The cell radius is 10 μm and the nuclear radius is 5 μm. The two peaks correspond to the two emissions from ^{212}Bi—6.05 MeV (36%) and 8.78 MeV (64%).

alpha-particle emissions that result in exactly one hit to the cell nucleus. Each time an individual alpha particle intersects the nucleus, the specific energy is determined and recorded. All of these single-event specific energies are then histogrammed to determine the single-event spectrum. Single-event spectra are often used in conjunction with analytical codes whereby the multi-event spectra are calculated by multiple convolutions of the single-event spectrum. Stinchcomb and Roeske [36] demonstrated that both methods produce the same multi-event spectra within the uncertainty of the calculation itself. An example of a single-event spectrum for ^{212}Bi distributed on the cell surface is shown in figure 7.2. A multi-event spectrum for 15 ^{212}Bi sources on the cell surface is shown in figure 7.3.

Microdosimetric spectra are often combined with a model of cell survival to estimate the survival following a particular type of irradiation. The surviving fraction following alpha particle irradiation is given by [35, 36, 50]:

$$S(\bar{z}) = \int_0^\infty f(z)e^{-z/z_o}\mathrm{d}z \tag{7.7}$$

where $f(z)$ is the multi-event specific energy spectrum representing the fraction of cells receiving specific energies between z and z+dz, z_o is the specific energy deposited within an individual cell that reduces the average cell survival to $1/e$, and $\exp(-z/z_o)$ is the fraction of cells that survive. The use of the exponential function assumes that there are no bystander effects and that intratrack interactions are dominant relative to intertrack interactions. When using the single-event spectrum, the relationship is given by [36]:

$$S(\bar{z}) = \exp[-\langle n\rangle\{1 - T_1(z_o)\}] \tag{7.8}$$

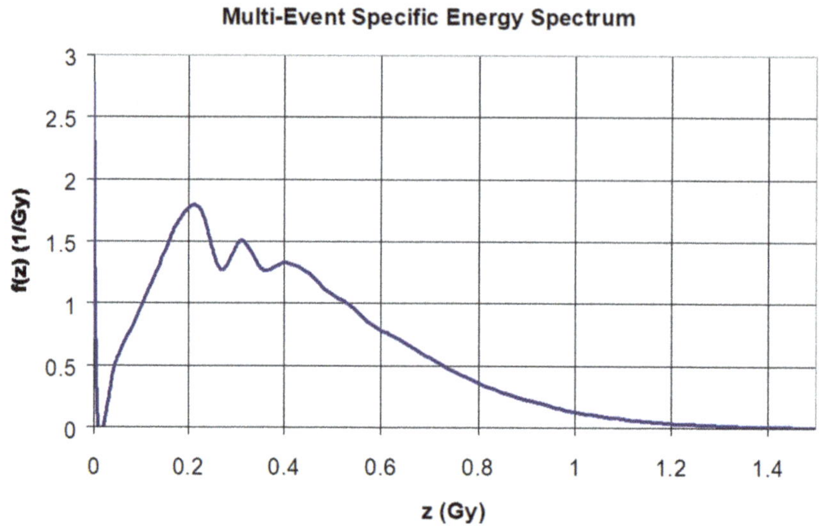

Figure 7.3. Multi-event specific energy spectrum for 15 alpha particle sources (^{212}Bi) distributed randomly on a cell surface with radius 10 μm. The nuclear radius is 5 μm. Note that a significant fraction of the cells receive zero hits to the cell nucleus.

where $<n>$ is the average number of hits to the cell nucleus, and $T_1(z_o)$ is the Laplace transform of the single-event spectrum. Note that the same value of z_o satisfies both equations (7.7) and (7.8).

It is important to point out that z_o is not equal to D_o, which is determined from the slope of the cell survival curve. Rather, it is a more fundamental quantity as D_o has folded into it not only the effects of the radiation, but also the effects of the source–target geometry. In the applications to be discussed, a recurrent theme is to use the observed surviving fraction to determine the inherent cell sensitivity—z_o.

7.2.2 Applications of microdosimetry to alpha-particle emitters

Microdosimetry has been used in a number of alpha-particle applications, particularly RIT. These applications can be broadly characterized as theoretical studies of simple cellular geometries, experimental analysis of cell survival following alpha-particle irradiation, and the microdosimetry of realistic geometries such as multi-cellular spheroids and bone marrow. The work in each of these categories will be discussed separately.

Fisher [51] was among the first to demonstrate the utility of microdosimetry for therapeutic applications. This particular report used an analytical code based on the previous work of Roesch [52], and suggested a number of geometries for RIT including sources distributed on and within individual cells, sources distributed within spherical clusters of cells, and sources located in cylinders (i.e., blood vessels) that deposited energy within spherical cell nuclei a short distance away. These geometries have served as the basis of those used in subsequent theoretical studies. In one such study, Humm [34] combined the physical calculation of specific energy (using Monte Carlo methods) with a model for cell survival to analyse the impact of stochastic energy deposition on the expected surviving fraction of a group of cells. In particular, two geometries were considered—cells located outside of a capillary and cells located within a tumour consisting of a uniform distribution of ^{211}At. The results of this analysis demonstrated that although the mean absorbed dose was similar for these two types of geometries, there was a significant variation in the expected cell survival due to the differences in the specific energy spectra. In particular, the fraction of cells without alpha-particle hits increased with distance from the capillary (due to the short range of the alpha particles) resulting in a bi-exponential cell survival curve. Initially, the slope of this curve was similar to that of a uniformly irradiated tumour. However, with increasing absorbed doses, the curve was less steep and asymptotically approached a value corresponding to the fraction of non-hit cells. Building upon the previous analysis, Humm and Chin [35] analysed the effects of cell nucleus size, binding fraction, cell volume fraction and nonuniform binding on calculated specific energy spectra. Their results demonstrated that the expected cell survival curves, produced by nonuniform distributions of alpha-particle emitting radiolabelled antibodies can depart significantly from the classical mono-exponential curves that were produced by a uniform, external source of alpha particles. In particular, although the inherent cell sensitivity (z_o) was held constant, the slope of the cell survival curve as a function of absorbed dose to the medium was highly dependent upon the source configuration. Those cases in which cells were more uniformly irradiated had steeper cell survival curves than simulations in which the distribution of alpha emitters was highly heterogeneous. Stinchcomb and Roeske [53] further studied the effects of cell size and shape on expected cell survival. In this study, the cell and nucleus were allowed to take on shapes ranging from spheres to ellipsoids where the ratio of the major-to-minor axis was varied from 1 to 5, while the volume of the nucleus was held constant. Separately, the size of the nucleus was varied while the nuclear shape was held constant. Using calculated specific energy spectra and a model of cell survival, they determined that the expected cell survival was not a strong function of the target shape, provided the volume was fixed. However, they noted significant variations in cell survival as the volume of the sensitive target was varied. In a study by Kvinnsland *et al* [54], microdosimetric spectra were calculated for a heterogeneous expression of cell surface antigens, as well as various diameters of cells and cell nuclei. Their results indicated that cell survival may be underestimated by two orders of magnitude if the mean values of any of these parameters are used. The largest contributor to this effect was the use of the average antigen concentration. All of these studies point out a number of features related to alpha-particle microdosimetry. In particular, they demonstrate the importance of accurately simulating the source/target geometry. Inaccuracies (such as using mean values) may impact the specific energy spectrum and subsequent estimation of cell survival.

Stinchcomb and Roeske [55] applied the previously developed methods of predicting cell survival to the analysis of experimentally produced cell survival curves. Two types of source configurations were considered. The first case consisted of cells irradiated in suspension by an ionic solution of ^{212}Bi. In this case, a small fraction of the ^{212}Bi accumulates on or within the cell. The second case consisted of ^{212}Bi chelated to DTPA—a geometry where the source remains in solution outside of the cell. By calculating the specific energy distribution and knowing the resultant cell survival for irradiation condition, the inherent cell survival (z_o) was determined using iterative methods. In this particular case, the authors observed a small difference in z_o for the two source geometries. These differences were attributed to using the average cell size (as opposed to the cell size distribution), and thus demonstrated the importance including the cell size distribution in future microdosimetric calculations. A microdosimetric approach was used by Larsen *et al* [19] to determine the inherent cell sensitivity of human glioma and melanoma cell lines. Both cells lines were grown in microcolonies and irradiated by ^{211}At labelled antibodies. A Monte Carlo method was used to calculate the single-event spectra of microcolonies modelled as an infinite plane. This analysis indicated that 1–2 hits to the cell nucleus reduced the surviving fraction to 37%. Furthermore, the microdosimetric analysis revealed that the sensitivity of individual cells was much greater than would be expected using a nonstochastic absorbed dose analysis. Charlton [56] made refinements to the Larsen [19] model by incorporating the effects of finite cluster sizes. The results of this work indicated that the absorbed dose was a maximum in the centre of the individual clusters and decreased by nearly a factor of 2 near the edges, dependent upon the size of the clusters. In particular, it was observed that for smaller clusters of cells (i.e., <25) the average number of passages for 37% survival was 1.26, which was smaller than the 1.62 estimated using the infinite plane assumption. This analysis indicates the importance in accurately modelling cell geometry and its effect in interpreting cell survival studies. Aurlien *et al* [57] compared the biological effects of ^{211}At radioimmunoconjugates to external gamma rays for osteosarcoma cells. Experiments were performed with cells in suspension containing a uniform solution of the alpha-particle emitting radioactivity. The distribution of cells sizes was measured and a Monte Carlo code was used to calculate the specific energy spectra. Their results indicated that when survival was plotted as a function of the average absorbed dose there was a significantly higher level of cell kill in osteosarcoma cells relative to bone marrow cells. Thus, this form of therapy may prove efficacious. In a departure from the previous studies, Charlton *et al* [58] applied a model of cell survival to *predict* cell survival following alpha-particle irradiation. Initially, the microdosimetric spectra were calculated for cells grown both as a monolayer and in suspension. Using these spectra, and a model of cell survival that took into account the specific energy deposited and the incident particle's LET, a theoretical cell survival curve was generated. A comparison of theoretical with experimental survival data indicated agreement within experimental uncertainties. This approach demonstrated that if knowledge of the cell response as a function of LET is available, it is possible to use microdosimetric spectra with a model of individual cell survival to accurately predict cell survival following alpha-particle irradiation.

The previous studies focused on combining microdosimetric spectra with biological models to analyse or predict survival of cells irradiated in suspension or as a monolayer. A logical extension of this work is the modelling of more complex geometries. One such geometry, multicellular spheroids, mimics small micrometastatic clusters and is therefore therapeutically relevant. However, because of the additional complexity, the computation is significantly more difficult and therefore fewer studies have been performed in this area. Kennel *et al* [59] considered cell survival following the irradiation of multicellular spheroids with an alpha-particle emitting radiolabelled antibody. Following irradiation, the spheroids were divided into three regions corresponding to an outer, middle and inner layer of cells. Cell survival was assessed individually for each layer. Additionally, a Monte Carlo code was developed to calculate specific energy spectra as a function of depth within the spheroid assuming the activity was concentrated on the outer 10 microns. Their results demonstrated a wide variation in the specific energy distribution in the three different levels of the spheroid. The average absorbed dose was a maximum near the outer spheroid surface and decreased rapidly with depth. Conversely, the fraction of cells receiving zero hits was a minimum near the surface and increased rapidly with depth. Additionally, they noted a qualitative correlation for which cells near the outer edges of the spheroids had a lower surviving fraction compared to those inside. Charlton [60] produced a calculational model for multicellular spheroids and simulated their subsequent survival. The distribution of alpha particle tracks throughout cells within the spheroid was calculated and combined with a cell survival model that takes into account the survival as a function of LET [61]. The results of this analysis indicated that cell survival decreased from 57% to 37% as the spheroid diameter increased from 75 to 225 microns for a uniform source with one decay per cell, and 50% packing. Longer ranged alpha-particle emitters increased the number of hits per cell for the larger spheroids. Increasing the packing fraction from 40% to 70% decreased the survival from 46% to 26% in 200 μm diameter spheroids, again with one decay per cell. This significant decrease in survival was due to the increased crossfire component of the absorbed dose as cells were brought closer together. In a different simulation, a small fraction of cells (20%) was assumed not to take up any activity due to lack of expression of a surface antigen. In this simulation, the unlabelled fraction did not significantly change the expected cell survival. In both of these studies, there is evidence of a highly nonuniform specific energy distribution that varies with distance from the spheroid surface. Thus, a single specific energy distribution is not representative of that through the entire tumour. Combining specific energy distributions with a model of cell survival may provide an overall measure of the therapeutic effectiveness, however, these cell survival models do not take into account second order processes such as bystander effects. These processes are more difficult to simulate and may play a significant role for these types of geometries. Chouin *et al* [62, 63] developed an analytical microdosimetric model, validated by comparison with Monte Carlo calculations, to determine cell survival from the microdosimetric spectra and the microdosimetric radiosensitivity (z_o). Isolated cells, as well as cells packed in a monolayer or a spheroid were considered. The model was applied to study the survival of two lymphoid cell lines (Ada and T2) irradiated with ^{213}Bi-radiolabelled monoclonal antibodies. The percentage of cell mortality, and of hit cells and hit nuclei, was represented as a function of the mean absorbed dose for both cell lines. It was observed that for mean absorbed doses below 0.01 Gy the percentage of mortality was higher than that of hit cells, which could be attributed to the bystander effect. For the Ada cells, with a higher radiosensitivity than T2 cells, the percentage of mortality was higher than that of hit nuclei. This suggests that indirect radiation effects may have to be considered to account for cell mortality. Roeske *et al* [64] determined the average number of hits to the cell nucleus to kill a tumour cell population using a tumour control probability model. They obtained that those cells with large nuclei, high radiosensitivity and in which the emissions occur in the nuclei would require a larger number of hits. Minguez *et al* [65] applied the model derived by Roeske *et al* [64] to the case of a nucleus embedded in a uniform distribution of ^{223}Ra atoms and determined the absorbed doses to eradicate a lesion allowing one to vary the cell nucleus size, the cell radiosensitivity and the lesion size. The comparison with clinically obtained absorbed doses suggested that higher ^{223}Ra activities may be necessary to improve treatment outcome. The ongoing refinement of these cell survival models is currently an area of active research [66–69].

Bone marrow is often the organ that limits the activity to administer in therapeutic applications. The dosimetry of bone marrow is difficult due to the complex geometry as well as the presence of tissue inhomogeneities. Thus, idealized models, as have been utilized in the previous studies, must be replaced by more realistic geometries. The work to date on estimating specific energy spectra for bone marrow has focused largely on using histological samples obtained from humans or animal models. While there has been considerable interest in bone marrow micro-dosimetry for protection purposes [70–73], this review will focus only on studies related to RIT. Akabani and Zaltusky [74] obtained histological samples of beagle bone marrow, and manually measured chord length distributions. Using a Monte Carlo program, they calculated the single-event spectra for both sources in the extracellular fluid and sources located on the surface of red marrow cells. These single-event distributions were combined with a model of cell survival. This analysis demonstrated that activity concentrated on the cell surface resulted in significantly higher levels of cell kill than activity in the extracellular fluid. The effect of LET on the survival of human haematopoietic stem cells in various geometries was studied by Charlton *et al* [75]. These geometries were determined from human marrow samples obtained from cadavers. The authors measured the distribution of distances between stem cells and fat cells. Microdosimetric spectra and cell survival were calculated for three different source/target geometries: (a) isolated cells labelled on the surface; (b) non-targeted distribution of decays in an extended volume; and (c) non-targeted decays in marrow with 36% of the marrow volume occupied by fat. Two different radionuclides were considered—^{149}Tb and ^{211}At. These simulations indicated that for targeted decays, ^{149}Tb was five times more effective than ^{211}At when compared on a hit-by-hit basis. This enhancement was due to the lower energy of ^{149}Tb resulting in a higher LET of the incident alpha particles. They also concluded that cell survival was a function of the position of decay relative to the cell nucleus. Using a similar model as Charlton [75], Utteridge *et al* [76] considered the risk of developing secondary malignancies (i.e., leukaemia) from alpha particles. This risk may be important in evaluating the future therapeutic application of alpha particles in patients who have an excellent prognosis. Three alpha-emitting radionuclides were considered based on the particle's relative range (short, medium and long). In this analysis, the authors calculated the fraction of cells that are hit and would survive (as these would potentially cause secondary malignancies). They determined that lowest fraction occurred for low energies and the highest fraction occurred for the highest energy alpha-particle emitter. Hobbs *et al* [77] simulated spheres that represented bone marrow cavities and placed ^{223}Ra on the trabecular bone surface or in the endosteal layer. The interior of the sphere was divided into voxels with the size of a cell and the energy deposited in each voxel determined by Monte Carlo calculations. Results showed a heterogeneous distribution of cellular absorbed dose, with the absorbed dose mainly deposited near the trabecular surface. Due to this heterogeneous distribution, a significant increase in the administered activity, which will result also in a proportional increase of the mean absorbed dose, would not substantially increase the risk of marrow toxicity.

Alpha-particle Monte Carlo codes have been used in a variety of theoretical and *in vitro* analyses. However, direct clinical application of microdosimetry has been limited due to a lack of knowledge of the source distribution as a function of time and location on the microscopic level. The previously described studies serve to provide valuable insight into the stochastics of alpha particle dosimetry that may be useful in interpreting the clinical results of patients treated with alpha-particle emitters.

7.3 Auger electron emitters

Auger electrons are emitted by any radionuclide that decays by either electron capture or internal conversion. Both of these decay processes result in an inner atomic electron vacancy, usually in the innermost K-shell. This vacancy initiates a cascade of inner atomic shell transitions in which a number of low energy Auger, Coster-Kronig and super Coster-Kronig electrons are liberated with emission energies corresponding to the differences in the binding shell energies involved in the transition. The total energy emitted following electron capture corresponds to the binding energy of a K (or L) shell electron for the appropriate atom. This energy is shared between all of the electrons and any fluorescent x-ray photons emitted. For this reason, the total energy emitted is small, and has often been disregarded in organ and total body dosimetry estimates. After internal conversion, the same Auger cascade ensues, with the difference that a potentially much higher 'conversion electron' from the K- (or L-) shell can be ejected with an energy corresponding to the difference between the nuclear excitation energy and the appropriate inner shell electron binding energy.

Radionuclides, which undergo electron capture and/or internal conversion, are extremely common. They constitute about half of the isotopes in the table of radionuclides, including many of the isotopes employed in routine diagnostic nuclear medicine studies e.g. ^{99m}Tc. ^{67}Ga, ^{123}I, ^{201}Tl, etc. The relevance of the soft electrons emitted as a consequence of electron capture or internal conversion went unnoticed until the studies by Ertl, Feinendegen and Heineger [78] and Hofer and Hughes [79]. These independent investigators discovered that when ^{125}I, which decays by electron capture followed by internal conversion, is directly incorporated into the cellular genome, via the thymidine analogue ^{125}I-iodiodeoxyuridine, the cell survival curves resemble the appearance of those produced by alpha particles. Estimates of the radiation absorbed dose revealed that intra-nuclear decays of ^{131}I deposited more absorbed dose to the cell nucleus than ^{125}I, yet investigators found ^{125}I was >12 times more lethal then ^{131}I on a per decay basis [80, 81]. This led radiobiologists to wonder how low energy and low LET electrons could exhibit high LET properties. The history of ^{125}I radiobiology has been reviewed by Sastry [82]. In brief, Charlton and Booz [83] simulated the atomic relaxation of ^{125}I following electron capture and internal conversion and revealed that on average 21 electrons are emitted per decay. The large number of electrons emitted, and the positive charge build-up in their wake, results in a massive local energy deposit within a few Angstroms of the decay site. Microscopic track structure simulations revealed the physical explanation for the high radiotoxicity of ^{125}I, when Charlton [84] showed that the density of ionisations around the decay site of ^{125}I is greater than in the core of an alpha-particle track. Since this time, improved calculations of Auger spectra for a number of electron capture and internal conversion decaying radionuclides have been generated [85, 86], more elaborate models of the DNA have been created [4, 87, 88], and better experimental data on strand break damage versus distance from the decay site has been measured on high resolution denaturing polyacrylamide gels [89, 90].

The inferences that can be drawn from this work are the following:

(1) The RBE of Auger electron sources is no different from any other low LET source, provided the decay does not occur adjacent to a critical cellular target [81].

(2) The range of effectiveness of Auger electron-emitting isotopes is extremely short and decay at or within one nanometre of a sensitive structure, such as the DNA, is subject to high LET effects [91].

(3) Since it is the pharmaceutical carrier which determines the subcellular location, the RBE for any Auger emitting radionuclide depends not only on the isotope, but also on the carrier molecule [92].

7.3.1 Monte Carlo simulation of Auger decays and their energy deposition

There is considerable stochastic variability in the way that energy is released from electron capture or internal conversion decaying radionuclides. This variability was first modelled by Monte Carlo simulation of the decay process of ^{125}I [83]. Since then, Auger electron spectra have been calculated for several other radionuclides [85, 86]. The number and distribution of energy between the electrons (and photons) emitted in an Auger cascade varies between individual disintegrations as governed by the inner shell transition probabilities. The number of permutations of possible de-excitation pathways through the atom can be hundreds, depending on the atomic number of the isotope. The way in which electron capture (or internal conversion) and the ensuing Auger cascade are computer simulated is described in detail by Charlton and Booz [83], Humm [93], and Howell [85]. In brief, a random number generator is used to select an initial inner shell vacancy according to the probabilities of capture within each of the respective orbitals K, L1, L2, L3, etc. The probabilities for all transitions to this vacancy are normalized to one, where each allowed transition is weighted by its respective probability according to published values for radiative [94] and non-radiative [95] transitions, determined from quantum mechanical calculations of oscillator strengths. A second random number is then used to select one of these transitions, according to their respective weights. The new electron occupancy of the orbitals is recorded and the next innermost shell vacancy is selected. This process then repeats until all of the vacancies have moved to the outermost shells (the lowest energy state), when further transitions are no longer possible.

For high Z elements e.g. tellurium following K-capture in iodine, the fluorescent yield is high (0.877). This means that 87.7% of the transitions to fill a K-shell vacancy emit characteristic x-rays and only 12.3% result in a K-shell Auger electron. This results in a variable total amount of electron energy released, which occurs contingent upon the photon transitions. One example of a possible atomic de-excitation pathway following electron capture for ^{123}I spectra is schematically depicted in figure 7.4.

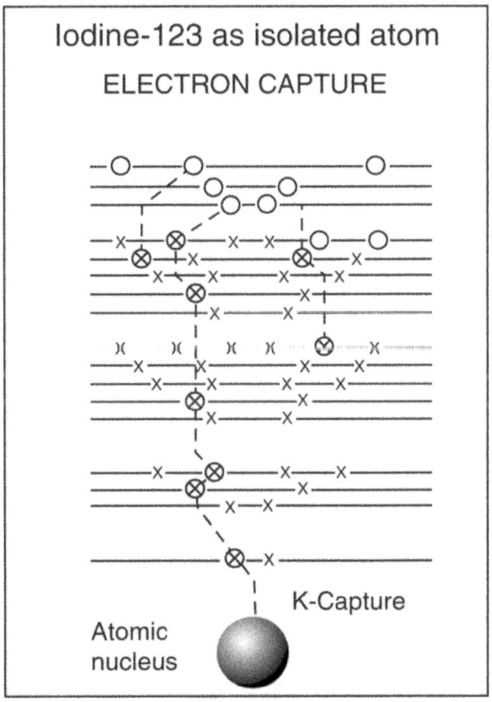

Figure 7.4. Schematic diagram of an ^{123}I atom undergoing electron capture. The crosses denote electron occupancy of each shell. A vacancy is represented by a empty circle, and a filled vacancy (as a consequence of a transition) by a circle containing a cross. Each transition moves a vacancy to a higher orbital until all vacancies occupy the outermost shells.

Simulation of 10,000 individual atomic de-excitations results in an output of individual Auger electron and x-ray spectra, containing the stochastics of the de-excitation process. These data can be averaged to yield an average Auger electron emission spectrum, an example of which is shown in figure 7.5. Note that this does not denote an emission spectrum from an individual decay, but rather a frequency weighted average of 10,000 simulated decays. Dosimetric estimates to different compartments of the cell have been performed using the average Auger electron spectrum, and the results are the same as when multiple individual Auger electron spectra are used. The stochastics of individual spectra is only pertinent when calculating the energy deposition along the DNA.

Charlton and Humm [96] performed Monte Carlo electron track simulations for 10,000 individual Auger electron spectra resulting from ^{125}I and other radionuclides [91]. They scored the coordinates of each energy deposition generated by the electron track structure code of Paretzke [97], relative to the origin of the site of decay placed at the position of a base in a computer-simulated geometric model of the DNA double helix. This model consisted of a central cylinder representing the bases surrounded by a concentric helical cylinder of pitch 0.36 nm, representing the sugar phosphate strands. The resulting energy depositions within opposing strand segments of this DNA model were then compared to experimental data on the radius of the strand break damage measured with DNA sequencing gels relative to the ^{125}I decay site by Martin and Haseltine [98]. By matching the theoretical profile of stochastic energy deposition versus distance from the site of ^{125}I decay, with the measured single strand break frequency profile, allowed a threshold energy for strand break damage to be estimated. A threshold energy of 17.5 eV for single strand break production provided the best agreement with experimental data. To improve statistics, the energy deposition was calculated for the same individual Auger spectrum 10 times, and it was found that the variability in local energy deposition

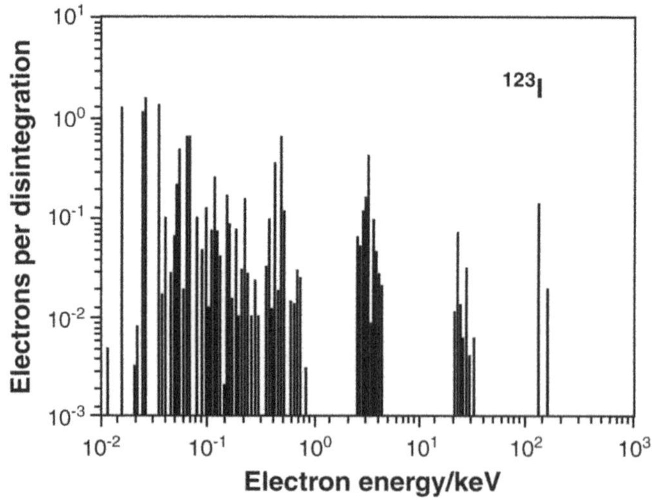

Figure 7.5. An example of an average electron energy spectrum from ^{123}I following 10,000 simulated decays.

was substantial and added yet another layer of stochastic variability in the data. In summary, variation in energy deposition results from the inherent variability on atomic de-excitation and second from the track structure of these emissions relative to the modelled target sites. Whereas these stochastics are not important for cellular level dosimetry estimates, except at very low numbers of decays per cell, they are important for predicting strand break damage. This is because any threshold model of strand break damage does not bear a linear relationship to the mean energy deposition.

The stochastic effects of Auger electron emitters and the high LET effects are relevant within a few nanometres of the decay site, since most Auger and Coster-Kronig electrons have ranges of this order. Therefore, for Auger sources that decay at the extracellular locations (as is the case for most diagnostic tracers used in nuclear medicine), concerns of high LET and the consequent radiotoxicity effects are unwarranted. However, the extreme short-range effects of Auger electron emitters do pose a new problem in assessing the dosimetry of novel radiopharmaceuticals. Humm *et al* [99] and Goddu *et al* [100] have recommended some guidelines for their evaluation, which requires the initial determination of the partition of the agent into extracellular, cell surface bound, intracellular, intra-nuclear and appended to the DNA. Cellular S-values for the former four compartments have been calculated by Goddu *et al* [101, 102] for a number of electron capture and internal conversion decaying radionuclides. The average radiation absorbed dose to the nucleus (or other compartment) can be determined by the piecewise addition of each contribution, upon experimental determination of the relative compartmental partitions.

7.3.2 Targeted therapy with Auger electron emitters

Auger electron emitters have been used in pre-clinical and clinical studies in a number of radiopharmaceutical forms designed to selectively deliver radiation to tumour cells. These include thymidine precursors [103–107], thymidine precursor prodrugs [108], minor DNA groove binding bisbenzimidazole ligands [109], triplet forming oligonucleotides [110], hormones [111], antibodies [112–115] and peptides [116, 117]. Reviews of emerging role of Auger targeting therapies can be found in Mariani *et al* [118] and Ku *et al* [119]. The most effective therapies target the Auger source directly to the DNA such as: iododeoxyuridine, Hoechst 33258 or triplex forming oligonucleotides. One drawback of these agents is the lack of tumour specificity, although attempts to attach these agents to tumour specific carriers, such as antibodies, as means of selective tumour targeting and intracellular access are currently under investigation. The potential therapeutic advantages of using Auger emitters for therapy with internalising antibodies are described by Daghighian *et al* [120] and Behr *et al* [121]. In brief, the short range of Auger electrons render them ineffectual in sterilizing the usual dose limiting tissues such as bone marrow, since they neither bind to nor are internalised by these cells. This was clinically demonstrated by Welt *et al* [114], who administered to patients activities in excess of 29 GBq (800 mCi) of ^{125}I-labelled A33 antibody and still did not reach absorbed dose-limiting toxicity. The cell cytoplasm acts to shield the nucleus and cellular genome from the low energy electron emissions. Antibody binding to cell surface antigen and eventual internalisation increases the spatial proximity of the Auger source to the targets of cell inactivation, thus giving rise to a significant geometric and hence dosimetric differential between tumour and normal cells. However, the deposition of energy from Auger electron emitters with antibody (or hormone) approaches to the tumour targeting do not result in the high LET effects observed with ^{125}I-iododeoxyuridine. With these approaches the Auger source is treated like a very soft β-source, such as tritium. As a consequence, for targeted Auger therapy to be successful, the radiolabelled carrier must reach every clonogenically potent tumour cell. This is a severe limitation. However, the theoretical rewards are enticing and it is this reason, which drives researchers to further pursue the enormous potential of high radiotoxicity associated with Auger emitters with limited toxicity to normal tissues.

7.4 Future directions

The number of radiopharmaceuticals which have alpha-particle emitters and are potentially usable for different therapies in molecular radiotherapy has increased in the last several years. Among those radiopharmaceuticals are ^{213}Bi-PSMA, ^{223}Ra-Cl_2 and ^{225}Ac-PSMA. There have been some applications of microdosimetry to clinical practice with those radiopharmaceuticals [65, 122] but they are limited for different reasons. One of them is that simulation of realistic geometries and activity distributions as a function of time at the cellular level is a challenging problem [37]. There are different techniques that can be used to determine the distribution of the radiopharmaceuticals at subcellular level from tissue samples. Among these techniques are microautoradiography, nanoscale secondary ion mass spectrometry, radioluminiscence microscopy and microdosimeters [123]. Despite the lack of uniformity in the distribution of the radiopharmaceuticals reported [124, 125], clinical results have proven the effectiveness of different radiopharmaceuticals [126, 127]. The response in targeted therapies with alpha particles could be satisfactorily explained from microdosimetry calculations if the energy deposition by alpha particles were the main cause for that response. However, biological effects, such as the bystander effect, DNA repair or immune mediated responses triggered by the tumour irradiation, such as the abscopal effect, have to be considered, so a more complex model would be needed to explain the response to the treatment [128].

Microdosimetry calculations can also be helpful in the study of therapeutic applications of radiopharmaceuticals with Auger electron emitting radionuclides. For instance, the cytotoxicity of ^{111}In-DTPA-hEGF [129] and ^{125}I-mIBG [130] has been studied, as has the potential use of ^{64}Cu-MM-302 in some treatments estimating a maximum tolerable activity to administer [131]. Microdosimetry calculations for different radiopharmaceuticals have shown that the highest lethality for Auger electrons occurs when they are incorporated into DNA [132]. For example, to obtain the same lethality for ^{64}Cu-ATSM with regard to ^{125}IUdR, the activity concentration per cell of ^{64}Cu-ATSM has to be 100 times higher than that of ^{125}I because of the lower accumulation of ^{64}Cu-ATSM in or near DNA [133].

Research aimed at therapeutic applications should be focused on internalisation of Auger electron emitters into cancer cells aimed at nuclear localisation [134]. Thus, for estimation of treatment response from microdosimetry calculations, knowledge of the radiopharmaceutical distribution would be essential in order to estimate the biological damage. Moreover, as with alpha particle emitters, in order to have a better estimation of that damage, it will be necessary to consider biological effects, such as the bystander effect or repair effects which have been proven to be influenced by the Auger electrons [119].

Calculated subcellular S-values with Monte Carlo codes have been reported to be comparable to those determined with analytical methods. However, Monte Carlo codes allow for more flexible models, for instance, regarding cell geometries and radiopharmaceutical distributions [129]. Currently, different Monte Carlo codes have been used and are available to perform microdosimetry calculations (e.g. Geant4, Geant4DNA, PENELOPE, GATE and MCNP) [135–138]. The results obtained using these codes have allowed an increased understanding of stochastic patterns of energy deposition by alpha particle and Auger electron emitters in both simple and complex geometries, and for different subcellular distributions. For the case of alpha particles, track structure models including delta and epsilon rays could lead to a better prediction of the biological damage [139]. For the case of Auger electrons, new Monte Carlo codes, such as the BriccEmis code [140], are improving the accuracy in the determination of the emission spectra of Auger electron emitting radionuclides, which has been shown to vary depending on the code used [135]. This variation will not affect substantially S-values calculated for the larger volumes, but it will have a significant impact at the nanoscale and it will affect the dose-point kernels in the first few hundred nanometres [119, 135]. The use of Monte Carlo codes can help to gain accuracy in calculations at the nanoscale [128]. DNA damage at the nanoscale should be correlated with damage at higher scales. Clustering of energy deposition at the nanoscale can produce clustered lesions which are harder to repair and whose frequency increases with the LET of the radiation [141, 142]. Some attempts have been made to include DNA repair processes in Monte Carlo simulations [129, 143].

An initial approach to estimate the therapeutic effectiveness within individual patients could be to study cell response to alpha particles or Auger electrons for tumour cells taken from patients. Next, using Monte Carlo codes, the activity distribution, cell geometry and cell response might be combined to predict cell survival. The results obtained might help to determine whether the treatment could be effective for the patient, or whether the activity to administer can be tailored for the individual patient. However, cell response *in vitro* may not correlate with tumour response *in vivo*. This approach could thus be refined with the use of the results reported in patients together with improved and updated radiobiological models [128]. Input from microdosimetry is required to understand and interpret the measured RBE values when based on common biological end points, such as cell survival fractions and double-strand break yields resulting from *in vivo* studies with alpha or Auger targeted molecules [144]. Microdosimetry could also be used to study toxicity effects caused by alpha particles or Auger electrons [145, 146], for example, the lung toxicity [147] or the bone marrow toxicity [77] caused by alpha particles. In pre-clinical studies, microdosimetry calculations could give a first approach of the biological damage caused by new radiopharmaceuticals with alpha particle or Auger electron emitting radionuclides. The spatial and temporal distributions of previously used molecules may be known, which will help to achieve a better prediction. Microdosimetry calculations from those distributions, together with the appropriate radiobiological models, may advance whether the new radiopharmaceutical could be effective and its development worthwhile.

References

[1] Rossi H H 1968 Microdosimetric energy distribution in irradiated matter *Radiation Dosimetry Vol.I Fundamentals* ed F H Attix and W C Roesch (New York: Academic)

[2] Charlton D E, Nikjoo H and Humm J L 1989 Calculation of initial yields of single- and double-strand breaks in cell nuclei from electrons, protons and alpha particles *Int. J. Radiat. Biol.* **56** 1–19

[3] Brenner D J 1990 Track structure, lesion development, and cell survival *Radiat. Res.* **124** S29–37

[4] Pomplun E 1991 A new DNA target model for track structure calculations and its first application to I-125 Auger electrons *Int. J. Radiat. Biol.* **59** 625–42

[5] Friedland W, Jacob P, Paretzke H G, Merzagora M and Ottolenghi A 1999 Simulation of DNA fragment distributions after irradiation with photons *Radiat. Environ. Biophys.* **38** 39–47

[6] Nikjoo H, O'Neill P, Terrissol M and Goodhead D T 1999 Quantitative modelling of DNA damage using Monte Carlo track structure method *Radiat. Environ. Biophys.* **38** 31–8

[7] de la Fuente L, Incerti S, Francis Z and Bernal M A 2018 Accounting for radiation induced indirect damage on DNA with the GEANT 4-DNA code *Phys. Med.* **51** 108–16

[8] Forster J C, Douglas M J J, Philips W M and Bezak E 2018 Monte Carlo simulation of the oxygen effect in DNA damage induction by ionizing radiation *Radiat. Res.* **190** 248–61

[9] Kellerer A M and Chmelevsky D 1975 Criteria for the applicability of LET *Radiat. Res.* **63** 226–34

[10] Makrigiorgos G M, Ito S, Baranowska-Kortylewicz J, Vinter D W, Iqbal A, Van den Abbeele A D, Adelstein S J and Kassis A I 1990 Inhomogeneous deposition of radiopharmaceuticals at the cellular level: experimental evidence and dosimetric implications *J. Nucl. Med.* **31** 1358–63

[11] Kozak R W, Atcher R W, Gansow O A, Friedman A M, Hines J J and Waldmann T 1986 Bismuth-212 labeled anti-Tac monoclonal antibody: alpha-particle-emitting radionuclides as modalities for radioimmunotherapy *Proc. Natl Acad. Sci.* **83** 474–8

[12] Kurtzman S H *et al* 1988 ^{212}Bismuth linked to an antipancreatic carcinoma antibody: model for alpha-particle-emitter radioimmunotherapy *J. Natl. Cancer Inst.* **80** 449–52

[13] Macklis R M, Kinsey B I, Kassis A I, Ferrara J L M, Atcher R W, Hines J J, Coleman C N, Adelstein S J and Burakoff S J 1989 Radioimmunotherapy with alpha-particle-emitting immunoconjugates *Science* **240** 1024–26

[14] Zalustky M R, McLendon R E, Garg P K, Archer G E, Schuster J M and Bigner D D 1994 Radioimmunotherapy of neoplastic meningitis in rats using an alpha-particle-emitting immunoconjugate *Cancer Res.* **54** 4719–25

[15] Zalutsky M R, Stabin M G, Larsen R H and Bigner D D 1997 Tissue distribution and radiation dosimetry of astatine-211-labeled chimeric 81C6, an alpha-particle-emitting immunoconjugate *Nucl. Med. Biol.* **24** 255–61
[16] Rotmensch J, Whitlock J L, Schwatrz J L, Hines J J, Reba R C and Harper P V 1997 *In vitro* and *in vivo* studies on the development of the alpha-emitting radionuclide bismuth 212 for intraperitoneal use against microscopic ovarian carcinoma *Am. J. Obstet. Gynecol.* **176** 833–41
[17] Kennel S J and Mirzadeh. S 1997 Vascular targeting for radioimmunotherapy with ^{213}Bi *Radiochim. Acta* **79** 87–91
[18] Larsen R H, Vaidyanathan G and Zalutsky M R 1997 Cytotoxicity of alpha-particle-emitting 5-[^{211}At]astato-2'-deoxyuridine in human cancer cell *Int. J. Radiat. Biol.* **72** 79–90
[19] Larsen R H, Akabani G, Welsh P and Zalutsky M R 1998 The cytotoxicity and microdosimetry of astatine-211-labeled chimeric monoclonal antibodies in human glioma and melanoma cells *in vitro Rad. Res.* **149** 155–62
[20] Couturier O *et al* 1999 Validation of ^{213}Bi-alpha radioimmunotherapy for multiple myeloma *Clin. Cancer Res.* **5** 3165s–70s
[21] Mulford D A, Scheinberg D A and Jurcic J G 2005 The promise of targeted alpha-particle therapy *J. Nucl. Med.* **46** 199S–204S
[22] Chang M Y, Seideman J and Sofou S 2008 Enhanced loading efficiency and retention of ^{225}Ac in rigid liposomes for potential targeted therapy of micrometastases *Bioconjug. Chem.* **19** 1274–82
[23] Chouin N, Lindegren S, Jensen H and Albertsson P 2012 Quantification of activity by alpha-camera imaging and small-scale dosimetry within ovarian carcinoma micrometastases treated with targeted alpha therapy *Q. J. Nucl. Med. Mol. Imaging* **56** 487–95
[24] Kim Y S and Brechbiel M W 2012 An overview of alpha targeted therapy *Tumour Biol.* **33** 573–90
[25] Chouin N, Lindegren S, Frost S H, Jensen H, Albertsson P, Hultborn R, Palm S, Jacobsson L and Back T 2013 *Ex vivo* activity quantification in micrometastases at the cellular scale using the alpha-camera technique *J. Nucl. Med.* **54** 1347–53
[26] Pandit-Taskar N, Larson S M and Carrasquillo J A 2014 Bone-seeking radiopharmaceuticals for treatment of osseous metastases. Part 1: α therapy with 223Ra-dichloride *J. Nucl. Med.* **55** 268–74
[27] Larson S M, Carrasquillo J A, Cheung N V and Press O W 2015 Radioimmunotherapy of human tumours *Nat. Rev. Cancer* **15** 347–60
[28] Kiess A P *et al* 2016 (2*S*)-2-(3-(1-Carboxy-5-(4–^{211}At-astatobenzamido)pentyl)ureido)-pentanedioic acid for PSMA-targeted alpha-particle radiopharmaceutical therapy *J. Nucl. Med.* **57** 1569–75
[29] Sattiraju A *et al* 2017 Alpha particle enhanced blood brain/tumor barrier permeabilization in glioblastomas using integrin alpha-v beta-3-targeted liposomes *Mol. Cancer Ther.* **16** 2191–200
[30] Sathekge M, Knoesen O, Meckel M, Modiselle M, Vorster M and Marx S 2017 ^{213}Bi-PSMA-617 targeted alpha-radionuclide therapy in metastatic castration-resistant prostate cancer *Eur. J. Nucl. Med. Mol. Imaging* **44** 1099–100
[31] Kratochwil C, Brutechseifer F, Rathke H, Hohenfellner M, Giesel F l, Haberkorn U and Morgenstern A 2018 Targeted α-therapy of metastatic castration-resistant prostate cancer with ^{225}Ac-PSMA-617: swimmer-plot analysis suggests efficacy regarding duration of tumor control *J. Nucl. Med.* **59** 795–802

[32] Sgouros G, Ballangrud A M, Jurcic J G, McDevitt M R, Humm J L, Erdi Y E, Mehta B M, Finn R D, Larson S M and Scheinberg D A 1999 Pharmacokinetics and dosimetry of an alpha-particle emitter labeled antibody: ^{213}Bi-HuM195 (Anti-CD33) in patients with leukemia *J. Nucl. Med.* **40** 1935–46

[33] Hall E J 2000 *Radiobiology for the Radiologist* (Philadelphia, PA: Lippincott, Williams and Wilkins)

[34] Humm J L 1987 A microdosimetric model of astatine-211 labeled antibodies for radioimmunotherapy *Int. J. Radiat. Oncol. Biol. Phys.* **13** 1767–73

[35] Humm J L and Chin L M 1993 A model of cell inactivation by alpha-particle internal emitters *Rad. Res.* **134** 143–50

[36] Stinchcomb T G and Roeske J C 1992 Analytic microdosimetry for radioimmunotherapeutic alpha emitters *Med. Phys.* **19** 1385–93

[37] Roeske J C, Aydogan B, Bardies M and Humm J L 2008 Small-scale dosimetry: challenges and future directions *Semin. Nucl. Med.* **38** 367–83

[38] Polig E 1978 The localized dosimetry of internally deposited alpha-emitters *Curr. Top. Rad. Res.* **13** 189–327

[39] Olko P and Booz J 1990 Energy deposition by protons and alpha particles in spherical sites of nanometer to micrometer diameter *Radiat. Environ. Biophys.* **28** 1–17

[40] Eckerman K F, Ryman J C, Taner A C and Kerr G D 1985 Traversal of cells by radiation and absorbed dose fraction estimates for electrons and alpha particles *Proc. of the Fourth Int. Radiopharmaceutical Dosimetry Symposium.* ed A T Schlafke-Stelson and E E Watson (Oak Ridge, TN: Oak Ridge Associated Universities) 67–86 CONF-851113

[41] Roeske J C and Hoggarth M 2007 Alpha-particle Monte Carlo simulation for microdosimetric calculations using a commercial spreadsheet *Phys. Med. Biol.* **52** 1909–22

[42] Attix F H 1986 *Introduction to Radiological Physics and Radiation Dosimetry* (New York: Wiley)

[43] Ziegler J F, Biersack J P and Littmark U 1985 *The Stopping and Range of Ions in Matter* (New York: Pergamon)

[44] Ziegler J F 1988 The stopping and range of ions in solids *Ion Implantation: Science and Technology* (San Diego, CA: Academic) pp 3–61

[45] Ziegler J F 1999 The stopping of energetic light ions in matter *Appl. Phys. Rev., J. Appl. Phys.* **85** 1249–72

[46] Bichsel H 1992 Stopping powers of fast charged particles in heavy elements *NIST Report IR-4550* (Springfield, VA: U.S. National Technical Information Service)

[47] ICRU 1993 Stopping Powers and Ranges for Protons and Alpha Particles *Report No. 49* (Bethesda, MD: International Commission of Radiation Units)

[48] Janni J F 1966 Calculation of energy loss, range, path length, straggling, multiple scattering and the probability of inelastic nuclear collisions for 0.1 to 1000 MeV protons (Technical Report No. AFWL-TR-65-150)

[49] Walsh P J 1970 Stopping power and range of alpha particles *Health Phys.* **19** 312–16

[50] Charlton D E and Sephton R 1991 A relationship between microdosimetric spectra and cell survival for high-LET irradiation *Int. J. Radiat. Biol.* **59** 447–57

[51] Fisher D R 1986 The microdosimetry of monoclonal antibodies labeled with alpha emitters *Proceedings of the Fourth International Radiopharmaceutical Dosimetry Symposium.* ed A T Schlafke-Stelson and E E Watson (Oak Ridge, TN: Oak Ridge Associated Universities) pp 26–36 CONF-851113

[52] Roesch W C 1977 Microdosimetry of internal sources *Rad. Res.* **70** 494–510
[53] Stinchcomb T G and Roeske J C 1995 Survival of alpha particle irradiated cells as a function of the shape and size of the sensitive target (nucleus) *Rad. Prot. Dosim.* **62** 157–64
[54] Kvinnsland Y, Stokke T and Aurilien E 2001 Radioimmunotherapy with alpha-particle emitters: Microdosimetry of cells with heterogeneous antigen expression and with various diameters of cells and nuclei *Rad. Res.* **155** 288–96
[55] Stinchcomb T G and Roeske J C 1994 Analysis of survival of C-18 cells after irradiation in suspension with chelated and ionic bismuth-212 using microdosimetry *Rad. Res.* **140** 48–54
[56] Charlton D E 1999 The survival of monolayers of cells growing in clusters irradiated by ^{211}At appended to the cell surfaces *Rad. Res.* **151** 750–53
[57] Aurlien E, Larsen R H, Akabani G, Olsen D R, Zalutsky M R and Bruland O S 2000 Exposure of human osteosarcoma and bone marrow cells to tumor-targeted alpha-particles and gamma-irradiation: analysis of cell survival and microdosimetry *Int. J. Radiat. Biol.* **76** 1129–41
[58] Charlton D E, Kassis A I and Adelstein S J 1994 A comparison of experimental and calculated survival curves for V79 cells grown as monolayers or in suspension exposed to alpha irradiation from ^{212}Bi distributed in growth medium *Rad. Prot. Dosim.* **52** 311–15
[59] Kennel S J *et al* 1999 Radiotoxicity of bismuth-231 bound to membranes of monolayer and spheroid cultures of tumor cells *Rad. Res.* **151** 244–56
[60] Charlton D E 2000 Radiation effects in spheroids of cells exposed to alpha emitters *Int. J. Radiat. Biol.* **76** 1555–64
[61] Charlton D E and Turner M S 1996 Use of chord lengths through the nucleus to simulate the survival of mammalian cells exposed to high-LET alpha-radiation *Int. J. Radiat. Biol.* **69** 213–17
[62] Chouin N, Bernardeau K, Davodeau F, Cherel M, Faivre-Chauvet A, Bourgeois M, Apostilidis C, Morgenstern A, Lisbona A and Bardies M 2009 Evidence of extra-nuclear sensitivity to alpha irradiation using a new microdosimetric model I. Presentation and validation of a new microdosimetric model *Radiat. Res.* **171** 657–63
[63] Chouin N, Bernardeau K, Bardies M, Faivre-Chauvet A, Bourgeois M, Apostilidis C, Morgenstern A, Lisbona A, Cherel M and Davodeau F 2009 Evidence of extra-nuclear sensitivity to alpha irradiation using a new microdosimetric model II. Application of the new microdosimetric model to experimental results *Radiat. Res.* **171** 664–73
[64] Roeske J C and Stinchcomb. T G 2006 The average number of alpha particle hits to the cell nucleus required to eradicate a tumour cell population *Phys. Med. Biol.* **51** N179–86
[65] Mínguez P, Roeske J C, Mínguez R, Gómez de Iturriaga A and Rodeño E 2018 Comparison of microdosimetry-based absorbed doses to control tumours and clinically obtained tumour absorbed doses in treatments with ^{223}Ra *Phys. Med. Biol.* **63** 105005
[66] Iyer M and Lehnert B E 2002 Alpha-particle-induced increases in the radioresistance of normal human bystander cells *Rad. Res.* **157** 3–7
[67] Huang C, Guatelli S, Oborn B M and Allen B J 2012 Microdosimetry for targeted alpha therapy of cancer *Comput. Math. Methods Med.* **2012** 153212
[68] Hobbs R F, Howell R W, Song H, Baechler S and Sgouros G 2014 Redefining relative biological effectiveness in the context of EQDX formalism : implications for alpha-particle emitter therapy *Rad. Res.* **151** 90–8
[69] Ma W, Wang X, Liu W, Ma H, Su Y, Yang Y, Liu N, Wang Y and Yang G 2019 A theoretical model for predicting and optimizing *in vitro* screening of potential targeted alpha-particle therapy drugs *Radiat. Res.* **191** 475–82

[70] Humphreys E R and Humm J L 1988 A Monte-Carlo approach to the microdosimetry of ^{224}Ra in murine compact and cancellous bone *Health Phys.* **54** 607–15

[71] Charlton D E, Utteridge T D and Beddoe A H 1996 Microdosimetry of haemopoietic stem cells irradiated by alpha particles from short-lived products of ^{222}Rn decays in fat cells and heamopoietic tissue *Int. J. Radiat. Biol.* **69** 585–92

[72] Charlton D E, Salmon P L and Utteridge T D 1998 Monte Carlo/numerical treatment of alpha-particle spectra from sources buried in bone and resultant doses to surface cells *Int. J. Radiat. Biol.* **73** 89–92

[73] Austin A L, Ellender M, Haines J W, Harrison J D and Lord B I 2000 Microdistribution and localized dosimetry of the alpha-emitting radionuclides ^{239}Pu, ^{241}Am and ^{233}U in mouse femoral shaft *Int. J. Radiat. Biol.* **76** 101–11

[74] Akabani G and Zalutsky M R 1997 Microdosimetry of astatine-211 using histological images: application to bone marrow *Rad. Res.* **148** 599–607

[75] Charlton D E, Utteridge T D and Allen B J 1998 Theoretical treatment of human haemopoietic stem cell survival following irradiation by alpha particles *Int. J. Radiat. Biol. Phys.* **74** 111–18

[76] Utteridge T D, Charlton D E and Allen B J 2001 Monte carlo modeling of the effects of injected short-, medium- and longer-range alpha-particle emitters on human marrow at various ages *Rad. Res.* **156** 413–18

[77] Hobbs R F, Song H, Watchman C J, Bolch W E, Aksnes A, Ramdahl T, Flux G D and Sgouros. G 2012 A bone marrow toxicity model for ^{223}Ra alpha-emitter radiopharmaceutical therapy *Phys. Med. Biol.* **57** 3207–22

[78] Ertl H H, Feinendegen L E and Heiniger H J 1970 Iodine-125, a tracer in cell biology: physical properties and biological aspects *Phys. Med. Biol.* **15** 447–56

[79] Hofer K G and Hughes W L 1971 Radiotoxicity of intranuclear tritium, 125 iodine and 131 iodine *Radiat. Res.* **47** 94–101

[80] Chan P C, Lisco E, Lisco S and Adelstein S J 1978 Cell survival and cytogenetic responses to 125I-UdR in cultured mammalian cells *Curr. Top. Radiat. Res.* **12** 426–35

[81] Hofer K G 1980 Radiation biology and potential therapeutic applications of radionuclides *Bull. Cancer* **67** 343–53

[82] Sastry K S 1992 Biological effects of the Auger emitter iodine-125: A review. Report No. 1 of AAPM Nuclear Medicine Task Group No. 6 *Med. Phys.* **19** 1361–70

[83] Charlton D E and Booz J 1981 A Monte Carlo treatment of the decay of ^{125}I *Radiat. Res.* **87** 10–23

[84] Charlton D E 1986 The range of high LET effects from ^{125}I decays *Radiat. Res.* **107** 163–71

[85] Howell R W 1992 Radiation spectra for Auger-electron emitting radionuclides: Report No. 2 of AAPM Nuclear Medicine Task Group No. 6 *Med. Phys.* **19** 1371–83

[86] Eckerman K F and Endo A (ed) 2008 *MIRD Radionuclide Data and Decay Schemes* (Reston, VA: The Society of Nuclear Medicine)

[87] Nikjoo H, Martin R F, Charlton D E, Terrissol M, Kandaiy S and Lobachevsky P 1996 Modelling of Auger-induced DNA damage by incorporated ^{125}I *Acta Oncol.* **35** 849–56

[88] Raisali G, Mirzakhanian L, Masoudi S F and Semsarha F 2013 Calculation of DNA strand breaks due to direct and indirect effects of Auger electrons from incorporated ^{123}I and ^{125}I radionuclides using the Geant4 computer code *Int. J. Rad. Biol.* **89** 57–64

[89] Lobachevsky P N and Martin R F 1996 DNA strand breakage by ^{125}I-decay in a synthetic oligodeoxynucleotide—2. Quantitative analysis of fragment distribution *Acta Oncol.* **35** 809–15

[90] Lobachevsky P N and Martin R F 2000 Iodine-125 decay in a synthetic oligodeoxynucleotide. I. Fragment size distribution and evaluation of breakage probability *Radiat. Res.* **153** 263–70
[91] Humm J L and Charlton D E 1989 A new calculational method to assess the therapeutic potential of Auger electron emission *Int. J. Radiat. Oncol. Biol. Phys.* **17** 351–60
[92] Narra V R, Howell R W, Harrapanhalli R S, Sastry K S and Rao D V 1992 Radiotoxicity of some iodine-123, iodine-125 and iodine-131-labeled compounds in mouse testes: implications for radiopharmaceutical design *J. Nucl. Med.* **33** 2196–201
[93] Humm J L 1984 The analysis of Auger electrons released following the decay of radioisotopes and photoelectric interactions and their contribution to energy deposition (KFA Report Juelich-Nr.1932, *PhD Thesis*, Germany, June 1984)
[94] Salem S I, Panossian S L and Krause R A 1974 Experimental K and L relative X-ray emission rates *At. Data Nucl. Data Tables* **14** 91–109
[95] Chen M H, Crasemann B and Mark H 1979 Relativistic radiationless transition probabilities fior atomic K and L shells *At. Data Nucl. Data Tables* **24** 13–37
[96] Charlton D E and Humm J L 1988 A method of calculating initial strand breakage following the decay of ^{125}I *Int. J. Radiat. Biol.* **53** 353–65
[97] Paretke H 1987 Radiation track structure *Kinetics of Non-homogeneous Processes* ed C R Freeman (New York: Wiley)
[98] Martin R F and Haseltine W A 1981 Range of radiochemical damage to DNA with decay of iodine-125 *Science* **213** 896–8
[99] Humm J L, Howell R W and Rao D V 1994 Dosimetry for Auger electron emitters: Report No.3 of AAPM Nuclear Medicine Task Group No.6 *Med. Phys.* **21** 1901–15
[100] Goddu S M, Howell R W and Rao D V 1996 Calculation of equivalent dose for Auger electron emitting radionuclides distributed in human organs *Acta Oncol.* **35** 909–16
[101] Goddu S M, Rao D V and Howell R W 1994 Cellular dosimetry: absorbed fractions for monoenergetic electron and alpha particle sources and S-values for radionuclides uniformly distributed in different cell compartments *J. Nucl. Med.* **35** 303–16
[102] Goddu S M, Rao D V and Howell R W 1994 Multicellular dosimetry for micrometastases: dependence of self-dose versus cross-dose to cell nuclei on type and energy of radiation and subcellular distribution of radionuclides *J. Nucl. Med.* **35** 521–30
[103] Bagshaw K D, Sharma S K, Southall P J, Boden J A, Boxer G M, Partridge T A, Antoniw P and Pedley R B 1991 Selective uptake of toxic nucleoside (^{125}IUdR) by resistant cancer *Br. J. Radiol.* **64** 37–44
[104] Mariani G *et al* 1996 targeting by intra-arterial infusion of 5-[^{123}I]iodo-2'-deoxyuridine in patients with liver metastases from colorectal cancer *J. Nucl. Med.* **37** 22s–5s
[105] Chiou R K, Dalrymple G V, Baranowska-Kortylewicz J, Holdeman K P, Schneiderman M H, Harrison K A and Taylor R J 1999 Tumor localization and systemic absorption of intravesical instillation of radio-iodinated iododeoxyuridine in patients with bladder cancer *J. Urol.* **162** 58–62
[106] Macapinlac H A, Larson S M, Daghighian F, Finn R, Zhang J, Humm J L, Squire O and Kemeny N 1996 I-125-5′-Iododeoxyuridine (IUdR) in the treatment of colorectal cancer metastatic to the liver: 1. Pilot clinical trial *J. Nucl. Med.* **37** 25s–9s
[107] Kassis A I, Adelstein S J and Mariani G 1996 Radiolabeled nucleoside analogs in cancer diagnosis and therapy *J. Nucl. Med.* **40** 301–19

[108] Baranowska-Kortylewicz J, Kortylewicz Z P, Hoffman D, Winoto A, Lai J and Dalrymple G V 1996 Colon-specific prodrugs of 5-radioiodo-2'-deoxyuridine *Acta Oncol.* **35** 959–64

[109] Karagiannis T C, Lobachevsky P N and Martin R F 2000 Cytotoxicity of an ^{125}I-labelled DNA ligand *Acta Oncol.* **39** 681–5

[110] Panyutin I G, Winters T A, Feinendegen L E and Neumann R D 2000 Development of DNA-based radiopharmaceuticals carrying Auger-electron emitters for anti-gene radiotherapy *J. Nucl. Med.* **44** 256–67

[111] DeSombre E R, Hughes A, Hanson R N and Kearney T 2000 Therapy of estrogen receptor-positive micrometastases in the peritoneal cavity with Auger electron-emitting estrogens—theoretical and practical considerations *Acta Oncol.* **39** 659–66

[112] Brady L W, Miyamoto C, Woo D V, Rackover M, Emrich J, Bender H, Dadparvar S, Steplewski Z, Koprowski H and Black P 1992 Malignant astrocytomas treated with iodine-125 labeled monoclonal antibody 425 against epidermal growth factor receptor: a phase II trial *Int. J. Radiat. Oncol. Biol. Phys.* **22** 225–30

[113] Derui L, Woo D V, Emrich J, Steplewski Z, Rodeck U, Herlyn D, Koprowski H, Miyamoto C and Brady L W 1992 Radiotoxicity of ^{125}I-labeled monoclonal antibody 425 against cancer cells containing epidermal growth factor receptor *Am. J. Clin. Oncol.* **15** 288–94

[114] Welt S, Scott A M, Divgi C R, Kemeny N E, Finn R D, Daghighian F, Germain J St, Richards E C, Larson S M and Old L J 1996 Phase I/II study of iodine 125-labeled monoclonal antibody A33 in patients with advanced colon cancer *J. Clin. Oncol.* **14** 1787–97

[115] Adelstein S J, Kassis A I, Baranowska-Kortylewicz J, van den Abbeele A D, Mariani G and Ito S 1991 Potential for tumor therapy with iodine-125 labelled immunoglobulin *Int. J. Rad. Appl. Inst.* B **18** 43–4

[116] Brom M, Joosten L, Oyen W J, Gotthardt M and Boerman O C 2012 Improved labelling of DTPA-and DOTA-conjugated peptides and antibodies with ^{111}In in HEPES and MES buffer *EJNMMI Res.* **2** 4

[117] Lever J R, Fergason-Cantrell E A, Carmack T L, Watkinson L D and Gallazi F 2019 Design, synthesis and evaluation of [^{111}In]-labeled DOTA-conjugated tetrapeptides having high affinity and selectivity for mu opioid receptors *Nucl. Med. Biol.* **70** 53–66

[118] Mariani G, Bodei L, Adelstein S J and Kassis A I 2000 Emerging roles for radiometabolic therapy of tumors based on auger electron emission *J. Nucl. Med.* **41** 1519–21

[119] Ku A, Facca V J, Cai Z and Reilly R M 2019 Auger electrons for cancer therapy—a review *EJNMMI Radiopharm. Chem.* **4** 27

[120] Daghighian F, Barendswaard E, Welt S, Humm J L, Scott A, Willingham M C, McGuffie E, Old L J and Larson S M 1996 Enhancement of radiation dose to the nucleus by vesicular internalization of iodine-125-labeled A33 monoclonal antibody *J. Nucl. Med.* **37** 1052–57

[121] Behr T M, Behe M, Lohr M, Sgouros G, Angerstein C, Wehrmann E, Nebendahl K and Becker W 2000 Therapeutic advantages of Auger electron-over beta-emitting radiometals or radioiodine when conjugated to internalizing antibodies *Eur. J. Nucl. Med.* **27** 753–65

[122] Gholami Y, Zhu X, Fulton R, Meikle S, El-Fakhri G and Kuncic. Z 2015 Stochastic simulation of radium-223 dichloride therapy at the sub-cellular level *Phys. Med. Biol.* **60** 6087–96

[123] Bavelaar B M, Lee B Q, Gill M R, Falzone N and Vallis K A 2018 Subcellular targeting of theranostic radionuclides *Front. Pharmacol.* **9** 996

[124] Herrera F G, Bourhis J and Coukos G 2017 Radiotherapy combination opportunities leveraging immunity for the next oncology practice *CA Cancer J. Clin.* **67** 65–85

[125] Suominen M I *et al* 2017 Radium-223 inhibits osseous prostate growth by dual targeting of cancer cells and bone microenvironment in mouse models *Clin. Cancer Res.* **23** 4335–46

[126] Parker C *et al* 2013 Alpha emitter radium-223 and survival in metastatic prostate cancer *N. Engl. J. Med.* **369** 213–23

[127] Kratochwil C, Bruchertseifer F, Giesel F L, Weis M, Verburg F A, Mottaghy F, Kokpa K, Apostolidis C, Haberkorn U and Morgenstern A 2016 ^{225}Ac-PSMA-617 for PSMA Targeted α-radiation therapy of metastatic castration resistant prostate cancer *J. Nucl. Med.* **57** 1941–4

[128] Hofmann W, Li W B, Friedland W, Miller B W, Madas B, Bardies M and Balásházy I 2020 Internal microdosimetry of alpha-emitting radionuclides *Radiat. Environ. Biophys.* **59** 29–62

[129] Cai Z, Pignol J P, Chan C and Reilly R M 2010 Cellular dosimetry of ^{111}In using Monte Carlo N-particle comupter code: comparison with analytic methods and correlation with *in vitro* toxicity *J. Nucl. Med.* **51** 462–70

[130] Shinohara A, Hanaoka H, Sakashita T, Sato T, Yamaguchi A, Ishioka N S and Tsushima Y 2018 Rational evaluation of the therapeutic effect and dosimetry of Auger electrons for radionuclide therapy in a cell culture model *Ann. Nucl. Med.* **32** 114–22

[131] Gaddy D F, Lee H, Zheng J, Jaffray D A, Wickham T J and Hendriks B S 2015 Whole-body organ-level and kidney micro-dosimetric evaluations of *64*Cu-loaded HER2/ErbB2-targeted liposomal doxorubicin (^{64}Cu-MM-302) in rodents and primates *EJNMMI Res.* **5** 24

[132] Cornelissen B and Vallis K A 2010 Targeting the nucleus: an overview of Auger-electron radionuclide therapy *Curr. Drug Discov. Technol.* **7** 263–79

[133] McMillan D D *et al* 2015 Validation of ^{64}Cu-ATSM damaging DNA via high-LET Auger electron emission *J. Rad. Res.* **56** 784–91

[134] Wang Y, Liu X, Nakamura K, Chen L, Rusckowski M and Hnatowich D J 2009 *In vivo* delivery of antisense MORF oligomer by MORF/carrier streptavidin nanoparticles *Cancer Biother. Radiopharm.* **24** 573–8

[135] Falzone N, Lee B Q, Fernández Varea J M, Kartsonaki C, Stuchbery A E, Kibedy T and Vallis K A 2017 Absorbed dose evaluation of Auger-electron emitting radionuclides: impact of input decay spectra on dose-point kernels and S-values *Phys. Med. Biol.* **62** 2239–53

[136] Di Maria S, Belchior A, Romanets Y, Paulo A and Vaz P 2018 Monte Carlo dose distribution calculation at nuclear level for Auger-emitting radionuclide energies *Appl. Radiat. Isot.* **135** 72 7

[137] Incerti S *et al* 2018 Geant4-DNA example applications for track structure simulations in liquid water: a report from the Geant4-DNA project *Med. Phys.* **45** 13048

[138] Lopez-Coello L I, Torres-Garcia E, Diaz-Sanchez L E, Oros-Pantoja E and Aranda-Lara L 2019 Differences in the S-value between male and female murine model for diagnostic, therapeutic and theragnostic radionuclides *Int. J. Rad. Appl. Instrum.* A **146** 61–5

[139] Friedland W and Kundrat P 2013 Track structure based modelling of chromosome aberrations after photon and alpha-particle irradiation *Mutat. Res.* **756** 213–23

[140] Lee B Q, Kibedy T, Stuchbery A E and Robertson K A 2012 Atomic radiations in the decay of medical radioisotopes: a physics perspective *Comput. Math. Methods Med.* **2012** 651475

[141] Bantsar A and Pszona S 2012 Nanodosimetry of ^{125}I Auger electrons *Int. J. Radiat. Biol.* **88** 895–8

[142] Hill M A 2018 Track to the future: historical perspective on the importance of radiation track structure and DNA as a radiobiological target *Int. J. Radiat. Biol.* **94** 759–68

[143] Carlson D J, Stewart R D, Semenenko V A and Sandison G A 2008 Combined use of Monte Carlo DNA damage simulations and deterministic repair models to examine putative mechanisms of cell killing *Radiat. Res.* **69** 47–59

[144] Li W B, Hofmann W and Friedland W 2018 Microdosimetry and nanodosimetry for internal emitters *Radiat. Res.* **115** 29–42

[145] Fakir H and Hofmann W 2006 Incorporation of microdosimetric concepts into a biologically-based model of radiation carcinogenesis *Rad. Prot. Dosim.* **122** 330–4

[146] Kassis A I 2011 Molecular and cellular radiobiological effects of Auger emitting radionuclides *Rad. Prot. Dosim.* **142** 241–47

[147] Truta-Popa L A, Hofmann W and Cosma C 2011 Prediction of lung cancer risk for radon exposures based on cellular alpha particle hits *Rad. Prot. Dosim.* **145** 218–23

Chapter 8

Monte Carlo calculations in therapeutic nuclear medicine

Michael Stabin and Habib Zaidi

8.1 Introduction

A general equation for dose to a target organ T from disintegrations can be given as [1]:

$$D_T = \frac{k \sum_S N_S \sum_i n_i \ E_i \ \phi_i(T \leftarrow S)}{m_T}$$

where D_T = absorbed dose in target organ T (Gy/or rad)
N_S = number of disintegrations in a source organ (MBq or μCi)
n_i = number of radiations with energy E_i emitted per nuclear transition
E_i = energy per radiation (MeV)
$\phi(T{\leftarrow}S)$ = fraction of energy emitted in a source that is absorbed in the target
m_T = mass of target region (kg or g)
k = some proportionality constant (Gy-kg MBq^{-1} s^{-1} MeV^{-1} or rad-g μCi^{-1} h^{-1} MeV^{-1})

Monte Carlo methods are frequently used to estimate values of $\phi(T{\leftarrow}S)$ for photon emissions. Values of ϕ for electron and alpha emissions are usually governed by simpler rules, i.e. $\phi(T{\leftarrow}S) = 1.0$, when the source and target are the same, and $\phi(T{\leftarrow}S) = 0.0$ for other cases, however, Monte Carlo methods have also been employed for beta emissions.

8.2 The Monte Carlo method

Monte Carlo methods, in general, involve the use of large quantities of randomly generated numbers to solve mathematical problems that do not lend themselves to analytical solutions. In radiation dosimetry, they may be applied to either internal or external sources of radiation. Figure 8.1 shows how Monte Carlo methods may be

doi:10.1088/978-0-7503-2694-0ch8

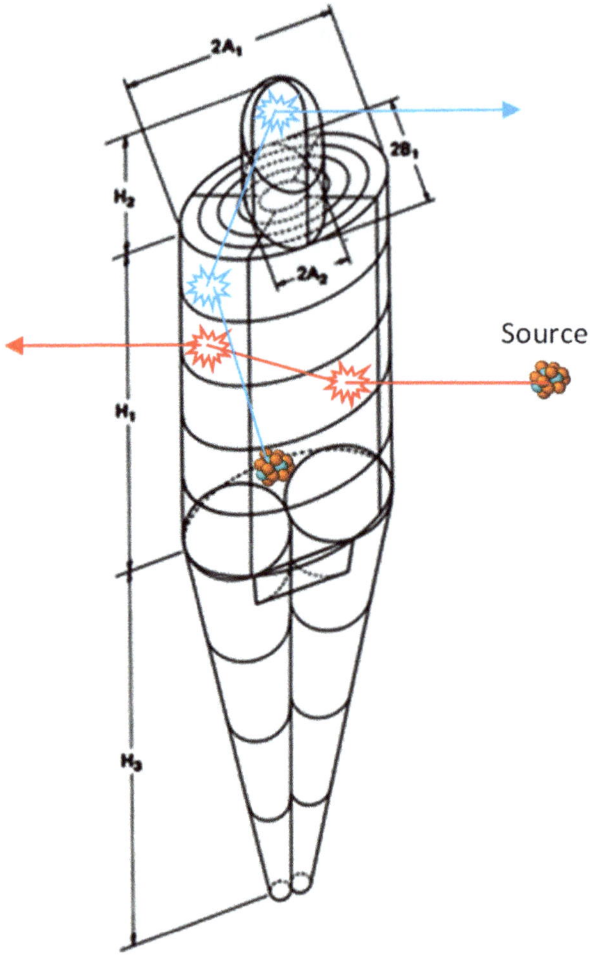

Figure 8.1. Visualization of Monte Carlo simulations of external and internal sources irradiating the ORNL mathematical phantom.

used for calculation of theoretical doses from external or internal sources of radiation. In this chapter, we will focus on internal sources of radiation.

The source, , represents a disintegration within a source region (S) in some location in the body. Sources are generally whole organs (liver, lung, etc). The first step is to select a random location within the source region, by generating random numbers and choosing an allowable x, y, and z within the limits of this region. For internal sources, emission of radiation is generally isotropic, so the next step is to randomly choose an orientation in 4π space. Then the photon is allowed to start moving in this direction, in small steps, i.e. distances. The size of each step is usually an adjustable variable. The smaller the step size, the better the simulation, but the longer that the simulation will take; thus optimization of this tradeoff is generally discovered by trial and error. Depending on the medium (soft tissue, lung or bone) and the energy of the particle, one can vary the step size from larger to smaller and find where smaller step sizes do not produce 'better' results. Along the pathway, the photon is allowed to take steps, and at each step, another random number is generated to predict whether an interaction, , has occurred, sampling from interaction probability functions for photoelectric, Compton, or pair production events (the latter of course only permitted for photons above 1.02 MeV). If a Compton event has occurred, yet another random number is generated to calculate the angle of scatter, and the photon begins a new 'random walk' in this new direction. For any type of interaction, some or all of the photon's energy is deposited at the interaction site, and this information is recorded, noting within which target region (T) it occurred. The photon continues its 'random walk' until it undergoes a photoelectric event, and all of its remaining energy is absorbed, or until it leaves the body simulated by the phantom. Photons leaving the body are generally emitted into an air environment, and are thought to have negligible backscatter towards the body, so further transport is neglected. Many photon histories (typically millions) are simulated for each source organ, and the average energy deposited in all of the target organs, as a fraction of the total energy emitted in the source, gives the estimate of $\phi(T \leftarrow S)$. For many target regions this can be successfully done with a reasonably large number of starting photons. With the estimate of $\phi(T \leftarrow S)$, values of the uncertainty are given, which are inversely proportional to the number of photon hits recorded. With more histories, better uncertainties can be obtained. Very small organs, however, like testes, ovaries, the thyroid, and others, may require more histories than, say, liver to obtain an acceptable level of uncertainty. A maximum level of 10% uncertainty is generally considered acceptable. Values of $\phi(T \leftarrow S)$ or $\Phi(T \leftarrow S)$ ('absorbed fraction' and 'specific absorbed fraction') may be reported as a function of starting photon energy, where $\Phi(T \leftarrow S) = \phi(T \leftarrow S)/m_T$. One must be very careful in using tables of absorbed fractions to be sure that the correct values are employed. Absorbed fractions are unitless, being some number between 0 and 1, whereas specific absorbed fractions (SAFs) will have units of 'per unit mass' (e.g. g^{-1}, kg^{-1}). Organs may have masses of hundreds of grams or kilograms, so use of the incorrect type of absorbed fractions may be wrong by a considerable amount.

With modern computing capabilities, even billions of histories can be generated, to drive uncertainties for smaller organs to acceptable levels. Cristy and Eckerman [2] developed SAFs for six adult/pediatric phantoms (Adult, 15-year-old, 10-year-old, 5-year-old, 1-year-old and newborn) and Stabin *et al* developed SAFs for models of the pregnant female at 3, 6, and 9 months' gestation [3]; these two phantoms were referred to as the 'ORNL phantom series', as the results were published as Oak Ridge National Laboratory (ORNL) documents (figure 8.2). The SAFs were generated using a code developed at ORNL named ALGAMP [4]. They used two methods to estimate values of $\phi(T \leftarrow S)$ for problematic organs:

1. The 'Reciprocal Dose Principle'—for photon emitters in a homogeneous medium, Lovenger [5] showed that for a given source and target region, $\phi(T \leftarrow S) = \phi(S \leftarrow T) = \phi(T \leftrightarrow S)$. The body is not totally homogeneous, being

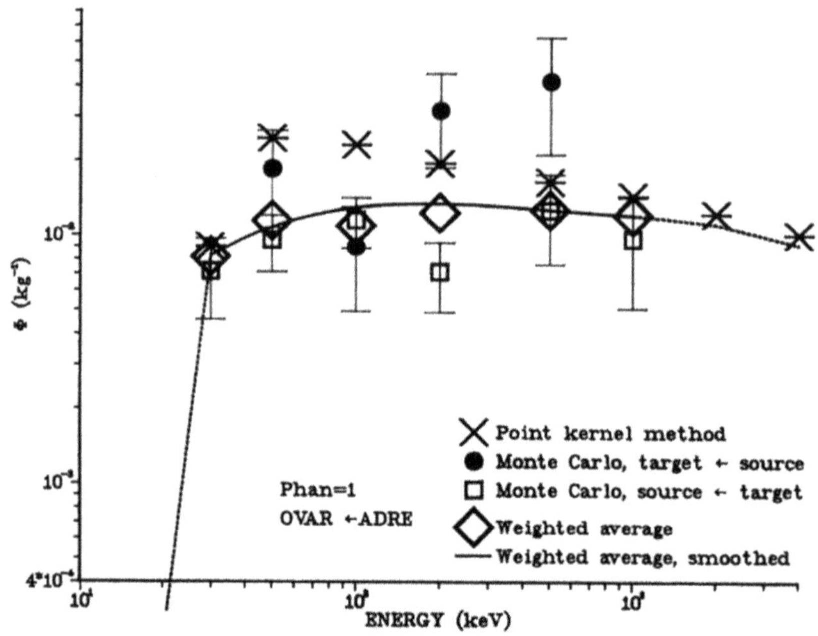

Figure 8.2. Plot of Monte Carlo estimates of ϕ vs energy for an organ pair (adrenals as source and ovaries as target) in 1-year-old ORNL mathematical phantom. Reprinted with permission from [2], ORNL-TM-8381, copyright ORNL.

primarily soft tissue (density $\cong$ 1.0 g cm^{-3}), but also containing lung tissue (density $\cong$ 0.3 g cm^{-3}) and bone (density $\cong$ 1.4 g cm^{-3}). The majority is soft tissue, so this approximation is generally reasonable. So, for example, while ϕ(thyroid←liver) may have a high uncertainty, ϕ(liver←thyroid) may reasonably be substituted. The authors gave also correction factors for some specific cases (e.g. source was lung or bone, or a target region was near the water/tissue interface).

2. Use of 'Point Kernel' estimates of $\phi(T \leftarrow S)$—this method assumes that the phantom is of uniform composition, and the equation here is applied:

$$\Phi(r) = \frac{\mu_{en}}{\rho} \cdot \frac{1}{4\pi r^2} \cdot e^{-\mu r} \cdot B(\mu r)$$

where

$\Phi(r)$ = point isotropic specific absorbed fraction at r (based on random selection of many points within the source and target organs),
μ_{en} = linear energy-absorption coefficient at the source energy,
μ = linear attenuation coefficient at the source energy,
ρ = density of medium,
$B(\mu r)$ = buildup factor, a factor representing the contribution of the scattered radiation to the energy absorption.

The assumption that $\phi(T \leftarrow S) = 1.0$ when $S = T$ is never absolutely correct; some leakage of electron energy occurs at the surface of all organs, regardless of size. For larger organs, this loss is very small, and neglecting the leakage produces no significant errors. Similarly electron leakage from an organ to an adjacent organ is usually very small. As computing power has increased, it has become common to estimate absorbed fractions for electrons as well as photons. The following plots show selected electron absorbed fractions, showing non-unity values for self-absorption and non-zero values for cross-irradiation (figure 8.3).

Monte Carlo simulations of electron transport have been used extensively in modeling of radiation transport with the skeleton. Most organs in the body are quite homogeneous; the skeleton, however, is a complex mixture of marrow and bone. The two source regions in the skeleton are the bone (trabecular or cortical) and the marrow; the two target regions are 'bone surfaces' (thin regions of dividing cells on the surfaces of trabecular and cortical bone) and red marrow. The skeleton also contains yellow marrow (in the adult this is the larger amount), but it is not of dosimetric interest. Red marrow is principally associated with trabecular, or 'spongy' bone. Electrons originating either in trabecular bone or marrow may cross several bone and marrow structures before depositing all of their energy. This is not easily handled in most Monte Carlo codes, as the regions have sizes of the order of

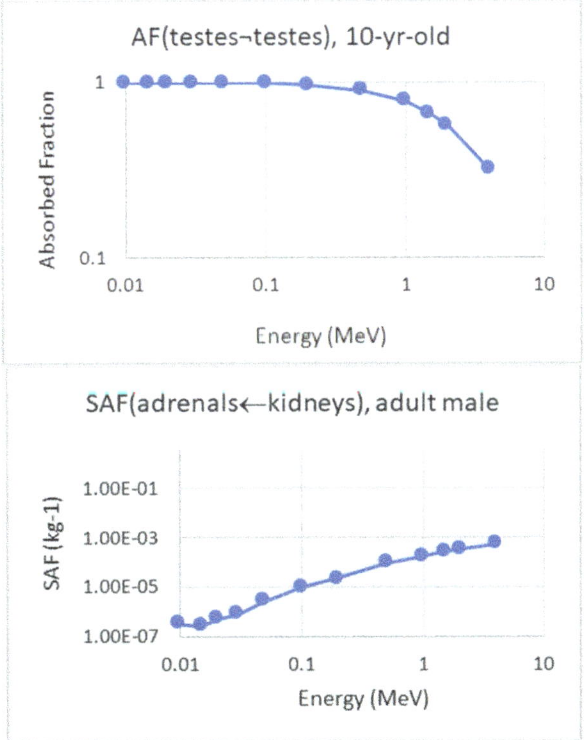

Figure 8.3. Top: self-absorbed fractions for testes in the 10-year-old phantom. Bottom: SAFs for kidneys irradiating adrenals in the adult male phantom.

micrometers, and obtaining good statistics for absorbed fractions is not easy. As will be discussed below, this is an area of significant inquiry, using specialized Monte Carlo methods.

8.3 Recent Monte Carlo efforts in nuclear medicine

For many years, the ORNL adult/pediatric phantom series and the ORNL pregnant female phantoms were the only specific absorbed fractions available to the nuclear medicine community to perform dose calculations for radiopharmaceuticals. In the 1990s, Stabin employed all of these absorbed fractions to develop a personal computer program called MIRDOSE [6]. Users could choose a radionuclide, one or more of the nine ORNL phantoms, and enter numbers of disintegrations for more than 20 source regions and obtain doses to more than 20 target regions. This automated and standardized dose estimates for users everywhere in the world. The name of the code was based on the idea that it calculated doses by the MIRD (Medical Internal Radiation Dose) (Committee of the Society of Nuclear Medicine) method. The MIRD Committee, however, vehemently objected to the name of the code, as it could imply that the committee developed it. Newer versions were thus renamed OLINDA [7] (Organ Level INternal Dose Assessment), with the same phantoms, but now using the RADAR (RAdiation Dose Assessment Resource) (Committee of the Society of Nuclear Medicine) calculational method (1). There were no other publicly available radiopharmaceutical dosimetry codes, although some researchers developed their own in-house software. In the late 1990s, a number of groups developed a number of phantoms and developed SAFs for use in internal dosimetry, using methods based on medical images to model organs and the whole body. The ORNL phantoms, sometimes referred to as 'stylized' phantoms, used combinations of geometric shapes to model organs and the whole body (figure 8.4).

Phantoms and associated SAFs have now been developed by numerous groups for various applications.

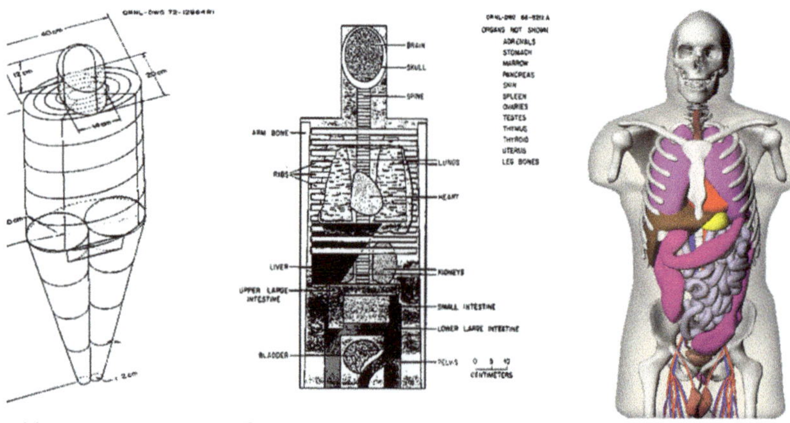

(a) (b) (c)

Figure 8.4. (a) Whole-body ORNL adult phantom, (b) internal organs, ORNL adult phantom, (c) whole-body and organs, RADAR realistic phantom.

8.4 Expansion of the field

A CRC handbook was published in 2010, summarizing the results of a US meeting in 2005 in which 64 authors from 13 countries presented information on anthropomorphic development [8] based on medical images of actual humans, too update the 'stylized' models developed at ORNL. There are now many hundreds of anthropomorphic phantoms that have been developed and used for various dosimetric analyses.

8.4.1 VIP man

The earliest entrant into the field was the 'VIP-Man' phantom, originally developed by Zubal *et al* [9] (figure 8.5).

With the broad availability of medical images (typically Computed Tomography (CT) scans), many other centers developed realistic phantoms, using a number of mathematical techniques, including simple (and very tedious) manual segmentation (associating groups of voxels with identifiable organs), using techniques such as Non-Uniform Rational B-Splines (NURBS), described in more detail below,

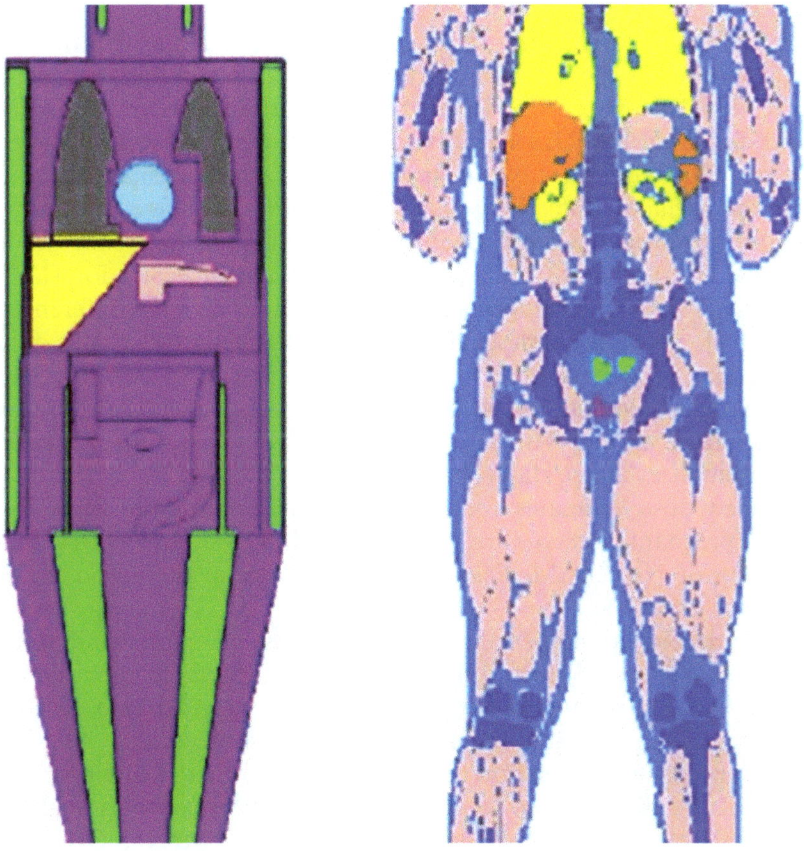

Figure 8.5. Zubal *et al* VIP-Man voxel phantom, right, compared to the Fisher–Snyder phantom, left.

polygon mesh, and other technologies. Various Monte Carlo codes have been employed by the authors to develop absorbed fractions for internal and external radiation sources in various applications, in medical imaging and therapy, general radiation protection, space exploration and others. In this chapter we will attempt to summarize the models presented in this symposium and the application addressed, with emphasis on the models that are having the most impact on routine use.

8.4.2 The GSF voxel phantoms

A series of phantoms were developed by the GSF (Research Center for Environment and Health) in Germany, based on CT images of living subjects. They created phantoms for male and female adults, and three other male and six female models, one of which was pregnant. Most were derived by manual segmentation of organ regions. The two adult models have been adopted by the International Commission on Radiological Protection (ICRP) for use in standard dose calculations for radiation workers [10].

8.4.3 RADAR anthropomorphic and animal models

The RADAR group developed a series of phantoms, starting with the adult male and female models developed by Segars [11] using a technology known as Non-Uniform Rational B-Splines (NURBS). Starting with CT images of an adult male and female, Segars fit NURBS surfaces to model individual organs. These surfaces are easily manipulated, either by linear scaling in any dimension, or by manually deforming the surfaces using individual points used to define the surfaces. The RADAR group scaled the adult phantoms to match the reference adult masses defined by the International Commission on Radiological Protection (ICRP) in its publication 89 [12], and developed phantoms for the other five reference ages, as with the ORNL phantom series. The ORNL adult/pediatric 'stylized' phantom series' individual phantoms were hermaphrodites, having both male and female organs in the one whole-body phantom, but of course not allowing cross-irradiation of male and female organs. The RADAR phantom series has a separate male and female phantom at each age (figure 8.6).

Standardized dose estimates for about 100 radiopharmaceuticals for adults, children, and pregnant women were subsequently developed by the RADAR group, for the Society of Nuclear Medicine [13, 14] and are available on the RADAR website for free download.

Segars also developed models of a rat and a mouse using micro-CT images [15]. RADAR scaled the one mouse and rat model to represent three mice and five rats of different mass, and calculated SAFs for the animal organs (figure 8.7). They also developed SAFs for a male and female beagle (unpublished data). All of the SAFs are available in the current version of OLINDA, for dose calculations.

Xu *et al* [16], in collaboration, developed three realistic phantoms representing the pregnant female at three stages of pregnancy, in parallel to the ORNL pregnant female series (figure 8.8).

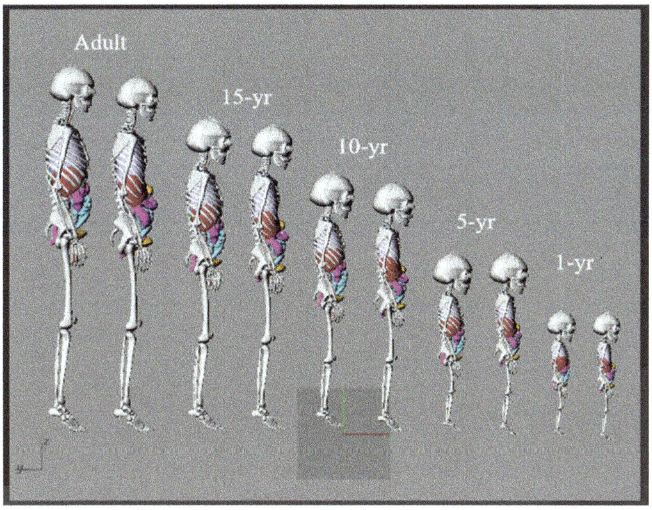

Figure 8.6. RADAR ICRP reference phantom series (newborns not shown).

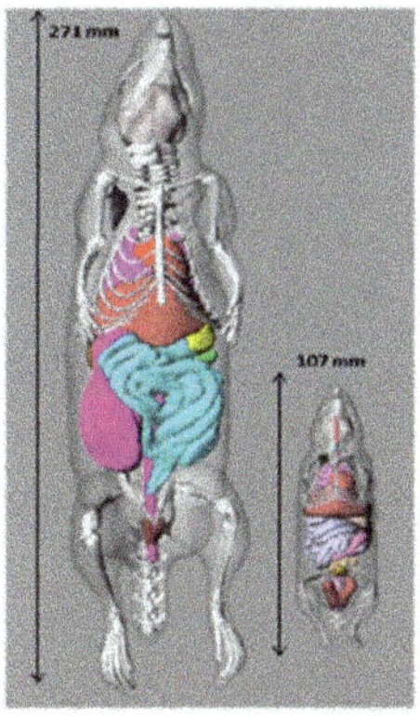

(a)

TABLE 1. Organ Masses in 3 Mouse Models

Organ	Organ mass (g) 25-g mouse	30-g mouse	35-g mouse
Brain	0.466	0.568	0.666
Heart	0.235	0.291	0.342
Stomach	0.055	0.069	0.082
Small intestine	1.74	2.12	2.49
Large intestine	0.583	0.709	0.830
Kidneys	0.302	0.374	0.432
Liver	1.74	2.15	2.57
Lungs	0.087	0.107	0.131
Pancreas	0.305	0.378	0.450
Skeleton	2.18	2.61	3.01
Spleen	0.111	0.136	0.157
Testes	0.160	0.197	0.228
Thyroid	0.014	0.016	0.020
Bladder	0.060	0.075	0.088
Body	24.11	29.80	35.27

(b)

TABLE 2. Organ Masses in 5 Rat Models

Organ	Organ mass (g) 200-g rat	300-g rat	400-g rat	500-g rat	600-g rat
Brain	1.57	2.32	3.16	3.93	4.54
Heart	1.80	2.64	3.55	4.39	5.28
Stomach	0.941	1.40	1.89	2.37	2.86
Small intestine	10.6	15.5	20.8	25.6	30.8
Large intestine	7.86	11.5	15.5	19.2	23.1
Kidneys	2.06	3.03	4.09	5.06	6.09
Liver	7.55	11.2	15.2	18.8	22.8
Lungs	0.594	0.884	1.21	1.50	1.82
Pancreas	0.368	0.535	0.732	0.908	1.10
Skeleton	15.3	22.0	29.2	35.2	38.7
Spleen	0.607	0.884	1.18	1.45	1.74
Testes	0.174	0.245	0.321	0.386	0.460
Thyroid	0.191	0.275	0.368	0.457	0.549
Bladder	0.475	0.682	0.916	1.12	1.34
Body	226	335	443	547	643

(c)

Figure 8.7. Segars *et al* mouse (MOBY) and rat (ROBY) NURBS models, (b) RADAR mouse model organ masses, (c) RADAR rat organ masses. Reprinted with permission from [17], copyright SNMMI.

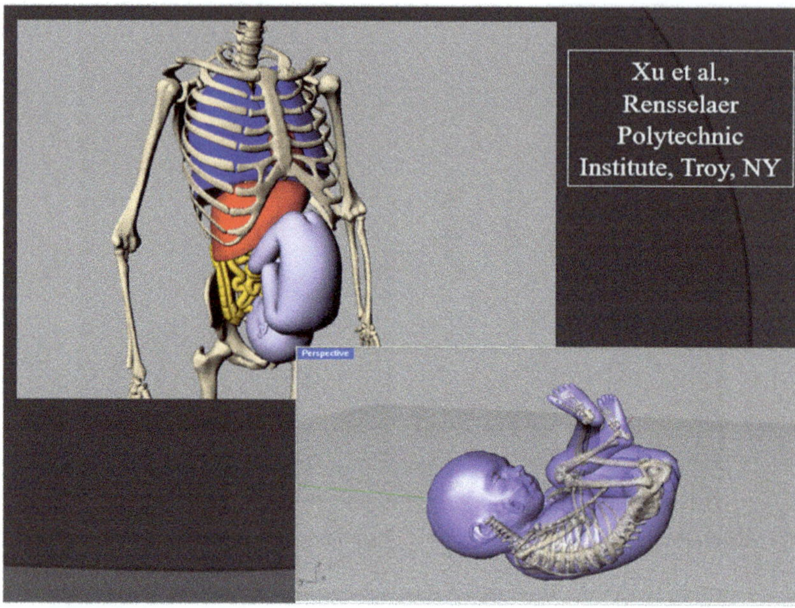

Figure 8.8. Example of Xu *et al* realistic pregnant female phantoms. Courtesy of Dr G Xu (Rensselaer Polytechnic Institute, NY, USA).

As noted above, NURBS models are very easy to deform; RADAR also deformed the reference phantoms (which represent the median, or 50th percentile, individual of a population), to represent individuals (mainly pediatric models) of 10th, 25th, 75th and 90th percentile individuals, and added some additional age phantoms, because of the interest in studying changes in CT dose in children with differing body size [18]. Dose estimates for various CT exams were provided as a function of age and body size. This has not been applied to any studies with nuclear medicine agents, however, all RADAR phantoms are available to the user community for either external or internal dose calculations. Figure 8.9 shows the front 3D views of 12 computational pediatric phantoms used in nuclear medicine dosimetry studies [19].

8.4.4 University of Florida phantom series

Researchers at the University of Florida developed a series of adult and pediatric anthropomorphic phantoms, and, as the RADAR group did, developed phantoms of different body shape and size (figure 8.10) [20].

The reference phantoms were also scaled to ICRP 89 reference values, and non-reference phantoms were scaled to model persons of different body habitus using published data on variations in human anatomy. The pediatric phantoms have been adopted by International Commission on Radiological Protection (ICRP) as the

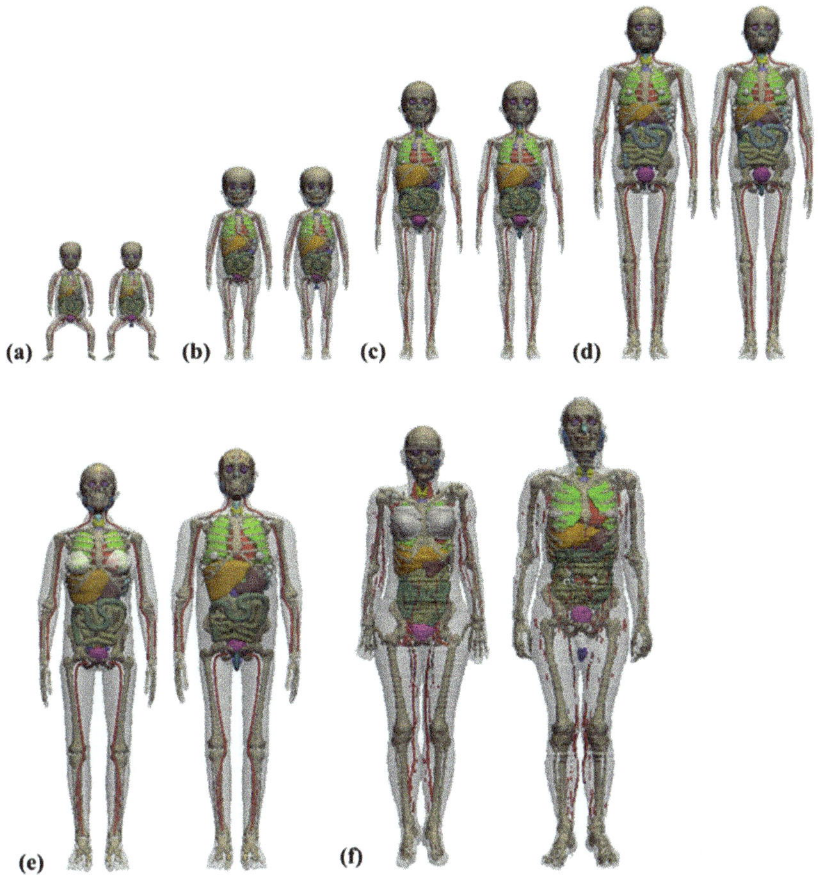

Figure 8.9. 3D visualization of computational phantoms of (a) newborn, (b) 1-year-old, (c) 5-year-old, (d) 10-year-old, (e) 15-year-old and (f) adult male (right) and female (left). Reprinted with permission from Xie *et al* [19] John Wiley & Sons. Copyright (2015) American Association of Physicists in Medicine.

international reference for calculations to radiation workers and members of the public, for basic radiation protection studies, but no data are currently available, and no dosimetry for nuclear medicine pharmaceuticals has been provided.

8.4.5 Other realistic phantoms

Other phantoms presented in the CRC handbook will be more briefly covered here, not because the work is not elegant, but because the phantom studies did not obtain international recognition as establishing standard SAFs and dose estimates, as have the RADAR and ICRP reference phantoms.

1. Kramer *et al* developed two adult computational phantoms to represent adult females and males called 'FAX' and 'MAX [21]. Manual segmentation

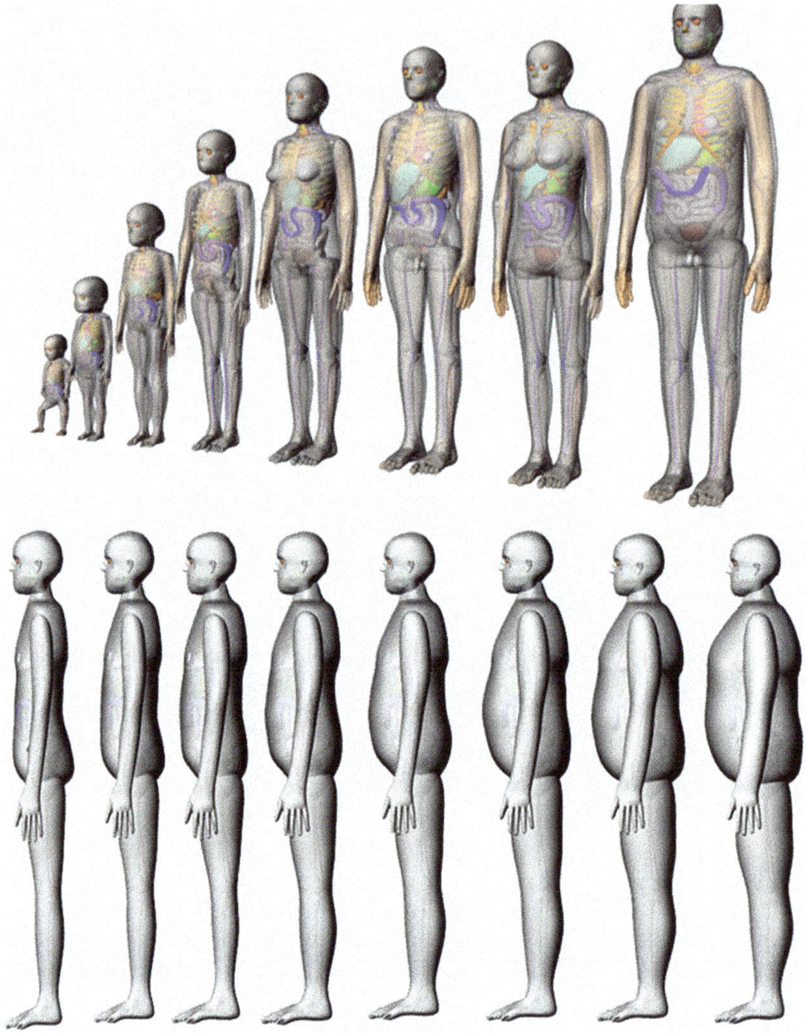

Figure 8.10. The UF adult and pediatric reference phantom series (top), and models of the adult male of different body habitus (bottom).

was used to identify about 30 internal organs; dose factors were provided for several external irradiation geometries.

2. A number of other studies cited in the CRC handbook describe efforts to produce models of individuals from different countries, particularly where members of the population are notably smaller than the ICRP 89 reference average values:
   - Researchers from the Japanese Atomic Energy Agency developed computational phantoms for adult males and females, a pregnant woman and several children of various different ages [22]. Absorbed fractions and Dose Factors were calculated.

- Two reference Korean phantoms were developed at the Hanyang University for a reference adult male and female Korean individuals [23].
- Voxel phantoms were developed to represent three Chinese adult males [24]; dose factors were given for several photon energies for external irradiation geometries.

3. A voxel phantom for a reference teenager, called ADELAIDE, was developed in a collaborative effort in Australia [25]. Effective doses for some CT examinations were calculated using Monte Carlo methods.
4. Na *et al* developed a series of female phantoms of differing breast size for use in 'virtual calibration' of lung counters [26].

8.5 Applications of the phantoms

Even with the earliest, very simple, spherical phantoms, the purpose for their development is to facilitate dose calculations in the human body. In the CRC handbook, Hintenlang *et al* described the use of physical phantoms for dosimetry, in which a physical object resembling the human body can be exposed to radiation fields, with radiation dosimeters placed inside the phantoms [27]. This is difficult to do for internal emitters, although an attempt was made by Poston *et al* to use physical measurements to validate SAFs calculated by Monte Carlo techniques (personal communication). Monte Carlo codes provide absorbed fractions for organ doses from internal or external exposures. Applications are numerous, but typical uses are:

- Calculations of internal dose from nuclear medicine procedures or internal intakes of radionuclides by workers.
- Calculations of external dose from medical irradiations—diagnostic (x-ray or CT) and therapeutic.
- Calculations of external dose to radiation workers or members of the public from industrial uses of radiation.
- Calculation of doses to individuals involved in space exploration.

It is difficult to imagine future phantom development can substantially improve on the current realistic models, but scientists are always looking for ways to improve the present state of the art. Models to address the issue of electron and alpha dose to bone and marrow are still developing; this was a topic considered outside the scope of this chapter.

Obviously, it was a huge advance in realism moving from the very simple 'body and all organs can be represented as spheres' to the first heterogeneous stylized phantoms. In the ICRP 2 calculations [28], the organs were simplified spheres, but also no cross-irradiation from one organ to another was possible. The dose estimates based on the Cristy/Eckerman 'stylized' phantoms were clearly better. However, the move from 'stylized' phantoms to the clearly more sophisticated and realistic image-based phantoms did not result in a significant change in overall dosimetry. The organs are closer together and overlap more, for example, kidneys and adrenals; however, effective doses in nuclear medicine procedures did not change substantially (figure 8.11) [29]:

		Adult	15-yr	10-yr	5-yr	1-yr
^{18}FDG	Old	1.86E-02	2.39E-02	3.58E-02	5.29E-02	8.90E-02
	New	1.73E-02	2.21E-02	3.26E-02	4.76E-02	7.83E-02
	Ratio	9.30E-01	9.25E-01	9.11E-01	9.00E-01	8.80E-01
^{11}C Acetate	Old	3.03E-03	3.77E-03	5.70E-03	8.60E-03	1.59E-02
	New	2.88E-03	3.72E-03	5.41E-03	8.20E-03	1.38E-02
	Ratio	9.50E-01	9.87E-01	9.49E-01	9.53E-01	8.68E-01
^{99m}Tc MAA	Old	1.17E-02	1.64E-02	2.35E-02	3.54E-02	6.56E-02
	New	1.10E-02	1.40E-02	2.21E-02	3.31E-02	5.85E-02
	Ratio	9.40E-01	8.54E-01	9.40E-01	9.35E-01	8.92E-01
111In Octreotide	Old	5.15E-02	6.76E-02	1.01E-01	1.49E-01	2.42E-01
	New	5.09E-02	6.34E-02	9.19E-02	1.30E-01	2.02E-01
	Ratio	9.88E-01	9.38E-01	9.10E-01	8.72E-01	8.35E-01
^{123}I BMIPP	Old	1.51E-02	1.92E-02	2.94E-02	4.44E-02	8.05E-02
	New	1.49E-02	1.94E-02	2.79E-02	4.08E-02	7.13E-02
	Ratio	9.87E-01	1.01E+00	9.49E-01	9.19E-01	8.86E-01

Figure 8.11. Effective doses for a variety of nuclear medicine pharmaceuticals using the 'Old' (Cristy/Eckerman stylized phantoms) and 'New' (RADAR NURBS) phantoms. FDG = Fluorodeoxyglucose, MAA = Macroaggregated Albumin, BMIPP = β-methyl-p-iodophenyl-pentadecanoic acid.

Three major Monte Carlo codes used for calculating SAFs are MCNP [30], EGS [31], and Geant [32]. The PEREGRINE code [33] has also been proposed for three-dimensional, computational dosimetry and treatment planning in radioimmunotherapy. Others have developed other, in-house, Monte Carlo codes that are not available for others to use.

The promise of this type of code is to provide three-dimensional, image-based dose calculations to the nuclear medicine physician which may be used in patient-specific treatment planning in radiopharmaceutical therapy (figure 8.12). The challenge, as noted above, is finding broad clinical acceptance for these approaches and a move away from the approach to therapy involving a uniform administration of activity to all patients, with no consideration of radiation dose.

8.6 Use of Monte Carlo methods to calculate absorbed doses in nuclear medicine therapy

Use of the OLINDA software, with patient-individualized biokinetics and adjustment to organ masses (using CT volumetrics) can probably provide dose estimates for planning of therapy with an accuracy of perhaps 10%–20% [35]. This article showed that the principal sources of uncertainty in an internal dose calculation are the numbers of disintegrations in source organs (depending on the effective half-time in source regions) and the mass of the target organs. These values vary markedly between individuals. Wahl studied the variability in these parameters in over a thousand patients treated with I-131 Bexxar [36]. As the figure below (figure 8.13) shows, the variability in the amount of activity needed to ensure a dose to the marrow of no more

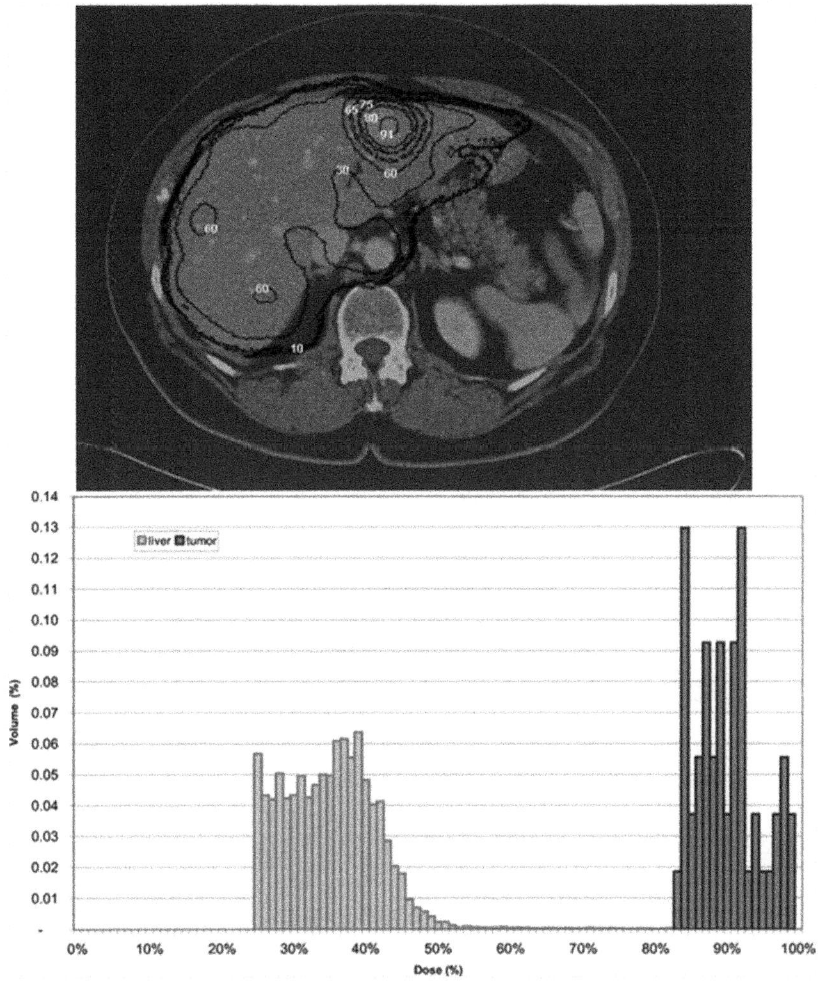

Figure 8.12. Example of 3D dose distributions. Reprinted with permission from Strigari *et al* [34] John Wiley & Sons. Copyright (2006) American Association of Physicists in Medicine.

than 75 cGy had a well-defined average, but values that varied by as much as a factor of 5, taking into consideration individual patient effective half-times and body sizes.

Many objections have been raised to performing patient-individualized calculations in nuclear medicine therapy, as we do for external beam therapy patients every day. Arguments include that it is too difficult, too expensive, that doses cannot be correlated with radiation effects, and other points. Stabin [37] in 2008 addressed all of these concerns and concluded that:

> *Treating all nuclear medicine patients with a single, uniform method of activity administration amounts to consciously choosing that these patients be treated with a lower standard of care than patients who receive radiation externally for cancer treatments.*

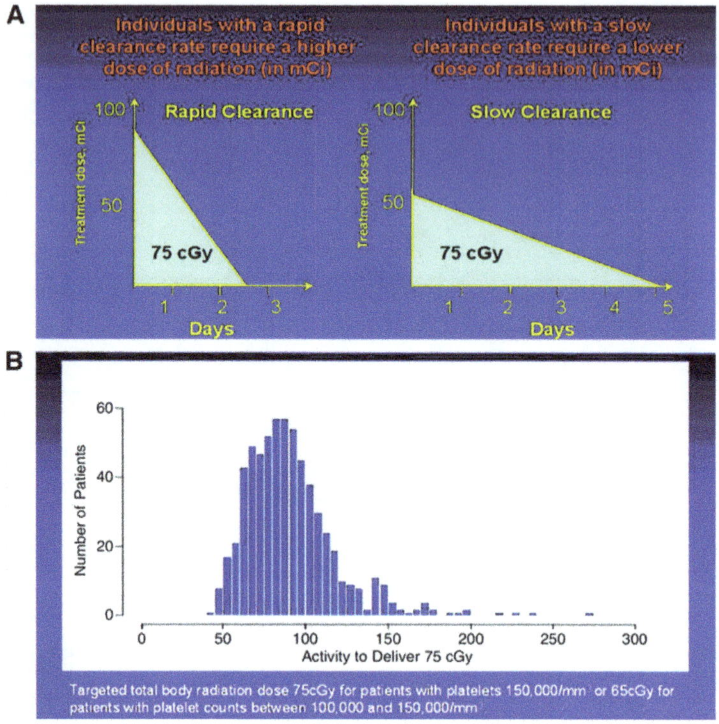

Figure 8.13. Variability in the amount of activity needed to give no more than 75 Gy to the marrow of Bexxar patients. Reprinted with permission from Wahl [36], copyright SNMMI.

There are now several computer codes on the market that can perform patient-individualized dose estimates at the voxel level, fusing patients' CT images with Single Photon Emission Computed Tomography (SPECT) or Positron Emission Tomography (PET) images from individual patients. Segmenting important organs and plotting and integrating the time-activity curves can provide three-dimensional, patient-individualized dose distributions and dose-volume histograms, as is routine in external beam therapy. These codes have been seen at technical meeting vendor areas, but are generally not documented in the literature. A few examples (not meant to be an exhaustive sampling) are shown in figure 8.14.

We have the tools, we have standardized methods for organ segmentation and kinetic analysis, the only thing hindering the use of patient-individualized dosimetry in nuclear medicine therapy is the willingness of the medical community to perform such calculations as standard practice. Were one to approach a medical physics department and ask them to stop performing individualized dosimetry for external beam therapy, they would quickly note that this would constitute malpractice. The only people on the planet that don't receive patient-individualized dosimetry are nuclear medicine therapy patients. We generate dose estimates for adults and children receiving CT scans, airline crews, nuclear workers, nuclear medicine doctors and technologists, just not for nuclear medicine therapy patients. Not only is this resulting in suboptimal therapy for the patients, when patients may need

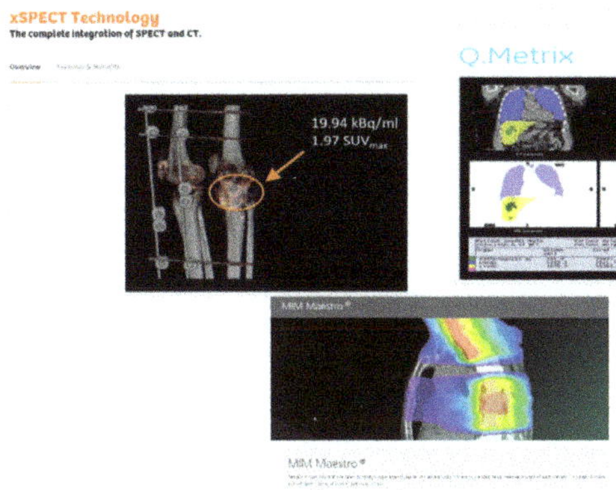

Figure 8.14. A few commercial three-dimensional, patient-individualized codes for nuclear medicine therapy planning. Courtesy of illustrated companies.

other therapies in the future, but also knowledge of previous therapy doses is essential to planning new therapies. It is time to change the historical practice of 'one dose fits all' approach to nuclear medicine therapy.

References

[1] Stabin M G and Siegel J A 2003 Physical models and dose factors for use in internal dose assessment *Health Phys.* **85** 294–310

[2] Cristy M and Eckerman K 1987 Specific absorbed fractions of energy at various ages from internal photons sources *ORNL/TM-8381 V1-V7* (Oak Ridge, TN: Oak Ridge National Laboratory)

[3] Stabin M, Watson E, Cristy M, Ryman J, Eckerman K, Davis J, Marshall D and Gehlen K 1995 Mathematical models and specific absorbed fractions of photon energy in the nonpregnant adult female and at the end of each trimester of pregnancy *ORNL Report ORNL/TM 12907* (Oak Ridge, TN: Oak Ridge National Laboratory)

[4] Warner G G, Craig A M J and ALGAM A 1968 Computer program for estimating internal dose for gamma-ray sources in a man phantom *ORNL-TM-2250* (Oak Ridge, TN: Oak Ridge National Laboratory)

[5] Loevinger R 1969 Distributed radionuclide sources *Radiation Dosimetry* ed F H Attix and E Tochilin vol III 2nd edn (New York: Academic) pp 51–90

[6] Stabin M G 1996 MIRDOSE: personal computer software for internal dose assessment in nuclear medicine *J. Nucl. Med.* **37** 538–46

[7] Stabin M G, Sparks R B and Crowe E 2005 OLINDA/EXM: the second-generation personal computer software for internal dose assessment in nuclear medicine *J. Nucl. Med.* **46** 1023–27

[8] Xu X G and Eckerman K F 2010 *Handbook of Anatomic Models for Radiation Dosimetry.* (Boca Raton, FL: CRC Press)

[9] Zubal I G *et al* 1994 Computerized three-dimensional segmented human anatomy *Med. Phys.* **21** 299

[10] ICRP 2009 Adult reference computational phantoms ICRP Publication 110 *Ann. ICRP* **39**

[11] Segars J P 2001 Development and application of the new dynamic NURBS-based Cardiac-Torso (NCAT) phantom *PhD Dissertation* The University of North Carolina

[12] ICRP 2003 Basic anatomical and physiological data for use in radiological protection: reference values. A report of age- and gender-related differences in the anatomical and physiological characteristics of reference individuals. ICRP Publication 89 *Ann. ICRP* **32** 5–265
[13] Stabin M and Siegel J A 2018 RADAR dose estimate report: a compendium of radiopharmaceutical dose estimates based on OLINDA/EXM Version 2.0 *J. Nucl. Med.* **59** 154–60
[14] Stabin M 2018 New generation fetal dose estimates for radiopharmaceuticals *J. Nucl. Med.* **59** 1005–6
[15] WP and Tsui B M W 2010 The MCAT, NCAT, XCAT and MOBY computational human and mouse phantoms *Handbook of Anatomic Models for Radiation Dosimetry*. ed X G Xu and K F Eckerman (Boca Raton, FL: CRC Press)
[16] XG, Shi C, Stabin M G and Taranenko V 2010 Pregnant female/fetus computational phantoms and the Latest RPI-P series representing 3, 6, and 9-month gestational periods *Handbook of Anatomic Models for Radiation Dosimetry*. ed X G Xu and K F Eckerman (Boca Raton, FL: CRC Press) pp 305–36
[17] Keenan M A, Stabin M G, Segars W P and Fernald M J 2010 RADAR realistic animal model series for dose assessment *J. Nucl. Med.* **51** 471–6
[18] Kost S D, Fraser N D, Carver D E, Pickens D R, Price R R, Hernanz-Schulman M and Stabin M G 2015 Patient-specific dose calculations for pediatric CT of the chest, abdomen and pelvis *Pediatr. Radiol.* **45** 1771–80
[19] Xie T, Lee C, Bolch W E and Zaidi H 2015 Assessment of radiation dose in nuclear cardiovascular imaging using realistic computational models *Med. Phys.* **42** 2955–66
[20] Lee *et al* 2010 The University of Florida pediatric phantom series *Handbook of Anatomic Models for Radiation Dosimetry* ed X G Xu and K F Eckerman (Boca Raton, FL: CRC Press) pp 199–220
[21] Kramer R, Khour H J, Vieira J W, Lima V J D M, Loureiro E C D M, Hoff G and Kawrakow I 2010 The FAX06 and the MAX 06 computational voxel phantoms *Handbook of Anatomic Models for Radiation Dosimetry* ed X G Xu and K F Eckerman (Boca Raton, FL: CRC Press) pp 163–98
[22] Saito K, Sato K, Kinase S and Nagaoka T 2010 Japanese computational phantoms: Otoko, Onago, JM, JM2, JF, TAR0, HANAKO, pregnant woman and deformable child *Handbook of Anatomic Models for Radiation Dosimetry* ed X G Xu and K F Eckerman (Boca Raton, FL: CRC Press) pp 221–54
[23] Lee C and Kim C H 2010 Korean computational phantoms: KMIRD, KORMAN, KORWOMAN, KTMAN-1, KTMAN-2 and HDRK-Man *Handbook of Anatomic Models for Radiation Dosimetry* ed X G Xu and K F Eckerman (Boca Raton, FL: CRC Press) pp 255–78
[24] Zhang B, Ma J, Zhang G, Liu Q, Qiu R and Li J 2010 Chinese voxel computational phantoms: CNMAN, VCH, and CVP *Handbook of Anatomic Models for Radiation Dosimetry* ed X G Xu and K F Eckerman (Boca Raton, FL: CRC Press) pp 279–304
[25] Caon M, Bibbo G and Pattison J E 2010 The ADELAIDE teenage female voxel computational phantom *Handbook of Anatomic Models for Radiation Dosimetry* ed X G Xu and K F Eckerman (Boca Raton, FL: CRC Press) pp 87–104
[26] Na Y H, Zhang J, Ding A and Xu X G 2010 Mesh-based and anatomically adjusted adult phantoms and a case study in virtual calibration of a lung counter for female workers *Handbook of Anatomic Models for Radiation Dosimetry* ed X G Xu and K F Eckerman (Boca Raton, FL: CRC Press) pp 347–76

[27] Hintenlang D E, Moloney W E and Winslow J 2010 Physical phantoms for experimental radiation dosimetry *Handbook of Anatomic Models for Radiation Dosimetry* ed X G Xu and K F Eckerman (Boca Raton, FL: CRC Press) pp 389–412
[28] ICRP 1959 Report of Committee II on permissible dose for internal radiation *ICRP Publication 2* (New York: Pergamon)
[29] Stabin M 2019 Personal communication
[30] Briesmeister J F 2000 MCNP – A General Monte Carlo N-Particle Transport Code, Version 4C, LA-13709-M, Los Alamos National Laboratory
[31] Bielajew A and Rogers D 1987 PRESTA: the parameter reduced electron-step transport algorithm for electron monte carlo transport *Nucl. Instrum. Methods.* **B18** 165–81
[32] Allison J, Amako K, Apostolakis J, Araujo H and Arce Dubois P *et al* 2006 Geant4 developments and applications *IEEE Trans. Nucl. Sci.* **53** 270–78
[33] Lehmann J *et al* 2005 Monte Carlo treatment planning for molecular targeted radiotherapy within the MINERVA system *Phys. Med. Biol.* Mar 7 **50** 947–58
[34] Strigari L, D'Andrea M, Ludovico Maini C, Sciuto R and Benassi M 2006 Biological optimization of heterogeneous dose distributions in systemic radiotherapy *Med. Phys.* **33** 1857–66
[35] Stabin M G 2008 Uncertainties in internal dose calculations for radiopharmaceuticals *J. Nucl. Med.* **49** 853–60
[36] Wahl R 2005 Tositumomab and 131I therapy in non-hodgkin's lymphoma *J. Nucl. Med.* **46** 128S–40S
[37] Stabin M G 2008 The case for patient-specific dosimetry in radionuclide therapy *Cancer Biother. Radiopharm.* **23** 273–84

Chapter 9

The OLINDA software package

Michael Stabin

9.1 Introduction

A general equation for dose to a target organ T from disintegrations can be given as:

$$D_T = \frac{k \sum_S N_S \sum_i n_i \; E_i \; \phi_i(T \leftarrow S)}{m_T}$$

where D_T = absorbed dose in target organ T (Gy/ or rad)
N_S = number of disintegrations in a source organ (MBq or μCi)
n_i = number of radiations with energy E_i emitted per nuclear transition
E_i = energy per radiation (MeV)
$\phi(T \leftarrow S)$ = fraction of energy emitted in a source that is absorbed in the target
m_T = mass of target region (kg or g)
k = some proportionality constant (Gy-kg MBq^{-1} s^{-1} MeV^{-1} or rad-g μCi^{-1} h^{-1} MeV^{-1})

This was formalized as

$$D = \sum_S N_S \times DF(T \leftarrow S)$$

where $DF(T \leftarrow S)$ includes all of the other terms shown above. This was published as the Radiation Dose Assessment Resource (RADAR) method of internal dosimetry [1]. The original absorbed fractions (ϕ), generated by Monte Carlo methods using mathematical 'phantoms' developed at Oak Ridge National Laboratory (ORNL). In the first version of OLINDA (Organ Level INternal Dose Assessment) code [2], the ORNL phantoms were used. The code was a successor to a personal computer program called MIRDOSE [3]. Users could choose a radionuclide, one or more of the nine ORNL phantoms, and enter numbers of disintegrations for more than 20 source

doi:10.1088/978-0-7503-2694-0ch9

regions and obtain doses to more than 20 target regions. This automated and standardized dose estimates for users everywhere in the world. The name of the code was based on the idea that it calculated doses by the MIRD (Medical Internal Radiation Dose) (Committee of the Society of Nuclear Medicine) method. The MIRD Committee, however vehemently objected to the name of the code, as it could imply that the committee developed it. Newer versions were thus renamed OLINDA, with the same phantoms, but now using the RADAR (RAdiation Dose Assessment Resource) (Committee of the Society of Nuclear Medicine) calculational method.

9.2 Advances in OLINDA v2

OLINDA 1 had over 800 radionuclides to choose from; OLINDA 2 has over 1000. The significant advance in OLINDA 2 was the use of the new RADAR ICRP 89 reference phantoms, based on medical images. The ORNL phantoms, sometimes referred to as ‘stylized’ phantoms, used combinations of geometric shapes to model organs and the whole body. In the 1990s, many groups developed new phantoms based on medical images of real persons, which significantly improved the realism of human body (anthropomorphic) models.

The RADAR group developed a series of phantoms, starting with the adult male and female models developed by Segars [4] using a technology known as Non-Uniform Rational B-Splines (NURBS). Starting with CT images of an adult male and female, Segars fit NURBS surfaces to model individual organs. These surfaces are easily manipulated, either by linear scaling in any dimension, or by manually deforming the surfaces using individual points used to define the surfaces. The RADAR group scaled the adult phantoms to match the reference adult masses defined by the International Commission on Radiological Protection (ICRP) in its Publication 89 [5], and developed phantoms for the other five reference ages, as with the ORNL phantom series. The ORNL adult/pediatric ‘stylized’ phantom series’ individual phantoms were hermaphrodites, having both male and female organs in the one whole body phantom, but of course not allowing cross-irradiation of male and female organs. The RADAR phantom series has a separate male and female phantom at each age.

Xu *et al* [6], in collaboration, developed three realistic phantoms representing the pregnant female at three stages of pregnancy, in parallel to the ORNL pregnant female series.

The original OLINDA code was documented in 2005 [2], and all details will not be repeated here. OLINDA 1 used the ORNL phantom series; OLINDA 2 kept all of the original ORNL phantoms, but incorporated the RADAR ICRP 89 reference phantoms—12 adult and pediatric phantoms and the three Xu *et al* pregnant female phantoms. It retained the ICRP 30 gastrointestinal (GI) tract model [7] and Cloutier *et al* bladder voiding model [8] (figures 9.1 and 9.2). Before OLINDA 2 was developed, the ICRP developed a new model of the GI tract, called the ICRP

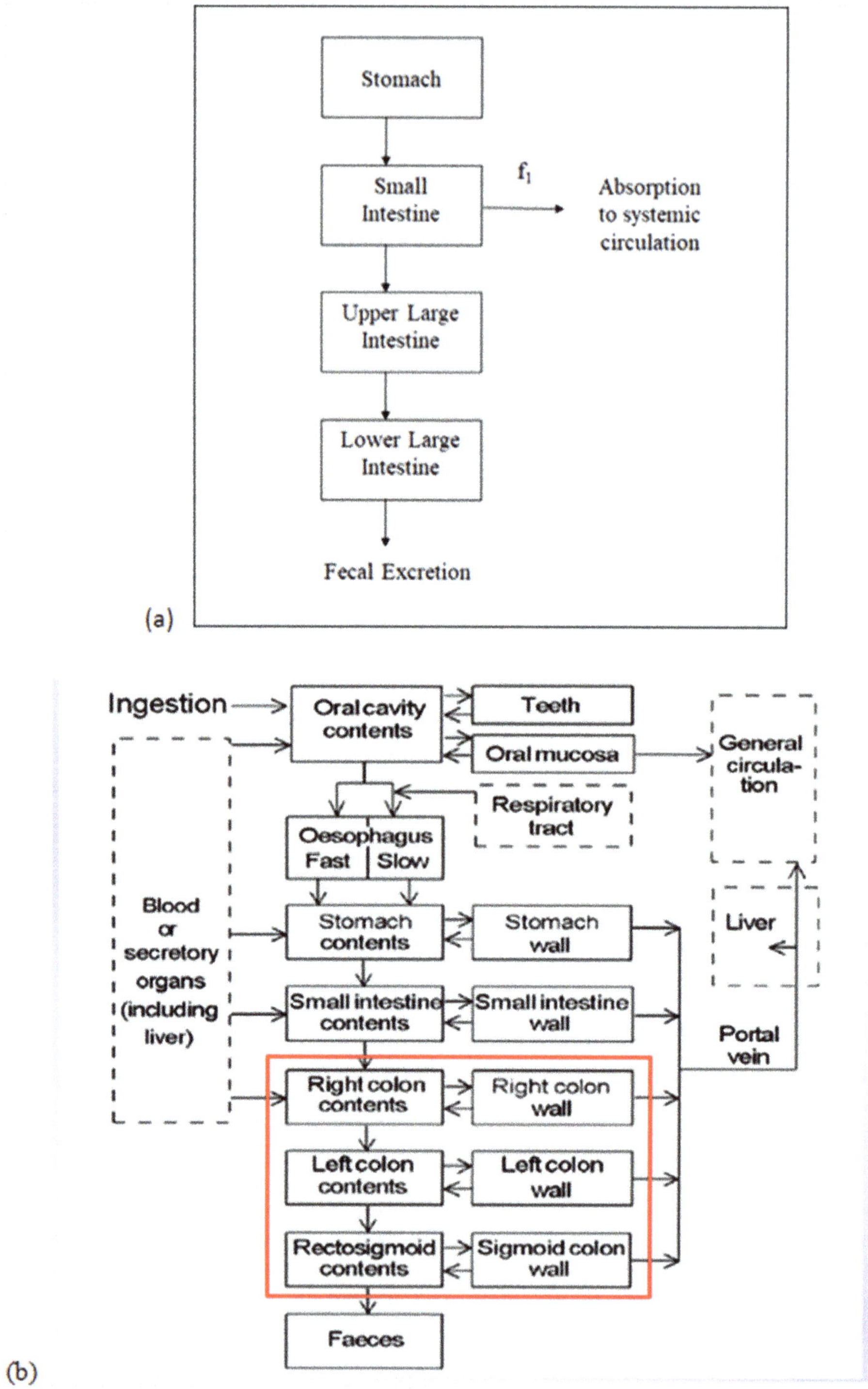

Figure 9.1. (a) The ICRP 30 and (b) ICRP 100 gastrointestinal tract models.

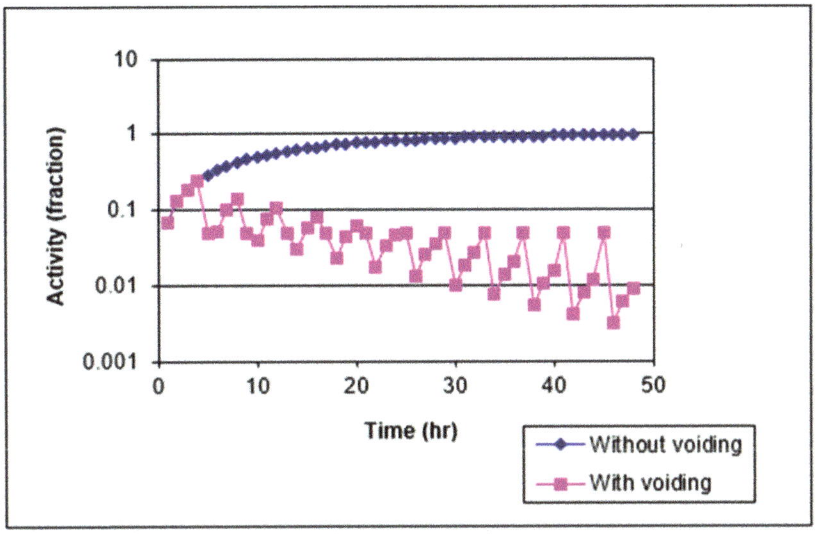

Figure 9.2. The Snyder *et al* dynamic urinary bladder voiding model.

Human Alimentary Tract (HAT) model. It redefined the upper and lower large intestines, which previously represented the ascending and transverse colon (upper) and descending and sigmoid colon (lower) to be the ascending and right half of the transverse colon (right) and left half of the transverse and the descending colon (left). This sigmoid colon was now represented separately. So, in the new RADAR reference phantoms, the Monte Carlo transport calculations were modified to represent these regions. Therefore, the ICRP 30 GI tract model may only be used with the ORNL phantoms and the ICRP HAT model can only be used with the RADAR ICRP 89 reference phantoms.

The transit times for all segments of the ICRP 30 and ICRP HAT model are shown in figure 9.3. Figure 9.4 shows a screen capture of the associated OLINDA 2 input screen for these models.

RADAR also developed realistic, image-based models of three mice and five rats, from models developed by Segars and Tsui [9]. They also developed SAFs for a male and female beagle (unpublished data). All of the SAFs are available in the current version of OLINDA, for dose calculations.

An important issue in the calculation of equivalent doses for alpha particles is the assignment of a radiation weighting factor (w_R). Traditionally, a factor of 20 has been applied in radiation protection [7]. Recent radiobiological evidence indicates that this value may be as low as 5 [10], or even 1 [11]. Similar arguments apply to the use of Auger emitters (for which literature values indicate a range of potential relative biological effectiveness (RBE)). Clearly more investigation and guidance from regulatory and international advisory bodies is needed. OLINDA/EXM

Segment	Transit time (h)
Stomach	1
Small Intestine	4
Upper Large Intestine	13
Lower Large Intestine	24
Total	42

	Adult Males	Adult Females	15-yr olds	10-yr olds	5-yr olds	1-yr olds	Newborns
Stomach	1.17	1.58	1.17	1.17	1.17	1.17	1.25
Small Intestine	4	4	4	4	4	4	4
Right Colon	12	16	11	11	11	10	8
Left Colon	12	16	11	11	11	10	8
Rectosigmoid	12	16	12	12	12	12	12
Total	41	54	39	39	39	37	33

Figure 9.3. Above, transit times in the ICRP 30 GI model, below, transit times in the ICRP HAT model.

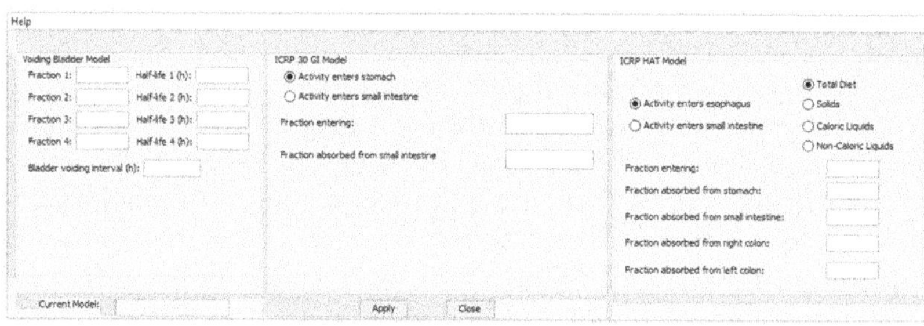

Figure 9.4. Data entry form for the dynamic urinary bladder model, ICRP 30 and ICRP 100 GI models.

assumes a w_R of 1 for photons and electrons and a value of 5 for alpha emissions, but these values can be varied by the user. These are the default values used, but OLINDA/EXM has a form in which these values, as well as the organ masses used in the ten whole body phantoms, can be varied. The latter is a useful feature in situations in which an organ is known to be much larger or smaller than those in the standard phantoms.

9.3 Implementation to develop standard dose estimates for radiopharmaceuticals

RADAR used OLINDA 2 were to calculate dose estimates for over 100 radiopharmaceuticals commonly used in nuclear medicine [12, 13]; they are available for free download from the RADAR website (www.doseinfo-radar.com). Sample tables are shown below.

Table 9.1. Radiation dose estimates (mSv/MBq) for ^{99m}Tc diethylenetriaminepentaacetic acid (DTPA). Normal renal function.

	Males				
	Adult	15-yo	10-yo	5-yo	1-yo
Adrenals	1.92E-03	2.44E-03	3.58E-03	5.12E-03	8.28E-03
Brain	8.68E-04	1.08E-03	1.78E-03	2.86E-03	4.92E-03
Esophagus	1.08E-03	1.38E-03	2.21E-03	3.51E-03	6.13E-03
Eyes	8.68E-04	1.08E-03	1.78E-03	2.86E-03	4.96E-03
Gallbladder Wall	1.45E-03	1.96E-03	3.18E-03	4.72E-03	5.70E-03
Left colon	1.80E-03	2.23E-03	3.23E-03	4.55E-03	7.52E-03
Small Intestine	2.67E-03	3.31E-03	5.22E-03	7.22E-03	9.78E-03
Stomach Wall	1.23E-03	1.63E-03	2.64E-03	3.71E-03	6.29E-03
Right colon	1.83E-03	2.37E-03	3.44E-03	4.88E-03	7.31E-03
Rectum	6.33E-03	8.14E-03	1.30E-02	1.63E-02	1.90E-02
Heart Wall	1.18E-03	1.49E-03	2.27E-03	3.38E-03	5.85E-03
Kidneys	4.56E-03	5.55E-03	7.64E-03	1.16E-02	1.74E-02
Liver	1.27E-03	1.62E-03	2.52E-03	3.72E-03	6.26E-03
Lungs	1.03E-03	1.34E-03	2.01E-03	3.08E-03	5.40E-03
Pancreas	1.45E-03	1.84E-03	2.80E-03	4.25E-03	6.96E-03
Prostate	8.33E-03	1.08E-02	3.37E-02	1.76E-02	2.02E-02
Salivary Glands	1.04E-03	1.32E-03	2.10E-03	3 38E-03	5.93E-03
Red Marrow	1.77E-03	2.16E-03	2.80E-03	3.50E-03	5.65E-03
Osteogenic Cells	2.71E-03	2.91E-03	3.83E-03	4.90E-03	8.03E-03
Spleen	1.36E-03	1.73E-03	2.70E-03	3.98E,-03	6.69E-03
Testes	2.48E-03	3.10E-03	7.44E-03	6.37E-03	8.39E-03
Thymus	1.03E-03	1.32E-03	1.98E-03	3.08E-03	5.45E-03
Thyroid	1.04E-03	1.33E-03	2.12E-03	3.40E-03	5.97E-03
Urinary Bladder Wall	6.67E-02	8.10E-02	1.19E-01	1.51E-01	1.75E-01
Total Body	1.76E-03	2.15E-03	3.34E-03	4.89E-03	7.19E-03
	Females				
	Adult	15-yo	10-yo	5-yo	1-yo
Adrenals	2.04E-03	2.18E-03	3.17E-03	4.77E-00	9.13E-02
Brain	1.07E-03	1.10E-03	1.78E-03	2.85E-06	7.13E,-03
Breasts	9.12E-04	9.75E-04	—	—	—
Esophagus	1.17E-03	1.42E-03	225E,-03	3.55E,48	1.77E-02
Eyes	1.07E-03	1.10E-03	1.78E-03	2.86E48	7.07E-03
Gallbladder Wall	2.13E-03	2.39E-03	3.71E-03	5.43E-06	5.41E-02
Left colon	2.56E-03	3.02E-03	4.70E-03	7.35E-06	2.63E,-02
Small Intestine	3.21E-03	3.55E,-03	5.21E,-03	7.69E-06	2.77E-02
Stomach Wall	1.55E-03	1.71E-03	270E,-03	3.85E,48	2.38E-02
Right colon	2.0E-03	2.25E,-03	3.44E-03	5.18E-C8	3.15E-02
Rectum	1.2E-02	1.74E-02	221E-02	2.90E-02	2.33E-02
Heart Wall	1.45E-03	1.50E-03	224E-03	3.36E-08	1.38E-02
Kidneys	5.07E-03	5.65E-03	7.50E-03	1.16E-02	6.87E-01
Liver	1.57E-03	1.64E-03	251E-03	3.77E-C6	4.54E-02
Lungs	1.3E-03	1.35E-03	200E-03	3.06E-06	1.21E-02

Ovaries	6.43E-03	8.69E-03	1.01E-02	1.74E-02	2.21E-02
Pancreas	1.85E-03	1.98E-03	296E-03	4.60E-03	3.62E-02
Salivary Glands	1.11E-03	1.34E-03	2.12E-03	3.39E-03	8.75E-03
Red 'Arrow	2.00E-03	1.95E-03	215E-03	2.56E-C6	5.59E-03
Osteogenic Cells	2.93E-03	2.75E-03	281E-03	2.95E-06	6.08E-03
Spleen	1.72E-03	1.82E-03	2.76E-03	4.11E-06	7.34E-02
Thymus	1.23E-03	1.34E-03	1.96E-03	3.06E-C6	1.04E-02
Thyroid	1.11E-03	1.34E-03	211E-03	3.38E-06	9.62E-03
Urinary Bladder Wall	7.1E-02	7.16E-02	1.03E-01	1.54E-01	8.24E-02
Uterus	1.32E-02	1.78E-02	283E-02	3.68E-02	2.77E-02
Total Body	2.25E-03	2.15E-03	3.24E-03	4.97E-03	1.61E-02

	Gender Average				
	Adult	15-yo	10-yo	5-yo	1-yo
Adrenals	1.98E-03	2.31E-03	3.37E-03	4.94E-03	4.86E,-03
Brain	930E-04	1.09E-03	1.78E-03	2.85E-03	2.85E-03
Breasts	912E-04	9.75E-04			
Esophagus	1.13E-03	1.40E-03	2.23E-03	3.53E-03	3.52E-03
Eyes	9.70E-04	1.09E-03	1.78E-03	2.86E-03	2.86E-03
Gallbladder Wall	1.79E-03	2.18E-03	3.44E-03	5.07E-03	4.98E-03
Left colon	2.18E-03	2.0E,-03	3.96E,-03	5.95E,-03	5.81E-03
Small Intestine	2.94E-03	3.43E-03	5.21E-03	7.46E-03	6.84E-03
Stomach Wall	1.39E-03	1.67E-03	2.67E-03	3.78E-03	3.74E-03
Right colon	1.95E-03	2.31E-03	3.44E-03	5.03E-03	4.86E-03
Rectum	9.55E-03	1.23E-02	1.76E-02	2.26E-02	2.0E-02
Heart Wall	132E-03	1.50E-03	2.25E,-03	3.37E-03	3.35E-03
Kidneys	4.81E-03	5.60E-03	7.57E-03	1.16E-02	1.15E-02
Liver	1.42E-03	1.0E-03	2.51E-03	3.74E-03	3.60E-03
Lungs	1.19E-03	1.34E-03	2.00E-03	3.07E-03	3.06E-03
Ovaries	6.43E-03	8.E0E-03	1.01E-02	1.74E-02	1.74E-02
Pancreas	1.65E-03	1.91E-03	2.88E-03	4.42E-03	4.35E-03
Prostate	8.33E-03	1.08E-02	3.37E-02	1.76E-02	1.25E-02
Salivary Glands	1.08E-03	1.33E-03	2.11E-03	3.38E-03	3.38E-03
Red Arrow	1.88E-03	2.05E-03	2.48E-03	3.03E-03	2.91E-03
Osteogenic Cells	2.82E-03	2.83E-03	3 32E-03	3.92E-03	3.22E-03
Spleen	1.54E-03	1.78E-03	2.73E,-03	4.05E,-03	3.99E-03
Testes	2.48E-03	3.10E-03	7.44E,-03	6.37E-03	5.04E-03
Thymus	1.16E-03	1.33E-03	1.97E-03	3.07E-03	3.06E-03
Thyroid	1.08E-03	1.33E-03	2.11E-03	3.39E,-03	3.38E-03
Urinary Bladder Wall	6.90E-02	7.63E-02	1.11E-01	1.52E-01	1.26E-01
Uterus	132E-02	1.78E-02	2.83E-02	3.68E-02	3.68E-02
Total Body	2.00E-03	2.15E-03	3.29E-03	4.93E-03	4.59E-03

Biokinetic data: ICRP 128 recommended model.
Physical models: RADAR ICRP 89 Reference Phantom Series.
Effective doses: ICRP 103 weighting factors.

Table 9.2. Embryo/fetus doses for selected radiopharmaceuticals.

Radiopharmaceutical	Estimated Fetal Dose (mGy/MBq*)			
	Early Pregnancy	3 months' Gestation	6 months' Gestation	9 months' Gestation
^{13}N Ammonia	2.3×10^{-3}	1.9×10^{-3}	2.0×10^{-3}	1.8×10^{-3}
^{18}F FDG	2.6×10^{-2}	1.9×10^{-2}	1.4×10^{-2}	6.9×10^{-3}
^{18}F Fluoride	3.1×10^{-2}	2.2×10^{-2}	8.6×10^{-3}	6.7×10^{-3}
^{32}P Phosphate	8.7×10^{-1}	1.7×10^{0}	2.3×10^{0}	2.3×10^{0}
^{67}Ga Citrate	9.4×10^{-2}	7.1×10^{-1}	1.1×10^{-1}	3.2×10^{-2}
^{123}I Hippuran	4.9×10^{-2}	3.3×10^{-2}	9.2×10^{-3}	6.6×10^{-3}
^{123}I IMP	2.2×10^{-2}	1.6×10^{-2}	1.1×10^{-2}	8.9×10^{-3}
^{123}I mIBG	2.2×10^{-2}	1.6×10^{-2}	8.2×10^{-3}	6.8×10^{-3}
^{123}I NaI (35% uptake)	2.3×10^{-2}	1.6×10^{-2}	9.3×10^{-3}	6.3×10^{-3}
^{123}I BMIPP	2.0×10^{-2}	1.5×10^{-2}	1.1×10^{-2}	9.6 x 10-3
^{125}I HSA	2.4×10^{-1}	1.6×10^{-1}	1.2×10^{-1}	1.0×10^{-1}
^{125}I mIBG	3.2×10^{-2}	2.0×10^{-2}	1.0×10^{-2}	9.1×10^{-3}
^{131}I Hippuran	9.8×10^{-2}	6.8×10^{-2}	2.0×10^{-2}	1.4×10^{-2}
^{131}I mIBG	1.2×10^{-1}	9.6×10^{-2}	7.3×10^{-2}	6.6×10^{-2}
^{131}I NaI (35% uptake)	7.6×10^{-2}	1.6×10^{-1}	2.9×10^{-1}	2.2×10^{-1}
^{111}In DTPA	9.4×10^{-2}	6.5×10^{-2}	2.1×10^{-2}	1.6×10^{-2}
^{111}In Pentetreotide	1.1×10^{-1}	7.8×10^{-2}	3.7×10^{-2}	3.0×10^{-2}
^{111}In Platelets	1.7×10^{-1}	1.3×10^{-1}	1.1×10^{-1}	9.8×10^{-2}
^{111}In RBCs	2.2×10^{-1}	1.6×10^{-1}	1.4×10^{-1}	1.2×10^{-1}
^{111}In WBCs	1.2×10^{-1}	9.3×10^{-2}	8.6×10^{-2}	8.0×10^{-2}
^{81m}Kr Gas	1.2×10^{-7}	1.4×10^{-7}	1.9×10^{-7}	2.0×10^{-7}
^{177}Lu DOTATATE [14]	1.9×10^{-2}	1.9×10^{-2}	2.0×10^{-2}	1.9×10^{-2}
^{82}Rb Chloride	9.5×10^{-4}	3.5×10^{-4}	6.3×10^{-4}	6.3×10^{-4}
^{153}Sm EDTMP [15]	2.0×10^{-2}	3.6×10^{-2}	1.8×10^{-2}	1.3×10^{-2}
^{89}Sr Chloride [16]	8.4×10^{-1}	2.1×10^{0}	2.3×10^{0}	2.3×10^{0}
^{99m}Tc Disofenin	1.9×10^{-2}	1.3×10^{-2}	7.8×10^{-3}	6.4×10^{-3}
^{99m}Tc DMSA	5.9×10^{-3}	8.8×10^{-3}	3.4×10^{-3}	2.2×10^{-3}
^{99m}Tc DTPA	1.7×10^{-2}	1.1×10^{-2}	3.8×10^{-3}	3.2×10^{-3}
^{99m}Tc DTPA aerosol	7.6×10^{-3}	5.1×10^{-2}	1.9×10^{-3}	2.1×10^{-3}
^{99m}Tc Glucoheptonate	1.6×10^{-2}	1.4×10^{-2}	4.8×10^{-3}	3.7×10^{-3}
^{99m}Tc ECD	1.4×10^{-2}	1.0×10^{-2}	4.9×10^{-3}	3.9×10^{-3}
^{99m}Tc HMPAO	1.0×10^{-2}	7.2×10^{-3}	4.3×10^{-3}	3.6×10^{-3}
^{99m}Tc MAA	3.2×10^{-3}	1.2×10^{-2}	2.6×10^{-3}	1.7×10^{-3}
^{99m}Tc MAG3	2.6×10^{-2}	1.8×10^{-2}	5.7×10^{-3}	4.1×10^{-3}
^{99m}Tc MDP	8.1×10^{-3}	9.1×10^{-3}	2.6×10^{-3}	2.1×10^{-3}
^{99m}Tc MIBI rest	1.7×10^{-2}	1.2×10^{-2}	6.9×10^{-3}	5.6×10^{-3}
^{99m}Tc MIBI stress	1.4×10^{-2}	9.5×10^{-3}	6.0×10^{-3}	4.8×10^{-3}
^{99m}Tc Pertechnetate	1.4×10^{-2}	6.2×10^{-2}	1.1×10^{-2}	5.8×10^{-3}
^{99m}Tc Red Blood Cells (in vivo)	7.0×10^{-3}	5.5×10^{-3}	2.2×10^{-3}	8.7×10^{-4}
^{99m}Tc Sulfur Colloid - normal	1.4×10^{-3}	2.1×10^{-3}	1.9×10^{-3}	1.9×10^{-3}
^{99m}Tc White Blood Cells	3.6×10^{-3}	2.7×10^{-3}	2.6×10^{-3}	2.4×10^{-4}
^{201}Tl Chloride [17]	8.2×10^{-2}	5.6×10^{-2}	5.0×10^{-2}	4.4×10^{-2}
^{133}Xe 5 min rebreathing	3.9×10^{-4}	3.7×10^{-4}	3.7×10^{-4}	3.6×10^{-4}

* mGy to the fetus per MBq administered to the mother. Conversion rad/mCi = mGy/MBq x 3.7. Shaded rows are nuclides for which placental crossover data were available.

References

[1] Stabin M G and Siegel J A 2003 Physical models and dose factors for use in internal dose assessment *Health Phys.* **85** 294–310

[2] Stabin M G, Sparks R B and Crowe E 2005 OLINDA/EXM: the second-generation personal computer software for internal dose assessment in nuclear medicine *J. Nucl. Med.* **46** 1023–7

[3] Stabin M G 1996 MIRDOSE: personal computer software for internal dose assessment in nuclear medicine *J. Nucl. Med.* **37** 538–46

[4] Segars W P 2001 Development and application of the new dynamic NURBS-based cardiac-torso (NCAT) phantom *PhD Thesis* University of North Carolina, Chapel Hill, NC

[5] ICRP. Publication 89 2002 Basic anatomical and physiological data for use in radiological protection: reference values *Ann. ICRP* **32** 1–277

[6] Xu X G, Taranenko V, Zhang J and Shi C 2007 A boundary-representation method for designing whole-body radiation dosimetry models: pregnant females at the ends of three gestational periods—RPI-P3, -P6 and -P9 *Phys. Med. Biol.* **52** 7023–44

[7] ICRP 1979 Limits for intakes of radionuclides by workers *ICRP Report No. 30* (New York: Pergamon)

[8] Cloutier R J, Smith S A, Watson E E, Snyder W S and Warner G G 1973 Dose to the fetus from radionuclides in the bladder *Health Phys.* **25** 147–61

[9] Segars P and Tsui B M W 2010 The MCAT, NCAT, XCAT and MOBY computational human and mouse phantoms *Handbook of Anatomical Models for Radiation Dosimetry* ed X G Xu and K F Eckerman (Boca Raton: CRC Press, Taylor and Francis)

[10] Feinendegen L E and McClure J 1996 Alpha-Emitters for Medical Therapy: Workshop of the United States Department of Energy *Radiat. Res.* **148** 195–201

[11] Fisher D R, Frazier M E and Andrews T K Jr. 1985 Energy distribution and the relative biological effects of internal alpha emitters *Radiat. Prot. Dosim.* **13** 223–7

[12] Stabin M G 2018 New-generation fetal dose estimates for radiopharmaceuticals *J. Nucl. Med.* **59** 1005–6

[13] Stabin M G and Siegel J A 2018 RADAR dose estimate report: a compendium of radiopharmaceutical dose estimates based on OLINDA/EXM Version 2.0 *J. Nucl. Med.* **59** 154–60

[14] Sandstrom M 2017 Personal communication. Uppsala University Hospital

[15] 1992 Radiation Internal Dose Information Centerpersonal communication. Oak Ridge, TN

[16] ICRP 1987 Radiation dose to patients from radiopharmaceuticals. Publication 53 *Ann. ICRP* **18** 1–377

[17] ICRP 2015 Radiation dose to patients from radiopharmaceuticals: a compendium of current information related to frequently used substances. Publication 128 *Ann. ICRP* **44** 7–321

Chapter 10

The three-dimensional radiobioligical dosimetry software package, 3D-RD

Robert F Hobbs and George Sgouros

10.1 Introduction

Radiopharmaceutical therapy (RPT) is currently enjoying a renaissance with the advent of new radiopharmaceuticals and new pharmaceutical company interest. Along with this new drug development and widespread use comes a desire to improve the existing therapies by moving towards personalized dosimetry-based treatment planning. Currently, radiopharmaceuticals are administered using fractionation and a fixed activity or mass-based paradigm consistent with other systemic therapy dosing, e.g. chemotherapy. RPT differs from these other systemic therapies with an obvious potential for a higher level of personalization of therapy. By administering a pre-therapeutic tracer amount of RPT, or of a companion diagnostic agent, and imaging the bio-distribution within the patient at multiple time points using 3D imaging (PET/CT or SPECT/CT), individual patient pharmacokinetics are measured. These pharmacokinetics, in turn, enable the calculation of a maximum tolerated activity (MTA) specific to each patient based on normal organ maximum tolerated dose (MTD) values. Similarly, imaging the therapeutic activity in the patient allows for retrospective dosimetry and establishing data for dose-response relationships.

Two methods for patient dosimetry exist: (1) the standard 'absorbed fraction' or 'MIRD' formalism which was developed by the Medical Internal Radiation Dose (MIRD) Committee [1, 2], and also reviewed in chapter 3 of this book. As indicated above, this method is fundamentally activity-based and relies on '*S* values' to convert time-integrated activity to absorbed dose values [3]. This procedure has been implemented in several software packages, most notably OLINDA/EXM [4]. To generate tables of *S* values for different radionuclides and source–target organ combinations, a standard model of human anatomy was adopted in which organ position, dimensions, and composition were mathematically defined. The radioactivity was assumed to be uniformly distributed throughout each source organ and the *S*

doi:10.1088/978-0-7503-2694-0ch10

values were defined as the mean absorbed dose to a target organ per unit time-integrated activity in a source organ. Recently, more realistic phantoms (voxelized, mesh or spline) have been developed to replace the original phantom consisting of simple geometric shapes, resulting in improved accuracy, especially for adjacent structures. However, since the position and size of tumors may vary within the body and since a standard model of human anatomy was adopted for generating the *S* value tables, tumors are not included in the published tables. A number of approaches have been developed for estimating the absorbed dose to tumors and the dose contributions between tumors and normal organs. The typical approximation is made by assuming that the tumor can be modeled as a sphere [5, 6] and using interpolation for simple spheres' self-dose obtained from Monte Carlo.

(2) The second method consists of Monte Carlo simulation of the individual patient's anatomical and activity distribution, a technique commonly known as 'voxelized dosimetry', either by direct Monte Carlo or by use of dose-kernel convolution methods. These latter have the important practical advantage of speed, but traditionally rely on the underlying assumption of a homogeneous medium. Point-kernel methods are now available that partially overcome this limitation by scaling the source-to-target distance to account for tissue inhomogeneity [7–9]. In contrast, Monte Carlo methods explicitly account for mass–density and atomic-number variations within the human body by simulating relevant particle transport events [10–18]. Either Monte Carlo-based method allows for direct tumor dosimetry within the patient's anatomy.

Both methods rely on: (1) accurate registration of time-ordered 3-dimensional data (SPECT/CT or PET/CT); (2) fitting of time-dependent data (either activity or dose rate); (3) Monte Carlo to convert activity and time-integrated activity to dose rate and absorbed dose, respectively. In the absorbed fraction methodology, the Monte Carlo results from phantoms are used, while in 'voxelized dosimetry' the Monte Carlo (or dose-kernel) is applied directly to the patient images.

The Three-Dimensional Radiobiological Dosimetry (3D-RD) software package combines both absorbed fraction and Monte Carlo-based dosimetry and also incorporates bio-effect modeling, including calculation of biological effective dose (BED), equivalent uniform dose (EUD) and sub-organ modeling. It evolved from 3D-ID (Three-Dimensional Internal Dosimetry) software, which was one of the first platforms for personalized dosimetry. That previous work introduced the concept of dose-volume histograms for internally administered radionuclides [10]. Over the course of its existence 3D-RD (and 3D-ID) has played a leading role in developing dosimetric techniques, including the first 3D prospective treatment planning [19], retrospective dose-response studies [20], and radiobiologically-driven rational therapy methodologies [21–23] as well as providing the gold standard benchmark for other 3D dosimetry platforms [9, 14].

10.2 The historical 3D-ID package

3D-ID was developed in response to the increasing use of RPT. Although point-kernel convolution as well as patient-specific Monte Carlo calculations had been implemented in external radiotherapy [24], no package was available to provide the platform and support structure required to carry out such calculations for patients treated with targeted radiopharmaceuticals. Likewise, although the fundamental schema of the MIRD methodology does not preclude patient-specific dosimetry calculations, translation of this schema had focused primarily on diagnostic uses of radionuclides in which standardized anatomy, uniform activity distribution and estimation of mean absorbed dose were deemed adequate in assessing the hazards involved in the use of radionuclides [3]. In RPT, administered activities greatly exceeding those used for diagnosis are combined with therapeutic radionuclides and tumor to non-tumor ratios that could easily lead to normal organ toxicity.

Up to that point, most of the effort in patient-specific 3-D dosimetry had been in the development of algorithms and computational techniques. Application of such techniques, however, required a software system that translates studies from multiple imaging modalities (CT or MR and PET or SPECT) into a single data format, providing the tools needed to draw contours around regions of interest (ROIs) for identifying source and target regions and outputs the absorbed dose calculations as images, isodose contours or dose-volume histograms [12]. Each of these steps required considerable user interaction and was input/output-intensive. 3D-ID was developed in order to provide a comprehensive and user-friendly framework in which to carry out the various and necessary steps to calculate absorbed dose rate or total dose to any user-defined target region.

In order to relate patient anatomy to radionuclide distribution, it is necessary to register the appropriate functional and anatomic image. The non-trivial issue of image registration was kept separate from the 3D-ID software system in order to keep the focus on the logistics of image handling, region-of-interest definition and absorbed dose calculation.

The 3D-ID software package consisted of a set of three modules sharing a common interface style [12] (figure 10.1). The first step provided an entry point into the software package for the diverse modality- and vendor-dependent image formats. Images in standard DICOM or Interfile3.3 were read in directly.

The second step provided tools to define regions-of-interest and to calculate source-to-target absorbed dose. Aligned multi-modality image sets were arranged in two rows corresponding to an anatomical modality (CT/MRI) and to a radionuclide modality (SPECT/PET) (figure 10.2). Images were displayed separately rather than fused or overlaid so that all available data from both modalities was preserved.

Contour drawing for region-of-interest definition was carried out on a side-by-side, dual modality display of individual 2D image slices, thus providing access to information from both imaging modalities (figure 10.3). Prior to drawing, the cursor appears over both images so that the user can easily identify corresponding points within the two images. Contours were drawn using either image and it was possible to switch back and forth between images during drawing. A number of drawing tools were provided: region growing which automatically generates a contour plot by using a percentage of the ratio between a user selected image value interior to the region and a selected background value outside the region [25]; edge following which is effective in images of high contrast; and standard copy and paste of individual

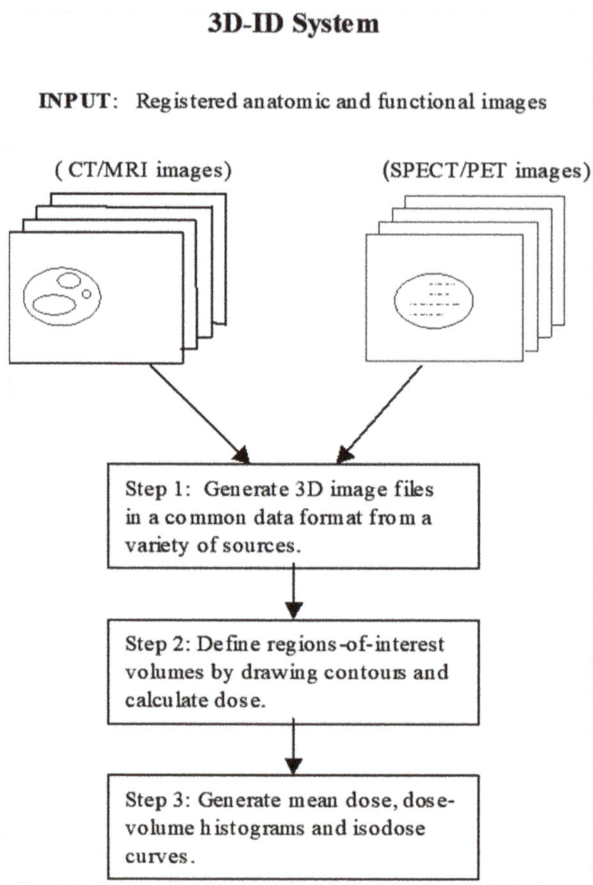

Figure 10.1. 3D-ID flowchart.

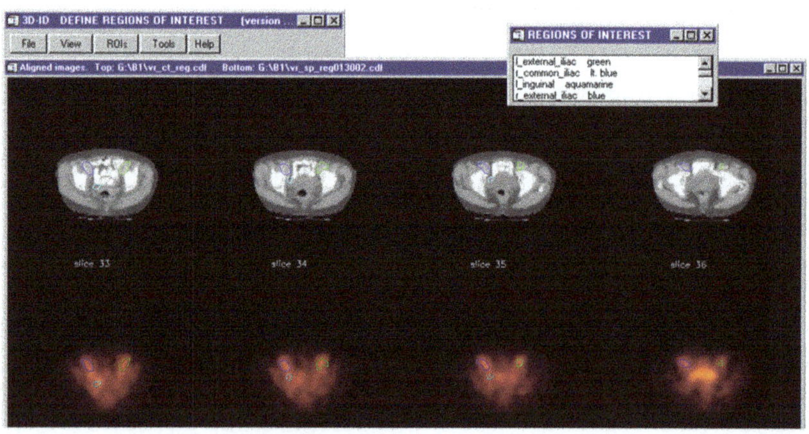

Figure 10.2. Multi-image display from 3D-ID showing four transverse slices from a set of CT-SPECT aligned images. The top row shows the CT with regions of interest drawn over three lesions identified as: right external iliac in blue, left external iliac in green, and right common iliac in light blue. The bottom row shows the corresponding I-131 SPECT slices with the lesion indicated.

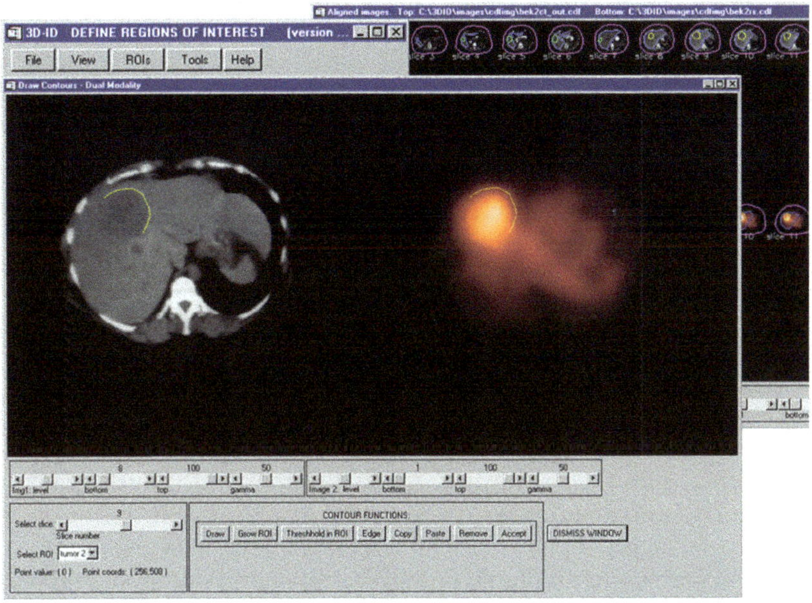

Figure 10.3. ROIs are defined by drawing individual contours using aligned image sets. As contours are drawn, they appear in the multi-slice, two-row, main display window (seen here behind the drawing window).

ROI contours. ROIs could be quantified in terms of total volume, pixel count, and magnitude and location of maximum value within the ROI.

Once ROI volumes had been defined, they could be identified as source or target for a particular dose calculation. A point-kernel file was also specified to complete input for the dose calculation.

The dose distribution for selected source and target combinations was calculated by convolving the point-kernel with the activity image. Convolution was implemented either by Fast Fourier Transform or by a table lookup procedure. In either case, the electron dose was deposited within the activity containing voxel for most human dosimetry calculations. If the electron range was significant relative to patient organ dimensions, an electron point-kernel could be added to the early portion of the photon point-kernel to account for the spatial distribution of the electron dose deposition.

Depending upon the size of the image files, the absorbed dose calculation could be a computationally intensive and time-consuming calculation. In such circumstances, batch mode execution of the absorbed dose calculation was possible.

The absorbed dose calculation yielded a 3-D array of absorbed dose values which are essentially images of absorbed dose distribution and could be viewed as ‘maps’ of the spatially varying absorbed dose in a target volume from a particular source volume or volumes. The modular characteristic of 3D-ID provided the flexibility to replace the absorbed dose calculation technique with other techniques such as point-kernel or Monte Carlo based.

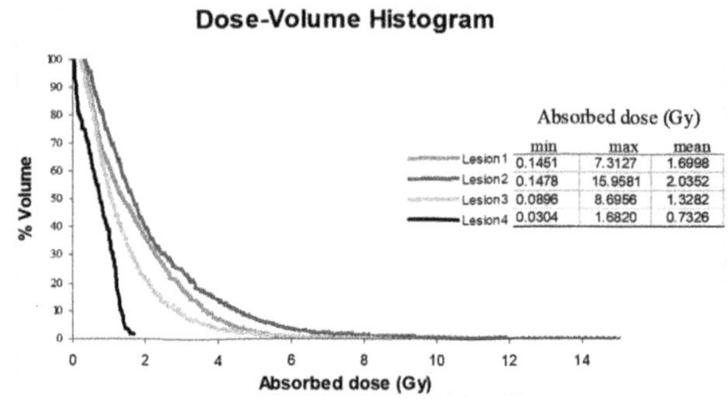

Figure 10.4. Sample integral absorbed dose-volume histogram showing distribution of absorbed dose over four lesions. Histograms were derived from I-124 PET studies in patients with thyroid carcinoma. Each curve represents an individual lesion seen in a single patient.

Once the calculation was performed, the final step in 3D-ID was the display and analysis of the 3D absorbed dose data set. This included: calculation of the mean absorbed dose to a target per unit time-integrated activity from a source (or sources); generation of absorbed dose-volume histograms of source-to-target regions in order to summarize the absorbed dose information in a clinically interpretable manner; and presentation of all image data in a summary display of anatomic, functional and absorbed dose distribution images.

If only a single time point of SPECT or PET was available, the absorbed dose in each target voxel could be scaled by a total source organ time-integrated activity that is typically obtained from time-ordered planar imaging kinetics. In this way, the spatial distribution of activity is assumed to be preserved throughout the study period. This is a first-order approximation which may be improved upon if additional (three-dimensional) pharmacokinetic data is obtained. Time-integrated activity estimation was external to the system.

The mean absorbed dose over a target volume was determined by taking the mean of all values in the dose array. Either integral or differential dose-volume histograms could be generated. In the former, the percent of volume receiving less than or equal to a particular absorbed dose value is plotted as a function of the absorbed dose value (figure 10.4). In the latter, the percent of volume receiving a particular absorbed dose range is plotted versus the dose range.

10.3 Radiobiological motivation FR 3D-RD

The objective of patient-specific voxelized dosimetry, such as that implemented by 3D-ID, is to better predict biologic effect. A further step toward this goal would couple the output described above with radiobiologic models that account for the spatial absorbed dose distribution and the rate at which it is delivered.

Unlike external beam, where fractional dose is determined by the physician, in RPT the patient's individual pharmacokinetics not only affect the absorbed dose per unit of administered activity to the various organs and tumors, but also the rate at which the dose is delivered. Just as different fractionation schemes result in different biological outcomes, so too does the patient-specific dose rate for the same absorbed dose affect biological outcome in RPT.

The variation in biological response to the same absorbed dose delivered at different dose rates or different number of fractions led to the concept of biologically effective dose (BED) (originally called the extrapolated tolerance dose, ETD) [26–28]. The BED is defined as the absorbed dose that is required to cause a given biological effect if the dose were delivered in infinitely small doses per fraction or, equivalently, at an infinitely low dose rate, thus providing a single standardized 'fractionation' equivalent. From this definition, the following relationship in the linear-quadratic formalism was established:

$$SF = \mathrm{e}^{-\alpha BED} = \mathrm{e}^{-\alpha D - G(T)\beta D^2} \tag{10.1}$$

By taking the natural logarithm of each side of the equation and solving for the BED, a formulation that is applicable to RPT [29, 30] is obtained:

$$\mathrm{BED} = D\left(1 + \frac{G(T)}{\alpha/\beta} \bullet D\right) \tag{10.2}$$

In the case of an organ or region of interest (ROI) whose absorbed dose is best modeled by a mono-exponentially decreasing time-dependent dose rate $\dot{D}(t)$ with rate constant λ of the form:

$$\dot{D}(t) = \dot{D}(0) \bullet e^{-\lambda t} \tag{10.3}$$

a simpler formulation may be obtained [29]:

$$\mathrm{BED} = D\left(1 + \frac{D}{\alpha/\beta} \bullet \frac{\lambda}{\lambda + \mu}\right) \tag{10.4}$$

Equation (10.4) is the most recognized form for the BED in RPT, but is only valid for organs or regions whose time-dependent dose rate is described by a mono-exponential function.

More complex forms of the BED have been derived for the multicomponent exponential dose rate curves that are typically observed in RPT [31, 32] and for more complex fits requiring numeric integration [33, 34].

The importance and validity of the BED concept was perhaps best illustrated by the study by Barone *et al* who found that kidney toxicity in peptide receptor radiation therapy correlated with the biological effective dose (BED) rather than with absorbed dose [35]. MIRD Pamphlet #20 demonstrated that this standardization was relevant and consistent with external beam absorbed dose values by comparing response to kidney dose normalized to BED from both the Barone study and external beam and showing that they are equivalent [36].

Another radiobiological consideration is that a single average dose value for a tissue or tumor may not adequately represent the dose distribution within the tissue or tumor volume and may therefore poorly predict biological outcome. Unlike in external-beam radiotherapy, where tumor dose is relatively uniform and normal-organ dose non-uniform (depending on beam placement and geometry), in RPT the distribution of the radiopharmaceutical in tumors is dependent on a number of factors (e.g., molecular weight, tissue penetration characteristics, vasculature, hypoxia) that will impact the activity and absorbed dose distributions and typically make them non-uniform. From a response standpoint, a tumor with a tumoricidal average absorbed dose may proliferate if a sub-volume of the tumor receives a sub-tumoricidal absorbed dose.

A dose-volume histogram (DVH) or a three-dimensional dose map is a more complete representation, but a quantity exists, the equivalent uniform dose (EUD), which provides a single value that is weighted in a way that accounts for the surviving fraction of tumor cells given the spatial distribution of absorbed dose within the tumor volume. For any dose distribution, the corresponding EUD is the absorbed dose in Gy, which, when distributed uniformly across the target volume, achieves the same survival fraction among clonogenic cells [37]. Since the concept of 'equivalent uniformity' is radiobiologically driven and more easily implemented with a linear model, most authors have extended the EUD by defining the equivalent uniform biological effective dose [13, 38–40], a concept first described by O'Donoghue [38]. In practice this distribution is composed of voxel values indexed by *i*:

$$e^{-\alpha EUD} = \frac{\sum_{i=1}^{N} e^{-\alpha \cdot BED_i}}{N} \tag{10.5}$$

Solving for the EUD yields:

$$\mathrm{EUD} = -\frac{1}{\alpha} ln\left(\frac{\sum_{i=1}^{N} e^{-\alpha \cdot BED_i}}{N}\right) \tag{10.6}$$

10.4 3D-RD package

As the radiobiological extension of 3D-ID, the basic modular aspect was retained, with the obvious addition of modules designed to convert absorbed dose to BED and to calculate EUD for tumors. In addition, new software was used as the basis for 3D-RD. The new Monte Carlo platform for the voxelized dosimetry became EGSnrc and the output from the Monte Carlo is processed using modular ROOT macros. The new procedure is as follows [13] (figure 10.5):

1. a series of longitudinal 3D SPECT/CT or PET/CT images are input.
2. The images are registered across time, preferably using CT-to-CT registration as a bridge for the emission images. Typically, the first time point is used as the reference frame. When the transmission image quality is insufficient, as when attenuation maps are provided instead of CT images at certain time points, then the emission images may be used for registration [41].

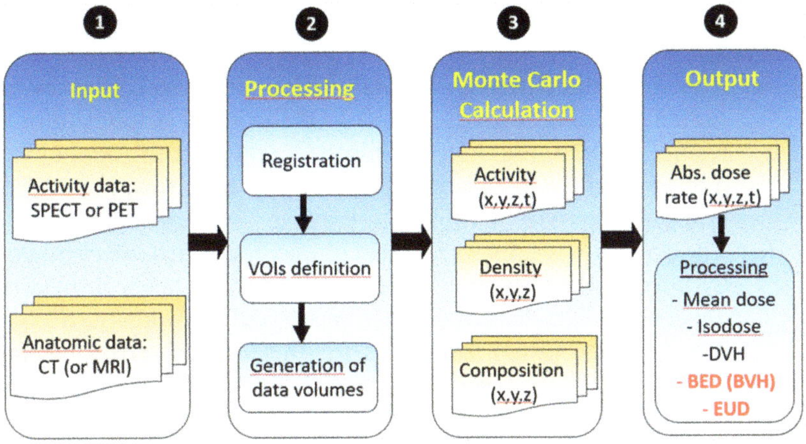

Figure 10.5. Schema of workflow for 3D-RD. In red are the radiobiological components.

3. Volumes of interest are delineated on the reference CT and a VOI map is created in the SPECT or PET voxel format and dimensions.
4. The reference CT image is used to create a density map and a composition map (i.e., tissue, lung, air, and bone) in the SPECT or PET voxel format and dimensions.
5. For each time point, the activity image, the matched density and composition images are used along with decay energy spectra files for (a) discrete energy photon emissions, (b) continuous energy spectrum electron emissions (beta particles) and (c) discrete energy electrons (e.g. Auger electrons) to perform a Monte Carlo calculation to estimate the energy deposition from radioactive decay of the activity in each voxel.
6. The Monte Carlo input files (density, composition, activity, VOI maps) and output files (beta electron energy, photon energy, discrete spectrum electron energy, i.e. Auger electrons) are read into the main 3D-RD program.
7. Analysis is performed at the VOI level:
   (a) Time-integrated activity is obtained for each VOI by fitting the most appropriate function (mono-exponential, bi-exponential, hybrid-trapezoid exponential, etc) and integrating the activity at each time point. The time-integrated activities are multiplied by external S values to obtain the absorbed dose values to the different organs.
   (b) Similarly, the dose rate is calculated at each time point on a VOI-basis by tallying the total energy deposited in the VOI and dividing by the VOI mass. The dose rate is then integrated to obtain the absorbed dose using the most appropriate function. Organ absorbed dose is compared between the two methods (a. and b.) for quality assurance.
   (c) The G-factor is calculated for each VOI and used with the α, β and μ radiobiological parameters specific to the VOI to obtain the VOI BEDs.

8. Analysis is performed at the voxel level, similarly to the organ (VOI)-level analysis:
   (a) Time-integrated activity is obtained for each voxel either by fitting the same functional form as for the VOI the voxel belongs to or, if the fit results in nonsensical or non-physical values, a trapezoid-exponential hybrid and then integrating the activity at each time point. A 3D time-integrated activity map is then created.
   (b) The dose rate is calculated at each time point on a voxel-by-voxel basis by tallying the total energy deposited in the voxel and dividing by the voxel mass. The dose rate is then integrated to obtain the absorbed dose using the functional form used for the VOI the voxel belongs to, or failing that, to a hybrid-exponential fit. A 3D absorbed dose map is created and the average voxel absorbed dose in each organ is compared to the whole organ mean absorbed dose for quality assurance.
   (c) The G-factor is calculated for each voxel and used with the α, β and μ radiobiological parameters specific to the VOI to obtain the voxel BEDs. A 3D BED map is then created.
   (d) For tumors, the EUD is calculated from the voxel BEDs.

The items of interest that have evolved since 3D-ID are as follows:

1. Both methods (absorbed fraction and Monte Carlo-based) are implemented. This is to provide double checks for consistency. The absorbed fraction method, though often touted as less precise, is in general within 5% of the voxelized methodology for large organs. Even for very large tumors and isotopes with significant photon absorbed dose, average tumor absorbed dose rarely exceeds 10%–15% differences [20].
2. The average organ absorbed dose is obtained by tallying the voxel energies and masses together in a single unit. This provides a value less susceptible to statistical and fitting error fluctuations than that obtained by averaging over voxel absorbed dose values calculated individually. While uncertainty in large regions is dominated by segmentation and activity quantification, curve fitting contributes substantially to the voxel-level absorbed dose calculation. For multiple time-point dosimetry (recommended for all except radioactive microspheres), fits to all voxel data are done automatically using a 'best fit' algorithm and a quality control metric to ensure that the area under the curve is sensical. However, with voxels numbering in the thousands, visual QC is not possible and often slight miss-registrations and fluctuations can lead to high artificial variability in the voxel absorbed doses. In particular, the area under the curve projected out to infinity can take on very large values. By tallying energy and the mass in the entirety of the region, this artificial variability is reduced.

Although originally designed as a research tool, given the widespread use of RPT and the growing interest in dosimetry and treatment planning, 3D-RD is now available through a service company, Radiopharmaceutical Imaging and Dosimetry

(RAPID), where the user-friendly interfaces common to 3D-ID are again available. New and upcoming developments provided through this company will be discussed in a later chapter.

10.5 Applications

Since its inception 3D-RD has played a significant role in advancing the field of RPT. Three examples of its application are presented here: the first case showed the feasibility of real-time prospective 3D dosimetry-based treatment planning, the second shows the importance of radiobiological considerations for dosimetry-based treatment planning in RPT and provides the framework for rational dosimetry-based patient-specific combinations of different modalities including external beam radiation therapy (EBRT) with RPT. The last example shows how anatomical modeling integrated with the absorbed fraction methodology can provide dosimetry for anatomical features smaller than the imaging resolution of the SPECT/CT or PET/CT.

1. A patient-specific three-dimensional Monte Carlo-based dosimetry software package (3D-RD) was used to perform treatment planning of ^{131}I therapy for an 11 year-old girl with differentiated papillary thyroid cancer with heavy lung involvement and cerebral metastases [19]. This was the first example of real-time fully three-dimensional prospective dosimetry-based treatment planning. Standard dosing, loosely based on hematological considerations, was considered to be potentially dangerous given the large amount of disease present in the lungs (see figure 10.6).

The patient received 92 MBq of ^{124}I, after which whole body PET/CT scans were performed at 1, 24, 48, 72, and 96 h. Using 3D-RD, Monte Carlo calculations for ^{131}I were performed on each image, with activity decay corrected to account for the difference in decay half-lives between ^{124}I (4.18 days) and ^{131}I (8.02 days) and normalized to unit (1 mCi) activity as soon as it became available. The resulting dose-rate images were co-registered over time, fitted and integrated over time to yield the mean absorbed dose to the lungs (figure 10.7) as well as the voxelized absorbed dose-volume histograms.

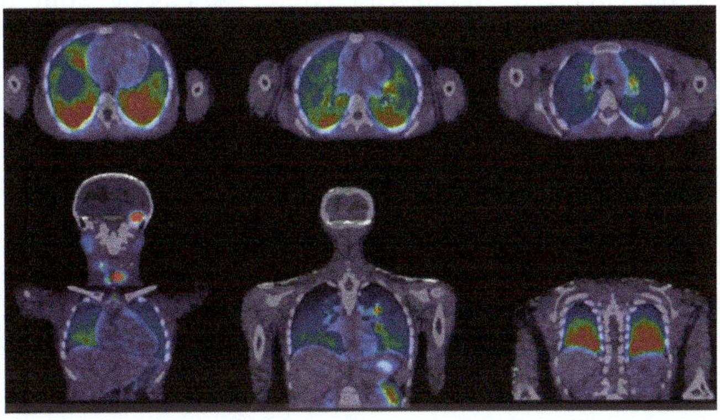

Figure 10.6. PET/CT at 24 h with uptake in lungs, thyroid remnant and brain lesions. This research was originally published in JNM [19]. Copyright SNMMI.

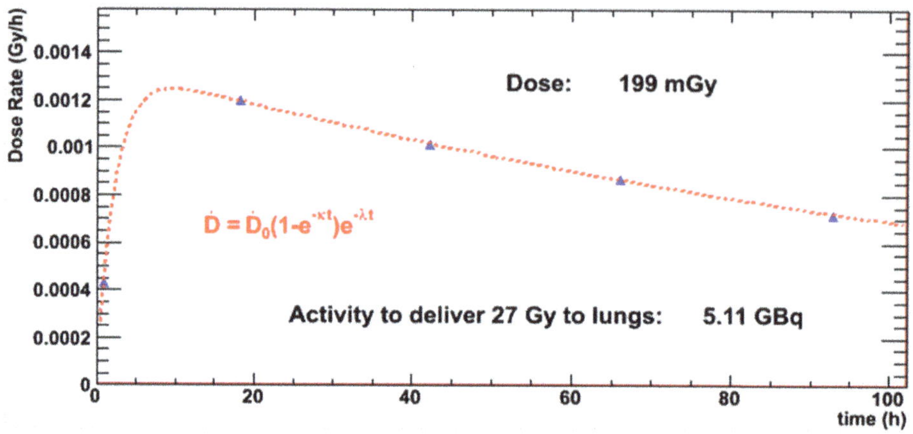

Figure 10.7. Dose-rate as a function of time in the lungs fit to a bi-exponential (uptake and decay) function.

Using a value of 27 Gy of maximum tolerated absorbed dose (MTD) for the lungs, a corresponding maximum tolerated activity (MTA) of 5.11 GBq was calculated. The absorbed fraction methodology was applied, using two source compartments, the lungs and the remainder of the body. *S* values were calculated for all five Christy–Eckermann phantoms (adult male, adult female, 15 year-old, 10 year-old, 5 year-old, 1 year-old). The results were plotted as a function of body mass, fit to a functional form and interpolated to the patient's body mass with the results predicting an MTA of 2.9 GBq. However, questions surrounding the validity of the assumptions of lung mass versus phantom whole body mass were raised, given the large amount of disease in the lungs and the subsequent increase in mass vis-à-vis a standard patient of the same body mass. Seven days after the tracer administration 5 GBq of ^{131}I was administered. Six months later, the patient was in remission with 50% and 90% reductions in the diameters of her two cerebral metastases, a dramatic reduction in TSH and no clinical evidence of pulmonary toxicity. The patient remains in remission 12 years out with no signs of toxicity. Retrospectively, the absorbed fraction method was revisited; now the phantom *S* values were plotted as a function of lung mass and the patient's lung mass used as the point of interpolation (figure 10.8).

The result was a predicted MTA consistent with the 3D-RD Monte Carlo calculation. This implementation proved the feasibility of real-time imaging-based Monte Carlo calculations as implemented in 3D-RD. A higher recommended administered activity than would have been administered with a favorable clinical outcome was obtained. In retrospect, the importance of using both methodologies with corroborating information was understood.

This demonstrated the feasibility and importance of patient-specific three-dimensional dosimetry-based treatment planning for RPT and ushered in the new era of voxelized dosimetry.

2. Effective cancer treatment generally requires combination therapy.

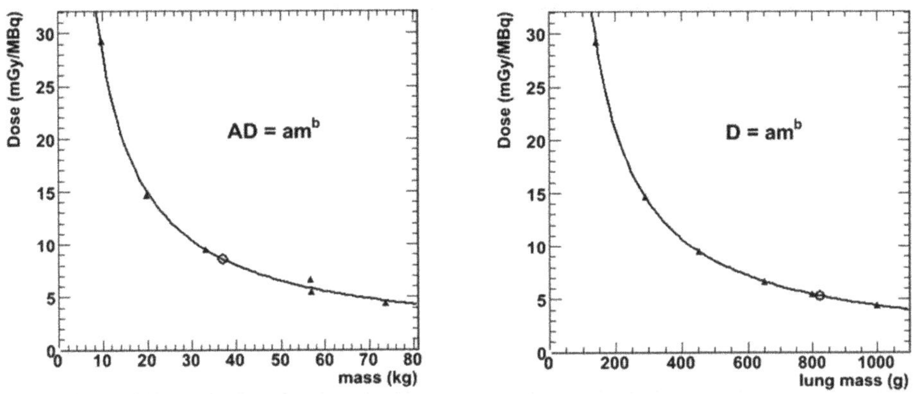

Figure 10.8. Lung-to-lung *S* values for the different Christy–Eckermann phantoms (triangles) as a function of phantom body mass (left) and phantom lung mass (right). The patient whole body mass and lung mass are shown by the open circle in both plots.

Currently many cancer therapies are designed with potential therapeutic synergy of several modalities while weighing the potential for combined or synergistic toxicity. These are designed using qualitative ad hoc patient class-based approaches. Combining methodologies that are administered on an ad hoc basis potentially treats most patients even less optimally than modalities administered individually. Alternatively, modalities that have potential intersecting toxicities are combined sequentially, with intervals of time designed to allow the patient to recover, as with radioimmunotherapy administered after chemotherapy. None of these approaches use quantitative patient-specific measures in their design. EBRT and RPT are both radiation modalities and have a common quantifiable language of damage and toxicity. This language, absorbed dose, must be translated properly from one modality to the other in order to combine effectively. Johns Hopkins Medical Institution has developed, tested and implemented a personalized dosimetry-based treatment planning methodology for combined RPT/EBRT therapy designed to provide therapeutic levels of absorbed dose while limiting toxicity to organs at risk using the 3D-RD platform.

The general principles of this combination were first described by Bodey *et al* [42, 43] The BED is a common radiobiological standard for both RPT and EBRT and can be used as a bridge to convert from one modality absorbed dose to the other. The equation for the BED in EBRT is given by:

$$\mathrm{BED} = D\left(1 + \frac{X}{\alpha/\beta}\right) \tag{10.7}$$

where X is the dose per fraction of EBRT. By setting this equation equal to equation (10.2), the conversion from RPT absorbed dose, D, to *EBRT* absorbed dose, $EQDX$, is obtained:

$$EQDX = \frac{D(\alpha/\beta + G(\infty)\bullet D)}{\alpha/\beta + X} \tag{10.8}$$

EQDX is the ICRU nomenclature used for external beam equivalent absorbed dose in x-gray fractions [44], termed 'equieffective dose'. Beyond the theoretical formalism, what 3D-RD enabled was actual application with the ability to provide a 3-D dose map in a reasonably short time frame and the technical know-how to overcome more complicated BED calculations on a voxel-by voxel basis [45] as well as characterized dead-time correction in the activity quantification for the high count rate. This application was first done retrospectively [46], before being applied prospectively for full patient-specific dosimetry-based combined ^{153}Sm-EDTMP-EBRT treatment of metastatic osteosarcoma in pediatric patients.

The rationale for combination was that even using a myeloablative regimen, the tumor absorbed dose to the patients was typically well below tumoricidal values [47, 48]. EBRT would similarly be limited to values of ~50 Gy for many of the tumors, often located in the pelvis or the spine and limited by the proximal spinal cord absorbed dose. As the toxicities are orthogonal, by combining the modalities a target tumoricidal absorbed dose (70 Gy of EQD2) could be safely achieved. The first patient was treated using the methodology outlined in the retrospective study: a baseline EBRT plan was designed in the treatment planning software and combined with the delivered RPT EQD2 map obtained from 3D-RD. The baseline EBRT plan was scaled by a factor *k*, obtained by considering the absorbed dose limit to the organs at risk, the maximum tolerated dose *MTD*, as the sum of the EQD2 values from both modalities:

$$k_i = \frac{MTD_i - (D_{RPT})_i}{(D_{EBRT})_i} \tag{10.9}$$

where D_{RPT} is the mean EQD2 delivered to a parallel organ at risk, indexed by *i*, and D_{EBRT} is the mean EQD2 to the same organ in the baseline plan. For a serial organ the values of k_i were calculated for each voxel indexed by *i*. The final scaling factor was taken as the lowest of the k_i values. This is illustrated in figure 10.9.

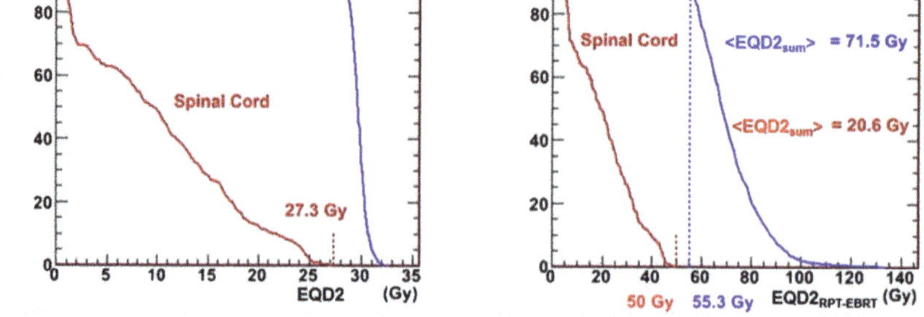

Figure 10.9. Integral DVHs of the baseline EBRT plan (left) and the combined scaled EBRT plan plus the RPT dose (right) for a case with the spinal cord as the organ at risk. Reprinted from [21], Copyright (2011) Elsevier Inc. All rights reserved.

After the first patient, a more truly prospective combined treatment planning protocol was designed:

(a) Stem cell collection for autologous transplant.
(b) CT simulation (CTsim) for EBRT planning/therapy.
(c) Administration of 1 mCi kg^{-1} of ^{153}Sm-EDTMP.
(d) SPECT/CT acquisition at 4, 24, and 48 h post-injection.
(e) 3D-RD EQD2 dose coefficient (absorbed dose per unit activity) map creation, including registration to CTsim.
(f) Import of RPT EQD2 map into EBRT treatment planning system as a weighted beam, creation of a combined EBRT-RPT plan.
(g) Administration of therapeutic ^{153}Sm-EDTMP (5–20 mCi kg^{-1}).
(h) SPECT/CT acquisition at 4, 24, and 48 h post-injection.
(i) 3D-RD delivered EQD2 map creation, registered to CTsim.
(j) Import of RPT EQD2 map into EBRT treatment planning system, adjustments to the EBRT portion of the combined EBRT-RPT plan based on actual delivered RPT EQD2.
(k) EBRT therapy.
(l) Autologous stem cell transplant when dosimetrically permissible.

In this approach, the 3D-RD pre-therapeutic EQD2 map was weighted as a beam and included in the forward treatment planning. The weighting had to be fixed as the planning system would systematically reduce the weight to near zero if left as a variable. But several options were tried, using a range of 5–20 mCi kg^{-1}, in combination with the beam setup to design a combined plan. The final EBRT was adjusted based on the actual delivered RPT EQD2. Only four patients were treated in all, the target lesion in all four patients received the target EQD2, delivering the tumoricidal dose that would not have been achievable with either single modality alone.

Unfortunately, from a clinical standpoint, the treatment was ineffective as the patients had disease that had progressed too much for effective treatment with widespread metastatic disease that could not be treated with the combination therapy.

This example stands as the first fully integrated RPT-EBRT patient-specific dosimetry-based treatment planning that can be expected to serve as a model for many future studies and applications.

3. Figure 10.10. Consecutive coronal [18F]-FDG PET slices for restaging of a lymphoma patient.
4. Three-dimensional voxelized dosimetry is currently becoming the norm in RPT dosimetry and is widely seen as superior to the MIRD absorbed fraction methodology based on phantom *S* values. However, there are limitations to the reliability and functionality of voxelized dosimetry that are often overlooked. In particular, voxelized dosimetry is limited by the resolution of the imaging modalities used to obtain the input information for the dosimetric calculations. In such instances, 3D-RD has resorted to using simple geometric volume simulations and Monte Carlo to ascribe absorbed dose. The example shown here is one where the concern of potential toxicity is the arterial wall for

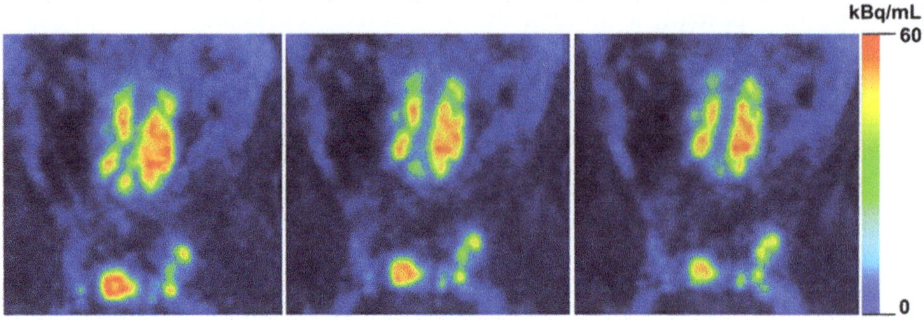

Figure 10.10. Consecutive coronal [18F]-FDG PET slices for restaging of a lymphoma patient. This research was originally published in [49] copyright SNMMI.

patients treated with ^{131}I-tositumomab for lymphoma [49]. Shown in figure 10.10 are ^{124}I PET images used for restaging in the treatment.

As can be divined from the images, the tumor is wrapped around the descending aorta, a pattern that was found in many of the treated patients. As radiation to vessel walls has been documented to potentially cause a multiplicity of negative events, such as stenosis, artherosclerosis and aneurysms [50–53], concern was expressed over the possibility of similar toxicities from this therapy. However, artery walls are too thin to be visualized in CT images, much less in the associated SPECT images used for activity quantification. The resolution to this problem was to apply the MIRD methodology [2] where the absorbed dose to a target region, r_T, is defined as:

$$D(r_T) = \sum_{r_S} S(r_T \leftarrow r_S) \bullet \tilde{A}(r_S) \tag{10.10}$$

where $\tilde{A}(r_S)$ is the time integrated activity (number of decays) in a source region and S is the S value associated with the source and target regions. A simple geometrical model of the aorta was used, with parametrization taken from the literature. Figure 10.11 illustrates this model. Several scenarios were used: the dimensions of the artery were varieds to represent either the aorta or one of the major arteries; additionally, the tumor enclosure of the artery was varied from 50% to 100%. The source regions that were considered were the artery content (blood), the tumor and the remainder of body. The model-specific S values were obtained by running GEANT4 Monte Carlo [54] for the different scenarios for the two source organs and the ^{131}I therapeutic isotope. The time-integrated activities were obtained using the standard activity quantification method and SPECT/CT imaging at five different time points for the tumor and using same time point planar images for the blood by quantifying the activity concentration in the heart. The remainder of body contribution was assumed to be from the photon component only. The results showed a relatively low range of BED to the artery walls, on the order of several Gy, well below assumed toxicity thresholds. However, by extrapolating to potential myeloablative regimens, such as those used by the University of Washington [55], the range of BED to the walls increased to the order of expected toxicity, such that in

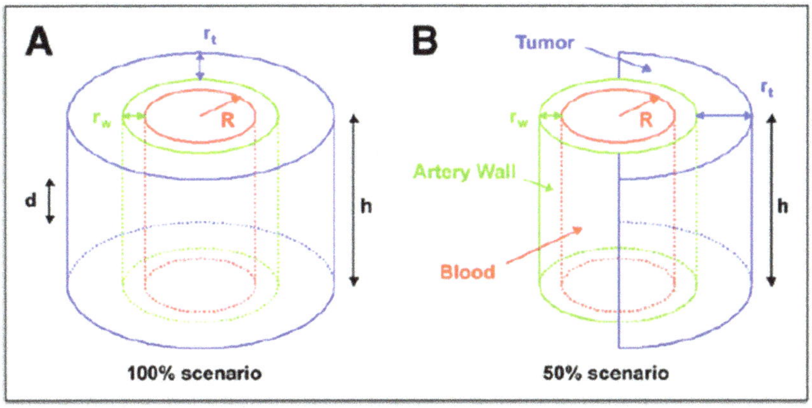

Figure 10.11. Geometrical model for the vessel walls. The artery wall is in green, the content (blood) in red and the tumor in blue. This research was originally published in [49] copyright SNMMI.

a myeloablative regimen, artery walls should be considered as potentially dose-limiting organs at risk.

This example reintroduces the importance and value of supplementing voxelized dosimetry with absorbed fraction dosimetry for regions that are not visualized or quantified with the clinical imaging modalities. Other examples implemented in 3D-RD include small-scale dosimetry for alpha-particle RPT [56, 57], a modality that will undoubtedly grow in popularity in the near future.

10.6 New and potential future uses

Two major areas of current development which will likely expand in the future are: (1) a commercialized version of 3D-RD, and (2) application to alpha-particle RPT (αRPT).

1. For most of its existence, 3D-RD has been a purely research tool. Recently, Drs Sgouros and Frey have created a dosimetry service company, Radiopharmaceutical Imaging and Dosimetry (Rapid) which provides dosimetry services for RPT. Rapid offers quantitative imaging and dosimetry consulting and analysis services and the software needed to implement them. The services offered to RPT developers include analysis and dosimetry of pre-clinical data, support for imaging trial design which includes determining the number and temporal spacing of the image acquisitions, image acquisition settings, description and analysis of phantom studies for site qualification and calibration, development of imaging manuals and procedures, centralized vendor-agnostic quantitative SPECT reconstruction, and standard phantom and full patient-specific 3D dosimetry calculations. With support from SBIR grants and contracts and internal funds, a desktop software application (RPTDose) based on the open-source 3D Slicer package (www.slicer.org) was developed. RPTDose (figure 10.12) that implements the full image analysis chain required to perform RPT dosimetry, generate dose maps, and output various radiobiologically-based dosimetric quantities.

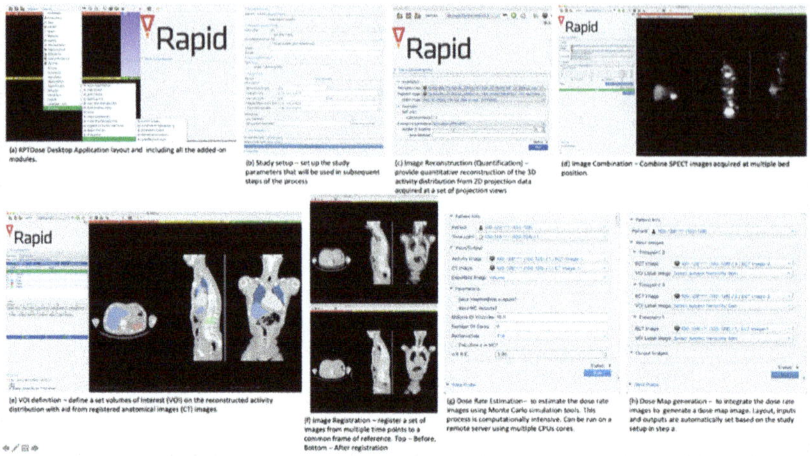

Figure 10.12. RPTDose desktop application and its modules.

Figure 10.12 shows the main modules in RPTDose including, study setup, image quantification, image combination, VOI definition, image registration and dose rate estimation and dose map calculation. RPTDose incorporates IRL and 3D-RD, where IRL (Iterative Reconstruction Library) is a vendor-neutral software package for quantitative reconstruction of SPECT images.

- Quantitative SPECT (QSPECT) reconstruction is performed using the ordered-subsets expectation-maximization (OS-EM) algorithm [58]. The iterative reconstruction models, and thus compensates for attenuation, scatter in the patient, and the collimator-detector system including septal penetration and scatter and characteristic x-ray. CT-based attenuation compensation is performed using an attenuation map generated from CT. Scatter is modeled using the effective source scatter estimation (ESSE) method. This method is exact for single scatter in a uniform object and includes approximations for non-uniform attenuators and multiple scatters. The collimator-detector response function (CDRF), which models the detection of photons emitted from a point source, is estimated by MC simulations (MCS) of point sources at various distances from the face of the collimator; these simulations included propagation of photons in the collimator and detector. The software includes the ability to reconstruct images of the activity distributions of multiple radionuclides from projection data acquired in an arbitrary number of energy windows while modeling contributions from each radionuclide to each energy window. These methods have been previously validated for a wide range of radionuclides (e.g., ^{99m}Tc, ^{201}Th, ^{131}I, ^{123}I, ^{111}In, ^{90}Y, ^{227}Th, ^{223}Ra, etc) using data from physical phantom studies and realistic MC simulated projection data [59–62].

The implementation of 3D-RD in RPTDose has been validated in comparison with the manual tool chain using data from an existing clinical trial for treatment of osteosarcoma planned using 3D-RD. The various intermediate steps (i.e., image reconstruction, dose rate estimation and dosimetry outputs) were validated for a set of five patients enrolled in the clinical study. Processing times using RPTDose and the research version of 3D-RD were compared; using RPTDose led to an ~80% reduction in user processing time.

In addition to the quantitative imaging and dosimetry services, Rapid has developed and validated a Standard-phantom, organ-based, MIRD-style [2] dosimetry tool, 3D-RD-S. The current version of the application is in the final stages of development; compiling the necessary documents for the 510k application for FDA clearance is in progress. 3D-RD-S calculations are based on ICRP-89 phantoms [63], ICRP-107 radionuclide decay data, and ICRP-133 specific absorbed fraction data [64]. It provides a set of 72 source regions and calculates doses to a set of 43 target tissues. It also supports calculation of tumor self-dose for spherical tumors with 5 compositions and 10 diameters ranging from 0.2 to 12 cm. The code supports calculation of dose from a radioactive nuclide and all of their radioactive daughters, allowing the user to assume that the daughters have the same, a scaled, or an independent distribution compared to their parents.

One unique feature of 3D-RD-S is that it is web based and inherently multi-user. Organizations or individuals can subscribe to 3D-RD-S. Organizations can have multiple members with different members having different roles and permissions. Data is organized hierarchically into projects, subjects, and datasets. A project is a collection of subjects whose dose calculations are based on the same radionuclide. A subject is an individual for which a dose calculation was performed. Each subject can have multiple datasets, which contain the input data and results from a dose calculation. The datasets can represent the results for a different set of input data or using different dose calculation methods, e.g., different curve fitting or integration of a time activity curve.

2. Alpha-particle RPT is a rapidly growing area of research for pre-clinical and clinical applications. Currently, neither absorbed fraction whole organ dosimetry nor Monte Carlo-based voxelized dosimetry methodologies are adequate to deal with the specificities and peculiarities of αRPT and they provide values that are inconsistent with observed pre-clinical or clinical toxicities, whether by over-predicting toxicity, as in the case of ^{223}Ra and the lack of bone marrow toxicity [57], or by under-predicting toxicity in the case of pre-clinical studies with ^{225}Ac [65]. Four reasons for these discrepancies exist; challenges that must be overcome in order to establish reliable alpha-particle dosimetry for correlative studies and potential dosimetry-based treatment planning:
    (a) The relative biological effect (RBE) for alpha-particles is poorly documented and not always well standardized. The RBE is a measure of the difference of the dose needed for alpha-particles (a high LET radiation) with respect to a low energy reference radiation for a specific biological endpoint. This reflects the fact that as a high LET radiation, alpha-particles do more biological damage per unit dose than low LET radiation. Mathematically it is the ratio of the reference radiation to the alpha-particle radiation. Essentially this comes down to quantifying the radiobiological parameters for a given tissue for

both alpha-particle radiation and the reference radiation. MIRD Pamphlet #22 recommends a value of 5 if the specific value is unknown [66]. However, studies often fail to use a same standard reference, such that much of the variability in the results comes down to a choice of reference.

(b) RPT activity uptake in normal organs are generally considered to be uniform. While the uptake may appear to be uniform at the organ level, it is driven by physiology and often confined to specific cell types or functional sub-units. For longer ranged radiation, this effect is often smoothed over to the point that the energy deposition and absorbed dose can still be assumed to be uniform. However, the range of energy depositing for alpha-particles is relatively short (50–100 μm), such that localized activity uptake results in localized absorbed dose distributions on a scale smaller than the resolving power of the SPECT/CT or PET/CT. Much like the example of the artery walls, small-scale modeling à la MIRD is necessary to associate activity and absorbed dose to the sub-organ regions.

(c) Many αRPT emitters have radioactive (often α-emitting) daughters. With the decay of the parent isotope, radiolysis most often occurs and the daughters are free to re-localize within the body. The absorbed dose to organs at risk must take into account not only the parent molecule, but the contributions from the daughters as well, often in different organs than those to which the parent localized.

(d) The high LET, localization and uncertainty of the dosimetry for αRPT results in low amounts of activity typically used for therapy. Administered activities for the FDA-approved clinical αRPT, Xofigo ($[^{223}Ra]RaCl_2$), range in the 100–200 μCi, an order of magnitude or more less than typical activities for diagnostic isotopes, such that the count rate in SPECT/CT is well below that needed for accurate quantification using current routinely available methodologies.

3D-RD and Rapid are leading the way in efforts to resolve these issues and develop solutions for αRPT dosimetry by addressing each of the above-mentioned challenges. By choosing a common ICRU reference radiation of EQD2, the RBE redefined as RBE2 eliminates much of the artificial variability [67]; the ICRU is poised to adopt the RBE2 as its official definition. The macro-to-micro methodology is the core of the small-scale dosimetry issue [56]; akin to the artery wall solution, it employs small-scale modeling coupled with *ex vivo* activity localization to provide small-scale (microscopic) dosimetry associated with organ-level (macroscopic) imaging. A complete αRPT dosimetry software will undoubtedly be forthcoming in 3D-RD.

Acknowledgments

This work was supported by NIH grants R01 CA62444, U01 CA58260 and P01 CA33049 and DOE grant DE-FG02–86ER-60407.

References

[1] Bolch W E *et al* 1999 MIRD pamphlet No. 17: the dosimetry of nonuniform activity distributions—radionuclide S values at the voxel level. Medical Internal Radiation Dose Committee. *J. Nucl. Med.* **40** 11S–36S

[2] Bolch W E, Eckerman K F, Sgouros G and Thomas S R 2009 MIRD pamphlet No. 21: a generalized schema for radiopharmaceutical dosimetry—standardization of nomenclature *J. Nucl. Med.* **50** 477–84

[3] Snyder W S, Ford M R, Warner G G and Watson S B 1975 *'S,' Absorbed Dose per Unit Cumulated Activity for Selected Radionuclides and Organs. MIRD Pamphlet No. 11* (Society of Nuclear Medicine)

[4] Stabin M G, Sparks R B and Crowe E 2005 OLINDA/EXM: the second-generation personal computer software for internal dose assessment in nuclear medicine *J. Nucl. Med.* **46** 1023–27

[5] Stabin M G and Konijnenberg M W 2000 Re-evaluation of absorbed fractions for photons and electrons in spheres of various sizes *J. Nucl. Med.* **41** 149–60

[6] Siegel J A and Stabin M G 1994 Absorbed fractions for electrons and beta particles in spheres of various sizes *J. Nucl. Med.* **35** 152–56

[7] Khazaee Moghadam M, Kamali Asl A, Geramifar P and Zaidi H 2016 Evaluating the application of tissue-specific dose kernels instead of water dose kernels in internal dosimetry: a Monte Carlo study *Cancer Biother. Radiopharm.* **31** 367–79

[8] Dieudonné A *et al* 2012 Study of the impact of tissue density heterogeneities on 3D abdominal dosimetry: comparison between dose Kernel convolution and direct Monte Carlo methods *J. Nucl. Med.* **54** 236–43

[9] Dieudonne A, Hobbs R F, Bolch W E, Sgouros G and Gardin I 2010 Fine-resolution voxel S values for constructing absorbed dose distributions at variable voxel size *J. Nucl. Med.* **51** 1600–7

[10] Furhang E E, Chui C S, Kolbert K S, Larson S M and Sgouros G 1997 Implementation of a Monte Carlo dosimetry method for patient-specific internal emitter therapy *Med. Phys.* **24** 1163–72

[11] Furhang E E, Chui C S and Sgouros G 1996 A Monte Carlo approach to patient-specific dosimetry *Med. Phys.* **23** 1523–29

[12] Kolbert K S *et al* 1997 Implementation and evaluation of patient-specific three-dimensional internal dosimetry *J. Nucl. Med.* **38** 301–8

[13] Prideaux A R *et al* 2007 Three-dimensional radiobiologic dosimetry: application of radiobiologic modeling to patient-specific 3-dimensional imaging-based internal dosimetry *J. Nucl. Med.* **48** 1008–16

[14] Botta F *et al* 2013 Use of the FLUKA Monte Carlo code for 3D patient-specific dosimetry on PET-CT and SPECT-CT images *Phys. Med. Biol.* **58** 8099–120

[15] Ferrer L, Chouin N, Bitar A, Lisbona A and Bardies M 2007 Implementing dosimetry in GATE: dose-point kernel validation with GEANT4 4.8.1 *Cancer Biother. Radiopharm.* **22** 125–29

[16] Yoriyaz H, dos Santos A, Stabin M G and Cabezas R 2000 Absorbed fractions in a voxel-based phantom calculated with the MCNP-4B code *Med. Phys.* **27** 1555–62

[17] Yoriyaz H, Stabin M G and dos Santos A 2001 Monte Carlo MCNP-4B-based absorbed dose distribution estimates for patient-specific dosimetry *J. Nucl. Med.* **42** 662–69

[18] Wilderman S J and Dewaraja Y K 2007 Method for fast CT/SPECT-based 3D Monte Carlo absorbed dose computations in internal emitter therapy *IEEE Trans. Nucl. Sci.* **54** 146–51
[19] Hobbs R F *et al* 2009 124I PET-based 3D-RD dosimetry for a pediatric thyroid cancer patient: real-time treatment planning and methodologic comparison *J. Nucl. Med.* **50** 1844–47
[20] Senthamizhchelvan S *et al* 2012 Tumor dosimetry and response for 153Sm-ethylenediamine tetramethylene phosphonic acid therapy of high-risk osteosarcoma *J. Nucl. Med.* **53** 215–24
[21] Hobbs R F *et al* 2011 A treatment planning method for sequentially combining radiopharmaceutical therapy and external radiation therapy *Int. J. Radiat. Oncol. Biol. Phys.* **80** 1256–62
[22] Hobbs R F *et al* 2013 Radiobiologic optimization of combination radiopharmaceutical therapy applied to myeloablative treatment of non-Hodgkin lymphoma *J. Nucl. Med.* **54** 1535–42
[23] Baechler S *et al* 2012 Three-dimensional radiobiological dosimetry of kidneys for treatment planning in peptide receptor radionuclide therapy *Med. Phys.* **39** 6118–28
[24] Mohan R *et al* 1988 A comprehensive three-dimensional radiation treatment planning system *Int. J. Radiat. Oncol. Biol. Phys.* **15** 481–95
[25] Erdi Y E *et al* 1997 Segmentation of lung lesion volume by adaptive positron emission tomography image thresholding *Cancer* **80** 2505–9
[26] Barendsen G W 1982 Dose fractionation, dose rate and iso-effect relationships for normal tissue responses *Int. J. Radiat. Oncol. Biol. Phys.* **8** 1981–97
[27] Fowler J F 1990 Radiobiological aspects of low dose rates in radioimmunotherapy *Int. J. Radiat. Oncol. Biol. Phys.* **18** 1261–69
[28] Dale R G 1996 Dose-rate effects in targeted radiotherapy *Phys. Med. Biol.* **41** 1871–84
[29] Dale R G 1985 The application of the linear-quadratic dose-effect equation to fractionated and protracted radiotherapy *Br. J. Radiol.* **58** 515–28
[30] Dale R G 1986 A graphical method to simplify the application of the linear-quadratic dose-effect equation to fractionated radiotherapy *Br. J. Radiol.* **59** 1111–15
[31] Howell R W, Goddu S M and Rao D V 1994 Application of the linear-quadratic model to radioimmunotherapy: further support for the advantage of longer-lived radionuclides *J. Nucl. Med.* **35** 1861–69
[32] Solanki J H *et al* 2017 Cellular response to exponentially increasing and decreasing dose rates: implications for treatment planning in targeted radionuclide therapy *Radiat. Res.* **188** 221–34
[33] Hobbs R F and Sgouros G 2009 Calculation of the biological effective dose for piecewise defined dose-rate fits *Med. Phys.* **36** 904–7
[34] Gustafsson J, Nilsson P and Gleisner K S 2013 On the biologically effective dose (BED)-using convolution for calculating the effects of repair: II. Numerical considerations *Phys. Med. Biol.* **58** 1529–48
[35] Barone R *et al* 2005 Patient-specific dosimetry in predicting renal toxicity with (90)Y-DOTATOC: relevance of kidney volume and dose rate in finding a dose-effect relationship *J. Nucl. Med.* **46** 99S–106S
[36] Wessels B W *et al* 2008 MIRD pamphlet No. 20: the effect of model assumptions on kidney dosimetry and response—implications for radionuclide therapy *J. Nucl. Med.* **49** 1884–99
[37] Niemierko A 1997 Reporting and analyzing dose distributions: a concept of equivalent uniform dose *Med. Phys.* **24** 103–10

[38] O'Donoghue J A 1999 Implications of nonuniform tumor doses for radioimmunotherapy *J. Nucl. Med.* **40** 1337–41

[39] Amro H, Wilderman S J, Dewaraja Y K and Roberson P L 2010 Methodology to incorporate biologically effective dose and equivalent uniform dose in patient-specific 3-dimensional dosimetry for non-Hodgkin lymphoma patients targeted with ^{131}I-tositumomab therapy *J. Nucl. Med.* **51** 654–59

[40] Dewaraja Y K *et al* 2010 ^{131}I-tositumomab radioimmunotherapy: initial tumor dose-response results using 3-dimensional dosimetry including radiobiologic modeling. *J. Nucl. Med.* **51** 1155–62

[41] Hobbs R F, Jentzen W, Bockisch A and Sgouros G 2013 Monte Carlo-based 3-dimensional dosimetry of salivary glands in radioiodine treatment of differentiated thyroid cancer estimated using ^{124}I PET *Q. J. Nucl. Med. Mol. Imag.* **57** 79–91

[42] Bodey R K, Evans P M and Flux G D 2004 Application of the linear-quadratic model to combined modality radiotherapy *Int. J. Radiat. Oncol. Biol. Phys.* **59** 228–41

[43] Bodey R K, Flux G D and Evans P M 2003 Combining dosimetry for targeted radionuclide and external beam therapies using the biologically effective dose *Cancer Biother. Radiopharm.* **18** 89–97

[44] Bentzen S M *et al* 2012 Bioeffect modeling and equieffective dose concepts in radiation oncology—terminology, quantities and units *Radiother. Oncol.* **105** 266–68

[45] Hobbs R and Sgouros G 2009 Calculation of the biological effective dose (BED) for Piece-wise defined dose-rate fits *Med. Phys.* **36** 904–7

[46] Hobbs R F *et al* 2010 A treatment planning method for sequentially combining radio-pharmaceutical therapy and external radiation therapy *Int. J. Radiat. Oncol. Biol. Phys.* **80** 1256–62

[47] Loeb D M *et al* 2009 Dose-finding study of ^{153}Sm-EDTMP in patients with poor-prognosis osteosarcoma *Cancer* **115** 2514–22

[48] Loeb D M *et al* 2010 Tandem dosing of samarium-153 ethylenediamine tetramethylene phosphoric acid with stem cell support for patients with high-risk osteosarcoma *Cancer* **116** 5470–78

[49] Hobbs R F *et al* 2010 Arterial wall dosimetry for non-Hodgkin lymphoma patients treated with radioimmunotherapy *J. Nucl. Med.* **51** 368–75

[50] Brennan S, Hann L E, Yahalom J, Oeffinger K C and Rademaker J 2008 Imaging of late complications from mantle field radiation in lymphoma patients *Radiol. Clin. North Am.* **46** 419–30

[51] Scott A S, Parr L A and Johnstone P A 2009 Risk of cerebrovascular events after neck and supraclavicular radiotherapy: a systematic review *Radiother. Oncol.* **90** 163–65

[52] Heidenreich P A *et al* 2007 Screening for coronary artery disease after mediastinal irradiation for Hodgkin's disease *J. Clin. Oncol.* **25** 43–9

[53] Hull M C, Morris C G, Pepine C J and Mendenhall N P 2003 Valvular dysfunction and carotid, subclavian, and coronary artery disease in survivors of hodgkin lymphoma treated with radiation therapy *JAMA* **290** 2831–37

[54] Agostinelli S *et al* 2003 GEANT4 - a simulation toolkit *Nucl. Instum. Methods Phys. Res.* A **506** 250–303

[55] Rajendran J G *et al* 2008 Myeloablative ^{131}I-tositumomab radioimmunotherapy in treating non-Hodgkin's lymphoma: comparison of dosimetry based on whole-body retention and dose to critical organ receiving the highest dose *J. Nucl. Med.* **49** 837–44

[56] Hobbs R F, Song H, Huso D L, Sundel M H and Sgouros G 2012 A nephron-based model of the kidneys for macro-to-micro alpha-particle dosimetry *Phys. Med. Biol.* **57** 4403–24

[57] Hobbs R F *et al* 2012 A bone marrow toxicity model for ^{223}Ra alpha-emitter radiopharmaceutical therapy *Phys. Med. Biol.* **57** 3207–22

[58] Hudson H M and Larkin R S 1994 Accelerated image reconstruction using ordered subsets of projection data *IEEE Trans. Med. Imaging* **13** 601–9

[59] Du Y, Tsui B M and Frey E C 2006 Model-based compensation for quantitative ^{123}I brain SPECT imaging *Phys. Med. Biol.* **51** 1269–82

[60] He B, Du Y, Song X, Segars W P and Frey E C 2005 A Monte Carlo and physical phantom evaluation of quantitative In-111 SPECT *Phys. Med. Biol.* **50** 4169–85

[61] Kadrmas D J, Frey E C and Tsui B M 1999 Simultaneous technetium-99m/thallium-201 SPECT imaging with model-based compensation for cross-contaminating effects *Phys. Med. Biol.* **44** 1843–60

[62] Song N, Du Y, He B and Frey E C 2011 Development and evaluation of a model-based downscatter compensation method for quantitative I-131 SPECT *Med. Phys.* **38** 3193–204

[63] ICRP 2003 *Basic Anatomical and Physiological Data for Use in Radiological Protection: Reference Values* vol 89 (Oxford: Pergammon Press)

[64] Bolch W E *et al* 2016 ICRP Publication 133: The ICRP computational framework for internal dose assessment for reference adults: specific absorbed fractions *Ann. ICRP* **45** 5–73

[65] Song H *et al* 2009 Radioimmunotherapy of breast cancer metastases with alpha-particle emitter ^{225}Ac: comparing efficacy with ^{213}Bi and ^{90}Y *Cancer Res.* **69** 8941–48

[66] Sgouros G *et al* 2010 MIRD Pamphlet No. 22 (abridged): radiobiology and dosimetry of alpha-particle emitters for targeted radionuclide therapy *J. Nucl. Med.* **51** 311–28

[67] Hobbs R F, Howell R W, Song H, Baechler S and Sgouros G 2014 Redefining relative biological effectiveness in the context of the EQDX formalism: implications for alpha-particle emitter therapy *Radiat. Res.* **181** 90–8

IOP Publishing

Monte Carlo Calculations in Nuclear Medicine (Second Edition)
Therapeutic applications
Habib Zaidi

Chapter 11

RAPID: radiopharmaceutical assessment platform for internal dosimetry

Bryan Bednarz, Abigail Besemer, Ian Marsh, David Adam and Joseph Grudzinski

11.1 Introduction

In radiopharmaceutical therapy (RPT), a radiolabeled biological molecule is engineered to preferentially enter or attach to cancer cells. The goal of the agent is to deliver a cytotoxic absorbed dose to cancer cells leading to radiation-induced cell death while sparing dose to healthy organs and tissues. Despite early clinical adoption of RPT nearly a century ago (e.g. $Na^{131}I$) [1], the incorporation of new RPT drugs into the armamentarium of oncologists during this time has been slow for a variety of reasons. However, in recent years the oncology community has seen a renaissance of RPT applications, with a growing number of drugs that are under development or have been approved for clinical use [2].

In general, there are five phases of RPT drug development: discovery and development, preclinical research, clinical research, regulatory review, and post-marketing safety monitoring [3]. The initial discovery and development process involves *in vitro* studies linking biology and chemistry to pinpoint appropriate biological targets and identify lead compounds. Once a suitable biological target and RPT drug pairing is identified, preclinical evaluation will commence. Preclinical testing in animals is done to assess *in vivo* specificity, biodistribution, pharmacokinetics and toxicology. Next, clinical testing is done in human subjects to evaluate safety and efficacy. An RPT drug candidate will be approved for clinical use once a regulatory body deems it safe and effective in humans. Following regulatory approval, the drug will likely be monitored for safety. RTP drug development is a complex and costly process whose success is associated with a high degree of uncertainty.

Through the RPT drug development process, a standardized radioactivity dosage is established which is intended to safely treat an average population of patients regardless of tumor burden or differences in rates of drug metabolism. Although it can be expected that the average patient will get a beneficial response from a treatment, the distribution of responses will be broad [4]. Interpatient variability in tumor burden and pharmacokinetics will lead to a wide range of normal organ and tumor doses from standardized dosage. Inevitably, a subset of patients that receive RPT will experience a severe toxicity. Moreover, a different subset of patients could tolerate more radioactivity which could lead to improved benefit. Because of potential adverse events, high cost, and safety concerns, predicting the efficacy and toxicity of any cancer treatment is a critical healthcare consideration. Such prediction would help to establish a clinical cost–benefit ratio of the therapy and optimal stratification of patients. For example, patients who are predicted to be less responsive to RPT may require additional therapy or a different type of therapy to improve treatment outcomes.

Through over 40 years of experience with external beam radiotherapy (EBRT), it is well-known that patient-specific absorbed dose is an excellent predictor of treatment efficacy and toxicity. A similar approach of determining patient-specific absorbed doses to achieve clinical objective(s) for each patient is starting to be considered for RPT. Several research studies have focused on the investigation of individual patient absorbed doses and their relationship with toxicity and treatment response providing strong evidence for the use of patient-specific dosimetry in RPT [5]. In particular, the integration of dosimetry in ^{90}Y microsphere treatment is notable, and the evidence of dose-effect relationships has been reported for hepatocellular carcinoma [6], neuroendocrine tumor metastases [7], colorectal cancer metastases [8], and for normal liver tissue [9, 10]. The status in ^{177}Lu- and ^{90}Y-peptide-receptor radiotherapy (PRRT) was recently reviewed [9], and tumor dose–response relationships have been demonstrated for both ^{177}Lu-DOTATATE (Lutathera™) and ^{90}Y-DOTATOC [11, 12]. For these drugs, patient-specific dosimetry is very important for reducing toxicity to the kidneys and bone marrow. Similarly, tumor dose–response relationships have also been demonstrated using ^{177}Lu-PSMA for patients with metastatic prostate cancer [13, 14]. For ^{177}Lu-PSMA, patient-specific dosimetry is very important for reducing toxicity to the kidneys, lacrimal glands, and salivary glands [15–17], making patient-specific dosimetry a requirement for the development of PSMA antigens [18]. Patient-specific dosimetry, which has the potential to provide substantial patient benefit, is mandated by the European Directive for every patient [19]. Currently, dosimetry is mandated by the FDA only during clinical trials, but because of an ever-growing body of supporting evidence, it is likely that the FDA will eventually mandate patient-specific dosimetry as a standard of care. Patient-specific dosimetry for RPT is becoming mainstream with many national scientific bodies such as the American Association for Cancer Research (AACR), Society of Nuclear Medicine and Molecular Imaging, (SNMMI), American College of Radiology (ACR), American Society for Radiation Oncology (ASTRO), and National Institutes of Health (NIH) making it a priority of their mission statements.

Accurate patient-specific 3D internal dosimetry is essential for establishing dose–response relationships. Given the unique dependencies that RPT drugs have on biology, chemistry, and physics, a complete understanding of dose–response relationships necessitates dosimetry at every phase of the RPT drug development process. Nano- and micro-dosimetry calculations at cellular and sub-cellular scales are needed to better understand how these drugs damage biological targets. Nano- and microdosimetry are particularly useful for drugs labeled with short ranged high-LET emitters such as alpha particles and Auger electrons. Preclinical dosimetry is needed to evaluate dose–response relationships in a controlled setting. And designing thoughtful clinical trials to ascertain these relationships will lead to more informed decision making once the RPT drug is approved for clinical use. Therefore, the development of multi-scale (ranging from cellular to organ level) dosimetry tools is warranted. While conventional organ-level dosimetry methods are still widely used in clinical practice, several research groups and a few companies have developed voxel-based Monte Carlo internal dosimetry packages (e.g. [20–31]). These platforms vary widely in their complexity, capabilities, and intended use. Notably, none of these platforms are designed for multi-scale dosimetry. As a result, the dosimetric impact of various parameters and methodologies implemented in these packages remains unclear and non-translatable. To this end, a multi-scale Monte Carlo internal dosimetry platform that calculates voxelized 3D dose distributions was developed (see figure 11.1). The platform can also implement basic dosimetry techniques commonly used in internal dosimetry such as MIRD S-value calculations so that differences between these techniques compared to advanced Monte Carlo calculations can be examined and quantified. The platform is called RAPID, which stands for Radiopharmaceutical Assessment Platform for Internal Dosimetry. The following sections describe the various features and implementation options available within RAPID.

11.2 Platform framework and implementation

The framework of RAPID is presented in figure 11.2. The essential parts of the framework include: imaging acquisition and import, image pre-processing, manual region of interest (ROI) contouring, ROI-based registration, Monte Carlo radiation transport, voxel-level absorbed dose rate (ADR) integration, radiobiological

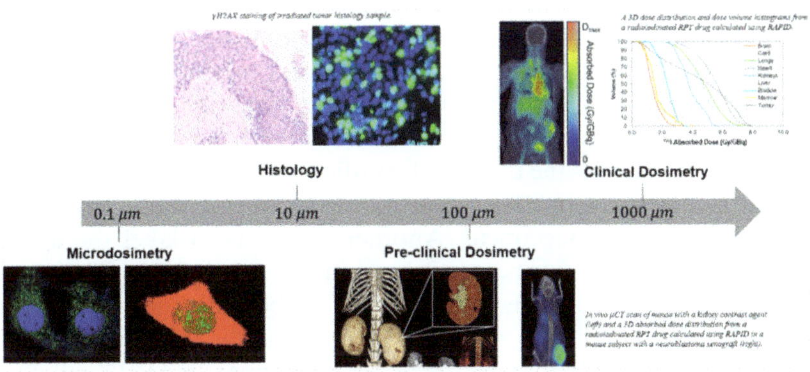

Figure 11.1. Overview of multi-scale functionality of RAPID.

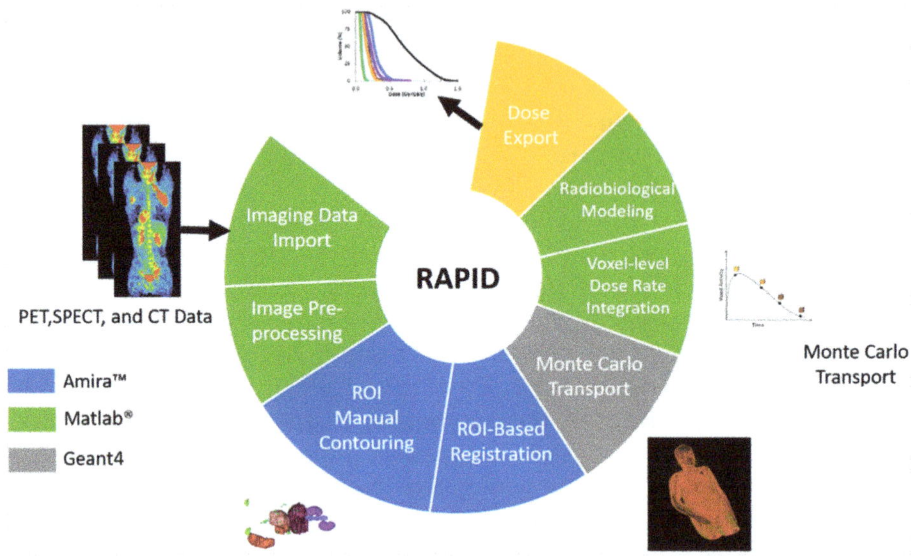

Figure 11.2. The RAPID framework and corresponding supporting software.

modelling, and dose statistics export. The following sub-sections will overview each of these parts in more detail.

11.3 Image acquisition

Patient-specific dosimetry is calculated based on serial quantitative SPECT/CT or PET/CT images acquired at multiple time points after injection of a radiopharmaceutical to map the patient's unique pharmacokinetics and anatomy. MRI image-scan can also be imported to better delineate the tumor volume and normal tissues compared with CT. The PET and SPECT must be in units of activity or activity concentration and all images must be in DICOM or raw binary format.

RAPID can compute dosimetry either retrospectively or prospectively (i.e. treatment planning). The dosimetry workflow is the same for both except for the image acquisition and the possible decay correction of the radionuclide activity when different radionuclides are used for imaging and therapy. Retrospective dosimetry is calculated with RAPID based on images acquired after injection of a therapeutic or diagnostic radiopharmaceutical to determine the doses that were received. Prospective dosimetry is calculated based on pre-treatment images acquired prior to therapy. If the therapeutic radionuclide emits photons or positrons, then pre-treatment SPECT or PET images can be acquired with a tracer amount of the therapeutic radiopharmaceutical. Another option is to label the agent with a different radionuclide that can be used as an imaging surrogate. These agents are typically called 'theranostic' as they can be used for both diagnostic and therapeutic applications. When the imaging surrogate radionuclide is different than the radionuclide that will be used for therapeutic dose calculations, RAPID will convert the imaging radionuclide activity to the desired therapeutic radionuclide activity by simply accounting for the differences in the decay rates. RAPID assumes that the imaging and therapeutic agents will have equivalent pharmacokinetics and that the absorbed dose scales linearly with the injection activity. For both prospective and retrospective dosimetry, the absorbed dose (AD) is normalized by the injected activity and the prescription dose is calculated in units of Gy/GBq of activity injected. For prospective dosimetry, the user can then compute the optimal therapeutic injection activity based on the desired tumor dose or the normal tissue maximum tolerated doses.

11.4 Image pre-processing

Following acquisition, the images are coregistered, resampled, rescaled, and masked. Images can also be manually corrected for partial volume effects (PVE) by the user. Currently, the image coregistration and resampling, as well as ROI contouring, is performed outside of RAPID using the AMIRA software (ThermoFisher Scientific, Waltham, MA).

First, the physiological PET or SPECT volumes, hereinafter referred to as nuclear medicine volumes, are coregistered to the anatomic CT (and MRI if available) at each time point using a global normalized mutual information affine transformation with 12 degrees of freedom (three rotational, three translational, three scaling, and three sheering). This type of registration is commonly used for the fusion of multimodality images [32, 33].

The image dataset at each imaging time point consists of a low-resolution nuclear medicine volume and a high-resolution CT volume. First, the low-resolution nuclear medicine volume is registered to high-resolution CT volume. The nuclear medicine volume is then resampled to match the resolution of the CT volume. This set of image volumes is used to draw contours and generate pharmacokinetic and ROI statistics (e.g. uptake, activity concentration, volume). Next, a dose grid resolution is generated at a resolution that is typically a factor of two to four of that of the CT resolution. Given that Monte Carlo dosimetry relies on accurate quantification of the activity distribution by the nuclear medicine volume and the material composition/density by the CT, this resampling approach has both advantages and disadvantages. Up-sampling the nuclear medicine volume is advantageous because ROIs can be more accurately delineated on the higher resolution CT and the simulation results in a finer dose deposition grid since energy deposition is scored on a scoring grid defined by the CT resolution. However, higher resolutions increase the simulation run time and uncertainty in the activity quantification is introduced when up sampling the nuclear medicine volume. Resampling can be performed using a variety of filter kernels. The default option is the Mitchell filter kernel which offers a good compromise between ringing and blurring artifacts.

If the pre-treatment nuclear medicine volumes were acquired using a different radionuclide than the therapeutic radionuclide, then the radioactivity of the imaging/tracer radionuclide is converted to the radioactivity of the therapeutic radionuclide at each time point by assuming the same injection activity and accounting for the difference in physical decay rates. The user can also apply partial volume corrections (PVCs) to the nuclear medicine volumes to correct for the apparent loss of activity in small regions due to the limited image resolution [34]. Currently, RAPID only supports spherical PVC corrections, which can be applied to the total absorbed dose in bulk tumor volumes. A mask contour is created based on the CT volume at each time point and is applied to the nuclear medicine volumes to remove any background activity outside the patient. This reduces the simulation time by eliminating extraneous sources.

11.5 Region of interest contouring

The tumors and normal tissues are contoured using anatomical and functional images. The user can choose to contour the ROIs manually or using a threshold method based on the CT HU, PET or SPECT radioactivity concentration, standardized uptake values (SUVs), or tumor-to-background ratios (TBRs).

11.6 Monte Carlo simulation

All simulations in RAPID are done using the MC code Geant4 version 9.6 [35]. Geant4 is a versatile object-oriented simulation toolkit that allows for the modeling of complex geometries, radiation sources, and detectors. Geant4 has been benchmarked for a variety of different medical physics applications [36–39]. RAPID provides the user with the flexibility to specify any voxelized geometry and source distribution under investigation. The input file permits the user to control the verbosity of the information provided in the simulation log file, the physics models that are used, the type of radioisotope being modelled, the size of the CT and nuclear medicine volume under consideration and the number of decays that are desired for each simulation.

The amount of information recorded in the simulation log file can be controlled by verbosity output flags in the input file. The minimal amount of information is provided when the verbose flag is set to zero. Increasing the value of the verbose flag increases the amount of information that is provided in the simulation log file. Detailed information regarding the overall run, events, and tracking/step parameters can be extracted from each simulation if desired. Verbosity flags can be especially useful for debugging.

The RAPID input file allows the user to specify the isotope that will be used for the simulation. The decay of the isotope is simulated using the G4RadioactiveDecay module [40] which allows the simulation of the decay of more than 2000 radioactive nuclei using the decay information from the ENSDF database [41]. The user selects the isotope to be simulated by specifying the atomic number, atomic mass, charge of the ion, and excitation energy. In addition, the user can also limit the nuclei in which radioactive decay will be applied. This is especially useful if the user wants to only simulate a single isotope that belongs to a multi-isotope decay chain.

The user can also select the electromagnetic (EM) physics models used to transport photons, electrons, hadron, and ions during the simulations. Currently, RAPID can instantiate three different EM physics constructors available in Geant4.

EM standard is the default EM constructor in Geant4. In addition, two other EM constructors can be used, EM Penelope and EM Livermore, which were designed for low energy applications that require higher accuracy for electron, hadron, and ion tracking. Each EM physics constructor has advantages and disadvantages, and the choice of the appropriate constructor should be evaluated on a case-by-case basis.

The CT and nuclear medicine volumes are used in the MC simulation to define the geometry and source distribution, respectively. The CT images are used to define the material composition and mass density of the simulation geometry. The entire range of HUs is segmented into 27 materials based on the human tissue compositions defined by Schneider *et al* [42]. Each HU is assigned a unique mass density according to the CT scanner specific calibration. If the scanner CT-to-density information is unknown, then the conversion defined in Schneider *et al* [42] is used as a default. The nuclear medicine volumes define the radionuclide activity or cumulated activity in each voxel. The decay point is sampled uniformly throughout each voxel.

RAPID computes the statistical relative error in the mean ADR on the voxel and ROI-level. The desired statistical uncertainty threshold is specified by the user at the whole body-, ROI-, or voxel-level. For most simulations, sufficient particles should be simulated such that the average ROI-level relative error (RE) is less than 1% and the average voxel-level RE within the patient volume is less than 10%. Note that the RE only refers to the statistical precision of the MC dose calculation and does not take into account other sources of uncertainty such as the activity quantification, material/density quantification, coregistration, resampling, kinetic fitting, or ROI contouring.

RAPID is deployable using the Docker platform which facilitates the code execution on high throughput computing systems (i.e. HTCondor) and is a perfect method for completing computationally intensive Monte Carlo absorbed dose calculations quickly [43]. Docker is a container solution that enables developers to construct a software environment in an exact way specific to a code including the base operating system, system libraries, environment variables, third party libraries, and a precise way in which to compile the application. Docker affords a key advantage in that the standardized computing environment ensures reproducible and consistent computation results across heterogeneous computing resources often managed by high throughput systems [44].

The multi-modal fusion of anatomic CT and physiological nuclear medicine images at each time point is an important aspect of patient-specific dosimetry. If the user is only interested in calculating average organ doses (like conventional model-based methods e.g. OLINDA) then further temporal registration is not needed. In this case the mean ADR for each ROI can be integrated to obtain a single mean AD. However, to obtain voxelized dose distributions, the ADR distribution at each time point must also be temporally coregistered to a common coordinate system so that the ADR can be integrated at the voxel-level. The temporal (4D) registration method depends on the imaging conditions and the specific clinical site.

For cases with reproducible and rigid immobilization during imaging and clinical sites with small variations in the intra- and inter-image growth/movement, simple temporal coregistration methods may be adequate. Manual [45, 46] and marker-based [47, 48] registration methods have been utilized as well as whole-body affine CT–CT sequential coregistration (WBAC) [26, 27, 47, 48]. Within RAPID, the WBAC method is implemented by registering the CT at each time point is to a single reference CT using normalized mutual information based affine CT–CT coregistration. The CT transformations are then applied to the ADR distributions at each corresponding time point and the ADRs are resampled to match the voxel coordinate system of the ADR at the reference time point. The AD is calculated on a voxel-basis over the entire image volume. An advantage to WBAC is that the ROIs only need to be contoured on the single reference PET/CT or SPECT/CT images.

For most 4D image sets in RPT, a single global affine coregistration may not be able to accurately register each region throughout the entire body. This is commonly due to positioning/setup variations during imaging, respiratory/cardiac motion, bladder/rectal filling, tumor shrinkage/growth, or patient weight loss/gain. For these cases, locally affine [49, 50] or deformable registration [51] may be more appropriate. Deformable methods using non-linear transformations have been applied for temporal registration in 3D internal dosimetry [28, 52]. To the best of our knowledge, locally affine registration techniques have not yet been utilized for 3D internal dosimetry calculations, but they may be an attractive alternative to more complicated deformable methods. The locally affine piecewise ROI–ROI coregistration (LAPRC) method used within RAPID is adapted from Pitiot *et al* [49] who proposed a piecewise affine registration algorithm where sub-images are extracted and coregistered independently using a hybrid affine/non-linear interpolation scheme. The LAPRC method is illustrated in figure 11.3. First, ROIs are generated from the PET/CT or SPECT/CT at each time point. Each ROI contour is then isolated and registered independently to the ROI at a single reference time point using a normalized mutual information based affine coregistration. The transformation used to register the ROIs is then replicated for the ADR distribution at each time point. The ADR distribution at each time point is then resampled so that the coregistered ADR distributions will have the same voxel coordinate system and the ADR can be integrated on a voxel-by-voxel basis. The AD is then calculated by integrating the ADR only for the voxels within each of the ROIs. One drawback of this method is that the ROI contours must be generated *a priori* at each time point. Another drawback is that full-body dose distributions cannot be calculated because the ADR is integrated only within the ROIs. However, LAPRC affords the calculation of ROI dose statistics and dose volume histograms (DVHs).

To summarize, the ADR distribution calculated at each time point needs to be temporally coregistered to calculate voxelized AD distributions. Within RAPID, this temporal coregistration can be performed using a whole-body affine CT–CT sequential coregistration (WBAC) or a locally affine piecewise ROI–ROI coregistration (LAPRC). Additionally, the platform offers the option to forgo the coregistration and integrate only the mean ADR for each ROI to obtain mean ADs for each ROI.

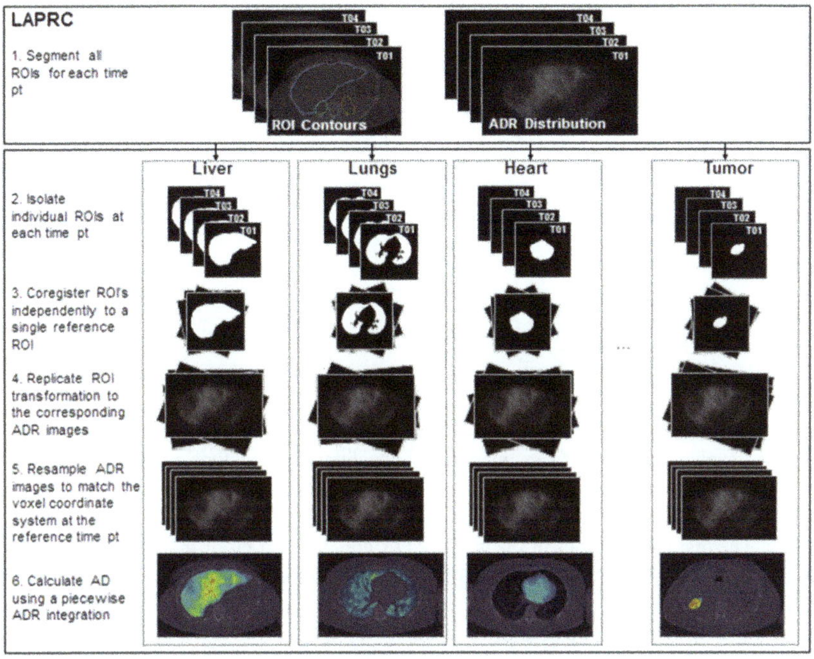

Figure 11.3. Detailed overview of the locally affine piecewise ROI–ROI coregistration (LAPRC) approach.

11.7 Kinetic fitting

The total AD for the treatment is calculated by integrating the voxelized ADR over all time. Because the shape of the pharmacokinetic curve depends on the TRT agent- and radionuclide-specific kinetics as well as the number and frequency of time points chosen to image and track the kinetics of the agent, a variety of different fitting methods have been employed to fit the pharmacokinetic curve [53]. The curve consists of three main regions: (1) pre-imaging the time from injection to the first imaging time point $t_0 \rightarrow t_1$; (2) imaging the time between the first and last time points ($t_1 \rightarrow t_f$); and (3) post-imaging—the time from the last time point to infinity ($t_f \rightarrow t_\infty$). Since imaging data is not available for the first and third regions, the ADR must be extrapolated from the imaging region. The fitting methods available in RAPID's pharmacokinetic fitting module are illustrated in figure 11.4.

The simplest method is to fit the curve with a single-, double-, or triple-exponential so that the time-integration can be performed analytically [28, 29, 54–56]. In general, multi-exponential fits are well suited for modelling both the initial rising uptake as well the decay portion of the curve. However, they may not always be practical because the curve fitting degrees of freedom (i.e. the difference between the number of measured image time points and the number of parameters in the fit) should be greater than one. Additionally, the ADR curve is not always well fit with an analytic exponential equation, especially at the voxel level (which may be noisy). Other functions have also been suggested to more realistically model the rising initial portion of the curve, such as a gamma variate with an exponential tail [57].

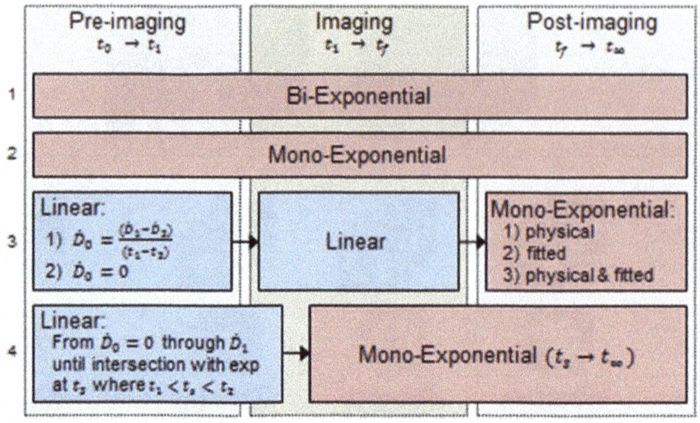

Figure 11.4. A summary of the kinetic fitting options available in RAPID. The curve consists of three main regions: (1) pre-imaging—the time from injection to the first imaging time point; (2) imaging—the time between the first and last time points; and (3) post-imaging—the time from the last time point to infinity. Imaging data is not available for the pre- and post-imaging regions so the ADR must be extrapolated from the imaging region. The first two methods fit the whole ADR curve with a bi- or mono-exponential fit. The third method fits the ADR linearly (e.g. trapezoidal) within the imaging region. The user has two options for linearly extrapolating within the pre-imaging region and three options for exponentially extrapolating within the post-imaging region. The first options are used as the default within RAPID. The fourth method fits the imaging region with a mono-exponential curve and fits the pre-imaging region linearly. The fitting method can be chosen automatically using the 'AutoFit' option which will fit the curve with a mono- or bi-exponential fit if the R^2 is above a user-defined threshold or else it will fit trapezoidally. The user can also manually force one of the four types of fit using the 'ForcedFit' option.

Alternatively, a piecewise linear approach can be used as shown in the third method in figure 11.4. Using this method, the curve can be fit linearly within the imaging region and integrated numerically using the trapezoidal rule [28, 47]. Numerical methods are advantageous because they avoid assumptions about the analytic shape of the curve and are also well suited for fitting non-monotonic curves. Additionally, trapezoidal integration results in the same mean doses regardless of whether the ADR is fit on the voxel-level or ROI-level because the summation and integration processes are interchangeable. Within the pre-imaging region, the ADR can be extrapolated back linearly from t_1 to t_0 assuming no initial uptake at the injection time so that $\dot{D}_0 = 0$ [55, 58] or by fitting a line from the first two time points so that $\dot{D}_0 = (\dot{D}_1 - \dot{D}_2)/(t_1 - t_2)$[28]. The linear extrapolation is advantageous because it can model uptake or decay. The impact of the pre-imaging fitting method decreases when the first image is acquired promptly after injection. Within the post-imaging region, the decay can be extrapolated exponentially with either the physical decay [47], the decay fitted from the last two time points [28], or a combination of both [47]. If there is uptake between the last two time points, then the fitted exponential decay constant will cause the integration to diverge. Thus, when uptake is present, an exponential fit that only accounts for physical decay is utilized [47]. By default, the pre-imaging region is extrapolated linearly back to t_0 from the first two time points and the post-imaging region is modelled as an exponential with a physical decay constant. The fourth method shown in figure 11.4 is also a hybrid approach that fits the imaging an imaging region with a mono-exponential curve and fits the pre-imaging region linearly. This option is available within RAPID to be consistent with methods used in previous publications [28, 55, 57].

The fits described above are applied using either the *ForcedFit* or *AutoFit* option in RAPID. The *ForcedFit* option applies either a mono-exponential, bi-exponential, or trapezoidal fit to all voxels or ROIs regardless of the number of image time points or the goodness of fit. Alternatively, the *AutoFit* option automatically fits the absorbed dose rate curve with a mono- or bi-exponential function if there are a sufficient number of image time points. *AutoFit* uses the exponential fit with the largest squared Pearson correlation coefficient (R^2) value. If there are not enough time points for curve fitting or the analytical equation does not fit the curve well (i.e., the R^2 value is less than a user-defined desired threshold), then a piecewise linear fit is used for that voxel or ROI. All kinetic fitting can be performed on the voxel- or ROI-level.

11.8 Output visualization and analysis

The dosimetry output can be visualized as 3D dose distributions overlaid on anatomical images, DVHs, and tables of ROI dose stats (i.e. minimum, mean, maximum, and standard deviation) and DVH stats (e.g. D_{99}, D_{95}, D_{90}). The 3D dose images can be output as binary files and the DVHs and tables can be output as .txt files which can then be imported into other visualization and plotting programs if desired. 3D dose distributions and DVHs can be generated for the ADR, AD, biological effective dose (BED) and radiobiological equivalent dose of 2 Gy (EQD2) and mean ROI values can be generated for the equivalent uniform dose (EUD), normal tissue complication probability (NTCP), and tumor control probability (TCP).

11.9 Benchmarking

A series of computational benchmarking experiments were conducted to compare dosimetry results in RAPID to other competing methods. Specific absorbed fractions (SAFs) in spherical tumors for a variety of radionuclides were calculated in RAPID and compared to spherical SAFs in the OLINDA/EXM software [54]. In addition, *S*-values for monoenergetic photon and electron sources in the Zubal phantom were compared to reference values. An additional series of benchmarking experiments were done to validate the microdosimetry module in RAPID. These experiments are summarized below and more details on the benchmarking work can be found in Besemer *et al* [59].

11.10 Spherical tumor phantom

In order to benchmark the tumor dosimetry SAFs were calculated for 1, 10, 100, and 1000 g unit-density spherical tumors with uniform activity distributions of β^- (^{177}Lu, ^{131}I, ^{153}Sm, and ^{90}Y), β^+ (^{18}F, ^{64}Cu, ^{89}Zr, ^{68}Ga, and ^{124}I), γ (^{111}I), Auger (^{125}I), and α (^{223}Ra, ^{211}At, and ^{213}Bi) emitting radionuclides. The sphere geometries were created out of 1 mm^3 voxels to ensure consistent voxel transport and scoring between the benchmark and patient CT geometries. The spherical sources were embedded in a

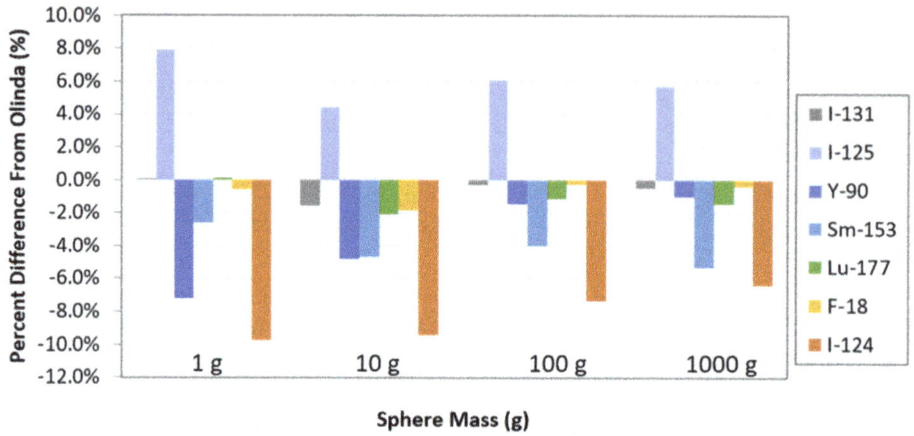

Figure 11.5. Percent differences between RAPID and OLINDA SAFs in the spherical phantoms.

semi-infinite unit-density voxelized medium. The SAFs calculated with RAPID were compared the references values for spheres of the same mass calculated by the OLINDA/EXM software.

The percent differences between the RAPID and OLINDA SAFs calculated in the spherical phantoms are shown in figure 11.5 for each radionuclide and sphere size. The *S*-values calculated with RAPID agreed well with the OLINDA/EXM reference values. The agreement was within 2% for ^{177}Lu, ^{131}I, ^{125}I, ^{18}F, and ^{64}Cu, within 3.5% for ^{211}At and ^{213}Bi, within 6.5% for ^{153}Sm, ^{111}In, ^{89}Zr, and ^{223}Ra, and within 9% for ^{90}Y, ^{68}Ga, and ^{124}I. The larger disagreement for the high energy beta emitters is likely due to the assumption in OLINDA/EXM that all betas are emitted with the average energy while Geant4 samples the entire energy spectrum.

11.11 Monoenergetic photon and electron *S*-values in the Zubal phantom

To benchmark the normal tissue dosimetry, *S*-values were calculated with RAPID and compared with reference values. SAFs for monoenergetic photons of 10, 100 keV, and 1 MeV and *S*-values for monoenergetic electrons of 0.935 MeV were calculated in the Zubal phantoms [60]. The Zubal phantom is a voxel-based representation of an adult male head and torso that consists of 33 segmented organs and structures. The SAFs and *S*-values calculated in this work were benchmarked against values previously published by Yoriyaz *et al* [61] and Chiavassa *et al* [27] who have developed similar MC-based dosimetry software. The calculations done by Yoriyaz *et al* used the MCNP-4B MC code [62] with the SCMS software interface and the calculations done by Chiavassa *et al* used the MCNPX MC code [62] with the OEDIPE software interface. In both cases, the Zubal geometry was simplified to only include air, lung, soft tissue, and bone with densities of 0.0012, 0.296, 1.04, and 1.40 g cm^{-3}, respectively. The tissue material compositions that were used were not mentioned in the previous publications. For the calculations in this work, the material compositions were chosen based on the tissues published by Schneider *et al* [42] that had the closest densities to the density values assumed above.

The 10 keV, 100 keV, and 1 MeV photon SAF values calculated with SCMS, OEdipe, and RAPID are shown in figure 11.6. The 935 keV electron *S*-values are shown in figure 11.7. Overall, the Zubal Phantom SAFs and *S*-values calculated with this platform agreed well with the reference values. All 935 keV electron *S*-values and 100 keV and 1 MeV photon SAFs differed by less than 5% from the reference values. Out of the 36 photon SAFs calculated for the 10 keV photons, 22 of them had an agreement within 5%, 12 agreed within 5%–10%, and the SAF(Liver ← Pancreas) and SAF(Pancreas ← Liver) agreed within 16%.

11.12 Spherical cellular phantom

Microdosimetry benchmarking was performed with the use of voxelized spheres to closely resemble the widely accepted MIRDCELL methodological formulation. The outside sphere had a radius of 5 μm and the inside spherical nucleus had a radius of 4 μm. Thus far, three isotopes have been investigated, a pure beta emitter ^{90}Y, an auger emitter ^{125}I, an alpha emitter, ^{223}Ra, and a positron emitter, ^{77}Br. Sub-cellular *S*-values were computed with RAPID and compared to MIRDCELL calculations. Four different source-to-target configurations were considered: nucleus to nucleus (*N*←*N*), nucleus to cytoplasm (*Cy*←*N*), cytoplasm to nucleus (*N*←*Cy*), and cytoplasm to cytoplasm (*Cy*←*Cy*). The percent differences between the RAPID and MIRDCELL *S*-values calculated in the spherical phantoms are shown in figure 11.8 for each radionuclide. Similar differences have been reported in the literature [63].

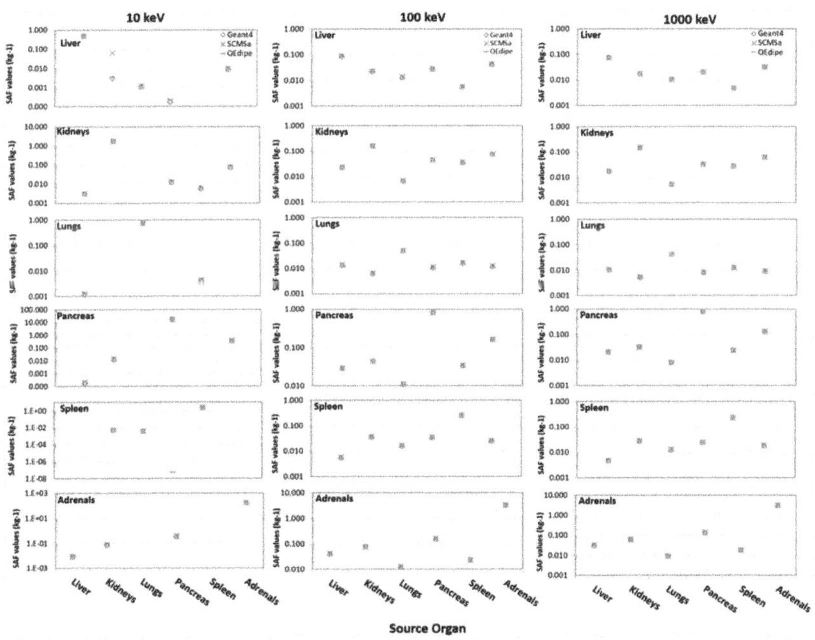

Figure 11.6. SAF values calculated with SCMS, OEdipe, and RAPID for a 10, 100 keV, and 1 MeV monoenergetic sources. Target organs are listed on the top left of each figure and the source organs are listed on the *x*-axis.

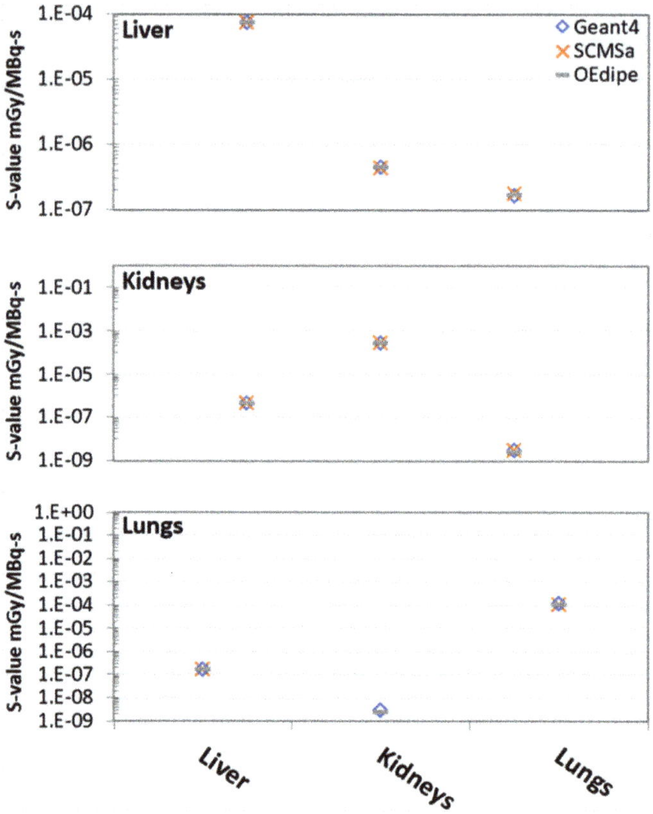

Figure 11.7. *S*-values for the 935-keV electron source. Target organs are listed on the top left of each figure and the source organs are listed on the x-axis.

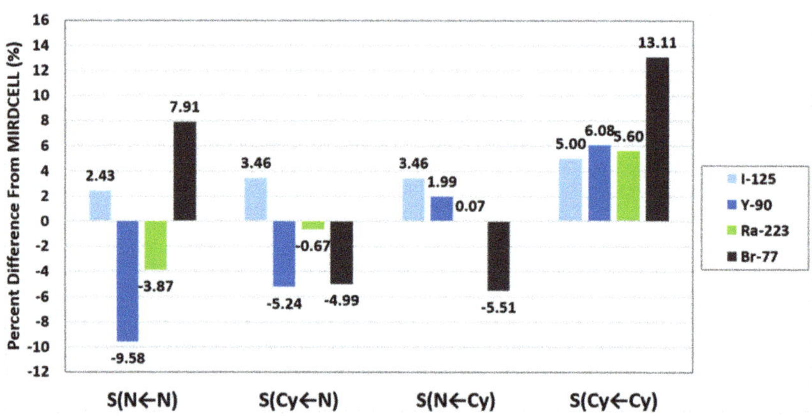

Figure 11.8. Percent difference in cellular *S*-values between RAPID and MIRDcell.

11.13 Applications of RAPID

As shown in figure 11.1, RAPID is a multi-scale dosimetry platform for RPT. To date, RAPID has been used to perform organ and tumor absorbed dose calculations in approximately 150 mice subjects, one rat subject, 5 canine subjects, and 15 human subjects. Microdosimetry calculations have been performed in multiple cell lines. Several isotopes have been investigated, and some are discussed in section 11.3. RAPID has been applied to drugs delivered intravenously, drugs delivered through embolization (e.g. ^{90}Y microspheres), and drugs delivered intrathecally.

11.14 Small-animal dosimetry

There is a compelling need to estimate absorbed doses to small animals from radionuclides used in preclinical studies, which has increased recently due in large part to the development of novel RPT agents. Most dose calculations performed by RAPID have been done in small animals. MicroPET/CT imaging is often used to characterize the biodistribution of RPT agents, thus providing useful information in the early stages of drug development. These microPET/CT images can then be used to estimate absorbed doses to organs or tumors to help evaluate the potential efficacy of RPT prior to being used on patients. Reference phantoms, such as MOBY [64], have been commonly used to represent the anatomy of the animals during the preclinical dosimetry process. However, it seems more prudent to utilize CT scans that are acquired during most preclinical PET studies to perform dosimetry on the actual anatomy of the animals under investigation, which is an essential feature of RAPID.

Bednarz *et al* [65] compared absorbed doses calculated with RAPID and absorbed doses calculated in a commonly used reference mouse phantom called MOBY. Four test mice containing different tumor xenografts underwent serial PET/CT imaging using a novel RPT agent NM404, which can be labeled with ^{124}I for imaging or ^{131}I for therapy. 3D dose distributions from ^{131}I-NM404 (Cellectar Biosciences, Florham Park, NJ) were calculated in the mice using RAPID. Mean organ doses in these four test mice were compared to mean organ doses derived by using two previously published ^{131}I S-values datasets in MOBY. In addition, mean tumor doses calculated in RAPID were compared to mean organ doses derived from unit density spheres. Large differences were identified between mean organ doses calculated in the test mice using RAPID and those derived in the MOBY phantom where percent errors in organs ranged between 0.3% and 333%. Overall, mass scaling improved agreement between MOBY phantom calculations and RAPID where percent errors were all less than 26%, except for the lung in which percent errors reached values of 48%. Percent errors in mean tumor doses in the test mice and unit density spheres were less pronounced, but still ranged between 8% and 23%. This work demonstrated the limitations of using pre-computed S-values in computational phantoms to predict organ doses in small animals from RPT procedures.

Morris *et al* [66] used RAPID to estimate organ and tumor doses from ^{131}I-CLR1404 (CLR-131) in mice with patient-derived xenografts. The group demonstrated selective distribution of CLR-131 (Cellectar Biosciences, Florham Park, NJ) and confirmed the therapeutic potential of the radiolabeled drug alone and together with moderate dose external beam radiation in murine models of head and neck cancer (HNC). Using PET/CT imaging of HNC tumor models treated with ^{124}I-CLR1404 (CLR-124), RAPID provided a dosimetric proof-of-principle for the capacity of CLR-131 to enable both normal tissue-sparing and tumor dose escalation. This works helped to inform an ongoing Phase 1 clinical trial that combines CLR-131 with reduced-dose external beam radiation in patients with loco-regional recurrence following prior HNC radiation.

In addition to HNC xenografts, RAPID has been used to provide a dosimetric evaluation of ^{131}I-CRL1404 in pediatric cancer xenografts. Marsh *et al* [67] evaluated the *in vivo* pharmacokinetics and estimated dosimetry for CLR-131 in mouse xenograft models of neuroblastoma, rhabdomyosarcoma, and Ewing sarcoma. Subject-specific tumor dosimetry for ^{131}I-CLR1404 presented a correlative relationship with tumor-growth delay after treatment. The group demonstrated that CLR-131 is a suitable RPT scaffold for dosimetry and therapy with potentially broad applicability in pediatric oncology. This work has helped to inform an ongoing multi-center dose escalation study in children and adolescents with relapsed or refractory malignant tumors called CLOVER-2 (NCT03478462).

More recently, Grudzinski *et al* used RAPID to perform a dosimetric evaluation of the theranostic pair $^{86/90}$Y-NM600 (Archeus Technologies, Madison, WI) in a variety of human cancer tumor models including lung, pancreatic, pediatric, lymphoid, skin, and prostate cancers. Absorbed doses delivered by ^{90}Y-NM600 per injected activity (Gy MBq^{-1}) were estimate with RAPID and were highest in the pancreatic and lymphoid cancer models at 2.72 ± 0.33 Gy MBq^{-1} and 2.67 ± 0.32 Gy MBq^{-1}, respectively. This work has helped to inform an ongoing single-center dose escalation study in dogs with naturally occurring metastatic melanoma (see section 11.15).

Finally, a proof-of-concept dosimetric evaluation was done in a rat that received an intrathecal injection of a ^{68}Ga PET imaging agent. The resulting dosimetry calculations are provided in figure 11.9. The rat was injected into the intrathecal space with 10.68 MBq of a Ga-68 labeled tracer and subjected to a dynamic 60 min PET/CT scan.

11.15 Companion animal dosimetry

The potential for companion (pet) species with spontaneously occurring tumors to serve as surrogates for preclinical development of cancer imaging technologies has become more apparent in the last decade. Until recently, this useful patient population for evaluating new therapeutics has been ignored by the RPT community. The value of the companion model centers around issues related to body size (including spatial target/normal anatomic characteristics), physical size and spatial distribution of metastasis, tumor heterogeneity, the presence of an intact syngeneic immune system and a syngeneic tumor microenvironment shaped by the natural evolution of the cancer [68]. Murine models generally do not replicate the size and

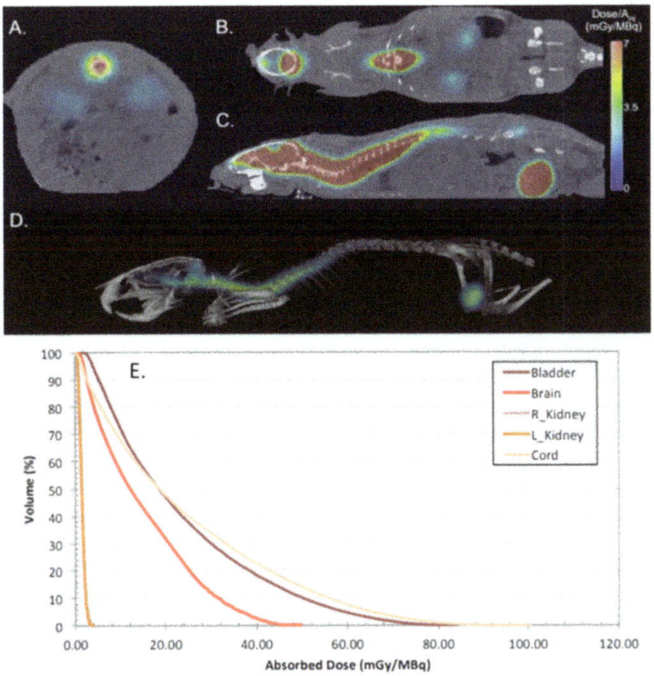

Figure 11.9. Dose color washes overlaid on the CT volume. The dose volume is overlaid on an (a) axial, (b) coronal, and (c) sagittal slice of the CT volume. The bottom panel (d) shows a 3D rendering of the dose volume overlaid on the CT volume which has a high-density threshold applied. (e) Cumulative DVHs are also displayed for several ROIs.

spatial distribution of human metastatic cancer and these factors strongly influence imaging and dosimetry.

To the best of our knowledge, the first use of companion animals to support the development of an RPT treatment is ongoing at the University of Wisconsin—Madison. Morris *et al* [69] discovered that distant metastatic sites serve as a nidus for immunosuppressive cells (e.g., Tregs), and these mediate systemic immunosuppressive effects that antagonize the *in situ* vaccination generated by EBRT—a phenomenon the team called concomitant immune tolerance (CIT). CIT is radiation sensitive, but it is infeasible to deliver EBRT to all distant metastatic sites in the patient. The team is using the RPT agent ^{90}Y-NM600 to deliver immunomodulatory radiation to overcome CIT at these distant sites.

Preclinical investigations in mice have revealed that there is a dosimetric window of ~2–5 Gy to treat CIT [70]. Delivering doses lower than 2 Gy will not control CIT, whereas delivering doses above 2–5 Gy could lead to unsafe levels of systemic immunosuppression. RAPID is being used in an ongoing canine trial to estimate how much activity is needed to deliver ~2 Gy of ^{90}Y-NM600 to distant metastatic sites. RAPID calculates 3D dose distributions from ^{90}Y-NM600 in each patient

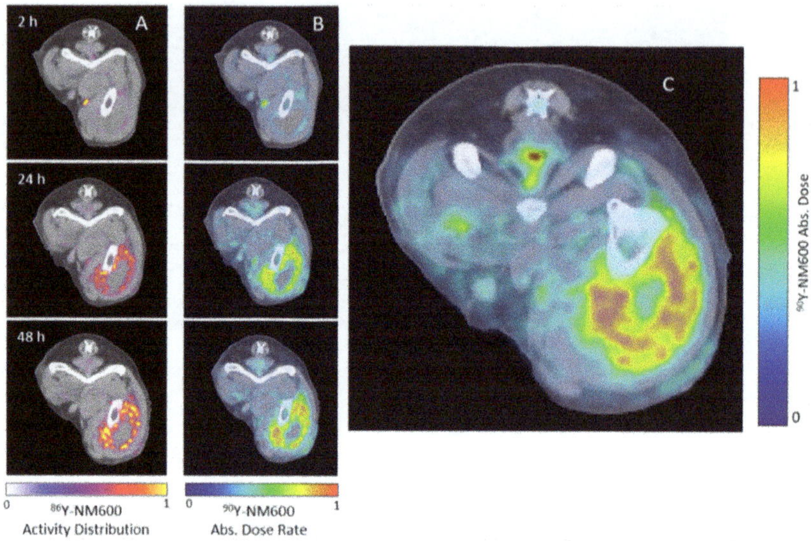

Figure 11.10. Multi-time point internal dosimetry study for a representative canine subject bearing a necrotic tumor adjacent to the right femur. Axial slices of the tumor with overlaid PET (a) and absorbed dose rate (b). Here you can see a clearly defined necrotic core at the center of the tumor. (c) The final 3D absorbed dose distribution calculated based on the temporal integration of the registered ADR volumes. All colorbars shown are normalized to the maximum value with the tumor ROI.

using multi-time point pre-treatment ^{86}Y-NM600 PET/CT scans. The activity needed to deliver a mean dose of 2 Gy to the primary tumor is then determined. To the best of our knowledge this is the first time that patient-specific 3D dosimetry has been used to prospectively plan RPT treatments in companion animals. Figure 11.10 provides imaging and dosimetry results in a canine that underwent a ^{90}Y-NM600 treatment.

11.16 Human dosimetry

A major goal of RAPID is to help physicians make more informed decisions about how they best treat RPT patients. Patient-specific dosimetry permits the administration of the therapeutic agent to be optimized to deliver the largest possible dose to the tumor(s) while keeping all other critical structures within their dose tolerances. Thus, the use of theranostic agents in radionuclide therapy is an important advancement toward the realization of personalized medicine.

There are many challenges that will need to be addressed before pre-treatment planning in RPT patients becomes a standard of care. Several challenges are non-technical. For example, mathematically speaking, increasing the number of imaging time points in an RPT patient should lead to more accurate dosimetry results. However, quantitative imaging scans required for dosimetry are costly and acquisition times in patients can be long, which places a fiscal and administrative burden on busy and resource-limited nuclear medicine departments. Furthermore, given that many RPT patients travel long distances to receive treatment, multi-time point imaging might require overnight lodging which may not be desirable or feasible for patients. Furthermore, even though nuclear medicine imaging has improved over the years, quantitative imaging for RPT is challenging since many therapeutic agents are not optimal for imaging [71]. Finally, despite several prior research studies demonstrating dose–response relationships for a variety of different RPT treatments, there is a burden of proof to show that RPT treatment planning is worth the time and fiscal investment to perform on a patient-by-patient basis. The merit of RPT treatment planning awaits large-scale clinical trials.

Given these challenges, RAPID has only been used to perform retrospective dosimetry in RPT patients. Besemer *et al* [72] investigated the impact of PET and MRI threshold-based tumor segmentation on TRT dosimetry in patients with primary and metastatic brain tumors treated with $^{124/131}$I-CLR1404. Historically, anatomical images from CT and MR have been used to manually segment tumor volumes. However, it is well known that interobserver variability in ROI contouring introduces one of the largest sources of imprecision in radiation therapy [73]. This type of variability could have a significant impact on RPT treatments given the heterogenous dose distributions involved in these treatments. Variability in the tumor volume could lead to uncertainties in dose estimates, resulting in inconsistencies in the prescription of administered activities if RPT treatment planning is used. These uncertainties could lead to challenges in establishing dose–response and dose–toxicity relationships, resulting in suboptimal tumor dosing and decreased treatment efficacy, as well as unexpected toxicities to normal tissues. In this study, it was shown that the tumor volume and the interpretation of the tumor dose is highly sensitive to the imaging modality, PET enhancement metric, and threshold level used for tumor volume segmentation. The large variations in tumor doses clearly demonstrate the need for standard protocols for multimodality tumor segmentation in RPT dosimetry.

In addition, Besemer *et al* [74] evaluated a pre-treatment dosimetry methodology in a triple-negative breast cancer patient who was uniquely enrolled in both a CLR-124 imaging clinical trial and a CLR 131 therapeutic dose escalation clinical trial. Retrospective dosimetry was calculated based on pre-treatment CLR 124 and CLR 131 2D planar images and compared with the therapeutic dose distribution calculated based on CLR 131 SPECT/CT images acquired after the therapeutic administration of CLR 131. This pilot study demonstrated that CLR 124 pre-treatment PET images had good concordance with CLR 131 3D SPECT/CT-based dosimetry (see figure 11.11).

Lastly, RAPID will be used to help facilitate the combination of EBRT and RPT in recurrent HNC patients. Preclinical studies using CLR-131 in murine model systems of HNC have been exceptionally promising [66]. Experimental data confirms preferential uptake and retention of CLR 131 in HNC xenografts and demonstrates that CLR 131 inhibits tumor growth. Additional tumor growth inhibition has been documented when CLR 131 is combined with EBRT [66]. Given these promising preclinical results, a clinical trial (NCT04105543) is underway at UW-Madison to investigate the efficacy of combining CLR 131 with EBRT with curative intent in treating relapsed HNC [75]. The premise of the trial is to

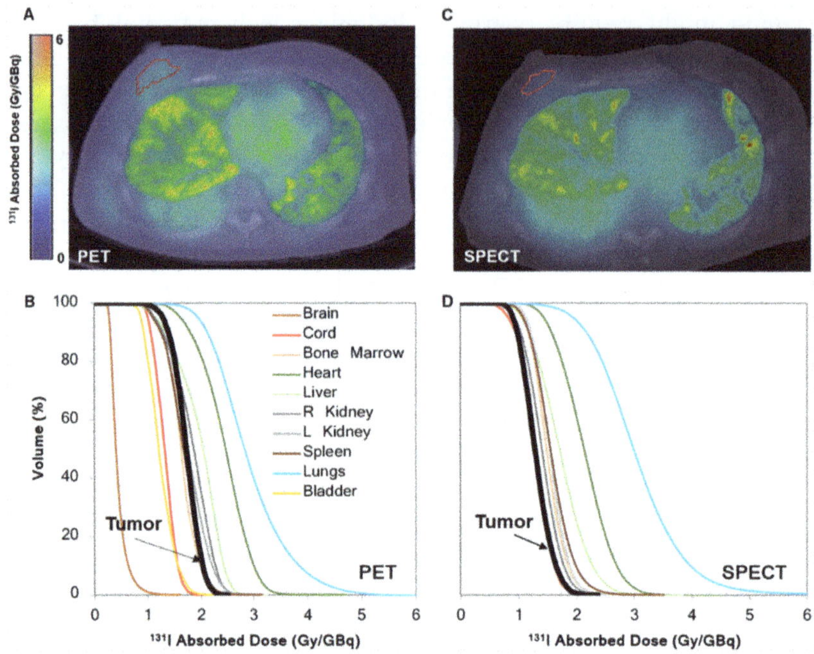

Figure 11.11. Axial dose color washes and DHVs for several ROIs calculated using PET/CT (a and b) and SPECT/CT (c and d).

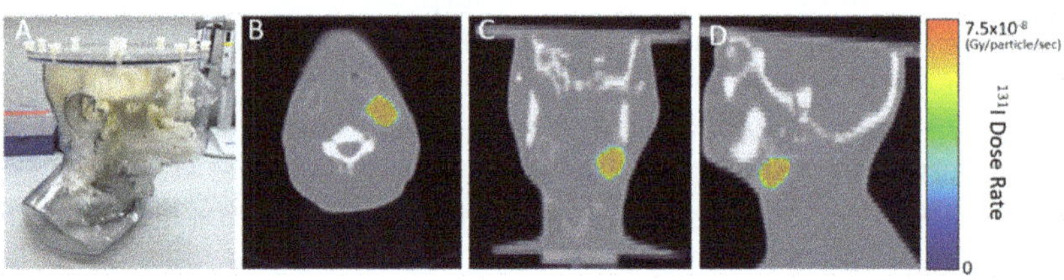

Figure 11.12. (a) A 3D printed tumor volume filled with ^{131}I inside a 3D printed head and neck phantom. (b–d) Dose rate difference between Monte Carlo calculations using SPECT/CT and ground truth uniform activity.

evaluate the combination of CLR 131 and EBRT in the treatment of recurrent HNC wherein the radiation dose deposited by the CLR 131 permits a dose reduction delivered by EBRT (e.g. standard of care 60–70 Gy will be targeted irrespective of radiation modality). The overall hypothesis of the clinical trial is that the dose reduction achieved in EBRT when given in combination with CLR 131 will be safe and tolerable while maintaining favorable tumor response rates and diminishing the adverse impact of radiation treatment on subject-specific symptoms, such as quality of life, salivary flow and swallowing function. To achieve the objectives of the trial, patient-specific dosimetry is needed to accurately characterize the absorbed dose to tumors and normal tissues from CLR 131. RAPID will use multi-time point SPECT/CT scans of CLR 131 to perform patient-specific dosimetry in these patients. An example of this process is provided in figure 11.12 which shows a dose calculation performed in a 3D printed head and neck phantom with fillable tumor and organ cavities that is being used for quality assurance for the trial.

11.17 Cellular dosimetry

Microdosimetric and nanodosimetric calculations are sometimes needed to understand and interpret the mechanisms of radiobiological effects from RPT. This is especially true for Auger emitting [76] and alpha emitting [77] RPT agents. Additionally, deterministic biological effects including tumor response and normal tissue toxicity for RPT do not correlate with mean absorbed dose metrics as they do for external beam radiotherapy [78]. Dose calculation methods that were developed for cellular level dosimetry for RPT (e.g. MIRDcell [79]) assume a spherical geometry and a uniform source distribution; however, previous work has demonstrated there to be a significant importance of including both a realistic geometric definition and source activity [80]. To this end, RAPID can also serve as a microdosimetry platform capable of importing confocal fluorescent microscopy Z-stack images (see figure 11.13(a)), transforming them into voxelized phantoms (see figure 11.13(b)), importing the radioactive compound distribution based upon confocal fluorescent microscopy, and then running transport simulations based upon the imported geometry, source distribution, and specified radionuclide. RAPID exploits single-electron scatter physics libraries as opposed to MIRDcell, which uses condensed history techniques. This feature in RAPID should help improve the understanding of *in vitro* radiobiology experiments and further elucidate microdosimetric properties not necessarily limited to targeted radionuclide compounds.

Adam *et al* [81] performed benchmarking with the use of voxelized spheres to resemble the widely accepted MIRDcell methodological formulation more closely. Three isotopes were investigated, a pure beta emitter ^{90}Y, an Auger emitter ^{125}I, and an alpha emitter, ^{223}Ra. Sub-cellular *S*-values were computed with RAPID and compared to MIRDcell. Good agreement was presented between RAPID and MIRDcell for

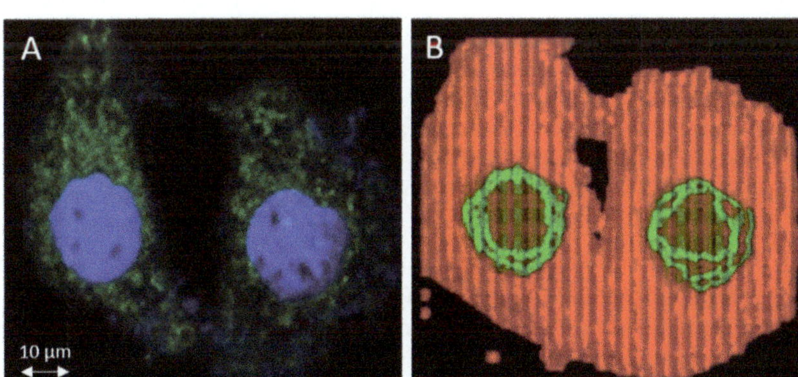

Figure 11.13. (a) Confocal fluorescent microscopy image of a human ovarian cancer cell (OvCar3) cell and (b) GEANT4 voxelized phantom generated from the microscopy image.

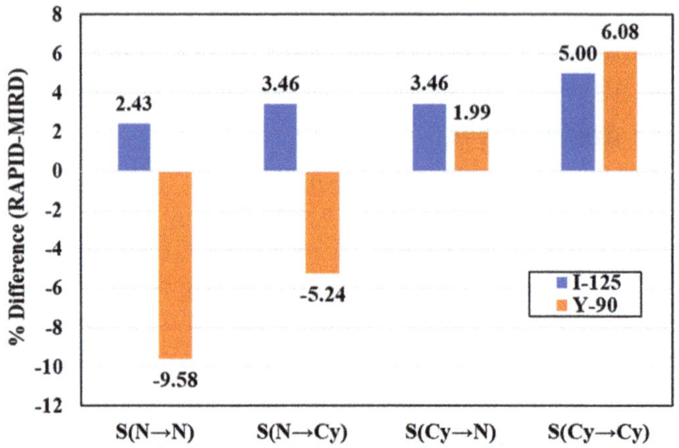

Figure 11.14. Percent differences between S(N←N), S(Cy←N), S(N←Cy), and S(Cy←Cy) calculated in a voxelized realistic OvCar cell and a voxelized sphere for ^{125}I and ^{90}Y.

spherical cell models. In addition, simulations were performed which compared the dose distributions between the realistic cellular geometry shown in figure 11.13 and a voxelized sphere geometry. Significant differences were calculated between the two geometries, as shown for ^{125}I and ^{90}Y sources plotted in figure 11.14.

11.18 Conclusions

RAPID is a Monte Carlo-based dosimetry platform designed to support preclinical and clinical research efforts intended to better understand the dosimetric and biological impacts of RPT. The multi-scale functionality of RAPID improves upon the limitations of existing dosimetry solutions for RPT that are not equipped to appropriately model the dosimetric and biological impacts of a growing number of emitters and conjugates that are currently approved or under development. This workflow includes quantitative image acquisition methods, pharmacokinetic profiling and modeling, micro- and macroscopic 3D dose calculations, and treatment response assessment. RAPID has been utilized to characterize the biodistribution and dosimetry of several radiopharmaceuticals both in the preclinical [66, 67, 82, 83] and clinical setting [72, 74].

RPT is rapidly becoming a mainstream modality with the development and approval of new drugs. Traditionally, RPT has been administered with fixed activities or empirical (e.g. mass-based) methods for determining administered activities following the approaches used in chemotherapy. Accompanying the growth of RPT in the clinic is a strong interest in the use of absorbed dose-based methodologies for treating patients with this modality. Optimizing patient treatments by planning to normal organ absorbed dose thresholds and/or tumoricidal doses using individual patient kinetics from 3D imaging (SPECT/CT and PET/CT) represents a safer and more effective therapy. Large professional organizations such as the American Association of Physicists in Medicine (AAPM), American Society for Radiation Oncology (ASTRO), the Society of Nuclear Medicine and Molecular

Imaging (SNMMI), and the National Cancer Institute (NCI) have all expressed a strong desire to move in this direction and it is expected for patient-specific treatment approaches to be the standard of care in the near future.

To this end, Voximetry Inc., a spin-off company from the University of Wisconsin-Madison founded in 2016, created a commercialized version of the RAPID workflow for clinical use called Torch^{TM}. Torch^{TM} is a software platform that is uniquely designed to compute absorbed dose on the voxel-level for patients undergoing radiopharmaceutical therapy (RPT). PET/CT or SPECT/CT volumes (e.g. DICOM) from multiple time points, along with externally generated organ contours (e.g. DICOM RT Structure Sets), are first imported into Torch^{TM}. Contours are then temporarily aligned on the voxel-level using GPU-accelerated deformable registration algorithms. If contours are only available for a limited number of time points, similar deformable registration algorithms are used for contour propagation. Next, Torch^{TM} performs automatic pharmacokinetic (PK) modeling in each voxel of each contour to help describe the rate of uptake and/or clearance of the radiopharmaceutical throughout the body over time. Then, Torch^{TM} analytically determines the time integrated activity within each voxel which is ultimately proportional to the number of radionuclide disintegrations that originate in each voxel. Lastly, radiation transport is performed either using GPU-accelerated convolution kernels (with or without density correction) or full transport GPU Monte Carlo. A 3D absorbed dose cloud is displayed, organ-level statistics are tabulated, and dose-volume histograms are generated. These display options support clinicians in prescribing injected activity or modifying ongoing cycles of RPT. The 3D absorbed dose volume (e.g. DICOM RT DOSE) can then be exported into other software for possible combination with other modalities (e.g. external beam radiotherapy). Figure 11.15 depicts the workflow that a Torch^{TM} user would follow to compute voxel-level dosimetry for patients being treated with RPT. The translation of RAPID into a fast and user-friendly software platform will allow sophisticated Monte Carlo RPT calculations to be performed in busy and resource limited clinical environments.

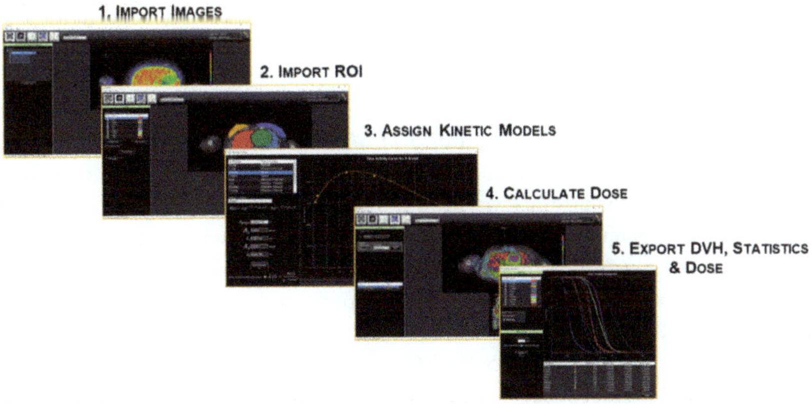

Figure 11.15. Screenshots from Torch^{TM} illustrating a treatment planning workflow.

References

[1] Sawin C T and Becker D V 1997 Radioiodine and the treatment of hyperthyroidism: the early history *Thyroid* **7** 163–76

[2] Dolgin E 2018 Radioactive drugs emerge from the shadows to storm the market *Nat. Biotechnol.* **36** 1125–27

[3] Le D T, Durham J N and Smith K N *et al* 2017 An immunogenic personal neoantigen vaccine for patients with melanoma *Nature* **12** 1–10

[4] Flux G D, Sjogreen Gleisner K and Chiesa C *et al* 2018 From fixed activities to personalized treatments in radionuclide therapy: lost in translation? *Eur. J. Nucl. Med. Mol. Imaging* **45** 152–4

[5] Strigari L, Konijnenberg M and Chiesa C *et al* 2014 The evidence base for the use of internal dosimetry in the clinical practice of molecular radiotherapy *Eur. J. Nucl. Med. Mol. Imaging* **41** 1976–88

[6] Chiesa C, Mira M and Maccauro M *et al* 2015 Radioembolization of hepatocarcinoma with ^{90}Y glass microspheres: development of an individualized treatment planning strategy based on dosimetry and radiobiology *Eur. J. Nucl. Med. Mol. Imaging* **42** 1718–38

[7] Chansanti O, Jahangiri Y and Matsui Y *et al* 2017 Tumor dose response in yttrium-90 resin microsphere embolization for neuroendocrine liver metastases: a tumor-specific analysis with dose estimation using SPECT-CT *J. Vasc. Interv. Radiol.* **28** 1528–35

[8] Van Den Hoven A F, Rosenbaum C E N M and Elias S G *et al* 2016 Insights into the dose-response relationship of radioembolization with resin ^{90}Y-microspheres: a prospective cohort study in patients with colorectal cancer liver metastases *J. Nucl. Med.* **57** 1014–19

[9] Cremonesi M, Ferrari M E and Bodei L *et al* 2018 Correlation of dose with toxicity and tumour response to ^{90}Y- and ^{177}Lu-PRRT provides the basis for optimization through individualized treatment planning *Eur. J. Nucl. Med. Mol. Imaging* **45** 2426–41

[10] Strigari L, Sciuto R and Rea S *et al* 2010 Efficacy and toxicity related to treatment of hepatocellular carcinoma with ^{90}Y-SIR spheres: radiobiologic considerations *J. Nucl. Med.* **51** 1377–85

[11] Ilan E, Sandstrom M and Wassberg C *et al* 2015 Dose response of pancreatic neuroendocrine tumors treated with peptide receptor radionuclide therapy using ^{177}Lu-DOTATATE *J. Nucl. Med.* **56** 177–82

[12] Pauwels S, Barone R and Walrand S *et al* 2005 Practical dosimetry of peptide receptor radionuclide therapy with ^{90}Y-labeled somatostatin analogs. *J. Nucl. Med.* **46** 92S–8S

[13] Hofman M S, Violet J and Hicks R J *et al* 2018 [^{177}Lu]-PSMA-617 radionuclide treatment in patients with metastatic castration-resistant prostate cancer (LuPSMA trial): a single-centre, single-arm, phase 2 study *Lancet Oncol.* **19** 825–33

[14] Violet J A, Jackson P and Ferdinandus J *et al* 2018 Dosimetry of Lu-177 PSMA-617 in metastatic castration-resistant prostate cancer: correlations between pre-therapeutic imaging and 'whole body' tumor dosimetry with treatment outcomes *J. Nucl. Med.* **60** 517–23

[15] Kabasakal L, AbuQbeitah M and Aygün A *et al* 2015 Pre-therapeutic dosimetry of normal organs and tissues of 177Lu-PSMA-617 prostate-specific membrane antigen (PSMA) inhibitor in patients with castration-resistant prostate cancer *Eur. J. Nucl. Med. Mol. Imaging* **42** 1976–83

[16] Delker A, Fendler W P and Kratochwil C *et al* 2016 Dosimetry for ^{177}Lu-DKFZ-PSMA-617: a new radiopharmaceutical for the treatment of metastatic prostate cancer *Eur. J. Nucl. Med. Mol. Imaging* **43** 42–51

[17] Herrmann K, Bluemel C and Weineisen M *et al* 2015 Biodistribution and radiation dosimetry for a probe targeting prostate-specific membrane antigen for imaging and therapy *J. Nucl. Med.* **56** 855–61
[18] Kratochwil C, Fendler W P and Eiber M *et al* 2019 EANM procedure guidelines for radionuclide therapy with ^{177}Lu-labelled PSMA-ligands (^{177}Lu-PSMA-RLT) *Eur. J. Nucl. Med. Mol. Imaging* **46** 2536–44
[19] Council Directive 2013/59/Euratom of 5 December 2013 Laying down basic safety standards for protection against the dangers arising from exposure to ionising radiation, and repealing directives 89/618/Euratom, 90/641/Euratom, 96/29/Euratom, 97/43/Euratom and 2003/122/Euratom
[20] Ljungberg M, Sjögreen K, Liu X, Frey E, Dewaraja Y and Strand S E 2002 A 3-dimensional absorbed dose calculation method based on quantitative SPECT for radionuclide therapy: evaluation for ^{131}I using Monte Carlo simulation *J. Nucl. Med.* **43** 1101–9
[21] Lu C C, Lin H H and Chuang K S *et al* 2014 Development and validation of a fast voxel-based dose evaluation system in nuclear medicine *Radiat. Phys. Chem.* **104** 355–9
[22] Kolbert K S, Sgouros G and Scott A M *et al* 1997 Implementation and evaluation of patient-specific three-dimensional internal dosimetry *J. Nucl. Med.* **38** 301–8
[23] Li T, Zhu L, Lu Z, Song N, Lin K-H and Mok G S P 2020 BIGDOSE: software for 3D personalized targeted radionuclide therapy dosimetry *Quant. Imaging Med. Surg.* **10** 160–70
[24] Tagesson M, Ljungberg M and Strand S E 1996 A Monte Carlo program converting activity distributions to absorbed dose distributions in a radionuclide treatment planning system *Acta Oncol.* **35** 367–72
[25] Wemple C A, Wessol D E and Nigg D W *et al* 2004 MINERVA—a multi-modal radiation treatment planning system *Appl. Radiat. Isotopes.* **61** 745–52
[26] Kost S D, Dewaraja Y K, Abramson R G and Stabin M G 2015 VIDA: a voxel-based dosimetry method for targeted radionuclide therapy using Geant4 *Cancer Biother. Radiopharm.* **30** 16–26
[27] Chiavassa S, Bardiès M and Guiraud-Vitaux F *et al* 2005 OEDIPE: a personalized dosimetric tool associating voxel-based models with MCNPX *Cancer Biother. Radiopharm.* **20** 325–32
[28] Marcatili S, Pettinato C and Daniels S *et al* 2013 Development and validation of RAYDOSE: a Geant4 based application for molecular radiotherapy *Phys. Med. Biol.* **58** 2491–508
[29] Grimes J, Uribe C and Celler A 2013 JADA: a graphical user interface for comprehensive internal dose assessment in nuclear medicine *Med. Phys.* **40** 072501
[30] Yoriyaz H, Stabin M G and dos Santos A 2001 Monte Carlo MCNP-4B-based absorbed dose distribution estimates for patient-specific dosimetry *J. Nucl. Med.* **42** 662–9
[31] Dewaraja Y K, Wilderman S J, Ljungberg M, Koral K F, Zasadny K and Kaminiski M S 2005 Accurate dosimetry in ^{131}I radionuclide therapy using patient-specific, 3-dimensional methods for SPECT reconstruction and absorbed dose calculation *J. Nucl. Med.* **46** 840–9
[32] Wyawahare M V, Patil P M and Abhyankar H K 2009 Image registration techniques: an overview *Int. J. Signal Proc. Image Proc. Pattern Recogn.* **2** 3
[33] Kessler M L 2006 Image registration and data fusion in radiation therapy *Br. J. Radiol.* **79** S99–108
[34] Jentzen W, Weise R and Kupferschläger J *et al* 2008 Iodine-124 PET dosimetry in differentiated thyroid cancer: recovery coefficient in 2D and 3D modes for PET(/CT) systems *Eur. J. Nucl. Med. Mol. Imaging* **35** 611–23

[35] Agostinelli S, Allison J and Amako K *et al* 2003 GEANT4—a simulation toolkit. *Nucl. Instrum. Methods Phys. Res. Sect. A Accel. Spectrom. Detect. Assoc. Equip.* **506** 250–303

[36] Faddegon B A, Kawrakow I, Kubyshin Y, Perl J, Sempau J and Urban L 2009 The accuracy of EGSnrc, Geant4 and PENELOPE Monte Carlo systems for the simulation of electron scatter in external beam radiotherapy *Phys. Med. Biol.* **54** 6151–63

[37] Aichambault L, Beaulieu L and Carrier J F *et al* 2003 Overview of Geant4 applications in medical physics *IEEE Nuclear Science Symp. Conf. Record* 3 1743–5

[38] Aso T, Kimura A and Tanaka S *et al* 2005 Verification of the dose distributions with GEANT4 simulation for proton therapy *IEEE Trans. Nucl. Sci.* **52** 896–901

[39] Thiam C O, Breton V, Donnarieix D, Habib B and Maigne L 2008 Validation of a dose deposited by low-energy photons using GATE/GEANT4 *Phys. Med. Biol.* **53** 3039–55

[40] Hauf S, Kuster M and Batič M *et al* 2013 Radioactive decays in Geant4 *IEEE Trans. Nucl. Sci.* **60** 2966–83

[41] Bhat M R 1992 Nuclear data for science and technology *Nuclear Data for Science and Technology.* Research Reports in Physics ed S M Qaim (Berlin, Heidelberg: Springer) pp 817–21

[42] Schneider W, Bortfeld T and Schlegel W 2000 Correlation between CT numbers and tissue parameters needed for Monte Carlo simulations of clinical dose distributions *Phys. Med. Biol.* **45** 459–78

[43] Younge A J, Pedretti K, Grant R E and Brightwell R 2017 A tale of two systems: using containers to deploy HPC applications on supercomputers and clouds *2017 IEEE Int. Conf. on Cloud Computing Technology and Science (CloudCom) 2017-December* (IEEE) 74–81

[44] Boettiger C 2015 An introduction to Docker for reproducible research *SIGOPS Oper. Syst. Rev.* **49** 71–9

[45] Kolbert K S, Hamacher K A, Jurcic J G, Scheinberg D A, Larson S M and Sgouros G 2001 parametric images of antibody pharmacokinetics in Bi213-HuM195 therapy of leukemia *J. Nucl. Med.* **42** 27–32

[46] Kolbert K S and Sgouros G 1998 Display and manipulation of multi-modality images for radiolabeled antibody therapy *Radiology* **209** 515

[47] Sgouros G, Kolbert K S and Sheikh A *et al* 2004 Patient-specific dosimetry for ^{131}I thyroid cancer therapy using ^{124}I PET and 3-dimensional-internal dosimetry (3D-ID) software *J. Nucl. Med.* **45** 1366–72

[48] Papavasileiou P, Flux G D, Flower M A and Guy M J 2001 An automated technique for SPECT marker-based image registration in radionuclide therapy *Phys. Med. Biol.* **46** 2085–97

[49] Pitiot A, Bardinet E, Thompson P M and Malandain G 2006 Piecewise affine registration of biological images for volume reconstruction *Med. Image Anal.* **10** 465–83

[50] Little J A, Hill D L G and Hawkes D J 1997 Deformations incorporating rigid structures *Comput. Vision Image Understand.* **66** 223–32

[51] Sotiras A, Davatzikos C and Paragios N 2013 Deformable medical image registration: a survey *IEEE Trans. Med. Imaging* **32** 1153–90

[52] Dewaraja Y K, Schipper M J and Roberson P L *et al* 2010 ^{131}I-tositumomab radioimmunotherapy: initial tumor dose-response results using 3-dimensional dosimetry including radiobiologic modeling *J. Nucl. Med.* **51** 1155–62

[53] Speer T 2012 *Targeted Radionuclide Therapy.* (Philadelphia, PA: Lippincott Williams & Wilkins)

[54] Stabin M G, Sparks R B and Crowe E 2005 OLINDA/EXM: the second-generation personal computer software for internal dose assessment in nuclear medicine *J. Nucl. Med.* **46** 1023–27

[55] Jentzen W, Freudenberg L, Eising E G, Sonnenschein W, Knust J and Bockisch A 2008 Optimized ^{124}I PET dosimetry protocol for radioiodine therapy of differentiated thyroid cancer *J. Nucl. Med.* **49** 1017–23
[56] Roberson P L, Amro H and Wilderman S J *et al* 2011 Bio-effect model applied to ^{131}I radioimmunotherapy of refractory non-Hodgkin's lymphoma *Eur. J. Nucl. Med. Mol. Imaging* **38** 874–83
[57] Graham M M, Peterson L M and Link J M *et al* 1997 Fluorine-18-fluoromisonidazole radiation dosimetry in imaging studies *J. Nucl. Med.* **38** 1631–36
[58] Hobbs R F and Sgouros G 2009 Calculation of the biological effective dose for piecewise defined dose-rate fits *Med. Phys.* **36** 904–7
[59] Besemer A E, Yang Y M, Grudzinski J J, Hall L T and Bednarz B P 2018 Development and validation of RAPID: a patient-specific Monte Carlo three-dimensional internal dosimetry platform *Cancer Biother. Radiopharm.* **33** 155–65
[60] Zubal I G, Harrell C R, Smith E O, Rattner Z and Gindi G 1994 Computerized three-dimensional segmented human anatomy *Med. Phys.* **21** 299–302
[61] Yoriyaz H, Dos Santos A, Stabin M G and Cabezas R 2000 Absorbed fractions in a voxel-based phantom calculated with the MCNP-4B code *Med. Phys.* **27** 1555–62
[62] Forster R A and Godfrey T N K 2006 MCNP—a general Monte Carlo code for neutron and photon transport *Monte-Carlo Methods and Applications in Neutronics, Photonics and Statistical Physics.* (Berlin: Springer) pp 33–55
[63] Hocine N, Chipana R and Sarda L 2020 Comparison of MCNPX and MIRDcell in assessing self-dose and cross-dose delivered to cell nuclei and the development of a realistic geometric model *Int. J. Radiat. Biol.* **96** 1008–16
[64] Segars W P, Tsui B M W, Frey E C, Johnson G A and Berr S S 2004 Development of a 4-D digital mouse phantom for molecular imaging research *Mol. Imaging Biol.* **6** 149–59
[65] Bednarz B, Grudzinski J and Marsh I *et al* 2018 Murine-specific internal dosimetry for preclinical investigations of imaging and therapeutic agents *Health Phys.* **114** 450–9
[66] Morris Z S, Weichert J P and Saker J *et al* 2015 Therapeutic combination of radiolabeled CLR1404 with external beam radiation in head and neck cancer model systems *Radiother. Oncol.* 116 504–9
[67] Marsh I R, Grudzinski J J and Baiu D C *et al* 2019 Preclinical pharmacokinetics and dosimetry studies of [^{124}I/^{131}I] CLR1404 for treatment of pediatric solid tumors in murine xenograft models *J. Nucl. Med.* **60** 1414–20
[68] Vail D M, LeBlanc A K and Jeraj R 2020 Advanced cancer imaging applied in the comparative setting *Front Oncol.* **10** 84
[69] Morris Z S, Guy E I and Werner L R *et al* 2018 Tumor-specific inhibition of *in situ* vaccination by distant untreated tumor sites *Cancer Immunol. Res.* **6** 825–34
[70] Grudzinski J J, Hernandez R and Marsh I *et al* April 2019 Preclinical characterization of $^{86/90}$Y-NM600 in a variety of murine and human cancer tumor models *J. Nucl. Med.* **118** 224808
[71] Li T, Ao E C I, Lambert B, Brans B, Vandenberghe S and Mok G S P 2017 Quantitative imaging for targeted radionuclide therapy dosimetry—technical review *Theranostics.* **7** 4551–65
[72] Besemer A E, Titz B and Grudzinski J J *et al* 2017 Impact of PET and MRI threshold-based tumor volume segmentation on patient-specific targeted radionuclide therapy dosimetry using CLR1404 *Phys. Med. Biol.* **62** 6008–25

[73] Riegel A C, Berson A M and Destian S *et al* 2006 Variability of gross tumor volume delineation in head-and-neck cancer using CT and PET/CT fusion *Int. J. Radiat. Oncol. Biol. Phys.* **65** 726–32

[74] Besemer A E, Grudzinski J J, Weichert J P, Hall L T and Bednarz B P 2019 Pretreatment CLR 124 positron emission tomography accurately predicts CLR 131 three-dimensional dosimetry in a triple-negative breast cancer patient *Cancer Biother. Radiopharm.* **34** 13–23

[75] 2019 CLR 131 Combined with radiation for head and neck cancer (Accessed May 3, 2020)

[76] Kassis A I 2004 The amazing world of Auger electrons *Int. J. Radiat. Biol.* **80** 789–803

[77] Sgouros G, Roeske J C and McDevitt M R *et al* 2010 MIRD pamphlet No. 22 (Abridged): radiobiology and dosimetry of α-particle emitters for targeted radionuclide therapy *J. Nucl. Med.* **51** 311–28

[78] O'Donoghue J A 1999 Implications of nonuniform tumor doses for radioimmunotherapy *J. Nucl. Med.* **40** 1337–41

[79] Vaziri B, Wu H, Dhawan A P, Du P and Howell R W 2014 MIRD pamphlet no. 25: MIRDcell V2.0 software tool for dosimetric analysis of biologic response of multicellular populations *J. Nucl. Med.* **55** 1557–64

[80] Šefl M, Incerti S, Papamichael G and Emfietzoglou D 2015 Calculation of cellular S-values using Geant4-DNA: the effect of cell geometry *Appl. Radiat. Isot.* **104** 113–23

[81] Adam D, Schweitzer N, Hoffman S and Bednarz B 2019 Development of a Monte Carlo-based microdosimetry platform for the analysis of targeted radionuclide therapy agents *in vitro Int. J. Radiat. Oncol.* **105** E642–3

[82] Grudzinski J, Marsh I and Titz B *et al* 2018 CLR 125 Auger electrons for the targeted radiotherapy of triple-negative breast cancer *Cancer Biother. Radiopharm.* **33** 87–95

[83] Grudzinski J J, Hernandez R and Marsh I *et al* 2019 Preclinical characterization of $^{86/90}$Y-Nm600 in a variety of murine and human cancer tumor models *J. Nucl. Med.* **60** 1622–8

Chapter 12

The VRAK software for internal radionuclide dosimetry

Price Jackson and Graeme O'Keefe

12.1 Scope of the task and application to clinical decision-making

Radiation dosimetry in nuclear medicine can have a variety of clinical applications ranging from assessment of the safety profile of diagnostic tracers at the population level up to prediction of tissue-level effects from high dose therapeutic radionuclide administrations in an individual. In either instance, the principles are the same: to infer an appropriate number of isotope disintegrations to the anatomy and use an understanding of radiation transport to translate that into regional absorbed dose or some metric of radiobiological effect. It is necessary to appreciate that this process should consider the many spatial and temporal uncertainties in both the imaging chain and the method of estimating pharmacokinetics from what is typically very sparse temporal data.

Image-based dosimetry, usually with multiple-time point quantitative SPECT imaging, provides a sequence of 3D snapshots of activity in time. In routine clinical practice or even the majority of clinical trials, it's often not possible to obtain more comprehensive measurements than a limited sequence of post-treatment images. If that data is to be used for clinical decision-making, it is also worthwhile to be able to derive meaningful measures of radiation dose without significant manual oversight. In some cases, it may be best to provide the output as a 3D dose volume which allows clinicians to directly measure volumes-of-interest on imaging workstations and in a manner that is familiar from reporting other diagnostic images. With that in mind, the aim of the voxel dosimetry code we've worked to build, VRAK (Voxel Registration and Kinetics), was to work directly in the 3D image space as an automated tool with the only manual measurements, delineation of organ and target VOIs, traceable to trained clinicians.

By automating the process, the barrier to request formal dosimetry for routine clinical cases is reduced to simply acquiring extra post-therapy quantitative SPECT images. Additionally, automated processes are inherently objective which are

doi:10.1088/978-0-7503-2694-0ch12

beneficial for the analysis of clinical trial data. The method can be shared across multiple centres without imparting bias for analyses conducted at different hospitals. With that in mind, the positive aspects of performing voxel dosimetry as an automated computer process can have concessions in terms of accuracy if users are not familiar with the limitations of the imaging chain and the programmatic steps involved in converting a sequence of 3D images into a single voxel dose volume.

Voxel dosimetry in nuclear medicine is fundamentally different from external beam radiation therapy where the planned radiation dose to a voxel of the simulation CT image is very nearly the same as the dose delivered by the treatment beams. In radiotherapy, MV-energy photon beams exhibit smooth changes in dose with depth. Additionally, dose gradients at the edge of treatment fields are smoothed by collimator leaf penumbra. As a result, computing dose at the native resolution of the CT images is possible and it is appropriate to expect that the planned dose to a given voxel will match the true physical quantity to that location within a few percent. In nuclear medicine, 3D dosimetry is constrained by the resolving power of our imaging systems. In practice, the short range of beta-emitting radionuclides—even more extreme with alpha- and auger-emitters—yields dose gradients that are generally much finer than even the resolution of clinical CTs which we use as the structural basis to score dose. And if the pattern of disintegrations used to perform the calculations is blurrier again due to the nature of the SPECT or PET imaging chains, it is necessary to apply a number of qualifications when reporting radiation dose to individual voxels in an image.

Much of this book relates to methods of computing radiation transport; many of which apply to estimation of radiation dose in the image space. It is actually a straightforward task with open-source Monte Carlo packages to convert a clinical CT scan into a 3D array of densities and material compositions and subsequently simulate an image-derived activity distribution. With GATE this is possible with a short macro script and the user need only handle the file conversion between DICOM and the compatible image types. Reasonable statistical accuracy can be achieved overnight on a desktop computer for the purposes of estimating mean dose to macroscopic volumes-of-interest. The material result, a three-dimensional map of radiation absorbed dose, is comparable in terms of its physical basis to an EBRT treatment plan; the structural heterogeneity is appropriately modelled and the photon/electron transport is precisely considered. However, the source itself is poorly characterised by the systematic loss in spatial fidelity from the imaging chain. We hope to at least capture the appropriate amount of energy deposition in tissue at a macroscopic level, but the finer patterns of dose heterogeneity are beyond our grasp. As a result, one should take care in interpreting voxel dose distributions in much the same manner that a nuclear medicine physician must rely on their expertise in untangling partial-volume effects and motion averaging when drawing conclusions from PET or SPECT images.

Fortunately, for an audience of molecular imaging professionals viewing voxel dose distributions through this correct lens often comes naturally as this is second nature in their day-to-day skillset. It is understood that peak uptake value for very small lesions is under-reported because of partial-volume effect compared to true physical tracer concentration. Further, this effect is less pronounced with increasing object size and negligible in regions with relatively uniform uptake. The ability to make these judgements is integral to accurately evaluating voxel dosimetry in nuclear medicine. Medical physicists may assist with providing quantitative methods to attempt to correct for these spatial pitfalls such as being familiar with recovery coefficients or other methods of resolution recovery to effectively *reconcentrate* the pattern of disintegrations or ionisation into a more accurate physical boundary.

12.2 Overview of voxel dose processing steps

As previously alluded to, dose processing can be divided into a series of discrete steps. While it is convenient to compute the collection as a single automated program, it is possible to interchange one or the other to suit specific applications or assist with troubleshooting challenging cases. In this case we are generally considering dosimetry based on a sequence of post-treatment quantitative SPECT/CT (qSPECT) images [1]. Each one is effectively a three-dimensional snapshot of the local dose rate in time. It is now possible to acquire quantitative SPECT images with therapeutic tracers directly using vendor-supported platforms [2]. This provides a conversion between the detected counts in a reconstructed image, the acquisition time and a known sensitivity factor to yield a distribution of activity; typically in the form of activity concentration (Bq ml^{-1}) as this is ubiquitous to PET imaging. Using image registration algorithms, we can align those volumes and track for each location how the activity concentration or local dose rate changes with time. The task is then to interpolate the predicted values for intermediate time points to yield a time integral. The final task, and much of the focus in this book and its preceding edition, is the computation of radiation dose based on a heterogeneous distribution of activity in tissue. This step may be computed at the outset, converting qSPECT images into estimated dose rate maps which is followed by co-registration and kinetics processing steps. Alternatively, it is mathematically comparable to align images, estimate pharmacokinetic time integral as disintegrations in each voxel and calculate radiation transport in the final instance. The latter has the advantage of requiring dose computation only once, which, if employing Monte Carlo methods, requires less time and avoids the summation of statistical noise from multiple time points. Using this framework, the final output image is a 3D dose volume with units of absorbed dose (Gy) which may be written as a conventional RT-Dose DICOM image or into pseudo-PET format which enables review and contouring on nuclear medicine reporting workstations.

Understanding the scope of voxel dosimetry as an image processing task, it is important to be familiar with traditional MIRD nuclear medicine techniques as this underpins the mathematical principles. Being able to evaluate a case through multiple different methods provides a quick sanity check when developing new protocols and troubleshooting potential errors. I frequently rely on OLINDA Sphere model dose factors to verify energy deposition per decay and check the efficiency of self-dose deposition as objects get very small in size [3]. This is the same case for being able to swap between methods for alignment, pharmacokinetics interpolation, and dose deposition. In each stage there are clinical or research situations where one technique may be more appropriate and certainly for performing developmental research, being familiar with a variety of methods provides a strong foundation to work on advances in the field.

12.3 Image co-registration

The VRAK routine uses the image registration package Elastix, written by Stefan Klein and Marius Staring [4]. Elastix is built around the Insight Toolkit (ITK) framework which includes a robust backend for image resampling as well as file reading and writing. ITK provides native image registration libraries which Elastix extends with additional transform operations, similarity metrics, and optimisation methods. For post-treatment dosimetry, the task involves alignment of multiple SPECT/CT image acquisitions. Over the length of one or more SPECT bed positions, rigid registration methods are unlikely to provide a reliable match for all relevant tissues across an image volume. As such, deformable registration algorithms are desirable to account for local displacements even if they are relatively small. The application of these methods does warrant consideration to achieve reliable performance. Many algorithms include a regularisation term which may be moderated to smooth out unnatural deformations. With Elastix, users may designate a 'bending energy penalty' as one component of the overall loss metric which may be weighted according to the anatomical challenge, image resolution, and scale of the output value for the similarity metric. There is also an option to assign rigidity penalties to labelled regions of an image which may present a more elegant solution for preserving the shape of rigid structures. In the case of VRAK, we achieved reliable image co-registration, free from unnatural warping effects, by weighting this bending energy penalty to approximately 50% of the similarity score (figure 12.1).

The similarity metric provides a single quantifiable value to evaluate whether the information contained in the fixed and moving image sets are aligned in space. In the simplest form, computing the sum of squared differences between voxel intensity

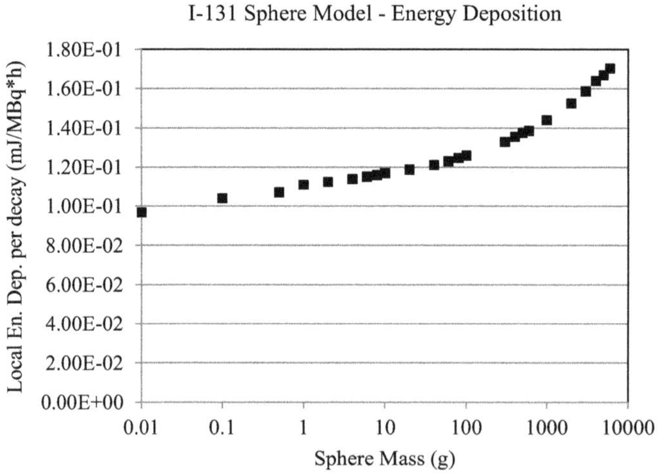

Figure 12.1. Transformation of sphere model dose factors for iodine-131 to determine local energy deposition as a proportion of joules per decay. Familiarity with other dose conversion techniques are helpful for validating Monte Carlo results and quickly appreciating cut-off sizes where different self- and cross-dose efficiency may warrant special consideration. Reproduced with permission, copyright AIP.

values may be used. This is well suited to matching CT images where anatomical structure is relatively stable in terms of physical density with time and image intensity values as Hounsfield Units are defined according to a standardised scale. For alignment of serial SPECT/CT or PET/CT images, we may rely on the inherent fusion between structural (CT) and functional (SPECT or PET) image components and perform registration on the CT data. The advantage here is that these structural images are acquired at a high resolution, the appearance of tissues are relatively stable. It is then possible to use the deformation vector field from matching the CT images to transform the functional volumes.

This method, however, is limited by the accuracy of the native SPECT/CT image fusion and may introduce artefacts if the patient shifts between the two scan components or, for example near the dome of the liver, if the CT is acquired with inconsistent breath holds between visits. Matching the CT which contains that single phase of the breathing cycle will shift the functional volume which is motion-averaged across the breathing phases. An alternative approach is to directly co-register serial functional images. One should consider that apparent pattern of tracer activity will change with time and as such a mutual information similarity metric is typically more appropriate. In practice, for alignment of ^{177}Lu-Octroetate or ^{177}Lu-PSMA-617 serial images, our experience shows it is more reliable to directly co-register SPECT images (figure 12.2). Having a first-hand understanding of the capabilities and limitations of the tools is important to ensure they are applied appropriately and be able to troubleshoot challenging cases. Many times if registration fails with one method, either CT- or SPECT-based, simply retrying with the other offers a quick solution. In the worst case, it is also possible to manually define any number of matching fiducial locations between images which may be used to anchor the alignment in specific areas of the image (figure 12.3).

In the intermediate stages of the dose processing framework, writing image files in DICOM format is not necessary. Data is more easily stored and reloaded using one of the stripped-down ITK formats (.nrrd, .mda, .nii, etc). These contain the

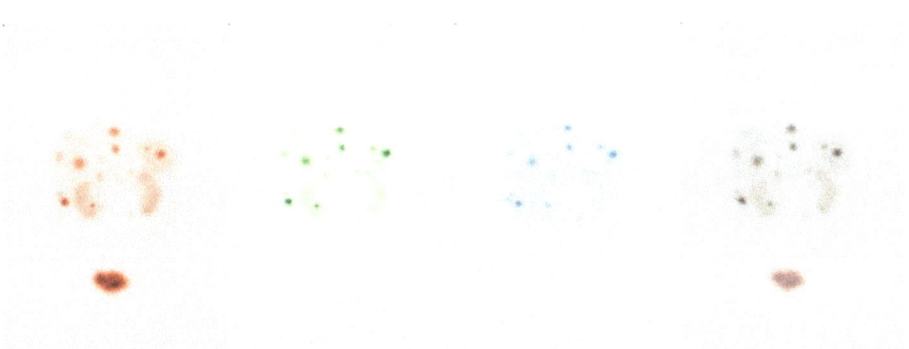

Figure 12.2. Alignment of images acquired at 4, 24, and 72 h post-administration (left-to-right in figure) shown as maximum-intensity-projections. Red/green/blue fusion of aligned images is shown on the right. This demonstrates the accuracy achievable with non-rigid registration algorithms, and when rendered from multiple angles provides a quick technique to identify potential alignment errors

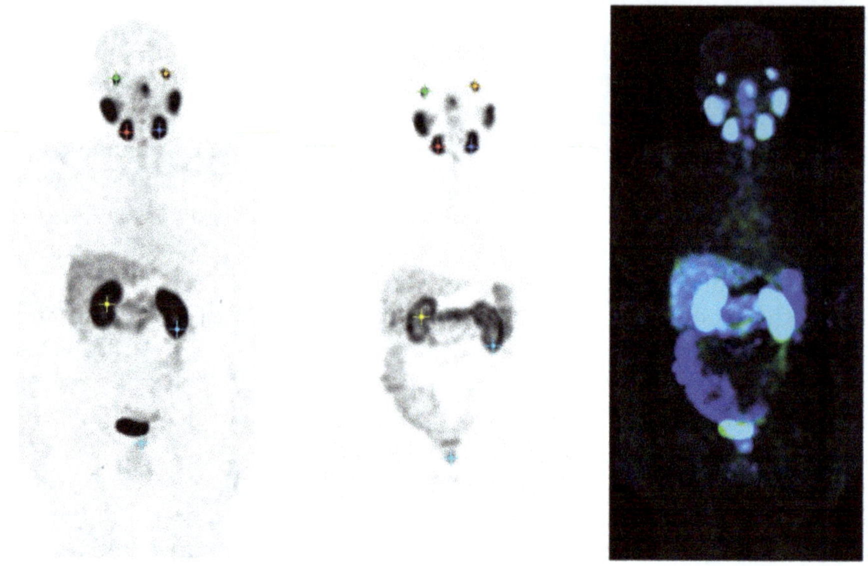

Figure 12.3. In some cases, manually annotated fiducial locations may assist with deformable co-registration. The alignment process optimises both the distance between matching points and the similarity metric used to evaluate the correspondence of intensity values in the overall image volume.

Table 12.1. Rigid and deformable registration in Elastix implemented in VRAK.

Parameter	Rigid value	Deformable value
Registration	MultiResolutionRegistration	MultiMetricMulti ResolutionRegistration
Transform	EulerTransform	BSplineTransform
Metric	AdvancedMattes MutualInformation	AdvancedNormalized Correlation, Transform BendingEnergyPenalty
Metric0RelativeWeight	n/a	0.5
Metric1RelativeWeight	n/a	0.5
NumberOfResolutions	3	3
Image Pyramid Schedule	8, 4, 1	8, 4, 2
GridSpacingSchedule	n/a	16, 8, 4
MaximumNumber OfIterations	2000, 500, 500	1500, 800, 600

necessary image voxel information with the flexibility to store values as integer or floating point format when larger intensity ranges are required. The images also store the basic spatial information required to align image data with physical geometry: voxel spacing, image origin, and the direction vector. These formats are ubiquitous to all image software based on the ITK backend. This includes Elastix, plastimatch [5], 3D Slicer [6], and a multitude of other developer-friendly medical imaging tools. In terms of modular design, it is straightforward to swap one package for another as long as the intermediate images are written in the same format. Subsequently no modification is needed for any of the downstream processing stages.

It is likely that in the near future, image alignment by convolutional neural network will become as accurate as traditional iteratively optimised methods [7]. By solving transform parameters directly, even in the current form they are extremely efficient, which presents new opportunities to implement in real-time. In any case, as long as it is possible to convert between medical image formats and the python data arrays that have become the standard for most neural network APIs [8, 9], it is a simple process to experiment with different registration methods in a total dose processing package.

12.4 Pharmacokinetics processing

Deriving an estimate of regional disintegrations, the number of decays per organ or in this case per voxel, is arguably the largest potential source of error in performing nuclear medicine dosimetry. The challenge is to derive a meaningful picture of the pharmacokinetic uptake and clearance between time of administration and the point when decays have effectively reduced to a rate of zero and in most clinical applications this must be done based on a limited sampling of static temporal information. It is not uncommon to rely on as few as three imaging time points to characterise this clearance information as each additional imaging study requires another visit from the patient, staff and camera resources, and (although small) typically an additional exposure from low-dose CT imaging. This may be further saddled with such practical restrictions as breaks in the working week with respect of the dates of isotope delivery. With these limitations in mind, it becomes particularly important to plan the timeline of imaging follow-up to best characterise the potential effective half-times in the tissues-of-interest; the combination of their biological and the isotope's physical half-lives.

The initial work in developing a pharmacokinetic algorithm sought to emulate the Exponential Modelling component of the OLINDA/EXM software package [3]. That allowed users to designate one-, two-, or three-phase exponential clearance curves. With three exponential phases, it is possible to describe periods of tracer uptake as well as fast and slow washout components which provides the flexibility to characterise the pharmacokinetics of most tissues with a single mathematical model. One of the challenges in performing voxelised dosimetry is utilising a routine that is generalisable and can be applied in the same manner for the variety of tissues in the human body. It's not practical to designate different equations in different organs or require modification of initial guesses when applying least-squares optimisers. Some voxelised curve-fitting approaches have aimed to fit multiple models (single- or multi-exponential) to each location and select the option that balances the best fit to data points with a penalty for the number of curve parameters [10]. This both aims to prevent overfitting, selecting a model with an unnecessary number of parameters, and reduces the likelihood of yielding a poorly optimised curve (local minimum result) by attempting the fit over multiple conditions. These are both important considerations in solving pharmacokinetics and while there are techniques to mitigate them, iterative optimisers may also have the risk of instability, potential for overfitting, and poor computational efficiency when applied over hundreds of thousands of voxels.

With the VRAK software package [11, 12], the aim was to develop an analytical method to replicate the behaviour of constrained iterative optimisers when applied to a limited number of data points. In practice, this has been applied for either two or three data points. The method for three points is described and it is possible to consider two points as a simplified case with only one observable clearance phase. For most locations, the technique applied is a simplified curve stripping method [13]. Starting from the last phase, the exponential line of slope may be solved to directly pass through the final two measurements. It is then possible to solve for a curve that represents the difference between this curve and the first measurement which depletes to near zero as it approaches the second measurement. This can similarly be applied to solve for a phase of uptake that rises from injection and peaks at the time of the first measurement. In a typical clinical image set, most regions of the image will display this clearance behaviour with highest activity on the first day, some intermediate level on the second, and again lower activity on delayed imaging. This method presents a straightforward function to mathematically interpret these changes and solve for the time integral which is the ultimate aim when estimating radiation exposure in the form of absorbed dose.

It is necessary, however, to accommodate instances where the uptake in a voxel does not display this typical behaviour. In some cases this may have a physiological basis, but—particularly in the voxel space—the presence of regions that display irregular patterns of uptake kinetics may also be related to either image alignment issues or statistical noise. Unfortunately, even with considered efforts to mitigate these errors, it is necessary to adapt to these conditions in a predictable fashion. In the VRAK algorithm, a quick evaluation of the data occurs at the outset and if necessary the data points are adjusted in a manner that replicates the behaviour of least-squares solvers by fitting a curve that approximately passes in between any outlier points. The direct curve-solving process can then complete as usual in an efficient and relatively predictable manner.

$$\text{Region activity} = A_1 e^{k_1 t} + A_2 e^{k_2 t} + A_3 e^{k_3 t}$$

Three activity concentration values are described as c_{1-3} and time points as t_{1-3}. As a preliminary step, the convention is taken to decay-correct all image data, $c_{i1-3} \rightarrow c_{1-3}$ (figure 12.4). This handles concentration values in pharmacological terms during fitting and while not a mathematical necessity, the decay-correction is straightforward to reverse when taking the final integral calculation and offers practical advantage to ensure that the pharmacological concentration does not continue to increase beyond the final imaging time point. This should be a reasonable assumption given judicious selection of imaging time points [14]. Specifically, it prevents the final integral from exceeding a value that would exceed the physical half-life of the isotope. These are curves which deplete more slowly than

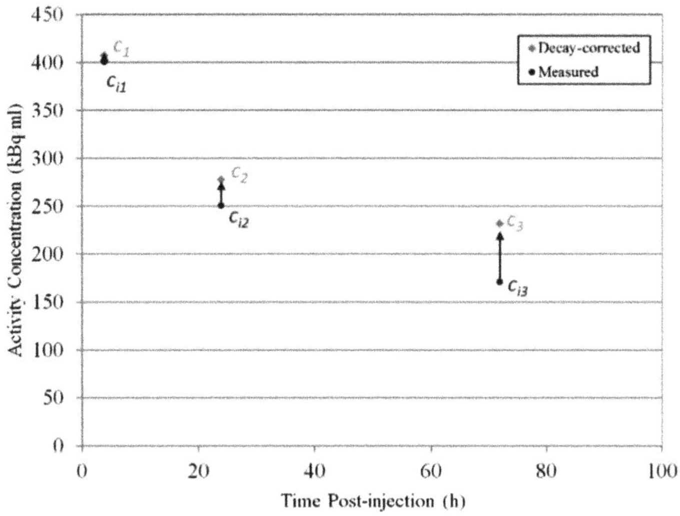

Figure 12.4. Decay-correcting initial measured data before fitting ensures algorithm does not fit increasing pharmacological concentration beyond final imaging time point. This is accounted for when calculating time-integrated activity and easily permits estimates from diagnostic/therapeutic isotope pairs such as $^{124}I/^{131}I$. Reproduced from [12] John Wiley & Sons. Copyright 2020 American Association of Physicists in Medicine.

the half-life of the radionuclide and it should be a fair assumption that the pharmacological concentration no longer increases after the final measurement; or if so, the effect is very subtle. If the highest activity measurement is detected at the final time point, fitting a slope very near zero, $k_3 \approx -\infty$, equates to only physical decay. Secondly, this provides a simple mechanism to adapt the algorithm for other isotope or diagnostic/therapeutic pairs, for example using sequential ^{124}I PET/CT imaging to predict radiation dose from ^{131}I radioactive iodine. The user only needs to provide the physical half-life of each tracer and the same algorithm may be applied to compute local time-integrated activity concentration (decays per unit volume) for the given quantity of activity administered. Different software tools can take the approach to solve the native (uncorrected) measurements with a constraint to limit the clearance parameters based on the isotope physical half-life which in principle yields a comparable result and has been shown to have limited effect when implemented on clinical data [12].

The ideal case can be described by a transient uptake phase between $t = 0$ and t_1 followed by distinct periods of rapid and delayed washout in the $t_{1\to 2}$ and $t_{2\to 3}$ time spans, respectively (figure 12.5). In clinical use of long half-life therapeutic isotopes, after initial uptake, one or two exponential clearance phases of activity from organs and tumours are widely described [15]. It is possible to fit a curve of exponential decay between two points as shown in figure 12.6 with the form:

$$A = A_0 e^{kt}$$

Solving for the linear fit, $y = mx + b$, by log transform of the activity values starting with the final phase, the kinetic parameter, k_3, is determined by:

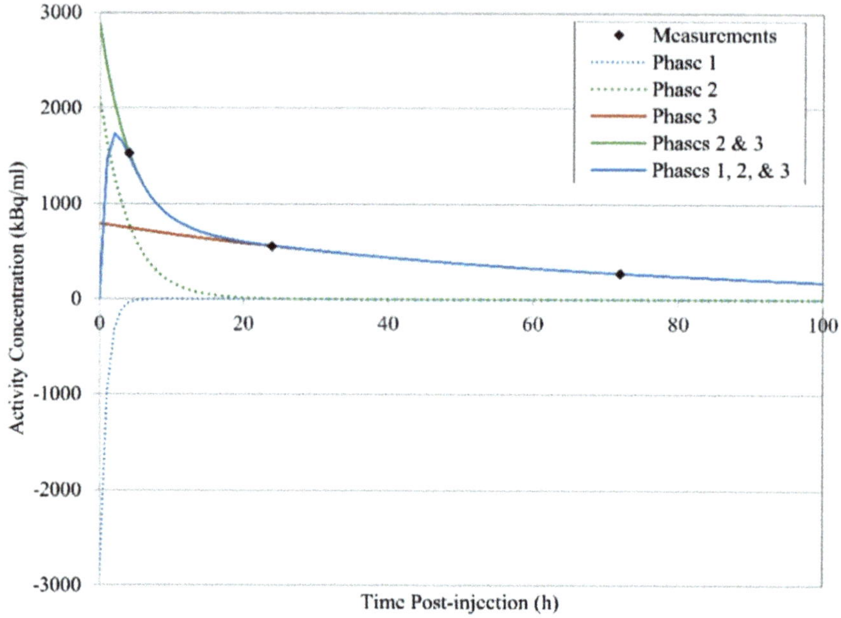

Figure 12.5. Illustration of typical time–activity curve comprised of three exponential phases that has been solved to pass through measured data points. Plots of the component exponential curves, one uptake and two clearance, are shown as dashed lines. The sum of phases 2 and 3 and all phases are shown as solid green and solid blue, respectively. Note that phase 1 and phase 2 effectively deplete as the curve nears measurement time points 1 and 2. Reproduced from [12] John Wiley & Sons. Copyright 2020 American Association of Physicists in Medicine.

$$k_3 = m = \frac{\ln(c_3) - \ln(c_2)}{t_3 - t_2}$$

And the amplitude parameter, A_3, is then:

$$A_3 = e^b = e^{\ln(c_2) - k_3 t_2}$$

As a first instance, this method provides the slope of the line representing the long-term retention phase. The preceding phase can be described as the difference between the slope described by the A_3, k_3 curve and the value at c_1 using another exponential decay equation that depletes as it approaches t_2. This very nearly builds a piecewise function from exponential terms. Solving by log transform requires non-zero values, so, for simplicity, we adjust the latter activity value, delta c_2, to be in the range of 1% of the measured value c_2 which will not appreciably affect the overall time-integral. The process may be more clearly illustrated by the mixture of curves in figure 12.7. The residual of the curve, Δc_1, is the difference between the slope of the late phase and the first measured activity value, c_1:

$$\Delta c_1 = c_1 - A_3 e^{k_3 t_1}$$

The sign of the delta value is evaluated: $\Delta c_2 = 0.01 \times c_2$ if Δc_1 is greater than zero. If not, the sign is reversed to $\Delta c_2 = -0.01 \times c_2$. That is, both delta values

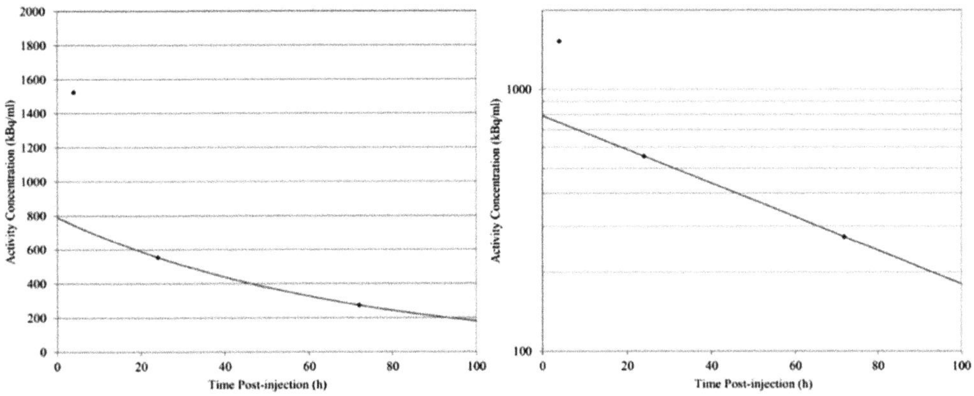

Figure 12.6. Standard (left) and logarithmic (right) plots of method to solve late-phase clearance based on exponential slope between measurement time points 2 and 3. Reproduced from [12] John Wiley & Sons. Copyright 2020 American Association of Physicists in Medicine.

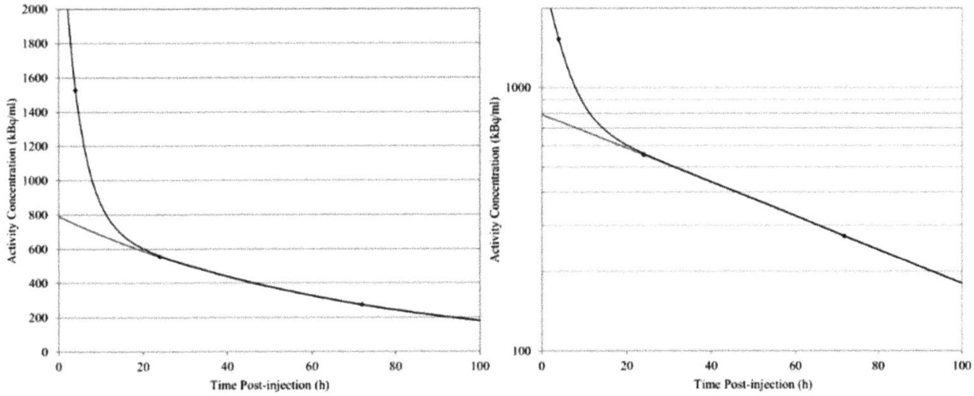

Figure 12.7. Method to solve phase 2, first clearance phase, based on the difference in the curve described by figure 12.2 and the measured activity at time point 1. Standard (left) and logarithmic (right) plots are shown to illustrate piecewise nature of the individual curve phases. Reproduced from [12] John Wiley & Sons. Copyright 2020 American Association of Physicists in Medicine.

should be positive or negative. There are conditions when a negative A_2 provides the most appropriate fit based on the measured data. Curve approximation then follows the previous method of solving for exponential slope to generate a phase that depletes as it approaches c_2:

$$k_2 = \frac{\ln(\Delta c_2) - \ln(\Delta c_1)}{t_2 - t_1}$$

$$A_2 = e^{\ln(\Delta c_1) - k_2 t_1}$$

Estimating the very early uptake kinetics requires an approximation when there are no intermediate points before the peak activity measurement. This is typical when only one image is available during the first day of administration. One approach is to infer the uptake pharmacokinetics based on a generic rate constant, k_1. For long half-life therapeutic isotopes, the majority of the time-integrated activity calculation is dictated by the late-phase retention. That is, if the physical half-life is several days or longer, only select tissues with significant initial uptake followed by rapid and sustained clearance could be appreciably impacted by the use of a generic uptake parameter. The notable exception is choosing a very small k_1 parameter, representing very rapid uptake, when employing the VRAK curve stripping technique. In this case, the presented algorithm may yield a curve that escalates dramatically at times close to $t = 0$ due to the nature of solving the line-of-fit for the fast clearance phase. With ^{177}Lu, a half-time of approximately 30 min, $k_1 = -1.3$, has been shown to offer reasonable agreement with other curve-fitting integrals over a range of different tissue types [3, 11]. An analysis of the influence of the generic rate parameter on time-integrated activity shows that over a range of half-times from 20 to 90 min, the effect is in the order of ±2% except in bladder and heart (blood pool) [12]. Finally, the amplitude, A_1, of the uptake phase is set such that concentration value passes through zero at $t = 0$. Here the term A_1 is simply the negative of the sum of A_2 and A_3.

In this manner, it is possible to analytically solve a three-phase exponential that very closely passes through three measured time points. This piecewise method, or a slightly modified version, can be applied wherever declining slope is detected between values of c_2 and c_3; that is, whenever some clearance is detected in the late phase of imaging as would be typical of most tracers and tissue types. The sequence of computations is depicted in the flowchart below with this common case designated by bold labels (figure 12.8).

With the solved time–activity curve, it is then possible to integrate the decay-corrected curve from $t_{0\to\text{inf}}$ including the physical half-life of the isotope in the form:

$$\tilde{A} = \frac{A_1}{\left(\frac{\ln 2}{t_{1/2}} + k_1\right)} + \frac{A_2}{\left(\frac{\ln 2}{t_{1/2}} + k_2\right)} + \frac{A_3}{\left(\frac{\ln 2}{t_{1/2}} + k_3\right)}$$

where $\tilde{A}$ is the time-integrated activity or total number of disintegrations in the region described by the time–activity curve, $t_{1/2}$ is the physical half-life of the therapeutic isotope, and the parameters A_{1-3} and k_{1-3} are the fitting values described previously. In most dosimetric applications, the final aim is to derive this time integral to approximate radiation absorbed dose.

As a side note, in some instances we may seek to determine a rudimentary estimate of absorbed dose two or even a single activity measurement. This may help to stratify patients into high- or low-dose categories and advise on suitability for further treatment cycles when considered alongside other areas of clinical presentation. If we are confident that tracer clearance follows a predictable trend, albeit with an unknown biological half-time, it is possible to determine a reasonable estimate of the time integral based on a single measurement at a carefully selected time. For example, if a single exponential clearance of Lu-177 is imaged at 48 h and we start estimating the curves across a range of biological half-times, many will yield very

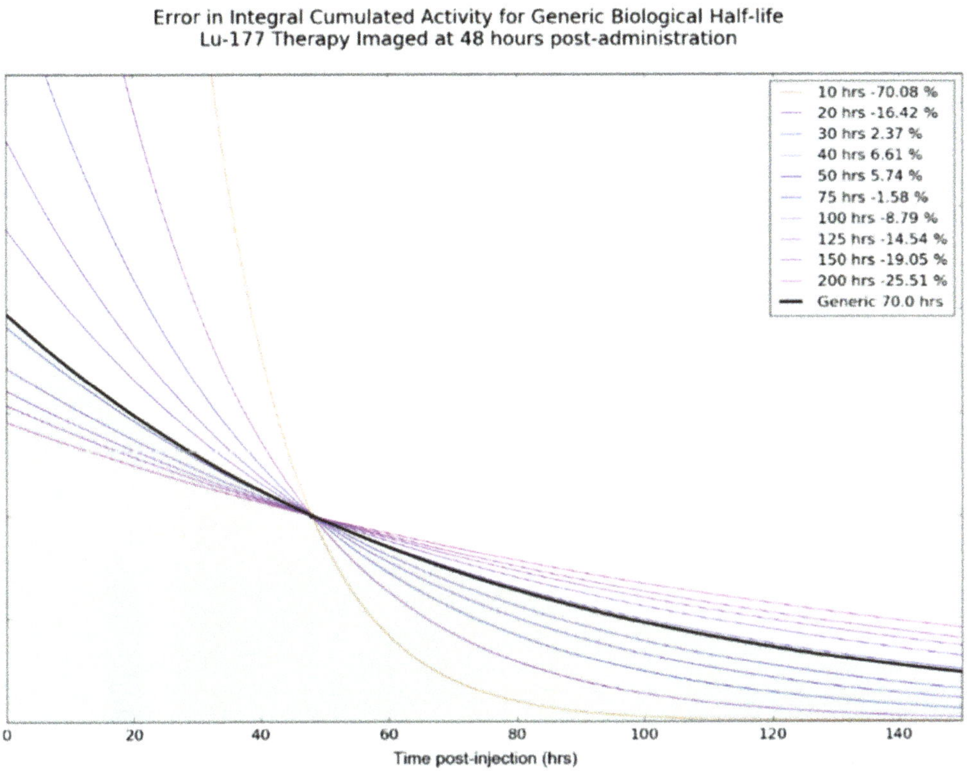

Figure 12.8. Illustration of potential pharmacokinetic curves based on a single measurement at 48 h for tissue with exponential biological clearance. The integral area defined by a generic biological half-time of 70 h is shown in green, a value that is within 10% of curves with biological half-lives ranging from 30 to 100 h.

similar time-integrals regardless of how accurately the clearance rate is predicted. That is, if the rate is overestimated, the area under the curve that precedes the measurement increases by approximately the same degree as the area that follows decreases [16]. By selecting this inflection point carefully and having some familiarity of the variability in observable pharmacokinetic ranges it is possible to minimise the potential error in estimated absorbed dose from multiple time points or even achieve a modest approximation from a single measurement altogether. In the example below, a generic biological half-time of 70 h imaged at 48 h yields an area under the curve within roughly 15% of the value for true half-times ranging from 20–125 h; comparable to the range observed in tumour for a variety of therapeutic radiopharmaceuticals (figure 12.9).

12.4.1 Handling of irregular data

When performing operations across many thousands of voxels that may be prone to noise or image alignment errors, there are likely to be challenging locations that warrant special consideration to yield curves and integrals which represent fair approximations of the underlying data. In instances where the highest activity is

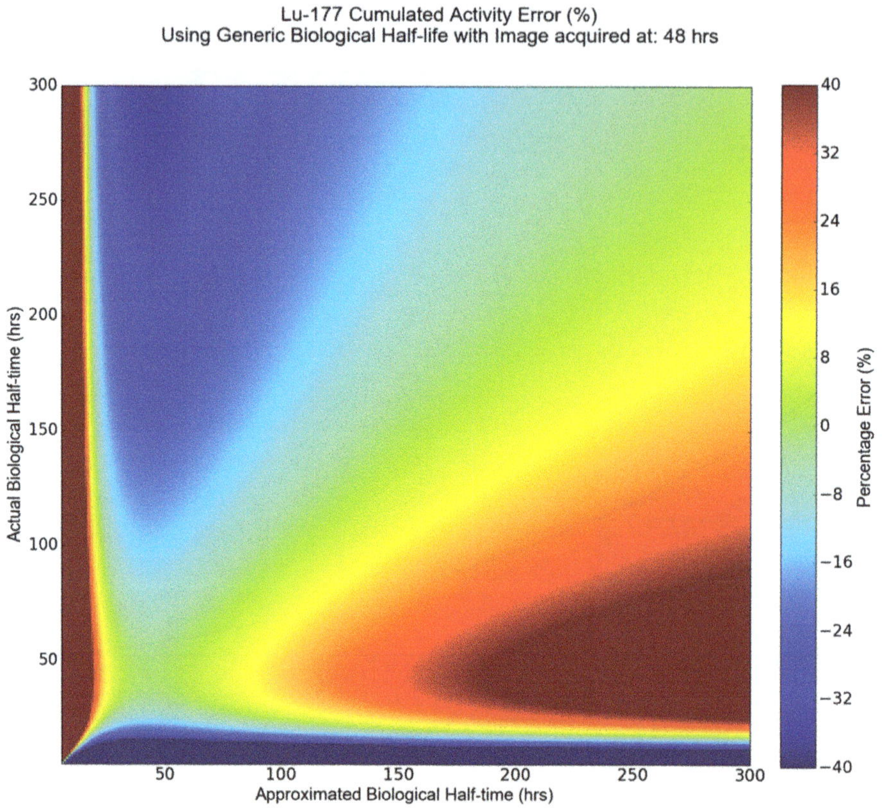

Figure 12.9. Heatmap illustrating error as a percentage when estimating area under the curve for models with a range of biological half-lives (y-axis) when using an approximated generic clearance parameter (x-axis). In this case shown for a measurement at 48 h with a ^{177}Lu radiopharmaceutical. A generic half-time in the order of 50 h is representative for a wide range of physiologically relevant clearance rates. These methods should be applied only when the variability in pharmacokinetics is well appreciated but may assist when serial measurements are not feasible.

detected at t_3, a reasonable convention is to assume that only physical decay occurs at time points beyond [17]; or for simplicity that there is very slight pharmacological clearance such that the effective half-time is determined by the physical half-life of the isotope, as illustrated in figure 12.10. In the proposed method, setting k_3 to a value near zero yields an equivalent computation when applied to decay-corrected data. The coefficient A_3 may be solved based on the negligible biological half-life and the delta from time of administration. There may be no need to fit multiple uptake phases in this instance ($A_2 = 0$). Instead, the biological half-time for a single uptake phase is solved such that it reaches a maximum as it nears c_3. There is scope to moderate the uptake kinetics based on the measured data points c_1 and c_2, as illustrated in the flowchart shown by figure 12.14. Nevertheless, the area under the curve for long half-life isotopes will be primarily dictated by the amplitude of uptake at measurement c_3.

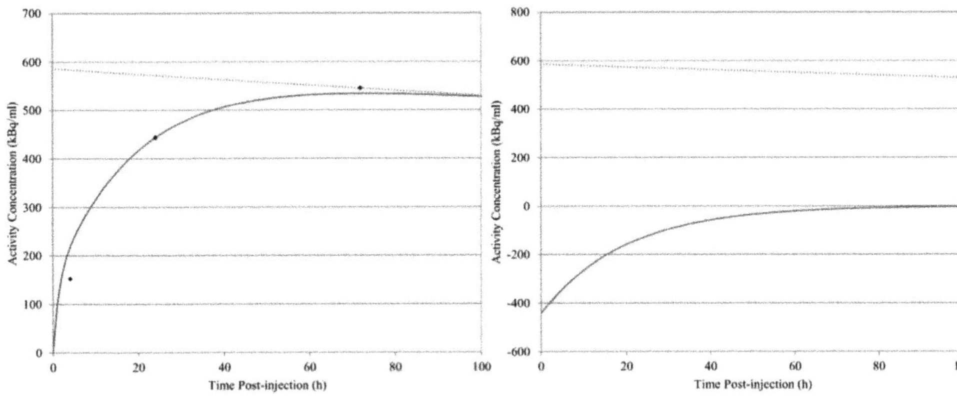

Figure 12.10. Time–activity curve approximated for data with greatest uptake measured at the final time point c_3. Physical decay is assumed beyond and a single uptake phase may be used to approximate the kinetics between $t = 0$ and c_3. The dotted line represents the late-phase retention beyond c_3. In the right plot, separate phases are plotted individually; uptake phase shown as a dashed line. The resulting time–activity curve is shown as a solid line against measured values on the left. Reproduced from [12] John Wiley & Sons. Copyright 2020 American Association of Physicists in Medicine.

The most apparent case where the data requires correction from the outset is when the trajectory of uptake or clearance changes multiple times across the measurement points. This may occur when a region initially displays a large activity concentration that appears to decrease at the time of the second image and subsequently increases again at the third. In image-based dosimetry, these instances are often artefactual; produced due to noise or small misalignments near the margin of high-uptake tissues such as kidneys and spleen. Here, the aim is to mitigate potential errors and yield a curve that approximates the behaviour of a least-squares fitting routine. In this methodology, the convention is taken that the final time point is most important to determine the time integral. The values of the first two points are moderated such that it is possible to apply a curve-solving methodology which balances error with respect to the early phase(s), as shown in figure 12.11.

Solving exponential slope between values that transition to very near zero can be problematic. This tends to occur for voxels in low-uptake regions where small changes in absolute activity concentration represents very large relative differences if one of the values is near zero. Although this type of measurement is largely influenced by scanner sensitivity in background regions and reconstructed noise, using those rate kinetics parameters leaves the potential to infer extremely high uptake in the very early time window and, as such, overestimate time-integrated activity. As a practical fix, the relative uptake can be initially compared across time points and adjusted if necessary—in this case c_3 is modified to 10% of the value at c_2 as shown in the preconditioning flow-chart, figure 12.13—to yield a more sensible curve and estimate of integral decays. Again, this correction is unlikely to be applied to consequential regions of an image.

Additionally, computation of two threshold c_2 values is used to define the limits for the slope between the values of c_1 and c_3. The first, $c_{2,\text{plateau}}$, is used to test

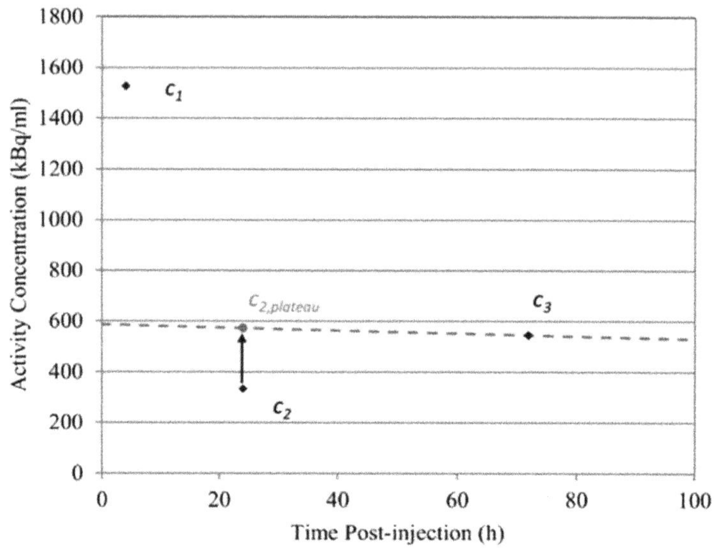

Figure 12.11. Example of serial activity measurements requiring conditional adjustment to solve analytical time–activity curve. This may occur due to image alignment issues or noise and is unlikely to represent true physiology. Here the activity declines at c_2 and increases again at c_3. This may be detected by conditional statement and the measurement c_2 is adjusted to match the slope for very near physical clearance between time points c_2 and c_3. The piecewise solving process may then proceed as normal. Reproduced from [12] John Wiley & Sons. Copyright 2020 American Association of Physicists in Medicine.

whether activity increases between time points 2 and 3. If the activity at c_1 is also very low, as may occur in some tissues that slowly accumulate tracer such as bowel in ^{177}Lu-PMSA therapy, no correction would be warranted. In other cases where activity initially declines from c_1 to c_2 only to increase again at the final measurement such as shown in figure 12.11, it is logical to assume that some element of the measurement is artefactual. Shifting the value of c_2 up to the plateau point where the exponential fit between c_2 and c_3 achieves minimal biological clearance; or effectively only physical decay from t_2 and passing through the measurement c_3. With this parameter solved, it is then possible to apply the piecewise methodology to solve the remaining phases that may appropriately pass through c_1.

When the concentration at c_2 appears as less than a linear interpolation between time points c_1 and c_3, a solution is to compute a value that falls on that line, $c_{2,\text{linear}}$, that may replace the original value if increasing slope is required to approximate the curve up to the final imaging time. The resulting curve is illustrated in figure 12.2 and the implementation into the algorithmic workflow can be traced in figure 12.14 and figure 12.15.

In practice these error handling methods are applied to only a small fraction of the total image space. They allow the pharmacokinetic algorithm to behave predictably, avoid large errors from dramatically over- or under-predicting certain regions, and yield output dose volumes that transition smoothly as the behavior changes from one path to another in the programmatic flow char.

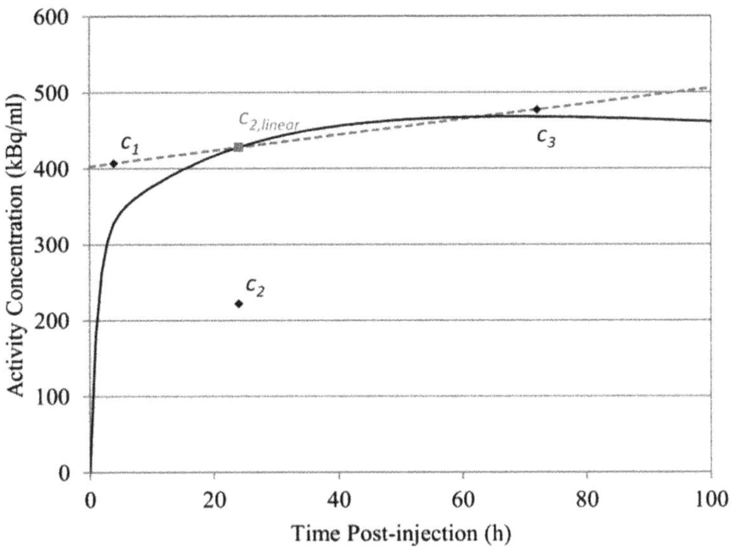

Figure 12.12. A second pattern of measurement that may be detected by conditional statement. By first shifting the measured activity at c_2 to the linear interpolation between c_1 and c_3, solving with the piecewise algorithm approximates the behaviour of a least-squares optimiser, yielding a curve that is a compromise of early measured data. Reproduced from [12] John Wiley & Sons. Copyright 2020 American Association of Physicists in Medicine.

12.4.2 Selection of generic uptake rate parameter (k_1)

For the common condition where fast- and slow-clearance components are observed it is sensible to implement a generic uptake rate in order to prevent unrealistically large activity extrapolations in the times between $t = 0$ and the first measurement. By setting the amplitude parameter A_1 equal to the negative sum of the A_2 and A_3 coefficients, the curve will necessarily pass through the origin at the time of administration and a generic k_1 coefficient should be applied which allows uptake to complete prior to the first measurement (typically in the first several hours) but mitigates the potential for large area under the curve in the initial moments post-administration. In this instance, an empirically chosen k_1 value of -1.3 (approximately 30 min half-time) appears to provide a reasonable approximation for a variety of tissue types. The results of experiments with other rate parameters for half-times of 20, 60, and 90 min yield time-integrated activity values within 2% in all instances, except in tissues with a particularly large component of early-phase clearance such as urinary bladder and heart (blood pool).

12.4.3 Activity and decays in units of concentration

With respect to the units of activity concentration and integrated decays, the MIRD committee outlines a standardised set of metrics and terminology which assist with many of the dosimetry-specific quantities that may be used in nuclear medicine [18]. The task group provides the best standard for reference in communications and general publication, however, there are some applications where it is sensible to

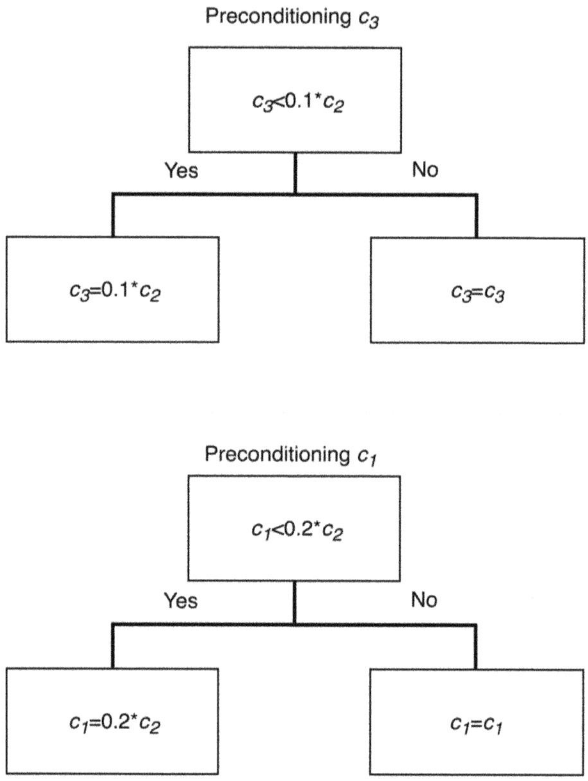

Figure 12.13. Initial preconditioning of values with very low measurements relative to the second time point. These tend to occur in low-uptake regions or where there are errors in image alignment and this may prevent calculation of very large exponential decay slopes. Reproduced from [12] John Wiley & Sons. Copyright 2020 American Association of Physicists in Medicine.

derive units which simplify the overall calculation. For PET images and pseudo-PET format as common to many quantitative SPECT file types, image activity is reported in terms of activity concentration, Bq ml^{-1}. This becomes a convenient quantity for radiation dosimetry. If we ignore the effects of cross dose between neighbouring voxels (a relatively insignificant effect for beta-emitters where imaging cannot resolve steep activity gradients) quantities in the form of Bq ml^{-1} will be approximately proportional to the local dose rate in terms of Gy per unit time. Conveniently, activity measures in the form of concentration can be readily resampled to match dose kernel or Monte Carlo grid dimensions without needing to correct for the change in voxel volume. Similarly, if the voxel dose factor accepts units in terms of time integrated decay concentration (the time integral of activity in Bq ml^{-1}, for example of the form Gy*ml/bq*h) it is possible to simply resample from a very high resolution voxel dose kernel to arbitrarily lower resolution kernels without affecting the integral dose conversion (though it is generally advisable to perform Monte Carlo for each dose kernel geometry or at minimum confirm the sum of dose for all voxels in the resampled kernel is equal to its original value).

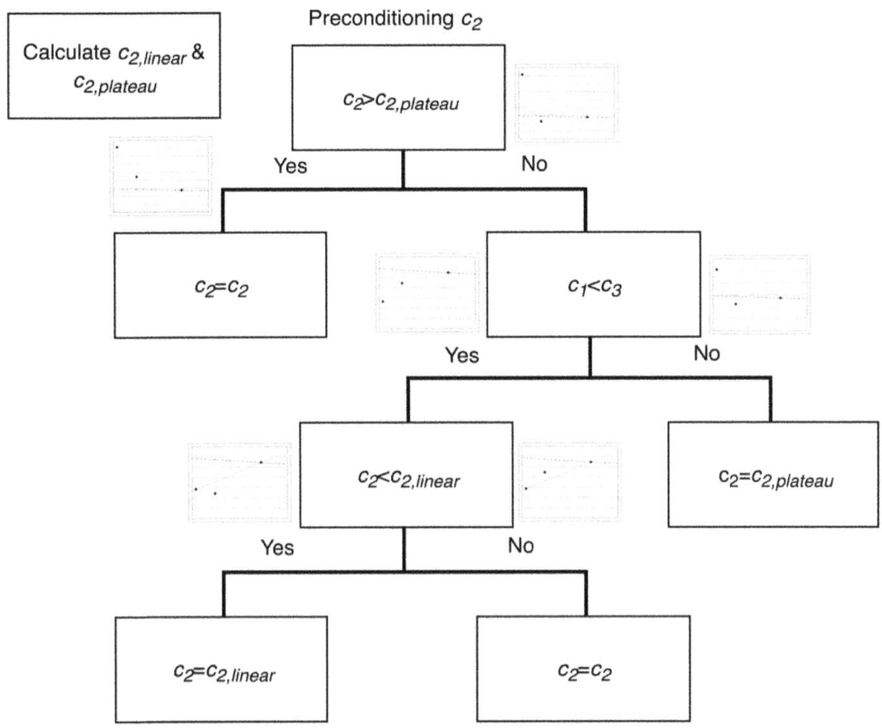

Figure 12.14. Calculation of comparison values $c_{2,\text{plateau}}$ and c_2, linear which may be employed to handle unrealistic cases (eg. high-low-high uptake). Reproduced from [12] John Wiley & Sons. Copyright 2020 American Association of Physicists in Medicine.

When taking the convention to decay-correct time–activity curves defined in units of Bq ml^{-1}, the values lose their association with a true physical quantity (decays per second per unit volume), however, they are still correctly proportional to the chemical concentration of the isotope species. While it may be sensible to compute the molecular concentration discretely when deriving dose estimates for different isotopes—for example determining the micrograms per millilitre of iodine when predicting the dose for ^{131}I therapy based on ^{124}I PET imaging—it is much simpler to denote that the activity concentration (Bq ml^{-1}) is decay-corrected and ensure that the appropriate isotope physical decay factors are applied to the resultant time activity curve which will yield an equivalent result.

Voxel pharmacokinetics estimation is likely the greatest source of uncertainty in the dose processing chain. It is also an area of biology that is frequently most unfamiliar to medical physicists and as such may stand to benefit from continued research efforts. With the VRAK processing algorithm, the voxelised results have been shown to be comparable to a number of previously developed methods including curves generated in OLINDA at the organ level and using another iterative solving method at the voxel level when compared in the same cohorts of patients [3, 11]. Importantly, we have also shown that the mean time integrals achieved from many curves computed at the voxel level are in close accordance with

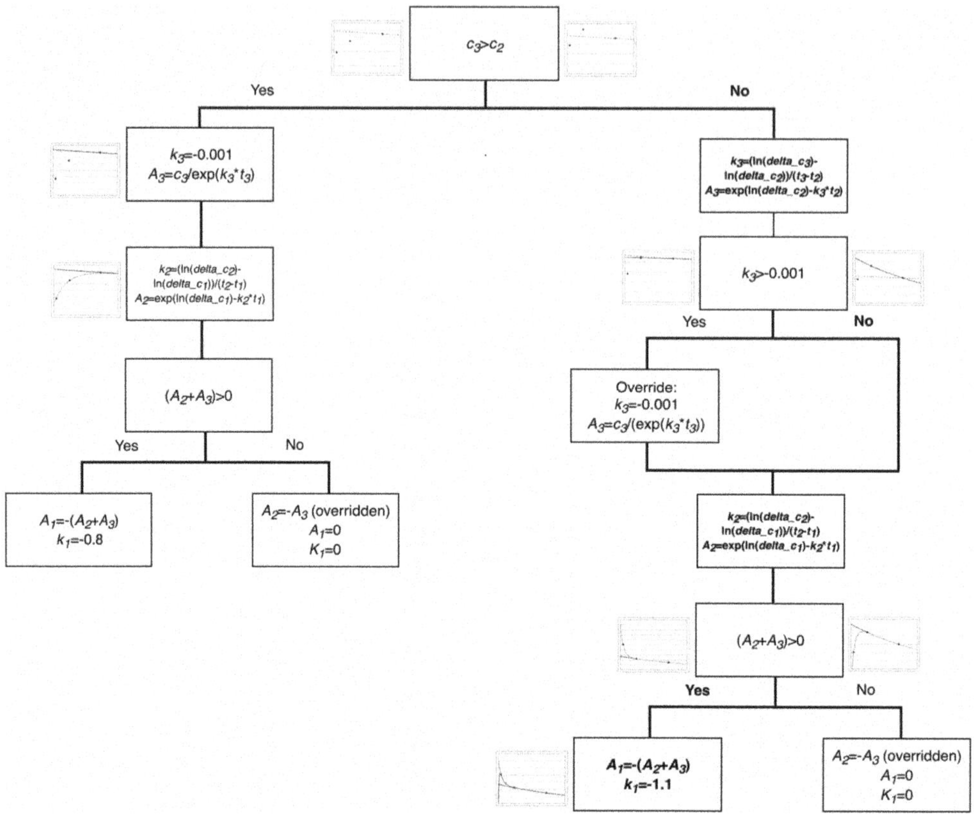

Figure 12.15. Primary flow-chart for solving time–activity curve as a two- or three-phase exponential decay curve. Note the common case with one uptake phase and two separate clearance components is denoted in bold. Reproduced from [12] John Wiley & Sons. Copyright 2020 American Association of Physicists in Medicine.

a single curve generated based on the mean activity for an organ over multiple time points. Figure 12.16 analyses the pharmacokinetics for tumour, bladder, kidney, and spleen in a ^{177}Lu-Octreotate therapy. One black curve is drawn for each voxel in the volume-of-interest with the mean of all voxel curves computed in blue. A single curve determined from the mean activity values is shown in red, closely aligning in terms of overall shape and integral area [19]. A comparison of time-integrated activity across a group of clinically relevant organs for a group of ^{177}Lu-Octreotate patients is given in table 12.2.

12.5 Voxel radiation transport estimation

The method outlined by the Committee on Medical Internal Radiation Dose (MIRD) initially considered physiological dosimetry based on the mean radiation absorbed dose to a selection of relevant tissues in the body [20]. The process for estimating radiation dose first involved solving for time integrated activity in each potential source organ and applying a set of dose factors to convert local

Table 12.2. Time integrated activity analysis applying the pharmacokinetic solving algorithm at the voxel and whole-organ level for a cohort of ^{177}Lu-Octreotate cases [19].

	Mean VOI activity concentration			Mean of voxel curves			Difference	
	MBq*h ml^{-1}		SD	MBq*h ml^{-1}		SD	Mean	St Dev
Marrow	2.92	±	1.57	3.07	±	1.65	5.7%	± 8.4%
Muscle	1.04	±	0.32	1.21	±	0.36	17.2%	± 4.6%
Lung	0.72	±	0.48	0.80	±	0.47	12.5%	± 5.4%
Heart	1.28	±	0.79	1.36	±	0.83	7.3%	± 7.1%
Stomach	11.83	±	8.40	12.29	±	9.22	4.2%	± 9.4%
Small bowel	3.21	±	1.57	3.59	±	1.41	16.8%	± 15.2%
Liver	25.49	±	30.44	26.45	±	30.25	6.6%	± 5.6%
Pancreas	29.47	±	43.57	29.04	±	42.79	0.1%	± 7.8%
Spleen	53.18	±	30.75	56.66	±	32.22	6.9%	± 3.3%
Kidney	33.14	±	11.50	35.67	±	11.76	8.2%	± 4.1%
Bladder	20.12	±	11.59	21.11	±	11.80	5.5%	± 7.5%
Tumour	324.53	±	136.17	317.5	±	133.3	−1.7%	± 7.2%

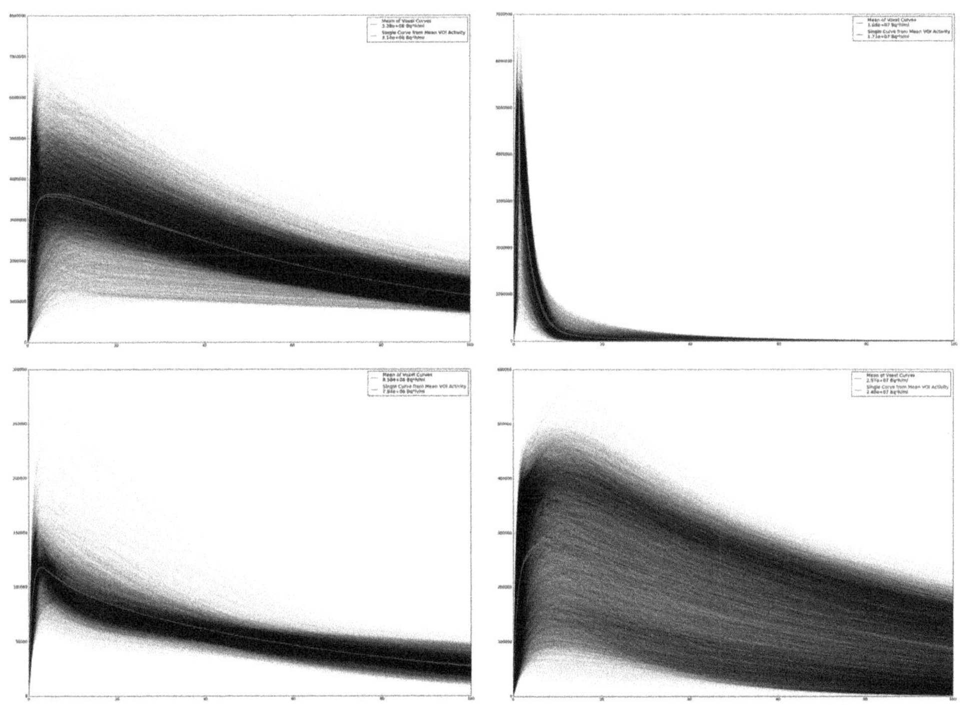

Figure 12.16. Voxelwise time–activity curves for tumour, bladder, kidney, and spleen (left to right, top and bottom rows). Individual voxels curves are denoted by black lines. Mean of all voxel curves computed in blue representing regional pharmacokinetics. Result is compared against single curve based on mean activity at each time point shown in red.

disintegrations into a quantity of self- and cross-dose to neighbouring tissues. Within each of these source regions, a uniform distribution of activity was assumed and simple geometric models were used to approximate the human anatomy. Monte Carlo was used to simulate that deposition of local radiation dose and transport into neighbouring tissues based on these static spatial arrangements. Improvements on the original MIRD methodology included corrections for the physical size of organs which would allow estimates to be more patient-specific, however, the dose factors in terms of the fraction of self- and cross-dose deposition would remain the same. Essentially, this always assumed the same physical distances between source and target tissues.

The assumption of uniform source organ activity and spatial configuration with respect to neighbouring tissues is a relatively subtle sources of error if the ultimate aim is to estimate organ doses as a single macroscopic quantity. For example, activity distributed very near a volume's surface will deposit a larger fraction of overall cross-dose, but the net effect (particularly with beta-emitting isotopes) is minor when considering the other sources of uncertainty in the dose estimation process. One notable omission when using standardised anthropomorphic phantoms is the ability to define tumours as a source region. They may be situated near radiosensitive tissues which warrants some consideration in terms of overall risk. Malignant tissues may also be irregularly shaped which can make it complicated to apply generic dose factors based on spherical objects [3].

Computing radiation transport in the voxel space addresses the shortcomings. Elsewhere in this chapter, a detailed discussion of some of the spatial considerations for deriving activity maps from clinical images is provided, however with that in mind computing radiation dose directly in the image space essentially follows the MIRD whole-organ framework only at a much more granular level. The image space is assumed to be constructed from many small, rectangularly-shaped regions rather than a collection of stylised geometric organs. In the functional image (SPECT or PET), the internal volume of each nearly cubic unit is assumed to be filled uniformly with the reported activity concentration (Bq ml^{-1}). That may be used to define a probability distribution for decays to originate from within its boundaries. Again, this follows the same standard MIRD source convention. Similarly, the structural CT component defines a physical density that uniformly fills the contents of each voxel in that image array. Look-up tables may be used to associate certain Hounsfield ranges with elemental compositions for human tissues in a patient-specific manner (figure 12.17) [21]. From there, using brute-force statistical methods via Monte Carlo simulation will determine the approximate self- and cross-dose from all of the potential source/target combinations as a single operation. This provides the basis to extend the original Monte Carlo simulations on anthropomorphic phantoms into a patient-specific form and with the ability to appreciate patterns of heterogeneous exposure at the sub-organ level.

To perform detailed voxel dose transport, the authors can recommend the open-source Monte Carlo package GATE [22], a macro-based extension of Geant4 [23]. EGS [24], Penelope [25], and MCNP [26] all offer comparable functionality and from previous head-to-head comparisons the differences in terms of short-range dose

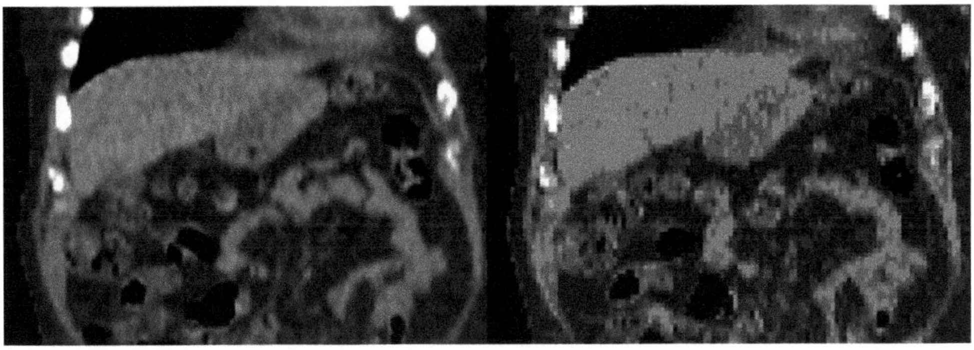

Figure 12.17. Original CT image (left) and density map used in Monte Carlo simulation with a binning threshold of 0.1 g cm^{-3}.

accuracy are very subtle [27]. Geant4 includes built-in radioactive decay tables which can generate the decay products of any clinically relevant isotope along with appropriate spectral information. It includes methods to construct regular voxelised geometry for the source distribution, physical phantom, and dose scoring array. As a practical benefit, it can interact with some of the standard ITK image types (or a very close variant) described in the registration section above making it relatively straightforward to link in with other stages of the dose processing framework. The main limitations of GATE are a reliance on linux-based operating systems for installation and the computational overhead to perform the simulations which prohibits processing cases in real-time. There is also the limitation the apparent distribution of tracer activity is of a much lower resolution—having been affected by the SPECT imaging chain—than the structural CT information. The implications in terms of accuracy are detailed in a later section, however, from a visual standpoint, it can lead to undesirable effects at the boundaries of significant density changes. This is most visible at the boundaries to air—both airways and body surface—where simulating beta-emission yields very high absorbed dose values owing to the short electron range in air (in comparison to photons) and a very low physical density. This is certainly an artefactual effect, as in practice there is no direct emission in air and the contribution from decays internal but very near the surface will be small. One simple solution is to perform radiation transport simulations as normal and clip the dose to certain density ranges as a post-processing step (see figure 12.18). Application of these methods should be considered with care and ensure that the clipped range is omitted from any subsequent dose analysis.

In GATE the overall process to define a voxel simulation macro involves the following steps: Setting a Hounsfield Material Generator which can be based on a default look-up table [21] and a reasonable density tolerance. A tolerance of 0.1 g cm^{-3} is illustrated in figure 12.17. The structural data is included as an Image Regionalized Volume which may be read in directly as a *.mhd ITK filetype with intensity values in CT Hounsfield Units. Radiation dose will be scored in a Dose Actor which can be set to match the size, resolution, and position of the input CT image. Finally, the source is defined as a 'voxel'

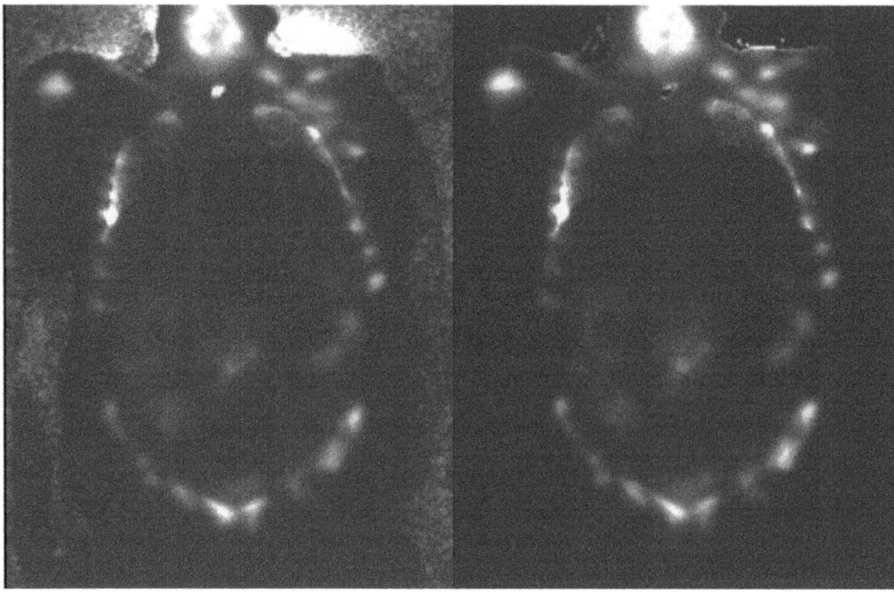

Figure 12.18. Original CT image (left) and density map used in Monte Carlo simulation with a binning threshold of 0.1 g cm^{-3}.

type. With the source defined, the activity map may be read in with the Interfile Reader method which is compatible with *.iff/*raw type images (note: this format is different from the interfile type associated with some PET raw data). It is possible to generate the raw file component by using an ITK writer to output to *.mhd/*.raw format and change elements of the mhd header format to match the iff form. Both are readable in plain text and it is a simple matter to script a conversion between the two. Note, the position of the voxel source reader is defined from the bottom corner while the CT phantom is situated based on its physical center. So a target volume defined at the origin (0, 0, 0) will be matched to a source offset by ½ of the x, y, and z extents ($-x$ extent/2, $-y$ extent/2, $-z$ extent/2). Finally, the source is defined to emit ion particles with the appropriate isotope atomic and mass numbers (Lu-177: 'ion 71 177 0 0') and an initial kinetic energy set to zero which yields only dose from the subsequent radioactive decay. In GATE the output dose is recorded as the sum per primary event, so rescaling the dose map involves multiplying by the ratio of total decays in the source image over the number of simulated primaries. The activity image is read in only to define the probability distribution. Dose may be output in *.mhd format and processed as any other ITK filetype for analysis of dicom conversion (figure 12.19).

In cases where tissues-of-interest are all of a similar density, it may be preferable to employ a dose kernel to quickly transform between a distribution of decays and an absorbed dose volume (figure 12.20). Voxelised geometry is regular array of spatial locations and therefore well suited to convolution operations. If we were to perform Monte Carlo using an activity distribution over a volume that was uniformly composed of a single material, the result would be nearly equivalent to one achieved by modelling the transport from a single source voxel to its neighbours

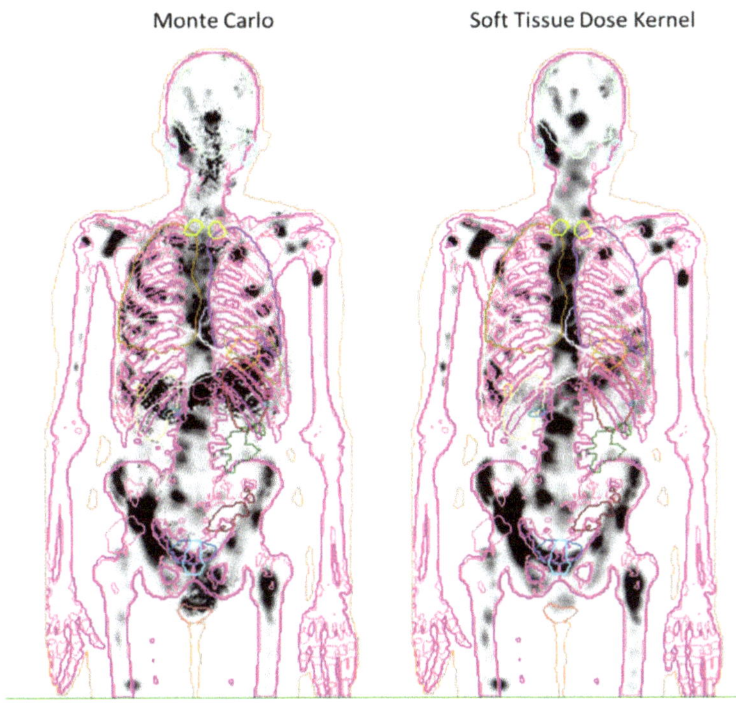

Figure 12.19. Maximum Intensity Projection of 3D dose image with radiation transport determined by Monte Carlo (left) and Voxel Dose Kernel used based on soft-tissue density. The overall appearance is similar with both methods, however dose to osseous lesions is estimated in the range of 15%–20% lower by accounting for physical density.

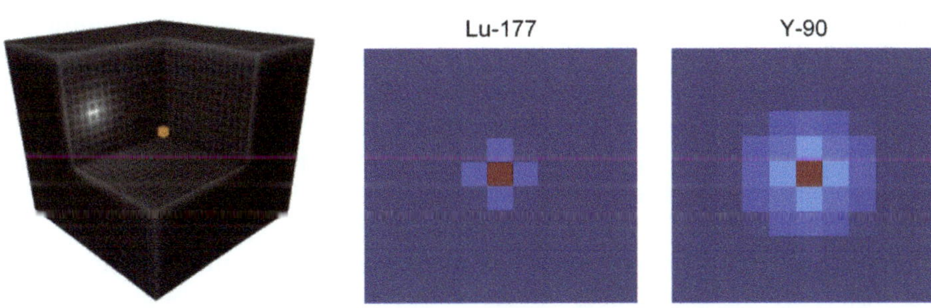

Figure 12.20. Illustration of Monte Carlo source geometry—a single source voxel in a larger scoring array—and examples illustrating the pattern of dose in a central slice of the resulting kernel for ^{177}Lu and the more energetic species ^{90}Y.

(self- and cross-dose) and using that function to approximate radiation transport for all potential source locations. This becomes an extremely efficient transform using with fast Fourier convolution and allows the translation between activity and absorbed dose in near real time. As many dosimetric applications of focus on soft tissue structures—tumour and most at-risk organs—the ability to quickly convolve dose represents a fair trade-off in comparison to patient-specific Monte Carlo.

Images are free from artefacts near the boundaries of airways and body surface and the range of error introduced with a simplified method is typically much smaller than the natural variability in dose between patients. With either method, one may confidently stratify cases into high- and low-dose categories and identify trends in terms of dose response (decreasing radiation exposure between cycles).

In practice it is beneficial to be confident with both methods as there are applications that are best-suited to the ability to precisely model tissue heterogeneity.

12.6 Discussion of limitations

Slow uptake in the bowel presents challenges both in terms of modelling the slow retention and the limitation in spatial accuracy achievable by deformable registration algorithms in the intestines. While radiation exposure to these tissues has limited correlation with onset of toxicity, it is of some consideration in PSMA therapies where delayed and significant uptake is consistently observed. Unfortunately, mapping the motion of bowel and matching locations over time involves essentially unfolding the tubular structure as a 3D surface and accounting for sliding motion in the interstitial space, far beyond the scope of the image intensity matching afforded by image registration algorithms. This does point in the direction of future development in analysis tools wherein using image recognition through neural networks or other advanced techniques will be able to label specific tissues [28, 29], improve temporal matching [7] and potentially assist with resolution recovery [2].

Resolution-related issues: apparent activity concentration moderated by the resolving power of the imaging system. In general, this will systematically reduce the observed concentration in lesions. The severity of this effect is dependent on object size, shape, and to a certain extent the local background activity. Potential solutions are to deconvolve the image volume based on an approximation of a scanner's Gaussian full-width half-maximum or another explicit function of image degradation. This may, however, introduce other unwanted artefacts such as negative values in some voxels and relies on an accurate model of detector response for a given isotope and collimator arrangement. Additionally, the interplay of activity foci within an image may yield non-isotropic effects when reconstructed by clinical imaging systems. This can be observed in phantoms where spherical objects are observed with an ovoid appearance or in clinical cases when a cluster of separate lesions which are distinguishable on diagnostic PET appear as a single smoothly accumulated mass (figure 12.21). In clinical cases effect is more nuanced than simple convolution with a point-spread function as it involves both the resolution of the gamma camera system and the nature of the OSEM reconstruction algorithm to allocate the observed count pattern from the tomographic projections into a smoothly optimised spatial matrix (figure 12.22). Siemens have demonstrated the potential to incorporate *a priori* boundary information into the reconstruction space to bias the placement of counts into osseous regions in ^{99m}Tc-MDP bone imaging based on the intensity values in the fused CT dataset. This method could be expanded by the use the margins of soft-tissue malignancies if they can be defined on previous contrast-enhanced CT or MR.

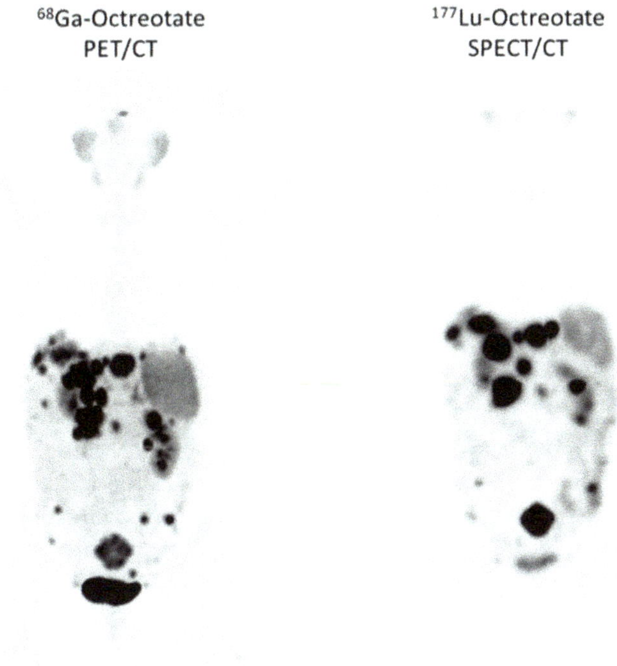

Figure 12.21. Image of the same patient prior to treatment with diagnostic PET/CT and during therapy with quantitative SPECT of ^{177}Lu. Image highlights the relative resolving power of gamma camera imaging with limitations in identifying lesion boundaries and accurately assigning activity concentration values.

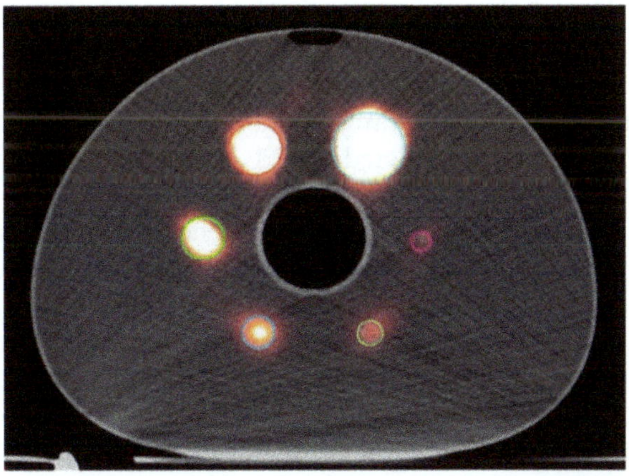

Figure 12.22. SPECT/CT image of Lu-177 in the spheres of a NEMA Image Quality phantom. All spheres are filled to the same homogeneous concentration of isotope. The ability to resolve these focal regions is highly dependent on their size.

In current practice, a rough guideline for use of voxel dosimetry is lesions with diameter greater than 2 centimetres and in organs with relatively homogeneous uptake. For volumes with smaller dimensions, it is possible to correct for partial-volume effect by taking an expanded contour near the approximate margins and subtracting a background concentration value based on an adjacent non-involved VOI. This approximates the total activity above background for a small lesion. The pharmacokinetics is calculated for the total activity in the region and a structural contour, either on CT or previous MRI, may be used to determine the physical size. If approximately round in shape, applying an appropriate OLINDA sphere dose factor determines radiation absorbed dose [3]. For intermediate sphere sizes, a power function may be used to interpolate between the values and determine a case-specific dose factor. Often a dose determined by this method will be similar to expanding a contour on the voxel dose map, subtracting the background adjacent dose and multiplying by the ratio of the expanded contour volume over the CT/MR-derived physical volume. Essentially, this reconcentrates the pattern of dose into the smaller volume while also accounting for the potential to overestimate by removing a background value. It does not fully account for the spatial effects of cross-dose escaping into neighbouring tissues, however, with beta-emitters this is a relatively subtle effect.

12.6.1 Considerations of radiobiology

Much of the dose computation in nuclear medicine is limited to computation of absorbed dose to a region, a single value representative of macroscopic effect. There are further spatial and temporal distinctions which account for the cellular repair over the duration of irradiation and the clustering of ionisation into subpopulations of cells while leaving others relatively free from the effects of radiation injury. This is encapsulated into the radiobiological metric Equivalent Uniform Dose (EUD). This was elegantly demonstrated by Konijnenberg *et al* using autoradiography of mouse renal tissue and MIRD 19 multi-region kidney model to demonstrate that the radiobiological effect from short-range beta-emission in somatostatin analogue therapy was much less than predicted based on the mean absorbed dose values used to derive toxicity thresholds from external beam radiotherapy [30–32]. The effect of millimeter-scale patterns of dose heterogeneity make it very difficult to infer dose constraints in radionuclide therapy based on EBRT population data. It may also explain some degree of variation in radiosensitivity between individuals as those with more heterogeneous uptake to malignancies will have inherently lower tumour control probability based on the same absorbed dose.

Much of the existing dose-response data is derived from experience with external beam radiation therapy. As mentioned at the outset, dose from MV-photon beams is relatively homogenous at this scale and delivered in a specific timeframe to optimize the difference in linear quadratic slope between tumour and healthy tissues; a function of radiosensitivity and cellular repair rates. In nuclear medicine, dose is delivered in a comparatively slow rate ranging in time from several days to a few weeks depending on isotope half-life. This yields dosimetric effects which are much more akin to low dose rate brachytherapy when applying equations of biological effective dose based on generic repair coefficients. These phenomena suggest that dose constraints in radionuclide therapy specifically for targeted therapies and potentially redefined for each isotope and targeting molecule which could potentially affect the pattern of dose at the tissue level.

12.7 Conclusions

Voxel dosimetry is vital to evaluate the patterns for radiation exposure in fine detail, but one of the main benefits is to enable more general use by automating much of the processing and becoming interactive for clinicians in a manner that is familiar. Many of the technical details (image co-registration, pharmacokinetics estimation, and radiation transport computation) have been addressed by numerous groups offering methods that achieve comparable ends with a high degree of accuracy. As we look ahead, a sustained effort to standardise methodologies and promote more routine clinical use—even with limited follow-up imaging—is necessary to build an understanding of dose-response models that are unique to radionuclide therapy and use them to inform future decision-making.

While many aspects of the process present finite limitations including the ability to routinely collect fine temporal sampling and the peak system resolution of scintillating gamma camera imaging systems, there is the potential to improve the overall utility and potential accuracy through intelligent resolution recovery, use of population pharmacokinetic models to labelled tissues, measurement through automated contouring afforded by emerging convolutional neural network tools.

With these potential shortcomings in mind, the authors strongly advocate for image-based dosimetry in nuclear medicine. The entirety of the process can be automated, which dramatically reduces the barrier for use in routine practice. Because volumes-of-interest are delineated on the end result, contours may be modified without reprocessing the regional pharamacokinetics or radiation transport. Voxel dosimetry can be approached as a sequence of modular steps which may be improved with time or substituted with purpose-specific algorithms to best consider image alignment, pharmacokinetics processing, and radiation dose convolution for individual applications. Looking to the future, the hope is that many of the known spatial limitations can be addressed with nuanced technical solutions in the imaging chain while the other aspects of dose processing, namely physical dose deposition and pharmacokinetics interpolation, have been addressed in a reliable manner.

References

[1] Beauregard J-M *et al* 2011 Quantitative ^{177}Lu SPECT (QSPECT) imaging using a commercially available SPECT/CT system *Cancer Imaging* **11** 56

[2] Vija A H 2013 *Introduction to xSPECT technology: evolving multi-modal SPECT to become context-based and quantitative.* Siemens Medical Solutions USA, Inc *Molecular Imaging, White Paper.*

[3] Stabin M G, Sparks R B and Crowe E 2005 OLINDA/EXM: the second-generation personal computer software for internal dose assessment in nuclear medicine *J. Nucl. Med.* **46** 1023–27

[4] Klein S *et al* 2009 Elastix: a toolbox for intensity-based medical image registration *IEEE Trans. Med. Imaging.* **29** 196–205

[5] Sharp G *et al* 2010 Plastimatch: an open source software suite for radiotherapy image processing *Proc. of the XVI'th Int. Conf. on the use of Computers in Radiotherapy (ICCR)*

[6] Pieper S, Halle M and Kikinis R 2004 *3D Slicer. in 2nd IEEE Int. Symp. on Biomedical Imaging: Nano to Macro (IEEE Cat No. 04EX821)* (IEEE)

[7] Yang X *et al* 2017 Quicksilver: fast predictive image registration–a deep learning approach *NeuroImage.* **158** 378–96

[8] Abadi M *et al* 2016 Tensorflow: a system for large-scale machine learning *12th USENIX Symp. on Operating Systems Design and Implementation (OSDI 16)*

[9] Paszke A *et al* 2019 Pytorch: An imperative style, high-performance deep learning library *Advances in Neural Information Processing Systems 32*

[10] Sarrut D *et al* 2017 Voxel-based multimodel fitting method for modeling time activity curves in SPECT images *Med. Phys.* **44** 6280–88

[11] Jackson P A *et al* 2013 An automated voxelized dosimetry tool for radionuclide therapy based on serial quantitative SPECT/CT imaging *Med. Phys.* **40** 112503

[12] Jackson P *et al* 2020 Rapid multi-exponential curve fitting algorithm for voxel-based targeted radionuclide dosimetry *Med. Phys.* **47** 4332–9

[13] Dunne A 1985 JANA: a new iterative polyexponential curve stripping program *Comput. Methods Programs Biomed.* **20** 269–75

[14] Siegel J A *et al* 1999 MIRD pamphlet techniques for quantitative radiopharmaceutical biodistribution data acquisition and analysis for use in human radiation dose estimates *J. Nucl. Med.* **40** 37S–61S

[15] Stabin M and Xu X G 2014 Basic principles in the radiation dosimetry of nuclear medicine *Semin. Nucl. Med.* **44** 162–71

[16] Rinscheid A *et al* 2020 Influence of sampling schedules on [^{177}Lu] Lu-PSMA dosimetry *EJNMMI Phys.* **7** 1–14

[17] Grassi E *et al* 2015 Quantitative comparison between the commercial software STRATOS® by Philips and a homemade software for voxel-dosimetry in radiopeptide therapy *Phys. Med.* **31** 72–9

[18] Bolch W E *et al* 2009 MIRD pamphlet a generalized schema for radiopharmaceutical dosimetry—standardization of nomenclature *J. Nucl. Med.* **50** 477–84

[19] Jackson P, Hofman M and Hicks R 2017 Does voxelised pharmacokinetics estimation for nuclear medicine dosimetry introduce a bias compared to whole organ methods? A comparison of post-treatment dosimetry in ^{177}Lu-octreotate therapy *Engineering & Physical Sciences in Medicine Conf.* (Hobart, Australia: Australasian College of Physical Scientists and Engineers in Medicine)

[20] Snyder W 1969 MIRD pamphlet estimates of absorbed fractions for monoenergetic photon sources uniformly distributed in various organs of a heterogeneous phantom *J. Nucl. Med.* **10** 1

[21] Schneider W, Bortfeld T and Schlegel W 2000 Correlation between CT numbers and tissue parameters needed for Monte Carlo simulations of clinical dose distributions *Phys. Med. Biol.* **45** 459

[22] Jan S *et al* 2004 GATE: a simulation toolkit for PET and SPECT *Phys. Med. Biol.* **49** 4543
[23] Agostinelli S *et al* 2003 GEANT4—a simulation toolkit *Nucl. Instrum. Methods Phys. Res., Sect. A.* **506** 250–303
[24] Hirayama H *et al* 2005 The EGS5 Code System (Menlo Park, CA: Stanford Linear Accelerator Center (SLAC))
[25] Salvat F, Fernández-Varea J M and Sempau J 2006 PENELOPE-2006: a code system for Monte Carlo simulation of electron and photon transport *Workshop Proc.* (Universitat de Barcelona)
[26] Waters L S 2002 MCNPX User's Manual (Los Alamos, NM: Los Alamos National Laboratory)
[27] Chauvin M *et al* 2017 Abstract ID: 155 OpenDose: a collaborative effort to produce reference dosimetric data with Monte Carlo simulation software. *Phys. Med* **42** 32–3
[28] Blanc-Durand P *et al* 2018 Automatic lesion detection and segmentation of ^{18}F-FET PET in gliomas: a full 3D U-Net convolutional neural network study *PLoS One.* **13** e0195798
[29] Jackson P *et al* 2018 Deep learning renal segmentation for fully automated radiation dose estimation in unsealed source therapy *Front. Oncol.* **8** 215
[30] Bouchet L G *et al* 2003 MIRD pamphlet absorbed fractions and radionuclide S values for six age-dependent multiregion models of the kidney. *J. Nucl. Med* **44** 1113–47
[31] Konijnenberg M *et al* 2007 Radiation dose distribution in human kidneys by octreotides in peptide receptor radionuclide therapy *J. Nucl. Med.* **48** 134–42
[32] Bentzen S M *et al* 2010 Quantitative Analyses of Normal Tissue Effects in the Clinic (QUANTEC): an introduction to the scientific issues *Int. J. Radiat. Oncol. Biol. Phys.* **76** S3–9

IOP Publishing

Monte Carlo Calculations in Nuclear Medicine (Second Edition)
Therapeutic applications
Habib Zaidi

Chapter 13

The Raydose software for image-based dosimetry in molecular radiotherapy

Emiliano Spezi, Nick Patterson and Salvatore Berenato

13.1 Introduction

Dosimetry involving unsealed sources of radioactivity in molecular radiotherapy (MRT) has long presented many challenges to dosimetry practice. Unlike external beam radiotherapy (EBRT), whereby the absorbed dose to target regions can be specified precisely, there are many other variables that impact the ability to perform dosimetry on those patients receiving MRT. Patient metabolism and disease status all play a significant role in understanding how a given administered activity of a radiopharmaceutical will present itself as a delivered dose to the target region. There has been a rapid expansion in the number of radiotherapeutics being utilised in clinics and interestingly there continues to be a lack of uniformity of approach [1] regarding how each radiotherapeutic is prescribed and how a dosimetry assessment is undertaken to determine the absorbed dose to tumour site or organs at risk (OAR). It is not uncommon for an MRT protocol to be developed during clinical trials to determine the maximum tolerated activity of a patient cohort, but this approach could result in sub-optimal treatment as the administered activity does not necessarily correlate with the absorbed dose to the target area. There is now a developing recognition of the need to move towards having clearer guidelines and methodologies for better understanding outcomes for MRT treatments [2] in the same manner as EBRT and some collaborative projects such as MetroMRT [3] and MRTDosimetry[1] [4] have already begun to make progress in this area. With this in mind, the Raydose research tool has been developed to perform dosimetry calculations using quantitative imaging and to calculate estimated absorbed doses to target areas or OAR using Monte Carlo (MC) methods. The MC engine with which Raydose has been developed is based on the Geant4 toolkit [5]. The Raydose application is capable of producing

[1] http://mrtdosimetry-empir.eu.

doi:10.1088/978-0-7503-2694-0ch13

patient-specific 3D voxel based absorbed dose maps for molecular radiotherapy treatments and will utilise all available post treatment nuclear medicine imaging scans (PET/SPECT) in order to make best efforts in accounting for the biokinetics of the radiopharmaceutical over the period of the treatment.

13.2 Raydose methodology and workflow

13.2.1 Input data and output data

DICOM [6] is the accepted standard for image communication in clinical care settings and is also widely supported by most research tools such as MATLAB[2] and Python[3]. Therefore, Raydose accepts all input data in the DICOM format and also generates all output in this format for reimportation into the users preferred analysis tool. The Grassroots DICOM library [7] along with the Insight Toolkit [8] and Visualisation Toolkit (VTK) [9] were used to both read, manipulate and write DICOM data. Great care was taken to ensure that the calculated absorbed dose map DICOM file, provided in the RTDOSE modality, was compliant with the DICOM standard and these files were verified to be compatible in research tools such as MATLAB and in clinical systems such as Velocity (Varian Medical Systems, Palo Alto, CA).

13.2.2 Patient geometry

The provision of a CT scan is a minimum requirement in the Raydose workflow as this is the image used to generate a three-dimensional volume (MC voxel detector) which aims to represent an anatomically correct representation of the patient. The MC voxel detector that is generated from the CT scans is the grid on which the absorbed dose is scored and can be created at the same size as the original CT resolution (e.g. 512 × 512) or can be downsampled to reduce the size of the scoring grid (e.g. 256 × 256 or 128 × 128). Downsampling the dose scoring grid has the advantage of increasing the speed of the simulation and limits consumption of computing memory but reduces the resolution of the generated absorbed dose maps. Of course, resampling the scoring grid is often of little consequence since the quantitative images are usually of a much lower resolution than the CT images.

This approach sets out to make a best effort regarding the accuracy of the dosimetry by ensuring the transport of radiation through an anatomically inhomogeneous and realistic phantom, both in terms of mass density and chemical composition. The Hounsfield Unit (HU) of each voxel is translated into a density by performing a multi-linear interpolation from a CT ramp provided by the user. In turn, different density ranges can be assigned to different material types. The chemical composition of the material assigned to each voxel is based on tissue substitute values outlined in [10, 11]. The current materials that are defined to simulate patient tissues are listed in table 13.1.

However, this is flexible, and a more refined list of materials can be included and mapped to the correct HU values.

---

[2] https://uk.mathworks.com/products/matlab.html.

[3] https://www.python.org.

Table 13.1. List of materials and density in g cm^{-3} used in Raydose for the dose calculation.

Material	Density (g cm^{-3})
Air	0.01225
Lung average	0.481
Adipose tissue	0.979
Breast	1.004
Water	1.043
Muscle	1.109
Liver	1.113
Trabecular bone	1.46
Dense bone	10.0

13.2.3 Quantitative activity images

Raydose utilises quantitative imaging representing the underlying activity distribution within the MC voxel phantom which is a key requirement in the production of the total absorbed dose map. Raydose will handle quantitative nuclear medicine images where the voxel units are presented as a concentration in absolute units of Bq mL^{-1}, but also if nuclear medicine data is provided where the voxel units are counts per second or simply counts per voxel, then a centre provided calibration factor can be provided to Raydose to calibrate the images in terms of an activity per voxel.

Each simulated event represents a radioactive decay which is then positioned within the patient MC voxel phantom depending on the underlying activity distribution. The decay products and all of the generated secondary particles are scored as they are transported through the patient phantom. The decay of the radionuclide is modelled using the G4RadioactiveDecay module [12] which is responsible for handling of the physics of radioactive decay and decay chains and allows for the simulation of more than 2000 radioactive nuclei by α, β^- and β^+ decay as well as electron capture. This model accounts for all emissions and spectra associated with the decay that are generated from evaluated empirical data or data-driven theoretical models, and so uses information derived from the Evaluated Nuclear Structure Data File (ENSDF) [13, 14]. The nuclear and atomic de-excitation of daughter nuclei in the decay chain is handled by the G4PhotonEvaporation class. Again, data from the ENSDF database along with theoretical models are used to derive branching ratios and energy level transition data.

After the decay of a nucleus is initiated, all charged particles and radiation from the initial decay and subsequent products are tracked as they are transported through the patient phantom. The energy deposition of each particle type is scored inside each voxel that is traversed and the absorbed dose is calculated and scored. Photon interactions such as photelectric effect, Compton scattering and Rayleigh scattering are simulated along with electron and other charged particle interactions using multiple scattering approaches which have been implemented in Geant4 using a model based on Lewis multiple scattering theory [15, 16].

13.2.4 Simulation and dose calculation

An outline of the simulation and calculation process implemented in Raydose is given in figure 13.1, however, it is best to present a specific example to illustrate how a total absorbed dose is determined. Raydose has been used in the study of single administration ^{177}Lu-DOTATOC with follow-up scans at 1, 4, 24, 44 and 72 h post-injection. These follow-up scans are to observe the pharmacokinetic distribution of the radiopharmaceutical and account for this in the dose calculation. A detailed look at the outcomes of one such study is presented in Case Study 3. The software requires DICOM files for the study to be presented in the folder format outlined in figure 13.2. A single CT image is required to generate the MC voxel phantom of the patient and any number of nuclear medicine scans can be provided to generate an understanding of the biokinetics of the radiopharmaceutical. The acquisition date and time of the scans are automatically extracted from the DICOM images in order to determine the time spacing between scans used in the generation of the time activity curves (TACs).

The first CT scan taken at $t = 1$ h post-injection, forthwith referred to as the 'reference CT' scan, is used to produce the patient-specific anatomy (Monte Carlo Voxel Phantom). Each of the successive nuclear medicine scans comes with its own associated CT scan which then must be registered, either rigidly or deformably, to the reference CT scan to account for any movement of the patient between scans. Once a registration map has been determined this can be applied to the accompanying nuclear medicine scan. This ensures that each of the five scans are in the same coordinate frame.

Each of the supplied nuclear medicine scans is utilised to generate an absorbed dose rate map $\dot{D}(t_n)$. The dose rate maps are generated from radioactive decays positioned within the MC voxel phantom according to the underlying activity distribution and the absorbed dose is scored within in each voxel. The dose rate maps are then normalised by the total number of simulated events (N_{events}).

For each voxel, a TAC $A_i(t)$, is generated using each of the provided scans. A representative image of such a TAC is shown in figure 13.3.

The TAC is used to determine the total number of decays originating from within each source voxel, i, between two successive scan times t_n and t_{n+1} and is calculated using

$$\widetilde{A}_i(t_n, t_{n+1}) = \int_{t_n}^{t_{n+1}} A_i(t)\mathrm{d}t \tag{13.1}$$

To determine the area under the function, $A_i(t)$ using the N scan points can theoretically be fitted with any function that is desired. In reality, in a standard nuclear medicine scan with resolution 256 × 256 and perhaps up to 100 slices, it would not be feasible to check the integrity of the fit for so many voxels [17]. As a default, the procedure is to compute the area between successive time points using a simple parallelogram method. The accumulated number of radioactive decays occurring within the entire patient volume (i.e. all voxels in the image) between two successive time points is expressed as

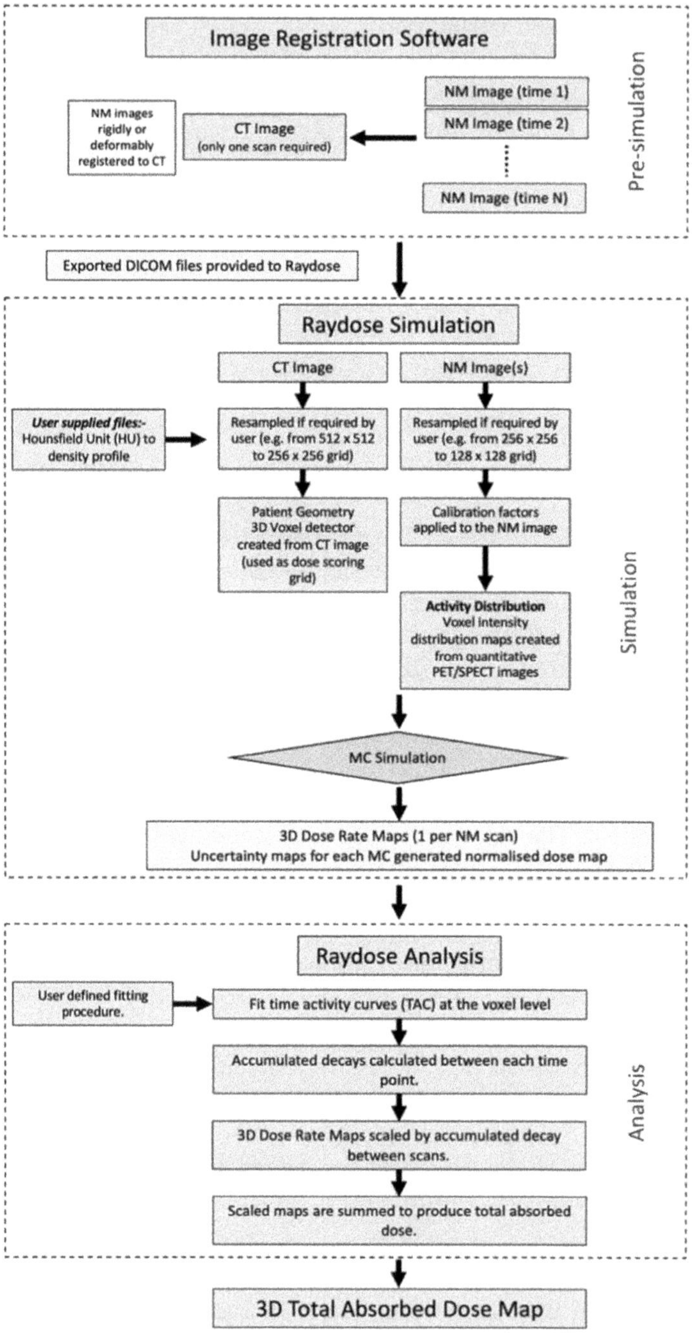

Figure 13.1. General workflow of Raydose in the production of a 3D total absorbed dose map.

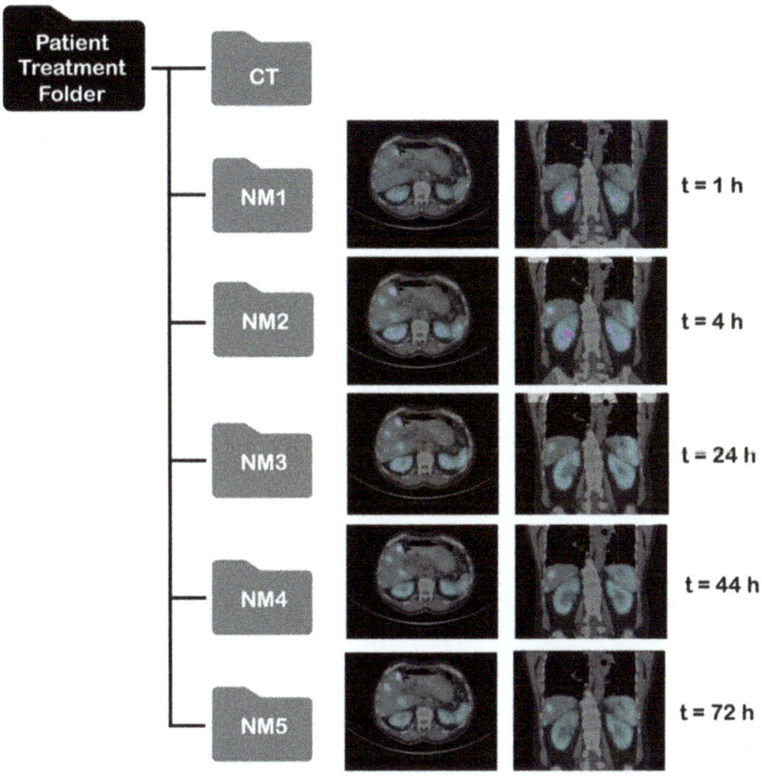

Figure 13.2. Overview of the directory structure of images that must be provided to Raydose in order to calculate the total absorbed dose (D) during a treatment regimen.

$$\widetilde{A}(t_n, t_{n+1}) = \sum_{i}^{N_{\text{voxels}}} \int_{t_n}^{t_{n+1}} A_i(t)\mathrm{d}t = \sum_{i}^{N_{\text{voxels}}} \widetilde{A}_i(t_n, t_{n+1}) \tag{13.2}$$

This quantity is required in order to scale each of the dose rate maps in order to obtain the total absorbed dose map between successive time points, such that

$$D(t_n, t_{n+1}) = \dot{D}(t_n)\bullet \widetilde{A}(t_n, t_{n+1}) \tag{13.3}$$

So, in the context of the ^{177}Lu-DOTATOC study, the total absorbed dose between the first and fifth scan would be calculated as

$$D = [\dot{D}(t_1)\bullet \widetilde{A}(t_1, t_2)] + [\dot{D}(t_2)\bullet \widetilde{A}(t_2, t_3)] + [\dot{D}(t_3)\bullet \widetilde{A}(t_3, t_4)] + [\dot{D}(t_4)\bullet \widetilde{A}(t_4, t_5)] \tag{13.4}$$

More generally, the total absorbed dose between the first and last provided nuclear medicine scan may be expressed as

$$D = \sum_{n}^{n=\text{Scans}} [\dot{D}(t_n)\bullet \widetilde{A}(t_n, t_{n+1})] \tag{13.5}$$

There is of course the question of how one should account for the dose delivered between the time the radiopharmaceutical is administered and the first

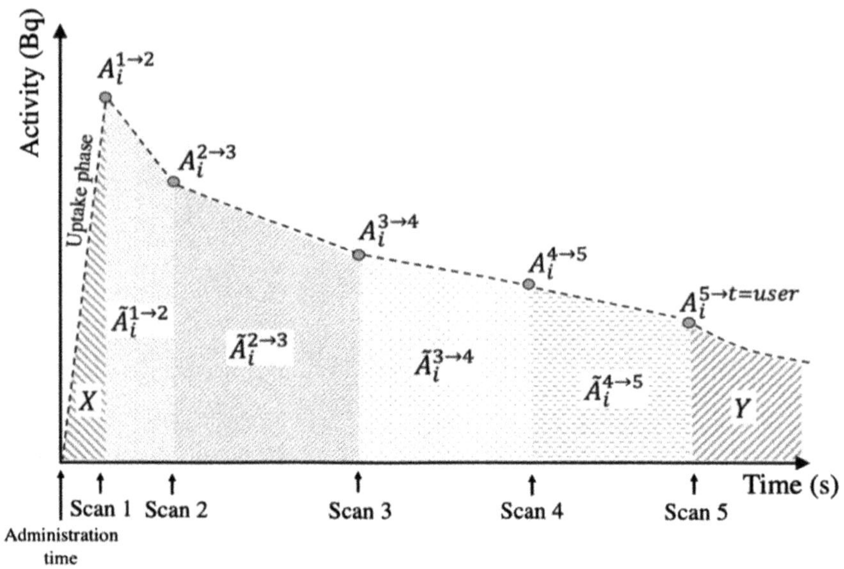

Figure 13.3. An illustrative Time Activity Curve (TAC) for a single voxel generated from multiple quantitative nuclear medicine scans. *X* and *Y* refer to accumulated decay occurring in the 'uptake' phase and the time after last scan, respectively.

scan point (see figure 13.3 area labelled 'X'). Also, the user may wish to make attempts to include the delivered dose after the last scan time (see figure 13.3 area labelled 'Y').

Raydose does include the option to estimate the delivered dose in the time between administration ($t = 0$ h) and the time of the first scan. The user is able to specify any functional form (linear, exponential or otherwise) that they wish in order to describe the uptake between injection and the time of the first scan. In doing so the delivered dose between the administration and the first scan time can be accounted for.

Furthermore, the user can account for the delivered dose after the last scan time point by specifying that the remaining activity in the last imaging scan point can be assumed to exponentially decay. The decay constant can reflect the decay constant of the radionuclide under investigation or could be user defined to reflect some other known quantity, such as biological clearance. The user can determine the absorbed dose after the last scan up to any time point they wish.

$$D = [\dot{D}(t_1)\bullet \widetilde{A}(t_0, t_1)] + [\dot{D}(t_1)\bullet \widetilde{A}(t_1, t_2)] + [\dot{D}(t_2)\bullet \widetilde{A}(t_2, t_3)] + \\ +[\dot{D}(t_3)\bullet \widetilde{A}(t_3, t_4)] + +[\dot{D}(t_4)\bullet \widetilde{A}(t_4, t_5)] + \left[\dot{D}(t_5)\bullet \int_{t_5}^{T} A(t_5)\bullet e^{-\lambda t}dt\right] \tag{13.6}$$

13.2.5 Uncertainties

The dose rate maps which are used in the calculation of the total absorbed dose inherently have an impact on the uncertainty surrounding the total dose calculated. Therefore, it is important to estimate the uncertainty of the scored dose generated in the MC simulation. A small number of simulated events will lead to large uncertainties in the scored voxel quantities, so it will always be preferable to run as many simulated events, N, as possible. There are two ways of assessing the uncertainty, the batch method or the event-by-event method. Both are methods that have been employed in the EGSnrc code and their implementation is discussed more widely elsewhere [18, 19].

The batch method involves splitting the simulated events or sample size (N) into a number of smaller batches, then the uncertainty on the average dose rate scored in a single voxel can be determined using:

$$s_{\bar{D}} = \sqrt{\frac{\sum_{i=1}^{N}\left(\dot{D}_i - \bar{\dot{D}}\right)^2}{N(N-1)}} \tag{13.7}$$

where N is the number of batches, $\dot{D}_i$ is the value of the dose rate in a specific voxel in the ith batch and D is the mean of the scored dose rate in a specific voxel over all batches. There are difficulties in using this approach, however. When splitting into batches the sample size is decreased and this can introduce large fluctuations into the uncertainty estimate by virtue of a small sampling size in each batch.

The event-by-event method does not require splitting the simulation into batches so mitigates this issue. The uncertainty on the mean scored dose rate in a given voxel can be calculated on the fly giving consideration to the full sample/simulation events, N, without having to split into batches and introducing the effect described earlier. The uncertainty in a given voxel can therefore calculated as:

$$s_{\bar{D}} = \sqrt{\frac{1}{N-1}\left(\frac{\sum_{i=1}^{N}\dot{D}_i^2}{N} - \left(\frac{\sum_{i=1}^{N}\dot{D}_i}{N}\right)^2\right)} \tag{13.8}$$

Raydose calculates this quantity for each dose rate map simulation. This helps to assess if the simulation has been run for long enough so that the uncertainty of the scored dose in regions of interest are sufficiently low. The uncertainty maps can be exported to DICOM format for assessment after the simulation has been performed.

13.2.6 Efficient sampling

It is imperative to ensure that efficient algorithms are used to ensure optimal speed performance when sampling decay events from the underlying distribution of activity from NM scans. If a SPECT or PET scan has a large number of voxels but is sparsely populated or has regions of very high and very low activity, a simple rejection algorithm is extremely slow and entirely dependent on the underlying

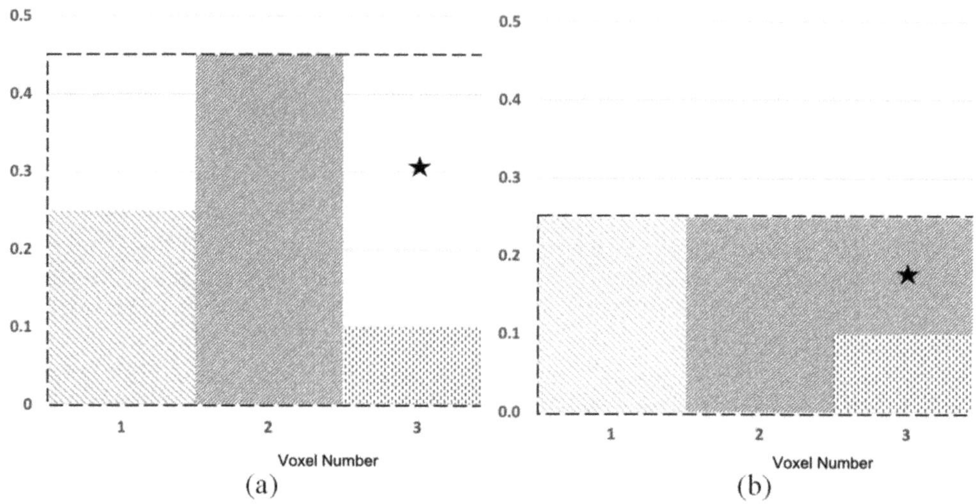

Figure 13.4. (a) Theoretical activity distribution of a 2 × 2 voxel grid (b) redistribution of activity according to the Vose-Alias method to ensure more efficient sampling from the activity distribution (the black dotted line represents the sampling space from which a voxel can be sampled).

activity distribution of the image. To demonstrate this inefficiency, consider a theoretical 2 × 2 image with each voxel containing an activity distribution to be sampled.

Figure 13.4(a) demonstrates such a scenario where the activity in each voxel is presented in a histogram and where activity bars have been normalised to one. The dashed black box represents the 'sampling-space' and the black star represents a randomly sampled point within that space. If the randomly sampled point falls within white space, which represents almost 50% the sampling space in this example, the sampled point would not lead to a sampled radioactive decay occurring within that voxel and is therefore wasteful of computational resources. This would be even more significant for discrete probability distribution containing millions of voxels as in a typical 3D NM scan. The problem becomes significantly worse for images where the activity between voxels varies significantly and where there are large voids in the sampling space, which are available for sampling but would not lead to an event being generated in the simulation.

Raydose implements the Vose-Alias method [20] which is far more efficient at sampling the underlying activity distribution, as demonstrated in figure 13.4(b). Here a new box is drawn which has a total area of 1. This new box becomes the sample space and the activity for each voxel in figure 13.4(a) can be cut and redistributes into any unfilled regions. A table must be constructed to keep track of where the activity distribution has been moved to and to which voxel it corresponds. However, it can be seen that when sampling from within the new box, every sampled point will lead to a sampled radioactive decay. For every random sample there is a guaranteed radioactive decay which will be appropriately positioned within the MC voxel phantom according to the underlying activity distribution.

13.3 Running Raydose

13.3.1 Graphical user interface

Raydose has been developed as a command line tool so that it can be run on large remote servers. However, a graphical user interface has been developed to allow the desktop user to easily create and run a simulation job, to run post analyses on the resulting simulation files and also to generate total absorbed dose maps. The Raydose user interface is shown in figure 13.5.

The typical parameters that must be set by the user in order to begin a simulation are the location of the input DICOM images and the factor by which these images are to be resampled. The isotope that is to be simulated must be specified by its atomic mass and number. If the user has been supplied with a calibration factor to convert nuclear medicine scan voxel counts to an absolute activity this must also be provided to the simulation. The user may also specify the number of threads that the application can use in order to complete the simulation.

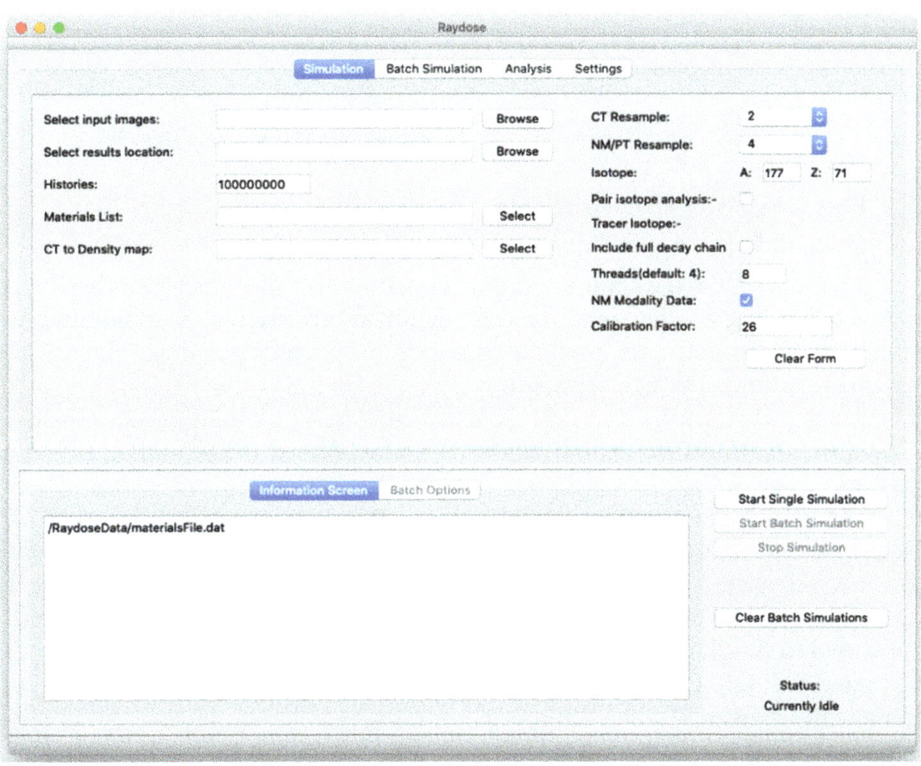

Figure 13.5. Main Raydose user interface allow for the user to setup the simulation parameters without having to utilise the command line.

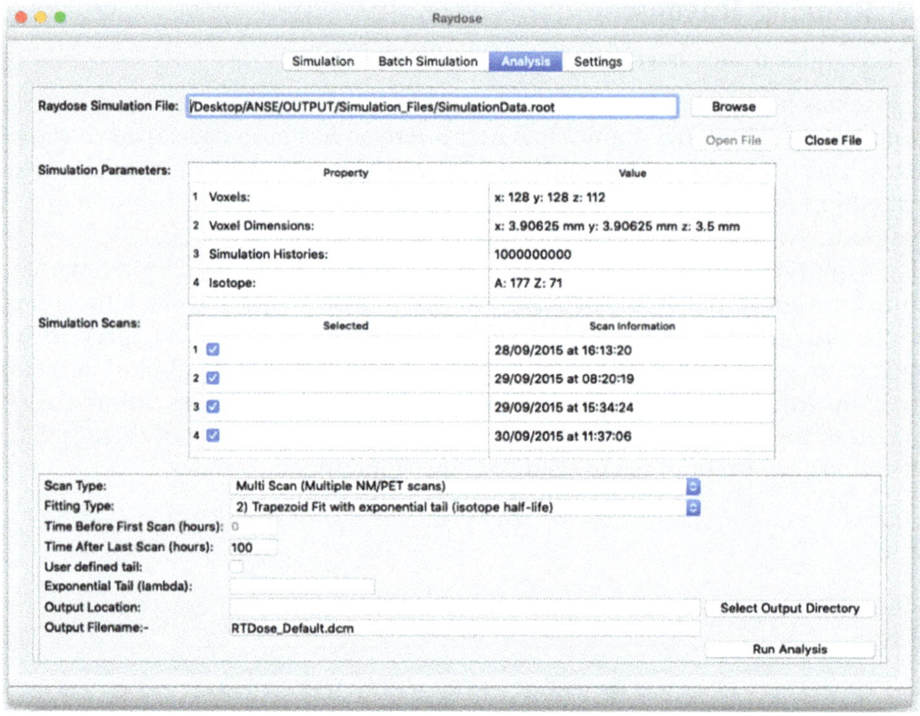

Figure 13.6. Analysis screen allowing the user to post-analyse a simulation file.

13.3.2 Post analysis of simulation data

After a simulation has been run and the dose rate maps are generated for each of the supplied patient scans, the simulation data is written to a file which can then be post-analysed according to the user's reconstruction requirements. The simulation file contains all information required to calculate total absorbed dose maps without having to reperform the MC simulations.

This is the view that the user is presented with when a simulation file is read in to the Raydose user interface to analyse the data and reconstruct a total absorbed dose to patient.

The simulation file contains the MC dose rate maps generated for each of the nuclear medicine scans, the CT density and materials data which defined the patient MC voxel phantom, the times that each of the scans were acquired along with statistical uncertainties of the MC simulations. The number of histories and the isotope used to simulate the dose rate maps is stored along with important data from the original CT DICOM images that will allow the exported RTDOSE file to be automatically registered when reimported into the users favoured analysis or clinical imaging software.

The types of analysis that can be employed to reconstruct the simulated dose rate maps into a total absorbed dose map are outlined in table 13.2.

Table 13.2. List of analysis types currently available in Raydose. The command line argument is also displayed for each analysis type.

Analysis	Description
Single scan	A single PET/NM image has supplied. The dose is calculated assuming exponential decay from been the time of scan out to infinity. The decay constant is that of the isotope simulated.
Single scan with limits	A single PET/NM image has been supplied. The dose is calculated assuming exponential decay from the scan time out to a user defined upper time limit after the acquired scan.
Multi-scan	Multiple PET/NM images are supplied. The dose is calculated assuming a linear fit between each scan point to determine the time activity curve (TAC) for each voxel in the 3D patient phantom.
Multi-scan exponential decay after last scan	Multiple PET/NM images are supplied which have been acquired at different time points post treatment. The dose is calculated assuming a linear fit between each scan point to determine the TAC for each voxel. An exponential tail is added after the last scan point with the relevant isotope decay constant.
Multi-scan exponential before first scan (no exponential tail)	Multiple PET/NM images are supplied which have been acquired at different time points post treatment. The dose is calculated assuming a linear fit between each scan point to determine the TAC for each voxel. However, a period of time BEFORE the first scan is considered, assuming that exponential uptake has occurred between the time of injection and the time of the first nuclear medicine scan.
Multi-scan with linear uptake (no exponential tail)	Multiple PET/NM images are supplied which have been acquired at different time points post treatment. The dose is calculated assuming a linear fit between each scan point to determine the TAC for each voxel. Linear uptake is assumed to have occurred between the time of injection and the time of the first nuclear medicine scan.
Multi-scan with exponential or linear uptake before first scan AND exponential tail.	As above, but with exponential tail after the patient last scan.

13.3.3 System requirements

Raydose is entirely capable of working on a standard desktop/notebook or on a large multiprocessor remote server. The underlying MC engine, GEANT4, includes the ability for the user to write multithreaded code and make use of multicore machines to increase the speed of simulations. However, there are many factors which determine the speed at which a dose calculation can be performed and it is not easy to state the length of time to complete a simulation since the parameters that affect it are:

- The size of the voxel grid on which dose is being scored. The smaller the grid the greater the requirements for the MC engine to track the particles as they cross geometrical boundaries.
- The radioisotope being simulated can affect the simulation speed depending on the complexity of the decay chain and also the types and energies of all subsequent decay radiation.
- The volume of interest to be included in the simulation. A whole abdomen SPECT or PET image will take many hours to simulate, whereas limiting a calculation to only radioactive decays emitted from regions of interest will significantly speed up the simulation.

13.4 Case studies

13.4.1 Quantitative ^{90}Y PET/CT-based dosimetry in liver radioembolization

This section describes the application of Raydose to the assessment of radionuclide therapy of hepatic lesions performed with ^{90}Y labelled microspheres. Selective Internal Radiation Therapy (SIRT) [21, 22], also commonly referred to as radio-embolization, delivers millions of radioactive microspheres into the hepatic artery of the patient which are carried into the arterioles and selectively lodge in the tumour microvasculature. Once the microspheres lodge within this microvasculature it is assumed that the microspheres remain static and *in situ* and the radioactive decays directly target the surrounding tumour tissue.

The study performed using Raydose is a treatment evaluation protocol which consists, as shown in figure 13.7, of an initial FDG-PET staging scan, a ^{90}Y-PET scan following SIRT administration and a further FDG-PET scan a period of 6 weeks following the SIRT treatment to evaluate outcomes. The intermediate ^{90}Y-PET scan is used to determine the biodistribution of the ^{90}Y microspheres but this imaging also plays a critical role in the dosimetry study. Using the ^{90}Y-PET scan Raydose is deployed to calculate the 3D dose distribution within the liver and to assess the dose delivered to the biological target volume (i.e. the tumour).

The ^{90}Y SIRT treatment allows for a relatively straightforward calculation of 3D absorbed dose at the voxel level since the radiopharmaceutical does not move over the course of the treatment regime. Essentially only one activity scan is required to deduce the biodistribution of the ^{90}Y microspheres within the liver and then they radioactively decay with a half-life of $T_{\frac{1}{2}} = 64.053$ h. Multiple follow-up scans to determine the biokinetics of the ^{90}Y radionuclide are not necessary.

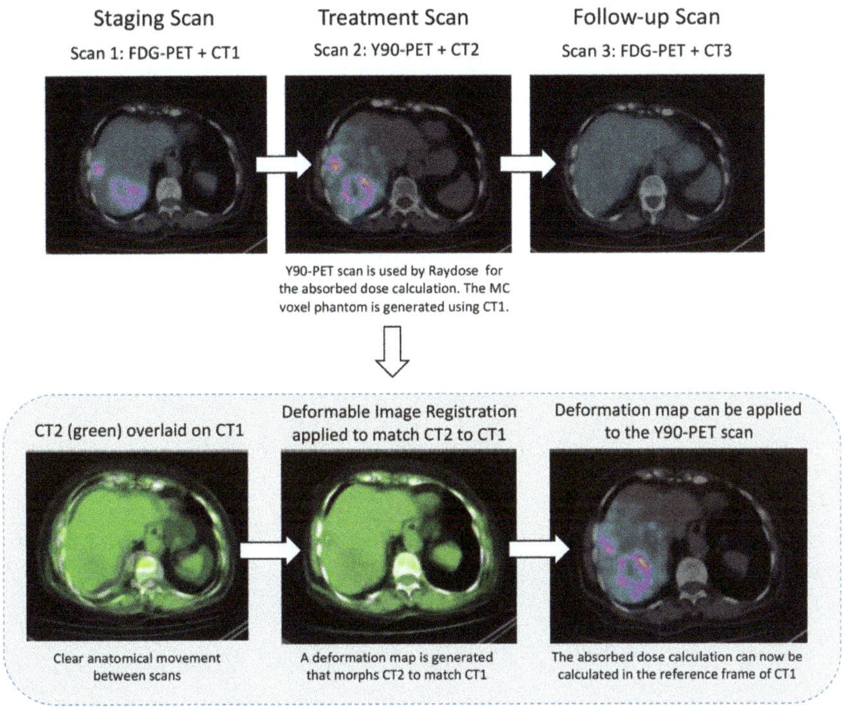

Figure 13.7. Outline of the image registration process for a ^{90}Y SIRT regimen.

For the ^{90}Y SIRT treatment simulation the first staging CT scan (CT1) is to generate the MC voxel phantom. The second ^{90}Y-PET scan is used by Raydose to calculate the absorbed dose to the tumour site. The third scan allows for follow-up assessment of the treatment outcomes to the areas of interest. To make comparisons between each scan it is essential that a single frame of reference is chosen. As the initial clinical evaluation is performed on the staging scan, the ^{90}Y-PET activity is mapped on to CT1 so that the absorbed dose delivered during the treatment regime can be calculated in the same frame of reference as the staging scan.

The same process is applicable to the follow-up scan so that the treatment results can be evaluated using any structures that the clinician may have outlined on the staging scan. When there is significant anatomical movement between staging and treatment scan like in the case presented in figure 13.7, a deformable image registration technique is applied to match the treatment CT (CT2) to the staging CT (CT1). The process generates a deformation map that can then be applied to the ^{90}Y-PET scan which in turn will register the PET image into the frame of reference of CT1. Please note, image registration must be done outside of Raydose and resampled images exported to file so that they can be used in the Raydose workflow.

The Raydose absorbed dose calculation can at this point be performed. The CT1 scan is used to generate the patient phantom and the newly registered PET data describes the activity distribution within the liver. The MC simulation is performed using the PET scan as the underlying intensity distribution. In the same way as is discussed earlier in this chapter a dose rate map is produced, normalised by the number of simulated events (or decays of ^{90}Y).

Once the dose rate maps are obtained from the MC simulation, they must be scaled to reflect the true number of decays of ^{90}Y occurring between user specified time periods. Similarly to figure 13.3, a time-activity curve must be generated for each voxel in order to determine the total number of decays occurring within each voxel and subsequently the total number of decays within the entire patient volume. The ^{90}Y-SIRT process is simplified given that there is only a single time-activity point and all voxels will have the same exponential decay and decay constant. The total number of decays occurring in the ith voxel between the scan time (t_0) and a user specified time (t_1) is given by:

$$\widetilde{A}_i(t_0, t_1) = \int_{t_0}^{t_1} A_i(t_0) \bullet \mathrm{e}^{-\lambda t} \mathrm{d}t \tag{13.9}$$

where $A_i(t_0)$ is the initial activity in the ith voxel.

The total number of decays occurring within the patient volume ($\widetilde{A}(t_0, t_1)$) is therefore given by:

$$\widetilde{A}(t_0, t_1) = \sum_{i}^{N_{\text{voxels}}} \widetilde{A}_i(t_0, t_1) \tag{13.10}$$

where N_{voxels} is the total number of voxels in the MC phantom.

The total absorbed dose map (D) is calculated using:

$$D(t_0, t_1) = \dot{D}(t_0) \bullet \widetilde{A}(t_0, t_1) \tag{13.11}$$

The result is a 3D absorbed dose map, such as shown in figure 13.8(a), provided by Raydose in RTDOSE DICOM that can be imported into any DICOM viewer or other analysis package. Figure 13.8(c) shows a contour around a liver lesion determined from the staging FDG-PET scan and figure 13.8(d) shows the evaluation follow-up FDG-PET scan. From the absorbed dose simulation, it is possible to calculate the cumulative dose-volume-histogram (DVH) inside the structure. In the DVH shown in figure 13.8(b) the maximum absorbed dose voxel dose is 152 Gy, the minimum voxel absorbed dose is 7 Gy and the mean average absorbed dose within the structure is 73 Gy.

13.4.2 Raydose phantom validation using ^{90}Y

Raydose has been used in a study aimed to investigate different computational approaches for ^{90}Y dosimetry [23]. In this case, full MC dose calculations were compared with (I) kernel convolution (II) local deposition algorithms and (III) MIRD analytical approaches. Absorbed dose measurements were also made using LiF:Mg,Cu,P thermoluminescent dosimeters (TLDs) placed in a homogenous solution of ^{90}Y. This allowed for direct comparisons to be made between the absorbed dose per beta decay (Gy/beta decay) in the TLD and the same quantity derived from the

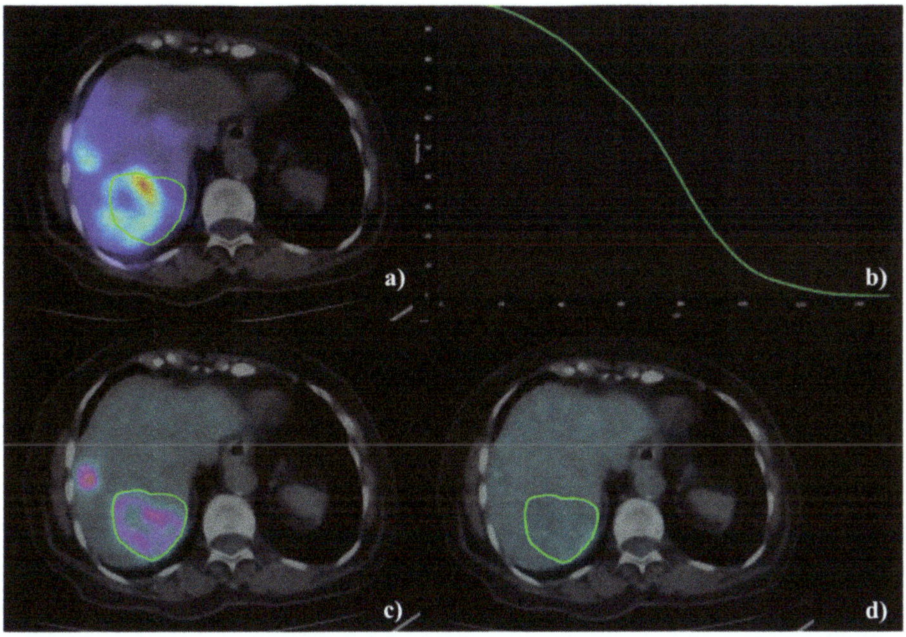

Figure 13.8. (a) Axial slice from a 3D total absorbed dose map calculated using MC methods within Raydose with a structure outlining a lesion drawn from the initial staging FDG scan. (b) Dose Volume Histogram (DVH) detailing delivered dose within the 3D structure, (c) initial FDG staging scan indicating location of one of the lesions of interest (d) a follow-up FDG scan deformably registered to CT1 evaluating uptake in the region of interest.

computed methods (I)–(III). The TLD acted as a gold standard which the computed methods could be benchmarked against. The absorbed dose calculations using methods (I)–(III) along with Raydose were performed using a solution of ^{90}Y within an anthropomorphic phantom incorporating lungs, liver, spine and a cylindrical lesion within the hepatic compartment, as shown in figure 13.9.

A solution of ^{90}Y filled both the liver background region with activity concentration of 0.89±0.04 MBq mL^{-1} and the cylindrical liver insert (volume of 19 mL) with an initial activity concentration of 5.5±0.3 MBq mL^{-1}, representing a 6:1 tumour to background ratio. The phantom was scanned on days 1, 4, 5, 6 and 12 post preparation until the time at which the activity concentration within the lesion insert fell to 0.31±0.02 MBq mL^{-1}.

It is worth noting that in figure 13.9 the image includes inhomogeneous activity distribution even though the two regions are known to have a uniform activity concentration. There are many attributable reasons for PET noise, however in ^{90}Y-PET imaging a significant factor is that the image is formed as a result of a very minor branch to the first 0+ excited state in the daughter ^{90}Zr nucleus. There is a subsequent rare electric monopole (E0) transition to the ^{90}Zr 0+ ground state resulting in a b^{+} emission via internal pair production [24]. Given the rarity of this decay mechanism there are detection inefficiencies which lead to the inhomogeneities in the activity distribution of the ^{90}Y solution.

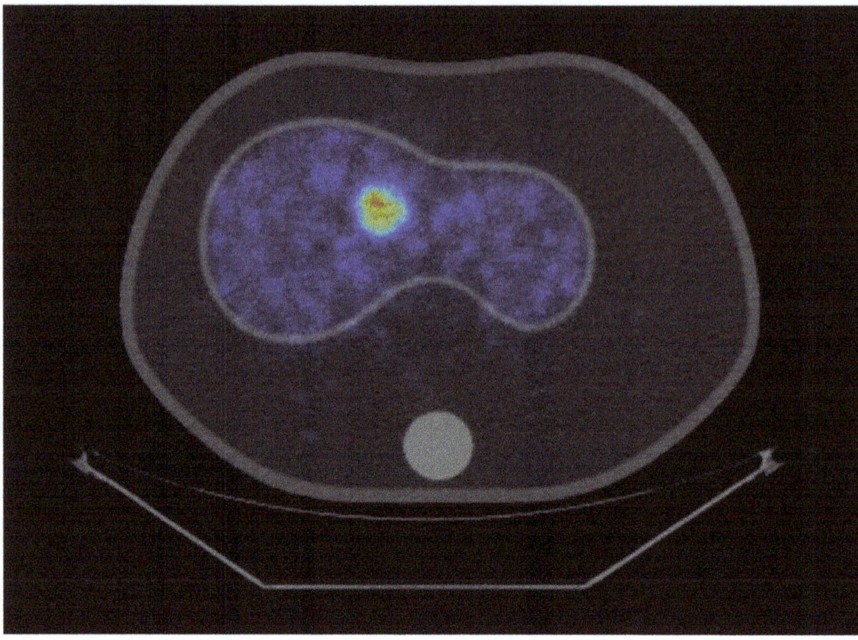

Figure 13.9. ^{90}Y-PET fused with CT of an anthropomorphic phantom incorporating a cylindrical lesion within the hepatic compartment.

For the purpose of studying each calculation method the average absorbed dose to the liver insert region was determined by considering the dose deposited in the proceeding 2.67 days (1 physical half-life of ^{90}Y). The results are presented in figure 13.10, with the vertical bars being a measure of the average absorbed dose and their standard deviations. Details of the calculation of the standard deviations can be found elsewhere [23]. The local deposition, MIRD and kernel convolution methods were in very good agreement with one another. Raydose along with MCNP4c [25], both of which are MC codes, were also in agreement which each other. Notably, the Raydose and MCNP4c packages produced absorbed doses that were systematically lower than the other three calculation methods. This was to be expected given that both Raydose and MCNP4c were capable of simulating the PMMA walls of the liver insert, thus accounting for attenuation of decays occurring in the surrounding liver background entering into the liver insert. What is apparent in the analysis is that as the activity of the insert decreases the quality of the dose calculation deteriorates and by day 12, the activity concentration is so low that the standard deviation of the calculated absorbed dose is too high to be useful in practice. Raydose was benchmarked against the gold standard TLD measurement by comparing the average absorbed dose to water. It was determined in that work that the TLD value was $(1.09 \pm 0.07) \times 10^{-11}$ Gy/beta decay and Raydose $(1.06 \pm 0.04) \times 10^{-11}$ Gy/beta decay, giving good agreement between the two within 3%. This gives confidence in the underlying performance of Raydose. There is no doubt that other methods such as local deposition or convolution methods will give similar

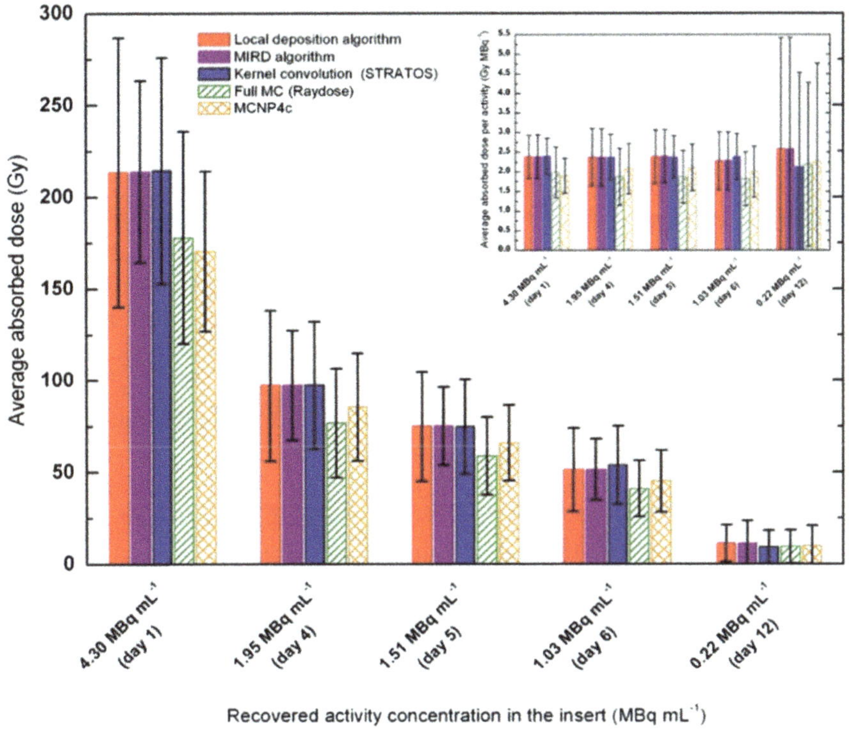

Figure 13.10. Comparison of absorbed doses of ^{90}Y within a cylindrical insert placed within the liver of an anthropomorphic phantom. Different dose calculation algorithms were used to compare average absorbed doses, namely (I) Local Deposition (LD) algorithm (red), (II) MIRD analytical approach (violet), (III) Kernel convolution using Philips Stratos (blue), (IV) full MC using Raydose (green stripe pattern), (V) MCNP4c (orange crosshatch pattern). The vertical bar chart displays the mean absorbed dose and its standard deviation [23]. Copyright (2017), the author(s). With permission of Springer.

results in a quicker time frame, but those methods do not deal with the underlying inhomogeneity of the patient anatomy which are not taken in to account in the same way as they are in full MC methods.

13.4.3 Quantitative ^{177}Lu SPECT/CT-based dosimetry in peptide receptor radionuclide therapy

This section describes the use of Raydose for ^{177}Lu SPECT/CT-based dosimetry in Peptide Receptor Radionuclide Therapy (PRRT). This is currently the most effective MRT treatment against the neuroendocrine tumours (NETs) but it is also used for prostate and pancreatic tumours. PRRT consists of a systemic administration of a targeted radionuclide in order to achieve important clinical outcomes such as improved quality of life, disease stabilization and improved overall survival.

The treatment protocol of this study involved the acquisition of five SPECT/CT scans using a Symbia T2 scanner (Siemens Medical, Germany) with a NaI:Tl detector. SPECT images were reconstructed with a 4.8 mm cubic voxel for body protocol and resulted on matrix size of 128 ×128 voxels on the transverse plane. Details of the validation of acquisition and reconstruction protocols can be found elsewhere [26, 27]. Patients were scanned at times of 1, 4, 24, 44 and 72 h after the first therapeutical administration of ^{177}Lu-radiolabelled derivatives DOTA-Tyr3-OCtreotide (^{177}Lu-DOTATOC). The mean administered activity for the full cohort was 4.27 ± 0.97 GBq, with activity ranging from 3.7 to 5.5 GBq.

In order for the dose calculation to be carried out in Raydose, the images must to be co-registered to the same frame of reference. This process helps minimising mismatches between the images taken at different time points due to internal organ movements, variations in patient repositions and respiratory motion. For this case study, a non-rigid registration algorithm is preferred to a rigid one as it provides the best performance [27, 28].

After all the images are co-registered to the same frame of reference (figure 13.11), Raydose is used to perform the dose calculation.

CT1 scan is used by the software to generate the material map (as described in section 13.2.2) and the registered SPECT are used to calculate the cumulative activity distribution between the different time points (as described in section 13.2.4).

For this study, the option 'Multi-scan exponential decay after the last scan point' (see section 13.3.2) is selected and a total of 10^8 histories per scan are simulated with a corresponding calculation time of ~24 h on a 20 core Dell PowerEdge rack server (Intel Xeon E5–2670 v2 @ 2.50 GHz) running Ubuntu 14.04 lTS and equipped with 64 GB RAM.

Raydose generates a dose rate map for each time point, normalised by the number of simulated decays of ^{177}Lu ($\dot{D}$). After their simulation, each dose rate map needed to be scaled to reflect the real number of decays between each time point. The total number of decays $\widetilde{A}$ which occurs between two scan points is computed using a parallelogram method, because it would not be feasible to check the fit curves for each voxel of the image.

After the last scan point, the number of decays is calculated with an exponential function. The decay constant can be automatically set by the code or it can be set by the user (as described in section 13.2.4). In this particular case study, the effective

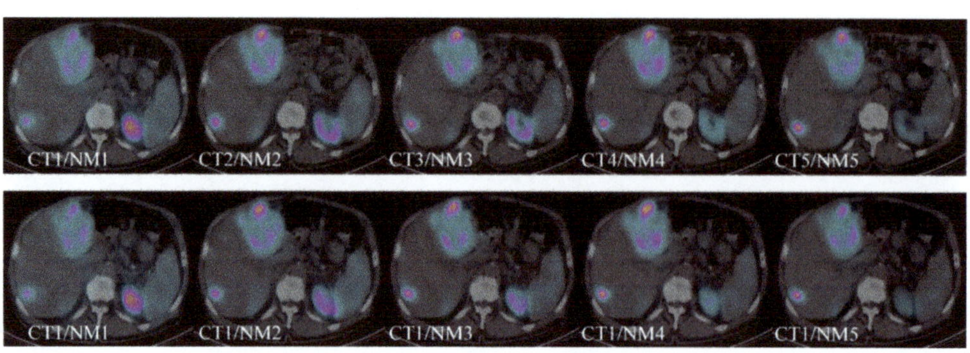

Figure 13.11. The first row shows the SPECT images registered with the corresponding CT image. The row below shows the SPECT images all non-rigidly co-registered with CT1.

decay constant for each volume of interest is calculated using an external spreadsheet software and manually set using the Raydose interface (figure 13.12).

For the last time point, the number of decays is calculated as shown in equation (13.1):

$$\widetilde{A}(t_5) = \int_{t_5}^{T} A(t_5)\bullet e^{-\lambda t} dt \tag{13.12}$$

where t_5 is the last scan point, $T = 1000$ h post-injection, λ is the decay constant, A is the activity and t is the time variable. The T parameter could also be set to infinite. The total absorbed dose map (D) is then calculated as the sum of the component from the trapezoidal integration for the first four time points and the last one equation (13.12):

$$D = [\dot{D}(t_1)\bullet \widetilde{A}(t_0, t_1)] + [\dot{D}(t_1)\bullet \widetilde{A}(t_1, t_2)] + [\dot{D}(t_2)\bullet \widetilde{A}(t_2, t_3)] \\ + [\dot{D}(t_3)\bullet \widetilde{A}(t_3, t_4)] + + [\dot{D}(t_4)\bullet \widetilde{A}(t_4, t_5)] + \left[\dot{D}(t_5)\bullet \int_{t_5}^{T} A(t_5)\bullet e^{-\lambda t} dt\right] \tag{13.13}$$

The result is a dose map provided in DICOM RTDOSE format and can be visualised with any DICOM compliant software package.

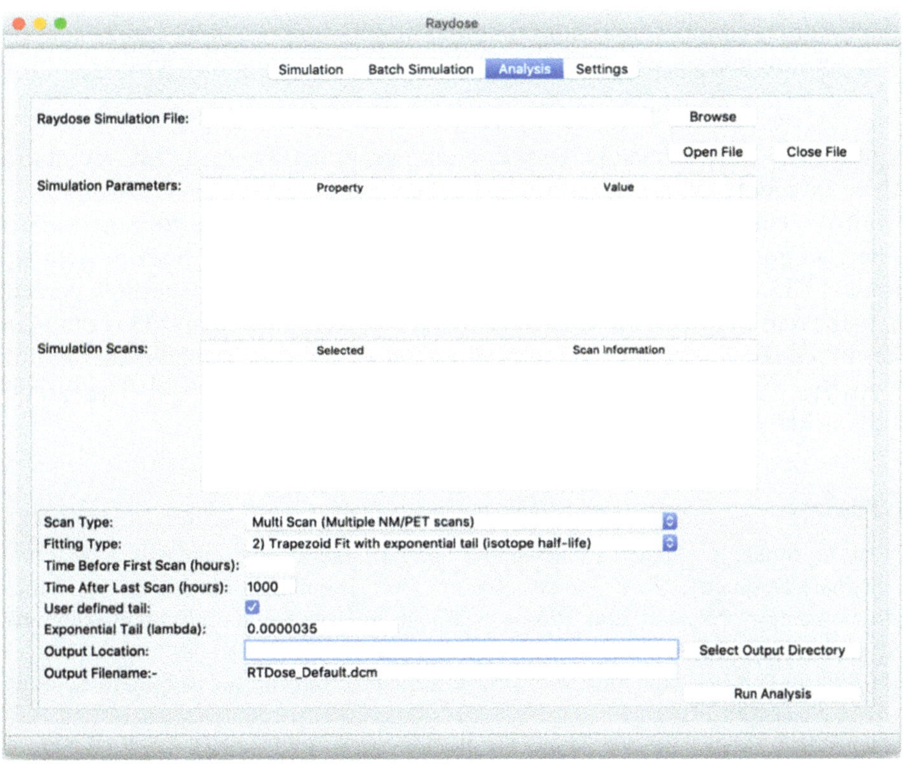

Figure 13.12. Screenshot of Raydose which shows the time after the last scan (in hour) and the user defined decay constant (in seconds).

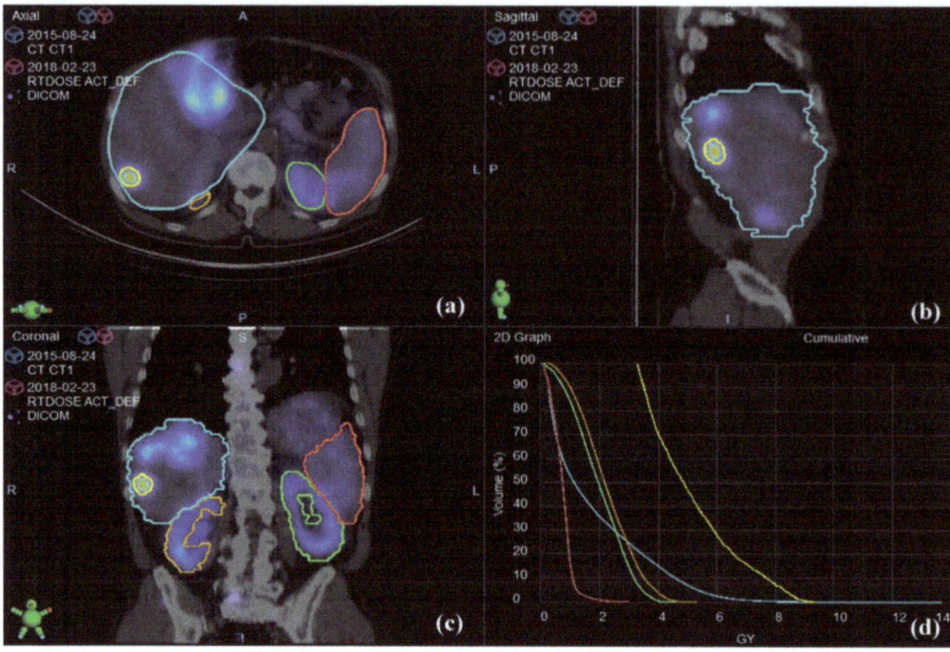

Figure 13.13. Dose map generated with the Raydose code overlapping CT1 (in greyscale). In the picture, the following volumes of interest are highlighted: left kidney (in green), right kidney (in orange), liver (in blue), spleen (in red) and a lesion (in yellow).

Figure 13.13 shows the absorbed dose map generated from the MC simulation in the three directions with organs at risk and lesions outlined.

From the calculated total absorbed dose map it is possible to generate a dose volume histogram inside the volume of interest as shown in the bottom right square of figure 13.13. In addition to the total dose map, Raydose generates percentage error maps and intermediate dose rate maps for each time point. This data can be further processed to estimate the total error associated with the dose calculation. In this case the total percentage error was less than 3% for every voxel. The estimation excludes other sources of error such as activity calibration.

References

[1] Flux G, Stokke C, Sjogreen Gleisner K, Spezi E, Aldridge M and Bacher K *et al* 2017 Internal Dosimetry Task Force Report On: Treatment Planning For Molecular Radiotherapy: Potential and Prospects 67 https://www.eanm.org/content-eanm/uploads/documents/EANM_2017_iDTF-Report_online.pdf

[2] McGowan D R and Guy M J 2015 Time to demand dosimetry for molecular radiotherapy? *Br. J. Radiol.* **88** 20140720

[3] MetroMRT. Metrology for Molecular Radiotherapy: EURAMET 2012 HLT11 https://euramet.org/research-innovation/search-research-projects/details/project/metrology-for-molecular-radiotherapy/

[4] MRTDosimetry 2019 *Metrology for Clinical Implementation in Molecular Radiotherapy* (Teddington: National Physics Laboratory) http://mrtdosimetry-empir.eu/wp-content/uploads/2019/08/15HLT06_Publishable_Report_M36_Final.pdf

[5] Agostinelli S, Allison J, Amako K, Apostolakis J, Araujo H and Arce P *et al* 2003 Geant4—a simulation toolkit *Nucl. Instrum. Methods Phys. Res., Sect.* A **506** 250–303

[6] Mildenberger P, Eichelberg M and Martin E 2002 Introduction to the DICOM standard *Eur. Radiol.* **12** 920–7

[7] Malaterre M GDCM Reference Manual 2008 [updated 2008. First] http://gdcm.sourceforge.net/gdcm.pdf

[8] McCormick M, Liu X, Jomier J, Marion C and Ibanez L 2014 ITK: enabling reproducible research and open science *Front Neuroinform.* **8** 13

[9] Schroeder W J, Martin K M and Lorensen B 2006 *The Vizualization Tookit* (New York: Kitware)

[10] White D R, Griffith R V and Wilson I J 1992 Report 46 Photon, electron, proton and neutron interaction data for body tissues *J. Int. Comm. Radiat. Units Meas.* **24** 11–3

[11] ICRU 1988 Report 44, Tissue substitutes in radiation dosimetry and measurement (Bethesda, MA: International Commission on Radiation Units and Measurements)

[12] Hauf S, Kuster M, Batic M, Bell Z W, Hoffmann D H H and Lang P M *et al* 2013 Radioactive decays in Geant4 *IEEE Trans. Nucl. Sci.* **60** 2966–83

[13] NNDC 2020 ENSDF: evaluated nuclear structure data file https://nndc.bnl.gov/ensdf/

[14] Tuli J K 1996 Evaluated nuclear structure data file *Nucl. Instrum. Methods Phys. Res., Sect.* A **369** 506–10

[15] Urban L 2005 A model for multiple scattering in GEANT4. CERN OPEN-2006-077

[16] Lewis H W 1950 Multiple scattering in an infinite medium *Phys. Rev.* **78** 526–9

[17] Sarrut D, Halty A, Badel J-N, Ferrer L and Bardiès M 2017 Voxel-based multimodel fitting method for modeling time activity curves in SPECT images *Med. Phys.* **44** 6280–8

[18] Kawrakow I, Mainegra-Hing E, Rogers D W O, Tessier F and Walters B R B 2020 *The EGSnrc Code System: Monte Carlo Simulation of Electron and Photon Transport* (Ottawa: National Research Council)

[19] Walters B, Kawrakow I and Rogers D 2003 History by history statistical estimators in the BEAM code system *Med. Phys.* **29** 2745–52

[20] Vose M D 1991 A linear algorithm for generating random numbers with a given distribution *IEEE Trans. Software Eng.* **17** 972–5

[21] Kennedy A 2014 Selective internal radiation therapy dosimetry *Future Oncol.* **10** 77–81

[22] Gulec S A and Fong Y 2007 Yttrium 90 microsphere selective internal radiation treatment of hepatic colorectal metastases *Arch. Surg.* **142** 675–82

[23] D'Arienzo M, Pimpinella M, Capogni M, De Coste V, Filippi L and Spezi E *et al* 2017 Phantom validation of quantitative Y-90 PET/CT-based dosimetry in liver radioembolization *EJNMMI Res.* **7** 94

[24] D'Arienzo M 2013 Emission of β+ particles via internal pair production in the 0^+—0^+ transition of ^{90}Zr: historical background and current applications in nuclear medicine imaging *Atoms.* **1** 2–12

[25] Briesmeister J F 2000 *MCNP: A General Monte Carlo N-Particle Transport Code.* Report No.: LA-13709-M. Los Alamos National Laboratory

[26] Grassi E, Fioroni F, Mezzenga E, Finocchiaro D, Sarti M and Filice A *et al* 2017 Impact of a commercial 3D OSEM reconstruction algorithm on the 177Lu activity quantification of

SPECT/CT imaging in a molecular radiotherapy trial *Radiol. Diag. Imaging* https://www.oatext.com/Impact-of-a-commercial-3D-OSEM-reconstruction-algorithm-on-the-177-activity-quantification-of-SPECT-CT-imaging-in-a-Molecular-Radiotherapy-trial.php#Article_Info

[27] Grassi E, Fioroni F, Berenato S, Patterson N, Ferri V and Braglia L *et al* 2018 Effect of image registration on 3D absorbed dose calculations in 177Lu-DOTATOC peptide receptor radionuclide therapy. *Phys. Med.* **45** 177–85

[28] Ward G, Ramasamy S, Sykes J R, Prestwich R, Chowdhury F and Scarsbrook A *et al* 2016 Superiority of deformable image co-registration in the integration of diagnostic positron emission tomography-computed tomography to the radiotherapy treatment planning pathway for oesophageal carcinoma *Clin. Oncol.* **28** 655–62

IOP Publishing

Monte Carlo Calculations in Nuclear Medicine (Second Edition)
Therapeutic applications
Habib Zaidi

Chapter 14

The VIDA software for targeted radionuclide therapy

Susan Kost, Jacob Hesterman and Keryn Gresco

14.1 Introduction

VIDA (Voxel-based Internal Dosimetry Application) is a 3D image-based dosimetry technique that employs the Geant4 Monte Carlo toolkit [1] to perform voxel-by-voxel absorbed dose calculations. 3D anatomical data (CT) is coupled with functional images (SPECT or PET) as input into the Monte Carlo simulation to perform particle transport based on the activity distribution at each time point; thus instantaneous dose rate maps are generated. The patient-specific total absorbed dose is determined by fitting the time-sequenced voxel dose rates to exponential functions and integrating over time. VIDA consists of two main components (figure 14.1), the Geant4 simulation code to generate 3D maps of voxel-level absorbed energy and a MATLAB-based custom exponential fitting tool to perform curve fitting and integration of dose for each voxel. VIDA was validated under several conditions including dose to uniform activity spheres and organ dose factors from various sources in an anthropomorphic phantom [2].

VIDA does not provide a method for quantitative SPECT image reconstruction. Quantitative estimation of activity from tomographic images requires attenuation and scatter correction and characterization of the collimator-detector response that is highly system specific. Use of VIDA for patient-specific dosimetry assumes the functional images modeling patient pharmacokinetics over time are representative of quantitative activity. Similarly, VIDA does not include a method for registration of time-sequenced images. Image registration must be performed prior to fitting the 3D dose rate maps by the exponential fitting tool.

14.2 Monte Carlo simulation

14.2.1 Detector geometry

Patient geometry is represented in the simulation by a 3D map of integer organ identification numbers. This approach allows patient anatomy to be represented by a

doi:10.1088/978-0-7503-2694-0ch14

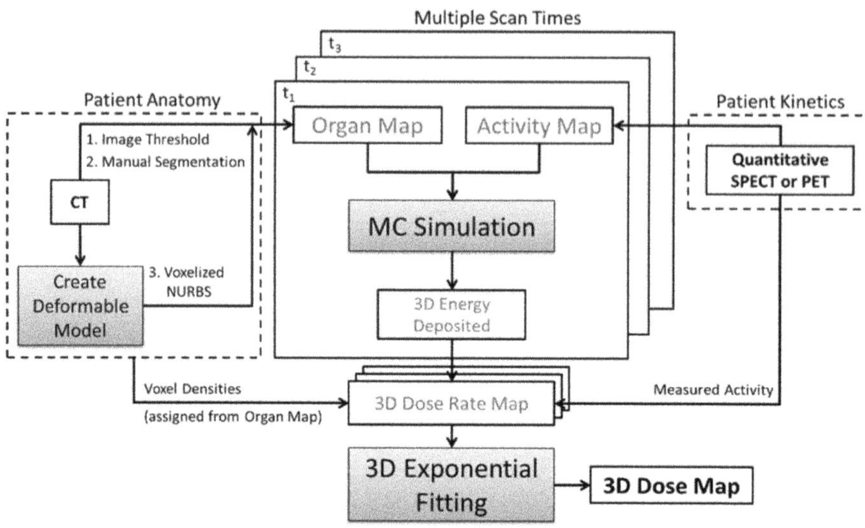

Figure 14.1. VIDA workflow with key inputs and application components. Reprinted from [2] with permission from Copyright © 2015, Mary Ann Liebert, Inc.

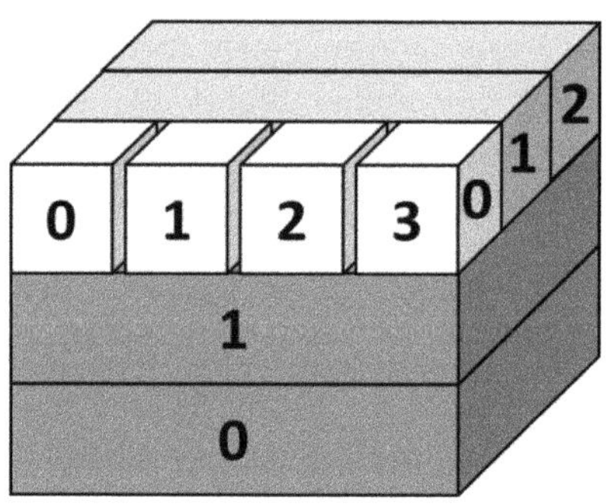

Figure 14.2. Schematic drawing of a nested parameterization in Geant4. Repetitions in each direction are assigned a copy number. Voxels in the geometry are identified by a unique combination of copy numbers representing the row, column and slice position.

deformable model derived from the NURBS-based XCAT phantom [3] that is digitized to a 3D organ map. Organ maps can also be created using conventional segmentation methods to delineate organs of interest and tumor volumes on the patient CT. For multiple time points, segmentation of each CT can be time-consuming and tissue maps of relevant materials can be quickly created for simulation using density thresholds.

Detector definition in VIDA consists of a nested parameterized 3D volume, as shown in figure 14.2. The physical parameterized volume is defined to match the patient image array size and voxel dimensions. Each voxel is assigned a tissue type by pairing a specific material with the integer identifier for each organ contained within the 3D organ map. The material definitions in VIDA include air for voxels surrounding the body contour and soft tissue, lung, adipose tissue, mammary tissue, cortical bone, red and yellow marrow, whole bone, and whole skull derived from data tabulated in International Commission on Radiation Units (ICRU) Publication 46 Appendix A [4].

VIDA includes several different material definitions for skeletal tissue and is capable of modeling the skeleton either whole bone representative of a mixture of cortical bone, trabecular bone, and marrow or with distinct bone and marrow spaces. The deformable NURBS model has cavities inside each bone so that cortical bone surfaces and marrow can be assigned specific materials. Similarly, if the skeleton is delineated by segmentation, hard bone and marrow voxels may be separately defined using specific cortical bone thicknesses and marrow cellularity values. If density thresholds are used to generate an organ map, skeletal tissue is designated as whole bone. A flag in the data input file controls how materials in the skeleton are assigned during simulation.

14.2.2 Particles and physics

Geant4 has the capability to model a wide range of particles and the physics processes governing their interactions with matter. Each particle type is associated with a unique set of processes. To accurately calculate the dose due to internal radioactive emissions, electromagnetic interactions for all decay products must be considered. The particle types used in VIDA are defined by the Geant4 decay physics class and include leptons, bosons, mesons, baryons and ions. The selected physics processes in the VIDA simulation model include decay of unstable particles, radioactive decay of unstable isotopes, and standard electromagnetic processes such as photoelectric absorption, Compton scattering, gamma conversion, coulomb scattering, and bremsstrahlung.

Geant4 contains a wide array of options for modeling electromagnetic physics processes. The standard electromagnetic physics package is optimized for electron and photon interactions at energies between 1 keV and 100 TeV. This model relies on parameterizations of atomic shell structure which can introduce error in particle interactions at lower energies. The physics processes implemented in VIDA (table 14.1) are a modified version of the standard electromagnetic package with selected options to increase accuracy when tracking electrons, hadrons, and ions at low energies.

VIDA includes a Compton scattering model that allows atomic de-excitation of the resulting ionized atom. The standard Compton scattering model in Geant4 ignores the binding energy when determining the kinetic energy of the electron based on the assumption that the energy of the recoil electron is large compared to the binding energy. This assumption becomes invalid in the low energy realm. The modified Compton scattering model in VIDA contains Doppler broadening due to the non-negligible influence of atomic shell effects at low interaction energies that may result in interactions with non-valance electrons that produce atomic shell vacancies.

Atomic relaxation processes induced by ionization events that leave the atom in an excited state are activated in VIDA. These processes include fluorescence and

Table 14.1. Electromagnetic processes in VIDA MC simulation.

Particle type	EM process	Model
Gamma	Photoelectric effect	*(G4PEEffectFluoModel)*
	Compton scattering	Klein-Nishina *(G4KleinNishinaModel)*
	Gamma conversion	Bethe-Heitler *(G4BetheHeitlerModel)*
e^- and e^+	Multiple scattering	L. Urban *(G4UrbanMscModel95)*
	Ionization	Möller (e^-e^-) Bhabha (e^+e^-) *(G4MollerBhabhaModel)*
	Bremsstrahlung	*(G4eBremsstrahlungModel)*
	e^+ e^- Annihilation (e^+ only)	*(G4eplusAnnihilation)*

Auger electron emission. Characteristic x-rays and Auger electrons are produced above the same threshold energy as secondary electrons and bremsstrahlung gammas, defined in the simulation as a production cut range (set to 1 mm for electrons, positrons, and gamma rays).

Geant4 pre-calculates and stores many interaction parameters for use in the simulation. These include the mean rate of energy loss below the secondary particle production cut, the total cross-section per atom for the ejection of a secondary of energy greater than the production cut and, given multiple processes providing energy loss for a given particle, the total continuous part of the energy loss. By default, these data are generated in 84 bins across the energy interval from 100 eV to 10 TeV. In order to increase the sampling across the range of energies encountered during particle transport of radioactive decay products used in therapeutic nuclear medicine, the energy interval in VIDA was set from 10 eV to 50 MeV divided into 220 bins.

One of the most critical parameters for accurate Monte Carlo simulation is the step size limit. The simulation must balance computational efficiency while also ensuring the step size is small enough that all relevant interaction cross-sections remain approximately constant during the step. Continuous energy loss imposes an additional limit on the step size due to the energy dependence of interaction cross-sections. Typically, tracking precision can be preserved while not compromising execution performance by limiting the step size to no more than 20% of the stopping range. At kinetic energies less than 1 MeV, this limit results in step sizes that are too short. Geant4 controls the step function with two parameters. At high energy, the step is limited by the ratio of step size to stopping range with a default value of 0.20. As the particle interacts and loses kinetic energy, the step size decreases gradually until it becomes lower than a specified cut-off. Below this range, the remaining distanced traveled is completed in a final step. A large portion of the energy imparted to tissue in internal dosimetry simulations comes from continuous energy loss from charged particles. In order to ensure accurate dose deposition, the final range cut-off was reduced in VIDA from the default value of 1.0 to 0.1 mm for electrons and positrons.

14.2.3 Primary event generation and scoring

Primary events in VIDA are generated voxel-by-voxel based on an activity map derived from quantitative SPECT or PET imaging. The simulation can be run to include events in the whole body or by selecting an organ or tumor volume as source. Within each voxel, events are assumed to be uniformly distributed, with randomly selected position and direction vectors.

VIDA employs the built-in radioactive decay module in Geant4 [5] to produce primary particles based on the branching ratios and decay energy of the radionuclide of interest. The radionuclide is defined as an ion with an atomic number and mass from the simulation input file. The decay library includes information on the nuclear half-life, nuclear structure of the parent and progeny, branching ratios, and energy of the decay structure taken from the Evaluated Nuclear Structure Data File [6] for each nuclide. For the case of nuclides that decay to excited isomers, the prompt de-excitation of the daughter nucleus via isomeric transition occurs based on photon evaporation files that include internal conversion coefficients. In the event the radionuclide decays to long-lived radioactive isotope that will be excreted by the body before its decay products are released (e.g., ^{99m}Tc to ground state ^{99}Tc with a half-life of 2.11×10^5 years), the radioactive progeny are killed by the simulation to prevent further particle transport.

VIDA employs a primitive scorer for deposited energy in the detector volume defined as an individual voxel. Energy deposition due to each primary event, including secondary particles and any subsequent gamma emission due to de-excitation of the nucleus, is tallied for each voxel. Once all events are tracked and scored, a 3D map of the total energy deposited is written out to memory to create a 3D dose rate map for voxel-level kinetics processing.

14.3 3D kinetics processing

The VIDA exponential fitting module (figure 14.3) for patient kinetics modeling consists of a MATLAB-based GUI for curve fitting by iterative least squares estimation. Nonlinear regression is performed using the Levenberg–Marquardt nonlinear least squares algorithm. The module has options for fit functions depending on patient uptake kinetics and the number of sequential scans. Instantaneous activity uptake, resulting in exponentially decaying dose rates, can be fit to a mono-exponential or bi-exponential function of the general form:

$$\dot{D}(t) = \dot{D}_1 e^{-\lambda_1 t} + \dot{D}_2 \mathrm{e}^{-\lambda_2 t}$$

Initial guesses for the rate components (D_1 and D_2) are derived from the time-sequence data and elimination constants (λ_1 and λ_2) from the physical half-life of the radionuclide of interest. The fitting algorithm automatically detects voxels with non-instantaneous uptake (i.e. tumor) and fits these data to a bi-exponential including the (0, 0) point of the form:

$$\dot{D}(t) = \dot{D}_0(\mathrm{e}^{-\lambda_1 t} - \mathrm{e}^{-\lambda_2 t})$$

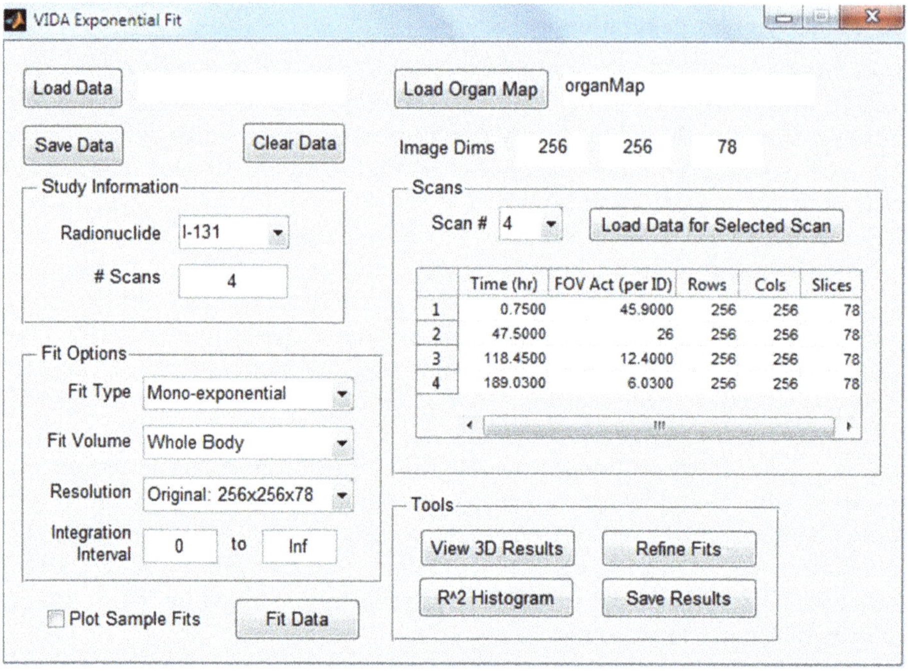

Figure 14.3. Graphical user interface of the exponential fitting tool in VIDA. Reprinted from [2] with permission from Copyright © 2015, Mary Ann Liebert, Inc.

Voxel dose is computed by integration of the fitted function based on a time interval supplied by the user, with a default range of zero to infinity.

The exponential fitting tool is user-driven with various options. The user may choose to fit the entire body in the image field of view or select specific organs and tumor volumes based on the segmented organ map. Also, the user may specify the resolution of the resulting dose map. A scaled map with voxel dimensions twice as large in the transverse plane is created by fitting a voxel dose curve to the data in the corresponding $2 \times 2 \times 1$ cell in the original array. This option decreases the overall processing time and may facilitate regression convergence when data are only available for a limited number of time points.

Evaluation tools are provided in the exponential fitting module to help the user assess and visualize results. A three-plane viewer plots raw data along with the fitted curve for the selected voxel (figure 14.4). The user can also plot a histogram of the coefficients of determination (R^2) and, if necessary, fits for voxels with R^2 values below a user-defined threshold may be refined by rerunning the regression using variable initial coefficient values.

VIDA was designed to generate dose rate maps at each time point as input for curve fitting and dose integration at the voxel level. Instantaneous dose rate calculation allows for tumor regression from response to therapy in the absorbed dose estimates. Radiobiological modeling also requires time-sequenced 3D absorbed dose rate images rather than the simulation of absorbed dose from a single map of

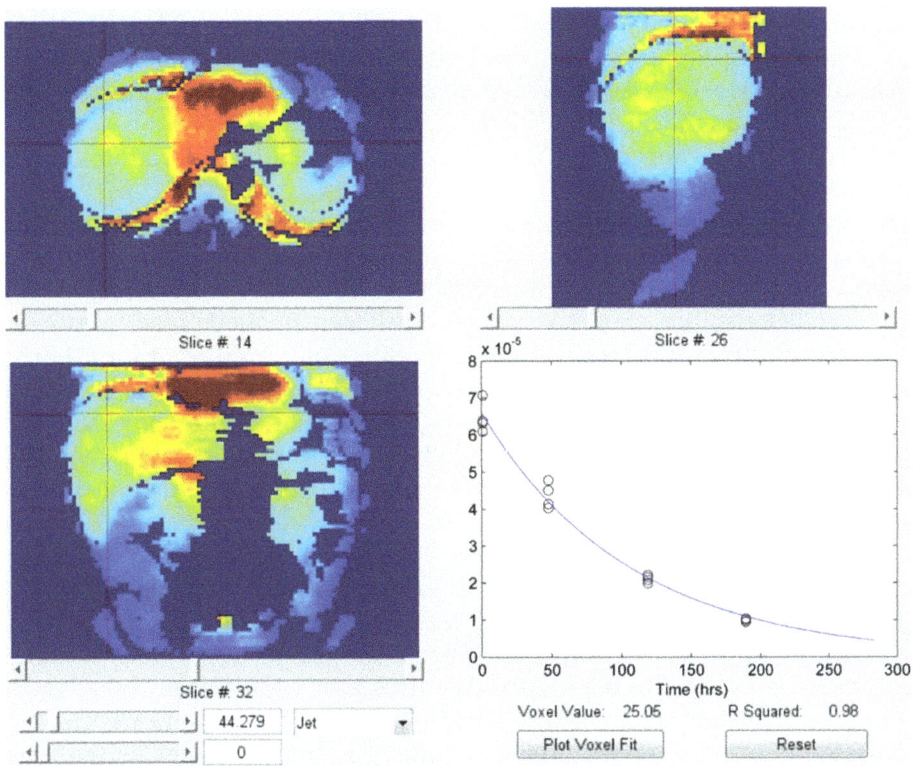

Figure 14.4. 3D viewer in VIDA exponential fitting tool to evaluate the 3D dose map and voxel fits.

cumulated activity. A similar approach of generating multiple dose rate maps has been implemented in other dosimetry studies [7–9]. If desired, a map of cumulated activity for Monte Carlo simulation input can be created using the exponential fitting tool to determine the 3D integrated dose with only one run.

14.4 Validation

VIDA simulation results were validated using two independent techniques. The first method applied a simple geometry of a spherical source in a semi-infinite scattering medium. The self-dose to spheres of various volumes was determined assuming a uniformly distributed activity. The second approach compared dose factors for selected target organs in a standard male reference phantom assuming uniform activity in the source organ.

Simulations were performed to determine absorbed fractions for self-absorption in unit-density spheres ranging in size from 10 to 1000 g. To be consistent with the organ maps used in patient studies, each sphere was 'voxelized' into an array of cells large enough to include a surrounding scattering medium. The spheres were modeled as a tissue-equivalent material surrounded by a semi-infinite scattering medium of water with tissue composition of the sphere taken from MIRD Pamphlets 3 and 8 [10, 11] to facilitate comparison to previously published results. Uniform activity was simulated for common therapeutic radionuclides including ^{90}Y, ^{131}I, ^{111}In and ^{177}Lu. One million events were tracked which resulted in absorbed energies with relative errors of less than 0.5%. Self-dose factors were calculated and compared the unit-density sphere model in OLINDA/EXM [12]. The differences between VIDA and OLINDA/EXM ranged from 0.4% to 5%, with the largest deviations occurring in the smallest sphere.

The relative difference in self-dose factors for the spheres improved with increasing mass, due to the increased likelihood of an electron completely depositing its energy within the sphere volume. The largest differences occurred for ^{90}Y, a nearly pure beta-emitter, most likely due to different methods used to sample the beta energy spectrum. OLINDA/EXM applies a single specific absorbed fraction (SAF) based on the average energy of the emission spectrum, whereas beta particles created in a decay event in Geant4 are assigned energies by sampling the β-Fermi-function. Thus, VIDA simulations include beta emissions with energy higher than the spectrum mean which have an increased possibility of exiting the sphere without being scored.

VIDA was validated in a more complex geometry by comparing dose factors for several organs of interest using the RADAR reference adult male phantom [13]. The phantom was rendered in voxel format with a resolution of $1.5 \times 1.5 \times 5.0$ mm. Any voxels in the array outside of the phantom were defined as air. Materials in Geant4 were changed to the densities and compositions of soft tissue, lungs and bone used to generate new specific absorbed fractions (SAFs) for OLINDA/EXM version 2.0 [14]. Simulations were performed for the same radionuclides as the uniform spheres (^{90}Y, ^{131}I, ^{111}In, and ^{177}Lu) with the source of radioactivity assigned to either the liver, spleen, or pancreas. Five million decay events were generated uniformly throughout the source organ, resulting in relative errors of less than 2% in the total energy deposited in each target volume. Target volumes included the three source organs and also lungs and kidneys to evaluate both self-dose and cross-organ doses to distant structures that may be of considered organs at risk from radionuclide therapies.

Deviations in the dose factors for organ self-dose ranged between 0.6% and 5%. The difference in cross-organ dose factors spanned from 0% to 9% for ^{131}I, ^{111}In and ^{177}Lu, with VIDA results being lower than OLINDA/EXM values for almost all source-target organ pairs. These radionuclides undergo beta decay, and the results are consistent with the self-dose factors for uniform spheres. As a pure beta-emitter with locally absorbed energy, cross-organ dose factors for ^{90}Y were excluded in the comparison due to low scoring statistics. The self-dose factors from VIDA are higher for ^{111}In than reported by OLINDA/EXM. ^{111}In decays by electron capture and locally deposited dose is governed by low energy Auger electrons and small yield internal conversion electrons. The difference in ^{111}In self-dose factors may be due to error in the interpolation by OLINDA/EXM of SAFs for the very low energy Auger electrons.

Results from VIDA were also compared to other groups that have performed simulations of internal radioactive sources using Geant4. Absorbed fractions of monoenergetic photons and electrons in ellipsoids were used to estimate the average energy deposited per disintegration of ^{131}I as a function of the generalized radius [15, 16]. They reported an energy deposited per disintegration for ellipsoidal sources with generalized radii of 1.34 cm (201 keV) and 1.68 cm of 201 and 206 keV, respectively, which are in excellent agreement with VIDA's results of 201 keV for the 10 g sphere and 205 keV for the 20 g sphere. The Geant4-based radionuclide therapy code GRNT [17] computed organ dose factors using liver as the source in the MIRD-5 phantom and found relative differences compared to RADAR reference data of approximately 5% for self-dose factors and 1%–10% for cross-organ dose factors to the kidneys, lungs, pancreas, and spleen for ^{131}I and ^{90}Y. These relative differences are similar to those between VIDA and OLINDA/EXM in the phantom study.

14.5 Application to intrathecal dosimetry

Intrathecal drug delivery has been used for over a century to support pain management [18]. Specific to nuclear medicine, intrathecal imaging techniques such as cisternography with ^{111}In-DTPA have long been employed for indications such as cerebrospinal fluid (CSF) leak and hydrocephalus [19]. Current efforts in nanomedicine and gene therapy have driven an expanded interest in intrathecal administration for delivery of antisense oligonucleotides and adeno-associated viruses [20–25]. Imaging, in particular SPECT and PET, is a critical tool for understanding both the pharmacokinetics and pharmacodynamics of such compounds and is also essential for optimization and evaluation of personalized drug delivery given the complexity of flow in the CSF space [26]. Along with interest in utilizing nuclear imaging for intrathecally delivered tracers and treatments comes a need to understand safety of delivering ionizing radiation through radiation dosimetry estimates.

Standard dosimetry models, such as OLINDA/EXM, are routinely used in the evaluation of new and existing radiotracers. These codes typically model radionuclide distribution based upon systemic circulation of radiotracer through intravenous injection and are not designed for intrathecal delivery. Utilization of voxel-based dosimetry estimation methods, specifically VIDA, allows the expansion of standard models to include dose delivered to the spinal cord regions and the radiation coming from the CSF compartment within the spine following an intrathecal injection.

VIDA has modeled the dosimetry following intrathecal injection using a modified XCAT phantom that includes CSF compartments [27]. MR images of healthy volunteers were used to segment spinal CSF and vertebrae into lumbar, lower thoracic, upper thoracic, and cervical regions and delineate brain ventricle regions. Spinal CSF volumes encompassed both the fluid and the spinal cord due to resolution limitations preventing the separation of the two regions. These regions were then incorporated into the adult male XCAT phantom and an organ map was created by voxelizing the modified phantom with resolution of 3.0 × 3.0 × 3.0 mm. 3D absorbed dose rate maps were generated by simulating ten million decay events of ^{99m}Tc in VIDA assuming a uniform distribution of activity within the source volume. Nine anatomical regions were defined as sources for activity including

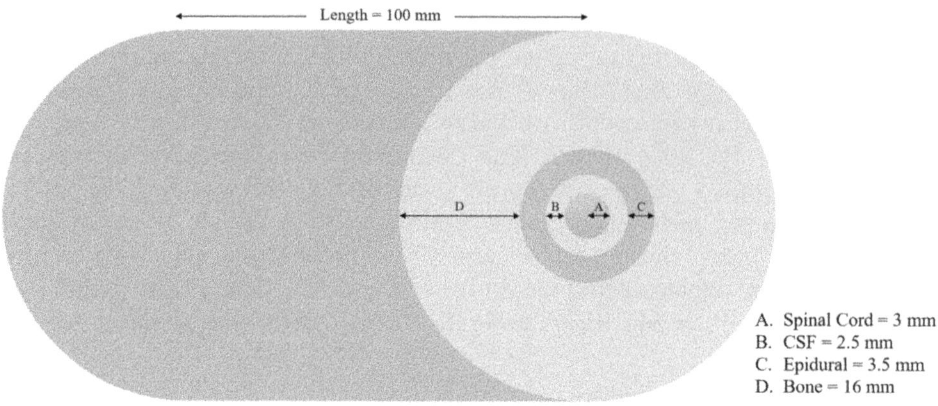

Figure 14.5. Cross-sectional view of cylindrical spine model for intrathecal dose estimates.

brain, brain ventricles, cervical, upper thoracic, lower thoracic, and lumbar CSF volumes, kidneys, and bladder contents.

To validate accuracy of the absorbed dose rate maps generated in VIDA, self-dose factors were generated for brain and kidneys and compared to OLINDA/EXM values. Phantom organ masses in OINDA/EXM were modified to be equal to those in the voxelized XCAT phantom prior to generation of the self-dose factors for comparison. Deviation between the two were found to be within 6%.

Absorbed dose rate maps were then multiplied by time integrated activity coefficient values for each source organ to generate patient-specific absorbed dose maps for example ^{99m}Tc patient data. Analysis of the dosimetry results indicated that CSF regions generally experience high levels of activity and it is important to evaluate spinal cord dose during the development of a new intrathecally-injected radiotracer.

VIDA has also been utilized to research the implications of intrathecal administration by specifically looking at the impact of separating the spinal cord and CSF into two compartments in calculating absorbed dose estimates [28]. A simple cylindrical model was generated to include spinal cord, CSF, and bone regions along with epidural space. A cross-section of the phantom is shown in figure 14.5 with regions (inner to outer) of spinal cord (radius = 3 mm), CSF (2.5 mm), epidural space (3.5 mm), and bone (16 mm). This model was voxelized at a resolution of 0.25 × 0.25 × 1.0 mm with a total length of 100 mm.

Dosimetry estimates were determined using VIDA considering the three different sources of radioactivity. The spinal cord and CSF as a combined compartment was simulated as the source region to reflect the assumptions made with the dosimetry estimates using the modified anthropomorphic model. Additionally, source activity in the CSF compartment only was simulated to generate the dose distribution expected from an intrathecal injection and a third simulation with activity in the epidural space was run to model a missed injection. It was found that when only the CSF region is the source of activity the dose to the spinal cord is about 85% less compared to the combined spinal cord and CSF compartment, and when the epidural space is the only source the dose to the spinal cord is about 94% less.

The intrathecal-extended dosimetry model described here is just one example of the utility of VIDA, and Monte Carlo-based codes in general, to extend nuclear medicine dosimetry estimates beyond conventional methods. The ability to generate voxel-level dosimetry for a user-defined phantom provides a vast number of scenarios in which more accurate dosimetry estimates are beneficial to both research and patient treatment. Some examples of this include applications to extended pediatric phantoms, and sub-organ dosimetry. For example, VIDA can be used to extend the options for dosimetry estimates for pediatric patients, utilizing an array of pediatric phantoms in age and body morphology [29] for dosimetry calculations which are more accurate compared to the standard infant, one-, five-, ten-, and fifteen-year old phantoms available in OLINDA/EXM. There is also the potential to use VIDA for sub-organ dosimetry through organ specific models that incorporate sub-regions and vasculature. This may be especially useful when looking at, for example, beta emitters which may be used in patient treatment.

References

[1] Agostinelli S, Allison J and Amako K A *et al* 2003 GEANT4—a simulation toolkit *Nucl. Instrum. Meth. A* **506** 250–303

[2] Kost S D, Dewaraja Y K, Abramson R G and Stabin M G 2015 VIDA: a voxel-based dosimetry method for targeted radionuclide therapy using Geant4 *Cancer. Biother. Radiopharm.* **30** 16–26

[3] Segars W P, Sturgeon G, Mendonca S, Grimes J and Tsui B M 2010 4D XCAT phantom for multimodality imaging research *Med. Phys.* **37** 4902–15

[4] White D R, Griffith R V and Wilson I J 1992 Report 46 Photon, electron, proton and neutron interaction data for body tissues *J. Int. Commn. Radiat. Units Measur.* **24** 11–3

[5] Hauf S, Kuster M, Batič M and Bell Z W *et al* 2013 Validation of Geant4-based radioactive decay simulation *IEEE Trans. Nucl. Sci.* **60** 2984–97

[6] Tuli J K 1996 Evaluated nuclear structure data file *Nucl. Instrum. Meth. A* **369** 506–10

[7] Dewaraja Y K, Wilderman S J, Koral K F, Kaminski M S and Avram A M 2009 Use of integrated SPECT/CT imaging for tumor dosimetry in I-131 radioimmunotherapy: a pilot patient study *Cancer. Biother. Radiopharm.* **24** 417–26

[8] Senthamizhchelvan S, Hobbs R F, Song H, Frey E C, Zhang Z, Armour E, Wahl R L, Loeb D M and Sgouros G 2012 Tumor dosimetry and response for ^{153}Sm-ethylenediamine tetramethylene phosphonic acid therapy of high-risk osteosarcoma *J. Nucl. Med.* **53** 215–24

[9] Sgouros G, Hobbs R F, Atkins F B, Van Nostrand D, Ladenson P W and Wahl R L 2011 Three-dimensional radiobiological dosimetry (3D-RD) with ^{124}I PET for ^{131}I therapy of thyroid cancer *Eur. J. Nucl. Med. Mol. Imaging* **38** 41–7

[10] Brownell G L, Ellet W H and Reddy A R 1968 MIRD pamphlet no. 3: absorbed fractions for photon dosimetry *J. Nucl. Med.* **27** 28–39

[11] Ellets W and Humes R 1972 *Absorbed Fractions for Small Volumes Containing Photon-Emitting Radioactivity MIRD Pamphlet No. 8.* (New York: Society of Nuclear Medicine)

[12] Stabin M G, Sparks R B and Crowe E 2005 OLINDA/EXM: the second-generation personal computer software for internal dose assessment in nuclear medicine *J. Nucl. Med.* **46** 1023–27

[13] Stabin M G, Xu X G, Emmons M A, Segars W P, Shi C and Fernald M J 2012 RADAR reference adult, paediatric, and pregnant female phantom series for internal and external dosimetry *J. Nucl. Med.* **53** 1807–13
[14] Stabin M G and Siegel J A 2018 RADAR dose estimate report: a compendium of radiopharmaceutical dose estimates based on OLINDA/EXM version 2.0 *J. Nucl. Med.* **59** 154–60
[15] Amato E, Lizio D and Baldari S 2009 Absorbed fractions for photons in ellipsoidal volumes *Phys. Med. Biol.* **54** N479
[16] Amato E, Lizio D and Baldari S 2010 Absorbed fractions for electrons in ellipsoidal volumes *Phys. Med. Biol.* **56** 357
[17] McKay E 2011 GRNT: A Geant4 code for radionuclide therapy dosimetry *Geant4 Workshop 2011* (Wollongong, Australia: Workshop Conducted at Wollongong University Centre for Medical Radiation Physics)
[18] Bottros M M and Christo P J 2014 Current perspectives on intrathecal drug delivery *J. Pain Res.* **7** 615–26
[19] Ziessman H A and O'Malley J P 2013 *Nuclear Medicine: The Requisites E-Book* (Philadelphia, PA: Elsevier Health Sciences)
[20] Leavitt B R *et al* 2020 Intrathecal drug delivery of antisense oligonucleotides in Huntington's disease: experience of Ionis/Roche RG6042 development programme and best practice considerations for real-world Use *Neurology* **94** 950
[21] Roberts T C, Langer R and Wood M J 2020 Advances in oligonucleotide drug delivery *Nat. Rev. Drug Discov.* **19** 673–94
[22] Gong Y *et al* 2019 Intrathecal adeno-associated viral vector-mediated gene delivery for adrenomyeloneuropathy *Hum. Gene Ther.* **30** 544–55
[23] Hocquemiller M *et al* 2016 Adeno-associated virus-based gene therapy for CNS diseases *Hum. Gene Ther.* **27** 478–96
[24] Fowler M J *et al* 2020 Intrathecal drug delivery in the era of nanomedicine *Adv. Drug Deliv. Rev.* **165** 77–95
[25] Hinderer C *et al* 2020 Translational feasibility of lumbar puncture for intrathecal AAV administration *Mol. Ther. Methods Clin. Dev.* **17** 969–74
[26] Coenen W *et al* 2019 Subject-specific studies of CSF bulk flow patterns in the spinal canal: implications for the dispersion of solute particles in intrathecal drug delivery *Am. J. Neuroradiol.* **40** 1242–49
[27] Hesterman J Y, Kost S D, Holt R W, Dobson H, Verma A and Mozley P D 2017 Three-dimensional dosimetry for radiation safety estimates from intrathecal administration *J. Nucl. Med.* **58** 1672–78
[28] Hesterman J Y, Segars P, Palmer K, Dobson H and Kost S 2021 Radiation dosimetry in translational intrathecal imaging *Proc. SPIE 11624, Visualizing and Quantifying Drug Distribution in Tissue V*
[29] Carver D E, Kost S D, Fraser N D, Segars W P, Pickens D R, Price R R and Stabin M G 2017 Realistic phantoms to characterize dosimetry in pediatric CT *Pediatr. Radiol.* **47** 691–700

Chapter 15

Novel approaches for voxel-based dosimetry

Tianwu Xie

15.1 Introduction

Targeted radionuclide therapy (TRT) is one of the most exciting development directions in nuclear medicine. By targeted delivery of radiopharmaceuticals to tumor cellular clusters and metastasized cells, TRT provides systemic therapy for malignant tumor and results in less collateral damage in normal tissues than external beam radiotherapy [1, 2]. With the rapid development of radiochemistry, biomedical engineering and cancer biology nowadays, in the last decays, the potential of nuclear medicine has grown markedly in clinics. More than 40 million nuclear medicine procedures are performed annually in worldwide. While patients gain acknowledged benefits from radiation-involved health technologies in nuclear medicine, they are also exposed to ionizing radiation irradiations and may suffer from potential health hazards. For example, the risk of skeletal and renal toxicity due to the emitted radiation from the radiopharmaceuticals is always an important concern for both patients and radiation oncologists when the great targeting capability of conjugated peptides with radionuclides were adored in clinical practices [3–5]. Therefore, accurate absorbed dose estimation and internal radiation dosimetry of radio-pharmaceuticals in the human body is essential for evaluating the response of targeted tumor and normal tissue to radiation exposures during radionuclide therapy and for optimizing of clinical procedures and minimizing of the risks of radiation-induced toxicity [6]. The accurate assessment of radiation dose in the region of interest (ROI) during radiotherapy is an important challenge for medical physicists [7]. The dosimetry scheme recommended by Medical Internal Radiation Dose Committee (MIRD) calculates the organ-level absorbed dose in the human body by using pre-calculated radionuclide *S*-values and time-integrated activities. It commonly assumes a homogeneous activity distribution within source organs and has been widely used in clinical practice [8]. While the MIRD organ-based radiation dosimetry approach does not consider patient-specific radiation activity

doi:10.1088/978-0-7503-2694-0ch15

distributions and individual organ anatomies, the voxel-based radiation dosimetry methods were developed to include the tissue heterogeneity and heterogeneous radiation activity in dose map calculations and provided more accurate absorbed dose estimations in clinics. This section reviews the state-of-the-art voxel-based dosimetry methods applied in radiotherapy and discusses the necessity of fast and accurate voxel-based radiation dosimetry methods for clinical TRT.

15.2 Dose kernel approaches

The dose point-kernel (DPK) method was first suggested by Berger *et al* [9] to include the heterogeneous radiation activity in dose calculations, and was later extended to dose voxel-kernel approach by introducing the voxel *S*-values (VSVs) [10]. These approaches commonly impose the assumption of homogeneous tissue and a fixed kernel for Fast Hartley Transform (FHT) or Fast Fourier Transform (FFT) for dose calculations. The radiation dose estimated by conventional dose kernel methods, either by DPK or VSV, considers the inhomogeneous radioactivity distribution while the tissue heterogeneities are usually ignored and, thus, has been limited for radiation dosimetry for tumor with homogeneous medium such as some lesions in the liver and brain. The DPK method defines a radial absorbed dose distribution surrounding an isotropic source point in homogeneous medium and requires continued calculation of spherical coordinates in Cartesian coordinates for the target voxels. The VSV method employs the MIRD schema at voxel level while the target and source regions are all defined in voxels and the voxel-based *S*-values are provided for dose calculation. It needs pre-calculated voxel-level *S*-value matrix for different isotopes at corresponding voxel resolutions. Therefore, the absorbed dose at voxel k can be calculated as:

$$D_k = \sum_{i=0}^{n} A_i \times VSV_{i \to k} \tag{15.1}$$

where A_i refers to the time-integrated activity at source voxel i and $VSV_{i \to k}$ is the tabulated voxel-level *S*-values with the given medium (water in most conditions). The VSV approach calculates voxel dose map by convoluting the time-integrated activity with the single medium–based VSV and provides better computational efficiency than the DPK method. However, most dose kernel methods do not consider the tissue heterogeneities, thus, their applications have been limited to tumor with homogeneous medium such as some lesions in the liver and brain. To overcome this shortcoming, Lee *et al* [11] proposed a multiple VSV approach for whole-body voxel-based personalized dosimetry in heterogeneous media with non-uniform activity distributions. Instead of utilizing a single VSV, the new approach introduces multiple VSVs for different mediums and densities.

Figure 15.1 shows the multiple VSV approach proposed by Lee *et al* [11] where the tabulated VSVs of Ga-68 point sources were pre-calculated by using Monte Carlo method for elemental compositions and densities of the air, lung, water and bone. The density-specific radiation dose map was calculated by convoluting the time-integrated radiation activity distributions with the VSVs. CT-based

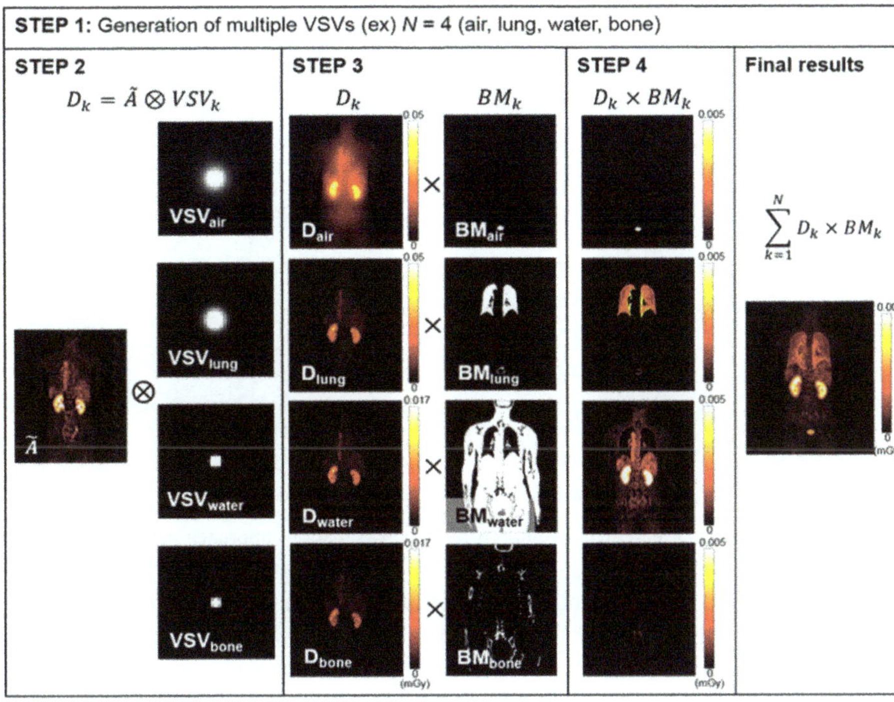

Figure 15.1. The multiple VSV approach proposed by Lee *et al* [11]. Reproduced with permission from [11]. Copyright SNMMI.

segmentation was performed to generate binary mask images for each density. Consequently, the binary mask image was multiplied by the generated density-specific dose map and the final voxel-level absorbed dose was calculated by:

$$D_{\text{voxel}} = \sum_{k=1}^{n} D_k \times (\text{Binary mask})_k \tag{15.2}$$

In the multiple VSV approach, the dose kernel size is carefully selected to cover the radiation dose range of adopted isotope to avoid dose underestimation while the absorbed dose was assumed to be mainly contributed by self-irradiations. The accuracy of calculated radiation dose map can be improved by increasing the number of VSV kernels. The computational burden of dose kernel method is commonly less than direct Monte Carlo calculation.

15.3 Direct Monte Carlo approach

Direct Monte Carlo simulation, which considers both the inhomogeneous distributions of radioactivity and tissue heterogeneity of individual patient for radiation dose calculations, is commonly considered the gold standard for implementing reliable dose map in clinical practice. Many Monte Carlo toolkits have been developed for radiation dosimetry studies, such as MCNP, Geant4, FLUKA,

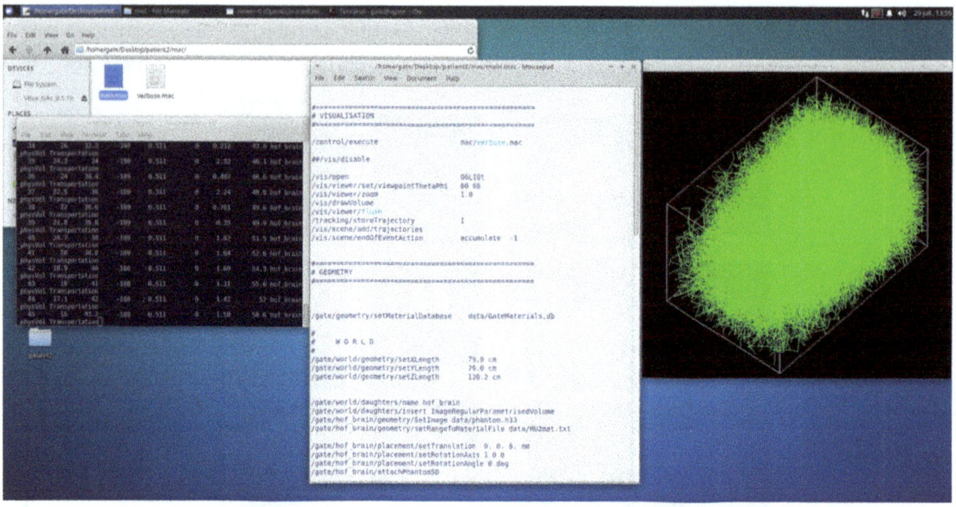

Figure 15.2. Illustration of the direct Monte Carlo-based calculations approach using GATE toolkit.

EGSnrc, GATE, etc. The Geant4 Application for Emission Tomography (GATE) v.8.1 simulation toolkit will be introduced in this section as an example for direct Monte Carlo simulation. GATE is advanced open source software developed by the international OpenGATE collaboration and dedicated to numerical simulations in medical imaging and radiotherapy. The standard electromagnetic physics package was used for simulating photon and electron interactions at each voxel. The voxel dose was obtained by using 'Dose Actor' mechanism. Owing to the long simulation time and extensive computational cost for the Monte Carlo simulation, a total of 10^8 actual events were performed while the statistical uncertainties of the voxel-level dose estimation was less than 0.5% for most voxels.

The PET images and CT images for the same patient were integrated in the input file of GATE for the simulation of voxelized source distribution in the human body. CT Hounsfield Units (HU) were converted into tissue densities and elemental weights for each voxel based on the published database while the PET image was used to determine the voxel radiation activity distribution in the human body. The voxels with the same HU value in the same organ were defined as a specific material for assigning tissue materials of simulations performed on high performance computing system equipped with AMD Ryzen 3960×24-Core 3.8 GHz CPU and 256 GB DDR4 RAM.

Figure 15.2 shows the dose map calculated using direct Monte Carlo approach in GATE under Ubuntu v20.04 operating system. The distribution of ^{18}F-FDG was obtained from PET scan. The dose detector was attached to the voxel-based human model for the collection of deposited voxelized absorbed dose in the human body. The individual tissue heterogeneity and radioactivity heterogeneity in clinical conditions were integrated in the simulations by defining the voxel-level activity distribution, tissue density and elemental weights based on CT images and PET images. The obtained absorbed dose map from direct Monte Carlo simulation can be used as a golden standard for creating the training set for the deep learning approach as long as ground truth for the development of novel radiation dosimetry is approached.

15.4 Deep learning approach

With the increasing processing power afforded by commercial GPUs and the development of more advanced algorithms, the technology of deep learning has gained huge attention in the field of medical image processing and outperformed some conventional approaches for image segmentation, registration and classification [12]. There also have been some applications for using deep learning methods for radiation dosimetry [13–15]. Mardani *et al* [16] reported a deep-learning based prediction approach for calculations of 3D dose volume for personalized inverse treatment planning. Nguyen *et al* [17] modified the U-net network model for dose prediction of the planning target volume and organs at risks. Ma *et al* [18] adopted deep convolutional neural network for isodose distribution prediction in radiation therapy. Kearney *et al* [13] implemented a 3D fully-convolutional neural network for radiation dose prediction. Lee *et al* [14] and Götz *et al* [15, 19] developed deep learning approaches for personalized internal dosimetry for ^{68}Ga- and ^{177}Lu-labelled radiotracers, respectively. Xue *et al* [20] reported a voxel-wise dosimetry prediction method for ^{177}Lu-PSMA I&T therapy using generative adversarial networks (GANs). Mirando *et al* [21] introduced deep-learning based voxel-level dosimetry tools for automated real-time patient-specific dose calculation for SPECT/CT imaging data of ^{177}Lu-DOTATATE. Azadeh *et al* [22] of Geneva University Hospital reported a novel deep learning approach for whole-body voxel-based internal dosimetry of ^{18}F using dose distribution kernels, and provided an accurate and robust dosimetry solution with a small training dataset. However, prohibitive computational cost of Monte Carlo calculations for generation of the training dataset was often required for deep-learning approaches and the implemented network usually could not be used for other isotopes.

In this section, we present a convolutional neural network for voxel-based radiation dosimetry for nuclear medicine and molecular imaging. The absolute radiation activity distribution provided by quantitative nuclear imaging, such as SPECT and PET, and tissue heterogeneity properties given by structure images, such as CT, are input into CNN for generation of 3D dose map. Figure 15.3 shows the geometry of the proposed network.

The suggested model was built upon the U-net architecture which consists of two contracting paths and an expanding path. The contracting path collapses the PET image and CT image down into a set of high-level features while the expanding path uses these generated features to produce the corresponding absorbed dose map. By skipping connections, outputs from early layers are concatenated to the input of subsequent layers to transfer missing information in the analysis path. Unlike conventional U-net architectures, the modified U-net model in figure 15.1 consists of encoder module and decoder module each with three resolution steps of 32, 64, and 128 feature maps. The analysis path is split into two streams, which serve as input to

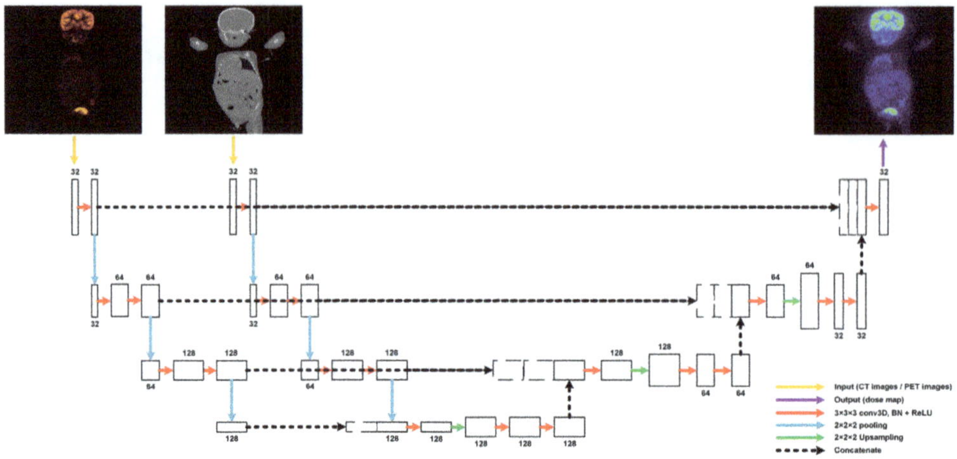

Figure 15.3. The volumetric architecture and the convolution layers and deconvolution layers of the suggested neural network.

CT image and PET image, respectively. The aim for processing different image modalities in separated streams is to disentangle radiation activity information and structure information (such as density) which otherwise would be fused in the analysis stage. In addition to better modeling relationships between different image modalities, we employed the dense connections in the network to improve the flow of information, alleviate the problem of vanishing gradient, introduce an implicit deep supervision into the network and reduce the risk of over-fitting when using small training datasets. The Monte Carlo simulation-based ground truth dose rate map was used as reference for network training. In the encoder module of the network, each layer contains two convolutions each followed by a rectified linear unit (ReLu), and then a max pooling with stride size of $2 \times 2 \times 2$. In the decoder module, each layer consists of an up-sampling with $2 \times 2 \times 2$ strides followed by two convolutions and a ReLu. For each convolutional layer, the size of all convolutional kernels is $3 \times 3 \times 3$. Shortcut connections between layers of equal resolution transfer the high-resolution features from the analysis paths of CT image and PET image to the synthesis path, respectively. The number of channels were doubled before max pooling to avoid bottlenecks. Batch normalization (BN) was introduced before each ReLU to normalize the batches with its mean and standard deviation, thus updating the global statistics in training. The architecture consists of 1,683,205 parameters in total and comprises 28 convolutional layers and a series of pooling options.

The input CT images and PET images and their corresponding dose rate map were used to train the network with the adaptive moment estimation (Adam) implementation of Keras [23] which computes adaptive learning rates for each parameter and keeps an exponentially decaying average of past gradients:

$$m_t = a_1 m_{t-1} + (1 - a_1) g_t \tag{15.3}$$

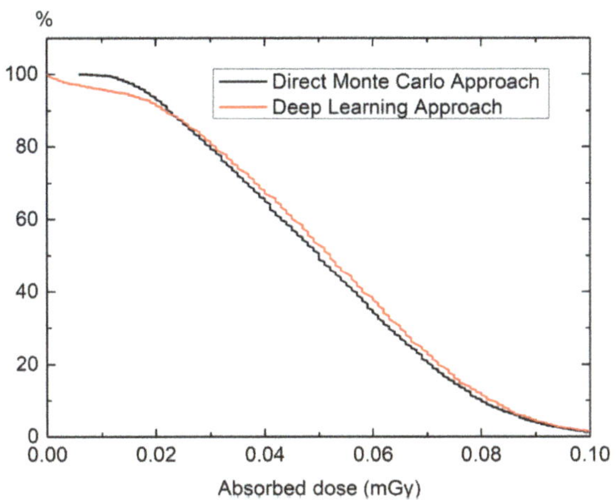

Figure 15.4. Comparison of DVHs of ROI estimated using direct Monte Carlo simulations and deep learning predictions.

$$v_t = a_2 v_{t-1} + \left(1 - a_2\right) g_t^2 \tag{15.4}$$

where m_t and v_t are estimates of the mean and the uncentered variance of the gradients, respectively. g_t is gradients at subsequent time step t and a_1 and a_2 are exponential decay rates with a_1, $a_2 \in [0, 1)$.

The CNN model was implemented using the open source Keras package [23] and ran on NVIDIA 2080Ti GPU with 256 GB of memory. To minimize the overheads and make maximum use of the GPU memory, the network was trained with a mini-batch size of 2 and for 1000 epochs. The learning rate was initialized as 10^{-3} and the network weight was initialized as normal distribution with zero mean and a standard deviation of 0.01. The dose voxel histograms (DVHs) estimated by using direct Monte Carlo simulation and deep learning approach for ^{18}F-FDG in region of interest (ROI) are presented and compared in figure 15.4. The differences for $D_{90\%}$, $D_{50\%}$ and $D_{20\%}$ of direct Monte Carlo simulation and deep learning approach are −4.35%, 4.00% and 2.86%, respectively.

15.5 Conclusions

Voxel-based dosimetry has several advantages over organ-based dosimetry, for example, it provides more accurate absorbed dose calculations in the human body because of the utilization of real activity distributions and individual anatomies from the real patient. The estimated whole-body dose distribution and 3-D dose map from voxel-based radiation dosimetry can provide heterogeneous dose information and be used for calculating dose-volume histograms for the ROI which can be visualized more conveniently.

Direct Monte Carlo approach provides the most reliable radiation dose estimates in heterogeneous media but suffers in the requirements of extensive computation resources. With the development of new VSV method and deep learning approach, fast voxel-based dosimetry can be conducted. The whole-body level voxel-based dose map can provide accurate dosimetric assessment and be helpful for TRT and personalized nuclear medicine. Further studies are required to optimize the CNN structure depending on the radionuclides and irradiation conditions. The development of a clinically feasible framework for accurate patient-specific radiation dosimetry is valuable for targeted radionuclide therapy, precision nuclear medicine and theranostic imaging.

References

[1] Dash A, Chakraborty S, Pillai M R A and Knapp F F Jr 2015 Peptide receptor radionuclide therapy: an overview *Cancer Biother. Radiopharm.* **30** 47–71

[2] Pool S E, Krenning E P, Koning G A, van Eijck C H, Teunissen J J, Kam B, Valkema R, Kwekkeboom D J and de Jong M 2010 *Preclinical and Clinical Studies of Peptide Receptor Radionuclide Therapy* (Amsterdam: Elsevier) pp 209–18

[3] De Jong M, Breeman W A, Kwekkeboom D J, Valkema R and Krenning E P 2009 Tumor imaging and therapy using radiolabeled somatostatin analogues *Acc. Chem. Res.* **42** 873–80

[4] Birn H, Spiegelstein O, Christensen E I and Finnell R H 2005 Renal tubular reabsorption of folate mediated by folate binding protein 1 *J. Am. Soc. Nephrol.* **16** 608–15

[5] Sandoval R M, Kennedy M D, Low P S and Molitoris B A 2004 Uptake and trafficking of fluorescent conjugates of folic acid in intact kidney determined using intravital two-photon microscopy *Am. J. Physiol.-Cell Physiol.* **287** C517–C526

[6] Strigari L, Konijnenberg M, Chiesa C, Bardies M, Du Y, Gleisner K S, Lassmann M and Flux G 2014 The evidence base for the use of internal dosimetry in the clinical practice of molecular radiotherapy *Eur. J. Nucl. Med. Mol. Imaging* **41** 1976–88

[7] Cremonesi M, Ferrari M, Di Dia A, Botta F, De Cicco C, Bodei L and Paganelli G 2011 Recent issues on dosimetry and radiobiology for peptide receptor radionuclide therapy *Q. J. Nucl. Med. Mol. Imaging* **55** 155

[8] Bolch W E, Eckerman K F, Sgouros G and Thomas S R 2009 MIRD pamphlet no. 21: a generalized schema for radiopharmaceutical dosimetry—standardization of nomenclature *J. Nucl. Med.* **50** 477–84

[9] Berger M J 1971 *Distribution of Absorbed Dose Around Point Sources of Electrons and Beta Particles in Water and Other Media* (Washington, DC: National Bureau of Standards)

[10] Bolch W E, Bouchet L G, Robertson J S, Wessels B W, Siegel J A, Howell R W, Erdi A K, Aydogan B, Costes S and Watson E E 1999 MIRD pamphlet no. 17: the dosimetry of nonuniform activity distributions—radionuclide S values at the voxel level *J. Nucl. Med.* **40** 11S–36S

[11] Lee M S, Kim J H, Paeng J C, Kang K W, Jeong J M, Lee D S and Lee J S 2018 Whole-body voxel-based personalized dosimetry: the multiple voxel S-value approach for heterogeneous media with nonuniform activity distributions *J. Nucl. Med.* **59** 1133–39

[12] Ronneberger O, Fischer P and Brox T 2015 *U-Net: Convolutional Networks for Biomedical Image Segmentation* (Berlin: Springer) pp 234–41

[13] Kearney V, Chan J W, Haaf S, Descovich M and Solberg T D 2018 DoseNet: a volumetric dose prediction algorithm using 3D fully-convolutional neural networks *Phys. Med. Biol.* **63** 235022

[14] Lee M S, Hwang D, Kim J H and Lee J S 2019 Deep-dose: a voxel dose estimation method using deep convolutional neural network for personalized internal dosimetry *Sci. Rep.* **9** 1–9

[15] Götz T I, Schmidkonz C, Chen S, Al-Baddai S, Kuwert T and Lang E 2020 A deep learning approach to radiation dose estimation *Phys. Med. Biol.* **65** 035007
[16] Mardani M, Dong P and Xing L 2016 Deep-learning based prediction of achievable dose for personalizing inverse treatment planning *Int. J. Radiat. Oncol. Biol. Phys.* **96** E419–20
[17] Nguyen D, Long T, Jia X, Lu W, Gu X, Iqbal Z and Jiang S 2019 A feasibility study for predicting optimal radiation therapy dose distributions of prostate cancer patients from patient anatomy using deep learning *Sci. Rep.* **9** 1–10
[18] Ma M K, Buyyounouski M, Vasudevan V, Xing L and Yang Y 2019 Dose distribution prediction in isodose feature-preserving voxelization domain using deep convolutional neural network *Med. Phys.* **46** 2978–87
[19] Götz T I, Lang E W, Schmidkonz C, Kuwert T and Ludwig B 2021 Dose voxel kernel prediction with neural networks for radiation dose estimation *Z. Med. Phys.* **31** 23–36
[20] Xue S, Gafita A, Afshar-Oromieh A, Eiber M, Rominger A and Shi K 2021 Voxel-wise prediction of post-therapy dosimetry for ^{177}Lu-PSMA I&T therapy using deep learning *J. Nucl. Med.* **61** 1424
[21] Mirando D, Dewaraja Y, Niedbala J, Wilderman S and Nelson A 2021 Integrating deep-learning based segmentation and 'fast' voxel-level dosimetry tools to enable automated real-time patient specific dosimetry *J. Nucl. Med.* **62** 75
[22] Akhavanallaf A, Shiri I, Arabi H and Zaidi H 2021 Whole-body voxel-based internal dosimetry using deep learning *Eur. J. Nucl. Med. Mol. Imaging* **48** 670–82
[23] Chollet F 2015 Keras: deep learning library for theano and tensorflow (https://keras.io/k7)

Chapter 16

The Monte Carlo method as a design tool in selective internal radiation therapy

Arnaud Dieudonné and Manuel Sánchez-García

16.1 Introduction to liver SIRT

16.1.1 Principle

Selective internal radiation therapy (SIRT), also known as trans-arterial radio-embolization, is a treatment of primary and secondary liver tumours. It consists in the injection of a radiolabelled agent, i.e. glass, resin, or polymer microspheres or lipiodol, via the hepatic artery, which is the main feeding vessel of liver tumours, while the parenchyma is mainly fed by the portal vein. This implies that the SIRT agent will preferentially feed the tumours over the liver parenchyma.

The liver is subdivided into two lobes (left and right), each one subdivided into segments (eight in total). They are all fed independently by a branch of the hepatic artery and have the same metabolic function (see figure 16.1). This parallel structure allows treating selectively by lobar or segmental injection, while preserving the remaining parenchyma.

Contrary to systemic therapies (i.e. molecular radionuclide therapy, peptide receptor radionuclide therapy etc.) SIRT particles are trapped in the capillaries, hence not entering the systemic flow and without metabolization. This implies that the irradiation follows the physical decay of the radionuclide, like brachytherapy with implanted seeds. The dosimetric implication is that a single *in vivo* assessment of biodistribution is required, via emission tomography (ET), i.e. single-photon or positron emission tomography (SPECT or PET).

16.1.2 Radionuclides and vendors

SIRT with ^{90}Y-labelled glass (Theraspheres, Boston Scientific) or resin microspheres (SIRSpheres, SIRTEX Medical) is widely used nowadays. Polymer microspheres labelled with ^{166}Ho (Quiremspheres, Quirem Medical) clinical applications are in

doi:10.1088/978-0-7503-2694-0ch16

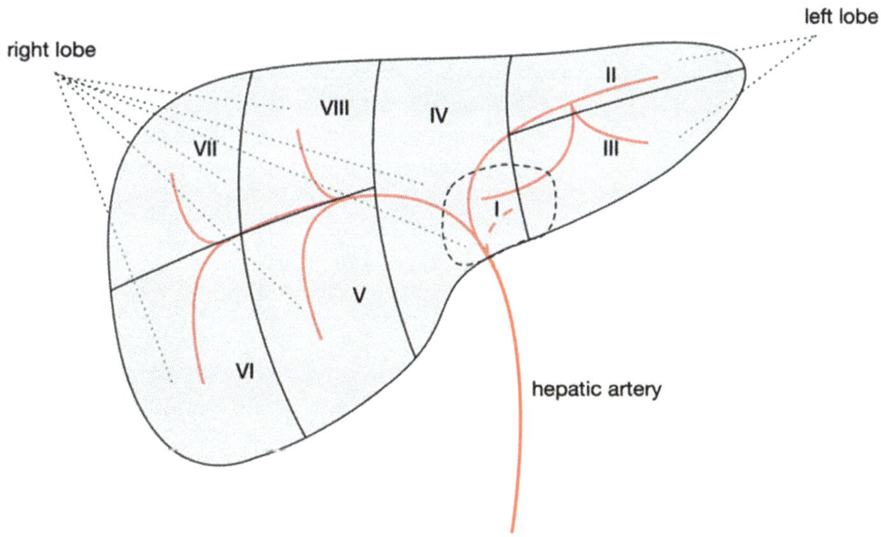

Figure 16.1. Liver segmentectomy and arterial branches.

Table 16.1. Description of materials characteristics for main SIRT agents.

	Manufacturer	Material	Diameter	Activity per sphere
^{90}Y-resin microspheres	SIRTEX	Resin	20–60 μm	50 Bq
^{90}Y-glass microspheres	BTG, Boston Scientific	Glass	20–30 μm	1250–2500 Bq
^{166}Ho-poly microspheres	Quirem	Polymer (L)-lactic acid (PLLA)	20–40 μm	450 Bq
^{131}I-lipiodol	n/a	Lipiodol	n/a	n/a
^{188}Re-lipiodol	n/a	Lipiodol	n/a	n/a

investigation [1], while ^{188}Re-lipiodol is being developed as replacement for ^{131}I-lipiodol [2].

These SIRT agents have all in common mechanical properties (i.e. size, density or viscosity, see table 16.1) making them appropriate to embolize arterioles whose diameter is between 40 and 110 μm. This, together with the use of β^- emitters (table 16.2), makes SIRT a locoregional therapy, which is why SIRT is sometimes referred to as intra-arterial brachytherapy [3].

16.1.3 Treatment phases

The SIRT procedure consists in driving a microcatheter from the femoral artery to the hepatic artery, under angiography, to inject the SIRT agent. Prior to this, a pre-therapeutic angiography is performed to identify the artery feeding the tumour and potential vessels feeding extrahepatic sites such as the gastric wall or the duodenum

Table 16.2. Radionuclides characteristics used in SIRT, data taken from MIRD [4] and NIST ESTAR Program [5].

	Physical half-life	β released energy per disintegration Δ CSDA range in water	Emission tomography characteristics
^{90}Y	64 h	933 keV 4 mm	SPECT of bremsstrahlung PET and pair creation of ^{90}Zr ($3.2\ 10^{-6}$)
^{166}Ho	27 h	665 keV 2.6 mm	SPECT of 81 keV γ emission (0.066)
^{131}I	192 h	182 keV 0.4 mm	SPECT of 365 keV γ emission (0.82)
^{188}Re	17 h	762 keV 3.1 mm	SPECT of 155 keV γ emission (0.16)

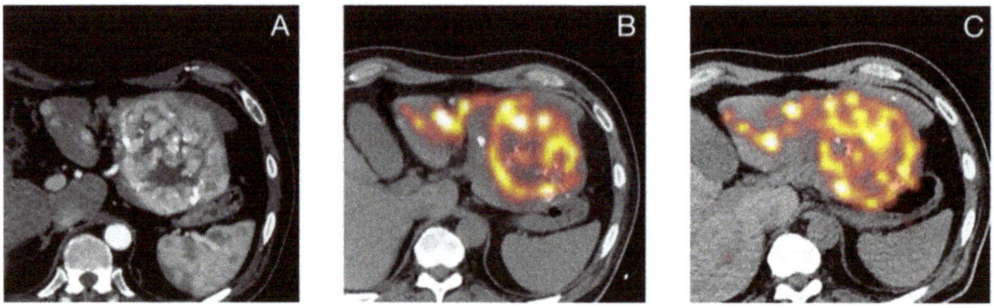

Figure 16.2. Transaxial images of a tumor lodged in the left liver lobe. (A) is a contrast-enhanced CT, (B) is a SPECT/CT of ^{99m}Tc macro-aggregated-albumin particles and (C) is a ^{90}Y-microspheres PET/CT. Fused ET images were smoothed for better rendering.

(figure 16.2). The injection site is determined to maximize the feeding of the tumour while minimizing the feeding of liver parenchyma.

Uptake in extrahepatic sites, mainly in lung and gastroduodenal area, can occur. To prevent any complication related to these unwanted uptakes, a scout dose of ^{99m}Tc labelled macro-aggregated albumin particles (^{99m}Tc-MAA) is injected mimicking the biodistribution of microspheres [6] to reveal any extrahepatic uptake [7]. ^{99m}Tc-MAA particles have the advantage of being highly available at low cost, and have the same mechanical behaviour as microspheres, even though they differ in nature [8].

The intrahepatic biodistribution of ^{99m}Tc-MAA is then measured with SPECT in order to calculate the absorbed dose to the tumour and the liver parenchyma and lungs, using methods that will be presented in sections 16.2 and 16.3. This pre-therapeutic dosimetry can then be used to personalize treatment planning of ^{90}Y microspheres [9, 10]. The scout dose can also be made of the same agent used for the therapeutic injection, with a low amount of radionuclide activity. This has been demonstrated to be superior to ^{99m}Tc-MAA for predicting the ^{166}Ho-microspheres biodistribution [11]. For ^{90}Y-microspheres, a phantom experiment has been conducted to investigate the feasibility of using such scout doses together with PET imaging [12].

Once the prescription activity is calculated, the treatment is delivered at the same injection site as the scout dose, generally one to two weeks after the pre-therapeutic angiography. Following the therapeutic procedure, post-therapeutic imaging through SPECT of PET is performed to confirm the absence of extrahepatic uptake and tumour targeting. Post-therapeutic dosimetry is performed based on this imaging to validate the conformity of the treatment to the pre-therapeutic procedure, requiring quantitative imaging of beta-emitting sources, the challenges of which will be discussed in section 16.4.

16.2 Dosimetry for SIRT

16.2.1 Application of MIRD formalism

According to the Medical Internal Radiation Dose (MIRD) formalism [13], the absorbed dose for a given target tissue r_T is calculated by equation (16.1)

$$D(r_T) = \sum_{r_S} \widetilde{A}(r_S) \times S(r_T \leftarrow r_S) \tag{16.1}$$

with $\widetilde{A}(r_S)$ being the time-integrated activity in source r_S obtained from the integration of $A(r_S, t)$, the activity in r_S over time.

Due to the absence of activity re-localization after injection in SIRT, $\widetilde{A}(r_S)$ can be calculated as

$$\widetilde{A}(r_S) = \frac{A(r_S, t = 0)}{\lambda} \tag{16.2}$$

with $A(r_S, t = 0)$ being the activity at time of therapeutic administration and λ the decay constant of the radionuclide. The measurement of $A(r_S, t = 0)$ through quantitative SPECT or PET will be developed in (16.4).

The value $S(r_T \leftarrow r_S)$ is itself given by

$$S(r_T \leftarrow r_S) = \sum_i \Delta_i \times \phi_i(r_T \longleftarrow r_S) \tag{16.3}$$

Δ_i being the energy released per disintegration of the ith transition, $\Delta_i = E_i \times Y_i$, where E_i is the particle energy of particles and Y_i the yield. $\phi_i(r_T \longleftarrow r_S)$ is the absorbed fraction between source and target, representing the transport of energy from source to target which, as we will see in (16.3), can be evaluated through Monte Carlo techniques.

Knowing this, equation (16.1) becomes:

$$D(r_T) = \frac{\sum_{r_S} A(r_S, t = 0) \times \sum_i \Delta_i \times \phi_i(r_T \longleftarrow r_S)}{\lambda \times M(r_T)} \tag{16.4}$$

$M(r_T)$ being the mass of the target.

This equation is formulated for structures, but it can be trivially extended to continuous activity distributions by replacing the sum in source structures with an integration over a continuous activity distribution, and the mass of the target structure with density at the target point, i.e.

$$D(x) = \frac{\iiint_{\Omega} A(s, t = 0) \times \sum_i \Delta_i \times \phi_i(s, x) \mathrm{d}s}{\lambda \times \rho(x)} \tag{16.5}$$

16.2.2 Treatment planning

The dosimetry planning consists in calculating the activity to administer $A(t = 0)$, prior to injection, as a function of the target dose $D(r_T)$.

This is done by first expressing the absorbed dose per unit of administered activity, also called absorbed dose coefficient [14], as a function of $a(r_T)$, the fraction of injected activity instead of $A(r_S, t = 0)$. $a(r_T)$ can be calculated directly from the scout images either by absolute or relative quantification. In the latter case, $a(r_T)$ is the ratio of the number of counts in r_T versus the sum of liver and lungs. In lungs, this value is also known as lung shunt fraction (LSF).

Second, the activity to administer would then be the result of the following equations:

$$\begin{cases} A(0) \times d(\text{tumor}) \geqslant D_T(\text{tumor}) \\ A(0) \times d(\text{liver}) < D_T(\text{liver}) \\ A(0) \times d(\text{lung}) < D_T(\text{lung}) \end{cases} \tag{16.6}$$

with $D_T(r_T)$, the targeted dose or dose constraint depending on if r_T is a tumor or an organ at risk.

16.2.3 Treatment verification

Following the intra-arterial administration of the SIRT agent, verification of the treatment delivery relies on post-therapeutic dosimetry. Post-therapy quantitative imaging allows for a direct measurement of the activity in $\mathrm{r_S}$ at the time of administration, while the dosimetry planning relies on absorbed dose coefficients with a prescribed activity. Knowing that the administered activity can differ from the prescribed activity, and that $\mathrm{a}(r_T)$ can differ between planning and the treatment phase, the direct measurement of $\mathrm{A}(r_S, t = 0)$ is essential for treatment verification.

16.3 Monte Carlo for energy transport

16.3.1 Calculation of absorbed fraction ϕ

One of the steps for calculating the absorbed dose consists in modelling the transport of the energy released by the isotope through the body tissues. Monte Carlo codes provide comprehensive tools for this purpose, simulating the transport of these particles and recording the energy lost and absorbed at each interaction site. In the MIRD scheme (see section 16.2), the results of energy transport are represented by the absorbed fractions. All combinations of sources and targets need to be computed, therefore complexity of the problem grows as $\mathcal{O}(N_s \times N_T)$, where N_s is the number of sources and N_T the number of targets.

The location of sources and targets can be derived from the specific anatomy of the patient through an x-ray computerized tomography (CT) and emission tomography (ET) scans or from a standardized anatomy and geometry. In the first case, sources and targets are the individual voxels conforming the CT or ET image. In the second case, numerical phantoms are used to represent the patient's anatomy and full organs are usually taken as sources and targets for dosimetry.

These phantoms go from the stylized phantoms developed by International Commission on Radiological Protection (ICRP) and at Oak Ridge National Laboratory in the 1980s and 1990s [15, 16] to the more modern phantoms based on Non-Uniform Rational Basis Splines (NURBS) that can be deformed to match patient characteristics like weight or body mass index [17] or the new standard voxelized phantoms introduced by ICRP [18].

If phantoms are used instead of the actual patient's geometry, absorbed fractions can be pre-calculated before the treatment. Since computation time is no longer a constraint, accurate algorithms, like full Monte Carlo simulations, can be used. However, the usage of phantoms not only imposes a standardized geometry but also a hypothesis like homogeneous intra-organ distribution. Indeed, the ever increasing computing resources at hospitals have made it so that in recent years the interest has shifted to performing energy transport on the actual anatomy of the patient being treated, as obtained from a CT scan, from the actual distribution of the radiopharmaceutical, as obtained through quantitative ET imaging.

In the following sections, several methods for energy transport and their applicability to SIRT are discussed.

16.3.2 Hypothesis of non-penetrating particles

One case met in SIRT where energy transport becomes trivial is the application of the MIRD formalism to pure β emitters when the source volume has dimensions much larger than the range of β particles.

The biodistribution of a radioisotope in units of activity (A in Bq), can be causally related to the total released energy (E) thanks to the radionuclide-specific constant $\frac{\Delta}{\lambda}$ (values on radionuclides used in SIRT can be found on table 16.3):

$$E = A \times \frac{\sum_i \Delta_i}{\lambda} = A \times \frac{\Delta}{\lambda} \tag{16.7}$$

If the distribution of activity in the source volume is considered homogeneous, all energy released in the volume is absorbed in that volume. That is to say, the absorbed fractions are simply reduced to equation (16.8).

Table 16.3. Values of Δ/λ for β components of radionuclides used in SIRT and self-absorbed fractions for spheres of different diameters. The values were calculated using the dose-point-kernel method in water.

	^{90}Y	^{166}Ho	^{131}I	^{188}Re
Δ/λ [10^{-9} J Bq^{-1}]	49.6	16.2	92.0	11.9
Density-scaled diameter of the sphere [g cm^{-2}]	Self-absorbed fraction			
0.5	0.39	0.51	0.29	0.44
1.0	0.63	0.71	0.31	0.64
1.5	0.74	0.79	0.32	0.73
2.0	0.80	0.83	0.33	0.78
2.5	0.84	0.86	0.34	0.81
3.0	0.87	0.87	0.34	0.83
4.0	0.90	0.90	0.35	0.86
5.0	0.92	0.91	0.36	0.87
10.0	0.96	0.94	0.41	0.91
20.0	0.98	0.96	0.49	0.93

$$\phi(r_T \leftarrow r_S) = \begin{cases} 1, & r_T = r_S \\ 0, & r_T \neq r_S \end{cases} \tag{16.8}$$

Thus, equation (16.4) can be expressed in terms of the energy emitted per unit of administered activity, Δ/λ, as:

$$D(r_T) = \frac{\Delta/\lambda}{M(r_T)} A(r_T, t = 0) \tag{16.9}$$

The non-penetrating particles assumption has been extensively used for ^{90}Y SIRT and is implemented in the partition model [19]:

$$D[\text{Gy}] = \frac{50 \times A[\text{GBq}]}{M[\text{kg}]} \tag{16.10}$$

The hypothesis of non-penetrating particles remains valid when the medium is considered infinite compared to the range of beta particles. When this is not the case or there is a significant photon component, one would rather rely on the calculation of self-absorbed fractions.

Self-absorbed fraction coefficients in water for spheres of different sizes and for SIRT radionuclides are presented in table 16.3. Since the medium is homogeneous, the coefficients have been calculated using a dose-point-kernel algorithm as we will describe later in this section. The kernel itself was generated using the Monte Carlo code MCNP6, scoring energy on concentric spherical shells 0.1 mm wide. Graphically, the behaviour can be seen in figure 16.3.

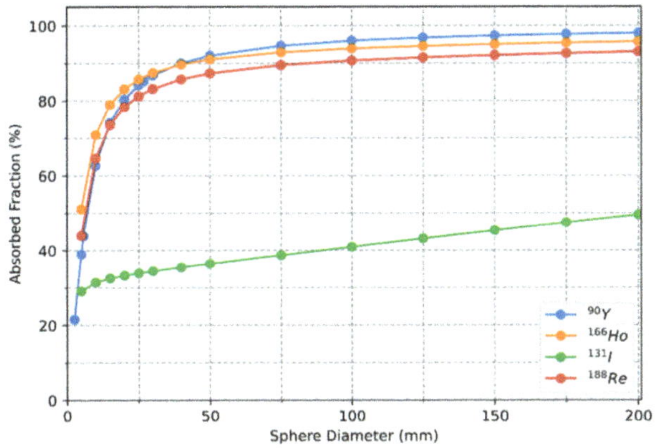

Figure 16.3. Graphical representation of self-absorbed fractions for spheres of different sizes. The values were calculated using the dose-point-kernel method in water.

16.3.3 Direct Monte Carlo

Monte Carlo is the gold standard in terms of accuracy for particle energy transport in internal dosimetry [20, 21]. Most standard Monte Carlo packages can be tuned to perform such calculations [22–24]. This requires carefully choosing the settings of the Monte Carlo code in case and providing ways to convert from CT scans and ET scans to geometry specification and source definition formats that are specific to each Monte Carlo software. To fulfil this need, several dedicated codes are available [25–28]. These dedicated programs may also introduce variance reduction techniques to speed up the computations from the general-purpose codes.

Both general-purpose programs, as well as dedicated ones, are explored in detail in other chapters of this book, including their use for energy transport calculations.

16.3.4 Kernel approach in homogenous medium

In SIRT, the fact that the main organ of interest, the liver, is mostly homogeneous in terms of density [29], allows the use of energy transport computations acceleration.

In a homogeneous medium, the absorbed fraction in equation (16.5) depends only on the distance between source and target. Using this fact and introducing ρ_m as the density of the medium, we define the isotope's dose point kernel for that medium as

$$d_m(\boldsymbol{x}) = \frac{\sum_i \Delta_i \times \phi_i(\boldsymbol{s} = \boldsymbol{0}, \boldsymbol{x})}{\rho_m} \tag{16.11}$$

Then, equation (16.5) takes the form of a convolution integral

$$D(\boldsymbol{x}) = \iiint \frac{A(\boldsymbol{s}, \boldsymbol{t} = \boldsymbol{0})}{\lambda} d_m(|\ \boldsymbol{s} - \boldsymbol{x}\ |)\mathrm{d}\boldsymbol{s} \tag{16.12}$$

The kernel, d_m, can be generated using Monte Carlo simulations [22, 23, 30, 31].

The formalism above is expressed for a continuous distribution of activity, however, in practice the biodistribution of activity is discretized from ET imaging.

The problem is thus divided in volume units, i.e. voxels. The usual approximation is to assume that the activity distribution is uniform within each voxel. Let *i,j,k* be the coordinates of a generic voxel, A_{ijk} be the initial activity, and Ω_{ijk} be the voxel's volume. Then the convolution integral becomes

$$D(\boldsymbol{x}) = \sum_{ijk} \frac{A_{ijk}}{\lambda} \iiint_{\Omega_{ijk}} \mathrm{d}(|\ \boldsymbol{x} - \boldsymbol{s}\ |)\mathrm{d}\boldsymbol{s} \tag{16.13}$$

To integrate the energy deposition over voxels, we express the absorbed dose in voxel *i,j,k* as:

$$D_{ijk} = \sum_{lmn} \frac{A_{lmn}}{\lambda} \iiint_{\Omega_{ijk}} \iiint_{\Omega_{lmn}} \mathrm{d}(|\ \boldsymbol{x} - \boldsymbol{s}\ |)\mathrm{d}\boldsymbol{s}\mathrm{d}\boldsymbol{x} \tag{16.14}$$

The result of the integrals depends only on the relative position of source and target voxels. Therefore, we can arbitrarily set the source voxel at coordinates 0,0,0 and define the dose voxel kernel (DVK), as

$$d_{i,j,k} = \iiint_{\Omega_{ijk}} \iiint_{\Omega_{000}} \mathrm{d}(|\ \boldsymbol{x} - \boldsymbol{s}\ |)\mathrm{d}\boldsymbol{s}\mathrm{d}\boldsymbol{x} \tag{16.15}$$

Thus, equation (16.15) becomes:

$$D_{ijk} = \sum_{lmn} \frac{A_{lmn}}{\lambda} .d_{i-l,\, j-m,\, k-n} \tag{16.16}$$

The dose-voxel-kernels (DVKs), also called voxel *S*-values [32], can be generated by different means:

(a) Monte Carlo simulations where the energy deposition is scored in a voxel matrix around a source voxel. In practice, the symmetry of the problem can be used as only one octant of the space is needed to score energy in (see figure 16.4). It should be noted that particles must be tracked to some extent outside the scoring octant as they can be backscattered into it.
(b) Numerical integration of equation (16.15). This requires performing a six-dimensional integral for each voxel in the DVK. An efficient method is Monte Carlo integration, where the points at which the integrand is evaluated are chosen randomly rather than on a regular grid.
(c) Analytical calculation from a fine resolution DVK [33] or by a parametric function [34], both approaches rely on pre-calculated Monte Carlo DVKs.

The advantage of equation (16.16) is that it can be implemented using a Fast Fourier Transform, making the DVK approach highly computationally efficient.

16.3.5 Kernel approach in heterogeneous medium

Handling a heterogeneous medium, such as lungs or liver/lung interface, requires slightly more sophisticated algorithms. To start, the notation of equation (16.5) can be simplified by introducing the concept of energy deposition kernel (EDK), defined as

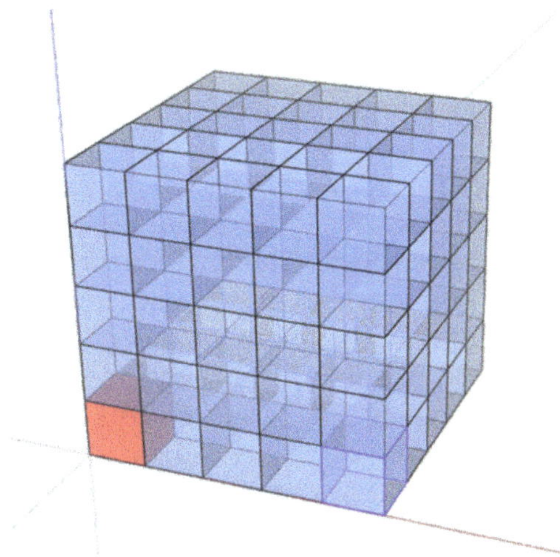

Figure 16.4. Illustration of a scoring geometry for DVK generation. The active voxel (red) is placed in one corner of the voxel matrix rather than at the center. The missing octants can be obtained from symmetry considerations. It should be noted that particles going outside the scoring geometry can still be backscattered into it.

$$h(\boldsymbol{x}) = \sum_i \Delta_i \times \phi_i(\boldsymbol{s}, \boldsymbol{x}) \tag{16.17}$$

Using this and defining the time-integrated activity $\widetilde{A}(\boldsymbol{s}) = A(\boldsymbol{s}, 0)/\lambda$, equation (16.5) becomes

$$D(\boldsymbol{x}) = \frac{\iiint \widetilde{A}(\boldsymbol{s})h(\boldsymbol{s}, \boldsymbol{x})\mathrm{d}\boldsymbol{s}}{\rho(\boldsymbol{x})} \tag{16.18}$$

The interpretation of this equation is straightforward: dose at a point x is computed as the energy density absorbed at that point divided by the mass density. In turn, energy density is computed as the sum over source locations s of the number of total decays $\tilde{A}(s)$ multiplied by the fraction of emitted energy at s that is absorbed at x ($h(s,x)$).

A simple method for handling heterogeneities is to make the homogeneous medium approximation only in the kernel but keep the local value of medium density. Thus, the energy deposition kernel for the medium, h_m, depends only on the distance between source and target points, so equation (16.18) becomes

$$D(\boldsymbol{x}) = \frac{\iiint \widetilde{A}(\boldsymbol{s})h_m(|\ \boldsymbol{s} - \boldsymbol{x}\ |)\mathrm{d}\boldsymbol{s}}{\rho(\boldsymbol{x})} \tag{16.19}$$

The integral becomes a convolution, so it is amenable to being solved with a Fast Fourier Transform and can be discretized using an energy voxel kernel like the DVK from section 16.3.4.

$$D_{ijk} = \frac{\sum_{lmn} \tilde{A}_{lmn} \cdot h_{i-l,\, j-m,\, k-n}}{\rho_{ijk}} \tag{16.20}$$

This works reasonably well for SIRT since we are dealing mostly with β^- emitters, where the range of electrons in tissue is limited, making energy transport depending mostly on short range heterogeneities. This method provides good estimates for average absorbed doses both for lung and liver in one single calculation. However, it shows its limitations at the interface between both organs [35].

A different strategy involves solving the superposition integral of equation (16.18), which requires determining h for each pair of $\boldsymbol{x}$ and $\boldsymbol{s}$. This can be done with Monte Carlo modelling, however, the advantage of the analytical resolution of the energy deposition would be lost. Another solution is to express h as a function of h_w, the energy deposition kernel in water [36, 37].

For β^- sources in a homogeneous medium, this can be achieved by scaling h_w with the effective medium density, λ_m, which is proportional to the relative mass density between the medium m, and water w [38].

$$\lambda_m = \kappa_m \frac{\rho_m}{\rho_w} \tag{16.21}$$

The energy deposition kernel in the medium would then be scaled from the water kernel using:

$$h_m = \lambda_m \lambda_m^2 h_w(\lambda_m \mid \boldsymbol{x} - \boldsymbol{s} \mid) \tag{16.22}$$

The κ_m factors describe how the behaviour in different media deviates from simple density scaling due to the medium composition. For low energy photons, a more complex methodology is needed to account for the photoelectric effect. Figure 16.5 shows the energy deposition kernels calculated for ^{90}Y beta emissions in bone, soft tissue and lung compared to a water kernel scaled by each medium's effective density.

The common approach in a heterogeneous medium is then to obtain h scaling h_w by the mean effective density $\boldsymbol{\lambda_{xs}}$ between the point s of energy release and the point **x** of energy deposition. In this approximation, h(**s**,**x**) can be expressed using the well-known expression from Ahnesjö *et al* [39]:

$$h_{het}(\boldsymbol{s}, \boldsymbol{x}) = \lambda(\boldsymbol{x}) \lambda_{xs}^2 h_w(\lambda_{xs} \mid \boldsymbol{x} - \boldsymbol{s} \mid) \tag{16.23}$$

where $\lambda(x)$ is the effective medium density at point x. Using this expression for the energy deposition kernel, we finally get:

$$D(\boldsymbol{x}) = \frac{\iiint \widetilde{A}(\boldsymbol{s}) \lambda(\boldsymbol{x}) \lambda_{xs}^2 h_w(\lambda_{xs} \mid \boldsymbol{x} - \boldsymbol{s} \mid) \mathrm{d}\boldsymbol{s}}{\rho(\boldsymbol{x})} \tag{16.24}$$

Finally, numerical integration of the equation above can be efficiently achieved using the collapsed cone algorithm [40, 41].

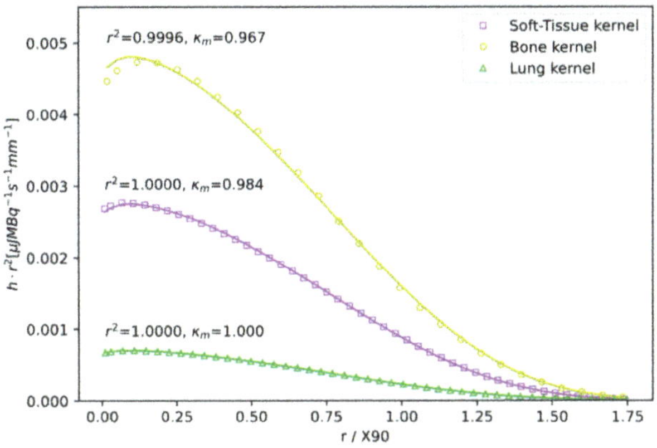

Figure 16.5. Scaled water kernels (solid lines) versus Monte Carlo generated dose point kernels (markers) in different media. k_m was fitted to the Monte Carlo kernels. The coefficients of determinations, r^2, are shown beside each curve. Kernels are displayed as function of the reduced radius (r/X90), where X90 is defined as the radius of the sphere in which 90% of the energy is absorbed. Copyright Institute of Physics and Engineering in Medicine. Reproduced from [40] by permission of IOP Publishing. All rights reserved.

16.3.6 Local deposition

The basic idea of local deposition for energy transport is to assume that all energy released at any given point is absorbed at that same point. Formally, this can be expressed using a Dirac's delta for the absorbed fraction in equation (16.5)

$$\phi_i(\mathbf{s}, \mathbf{x}) = \delta(\mathbf{s} - \mathbf{x}) \quad \forall \quad i \tag{16.25}$$

Then, the integral is trivially solved and dose at any given point can be computed by multiplying the number of decays by the mean released energy per decay and dividing the result by the local density:

$$D(\mathbf{x}) = \frac{A(\mathbf{s}, t = 0)\sum_i \Delta_i}{\lambda \rho(\mathbf{x})} = \frac{A(\mathbf{s}, t = 0)}{\rho(\mathbf{x})}\frac{\Delta}{\lambda} \tag{16.26}$$

The local deposition approximation can be applied in several situations. The first being when charged particle equilibrium is met. This happens when the number of charged particles leaving a volume is equal to the number entering it. In SIRT this holds when the biodistribution of activity is uniform on the patient over distances much larger than the range of the emitted particles from the isotope. In this case, at each point, the energy released that is absorbed in the surroundings is compensated by energy released in the surroundings that is absorbed locally.

A second situation where local deposition may be applied is when the range of emitted particles is much smaller than the voxel size. This may be the case for the beta component of ^{131}I and, to a lesser extent, ^{166}Ho. It should be noted, though, that the β– component of ^{131}I represents only about one third of the total released energy.

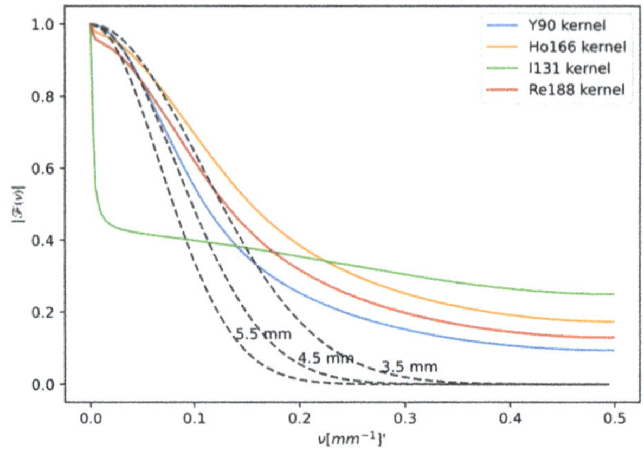

Figure 16.6. Comparison of dose-point-kernels to Gaussian PSFs for the isotopes relevant in radioembolization. The comparison is performed in the Fourier space for clarity.

Another application surfaces when we consider the finite spatial resolution of the imaging systems. Under the assumption that the point spread function (PSF) representing the system's resolution is stationary, the measured activity biodistribution is the convolution of the actual one with the system's PSF. On the other hand, it has been shown that (see section 16.3) for a homogeneous medium, the absorbed dose can be expressed as the convolution of the actual activity biodistribution with a dose point kernel. Formally:

$$\begin{aligned} D(\boldsymbol{x}) &= \widetilde{A} \otimes d \\ \widetilde{A}_{\text{measured}} &= \widetilde{A} \otimes psf \end{aligned} \tag{16.27}$$

Or, in the Fourier space:

$$\begin{aligned} \mathcal{F}(D(\boldsymbol{x})) &= \mathcal{F}(\widetilde{A}) \times \mathcal{F}(d) \\ \mathcal{F}(\widetilde{A}_{\text{measured}}) &= \mathcal{F}(\widetilde{A}) \times \mathcal{F}(psf) \end{aligned} \tag{16.28}$$

The formal similarity between both equations is apparent. If the system's PSF is close enough to the isotope's DPK, the argument can be made that the imaging system already mimics the energy transport and local deposition can be applied.

Monte Carlo techniques can be used to verify the validity of this hypothesis, by comparing Monte Carlo generated DPKs for the isotopes used in SIRT with Gaussian PSFs. Figure 16.6 illustrates such a comparison, where the curves are visually compared in the Fourier space for presentation clarity.

From the figure, the ^{90}Y kernel matches the 4.5 mm Gaussian for frequencies up to 0.1 mm^{-1}, which corresponds to the Nyquist frequency for voxels of 5 mm side. For higher frequencies, i.e. smaller voxels, the two curves start to diverge. State-of-the-art PET systems are capable of 4–5 mm resolutions, justifying the use of local deposition for dosimetry calculations in ^{90}Y when imaged with such systems [42–44].

On the opposite extreme, ^{131}I does not match any of the Gaussians, due to the combination of very low energy beta particles, that preserve high frequencies in the activity biodistribution and the relative importance of the photon component, which contributes mostly to the low frequencies in the curve.

^{166}Ho and ^{188}Re do not have as dramatic a departure from the Gaussian behaviour as ^{131}I but significant differences are apparent, with peaks at low frequencies coming from the photon component and relatively flat curves at high frequencies due to the beta component.

16.3.7 Micro-scale biodistribution

All aforementioned approaches consider the unit of volume to be the voxel of CT or ET images. The implicit assumption is made that activity distribution within a voxel is uniform, whilst the microspheres travel within the blood vessels and induce dose inhomogeneities at the microscopic level varying according to the concentration of microspheres/clusters per volume unit. This results in a broader distribution of dose values, leading to a potential lower biological effect for a given average absorbed dose (at the macroscopic scale) [45–47]. In order to have a 3D dosimetry approach taking into account the non-uniform distribution of activity from clusters of microspheres, the role of Monte Carlo is two-fold: to simulate the random deposition of heterogeneous distribution of microspheres as clusters and generating dose-point-kernels for these cluster sources [45, 48]. To this purpose, data published from microscopic microspheres distribution in human or animals *ex vivo* samples are available (see table 16.4).

Table 16.4. List of studies and their notable results on microscopic microspheres distribution assay.

Reference	Type	Subjects	Diameter	Number of infused microspheres	Cluster population median mean max
Pillai *et al* [45]	polystyrene microspheres (Polysciences Inc. USA)	Rabbits with hepatic VX2 tumours	27 μm	1.5×10^7 3×10^7	2 3.3–3.6 25+ 3 4.4–8.1 25+
Campbell *et al* [46]	resin microspheres (SIRTEX Medical)	Patient with a liver tumour	32 μm	6×10^7	5 ? 98
Pasciak *et al* [49]	glass microspheres (BTG International)	Pigs	20–30 μm[a]	2.20×10^5 to 8.8×10^6	2.04–2.60 4.73–5.02 ?

[a]BTG package insert data

16.4 Monte Carlo for quantitative imaging of beta emitters

Quantification of the distribution of activity is a pre-requisite for computing absorbed dose. Indeed, while the anatomic information required for energy deposition can be standardized (see section 16.3), activity quantification requires an *in vivo* assay through 3D imaging. Common nuclear medicine imaging devices are amenable to quantification for SIRT radionuclides such as ^{90}Y, ^{166}Ho, ^{188}Re and ^{131}I. Their use is, however, not absent of challenges. The particularity of these radionuclides is the presence of a strong β^- component, sometimes accompanied by a very low photon component (see table 16.2). However, imaging devices in nuclear medicine departments are generally optimized for gamma emitters, such as ^{99m}Tc, ^{123}I and ^{111}In for SPECT, and positron emitters ^{18}F, ^{68}Ga for PET. The difference is significative since SPECT and PET reconstruction for quantification purposes needs to take into account the physical effects altering the acquired signal, through the correction of attenuation, scattering, non-stationary spatial resolution, and random events and time-of-flight for PET.

16.4.1 Bremsstrahlung and scatter modelling for SPECT

The SPECT acquisition of β^- emitters is the sum of the contributions of multiple components due to the generation of bremsstrahlung photons inside the patient and the photons scattering. Monte Carlo simulations can be used to model the processes taking places at the emission point, inside the patient and the detectors. This has proven useful in characterizing the different sources of photons and the influence of bremsstrahlung on the imaging performance for different camera set-ups.

The different sources contributing to the photon spectrum can be detailed as follows [50, 51]:

1. geometric γ photons;
2. scattered γ photons[1]
3. geometric bremsstrahlung photons;
4. scattered bremsstrahlung photons
5. backscattered γ photons[1]
6. backscattered bremsstrahlung photons;
7. secondary bremsstrahlung photons issued from secondary electrons;
8. characteristic photons from interactions of bremsstrahlung photons with the collimator lead.

The relative importance of these components will greatly depend on the radionuclide, the relative yield of β^- emissions compared to the γ emissions, and the energy range of the collimator, i.e. low, medium, or high.

Heard *et al* [51] used the EGSnrc Monte Carlo code to optimize energy window and collimator set-ups on an ADAC Forte gamma camera for ^{90}Y. For a β^- source, most energy loss events do not produce photons contributing to the final image, therefore a full simulation accumulating enough statistics in the detectors would be

[1] For gamma emitters (^{131}I, ^{166}Ho, ^{188}Re).

prohibitively expensive. Instead, they simulated ^{90}Y point sources in water to generate probability distribution functions (PDFs) of bremsstrahlung photon energy and spatial distribution. In their work, they found that only 1.85% of beta decays resulted in bremsstrahlung photons that entered the acquisition energy window (50–300 keV) and contributed to the final image. Then, these PDFs were used as extended photon sources placed around the original β^- ones. Finally, photon interactions were simulated with EGSnrc up to the gamma-camera detector. With this technique, they were able to compare the performance of low and medium-energy-general-purpose collimators (LEGP and MEGP) for various window set-ups, from 50 to 300 keV. As expected, the MEGP collimator outperformed LEGP for limiting backscatter and septal penetration, which resulted in a contrast recovery 2–20 times greater with the MEGP collimator, despite a LEGP sensitivity five times greater.

Rather than converting ^{90}Y into a photon distribution with photon source PDFs Takahashi *et al* [52] integrated the PDFs directly into the MCEP Monte Carlo code [53]. They simulated an acquisition of a phantom with spheres filled with ^{90}Y at various levels of concentration of ^{90}Y (74 to 2246 kBq mL^{-1}) on a two-headed SPECT camera with a NaI crystal 9.4 mm thick and a high-energy-general-purpose (HEGP) collimator. They found that 5.3% of beta decays resulted in photons with energy larger than 10 keV. After photoelectric and Compton interactions in the phantom, only 0.36% of the decays resulted in photons in the 105–195 keV window, which is typically the energy range of most SPECT radionuclides. Out of the photons in the energy window, around 50% were scattered at least once (25% in the phantom, and 25% in the collimator) and about 50% did not interact with the water or the collimator, thus preserving the information on the original position of the source. Therefore, the signal is not clean within the energy windows, thus the need for accurate scattering modelling in the reconstruction. This is vastly different from the radionuclides for which commercial reconstruction software is optimized.

Regarding the validity of this technique,[2] Rault *et al* [54] generated PDFs for their use with the GATE simulation software and they also compared them to actual measurements in order to evaluate the accuracy of bremsstrahlung simulations. They observed a large overestimation of low energy photons fluence (below 75 keV), but there still resulted a good concordance between simulated and measured point spread function with a medium energy collimator on a PRISM 3000 XP+ SPECT camera (22.4 versus 23.5 mm), showing that the low energy photons contribute little to the final image.

Accurate estimations of bremsstrahlung photons remain relevant even for isotopes like ^{188}Re which has a γ emission of 155 keV at a relatively low intensity of 16% together with the main β^- decay (maximum energy of 2.1 MeV). Indeed, the 155 keV peak is contaminated by scatter from higher energy γ emissions and bremsstrahlung photons. In this study, Monte Carlo simulations were performed to evaluate the performances of scatter correction with a triple energy window

[2] Interestingly, on a lower energy window (76–105 keV), around 5% of the photons hitting the detector were formed in the collimator, i.e. lead characteristic photons.

technique. For that purpose, acquisition of a sphere (1 cm radius) and a point source were simulated with Monte Carlo code GATE [55]. The results allowed validation of the usage of the triple energy window technique for ^{188}Re.

16.4.2 SPECT reconstruction

Quantitative reconstruction requires one to properly model the contribution within the imaging energy window of the photons having undergone any interaction. This is possible in iterative algorithms, by building forward projectors accordingly. The complexity derived from the fact that we have a continuous spectrum, invalidates scatter correction methods that were developed for simple gamma emissions, i.e. discrete spectrum.

For ^{90}Y, which is a pure β^- emitter, Minarik *et al* [56] have implemented a scatter correction based on the effective source scatter estimation (ESSE [57]) and a collimator depth-response (CDR) compensation for General Electric Discovery MH camera. They used the Monte Carlo code SIMIND to generate both ESSE and CDR kernels. Rong *et al* [58] then developed the method further by implementing a multiple energy range (MER) schema, as opposed to single energy range (SER), where they modelled bremsstrahlung photons of different energy ranges to take into account the photon beam hardening. At each range, they applied attenuation correction, CDR compensation and ESSE into an OS-EM iterative reconstruction. They managed to show improved activity recovery with MER compared to SER on a physical phantom experiment that was performed on a Philips Precedence camera, using an HEGP collimator with an energy window between 100 and 500 keV.

Alternatively, Elschot *et al* [59] applied a full Monte Carlo based SPECT reconstruction method (MC-SPECT) on patient data. Their method used their UMCS Monte Carlo code to generate a system's matrix including scatter of bremsstrahlung photons individualized for the patient's geometry. For CDR they used kernels generated using the MCNPX Monte Carlo by introducing the blurring effect of β^- particles. All modelling was energy dependent as proposed by Rong *et al* [60]. They validated the method on a NEMA Body phantom acquisition and achieved similar performance between MC-SPECT and ^{90}Y-PET.

Besides pure β^- emitters, accurate modelling of bremsstrahlung photons remains relevant of isotopes, like ^{166}Ho, which is a β^- emitter (maximum energy of 1.9 MeV), but also has a γ emission of 81 keV, intensity of 6.6%, that is suitable for SPECT imaging. Elschot *et al* [61] showed how full Monte Carlo modelling significantly improved the quantification performances of SPECT with respect to a standard reconstruction method using analytical modelling. On a NEMA Body phantom acquisition they showed a 70% increase of contrast recovery for spheres over 25 mm of diameter and a reduction of 50% in counts detected in the lung insert, void of activity, while no noise reduction was observed in the background region. They achieved these results using the Monte Carlo code UMCS to model the CDR energy dependency, the collimator-detector degrading effect and scatter for the γ and bremsstrahlung emission of ^{166}Ho. They used 8 energy bins from 75 to 2000 keV for an acquisition with a 75–87 keV window and medium energy collimator. MCNPX was used to generate CDR for each energy bin.

Finally, the same Monte Carlo method was extended by van Rooij *et al* [62] to model the crosstalk in dual isotope SPECT for simultaneous acquisition of ^{99m}Tc-phytate for healthy liver delineation and ^{166}Ho-microspheres as a pre-treatment dose.

16.4.3 Bremsstrahlung and scatter modelling for PET

PET imaging of ^{90}Y is possible thanks to internal pair creation at the 0^+–0^+ transition of ^{90}Zr following the β^- decay with a branching ratio of 31.86×10^{-6} [63]. The PET detection characteristics have been studied and the different components contributing to the detection of coincidences are listed below [64]:

1. Annihilation photons;
2. scattered annihilation photons;
3. 511 keV bremsstrahlung;
4. ^{176}Lu photons (LSO).

Strydhorst *et al* [65] have evaluated the sources of quantification error of a Siemens Biograph mCT PET using the GATE simulation software. They have simulated the bremsstrahlung production, as well as the internal pair production leading to positron emission, as well as the background activity of ^{176}Lu inherent of the LSO crystals of the PET system. Bremsstrahlung PDFs were first generated around a point source up to 12 mm from the source and up to 2000 keV. They simulated an acquisition of a NEMA Body phantom with filled spheres. They showed a comparable contrast recovery coefficient compared to ^{18}F acquisitions, but overestimated the performances compared to experiment with the same PET system [64]. These simulations showed the comparable impact of bremsstrahlung photons and ^{176}Lu decay photons to true and random coincidences.

In the aforementioned study by Takahashi *et al* [52] SPECT and PET detection characteristics towards, respectively, the emission of bremsstrahlung and positron of ^{90}Y were compared. While the bremsstrahlung emission abundance is 1500 times greater than that of the annihilation photons, the better efficiency of PET detection compared to SPECT leads to a comparable number of events contributing to the image. They concluded that PET imaging performance was superior to that of SPECT, but without considering the ^{176}Lu decay photons in the PET simulations.

References

[1] Klaassen N J M, Arntz M J, Gil Arranja A, Roosen J and Nijsen J F W 2019 The various therapeutic applications of the medical isotope holmium-166: a narrative review *EJNMMI Radiopharm. Chem.* **4** 19

[2] Bouvry C, Palard X, Edeline J, Ardisson V, Loyer P, Garin E and Lepareur N 2018 Transarterial radioembolization (TARE) agents beyond ^{90}Y-microspheres *Biomed Res. Int.* **2018** 1435302

[3] Morgan B, Kennedy A S, Lewington V, Jones B and Sharma R A 2011 Intra-arterial brachytherapy of hepatic malignancies: watch the flow *Nat. Rev. Clin. Oncol.* **8** 115–20

[4] Eckerman K F and Endo A 2007 *MIRD: Radionuclide Data and Decay Schemes* vol 50 (Reston, VA: Society of Nuclear Medicine)

[5] Berger M J, Coursey J S, Zucker M A and Chang J 2005 ESTAR, PSTAR, and ASTAR: computer programs for calculating stopping-power and range tables for electrons, protons, and helium ions (version 1.2.3) http://physics.nist.gov/Star

[6] Lau W Y, Leung W T, Ho S, Leung N W Y, Chan M, Lin J, Metreweli C, Johnson P and Li A K C 1994 Treatment of inoperable hepatocellular carcinoma with intrahepatic arterial yttrium-90 microspheres: a phase I and II study *Br. J. Cancer* **70** 994–9

[7] Bailey J J, Dewaraja Y, Hubers D, Srinivasa R and Frey K A 2017 Biodistribution of ^{99m}Tc-MAA on SPECT/CT performed for ^{90}Y-radioembolization therapy planning: a pictorial review *Clin. Transl. Imaging* **5** 473–85

[8] Van de Wiele C, Maes A, Brugman E, D'Asseler Y, De Spiegeleer B, Mees G and Stellamans K 2012 SIRT of liver metastases: physiological and pathophysiological considerations *Eur. J. Nucl. Med. Mol. Imaging* **39** 1646–55

[9] Lau W-Y *et al* 2012 Patient selection and activity planning guide for selective internal radiotherapy with yttrium-90 resin microspheres *Int. J. Radiat. Oncol. Biol. Phys.* **82** 401–7

[10] Salem R *et al* 2019 Clinical and dosimetric considerations for Y90: recommendations from an international multidisciplinary working group *Eur. J. Nucl. Med. Mol. Imaging* **46** 1695–704

[11] Reinders M T M, Smits M L J, van Roekel C and Braat A J A T 2019 Holmium-166 microsphere radioembolization of hepatic malignancies *Semin. Nucl. Med.* **49** 237–43

[12] Khazaee M, Kamali-Asl A, Geramifar P and Rahmim A 2017 Low-dose ^{90}Y PET/CT imaging optimized for lesion detectability and quantitative accuracy *Nucl. Med. Commun* **38** 985–97

[13] Loevinger R, Budinger T F and Watson E E 1991 *MIRD Primer for Absorbed Dose Calculations* (New York: Society of Nuclear Medicine)

[14] Bolch W E, Eckerman K F, Sgouros G and Thomas S R 2009 MIRD pamphlet No. 21: a generalized schema for radiopharmaceutical dosimetry—standardization of nomenclature *J. Nucl. Med.* **50** 477–84

[15] ICRP 1975 *Report of the Task Group on Reference Man* ICRP Publication 23

[16] Cristy M and Eckerman K F 1987 *Specific Absorbed Fractions of Energy at Various Ages from Internal Photon Sources: 7 ORNL/TM-8381 V1–V7* (Oak Ridge, TN: Oak Ridge National Laboratory)

[17] Segars W P, Sturgeon G, Mendonca S, Grimes J and Tsui B M W 2010 4D XCAT phantom for multimodality imaging research *Med. Phys.* **37** 4902–15

[18] Menzel H G, Clement C and DeLuca P 2009 Realistic reference phantoms: an ICRP/ICRU joint effort *Ann. ICRP* **39** 3–5

[19] Ho S, Lau W Y, Leung T W, Chan M, Ngar Y K, Johnson P J and Li A K 1996 Partition model for estimating radiation doses from yttrium-90 microspheres in treating hepatic tumours *Eur. J. Nucl. Med.* **23** 947–52

[20] Andreo P 1991 Monte Carlo techniques in medical radiation physics *Phys. Med. Biol.* **36** 861–920

[21] Zaidi H 1999 Relevance of accurate Monte Carlo modeling in nuclear medical imaging *Med. Phys.* **26** 574–608

[22] Botta F *et al* 2011 Calculation of electron and isotopes dose point kernels with fluka Monte Carlo code for dosimetry in nuclear medicine therapy *Med. Phys.* **38** 3944
[23] Ferrer L, Chouin N, Bitar A, Lisbona A and Bardiès M 2007 Implementing dosimetry in GATE: dose-point kernel validation with GEANT4 4.8.1 *Cancer Biother. Radiopharm.* **22** 125–9
[24] Waters L S, McKinney G W, Durkee J W, Fensin M L, Hendricks J S, James M R, Johns R C and Pelowitz D B 2007 *The MCNPX Monte Carlo Radiation Transport Code AIP Conf. Proc.* vol 896 (Melville, NY: AIP Publishing LLC) pp 81–90
[25] Kost S D, Dewaraja Y K, Abramson R G and Stabin M G 2015 VIDA: a voxel-based dosimetry method for targeted radionuclide therapy using Geant4 *Cancer Biother. Radiopharm.* **30** 16–26
[26] Chiavassa S, Aubineau-Lanièce I, Bitar A, Lisbona A, Barbet J, Franck D, Jourdain J R and Bardiès M 2006 Validation of a personalized dosimetric evaluation tool (Oedipe) for targeted radiotherapy based on the Monte Carlo MCNPX code *Phys. Med. Biol.* **51** 601–16
[27] Prideaux A R, Song H, Hobbs R F, He B, Frey E C, Ladenson P W, Wahl R L and Sgouros G 2007 Three-dimensional radiobiologic dosimetry: application of radiobiologic modeling to patient-specific 3-dimensional imaging-based internal dosimetry *J. Nucl. Med.* **48** 1008–16
[28] Besemer A E, Yang Y M, Grudzinski J J, Hall L T and Bednarz B P 2018 Development and validation of RAPID: a patient-specific monte carlo three-dimensional internal dosimetry platform *Cancer Biother. Radiopharm.* **33** 155–65
[29] Kennedy A, Dezan W and Weiss A 2011 Patient specific 3D image-based radiation dose estimates for ^{90}Y microsphere hepatic radioembolization in metastatic tumors *J. Nucl. Med. Radiat. Ther.* **2** 1–8
[30] Simpkin D J and Mackie T R 1990 EGS4 Monte Carlo determination of the beta dose kernel in water *Med. Phys.* **17** 179–86
[31] Mainegra-Hing E, Rogers D W O and Kawrakow I 2005 Calculation of photon energy deposition kernels and electron dose point kernels in water *Med. Phys.* **32** 685
[32] Bolch W E *et al* 1999 MIRD pamphlet No. 17: the dosimetry of nonuniform activity distributions radionuclide S values at the voxel level. Medical Internal Radiation Dose Committee *J. Nucl. Med.* **40** 11S–36S
[33] Dieudonné A, Hobbs R F, Bolch W E, Sgouros G and Gardin I 2010 Fine-resolution voxel S values for constructing absorbed dose distributions at variable voxel size *J. Nucl. Med.* **51** 1600–7
[34] Amato E, Minutoli F, Pacilio M, Campenni A and Baldari S 2012 An analytical method for computing voxel S values for electrons and photons *Med. Phys.* **39** 6808–17
[35] Dieudonné A, Hobbs R F, Lebtahi R, Maurel F, Baechler S, Wahl R L, Boubaker A, Le Guludec D, Sgouros G and Gardin I 2013 Study of the impact of tissue density heterogeneities on 3-dimensional abdominal dosimetry: comparison between dose kernel convolution and direct Monte Carlo methods *J. Nucl. Med.* **54** 236–43
[36] Ahnesjö A 1989 Collapsed cone convolution of radiant energy for photon dose calculation in heterogeneous media *Med. Phys.* **16** 577
[37] Ahnesjö A and Aspradakis M M 1999 Dose calculations for external photon beams in radiotherapy *Phys. Med. Biol.* **44** R99–155
[38] Cross W G 1968 Variation of beta dose attenuation in different media *Phys. Med. Biol.* **13** 611–8

[39] Ahnesjö A, Andreo P and Brahme A 1987 Calculation and application of point spread functions for treatment planning with high energy photon beams *Acta Oncol.* **26** 49–56

[40] Sanchez-Garcia M, Gardin I, Lebtahi R and Dieudonné A 2014 A new approach for dose calculation in targeted radionuclide therapy (TRT) based on collapsed cone superposition: validation with ^{90}Y *Phys. Med. Biol.* **59** 4769–84

[41] Sanchez-Garcia M, Gardin I, Lebtahi R and Dieudonné A 2015 Implementation and validation of collapsed cone superposition for radiopharmaceutical dosimetry of photon emitters *Phys. Med. Biol.* **60** 7861–76

[42] Ljungberg M, Frey E, Sjögreen K, Liu X, Dewaraja Y and Strand S-E 2003 3D absorbed dose calculations based on SPECT: evaluation for 111-In/90-Y therapy using Monte Carlo simulations *Cancer Biother. Radiopharm.* **18** 99–107

[43] Pasciak A S, Bourgeois A C and Bradley Y C 2014 A comparison of techniques for ^{90}Y PET/CT image-based dosimetry following radioembolization with resin microspheres *Front. Oncol.* **4** 121

[44] Chiesa C *et al* 2012 A dosimetric treatment planning strategy in radioembolization of hepatocarcinoma with ^{90}Y glass microspheres *Q. J. Nucl. Med. Mol. Imaging* **56** 503–8

[45] Pillai K M, McKeever P E, Knutsen C A, Terrio P A, Prieskorn D M and Ensminger W D 1991 Microscopic analysis of arterial microsphere distribution in rabbit liver and hepatic VX2 tumor *Sel. Cancer Ther.* **7** 39–48

[46] Campbell A M, Bailey I H and Burton M A 2000 Analysis of the distribution of intra-arterial microspheres in human liver following hepatic yttrium-90 microsphere therapy *Phys. Med. Biol.* **45** 1023–33

[47] Walrand S, Hesse M, Jamar F and Lhommel R 2014 A hepatic dose-toxicity model opening the way toward individualized radioembolization planning *J. Nucl. Med.* **55** 1317–22

[48] Pasciak A S, Bourgeois A C and Bradley Y C 2016 A microdosimetric analysis of absorbed dose to tumor as a function of number of microspheres per unit volume in ^{90}Y radioembolization *J. Nucl. Med.* **57** 1020–6

[49] Pasciak A S *et al* 2019 The number of microspheres in Y90 radioembolization directly affects normal tissue radiation exposure *Eur. J. Nucl. Med. Mol. Imaging* **47** 816–27

[50] Uribe C F, Esquinas P L, Gonzalez M and Celler A 2016 Characteristics of bremsstrahlung emissions of ^{177}Lu, ^{188}Re, and ^{90}Y for SPECT/CT quantification in radionuclide therapy *Phys. Medica* **32** P691–700

[51] Heard S, Flux G D, Guy M J and Ott R J 2004 Monte Carlo simulation of ^{90}Y bremsstrahlung imaging *IEEE Symp. Conf. Record Nuclear Science.* vol 6 (Piscataway, NJ: IEEE) 3579–83

[52] Takahashi A, Himuro K, Yamashita Y, Komiya I, Baba S and Sasaki M 2015 Monte Carlo simulation of PET and SPECT imaging of ^{90}Y *Med. Phys.* **42** 1926–35

[53] Uehara S 1986 The development of a Monte Carlo code simulating electron-photon showers and its evaluation by various transport benchmarks *Nucl. Instruments Methods Phys. Res. Sect. B Beam Interact. with Mater. Atoms* **14** 559–70

[54] Rault E, Staelens S, Van Holen R, De Beenhouwer J and Vandenberghe S 2010 Fast simulation of yttrium-90 bremsstrahlung photons with GATE *Med. Phys.* **37** 2943–50

[55] Jan S *et al* 2004 GATE: a simulation toolkit for PET and SPECT *Phys. Med. Biol.* **49** 4543–61

[56] Minarik D, Sjögreen Gleisner K and Ljungberg M 2008 Evaluation of quantitative ^{90}Y SPECT based on experimental phantom studies *Phys. Med. Biol.* **53** 5689–703

[57] Farncombe T H, Gifford H C, Narayanan M V, Pretorius P H, Frey E C and King M A 2002 Assessment of scatter compensation strategies for 67Ga tumor SPECT using numerical observers and human LROC studies *Nuclear Science Symp. Conf. Record* vol 2 (Piscataway, NJ: IEEE) 1035–9

[58] Rong X, Du Y and Frey E C 2012 A method for energy window optimization for quantitative tasks that includes the effects of model-mismatch on bias: application to Y-90 bremsstrahlung SPECT imaging *Phys. Med. Biol.* **57** 3711–25

[59] Elschot M, Lam M G E H, van den Bosch M A A J, Viergever M A and de Jong H W A M 2013 Quantitative Monte Carlo-based ^{90}Y SPECT reconstruction *J. Nucl. Med.* **54** 1557–63

[60] Rong X, Du Y, Ljungberg M, Rault E, Vandenberghe S and Frey E C 2012 Development and evaluation of an improved quantitative ^{90}Y bremsstrahlung SPECT methodMed. Phys. **39** 2346–58

[61] Elschot M, Smits M L J, Nijsen J F W, Lam M G E H, Zonnenberg B A, van den Bosch M A A J, Viergever M A and de Jong H W A M 2013 Quantitative Monte Carlo-based holmium-166 SPECT reconstruction *Med. Phys.* **40** 112502

[62] van Rooij R, Braat A J A T, de Jong H W A M and Lam M G E H 2020 Simultaneous ^{166}Ho/^{99m}Tc dual-isotope SPECT with Monte Carlo-based downscatter correction for automatic liver dosimetry in radioembolization *EJNMMI Phys.* **7** 13

[63] Selwyn R G, Nickles R J, Thomadsen B R, DeWerd L A and Micka J A 2007 A new internal pair production branching ratio of ^{90}Y: the development of a non-destructive assay for ^{90}Y and ^{90}Sr *Appl. Radiat. Isot.* **65** 318–27

[64] Carlier T, Willowson K P, Fourkal E, Bailey D L, Doss M and Conti M 2015 ^{90}Y -PET imaging: exploring limitations and accuracy under conditions of low counts and high random fraction *Med. Phys.* **42** 4295–309

[65] Strydhorst J, Carlier T, Dieudonné A, Conti M and Buvat I 2016 A gate evaluation of the sources of error in quantitative ^{90}Y PET *Med. Phys.* **43** 5320

IOP Publishing

Monte Carlo Calculations in Nuclear Medicine (Second Edition)
Therapeutic applications
Habib Zaidi

Chapter 17

The Monte Carlo method as a design tool in boron neutron capture therapy

Leyla Moghaddasi

17.1 Introduction

The conceptual realization of the use of alpha emitters for cancer treatment dates backs to early 1900. However, the clinical implication of these radionuclides for a targeted approach has been slow, with first proposal of localized neutron capture therapy being made by Locher in 1936 [1], four years after the discovery of the neutron [2], and more recently, targeted alpha therapy.

Boron Neutron Capture Therapy (BNCT) is a binary type of radiotherapy which is envisioned as a two-step procedure involving, first, the administration of a ^{10}B agent which is taken up preferentially by malignant cells, and second, irradiation of the target volume with a beam of epithermal neutrons. This technique is based on a nuclear reaction in which a thermal neutron is captured by a stable ^{10}B nucleus, resulting in an emission of a high linear energy transfer (LET) alpha particle (~1.47–1.77 MeV), a recoiling ^{7}Li nucleus (~0.84–1.02 MeV), and a 478 keV gamma ray.

Developments in x-ray External Beam Radiation Therapy (EBRT) in terms of increased precision of dose delivery over the last two decades have failed to offer improved prognosis for malignancies with certain aggressive features such as hypoxia and infiltrative growth. BNCT, in theory, proposes an attractive alternative for such malignancies. However, the successful implementation of this technique is subject to several limitations which will be elaborated in this chapter.

17.2 Nuclear reactions

A list of possible reactions is given in equation (17.1). The track length of alpha particles and ^{7}Li nuclei range from 4 to 7 μm (LET rate of about 240 keV μm^{-1}).

doi:10.1088/978-0-7503-2694-0ch17

$$
\begin{aligned}
&^{1}n + {}^{10}\mathrm{B} \rightarrow {}^{11}\mathrm{B}^{*}\\
&^{11}\mathrm{B}^{*} \rightarrow {}^{7}\mathrm{Li} + {}^{4}He(6.1\%\ yield)\\
&^{11}B^{*} \rightarrow {}^{7}Li + {}^{4}He + \gamma(93.9\%\ yield)
\end{aligned}
\tag{17.1}
$$

Figure 17.1 shows the cross-section for the $^{10}\mathrm{B}(n,\alpha)^{7}\mathrm{Li}$ neutron capture reaction at different neutron incident energies. The data have been obtained from the Experimental Nuclear Reaction Data (EXFOR) library, compiled by the International Network of Nuclear Reaction Data Centers (NRDC) coordinated by the IAEA Nuclear Data Section (NDS), and contains the most comprehensive compilation of experimental nuclear reaction data [3].

It is evident that the cross-section for this reaction is high in the thermal energy range ($E < 1$ eV) of neutrons. However, the penetration depth for thermal neutrons is limited and a beam designed such that most of the neutrons have energies in the thermal range is only applicable for surface tumors, such as melanoma [4]. Epithermal neutrons (1 eV–10 keV) are more useful for clinical purposes as they lose energy while traversing the medium and will undergo boron neutron capture at larger depths. Epithermal neutron beams designed for BNCT are predominantly produced by fission reactors and accelerators, and therefore have broad spectra and contain unwanted fast neutrons ($E > 10$ keV) and gamma contamination as a result of neutron interactions with beam shaping assembly (BSA) components. IAEA-TECDOC-1223 defines an optimal neutron spectrum for BNCT to contain high thermal and epithermal and minimal fast neutron components [5].

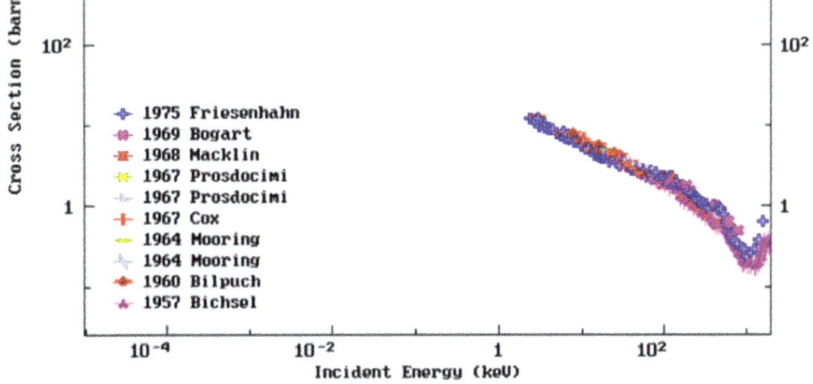

Figure 17.1. $^{10}\mathrm{B}(n,\alpha)^{7}\mathrm{Li}$ cross-sections extracted from IAEA EXFOR library.

The slowing down of a neutron beam traversing a human body is a complex and dynamic process. Neutrons undergo a chain of nuclear reactions with biological tissue resulting in dose deposition in normal tissue. These reactions primarily consist of neutron reactions with hydrogen nuclei in forms of elastic scattering and neutron absorption reaction, equation (17.2):

$$\begin{aligned} {}^{1}n + {}^{1}\mathrm{H} &\rightarrow {}^{1}\mathrm{H} + {}^{1}n \\ {}^{1}n + {}^{1}\mathrm{H} &\rightarrow {}^{2}\mathrm{H} + \gamma \end{aligned} \qquad (17.2)$$

and nitrogen nuclei in the human body, equation (17.3):

$$\begin{aligned} {}^{1}n + {}^{14}\mathrm{N} &\rightarrow {}^{1}\mathrm{H} + {}^{14}\mathrm{C} \\ {}^{1}n + {}^{14}\mathrm{N} &\rightarrow {}^{14}\mathrm{N} + {}^{1}n \end{aligned} \qquad (17.3)$$

The cross-sections for neutron–proton and neutron–nitrogen obtained from IAEA Evaluated Nuclear Data File (ENDF) database [3] are plotted in figure 17.2. In addition, recoiled or generated charged particles and gamma emissions by nuclear reactions can induce new nuclear reactions and electromagnetic cascades which further contribute to the absorbed dose along and lateral to the neutron beam passage [6]. Due to complexity of neutron interactions with biological tissue, as described above, Monte Carlo (MC) simulation method using a stepwise particle tracking technique is the only reliable means for BNCT application.

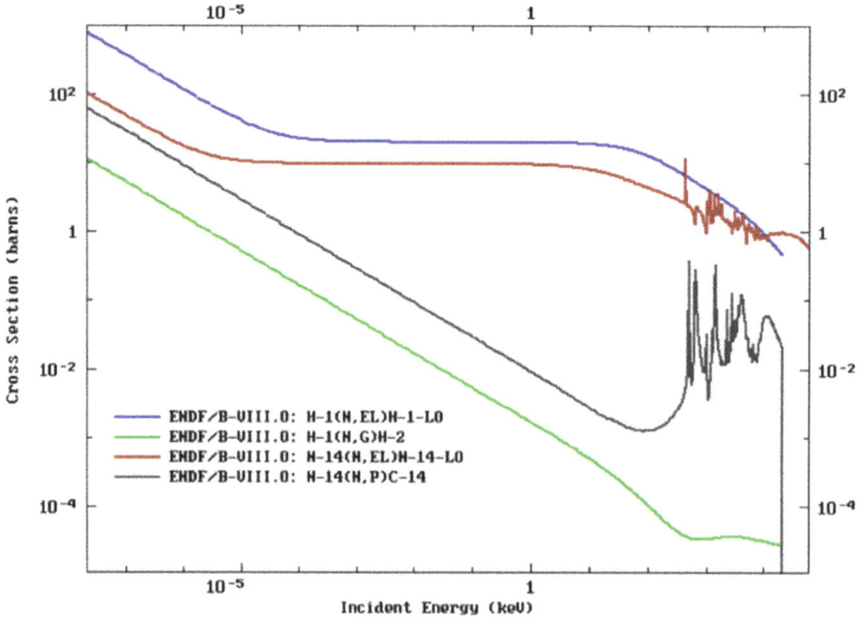

Figure 17.2. Energy dependence of neutron–hydrogen and neutron–nitrogen elastic scattering and neutron absorption reaction cross-sections, extracted from IAEA ENDF database.

17.3 Selective-localizing boron compounds

The success of BNCT relies on the occurrence of boron neutron capture reaction within tumor cells while minimizing normal tissue toxicity. Therefore, the design of BNCT boronated agent must aim to meet two main requirements: sufficient concentration of ^{10}B in tumor cells (i.e. 20–50 $\mu g\ g^{-1}$ or 10^9 atoms/cell [7]); and differential uptake of boron in tumor and normal cells (tumor-to-blood boron concentration ration of ~3) which rapidly clears from normal tissue and resides long enough in tumor cells for neutron irradiation .

The two main boron delivery agents that are used in clinical trials do not possess intrinsic selective uptake and attain higher tumor boron concentration by physiological or metabolic means. These agents are:

- Sodium mercaptoundecahydro-closo-dodecaborate ($Na_2B_{12}H_{11}SH$) or commonly named BSH; Preferential tumor uptake using BSH is physiological, utilizing compromise of blood–brain barrier (BBB) integrity in tumor cells to diffuse through leaky capillaries endothelial cell junctions.
- (L)-4-Dihydroxy-borylphenylalanine or BPA; BPA uptake is metabolic, hence, higher concentration in neoplastic cells can be attained.

Neither of these agents fulfill the biochemical criteria for an effective boronated agent for the clinical application. Extensive animal studies were conducted to investigate the potential gain from the combination of BSH or BPA with the infusion of a other drugs, such as a hyperosmotic solution of mannitol [8, 9] and specific amino acid molecules, L-DOPA [10] to disrupt the BBD (BBB-D) using different routes of administration, intravenous (i.v.) versus the intra-carotid (i.c.). Improved tumor uptake, while far from optimal, was obtained using an adjuvant BBB-D infusion through i.c. administration route.

The quest for new boron delivery agents with suitable selectivity and biodistribution for clinical use has been the subject of considerable research, resulting in synthesis of hundreds of new BNCT agents [11]. Of particular interest are targeted boron delivery agents in which the boronated compound is incorporated with tumor targeting molecules such as natural and unnatural amino acids, peptides and monoclonal antibodies which have the ability to recognize tumor-associated epitopes for improved selectivity [12–15]. Recent advances in synthesis technique have allowed for development of nanomaterial delivery systems. Boron delivery nanomaterials are an advanced type of targeted boron delivery agent utilizing the enhanced permeability and retention (EPR) effects of nanomaterials alongside selectivity effects intrinsic to targeted boron delivery agents [12, 16]. As promising as the concept of targeted boron delivery agents may seem, unfavorable histopathological characteristics of malignancies treated by BNCT, most notably extensive inter- and intra-tumor heterogeneity of glioblastoma multiforme (GBM), represent a significant hurdle for the efficacy of the drug delivery. The comprehensive review papers by Barth *et al* [8, 16, 17], Kuan *et al* [12] and Allen *et al* [18] can be referred to for further information on current status of boronated agents for BNCT.

Despite great effort expended by chemists and biologists over the last five decades, BSH and BPA, despite their known shortcomings, remain the only two agents for clinical use. The set of requirements for a suitable boron agent is a challenging goal to achieve. These agents must deliver sufficiently large number of ^{10}B atoms to induce lethal damage through boron neutron reaction, selectively to the entire population of tumor cells. Furthermore, they must sustain in tumor cells for large enough time to complete irradiation, while depleting rapidly from surrounding normal tissue to prevent toxicity. Biodistribution and toxicologic evaluations of the new generation of boron compounds have not demonstrated convincing data to warrant human trials and the pursuit of either new or more effective boronated agents or enhanced version of current agents remains an active research topic in the field of BNCT.

17.4 Radiobiological considerations

Targeted radiation therapy is often proposed for certain malignancies with aggressive features and inherent complexities, such as diffusive infiltration, hypoxia and intrinsic radioresistance. Despite major advances in x-ray external beam radiation radiotherapy (EBRT) in the last two decades, EBRT approach is largely ineffective once the tumor has metastasized and evolved into later stages, which for certain malignancies occurs before any symptoms have appeared. The eradication of dispersed tumor cells, or of a subpopulation of tumor stem cells, requires a targeted approach that is potent enough to sterilize individual tumor cells and microscopic clusters regardless of their low oxygen tension and radioresistance, while maintaining an acceptable toxicity profile [19]. From a radiobiological perspective, the BNCT modality has several advantages over EBRT [20]:

- The treatment is delivered by high LET particles. The densely ionizing track by high LET radiation and short path length of the emitted positively charged particles (4–7 μm) induce clustered damage to cellular DNA, increasing the probability of yielding DNA double strand (dsb) within the same cell (~15 μm). As the repair of sublethal and potentially lethal damage is far less efficient for dsb, such damage increases the possibility to overcome tumor cell intrinsic radioresistance and aggressive features and offers a higher tumor kill/sterilization compared to x-ray radiotherapy.
- As BNCT is a binary modality, it has the potential to selectively deliver localized dose to tumor cells (providing that substantially higher concentrations of ^{10}B can be delivered and retained in tumor cells) dispersed in normal tissue, while minimizing normal tissue toxicity. This makes BNCT an attractive modality for tumors which grow infiltratively rather than by forming a solid tumor [21].
- Charged particulate radiation interacts with biological tissue predominantly through direct interaction of incident particles or secondary electrons from ionized atoms with DNA, rather than producing free radicals which in return cause DNA damage. As a result, the cell killing process in BNCT is less sensitive to oxygen pressure compared to x-ray radiotherapy, which makes it advantageous for tumors with extensive hypoxia [22].

17.4.1 Biological dosimetry

The dosimetry for BNCT is more complicated than other radiotherapy modalities. Epithermal neutron beams designed for BNCT have broad spectra which differs from center to center and contain fast neutron and gamma contaminations which is also variable, depending on the beam production lines at different centers. That is, the dose arising from BNCT consists of contribution from different LET radiation quality components with varying inter-center ratios. BNCT treatment planning protocols, i.e. Harvard-MIT [23] and IAEA-TECDOC-1223 [24], recommend that four major dose components should be considered when an epithermal neutron spectra interact with biological medium:

- Boron dose (D_B) from thermal neutron capture through ^{10}B (n,α) ^{7}Li reaction (which is the sum of doses deposited by ^{7}Li and alpha particles);
- Thermal neutron/proton dose (D_P) as a result of thermal neutron capture by nitrogen atoms in biological tissue: ^{14}N (n,p) ^{14}C;
- Gamma dose (D_γ) from neutron capture by hydrogen atoms as well as from gamma contamination in the incident neutron beam;
- Fast neutron dose (D_n) which is due to elastic and inelastic neutron scattering (e.g. ^{1}H (n, n') ^{2}H) and several possible nuclear reactions (e.g. ^{16}O (n, α) ^{13}C).

Refer to section 17.2 for detailed discussion on nuclear reactions and corresponding cross-sections.

In current BNCT practices, to account for the variability of radiobiological effects of different physical dose components arising from a neutron beam spectrum, each physical dose component is weighted by its respective Relative Biological Effectiveness (RBE) factor. The concept of RBE was introduced to quantitatively evaluate the quality of a radiation beam in terms of biological damage it produces. RBE is the ratio doses from a reference x-ray beam and a test radiation beam, where both beams result in the same biological endpoint [25].

Due to inhomogeneity of boron biodistribution and short range of ^{7}Li and alpha particles, the concept of absorbed dose for boron dose component (D_B) is not applicable. Consequently, RBE for boron dose component cannot be determined. However, the product of RBE and boron compound factor (boron carrier dependant), referred to as Compound Biological Effectiveness (CBE) can be determined for a certain tissue type and experimental set-up [24].

Note that, it is not possible to assign a single RBE to a certain beam quality. RBE is a complex factor which is dependent upon several parameters including LET, cancer site, microenvironment status (e.g. hypoxia), fractionation, biological endpoint and so forth [26]. Therefore, RBE and CBE values for BNCT dose components, experimentally determined for a given treatment site and regimen, are only valid for the conditions for which they have been acquired.

The total BNCT dose is then expressed as the sum of RBE-weighted components with a unit of RBE-Gy calculated as:

$$D_{\mathrm{tot}} = C_B \bullet D_B + w_n \bullet D_n + w_P \bullet D_p + w_\gamma \bullet D_\gamma \qquad (17.4)$$

Table 17.1. The CBE and RBE factors for x-rays and the different BNCT dose components for brain.

	w_γ	w_n	w_P	C_B
Brain tissue	1	3.2	3.2	1.3
Tumor tissue	1	3.2	3.2	3.8

where D_B, D_n, D_P and D_γ are boron, fast neutron, thermal/proton and gamma dose, respectively. C_B, w_n, w_P and w_γ denote CBE for boron dose, and RBE for fast neutron, thermal/proton and gamma dose, respectively.

Table 17.1 lists the CBE and RBE values used for clinical use for high grade gliomas in Japan, United States and Sweden [27–29]. These values were determined from the results of *in vitro* and *in vivo* clonogenic assays of rat 9L gliosarcoma in Brookhaven National Laboratory by Coderre research group [30, 31]. Different factors may be reported in literature for the same malignancy, which is then again due to the different clinical conditions, e.g. the choice and dose of boron agent.

The BPA-based CBE and RBE (for fast neutron and thermal dose components or collectively referred to as proton dose) values for melanoma derived from thermal neutron based BNCT of cutaneous melanoma in patients [32] and epithermal BNCT in dogs were 2.5 and 3.0, respectively [31, 33]. These factors were later supported by the results of *in vivo* clonogenic assays of three human melanoma cell lines [34]. The data for CBE and RBE factors for other tumor sites (e.g. head and neck and liver), while scarce, can be found in publications from centers that provide BNCT for those sites.

17.5 Neutron sources for NCT

17.5.1 Neutron beam specifications

The two main NCT beam characteristics are beam intensity and beam quality. Beam intensity, or more importantly epithermal (0.5 eV–10 keV) beam intensity (φ_{epi}), is the main factor affecting the length of irradiation time. Beam quality, which is a measure of epithermal beam purity and divergence, is determined by four parameters, namely, fast neutron ($\dot{D}_{fast}/\varphi_{epi}$) and gamma ray ($\dot{D}_\gamma/\varphi_{epi}$) components, thermal to epithermal flux ratio ($\varphi_{thermal}/\varphi_{epi}$) and total neutron current to flux ratio (J/φ) which is the fraction of neutrons that are moving in the forward beam direction (indicative of beam divergence). Minimum performance specification of an epithermal neutron beam required for a successful NCT, as recommended by IAEA-TECDOC 1223 [24] are listed in table 17.2.

17.5.2 Neutron sources

The initiation of clinical studies on BNCT dates to 1951 at the Brookhaven National Laboratory (BNL) using fission reactor-produced thermal neutron beams.

Table 17.2. BNCT epithermal neutron beam performance criteria as recommended by IAEA-TECDOC 1223.

BNCT beam port parameter	φ_{epi} (n cm^{-2}.s)	$\varphi_{thermal}/\varphi_{epi}$	$\dot{D}_{fast}/\varphi_{epi}$ (Gy.cm^2)	$\dot{D}_{\gamma}/\varphi_{epi}$ (Gy.cm^2)	J/φ
Recommendations	$>10^9$	< 0.05	2×10^{-13}	2×10^{-13}	>0.7

Thereafter, a BNCT facility was developed at Massachusetts Institute of Technology (MIT) in 1958. The short penetration of thermal neutron beams forced the MIT research team to employ an intra-operative technique with unflattering outcome, primarily due to intrinsic complications of the technique. In the early 1990s, following a 25-years pause in BNCT research due to lack of evidence of effectiveness, the use of epithermal neutron beams with superior penetration (~8 cm) compared to thermal neutron beams (~3–4 cm) was proposed.

Sources of epithermal neutrons for BNCT facilities have since included nuclear reactors, californium (^{252}Cf) sources, accelerator-based (AB) neutron sources.

17.5.2.1 Nuclear reactors

The first epithermal neutron reactor was installed and tested at the Brookhaven Medical Research Reactor [35]. This was rapidly succeeded by other epithermal reactors being constructed or modified to produce epithermal neutron beams across US, Japan, Europe, Argentina and Asia. Three main approaches have been taken in the design of epithermal neutron reactors for BNCT facilities: conversion or modification of existing thermal reactors, fast reactors, fission converters.

Conversion of an existing reactor involves using spectrum shifting or filtering to produce neutron spectra suitable for NCT. Spectrum shifting, used in reactors with a large irradiation facility, moderates the fast neutrons directly leaking from the core to an appropriate epithermal spectrum. Filtering technique filters neutrons of undesired energies, resulting in a lower flux-to-power ratio as compared to the spectrum shifting technique [5, 17].

Fast reactors use highly enriched uranium (HEU) fuel which produces fast neutrons as the initial source of neutrons. Therefore, compared to thermal reactors, far larger fission converters can be achieved using fast reactors in a much more compact facility. On the other hand, the limited availability of HEU has limited the use of such reactors to few facilities [35, 36].

Fission converters convert a reactor's thermal neutrons to higher energy fission neutrons which will then be moderated or filtered to produce a suitable neutron spectrum. The flux-to-power ratio of fission converters are superior to thermal reactors, allowing a shorter irradiation time for BNCT. MIT research reactor (MITR) with a fission converter is the most advanced reactor-based facility for epithermal neutron beam production [37, 38]. Detailed performance characteristics for different types of fission reactors are summarized in the publication by Harling *et al* [39].

17.5.2.2 ^{252}Cf sources

The use of californium (^{252}Cf) radioactive isotope has been proposed as a possible neutron source for BNCT [40]. Although ^{252}Cf can provide compact BNCT facility, the proposal has not been implemented in clinical BNCT; the reason being, the need for frequent replacement of the source due its short half-life (2.6 years), the cost of isotope and failure to produce an intense enough beam in a reasonable treatment time [5, 18].

17.5.2.3 Accelerator-based neutron source

Fission reactor-based neutron sources have been the predominant approach and research topic in the field of clinical BNCT from the early 1990s until recently. While a great deal of progress had been made using nuclear reactors, clinical application of BNCT remained limited, partially due to concerns rightfully raised by the public and government agencies regarding the presence of nuclear reactors at the site of hospitals. Accelerator-based neutron sources presented an attractive alternative to nuclear reactors for several reasons: (1) low radiation hazard as they can be turned off when the source is not needed; (2) the cost is significantly lower for AB-BNCT; (3) there is a long history of clinical experience with accelerators in radiotherapy departments; and (4) accelerator systems can be designed to produce superior epithermal beam quality compared to reactors. The development and implementation of ABNS has been of interest to several groups [41–44], however, the research accelerated substantially after the Fukushima nuclear disaster in 2011 resulted in the closure of Japanese nuclear reactors [11].

Neutrons in accelerator-based systems are produced through neutron-producing reactions induced by accelerated charged particles with sufficiently high energy to overcome the given nuclear reaction's threshold energy. The most popular AB-BNCT reactions are: ^{7}Li (p, n)^{7}Be with the threshold energy of ~1.88 MeV (i.e. a 1.88 MeV proton projectile can eject a neutron with kinetic energy of zero from a lithium target); and ^{9}Be (p, n)^{9}B with the threshold energy of ~2.06 MeV. Examples of application of proton-lithium nuclear reaction are IBA AB-BNCT system [45] and nuBeam manufactured by Neutron Therapeutics and installed in Helsinki University Hospital in Finland [46], and proton-beryllium nuclear reaction is iBNCT developed by University of Tsukubai [47]. Regardless of the neutron-producing reaction used, the emitted neutrons must still be moderated or filtered to an energy suitable for clinical application. Table 17.3 lists threshold (E_{th}) and bombarding energies (E_i), neutron production current, fraction of neutrons with energy less than 1 MeV, and maximum and minimum energy of emitted neutrons for neutron-producing reactions used in AB-BNCT [48].

In recent years, several AB-BNCT facilities, mostly in Japan, have been constructed and commissioned and more in different status of implementation. Table 17.4, adopted from the Kiyanagi *et al* report, lists AB-BNCT facilities across the world and their present status of implementation as of the date of publication [49].

Table 17.3. Threshold (E_{th}) and bombarding energies (E_i), neutron production current, fraction of neutrons with energy less than 1 MeV, and maximum and minimum energy of emitted neutrons for neutron-producing reactions used in AB-BNCT [48].

Reaction	E_{th} (MeV)	E_i (MeV)	n/mA	$E_n < 1$ MeV(%)	$E_{n,max}$ (keV)	$E_{n,min}$ (keV)
$^{7}Li(p,n)^{7}Be$	1.880	1.88	0	100	30	30
		1.89	6.3×10^{9}	100	67	0.2
		2.5	9.3×10^{11}	100	787	60
$^{9}Be(p,n)^{9}B$	2.057	2.5	3.9×10^{10}	100	574	193
$^{9}Be(d,n)^{10}B$	0	1.5	3.3×10^{11}	50	4279	3874
$^{13}C(d,n)^{14}N$	0	1.5	1.9×10^{11}	70	6722	5616
$^{12}C(d,n)^{13}N$	0.327	1.5	6.0×10^{10}	80	1.188	707
$d(d,n)^{3}He$	0	0.2	1.1×10^{9}	0	3054	2074

Table 17.4. The status of AB-BNCT facilities worldwide in 2019 [49].

Location	Facility	Target	Max neutron energy (MeV)	Designed current (mA)	Present current (mA)	Present status
Japan	Kyoto University	Be	28	1	1	Clinical trial
	Tohoku Hospital	Be	28	1	1	Clinical trial
	Tsukuba University	Be	6	5	<2	Commissioning
	National Cancer Center	Solid Li	1	20	12	Commissioning
	Kansai BNCT Medical Center	Be	28	1		Commissioning
	Edogawa Hospital	Solid Li	1	20		Constructing
	Nagoya University	Solid Li	1	15		Commissioning
Russia	Budker Institute	Solid Li	1	10	3	Developing
Finland	Helsinki University Hospital	Solid Li	1	30	20	Constructing
Israel	SARAF	Liquid Li	1	20	1–2	Developing

Argentina	CNEA	Be	6	30	<1	Constructing
Italy	Legnaro INFN	Be	2	30		Developing
Korea	A-BNCT	Be	8	8		Constructing
China	Xiamen	Solid Li	1	10		Developing

17.6 Monte Carlo optimization for BNCT

17.6.1 Neutron beam design and specification

The production and delivery of an epithermal neutron beam with minimal fast neutron and gamma contamination and an intensity sufficiently high to be suitable for clinical application is complicated. Furthermore, optimal beam quality for different treatment sites are different. Neutrons depth of penetration and interaction in different types of tissues varies and therefore the design of a beam facility should be optimized for each cancer site. The beam design for a BNCT facility involves optimization of several inter-dependent and often competing parameters from neutron source through beamline and out of beam-shaping assemblies (BSAs) to the patient surface. Neutrons emerging from the source are moderated or filtered, transported, and collimated through BSA. Generally smaller BSAs are desired because a higher neutron flux per mA can be generated as fewer neutrons are lost through neutron capture reactions with materials of BSA. For example, the optimization of BSA for AB-BNCT systems include the choice of a neutron-producing reaction with the lowest bombarding energy and at the same time low-energy neutron spectrum which requires a smaller moderation volume, consequently smaller ABS, see table 17.3. A great deal of research work has been carried out in an effort to maximize achievable useful neutron flux by modifying accelerator components in AB (e.g. [50–52]) and nuclear reactor (e.g. [37, 38, 43, 44, 53]) BNCT facilities.

Monte Carlo particle transport codes are the only robust and reliable method for design of a beam facility as well as treatment planning aspect of BNCT. A variety of potential configurations are modelled to determine the optimal design. MC toolkits, e.g. general MC *n*-particle (MCNP/MCNPX [54]) and geometry and tracking (GEANT4 [55]) transport codes which include libraries for hadronic processes are considered for BNCT simulations.

The MC MCNP transport code MCNP is a simulation code developed at the Los Alamos National Laboratory that has been intensively benchmarked [56, 57]. This MC code has been widely used for BNCT application in both clinical and research capacities. BNCT treatment planning systems (TPS) for neutron dose calculations are generally based on MCNP, and currently MCNPX transport code [58–61]. MCNP MC code was used to design M67 epithermal neutron beam at the MIT Nuclear Reactor Laboratory. The calculations were extensively validated against measurements, both free in-air (to obtain the beam performance parameters in

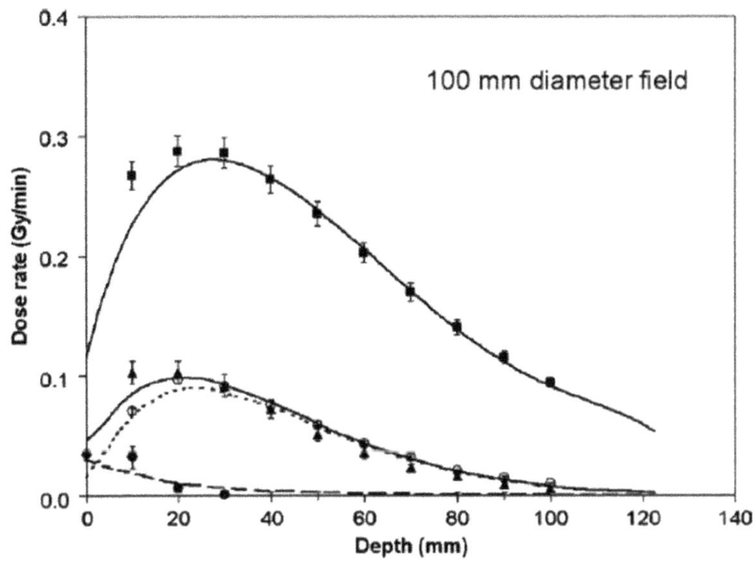

Figure 17.3. Measured (markers) and calculated (lines) gamma (■), total neutron (▲), thermal neutron including both boron dose (○) and fast neutron (●) dose rates versus depth in phantom from MIT fission converter based epithermal neutron beam, data adopted from [63]. Reproduced from [62], copyright (2002) with permission from Elsevier.

table 17.2), and at different depths in an ellipsoidal head phantom. M67 measured physical dose rates, i.e. not scaled by RBE, at different depths in the ellipsoidal phantom are graphed in for all four dose components as shown in figure 17.3 [62].

Design and characterization of the MIT fission converter based epithermal neutron beam was performed using MCNP code and acceptable agreement with measured spectra and in-phantom dose distributions were obtained [38, 63–65].

Likewise, MNCPX was used to simulate radiation transport through the core, epithermal column and an anthropomorphic phantom for optimization of a reactor based BNCT facility, TAPIRO in Italy [66] and THOR in Taiwan [67], see for calculated RBE-weighted dose rates as a function of depth in phantom from THOR.

MC calculations have been performed using MCNPX code to study the optimization of different aspects of AB-BNCT facilities, e.g. [68–74] (figure 17.4).

Geant4 is an open source and powerful MC toolkit capable of simulating complex experimental set-ups and offers an alternative to MCNP/MCNPX for BNCT. This MC toolkit, which is the result of an ongoing worldwide collaboration, has been developed to precisely track the passage of radiation through matter for a diverse range of particles and interaction processes over an extended energy range. While not as well-verified as MCNP for this specific application, the suitability of Geant4 for BNCT simulation has been demonstrated in recent years [75–77] (figure 17.5).

A Geant4 BNCT beam model was developed and validated against reported measured and MCNP-calculated data [76]. The published cyclotron-produced neutron spectrum for BNCT-enhanced FNT at the University of Washington

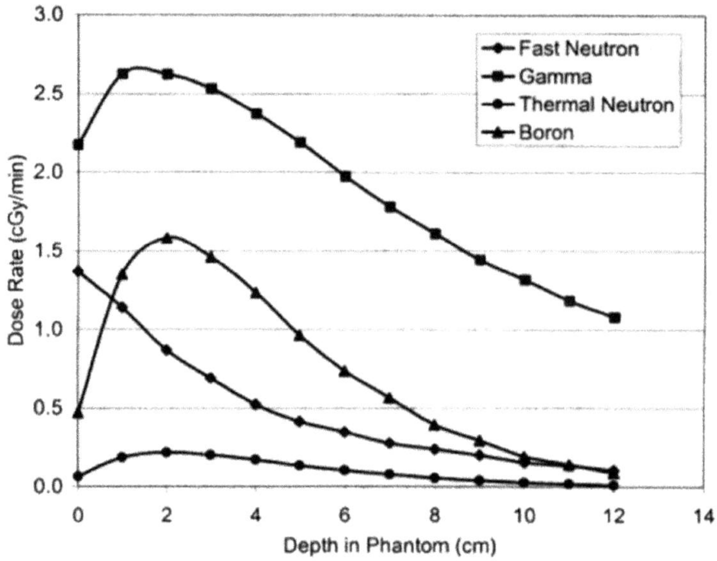

Figure 17.4. Physical dose rate components as a function of depth in the ellipsoidal head phantom from M67 epithermal neutron beam at the MIT Nuclear Reactor Laboratory. Gamma and fast neutron doses were measured directly, thermal neutron dose was calculated from measured thermal neutron flux, and boron dose were derived from measured neutron flux assuming a ^{10}B concentration of 15 μg g^{-1} in^{-1} normal tissue. Reproduced from [63]. Copyright IOP Publishing Ltd. All rights reserved.

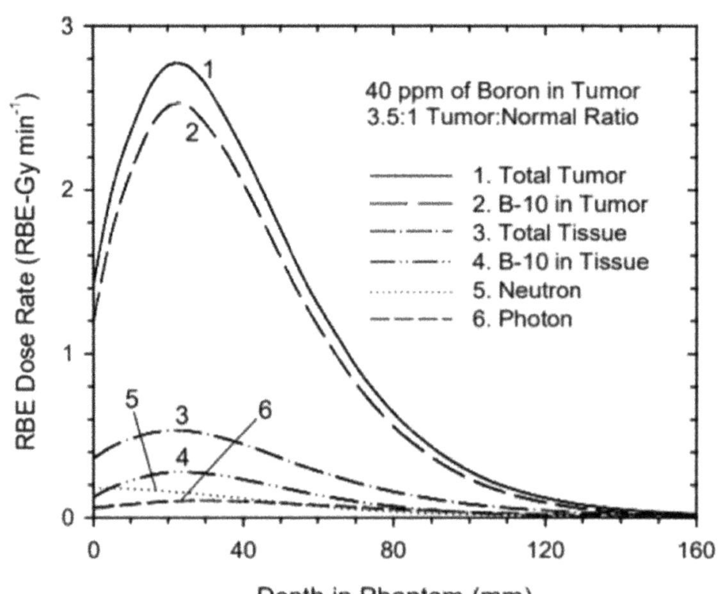

Figure 17.5. RBE-weighted dose rate components as a function of depth from THOR epithermal neutron beam. ^{10}B concentration of 11.4 μg g^{-1} in normal tissue and 40 μg g^{-1} in tumor, i.e. 3.5:1 tumor-normal ^{10}B concentration ratio. Reproduced from [66]. Copyright 2005, Oxford University Press

(UW) [78] was modelled. The simulation set-up was designed to represent the measurement/calculation set-up in the study of Nigg *et al* [78]. For this application, to transport a BNCT beam including neutrons down to thermal energies, the Geant4 neutron high precision data libraries G4NDL4 (valid for 0 eV–20 MeV) for elastic, inelastic and neutron capture procedures were used for modelling low energy neutrons transport and G4EmStandardPhysics_option4 was used to simulate photons and electrons. The tracks of ^{7}Li and alpha particles from the boron capture reaction, gammas resulting from the neutron capture reaction and all other particles and nuclear fragments from fast neutron reactions and thermal neutron capture were individually tracked in the simulation. Geant4-calculated percentage depth doses for total and boron doses (sum of ^{7}Li and alpha particles dose) were compared favourably with measured data in the study of Nigg *et al*, see figure 17.6 [76]. The beam model was then used to conduct sensitivity studies to quantitatively analyse the impact of the neutron beam spectrum and tumor boron concentration on the boron to non-boron dose ratio. A more recent study evaluated the validity of a number of Geant4 physics lists for BNCT application by comparing in-phantom boron and non-boron depth dose calculations with the results from FLUKA code [75]. Neutron high precision physics lists were found to be in best agreement with FLUKA calculations.

17.6.2 Treatment optimization

The success of BNCT depends on sufficiently high ratio of boron to non-boron dose component which is subject to surmounting two major challenges: (1) generating an optimal epithermal beam to deliver thermal neutrons to the tumor site with acceptable intensity and limited contamination; and (2) synthesizing an effective boron agent that could selectively deliver at least approximately 10^{9} of ^{10}B atoms to tumor cells. The design of an optimal BNCT epithermal beam, consisting of MC optimization of the beam spectrum followed by experimental characterization, was discussed in section 17.6.1. This section addresses treatment design and optimization to overcome the second key challenge for possible treatment of cancer sites by BNCT is discussed.

Treatment optimization generally involves Monte Carlo calculation of radiation transport in simulated phantoms or tumors in given cancer sites to evaluate BNCT outcome for various treatment configurations (e.g. fractionations, boron agent and mode of delivery and other treatment planning parameters).

To investigate the potential of BNCT for several cancer sites other than commonly-considered malignancies (i.e. high-grade glioma and melanoma), BNCT depth dose distributions for all dose components calculated by MCNP MC code in the ICRU computational model of ADAM & EVA [79] were evaluated in terms of depth of penetration, lethal tumor dose and normal tissue tolerance [80]. In this study, the epithermal neutron beam spectrum from a TRIGA-II reactor was used as the neutron source. In a study of similar nature, MCNP5/X was used to simulate particle transport through Oak Ridge National Laboratory (ORNL), using epithermal neutron beam spectrum from a TRIGA-II reactor as source with

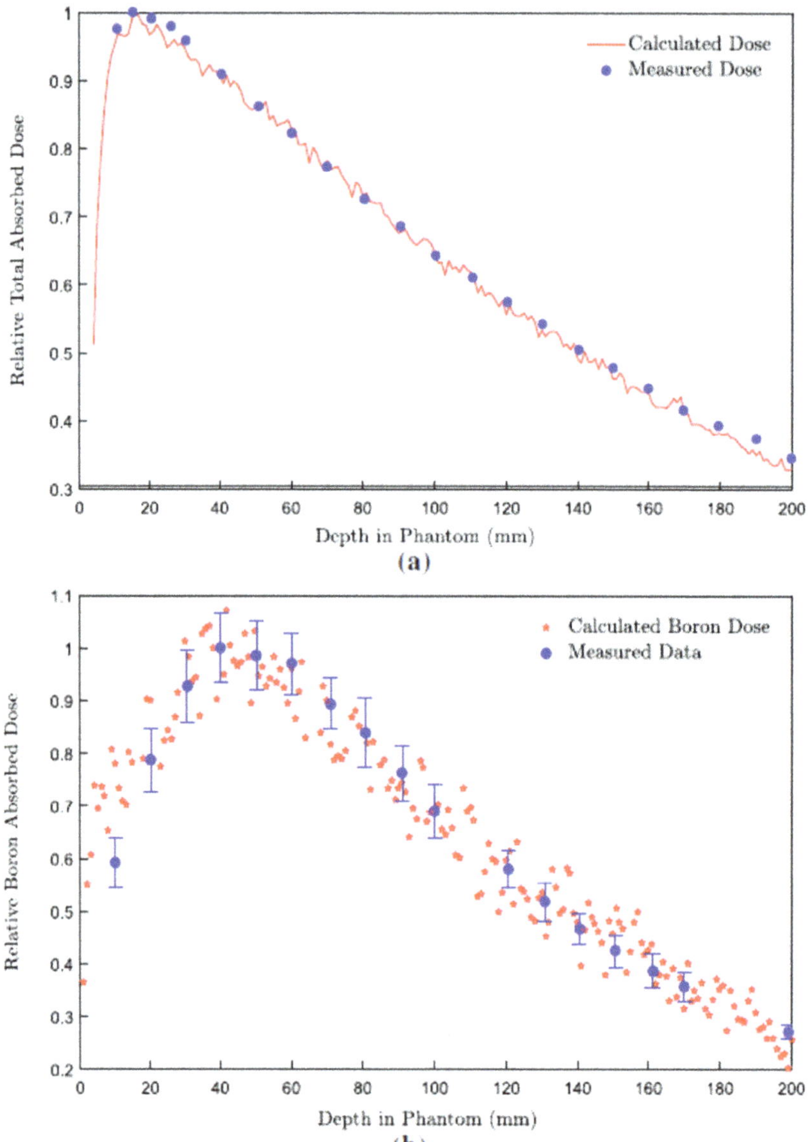

Figure 17.6. Geant4-calculated normalized absorbed total and boron dose as compared to measured data (reproduced from Nigg *et al* data) per million (ppm) ^{10}B. Reproduced from [76] Copyright 2018, Australasian College of Physical Scientists and Engineers in Medicine. With permission of Springer.

rectangular or elliptical fields, to calculate dose in the healthy lung tissue and lung tissue with metastasis spread and several organs at risk. Only the total physical dose was reported (i.e. the four dose components were not differentiated) [81]. The results of this study [81] indicated the feasibility of BNCT for lung cancer which supported the findings reported by Matsumoto [80].

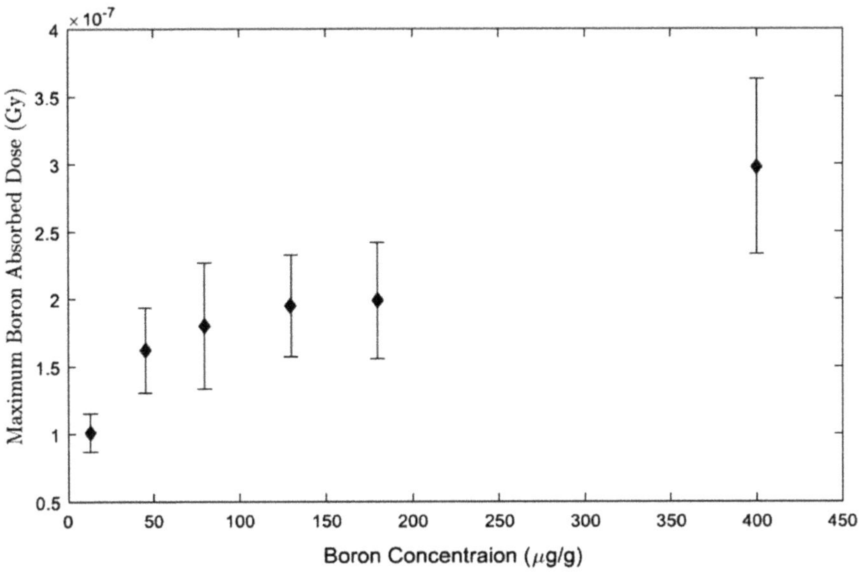

Figure 17.7. Maximum boron dose versus ^{10}B concentration in a cylindrical water phantom. Reproduced from [76] Copyright 2018, Australasian College of Physical Scientists and Engineers in Medicine. With permission of Springer.

The effect of varying concentrations of ^{10}B in brain tumor, blood and normal brain cells in the tally of neutron collisions and capture reactions was investigated in a MCNPX MC radiation transport through a simulated anthropomorphic phantom representing a human brain with realistic composition and tumor allocation based on 3D probabilistic glioblastoma multiforme location atlas [82]. The study focused on neutron transport dynamics and its interplay with chemical constituents of the brain with varying concentration of boron in different tissues and not the biology of ^{10}B uptake. In a more generic platform, another study investigated the sensitivity of boron dose to ^{10}B concentration using Geant4 MC particle transport simulation through a cylindrical water phantom using an epithermal neutron beam spectrum from a nuclear reactor as the source of neutrons [76]. It was demonstrated that enhancement of tumor cell to normal tissue boron concentration above a standard value may not translate to increased boron to non-boron dose ratio, see figure 17.7.

In ongoing research by our group, we developed an integrated radiobiological platform containing a micro-scale GBM model, taking into account cellular oxygen status, radiation sensitivity and microscopic extension models. The treatment modelling framework, originally designed for x-ray external radiotherapy (EBRT), is basically a hybrid model, integrating a Geant4 MC dose calculation in an *in silico* GBM with an analytical cell survival algorithm to study the radiation effects in the individual [83, 84]. Thereafter, the GBM model was extended to incorporate BNCT (i.e. simulation of the brain after boron agent administration with varying ^{10}B concentrations in tumor clonogenic cells, including infiltrative cells, and normal brain cells) to investigate the effect of clinical target volume (CTV) extensions as well as BNCT efficacy as compared to x-ray EBRT in terms of GBM cell survival [20]. The GBM model in this work consisted of cells with homogeneous radiosensitivities and isotropic distribution of histological infiltration. The BNCT treatment modelling framework was further expanded by including a more biologically plausible semi-realistic model of GBM that exhibits heterogeneous radiosensitivity and anisotropic infiltration distributions to allow comparison of BNCT treatment to the non-uniform oxygen distribution of EBRT (paper under revision by *Medical Physics*). The model served as a quantification tool to estimate the reduction in survival fraction due to extension of the CTV for three infiltration distributions. The results of this study suggested that the reduction in tumor cell survival fraction, which was considerably lower than that for x-ray EBRT for all microscopic extension distributions, may not justify the increased normal brain exposure to high LET radiation. In addition, it was demonstrated that BNCT is efficacious for GBM within the beam in terms of cell kill efficacy. Any increase in the dose and the CTV margin extension should be subject to further improvement of boron microdistribution and the neutron beam spectrum and should be evaluated with care.

Recent development in Geant4 MC code to include processes for the modeling of biological damage induced by ionizing radiation at the DNA scale will extend the scope of MC for treatment optimization, particularly for treatment modalities using charged particles which induce DSB. At present, the available MC software, capable of simulating particle track structure, do not support boron material and neutron interactions and therefore the development of more accurate cell survival models is subject to further developments of Geant4-DNA [85] or RITRACKS [86]. Nevertheless, the enormity of CPU/GPU time associated with such features make these techniques only suitable at research capacity with current computational technology.

17.7 Clinical studies: history and current status

The discussion in this chapter thus far has demonstrated that BNCT is a promising, but complex, treatment modality. Whether BNCT is an attractive treatment modality for malignancies of certain types, ultimately depends on the results of controlled randomized clinical trials aimed to assess the BNCT safety and efficacy.

Earlier clinical studies of BNCT were primarily focused on the treatment of patients with high grade gliomas that represent one of the most aggressive and devastating tumors in neuro-oncology. Glioblastoma multiforme, classified as grade V gliomas according to the World Health Organisation (WHO) [87], is the most common primary intracranial neoplasm. The prognosis for GBM patients remains poor, despite major advances in x-ray EBRT in the last two decades. According to a recent report, the median survival time (MST) for GBM patients was found to be 9.2 months using concurrent chemo-radiotherapy in a cohort of 1479 GBM patients [88]. The MST showed a mild improvement, 17 months, in a population of 233 GBM patients managed with adjuvant radiation therapy and temozolomide [89]. Further down the path, the application of BNCT was expanded to the treatment of recurrent tumors of head and neck (H&N) region, metastatic diseases (e.g. lung and prostate) and malignancies difficult to treat with conventional EBRT due to sever normal tissue toxicity, such as melanoma of the genital region.

17.7.1 Clinical trials: brain tumors

Several clinical studies for the treatment of brain tumors, primarily high-grade gliomas, using BNCT have been conducted in Japan, USA, Finland and Sweden (table 17.5). The earliest clinical responses of malignant brain tumors to BNCT using epithermal neutron beam were reported by the Hatanaka and Nakagawa group [90–92]. However, the results of this cohort, as impressive as to suggest a cure for GBM [90], should be interpreted with care [11, 93]. The reason is that survival data were presented from mixed histopathologies and there seems to be an exclusion bias which requires statistical re-examination of their data. Other Japanese neurosurgical institutions also conducted clinical trials to test modified BNCT treatment protocols in terms of tumor control, and overall survival by extension, and adverse effects, e.g. radiogenic necrosis. Barth and his research team conducted extensive *in vivo* animal studies including using F98 glioma cells implanted into rat brain and demonstrated significant enhancement of the MSTe using an x-ray boost after BNCT [7]. Clinical validation of combination therapy with x-ray EBRT and BNCT for newly diagnosed GBM patients was investigated by the Kawabata and Miyatake research team [94–96]. Therapeutic gain was obtained using BNCT in both protocols (BNCT only and BNCT+XRT) where MST was improved from 10.3 (7.4–13.2) for historical control group to 15.6 (12.2–23.9) for BNCT group.

Other publications from Japanese groups investigated the therapeutic gain, in terms of overall survival and safety, and dose distributions in GBM patients who were treated with either intra-operative BNCT (IO-BNCT) or external beam NCT (EB-BNCT) [97–99]. No significant difference was found between MSTs using IO-BNCT (= 23.3 months) and EB-BNCT (= 27.1 months) in Yamamoto's cohort [97]. Median survival times for BSH-mediated IO-BNCT for high grade gliomas in Nakagawa [92] and GBM in Kageji [98] reports were 22.9 and 19.5 months, respectively. In a group of eight GBM patients treated by EB-BNCT and boron agent dose similar to protocols used by Kageji, an MST of 27.9 months was achieved with treatment failure occurring predominantly in-field [100]. This result brings to attention the major limitation of BNCT which is insufficient or heterogenous microdistribution of boron compounds as discussed in section 17.3.

The MSTs for the Swedish [101, 102] and Finnish [103] subjects were 14.2 and 13 months, respectively, for GBM patients treated by BNCT, a duration not unlike that seen with resection and standard x-ray EBRT. The adverse effects, including various toxicities as well as quality of life parameters were comprehensively investigated and quantified by the Swedish team [102]. Two major reports from an American cohort are from BNCT trials at BNL and MIT conducted beginning in the early 1990s (1994 and 1996) to 1999. Phase I and II BNCT trials for patients with GBM at BNL were conducted for patients with GBM with two objectives: evaluation of the safety of BNCT using one, two or three irradiation fields and a dose escalation scheme; and

Table 17.5. BNCT clinical studies for malignant brain tumors.

References	Country	N	Histology	Boron agent (mg/kg)	Clinical outcome: OS[d], MST[e]	Adverse effects	Period
Nakagawa and Hatanaka [91]	Japan	60	GBM	BSH (60–80)	2Y: 12%	Radiogenic necrosis for recurrent cases	1968–1995
		39	Anaplastic Astrocytoma		2Y: 56%		
		13	Low-grade Astrocytoma		2Y: 62%		
Nakagawa [92]	Japan	89[a]	GBM	BSH (60–80)	3Y: 9%	Radiogenic necrosis and edema	1978–1997
Kawabata [94, 95]	Japan	21	Newly diagnosed GBM	BSH (100) + BPA (700)	2Y: 25%	No acute or severe toxicity	2002–2007
Yamamoto [97]	Japan	15	Newly diagnosed GBM	BSH (5g/body) + BPA (250) for EB-BNCT	2Y: 53.3%	Orbital swelling, post-epileptic brain swelling	1998–2007
Kageji [98]	Japan		GBM	BSH (100) + BPA (250) for EB-BNCT	2Y: 19.9%	Radiogenic necrosis	2001–2004
Diaz [104]	USA/BNL	53	GBM	BPA (250–330)	3Y: 4%–12% MST: 11.9–14.8 M	CNS toxicity for MTD>35 RBE-Gy	1994–1999
Busse	USA/MIT	18	GBM	BPA (250–350)	MST: 13 M	Toxicities classified to 5 grades, e.g. ICP[b], ARDS[c]	1996–1999

(Continued)

Table 17.5. (*Continued*)

References	Country	N	Histology	Boron agent (mg/kg)	Clinical outcome: OS[d], MST[e]	Adverse effects	Period
Capala [101]	Sweden	17	GBM	BPA (900)	Insufficient statistics	Seizures, thrombosis, abdominal pain	2001–2002
Henriksson [102]	Sweden	30	GBM	BPA (900)	MST: 14.2 M	Seizures, thrombosis, abdominal and mucous issues	2001–2003
Joensuu [103]	Finland	18	GBM	BPA (290–400)	MST: 13 M		1999–2001

[a]23 patients were excluded for reasons including different treatment protocol, imaging technique, and death within 3 months of BNCT due to rapid tumor enlargement/infiltration.
[b]ICP: increase in intracranial pressure;
[c]ARDS: Adult respiratory distress syndrome,
[d]OS: overall survival;
[e]MST: Median survival time.

to assess the tumor control in terms of MST and time to tumor progression [104, 105]. The MST ranged from 11.9 to 14.8 months with statistically significant trend toward lower MSTs for three irradiation fields as compared to one or two fields [104]. The dose escalation scheme with minimum tumor dose (MTD) of more than 35 RBE-Gy did not result in improved three-year survival with statical significance; 12% versus 4% for the group treated with MTD less than 35RBE-Gy [104]. The primary objective of phase I trial at Harvard-MIT was evaluation of normal tissue tolerance with tumor response coming as the second endpoint [27]. The MST for the subjects in this series was 13 months, however, the project did not achieve statistical significance to determine an MTD for CNS tolerance.

17.7.2 Clinical trials: head and neck

Up until the early 2000s, post-epithermal BNCT clinical studies mostly concentrated on malignant brain tumors and melanomas. Later, certain cancers of head and neck regions were considered as a suitable candidate for BNCT. Head and neck cancers are invasive carcinomas of the mucosal epithelium involving the complex anatomy of this area. The management of this malignancy whilst conserving functionality of organ at risks in such complex anatomy is difficult [106]. Furthermore, intrinsic radioresistance of cancer stem cells, genetic heterogeneity, hypoxia, radio-chemo resistance for locally recurred disease are additional factors contributing to poor prognosis in head and neck patients after conventional therapy [107]. BNCT, delivering selectively localized high LET radiation, presents a promising alternative approach for this malignancy.

Clinical studies to investigate the efficacy and safety of BNCT for different strands of cancers in head and neck regions were initiated in Japan for a small group of six patients with locally recurred head and neck cancer [108]. Encouraged by improved tumor control and minor side effects, further clinical trials were designed for a larger group of patients. The Kankaanranta research group reported on the clinical outcome of their phase I/II trial in Finland including 30 patients with inoperable, irradiated, locally recurred head and neck cancer treated by BNCT [109]. The next BNCT trial at the same facility in Finland adopted a more specific patient inclusion criteria, addressing only recurrent laryngeal cancer cases [110]. Although the results of this study should be considered with care due to the rather small size of the series, the feasibility of BNCT for recurred laryngeal cancer was verified with more favourable results for smaller tumor sizes [110]. The most recent clinical study from Finnish BNCT facility, the largest cohort reported to date, investigated the efficacy of BNCT for 79 patients with recurrent squamous cell carcinoma (SCC) and possible prognostic factors, e.g. minimum tumor dose and volume, of treatment response [111].

Similarly and during the same period, Suzuki's team conducted a BNCT trial including patients with unresectable advanced or recurrent head and neck cancers [112–114]. In terms of efficacy, the one- and two-year overall survival rates were 43.1% and 24.2%, respectively. As for safety, other than manageable acute toxicities (i.e. listed in table 17.6), six major BNCT-related toxicities, three deaths and three carotid artery hemorrhages were observed. Based on these results, the study warranted further

Table 17.6. BNCT clinical studies for malignancies of head and neck regions.

References	Country	N	Histology	Boron agent (mg/kg)	Clinical outcome: OS[c] MST[d]	Adverse effects	Period
Kato [108]	Japan	6	rHN[a] (3 SCC[b])	BPA (250) + BSH (5g/body)		Transient mucositis, minor alopecia	2001–2004
Suzuki [113]	Japan	62	rHN:49; HN:13. various histology	Various drug and dose	2Y: 24%, MST: 10.1 M	Hyperamylasemia; mucositis; fatigue; death (3/62) due to toxicity	2001–2007
Kankaanranta [109]	Finland	30	rHN (24 SCC)	BPA (400)	2Y: 30%, MST: 13 M	Oral mucositis; fatigue; pneumonia	2003–2008
Wang [115]	China	17	rHN (11 SCC)	BPA (400)	2Y: 47%	laryngeal edema; carotid haemorrhage; mucositis; cranial neuropathy	2010–2013
Wang [116]	China		rHN	BPA (400)		Same as above	2014–ongoing
Haapaniemi [110]	Finland	9	rLaryngeal	BPA (400)	MST: 13.3 M	Stomatitis; fatigue; oral pain	2006–2012
Koivunoro [111]	Finland	79	rHN-SCC	BPA (400)	2Y: 21%; MST: 25 M	Not reported	2003–2012

[a]rHN: recurrent head and neck;
[b]SCC: squamous cell carcinoma;
[c]OS: overall survival;
[d]MST: Median survival time.

controlled clinical trials due to low incidence of serious morbidity/mortality. However, the group has not reported any follow-up clinical studies to date.

More recently, the Wang research group designed a phase I/II trial investigate the efficacy and safety of 2-fractions BNCT for irradiated, recurrent head and neck cancer in a cohort of 17 patients [115]. The efficacy was evaluated in terms of overall survival and time to progression. The long follow-up period (median of 19.7 months) in this trial allowed evaluating both acute and late toxicities. In the light of the experience from this trial and published trials, indicating locoregional recurrence as the major cause of treatment failure, a subsequent clinical trial was designed by this group to investigate combination of BNCT with image-guided x-ray therapy with planning target volume extensions to ensure coverage of tumor and its microscopic extension [116].

The question is whether BNCT can propose a superior treatment outcome, both tumor response and BNCT-related toxicities, compared to conventional radiotherapy for recurrent head and neck cancer patients. The data summarized in table 17.6 for recurrent head and neck cancer indicate comparable tumor response (e.g. MST of ranging from 10–25 months) and treatment related acute and late toxicities. The difference can be attributed to varying histologies, treatment protocols, statistics etc. Wald *et al* reported treatment outcome for a cohort of 61 patients with recurrent head and neck cancer treated with salvage surgery followed by intra-operative electron therapy [107]. The median survival rate in this series was 15.9 months and two-year overall survival rate was 42%, a response not unlike what has been achieved by BNCT protocol. Unlike BNCT trials, the incidence of major toxicities (Grade 5) was minimal in this group of patients with only one case of carotid rapture [107]. Failing development of optimal boron agents with high selectivity and administration techniques to deliver ^{10}B homogeneously to tumor cells, BNCT cannot deliver to its capacity and make a significant difference [11].

17.8 Conclusion

The rationale for BNCT is its ability to deliver a localized high LET absorbed dose to tumor cells containing a sufficient number of ^{10}B atoms, while minimizing the dose to normal tissue cells which contain a fewer number of boron atoms. Therefore, an effective and nontoxic BNCT depends on the selective uptake of a boron compound by tumor cells or, in other words, on the ratio of tumor to normal tissue boron concentration as well as the number of thermal neutrons arriving at the tumor site. As technology advances, so does our ability to circumvent roadblocks in our way to an optimized BNCT. The introduction of AB-BNCT has tackled BNCT issues related to epithermal neutron beam performance, and more importantly radiation safety concerns regarding reactor-based facilities. The future prospect of BNCT relies on the development of a boron delivery agent with optimal selectivity and microdistribution. Only then we can expect the new BNCT clinical trial restart and hopefully produce results which can convince physicians of the promise of this modality. Until then, the jury is out.

References

[1] Locher G L 1936 Biological effects and therapeutic possibilities of neutrons *Am. J. Roentgenol. Radium Therapy.* **36** 1–13

[2] Chadwick J 1932 The existence of a neutron *Proc. R. Soc. A.* **136** 692–708

[3] Otuka N *et al* 2014 Towards a more complete and accurate experimental nuclear reaction data library (EXFOR): int. collaboration between nuclear reaction data centres (NRDC) *Nucl. Data Sheets.* **120** 272–76

[4] Menéndez P *et al* 2009 BNCT for skin melanoma in extremities: updated Argentine clinical results *Appl. Radiat. Isot.* **67** S50–53

[5] IAEA-TECDOC-1223 2001 *Current Status of Neutron Capture Therapy.* (Vienna: Int. Atomic Eenergy Agency)

[6] Seki R *et al* 2017 Physics of epi-thermal boron neutron capture therapy (epi-thermal BNCT) *Radiol. Phys. Technol.* **10** 387–408

[7] Barth R F *et al* 2004 Combination of boron neutron capture therapy and external beam radiotherapy for brain tumors *Int. J. Radiat. Oncol. Biol. Phys.* **58** 267–77

[8] Barth R F 2009 Boron neutron capture therapy at the crossroads: challenges and opportunities *Appl. Radiat. Isot.* **67** S3–6

[9] Barth R F *et al* 2000 Boron neutron capture therapy of brain tumors: enhanced survival and cure following blood–brain barrier disruption and intracarotid injection of sodium borocaptate and boronophenylalanine *Int. J. Radiat. Oncol. Biol. Phys.* **47** 209–18

[10] Capuani S *et al* 2008 L-DOPA preloading increases the uptake of borophenylalanine in C6 glioma rat model: a new strategy to improve BNCT efficacy *Int. J. Radiat. Oncol. Biol. Phys.* **72** 562–67

[11] Barth R F and Grecula J C 2020 Boron neutron capture therapy at the crossroads—where do we go from here? *Appl. Radiat. Isot.* **160** 109029

[12] Hu K *et al* 2020 Boron agents for neutron capture therapy *Coord. Chem. Rev.* **405** 213139

[13] Stretch J *et al* 1992 Boronated monoclonal antibody for neutron capture therapy of melanoma: specificity retention shown by autoradiography of tissue sections *Antib. Immunoconjug. Radiopharm.* **5** 261–65

[14] Tamat S R, Moore D E and Allen B J 1989 Determination of the concentration of complex boronated compounds in biological tissues by inductively coupled plasma atomic emission spectrometry *Pigment Cell Res.* **2** 281–85

[15] Yang W *et al* 2009 Boron neutron capture therapy of EGFR or EGFRvIII positive gliomas using either boronated monoclonal antibodies or epidermal growth factor as molecular targeting agents *Appl. Radiat. Isot.* **67** S328–331

[16] Barth R F, Mi P and Yang W 2018 Boron delivery agents for neutron capture therapy of cancer *Cancer Commun.* **38** 35

[17] Barth R F *et al* 2012 Current status of boron neutron capture therapy of high grade gliomas and recurrent head and neck cancer *Radiat. Oncol.* **7** 1–21

[18] Allen B J 2006 Internal high linear energy transfer (LET) targeted radiotherapy for cancer *Phys. Med. Biol.* **51** R327

[19] Sgouros G *et al* 2010 MIRD Pamphlet No. 22 (abridged): radiobiology and dosimetry of α-particle emitters for targeted radionuclide therapy *J. Nucl. Med.* **51** 311–28

[20] Moghaddasi L and Bezak E 2017 Development of an integrated Monte Carlo model for glioblastoma multiforme treated with boron neutron capture therapy *Sci. Rep.* **7** 1–14

[21] Unkelbach J *et al* 2014 Radiotherapy planning for glioblastoma based on a tumor growth model: implications for spatial dose redistribution *Phys. Med. Biol.* **59** 771

[22] Barendsen G *et al* 1966 The effect of oxygen on impairment of the proliferative capacity of human cells in culture by ionizing radiations of different LET *Int. J. Radiat. Biol. Relat. Stud. Phys. Chem. Med.* **10** 317–27
[23] Zamenhof R *et al* 1996 Monte Carlo-based treatment planning for boron neutron capture therapy using custom designed models automatically generated from CT data *Int. J. Radiat. Oncol. Biol. Phys.* **35** 383–97
[24] IAEA 2001 *Current Status of Neutron Capture Therapy*. IAEA TECDOC 1223 (Vienna: International Atomic Energy Agency)
[25] Marcu L, Bezak E and Allen B 2012 *Biomedical Physics in Radiotherapy for Cancer.* (London: Springer)
[26] Hopewell J W *et al* 2012 Boron neutron capture therapy: application of radiobiological principles *Neutron Capture Therapy.* (Berlin: Springer) pp 329–58
[27] Busse P M *et al* 2003 A critical examination of the results from the Harvard-MIT NCT program phase I clinical trial of neutron capture therapy for intracranial disease *J. Neuro-Oncol.* **62** 111–21
[28] Henriksson R *et al* 2008 Boron neutron capture therapy (BNCT) for glioblastoma 4 multiforme: a phase II study evaluating a prolonged high-dose of 5 boronophenylalanine (BPA) at the Studsvik facility in Sweden *Radiother. Oncol.* **88** 288
[29] Miyatake S-I *et al* 2005 Modified boron neutron capture therapy for malignant gliomas performed using epithermal neutron and two boron compounds with different accumulation mechanisms: an efficacy study based on findings on neuroimages *J. Neurosurg.* **103** 1000–9
[30] Coderre J A *et al* 1993 Derivations of relative biological effectiveness for the high-LET radiations produced during boron neutron capture irradiations of the 9L rat gliosarcoma *in vitro* and *in vivo Int. J. Radiat. Oncol. Biol. Phys.* **27** 1121–29
[31] Coderre J A and Morris G M 1999 The radiation biology of boron neutron capture therapy *Radiat. Res.* **151** 1–18
[32] Fukuda H *et al* 1994 Boron neutron capture therapy of malignant melanoma using ^{10}B-paraboronophenylalanine with special reference to evaluation of radiation dose and damage to the normal skin *Radiat. Res.* **138** 435–42
[33] Gavin P R *et al* 1992 Large animal model studies of normal tissue tolerance using an epithermal neutron beam and borocaptate sodium *Boron Neutron Capture Therapy.* (Berlin: Springer) pp 197–209
[34] Rossini A *et al* 2010 Determination of relative biological effectiveness factors for human melanoma cell lines for the improvement of boron neutron capture therapy (BNCT) *Proc. of the 14th Int. Congress on Neutron Capture Therapy 14-ICNCT* (Buenos Aires: National Atomic Energy Commission CNEA)
[35] Burian J *et al* 2004 Report on the first patient group of the phase I BNCT trial at the LVR-15 reactor *Int. Cong. Ser.* (Amsterdam: Elsevier)
[36] Rosi G *et al* 2005 Role of the Tapiro fast research reactor in neutron capture therapy in Italy calculations and measurements research reactor utilization, safety, decommissioning, fuel and waste management IAEA-CN-100/97
[37] Harling O *et al* 2002 The fission converter-based epithermal neutron irradiation facility at the Massachusetts Institute of Technology reactor *Nucl. Sci. Eng.* **140** 223–40
[38] Kiger W III, Sakamoto S and Harling O 1999 Neutronic design of a fission converter-based epithermal neutron beam for neutron capture therapy *Nucl. Sci. Eng.* **131** 1–22
[39] Harling O K 2009 Fission reactor based epithermal neutron irradiation facilities for routine clinical application in BNCT—Hatanaka memorial lecture *Appl. Radiat. Isot.* **67** S7–11

[40] Yanch C *et al* 1990 *Comparison of Dose Distributions with B-10 Augmentation Near Linear Sources of Cf-252 Obtained by Monte Carlo Simulation and by Experimental Measurement Progress in Neutron Capture Therapy* ed B J Allen, D E Moore and B V Harrington (Sydney: Plenum)

[41] Blackburn B W, Yanch J C and Klinkowstein R E 1998 Development of a high-power water cooled beryllium target for use in accelerator-based boron neutron capture therapy *Med. Phys.* **25** 1967–74

[42] Beynon T *et al* 2002 *Status of the Birmingham accelerator based BNCT facility.* Research and Development in Neutron Capture Therapy *Proc. of 10th Int. Congress on Neutron Capture Therapy (Essen, Germany, 8–13 September 2002)* Ed W Sauerwein *et al* (Milan: Monduzzi Editore) pp 225–8

[43] Zimin S and Allen B J 2000 Study of moderator thickness for an accelerator-based neutron irradiation facility for boron neutron capture therapy using the ^{7}Li (p, n) reaction near threshold *Phys. Med. Biol.* **45** 59

[44] Zimin S and Allen B 1998 Evaluation of useful neutron flux for accelerator boron neutron capture therapy using the ^{7}Li (p, n) reaction *Australas. Phys. Eng. Sci. Med.* **21** 193–99

[45] Forton E *et al* 2009 Overview of the IBA accelerator-based BNCT system *Appl. Radiat. Isot.* **67** S262–265

[46] *A Compact Neutron Source Designed for the Hospital Environment.* Neurotherapeutics, available online at https://www.neutrontherapeutics.com/technology/

[47] Kumada H *et al* 2019 Beam performance of the iBNCT as a compact linac-based BNCT neutron source developed by University of Tsukuba *AIP Conf. Proc.* (Melville, NY: AIP Publishing LLC)

[48] Kreiner A J 2012 Accelerator-based BNCT *Neutron Capture Therapy.* (Berlin: Springer) pp 41–54

[49] Kiyanagi Y *et al* 2019 Status of accelerator-based BNCT projects worldwide *AIP Conf. Proc.* (AIP Publishing LLC)

[50] Burlon A *et al* 2005 Optimization of a neutron production target and a beam shaping assembly based on the ^{7}Li (p, n) 7Be reaction for BNCT *Nucl. Instrum. Methods Phys. Res., Sect. B.* **229** 144–56

[51] Burlon A and Kreiner A 2008 A comparison between a TESQ accelerator and a reactor as a neutron sources for BNCT *Nucl. Instrum. Methods Phys. Res., Sect. B.* **266** 763–71

[52] Bleuel D *et al* 1998 Designing accelerator-based epithermal neutron beams for boron neutron capture therapy *Med. Phys.* **25** 1725–34

[53] Cerullo N *et al* 2002 An irradiation facility for Boron Neutron Capture Therapy application based on a radio frequency driven D–T neutron source and a new beam shaping assembly *Rev. Sci. Instrum.* **73** 3614–18

[54] Briesmeister J F 2000 *MCNPTM-A General Monte Carlo N-Particle Transport Code.* (Los Alamos, NM: Los Alamos National Laboratory) 2 Version 4C, LA-13709-M

[55] Agostinelli S *et al* 2003 GEANT4—a simulation toolkit *Nucl. Instrum. Methods Phys. Res., Sect. A.* **506** 250–303

[56] Gierga D P and Adams K 1999 *Electron/Photon Verification Calculations using MCNP4B* (Los Alamos, NM: Los Alamos National Laboratory)

[57] Whalen D J *et al* 1991 *MCNP: Neutron Benchmark Problems.* (Los Alamos, NM: Los Alamos National Laboratory)

[58] Carolan M *et al* 1996 Validation of Monte Carlo dose planning calculations by epithermal beam dose distribution measurements in phantoms *Cancer Neutron Capture Therapy.* (Berlin: Springer) pp 303–10
[59] Wallace S, Allen B and Mathur J 1996 Monte Carlo neutron photon treatment planning calculations *Cancer Neutron Capture Therapy.* (Berlin: Springer) pp 295–302
[60] Nievaart V, Daquino G and Moss R 2007 Monte Carlo based treatment planning systems for boron neutron capture therapy in Petten, the Netherlands *J. Phys. Conf. Ser.* **73** 012012
[61] Nigg D W 2003 Computational dosimetry and treatment planning considerations for neutron capture therapy *J. Neuro-Oncol.* **62** 75–86
[62] Palmer M, Goorley J T, Kiger W S, Busse P M, Riley K J, Harling O K and Zamenhof R G *et al* 2000 Treatment planning and dosimetry for the Harvard-MIT phase-I clinical trial for cranial BNCT *Int. J. Radiat. Oncol. Biol. Phys.* **53** 1361–79
[63] Riley K, Binns P and Harling O 2003 Performance characteristics of the MIT fission converter based epithermal neutron beam *Phys. Med. Biol.* **48** 943
[64] Riley K J 2003 Construction and characterization of a fission converter based epithermal neutron beam for BNCT *PhD Thesis* MIT
[65] Sakamoto S, Kiger W III and Harling O 1999 Sensitivity studies of beam directionality, beam size, and neutron spectrum for a fission converter-based epithermal neutron beam for boron neutron capture therapy *Med. Phys.* **26** 1979–88
[66] Nava E *et al* 2005 Monte Carlo optimisation of a BNCT facility for treating brain gliomas at the tapiro reactor *Radiat. Prot. Dosim.* **116** 475–81
[67] Tung C *et al* 2004 Characteristics of the new THOR epithermal neutron beam for BNCT *Appl. Radiat. Isot.* **61** 861–64
[68] Kreiner A J *et al* 2016 Present status of accelerator-based BNCT *Rep. Pract. Oncol. Radiother.* **21** 95–101
[69] Rahmani F and Shahriari M 2010 Hybrid photoneutron source optimization for electron accelerator-based BNCT *Nucl. Instrum. Methods Phys. Res., Sect. A.* **618** 48–53
[70] Chegeni N *et al* 2017 Optimization of the photoneutron target geometry for e-accelerator based BNCT *Electron. Phys.* **9** 4590
[71] Rahmani F and Shahriari M 2011 Beam shaping assembly optimization of Linac based BNCT and in-phantom depth dose distribution analysis of brain tumors for verification of a beam model *Ann. Nucl. Energy.* **38** 404–9
[72] Rahmani F *et al* Design of photon converter and photoneutron target for high power electron accelerator based BNCT *Appl. Radiat. Isot.* **106** 45–8
[73] Minsky D M and Kreiner A J 2014 Beam shaping assembly optimization for ^{7}Li (p, n) ^{7}Be accelerator based BNCT *Appl. Radiat. Isot.* **88** 233–37
[74] Culbertson C N *et al* 2004 In-phantom characterisation studies at the Birmingham Accelerator-Generated epIthermal Neutron Source (BAGINS) BNCT facility *Appl. Radiat. Isot.* **61** 733–38
[75] Chen Z *et al* 2019 Comparison of BNCT dosimetry calculations using different GEANT4 physics lists *Radiat. Prot. Dosim.* **187** 88–97
[76] Moghaddasi L and Bezak E 2018 Geant4 beam model for boron neutron capture therapy: investigation of neutron dose components *Australas. Phys. Eng. Sci. Med.* **41** 129–41
[77] Ribeiro R M and Souza-Santos D 2017 Comparison of the neutron ambient dose equivalent and ambient absorbed dose calculations with different GEANT4 physics lists *Radiat. Phys. Chem.* **139** 179–83

[78] Nigg D W *et al* 2000 Modification of the University of Washington Neutron Radiotherapy Facility for optimization of neutron capture enhanced fast-neutron therapy *Med. Phys.* **27** 359–67

[79] ICRU 1992 Report 48, Computational models in therapy, diagnosis and protection *Int. Commission on Radiation Units and Measurements*

[80] Matsumoto T 2007 Monte Carlo simulation of depth–dose distribution in several organic models for boron neutron capture therapy *Nucl. Instrum. Methods Phys. Res., Sect. A.* **580** 552–57

[81] Krstic D *et al* 2014 Monte Carlo calculations of lung dose in ORNL phantom for boron neutron capture therapy *Radiat. Prot. Dosim.* **161** 269–73

[82] Chin M and Spyrou N 2009 A detailed Monte Carlo accounting of radiation transport in the brain during BNCT *Appl. Radiat. Isot.* **67** S164–167

[83] Moghaddasi L, Bezak E and Harriss-Phillips W 2015 Evaluation of current clinical target volume definitions for glioblastoma using cell-based dosimetry stochastic methods *Br. J. Radiol.* **88** 20150155

[84] Moghaddasi L, Bezak E and Harriss-Phillips W 2016 Monte-Carlo model development for evaluation of current clinical target volume definition for heterogeneous and hypoxic glioblastoma *Phys. Med. Biol.* **61** 3407

[85] Incerti S *et al* 2010 The geant4-dna project *Int. J. Model. Simul. Sci. Comput.* **1** 157–78

[86] Plante I and Cucinotta F A 2011 Monte-Carlo simulation of ionizing radiation tracks *Application of Monte Carlo Methods in Biology, Medicine and Other Fields of Science.* (Rijeka: InTech) pp 315–56

[87] Ostrom Q T *et al* 2015 CBTRUS statistical report: primary brain and central nervous system tumors diagnosed in the United States in 2008–12 *Neuro-Oncology.* **17** iv1–62

[88] Kole A J *et al* 2016 Concurrent chemoradiotherapy versus radiotherapy alone for ‘biopsy-only’ glioblastoma multiforme *Cancer.* **122** 2364–70

[89] Jayamanne D *et al* 2018 Survival improvements with adjuvant therapy in patients with glioblastoma *ANZ J. Surg.* **88** 196–201

[90] Hatanaka H and Nakagawa Y 1994 Clinical results of long-surviving brain tumor patients who underwent boron neutron capture therapy *Int. J. Radiat. Oncol. Biol. Phys.* **28** 1061–66

[91] Nakagawa Y and Hatanaka H 1997 Boron neutron capture therapy: clinical brain tumor studies *J. Neuro-Oncol.* **33** 105–15

[92] Nakagawa Y *et al* 2003 Clinical review of the Japanese experience with boron neutron capture therapy and a proposed strategy using epithermal neutron beams *J. Neuro-Oncol.* **62** 87–99

[93] Linz U 2008 Boron neutron capture therapy for glioblastoma: is it worth pursuing? *Technol. Cancer Res. Treat.* **7** 83–8

[94] Kawabata S *et al* 2011 Phase II clinical study of boron neutron capture therapy combined with x-ray radiotherapy/temozolomide in patients with newly diagnosed glioblastoma multiforme—study design and current status report *Appl. Radiat. Isot.* **69** 1796–99

[95] Kawabata S *et al* 2009 Survival benefit from boron neutron capture therapy for the newly diagnosed glioblastoma patients *Appl. Radiat. Isot.* **67** S15–18

[96] Kawabata S *et al* 2003 The early successful treatment of glioblastoma patients with modified boron neutron capture therapy *J. Neuro-Oncol.* **65** 159–5

[97] Yamamoto T *et al* 2009 Boron neutron capture therapy for newly diagnosed glioblastoma *Radiother. Oncol.* **91** 80–4

[98] Kageji T *et al* 2011 Long-survivors of glioblatoma treated with boron neutron capture therapy (BNCT) *Appl. Radiat. Isot.* **69** 1800–2
[99] Kageji T *et al* 2014 Correlation between radiation dose and histopathological findings in patients with gliblastoma treated with boron neutron capture therapy (BNCT) *Appl. Radiat. Isot.* **88** 20–2
[100] Matsuda M *et al* 2009 Dose distribution and clinical response of glioblastoma treated with boron neutron capture therapy *Appl. Radiat. Isot.* **67** S19–21
[101] Capala J *et al* 2003 Boron neutron capture therapy for glioblastoma multiforme: clinical studies in Sweden *J. Neuro-Oncol.* **62** 135–44
[102] Henriksson R *et al* 2008 Boron neutron capture therapy (BNCT) for glioblastoma multiforme: a phase II study evaluating a prolonged high-dose of boronophenylalanine (BPA) *Radiother. Oncol.* **88** 183–91
[103] Joensuu H *et al* 2003 Boron neutron capture therapy of brain tumors: clinical trials at the Finnish facility using boronophenylalanine *J. Neuro-Oncol.* **62** 123–34
[104] Diaz A Z 2003 Assessment of the results from the phase I/II boron neutron capture therapy trials at the Brookhaven National Laboratory from a clinician's point of view *J. Neuro-Oncol.* **62** 101–9
[105] Chanana A D *et al* 1999 Boron neutron capture therapy for glioblastoma multiforme: interim results from the phase I/II dose-escalation studies *Neurosurgery* **44** 1182–93
[106] Reid P *et al* 2020 Influence of the human papillomavirus on the radio-responsiveness of cancer stem cells in head and neck cancers *Sci. Rep.* **10** 1–8
[107] Wald P *et al* 2019 Intraoperative electron beam radiotherapy for locoregionally persistent or recurrent head and neck cancer *Head Neck* **41** 2148–53
[108] Kato I *et al* 2004 Effectiveness of BNCT for recurrent head and neck malignancies *Appl. Radiat. Isot.* **61** 1069–73
[109] Kankaanranta L *et al* 2012 Boron neutron capture therapy in the treatment of locally recurred head-and-neck cancer: final analysis of a phase I/II trial *Int. J. Radiat. Oncol. Biol. Phys.* **82** e67–75
[110] Haapaniemi A *et al* 2016 Boron neutron capture therapy in the treatment of recurrent laryngeal cancer *Int. J. Radiat. Oncol. Biol. Phys.* **95** 404–10
[111] Koivunoro H *et al* 2019 Boron neutron capture therapy for locally recurrent head and neck squamous cell carcinoma: an analysis of dose response and survival *Radiother. Oncol.* **137** 153–58
[112] Suzuki M *et al* 2006 Impact of intra-arterial administration of boron compounds on dose–volume histograms in boron neutron capture therapy for recurrent head-and-neck tumors *Int. J. Radiat. Oncol. Biol. Phys.* **66** 1523–27
[113] Suzuki M *et al* 2014 Boron neutron capture therapy outcomes for advanced or recurrent head and neck cancer *J. Radiat. Res.* **55** 146–53
[114] Suzuki M 2020 Boron neutron capture therapy (BNCT): a unique role in radiotherapy with a view to entering the accelerator-based BNCT era *Int. J. Clin. Oncol.* **25** 43–50
[115] Wang L-W *et al* 2016 Fractionated boron neutron capture therapy in locally recurrent head and neck cancer: a prospective phase I/II trial *Int. J. Radiat. Oncol. Biol. Phys.* **95** 396–403
[116] Wang L-W *et al* 2018 Clinical trials for treating recurrent head and neck cancer with boron neutron capture therapy using the Tsing-Hua Open Pool Reactor *Cancer Commun.* **38** 1–7

Chapter 18

Monte Carlo-based dosimetry of Auger electrons

Nadia Falzone

18.1 Introduction

If there ever was an underdog amongst the forms of particulate radiation used for molecularly targeted radionuclide therapy (MRT) in nuclear medicine, Auger electrons (AE) would fit the criteria. First described almost 100 years ago by Meitner in 1922 [1] while studying radioactive decay and Auger in 1923 [2], by the excitation of atoms by x-ray bombardment, their utility as potential cancer therapeutics was only expounded some 50 years later [3].

Auger electrons used for clinical application arise either from the decay of unstable high atomic number elements by electron capture (EC) and/or internal conversion (IC) or are induced by x-ray exposure (table 18.1). Each of these processes result in the formation of an orbital electron vacancy. During EC, an inner (K-shell) orbital electron is absorbed by a proton-rich nucleus, forming a neutron and neutrino thus maintaining atomic mass. The neutrino is subsequently emitted and de-excitation of the daughter nucleus (if in an excited state) occurs either via gamma (γ)-ray emission or IC. The primary electron vacancy is filled by a higher orbital electron of lower energy (i.e. L-shell). The difference in the electronic binding energy of the two orbitals is either emitted as a characteristic x-ray or is transferred as kinetic energy to a secondary electron of lower binding energy, ejecting the electron in the process. This secondary ejected electron is referred to as an Auger electron when the transition occurs within the same principal energy level. Coster–Kronig transitions are special cases of the Auger process where the initial vacancy is filled by an electron from a higher subshell of the same shell. Conversely, if the emitted Auger electron originates from the same shell, it is referred to as a Super Coster–Kronig transition. The resulting vacancies from the AE emission produce a cascade of Auger electrons with energies strictly governed by allowable transitions [4].

The electron vacancy in the case of IC occurs due to the electromagnetic interaction of the excited nucleus with one of the orbital electrons resulting in the

doi:10.1088/978-0-7503-2694-0ch18

Table 18.1. Summary of Auger electron emitting radionuclides used as theragnostic agents. Yield is the number of radiative species released per decay (/nt).

Per decay	$^{67}Ga_{(A)}$	$^{80m}Br^{\dagger}_{(B)}$	$^{89}Zr_{(B)}$	$^{90}Nb_{(B)}$	$^{99m}Tc_{(A)}$	$^{111}In_{(A)}$	$^{119}Sb_{(A)}$	$^{123}I^{*}_{(A)}$	$^{125}I_{(A)}$	$^{195m}Pt_{(A)}$	$^{201}Tl_{(A)}$
Half-life (days); Decay mode	3.26; EC[e]	0.18; IT[f]	3.27, EC β^+	0.61; EC β^+	0.25; IT β^-	2.80; EC	1.59; EC	0.55;EC	59.40; EC	4.02; IT	3.04; EC
Yield of AE[a] & CK[b] e^-	4.96	9.60	9.45	8.77	4.41	7.43	23.68	13.71	23.01	36.55	20.91
Number of AE and CK e^-	9	9	11	11	22	14	13	13	13	14	15
Total AE & CK energy (keV/nt)	6.64	7.87	4.37	4.42	0.94	6.88	8.86	7.23	1.20	23.10	15.00
Total energy released (keV/nt)	196	85.9	126E04	462E04	1.43	441	49.3	201	621	262	139
(p/e)[c] ratio	4.50	0.39	11.36	9.10	7.90	11.10	0.91	6.10	2.20	0.42	2.00
SpA[d] ($Bq\ kg^{-1}$)	2.18E19	3.24E20	1.65E19	8.74E19	1.93E20	1.54E19	2.53E19	7.05E19	6.46E17	6.13E18	7.88E18

A: SPECT; B: PET. $^{\dagger 80m}Br$—decays to ^{80}Br (yield 1); $^{*123}I$ decays to ^{123}Te and ^{123m}Te, with half-lives too long to play role in mTRT.
[a]Auger electron;
[b]Coster–Kronig electron;
[c]Ratio of penetrating to non-penetrating ionizing radiation,
[d]Specific activity,
[e]Electron capture,
[f]Isomeric transition.

de-excitation of the unstable nucleus, while sufficient energy is imparted to eject the electron from its inner shell orbit. As is the case with γ-decay, the atomic number is maintained and no transmutation of one element to another takes place. Auger electron- and γ-ray emission are competing processes during IC. The atomic transition rates, and whether γ-ray or AE-emission is dominant, depends on the atomic number, the electron shells involved, and the electron configuration of the atom [5]. Similarly, photoelectrons and AE can be generated from the irradiation of metal-based nanoparticles [6–8].

In all the aforementioned processes, the relaxation of the inner-shell vacancy results in a cascade of atomic transitions following the filling of the initial vacancy with higher subshell electrons, consequently a shower of low energy AE and characteristic x-ray are emitted. The majority of the emitted AE have energies of a few eV with deposition over short (nm–μm) distances in tissues, although K-shell transitions can have energies higher than 25 keV, and up to 100 keV, but with yields per decay of less than 0.1. The extremely short range of AE yields a high linear energy transfer (LET), with electron energies below 1 keV peaking at around 26 keV μm^{-1}, resulting in a very high ionization density around the point of decay [9]. It is this observation that AE-emitters can be highly radiotoxic when they decay in close proximity to cell sensitive structures such as the DNA [10–13] or the cell surface [14] that make them desirable as therapeutic agents. However, this very characteristic has also hampered the clinical translation of most AE-emitting radiotherapeutics. This dichotomy arises as AE-emitting radionuclides require targeted binding to the cell membrane or cellular internalization and translocation to cell sensitive organelles or the nucleus to be effective. Unlike high LET (60–230 keV μm^{-1}) α-emitters used for therapy that deposit extremely high energies (several MeV) over a few cell diameters (50–100 μm) [15]. Or the longer range β-emitters (2–10 mm) that have kinetic energies ranging from tens of keV to several MeV. Although lower LET (~0.2 keV μm^{-1}), β-emitters are more widely used in MRT as the particles can target neighboring cells thus extending their toxicity.

18.2 Auger electron molecularly targeted radionuclide therapy

During the decay of AE-emitting radionuclides, AE are emitted as well as γ-rays, x-rays and IC electrons, with energies ranging from a few eV to hundreds of keV. As a consequence, energy deposition in soft tissue occurs at a few nm to tens of μm for AE and IC electrons, while the more penetrating γ-rays and x-rays, deposit energy at cm distances apart from that for ultra-soft x-rays where energy is deposited at a few μm [9]. When evaluating AE-emitting radionuclides as potential theranostic agents, several aspects must be considered. The therapeutic efficacy of a radionuclide depends on the total number of electrons emitted per decay (including AE, CK and internal conversion electrons (IC) as well as β-particles) along with the total energy released. If the total energy released is carried away by fewer electrons, these higher energy electrons would have longer ranges, that would enable targeting of the nucleus from the cytoplasm or cell surface [16]. Another key factor is the physical half-life ($T_{p½}$). As the maximum theoretical specific activity of a radionuclide is inversely proportional to $T_{p½}$, a too long $T_{p½}$ may result in the redistribution of the therapeutic agent before sufficient decays have occurred to cause lethal damage. It is also important to consider the ratio of penetrating (x- and γ-ray) to non-penetrating (electron or β-particle) forms of ionizing radiation (p/e) because the moderate-high energy but low LET γ-emissions from some radionuclides could irradiate and potentially kill non-targeted normal cells. It has therefore been proposed that an ideal therapeutic radionuclide should have a (p/e) ratio of 2 [17]. Conversely, for imaging purposes a high proportion of γ emissions are desirable, which poses a trade-off between sparing healthy tissue surrounding the target region and providing adequate absorbed dose to the target region.

The most critical point to consider, however, when using AE-emitting radionuclides for MRT is to increase their intra-nuclear accumulation to maximize therapeutic potential due to the very short range of the Auger and CK electrons [18]. Since the dimensions of the different DNA condensation states (e.g., chromatin fibers, nucleosomes and double-stranded DNA) are all within the range of typical Auger and CK electrons, nucleus incorporation leads to extreme radiotoxicity [19], resembling high-LET radiation with relative biological effectiveness (RBE) similar to that achieved by alpha-emitting radionuclides [20]. In contrast, radionuclides bound outside the cellular nucleus, e.g. in the cytoplasm, on the cellular membrane or extra-cellularly, do not produce severe lethal effects, and have RBE values comparable to those observed for low-LET radiation (e.g., x-rays) [21]. This is the consequence of the precipitous drop in energy deposition as a function of distance from the decay site. However, recent studies have challenged this premise as it was noted that the membrane is a more sensitive target than the cytoplasm for the dense ionization produced by AE [22]. Furthermore, Auger-induced bystander effects may be of consequence in MRT [23–25]. Several strategies have been proposed to achieve the optimal localization of the radionuclides with respect to the sensitive targets in the cells, while other studies concluded that nuclear accumulation is not a prerequisite. Regardless of the targeting strategy adopted, the biological response following radionuclide incorporation relies mostly on the dose absorbed.

18.3 Dosimetry of auger electrons

The Medical Internal Radiation Dose (MIRD) formalism is the gold standard when estimating the internal absorbed dose from radiopharmaceuticals [26]. The true utility of the MIRD schema lies therein that it can equally be applied to determine dose at organ, sub-organ or cellular scales [27]. Photons (γ- and x-rays) are the main contributors of dose at an organ level when the target organ differs from the source organ. For shorter-range electrons, the source and target organs mostly overlap as energy is locally deposited. In the case of AE, energy deposition is at a cellular level where the source can be the cell membrane, cytoplasm, sub-cellular organelles, or the nucleus. The primary contribution of cytotoxic effects from AEs are mainly considered from radioactive decays in the cellular nucleus. Since the energy deposition of AE-emitters is concentrated at a cellular scale, dosimetry is most effective if performed on a spatial scale comparable to the penetration range of the emitted radiation and size of the target structures. Hence, Monte Carlo (MC) simulation utilized in dosimetric calculations in this chapter, will predominately focus on cellular dosimetry.

18.3.1 Deconstructing the MIRD formulism for Auger electron dosimetry

The MIRD schema defines the mean absorbed dose $D(r_T,\ T_D)$ to a target region r_T, calculated over a dose integration period (T_D), from a radionuclide distributed uniformly within a source region (r_S) as;

$$D\ (r_T, T_D) = \sum_{r_S} \widetilde{A}\,(r_S, T_D)\ S\ (r_T \leftarrow r_S)\ (\text{Gy})$$

where, the time integrated cumulated activity in the source region is, $\widetilde{A}\,(r_S, T_D)$ and $S(r_T \leftarrow r_S)$ is the absorbed dose in the target region r_T per unit cumulated activity in the source region r_S, which can equally be applied to determine the dose at an organ or cellular scale [28].

$$S(r_T \leftarrow r_S) = \frac{\sum_i n_i E_i \phi_i(r_T \leftarrow r_S)}{m_T}\ (\frac{\text{Gy}}{\text{Bqs}})$$

with n_i the number of disintegrations, E_i the energy of the radionuclide emissions and $\phi_i(r_k \leftarrow r_h)$ the fraction of the energy emitted by the source organ that is absorbed by the target organ, which depends on the amount of radiation energy reaching the target organ (tissue and distance attenuation), the volume and composition of the target organ, the anatomical relationship of the source–target pair and the type and energy of the ith emission [29].

18.3.2 Cumulated activity

The cumulated activity $\widetilde{A}\,(r_S, T_D)$ is simply the sum or accumulation of all nuclear transitions in the source region, during a specified time interval and is written as;

$$\widetilde{A}\,(r_S, T_D) = \int A(r_S, T_D)\mathrm{d}t$$

The activity in a volume of interest (either at the organ or voxel level) is determined at specific time points by serial PET or SPECT imaging in the case of clinical or pre-clinical assessment. Alternatively, the activity can be determined directly in pre-clinical studies by harvesting organs of interest at specific time points and counting the activity in a suitable γ-counter. Similarly, activity in cells can be determined at specific time points using cellular fractionation protocols which disrupts the cell membrane [30]. The resulting homogenate is then separated by centrifugation in different buffers and the sub-cellular components isolated and activity associated with each fraction quantified. Cumulated activity can then be determined by fitting the resulting time-activity curve (TAC) by a one-phase, two-phase or multi-phase exponential decay model and numerically integrating the function [31].

18.3.3 *S*-values

The MIRD formulism has been extended to provide a compendium of cellular *S*-values of radionuclides for several cell/nucleus radii and source–target arrangements [28]. MIRD cellular *S*-values are based on Cole's analytic range-energy expression calculations with improvements by Howell and Goddu [32, 33]. A recent addition is a software applet, MIRDcell [34] providing *S*-values for single cells, monolayers and cell clusters. A limitation of this software is the use of unit density spherical geometries allowing only uniform activity distributions and a semi-analytical radiation transport model adopting the continuous-slowing-down approximation (CSDA) [35]. The use of theoretical methods, such as the CSDA approximation becomes problematic for micro-scale dosimetry as the discrete and stochastic nature of energy loss processes associated with straggling, scatter and the finite range of secondary electrons cannot be neglected at the sub-cellular scale [36]. Several authors using full MC simulation to determine *S*-values have confirmed the discrepancies with MIRD tabulations and MIRDcell [19, 35, 37, 38], which can lead to sizeable differences in dosimetry calculations.

S-value estimates, depends on the physical properties of the radionuclide (i.e. emission spectra and yield), the size and geometry of the source and target regions as well as the spatial distribution of the radionuclide in these regions (i.e. uniform versus heterogeneous) [19, 39, 40]. This will be discussed in detail in the following paragraphs.

18.3.4 Emission spectra

MC simulation of AE-emitters requires accurate data of the radionuclide emission spectra and AE cascade with associated energies and yields. These data have largely been provided by empirical and semiempirical models as well as Monte Carlo (MC) simulation [41–43], with a list of worldwide databases provided in a comprehensive review by Nikjoo *et al* [44]. Traditionally, the scientific community refers to the MIRD monograph which includes up-to-date tabulations of radiation spectra for 333 radionuclides as the database for AE-emitter emission spectra [9]. Alternatively, the electronic database of ICRP publication 107 is consulted with information on the half-lives, decay chains, yields and energies of radiations emitted in nuclear transformations of 1252 radionuclides of 97 elements [45]. These documents are based on the evaluated nuclear structure data files (ENSDF) which form part of the computer database of the National Nuclear Data Center at Brookhaven National Laboratory.

In the calculation of atomic relaxation spectra emitted after electron capture (EC) or internal conversion (IC), several assumptions were made in the MIRD tabulations partly due to computational limits at the time and the paucity of experimental data. These include the use of binding energies of neutral atoms and neglect of spectator vacancies that could potentially lead to inaccurate estimates of the emissions of low-energy x-rays and AE. An improved computational route is therefore required to address the limitations of the MIRD committee data as suggested by the International Nuclear Data Committee of the International Atomic

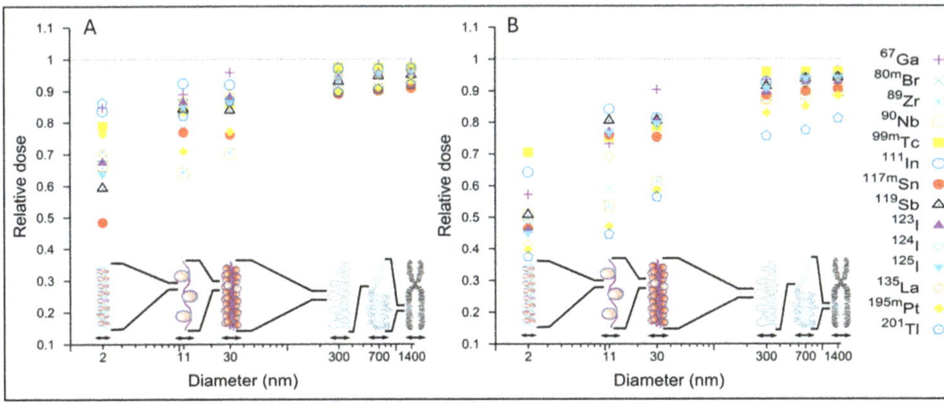

Figure 18.1. Relative dose given as the ratio of energy deposition in spheres with diameters representing different DNA condensation states, simulated using emission spectra for condensed as well as isolated-atom approximations from the stochastic atomic relaxation cascade model [50] were compared with energy deposition simulated using MIRD RADTABS emission spectra (figure adapted from [54], copyright IOP Publishing Ltd, all rights reserved).

Energy Agency [46]. For a comprehensive overview of the inadequacies in existing decay data refer to [47].

It is evident that experimental data is needed to confirm the validity of the calculations especially since experimental observation of AE energies and yields differ significantly from data provided by published libraries, i.e. Evaluated Atomic Data Library (EADL) [48]. The last decade has seen vast advancements in experimental techniques in atomic physics that could potentially address some of the limitations and accuracy of theoretical models [36]. Particularly, how the local environment affects assumptions made regarding the charge neutralisation of valence vacancies during a cascade. These assumptions could potentially lead to inaccurate calculation of the emissions of low-energy x-rays and AE that have significant implications for AE MC simulations in MRT.

Recently Lee and co-workers [49, 50] published a stochastic atomic relaxation cascade model which requires only the decay data of a radionuclide in the ENSDF format, to simulate the AE and x-ray cascade of elements up to fermium, $Z = 100$. The decay data provided is analysed *in situ* to produce the initial vacancy distribution, which determines the initial atomic state before the cascade simulation commences. As there is no consensus in literature towards the fate of valence vacancies produced during the cascade, both a condensed-phase approach where charge transfer between the environment and the valence shell is allowed, leaving the atom completely neutralised at the end of the cascade process, and an isolated-atom approach which leaves the atom in an ionised state were considered. Major differences in the emission spectra for the two approximations are found at energies below 100 eV. These differences resulted in significant underestimation of dose deposited at sub-cellular dimensions (figure 18.1) when comparing MC simulation of either condensed or isolated-atom approximation derived emission spectra with that of MIRD RADTABS [9].

Interestingly, experimental data using synchrotron irradiation to excite the C, N, and O K-edges of a target molecule along with computational modelling accounting for chemical environmental effects on excitation, ionization, and fragmentation of the molecule has shown ultra-fast charge redistribution from within the molecule [51, 52]. These experimental observations along with pulse-laser technology used to study sub-femtosecond charge migration after core shell ionisation [53] supports the condense-phase approach allowing the replenishment of valence electrons when calculating Auger electron spectra, an approach adopted in the AAPM report no. 2 [43]. It is thus clear that improved computational routes are needed to determine energies and yields of low energy AE and x-rays, in turn this would permit the validation of computer models used to calculate AE spectra.

18.3.5 Source–target geometry

It is common practice to model cell geometries with simplified regular models, i.e. where the cell model consists of two homogenous spheres of liquid water (mass density $\rho = 1$ gcm^{-3}), representing the cell and its nucleus. However, for detailed MC simulation appropriate biological models that reflect the 3-D architecture of cells and micrometastasis are required. Recently, Tamborino and co-workers used polygonal meshes (PM) to accurately render cellular and organelle geometry [55]. *S*-values obtained by MC simulation in Geant4–10.03 in this refined geometry were compared with spheres (MIRDcell) [34] and truncated cone-shaped constructive solid geometry (using MCNP6.1). They concluded that the heterogeneity of cells morphology and sub-cellular distribution of the radionuclide were the major contributing factors affecting the absorbed dose. This is an interesting observation as subsequent clonogenic survival assays show that ^{177}Lu-DOTATATE may be more effective than that suggested by average spherical cell dosimetry despite the 3x lower absorbed dose predicted using the PM models compared with MIRDcell.

Although ^{177}Lu and ^{161}Tb are medium range β-emitters, ^{161}Tb emits a higher percentage of conversion and Auger electrons. Hindié *et al* performed MC simulations using the MC code CELLDOSE in spheres with diameters ranging from 10 mm to 10 μm. ^{161}Tb delivered almost twice the dose of ^{177}Lu in 100 μm sphere and four times the dose in a 10 μm sphere. These simulations suggest that ^{161}Tb may be effective not only for treating larger tumors but also micrometastasis.

Several authors have noted that the proximity of the cell membrane and cytoplasm to the nucleus as well as non-homogenous source distribution can significantly influence cellular *S*-values [19, 40, 56, 57]. Falzone *et al* extended this observation to asymmetric cell-nucleus arrangements in multicellular tumor spheroids as a model of micrometastasis [55]. A random close-packed algorithm was used to model the variation of cellular and nucleus radii of cells and symmetrical or a-symmetrical cell-nucleus arrangements, and event-by-event MC simulation with PENELOPE based on the actual radionuclide distribution as informed by micro-autoradiograpy. They noted that *S*-values were greatly affected by the variation of cell volume and hence dose may be significantly under- or overestimated when variation in cell volume is ignored.

18.3.6 Radionuclide spatial distribution

A major advantage for MRT is that many of the AE-emitting radionuclides also emit γ-rays suitable for single-photon emission computed tomography (SPECT) or positron emission tomography (PET), in the case where annihilation photons are emitted. It is thus possible to visualize drug distribution and tumour targeting in patients by means of PET or SPECT. However, visualisation of cellular and sub-cellular AE-distribution as a means to inform MC simulation, remains extremely challenging. In most instances, $\sum_j y_j E_j \phi_j(\mathrm{T} \leftarrow \mathrm{S})$ is simulated assuming that the activity is uniformly distributed within the nucleus (N←N), the cytoplasm (N←Cy) or the cell surface (N←CS). The absorbed dose to the nucleus per unit cumulated activity in the cell *S*(N←C), is then the sum of the contributions *S*(N←N), *S*(N←Cy) and *S*(N←CS). This assumption is extremely problematic for accurate *S*-value simulation, especially as cell targeting is mainly directed to membrane receptors due to accessibility, relying on receptor-mediated endocytosis for cellular internalization and redistribution of the construct to sensitive intracellular structures or the DNA. Several authors have noted that intracellular radionuclide distribution can be extremely non-homogeneous [20, 55], leading to inaccurate estimation of *S*-values and hence absorbed dose. A recent review by Bavelaar and co-workers summarizes some of the experimental techniques currently employed to elucidate the cellular distribution of AE-emitters, including cellular fractionation, micro-autoradiography, electron microscopy combined with micro-autoradiography, photoresist autoradiography, nanoscale secondary ion mass spectrometry (nano-SIMS), radioluminescence microscopy and microdosimeters [58]. Most microscopy techniques are limited by the diffraction threshold of optical microscopy and cannot achieve the sub-cellular resolution required for quantitative assessment of radionuclide distribution. On the other hand, electron microscopy, photoresist autoradiography and nano-SIMS provide exquisite nm range resolution but are ablative techniques by nature, and not suited for live cell imaging. An interesting combination of transmission electron microscopy (TEM) and micro-autoradiography achieves nm-resolution with the potential to quantitate radionuclide distribution. A voxelised dose map calculated from an MC simulated dose point kernel (DPK) of the radionuclide is overlaid onto the TEM image (figure 18.2). Extremely high doses ($>10^3$ Gy), are achieved at the point of decay, indicated by the developed silver coil, however, due to the lack of energetic AE in the ^{111}In spectra, very-low dose per voxel (<0.01 Gy) is noted throughout the remainder of the section (unpublished data, Falzone *et al*).

18.4 Applications of Auger electron MC dosimetry

It is postulated that Auger electron (AE)-emitting radionuclides that bind or intercalate into DNA are ideal for MRT of single cells, limited volume disseminated cancer and micrometastases [59, 60]. This hypothesis is based on the high linear-energy transfer of low-energy AE that deposit energy within a few cubic nanometers of the decay site. The localized absorption of electrons results in complex irreparable

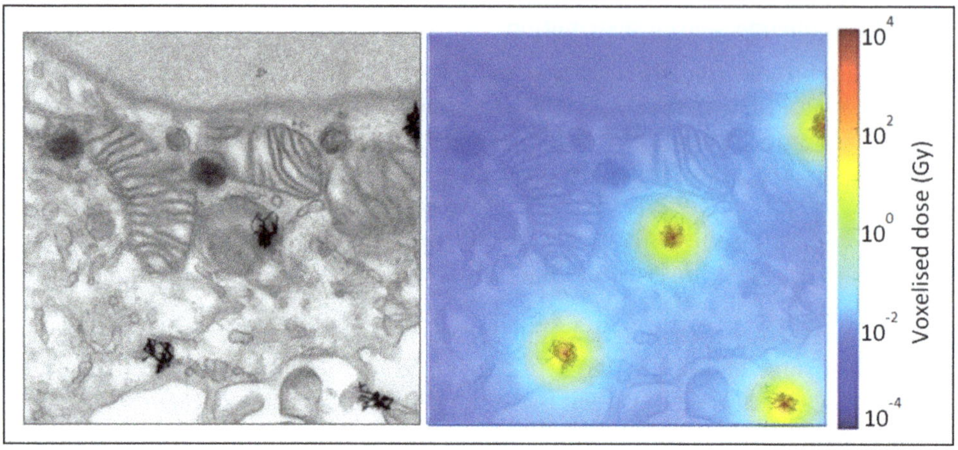

Figure 18.2. ^{111}In voxel dose overlaid onto a TEM micrograph, showing the sub-cellular distribution of the construct. The range of the AE extends over the dimensions of the mitochondria shown in the insert, resulting in a dose deposition $>10^2$Gy/voxel.

DNA damage. This feature endows AE-emitting MRT agents with an enhanced RBE when compared to α- or β-emitting radionuclides [61]. To advance our understanding of the radiobiology of AE-emitting therapeutics, MC simulation, incorporating accurate decay spectra and accounting for spatial distribution is a valuable source for dosimetry.

18.4.1 In silico

There is an increased probability of photoelectric interaction resulting in the emission of low-energy and short-range secondary electrons when high atomic number (*Z*) nanoparticles (NPs) are activated by photons of keV to MeV energies, electrons, neutrons or fast ions (>50 MeV amu^{-1}) [7, 62]. The NPs act as radiosensitizers and amplify the effects from external MeV beams, brachytherapy sources, and molecules labeled with radioisotopes. Van der Heuvel [6] demonstrated by classic MC simulation that the dose enhancement (DE) changes with energy but more importantly that there is a strong 1:1 correlation between dose and concentration of the NPs. This theme has extensively been explored for gold nanoparticles (GNPs) [63, 64]. However, Gholami proposed the use of radiolabelled superparamagnetic iron oxide (SPIO) NPs to maximize their dose enhancing effect by increasing the interaction probability due to the close proximity of the radiation source to the NPs [65, 66]. They noted that for all the radio-SPIO NP sources evaluated, a DE of 15%–20% was achieved at SD <20 nm, which corresponds to the range of electrons in the peak of the secondary energy spectra.

18.4.2 *In vitro*

Several β-emitters used in clinical practice, i.e. ^{177}Lu, ^{67}Ga, ^{64}Cu have non-negligible AE yields. Tamborino *et al* explored the influence of different cell morphologies and sub-cellular activity distribution of [^{177}Lu]-DOTA-[Tyr3]octreotate (^{177}Lu-DOTATATE) on self-dose using Geant4–10.03 [67]. ^{177}Lu-DOTATATE targets the somatostatin receptor type 2 (SST2) and upon binding, SST2 is recycled to the plasma membrane or is relocated to the trans-Golgi-network [68]. If the receptor agonist pair translocates to the Golgi, the shorter-range AE emitted during the decay of ^{177}Lu, would contribute significantly to the nuclear dose. This was confirmed by simulation, where activity distribution in the Golgi resulted in a greater nuclear dose (up to 149% in certain cell geometries) compared to a homogeneous cytoplasmic distribution of activity. This could in-part explain the therapeutic success of ^{177}Lu-DOTATATE [69].

Prostate-specific membrane antigen (PSMA) targeting in men with castration-resistant metastatic prostate cancer (mPCa) is an interesting option for AE MRT, especially as membrane and peri-nuclear localization of a PSMA-targeting small molecule was observed in PSMA-expressing PC cells [70]. Development of an AE-emitting construct targeting PSMA could potentially alleviate some of the dose limiting bone marrow toxicity and xerostomia observed from α- and β-particle emitting therapeutics, while targeting micro-metastatic disease.

18.4.3 Pre-clinical

Shen *et al* [71] recently tested the hypothesis that the AE-emitting TRT, ^{125}I-DCIBzL is effective in treating mPCa in a murine PC model. They observed that ^{125}I-DCIBzL significantly delayed development of detectable metastatic disease and improved survival with no long-term toxicities for up to 12 months. This observation was backed up by MC modelling in Geant4 using a published nephron model of the kidneys simulating emissions originating on the surface of the tubules or in the cytoplasm. The energy was tabulated both at the cellular and sub-cellular (nuclear) level. Results were validated using the MIRDCell 2.0 software and comparing simple cell-sized spheres. Interestingly, they noted that only 0.8% of the energy emitted by ^{125}I decay on the surface of the tubule, will be deposited in the nucleus, while the remainder is deposited in the cytoplasm (34%) or in the tubule lumen. The consequence of this for clinical transition is that the absorbed dose in the glomerular cell nuclei and proximal tubule nuclei will be negligible even though the average absorbed dose on an organ scale is high. Apart from the kidneys, dose limiting toxicity to the salivary glands is of consequence from α- or β-emitting PSMA targeting agents However, for ^{125}I-DCIBzL the average absorbed dose coefficient to the salivary glands was relatively low (12 mGy MBq^{-1}). In contrast, for mPCa tumor cells, energy deposition in the nucleus resulted from peri-nuclear decay of^{125}I-DCIBzL (38%), cell membrane (5%), cytoplasmic (8%) activity, resulting in an average nuclear absorbed dose of 35% i.e. 100 Gy for an injected activity of 111 MBq. MC simulation in these cases is extremely useful in predicting the toxicity to organs normally at risk versus efficacy of the AE-emitting construct and provides an important evaluation of the clinical translation of the agent.

18.4.4 Clinical

There is a paucity in clinical studies using AE-emitting MRT and very few fully utilizing dosimetry in planning or subsequent repeat treatments [72–75]. Full MC simulations for whole-body dosimetry in nuclear medicine would be desirable, especially considering that the MIRD approach is based on a homogeneous source activity distribution. Currently, there is no commercially available software that implements full event-by-event MC simulation for nuclear medicine MRT dosimetry. Notwithstanding, this is an active field of research, with several studies evaluating full stochastic MC simulations compared with analytical approaches like DPK convolution and local energy deposition [76–79].

Accurate mapping of the spatial distribution of radionuclides using PET or SPECT/CT is hampered by the resolution achievable in clinical scanners. This in turn gives rise to significant partial volume effects (PVEs) leading to an underestimation of the radioactivity concentration and hence, dose. There have been several advances in quantitative reconstruction algorithms that have increased the spatial resolution of PET/CT and SPECT/CT systems [80–82]. Recently, Marquis *et al* implemented a hybrid kernelised expectation maximisation (HKEM) algorithm in STIR to obtain SPECT images with PET-equivalent resolution [83]. The two-fold increase in resolution led to an improvement of volume resolution of a factor of 8 (i.e. 2^3). This is of particular importance as many metastatic lesions and malignant nodes are of a size that is unresolvable by SPECT.

In addition, for organs at risk, like the kidneys it is imperative that dose heterogeneity is addressed for accurate dose estimation especially since repeat administrations of MRT is planned around the maximal tolerated dose to organs at risk (OAR) [84]. This was highlighted in a Phase 1 trial in 50 patients with SST2-positive tumors treated with repeated administration to a total of 20–160 GBq ^{111}In-DTPA-octreotide at two-weekly intervals [74]. The kidney dose was estimated as 0.45 mGy MBq^{-1}, corresponding to doses ranging from 9 to 72 Gy in total, with the latter almost three times the accepted dose limit for external beam radiation. However, no renal toxicity was observed in any of the patients treated. Using MC simulation, Konijnenberg *et al* confirmed that distribution of ^{111}In-DTPA-octreotide in the renal tubules combined with the very short range of the AEs emitted by ^{111}In, may protect the more radiation sensitive glomeruli [85]. MC simulation can be used in this way to inform clinical management of patients.

18.5 Conclusion

Clinical dosimetry is still some way from incorporating absorbed dose at the scale at which dose deposition occurs with AE-emitting radionuclides. In this regard MC modelling plays an essential part in characterising the dose effect relationship of AE-emitters. Only by implementing this knowledge at a clinical level will AE-emitters receive their place amongst the particulate radiation used for MRT as truly precision systemic therapies targeting disseminated tumour cells and occult micrometastases.

Acknowledgements

With special thanks to Professor Katherine A Vallis for her continued support of my work over the years, Professors Sverre Myhra and José M Fernández-Varea and Dr Boon Quan Lee for many insightful conversations about experimental techniques, MC simulations and AE emission spectra.

References

[1] Meitner L 1922 Über die Entstehung der β-Strahl-Spektren radioaktiver Substanzen *Z. Phys.* **9** 131–44

[2] Auger P 1923 Sur les rayons beta secondaries produits dans gaz par des rayons X *CR Acad. Sci.* **177** 169–71

[3] Howell R W 2008 Auger processes in the 21st century *Int. J. Radiat. Biol.* **84** 959–75

[4] Lee B Q, Nikjoo H, Ekman J, Jonsson P, Stuchbery A E and Kibedi T 2016 A stochastic cascade model for Auger-electron emitting radionuclides *Int. J. Radiat. Biol.* **92** 641–53

[5] Lee B Q, Kibédi T, Stuchbery A E and Robertson K A 2012 Atomic radiations in the decay of medical radioisotopes: a physics perspective *Comput. Math. Methods Med.* **2012** 651475

[6] Van den Heuvel F, Locquet J-P and Nuyts S 2010 Beam energy considerations for gold nano-particle enhanced radiation treatment *Phys. Med. Biol.* **55** 4509–20

[7] Kuncic Z and Lacombe S 2018 Nanoparticle radio-enhancement: principles, progress and application to cancer treatment *Phys. Med. Biol.* **63** 02TR01

[8] Laprise-Pelletier M, Simão T and Fortin M-A 2018 Gold nanoparticles in radiotherapy and recent progress in nanobrachytherapy *Adv. Healthc. Mater.* **7** e1701460

[9] Eckerman K F and Endo A 2008 S of NM (1953–); MIRD Committee: Radionuclide Data and Decay Schemes (Reston, VASociety of Nuclear Medicine)

[10] Panyutin I G and Neumann R D 1997 Radioprobing of DNA: distribution of DNA breaks produced by decay of ^{125}I incorporated into a triplex-forming oligonucleotide correlates with geometry of the triplex *Nucleic Acids Res.* **25** 883–7

[11] Lobachevsky P N, White J, Leung M, Skene C, White J and Martin R F 2008 Plasmid breakage by ^{125}I-labelled DNA ligands: effect of DNA-iodine atom distance on breakage efficiency *Int. J. Radiat. Biol.* **84** 991–1000

[12] Karamychev V N, Reed M W, Neumann R D and Panyutin I G 2000 Distribution of DNA strand breaks produced by iodine-123 and indium-111 in synthetic oligodeoxynucleotides *Acta Oncol.* **39** 687–92

[13] Freudenberg R, Runge R, Maucksch U, Berger V and Kotzerke J 2014 On the dose calculation at the cellular level and its implications for the RBE of ^{99m}Tc and ^{123}I *Med. Phys.* **41** 62503

[14] Paillas S *et al* 2016 Localized irradiation of cell membrane by Auger electrons is cytotoxic through oxidative stress-mediated nontargeted effects *Antioxid. Redox Signal.* **25** 467–84

[15] Sgouros G, Roeske J C, McDevitt M R, Palm S, Allen B J, Fisher D R, Brill A B, Song H, Howell R W and Akabani G 2010 MIRD Pamphlet No. 22 (abridged): radiobiology and dosimetry of α-particle emitters for targeted radionuclide therapy *J. Nucl. Med.* **51** 311–28

[16] Govindan S V, Goldenberg D M, Elsamra S E, Griffiths G L, Ong G L, Brechbiel M W, Burton J, Sgouros G and Mattes M J 2000 Radionuclides linked to a CD74 antibody as

therapeutic agents for B-cell lymphoma: comparison of Auger electron emitters with beta-particle emitters *J. Nucl. Med.* **41** 2089–97

[17] Bernhardt P, Forssell-Aronsson E, Jacobsson L and Skarnemark G 2001 Low-energy electron emitters for targeted radiotherapy of small tumours *Acta Oncol.* **40** 602–8

[18] Ku A, Facca V J, Cai Z and Reilly R M 2019 Auger electrons for cancer therapy—a review *EJNMMI Radiopharm. Chem.* **4** 27

[19] Falzone N, Fernandez-Varea J M, Flux G and Vallis K A 2015 Monte Carlo evaluation of Auger electron-emitting theranostic radionuclides *J. Nucl. Med.* **56** 1441–6

[20] Kassis A I 2008 Therapeutic radionuclides: biophysical and radiobiologic principles *Semin. Nucl. Med.* **38** 358–66

[21] Howell R W, Rao D V, Hou D Y, Narra V R and Sastry K S 1991 The question of relative biological effectiveness and quality factor for auger emitters incorporated into proliferating mammalian cells *Radiat. Res.* **128** 282–92

[22] Pouget J-P *et al* 2008 Cell membrane is a more sensitive target than cytoplasm to dense ionization produced by auger electrons *Radiat. Res.* **170** 192–200

[23] Akudugu J M, Azzam E I and Howell R W 2012 Induction of lethal bystander effects in human breast cancer cell cultures by DNA-incorporated Iodine-125 depends on phenotype *Int. J. Radiat. Biol.* **88** 1028–38

[24] Pouget J-P, Georgakilas A G and Ravanat J-L 2018 Targeted and off-target (Bystander and Abscopal) effects of radiation therapy: redox mechanisms and risk/benefit analysis *Antioxid. Redox Signal.* **29** 1447–87

[25] Prise K M and O'Sullivan J M 2009 Radiation-induced bystander signalling in cancer therapy *Nat. Rev. Cancer* **9** 351–60

[26] Bolch W E, Eckerman K F, Sgouros G and Thomas S R 2009 MIRD pamphlet No. 21: a generalized schema for radiopharmaceutical dosimetry—standardization of nomenclature *J. Nucl. Med.* **50** 477–84

[27] Thierens H M, Monsieurs M A, Brans B, Van Driessche T, Christiaens I and Dierckx R A 2001 Dosimetry from organ to cellular dimensions *Comput. Med. Imaging Graph.* **25** 187–93

[28] Goddu S M 1997 MIRD Cellular S values: Self-Absorbed Dose Per Unit Cumulated Activity for Selected Radionuclides and Monoenergetic Electron and Alpha Particle Emitters Incorporated into Different Cell Compartments (Reston, VA: Society of Nuclear Medicine)

[29] Stabin M G and Siegel J A 2018 RADAR dose estimate report: a compendium of radiopharmaceutical dose estimates based on OLINDA/EXM version 2.0 *J. Nucl. Med.* **59** 154–60

[30] Alberts B, Johnson A, Lewis J, Raff M, Roberts K and Walter P 2002 *Molecular Biology of the Cell.* ed D Drubin (New York: Garland Sciences)

[31] Li T, Ao E C I, Lambert B, Brans B, Vandenberghe S and Mok G S P 2017 Quantitative imaging for targeted radionuclide therapy dosimetry—technical review *Theranostics* **7** 4551–65

[32] Howell R W, Rao D V and Sastry K S 1989 Macroscopic dosimetry for radioimmunotherapy: nonuniform activity distributions in solid tumors *Med. Phys.* **16** 66–74

[33] Goddu S M, Howell R W and Rao D V 1994 Cellular dosimetry: absorbed fractions for monoenergetic electron and alpha particle sources and S-values for radionuclides uniformly distributed in different cell compartments *J. Nucl. Med.* **35** 303–16

[34] Vaziri B, Wu H, Dhawan A P, Du P and Howell R W 2014 MIRD pamphlet No. 25: MIRDcell V2.0 software tool for dosimetric analysis of biologic response of multicellular populations *J. Nucl. Med.* **55** 1557–64

[35] Emfietzoglou D, Kostarelos K, Hadjidoukas P, Bousis C, Fotopoulos A, Pathak A and Nikjoo H 2008 Subcellular S-factors for low-energy electrons: a comparison of Monte Carlo simulations and continuous-slowing-down calculations *Int. J. Radiat. Biol.* **84** 1034–44

[36] Howell R W 2020 Advancements in the use of Auger electrons in science and medicine during the period 2015–2019 *Int. J. Radiat. Biol.* 1–26

[37] Cai Z, Kwon Y L and Reilly R M 2017 Monte Carlo N-particle (MCNP) modeling of the cellular dosimetry of ^{64}Cu: comparison with MIRDcell S values and implications for studies of its cytotoxic effects *J. Nucl. Med.* **58** 339–45

[38] Brivio D, Nguyen P L, Sajo E, Ngwa W and Zygmanski P 2017 A Monte Carlo study of I-125 prostate brachytherapy with gold nanoparticles: dose enhancement with simultaneous rectal dose sparing via radiation shielding *Phys. Med. Biol.* **62** 1935–48

[39] Siragusa M, Baiocco G, Fredericia P M, Friedland W, Groesser T, Ottolenghi A and Jensen M 2017 The COOLER code: a novel analytical approach to calculate subcellular energy deposition by internal electron emitters *Radiat. Res.* **188** 204–20

[40] Šefl M, Incerti S, Papamichael G and Emfietzoglou D 2015 Calculation of cellular S-values using Geant4-DNA: the effect of cell geometry *Appl. Radiat. Isot* **104** 113–23

[41] Charlton D E and Booz J 1981 A Monte Carlo treatment of the decay of ^{125}I *Radiat. Res.* **87** 10–23

[42] Pomplun E, Booz J and Charlton D E 1987 A Monte Carlo simulation of Auger cascades *Radiat. Res.* **111** 533–52

[43] Howell R W 1992 Radiation spectra for Auger-electron emitting radionuclides: report No. 2 of AAPM Nuclear Medicine Task Group No. 6 *Med. Phys.* **19** 1371–83

[44] Nikjoo H, Emfietzoglou D and Charlton D E 2008 The Auger effect in physical and biological research *Int. J. Radiat. Biol.* **84** 1011–26

[45] Eckerman K and Endo A 2008 ICRP Publication 107. Nuclear decay data for dosimetric calculations *Ann. ICRP* **38** 7–96

[46] Nichols A L, Qaim S M and Noy R C 2011 *Summary Report of the Technical Meeting on Intermediate-Term Nuclear Data Needs for Medical Applications: Cross Sections and Decay Data* (Vienna: International Atomic Energy Agency, IAEA)

[47] Nichols A L 2015 Recommended decay data and evaluated databases—international perspectives *J. Nucl. Sci. Technol.* **52** 17–40

[48] Alotiby M, Greguric I, Kibedi T, Lee B Q, Roberts M, Stuchbery A E, Tee P, Tornyi T and Vos M Measurement of the intensity ratio of Auger and conversion electrons for the electron capture decay of (125)I *Phys. Med. Biol.* **63** 06NT04

[49] Lee B Q, Kibédi T, Stuchbery A E and Robertson K A 2012 Atomic radiations in the decay of medical radioisotopes: a physics perspective *Comput. Math. Methods Med.* **2012** 651475

[50] Lee B Q, Nikjoo H, Ekman J, Jönsson P, Stuchbery A E and Kibédi T 2016 A stochastic cascade model for Auger-electron emitting radionuclides *Int. J. Radiat. Biol.* **92** 641–53

[51] Bolognesi P *et al* 2019 Core shell investigation of 2-nitroimidazole *Front. Chem.* **7** 151

[52] Boudjemia N, Jänkälä K, Gejo T, Nagaya K, Tamasaku K, Huttula M, Piancastelli M N, Simon M and Oura M 2019 Deep core photoionization of iodine in CH_3I and CF_3I molecules: how deep down does the chemical shift reach? *Phys. Chem. Chem. Phys.* **21** 5448–54

[53] Kuleff A I, Kryzhevoi N V, Pernpointner M and Cederbaum L S 2016 Core ionization initiates subfemtosecond charge migration in the valence shell of molecules *Phys. Rev. Lett.* **117** 93002
[54] Falzone N, Lee B Q, Fernandez-Varea J M, Kartsonaki C, Stuchbery A E, Kibedi T and Vallis K A 2017 Absorbed dose evaluation of Auger electron-emitting radionuclides: impact of input decay spectra on dose point kernels and S-values *Phys. Med. Biol.* **62** 2239–53
[55] Falzone N, Lee B Q, Able S, Malcolm J, Terry S, Alayed Y and Vallis K A 2019 Targeting micrometastases: the effect of heterogeneous radionuclide distribution on tumor control probability *J. Nucl. Med.* **60** 250–8
[56] Salim R and Taherparvar P 2019 Monte Carlo single-cell dosimetry using Geant4-DNA: the effects of cell nucleus displacement and rotation on cellular S values *Radiat. Environ. Biophys.* **58** 353–71
[57] Arnaud F X, Paillas S, Pouget J P, Incerti S, Bardiès M and Bordage M C 2016 Complex cell geometry and sources distribution model for Monte Carlo single cell dosimetry with iodine 125 radioimmunotherapy *Nucl. Instruments Methods Phys. Res. Sect.* B **366** 227–33
[58] Bavelaar B M, Lee B Q, Gill M R, Falzone N and Vallis K A 2018 Subcellular targeting of theranostic radionuclides *Front. Pharmacol.* **9** 996
[59] Zanzonico P 2011 Cell-level dosimetry and biologic response modeling of heterogeneously distributed radionuclides: a step forward *J. Nucl. Med.* **52** 845–7
[60] Gill M R, Falzone N, Du Y and Vallis K A 2017 Targeted radionuclide therapy in combined-modality regimens *Lancet Oncol.* **18** e414–23
[61] Fourie H, Nair S, Miles X, Rossouw D, Beukes P, Newman R T, Zeevaart J R, Vandevoorde C and Slabbert J 2020 Estimating the relative biological effectiveness of Auger electron emitter ^{123}i in human lymphocytes *Front. Phys.* **8** 519
[62] Lacombe S, Porcel E and Scifoni E 2017 Particle therapy and nanomedicine: state of art and research perspectives *Cancer Nanotechnol.* **8** 9
[63] Jayarathna S, Manohar N, Ahmed M F, Krishnan S and Cho S H 2019 Evaluation of dose point kernel rescaling methods for nanoscale dose estimation around gold nanoparticles using Geant4 Monte Carlo simulations *Sci Rep.* **9** 3583
[64] Ahn S H, Chung K, Shin J W, Cheon W, Han Y, Park H C and Choi D H 2017 Study on dependence of dose enhancement on cluster morphology of gold nanoparticles in radiation therapy using a body-centred cubic model *Phys. Med. Biol.* **62** 7729–40
[65] Gholami Y H, Maschmeyer R and Kuncic Z 2019 Radio-enhancement effects by radiolabeled nanoparticles *Sci Rep.* **9** 14346
[66] Maschmeyer R T, Gholami Y H and Kuncic Z 2020 Clustering effects in nanoparticle-enhanced β(−) emitting internal radionuclide therapy: a Monte Carlo study *Phys. Med. Biol.* **65** 125007
[67] Tamborino G, De Saint-Hubert M, Struelens L, Seoane D C, Ruigrok E A M, Aerts A, van Cappellen W A, de Jong M, Konijnenberg M W and Nonnekens J 2020 Cellular dosimetry of [^{177}Lu]Lu-DOTA-[Tyr(3)]octreotate radionuclide therapy: the impact of modeling assumptions on the correlation with *in vitro* cytotoxicity *EJNMMI Phys.* **7** 8
[68] Csaba Z, Peineau S and Dournaud P 2012 Molecular mechanisms of somatostatin receptor trafficking *J. Mol. Endocrinol.* **48** R1–12
[69] Strosberg J *et al* 2017 Phase 3 trial of ^{177}Lu-dotatate for midgut neuroendocrine tumors *N. Engl. J. Med.* **376** 125–35

[70] Kiess A P, Minn I, Chen Y, Hobbs R, Sgouros G, Mease R C, Pullambhatla M, Shen C J, Foss C A and Pomper M G 2015 Auger radiopharmaceutical therapy targeting prostate-specific membrane antigen *J. Nucl. Med.* **56** 1401–7
[71] Shen C J *et al* 2020 Auger radiopharmaceutical therapy targeting prostate-specific membrane antigen in a micrometastatic model of prostate cancer *Theranostics* **10** 2888–96
[72] Vallis K A *et al* 2014 Phase I trial to evaluate the tumor and normal tissue uptake, radiation dosimetry and safety of ^{111}In-DTPA-human epidermal growth factor in patients with metastatic EGFR-positive breast cancer *Am. J. Nucl. Med. Mol. Imaging* **4** 181–92
[73] Limouris G S, Chatziioannou A, Kontogeorgakos D, Mourikis D, Lyra M, Dimitriou P, Stavraka A, Gouliamos A and Vlahos L 2008 Selective hepatic arterial infusion of In-111-DTPA-Phe1-octreotide in neuroendocrine liver metastases *Eur. J. Nucl. Med. Mol. Imaging* **35** 1827–37
[74] Valkema R *et al* 2002 Phase I study of peptide receptor radionuclide therapy with [In-DTPA] octreotide: the Rotterdam experience *Semin. Nucl. Med.* **32** 110–22
[75] Li L, Quang T S, Gracely E J, Kim J H, Emrich J G, Yaeger T E, Jenrette J M, Cohen S C, Black P and Brady L W 2010 A Phase II study of anti-epidermal growth factor receptor radioimmunotherapy in the treatment of glioblastoma multiforme *J. Neurosurg.* **113** 192–8
[76] Botta F *et al* 2013 Use of the FLUKA Monte Carlo code for 3D patient-specific dosimetry on PET-CT and SPECT-CT images *Phys. Med. Biol.* **58** 8099–120
[77] Papadimitroulas P 2017 Dosimetry applications in GATE Monte Carlo toolkit *Phys. Med.* **41** 136–40
[78] Sarrut D *et al* 2014 A review of the use and potential of the GATE Monte Carlo simulation code for radiation therapy and dosimetry applications *Med. Phys.* **41** 64301
[79] Pérez P and Valente M 2019 DOSIS: an integrated computational tool for patient-specific dosimetry in nuclear medicine by Monte Carlo and dose point kernel approaches *Appl. Radiat. Isot.* **150** 135–40
[80] Wang X, Zhou L, Wang Y, Jiang H and Ye H 2021 Improved low-dose positron emission tomography image reconstruction using deep learned prior *Phys. Med. Biol.* **66** 115001
[81] Bal H, Kiser J W, Conti M and Bowen S L 2021 Comparison of maximum likelihood and conventional PET scatter scaling methods for ^{18}F-FDG and ^{68}Ga-DOTATATE PET/CT *Med. Phys.* **48** 4218–28
[82] Arabi H, AkhavanAllaf A, Sanaat A, Shiri I and Zaidi H 2021 The promise of artificial intelligence and deep learning in PET and SPECT imaging *Phys. Med.* **83** 122–37
[83] Marquis H, Deidda D, Gillman A, Willowson K P, Gholami Y, Hioki T, Eslick E, Thielemans K and Bailey D L 2021 Theranostic SPECT reconstruction for improved resolution: application to radionuclide therapy dosimetry *EJNMMI Phys.* **8** 16
[84] Willowson K P, Eslick E, Ryu H, Poon A, Bernard E J and Bailey D L 2018 Feasibility and accuracy of single time point imaging for renal dosimetry following ^{177}Lu-DOTATATE ('Lutate') therapy *EJNMMI Phys.* **5** 33
[85] Konijnenberg M W, Bijster M, Krenning E P, De and Jong M 2004 A stylized computational model of the rat for organ dosimetry in support of preclinical evaluations of peptide receptor radionuclide therapy with ^{90}Y, ^{111}In, or ^{177}Lu *J. Nucl. Med.* **45** 1260–9

Chapter 19

Monte Carlo models and methods of skeletal and bone marrow dosimetry

Wesley Bolch

19.1 Introduction

In this chapter, we consider one of the most anatomically complex and variable tissues of the human body—the skeleton. The tissues of the human skeleton are inherently radiosensitive given that they contain stem cells whose primary function is to continually replenish the supply of blood cells which support CO_2/O_2 exchange, immune response, and blood clotting. The life-span of erythrocytes is ~120 days, and the life-span of granulocytes, lymphocytes, and platelets are 7–30 h, 4.4 years, and 9–12 days, respectively [1]. Thus, there is constant cell division and differentiation occurring within the marrow tissues. Risks of marrow irradiation include induction of leukemia and other blood cancers and, at higher doses seen in radiopharmaceutical therapy (RPT), marrow toxicity which can result in either transitory or potentially permanent declines in circulating blood cell concentrations.

Skeletal dosimetry is further complicated by the fact that these tissues are distributed throughout the body amongst individual bones that vary widely both macroscopically (shape, size, cortical thickness, internal volumes) and microscopically (fraction of internal volumes filled by bone, marrow hematopoietic cells, and marrow fat cells). While dose estimates to hematopoietically active (or red) bone marrow are needed for dose-response modeling of marrow toxicity in RPT, one typically assigns that dose as a single value in mGy or Gy. This reported value in fact is an average value of the active marrow dose seen in each of the individual bones of the skeleton, each of which probably never receives this skeletal-averaged dose. Consequently—as will be seen throughout this chapter—skeletal dosimetry models in internal dosimetry fundamentally are built from a collection of bone site-specific models (e.g., cranium, ribs, lumbar vertebrae, etc) from which skeletal-averaged doses are derived.

doi:10.1088/978-0-7503-2694-0ch19

It will also be seen in this chapter that the majority of skeletal tissue dose estimates reported in therapy nuclear medicine have heavily relied upon 'reference' models coming from radiological protection. While some patient-specific adjustments can be made to reference model marrow doses—such as proportional scaling by total body or lean body mass of the patient and reference computational phantom—more refined adjustments to these models for patient-specific dosimetry have only been pursued within the last ~15 years to include MR-methods for assessing marrow cellularity, and dual-energy CT methods for assessing patient-specific bone volume fractions.

19.1.1 Anatomy of the skeletal system

The entire skeleton in an adult consists of 200 distinct bones divisible in four classes: long, short, flat, and irregular bones [2]. These 200 bones form a rigid framework to which the softer tissues and organs of the body are attached. Microscopically, the human skeletal system is a complex structure that can be classified into two distinct types: trabecular bone and cortical bone (figure 19.1). The trabecular bone (also called cancellous bone) is a porous, spongy structure found in the interior of the flat bones and at the ends of the long bones. It is formed of a complex network of bone trabeculae (osseous plates and rods) and tissue cavities. Each cavity is lined by a layer of osteogenic cells called the endosteum and is filled with marrow that is either hematopoietically active (red marrow) or inactive (yellow marrow). The composite regions of bone trabeculae and bone marrow is called spongiosa, and in whole-body computational human phantoms, this region is presented as a homogeneous mixture of these two tissue types. The cortical bone is a hard, dense structure also called compact bone. It is found in the shafts of the long bones and in the outer cortex of all

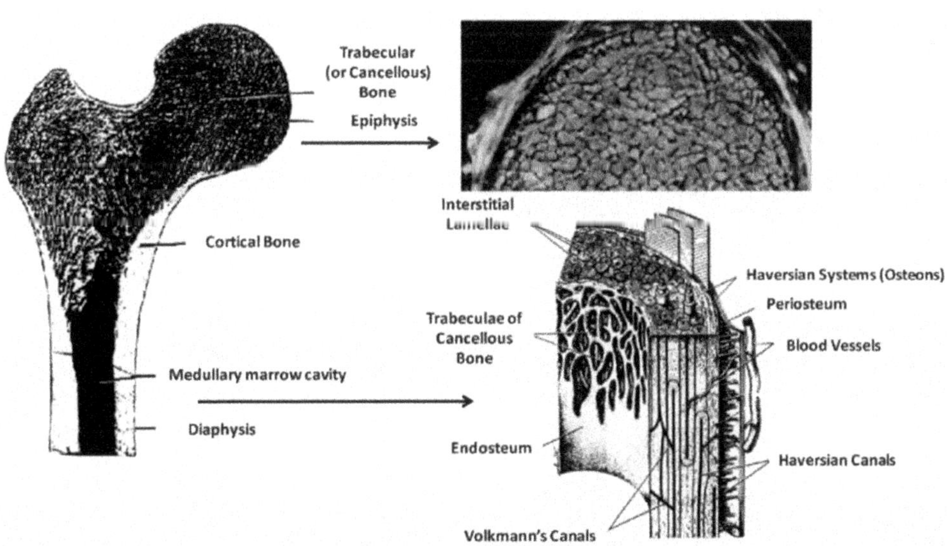

Figure 19.1. Diagram of the upper half of a femur showing the trabecular (or cancellous) regions of the femoral head (NMR image) and neck epiphysis, and the cortical bone and medullary cavity regions of the diaphysis.

bones. The dominant microstructure of the cortical bone is the osteon, which is formed by a central Haversian canal lined with a layer of endosteum and surrounded by bone lamellae. Typical dimensions of the microstructures of an adult trabecular and cortical bone are given in table 19.1.

Radiosensitive cells in the human skeletal system have been identified as (a) hematopoietic cells present in bone marrow, (b) endosteal cells lying close to bone surfaces, and (c) epithelial cells close to bone surfaces in the air sinuses of the skull [3]. Dosimetrically important regions in therapeutic nuclear medicine are considered to be the hematopoietic stem and precursor cells in the marrow cavities and the osteogenic cells on the surface of the bone trabeculae. Doses to epithelial cells lining ethmoid, sphenoid, mandibular, and maxillary sinuses from alpha emitters in bone or in gaseous form [4–6] are generally not applicable for nuclear medicine dosimetry.

19.1.2 Overview of the dosimetry of the skeletal system

Throughout the development of therapeutic radiopharmaceuticals for systemic and targeted cancer treatment, the skeletal system, and more precisely the active bone marrow, has been identified as one of the main dose-limiting organs [8–11]. Dose calculations in nuclear medicine are principally based on the MIRD methodology [12]. In a practical approach, the MIRD schema decomposes the average absorbed dose to a target tissue region r_T from a collection of tissue-specific radiopharmaceutical source region r_S into a 'biological/physiologic' term, $\widetilde{A}(r_S)$, and an 'anatomic/physics' term, $S(r_T \leftarrow r_S)$:

$$D(r_T) = \sum_{r_S} \widetilde{A}(r_S)\, S(r_T \leftarrow r_S) \tag{19.1}$$

where $\widetilde{A}(r_S)$ is the total number of decays (or time-integrated activity) of the radionuclide in the source region r_S, $S(r_T \leftarrow r_S)$ is the S value or the mean absorbed

Table 19.1. Typical geometrical dimensions of trabecular and cortical bone microstructure in the adult [7].

Bone morphometry feature	Typical dimensions (mm)
Trabecular bone	
Trabeculae in parietal bone	511
Trabeculae in cervical vertebrae	279
Trabeculae in femoral head	232
Marrow cavity in parietal bone	389
Marrow cavity in cervical vertebrae	910
Marrow cavity in femoral head	1157
Cortical bone	
Osteon length (between Volkmann canals)	2500
Concentric lamellae thickness	7
Osteon diameter	200–400
Haversian canal diameter	20–200

dose in the target per time-integrated activity in the source region, and the summation runs over all potential source regions containing the radiopharmaceutical. The time-integrated activity is generally determined on a per patient basis from planar imaging (scintigraphy), SPECT/CT or PET/CT, or blood activity counting (see chapter 2). S values are tabulated for specific radionuclides and source/target combinations. The calculation of an S value requires knowledge of the radiations emitted by the radionuclide and information on the distribution of absorbed energy within target regions of interest in the anatomical model for monoenergetic sources distributed uniformly within source regions of the model. Such data are referred to as absorbed fractions of energy. The calculation of absorbed fractions of energy requires the use of a geometric model of the anatomy of interest (source and target regions as well as the surrounding tissue media) and a particle transport code. For most source–target regions in the body, both models are well differentiated and anatomical models are based on either stylized (mathematical) anatomic models, voxel-based anatomic models, or more recently, mesh-based anatomic models (see chapter 3). Particle transport may be performed via the Monte Carlo method (see chapter 1) or dose-point kernel algorithms (see chapter 4).

Macroscopically, the skeletal system is well defined throughout the body, and its geometry can be modelled using simple mathematical shapes [13–15]. However, it is an inhomogeneous region with radiosensitive tissues embedded within a dense bone matrix defined at the microscopic level. Therefore, a different approach must be employed to incorporate both the macroscopic and microscopic anatomy, and because directly-ionizing and indirectly-ionizing radiations have widely different interaction ranges, they are separately considered in the dosimetry modelling.

There are three approaches that have been developed and used to calculate the energy deposited by indirectly-ionising radiations in the sensitive regions of the skeletal system. These three methods are all based on the use of a homogeneous macroscopic model of the skeletal system (uniform mixture of bone and soft tissue with density 1.4 g cm^{-3}) within mathematical anthropomorphic phantoms [13, 14] that is coupled to a photon transport Monte Carlo code. The first method was developed by Snyder *et al* for the tabulation of dose conversion factors for various radionuclides [16], and was subsequently used in MIRD Pamphlets Nos. 5 [13] and 11 [17]. Snyder *et al* calculated the energy deposited by photon interactions in the macroscopic skeletal model, and partitioned it by target mass to generate the energy deposited in the sensitive tissues of the skeletal system. This approach assumed that the range of secondary electrons was smaller than the marrow cavity sizes and equivalence in the mass absorption coefficients for marrow and bone. In an effort to improve upon these assumptions, Eckerman presented a new approach in which photon fluence calculated in a homogenous skeletal region of the anthropomorphic phantom is coupled with dose per unit fluence response factors derived considering the creation and transport of the secondary electrons in a microscopic model of the skeletal region [14, 18, 19]. This second approach was used in the tabulation of specific absorbed fractions of energy (absorbed fraction of energy per unit mass of target region) published in ORNL TM-8381 [14], which were subsequently used in the MIRDOSE code [20]. A similar approach was proposed by Bouchet *et al* consisted of calculating the energy of the secondary electrons created by photon interactions in the macroscopic model of the skeletal system and coupling them to electron absorbed fractions of energy calculated using a microscopic model of the skeletal system [21]. The last two methods differed principally in the use of two different microscopic models of the skeletal system that will be presented and compared in subsequent sections.

For directly-ionising radiations, there are essentially two approaches to calculate the energy deposited in the sensitive regions of the skeletal system. The first is to focus the modelling on closely representing the anatomical microstructures of the skeletal system and to couple the model to a Monte Carlo particle transport code. However, because of its intricate anatomy and its complex microstructure, it is difficult to exactly model the skeletal system without using three-dimensional imaging. The second approach is to model the particle transport in such a way that the traversed distance across a given microscopic region (bone trabeculae or marrow cavity for example) by a particle matches exactly the real path of the particle in the skeletal microstructure. This approach initially was proposed by the research group of Spiers [7, 22–28] and subsequently was used in several skeletal dosimetry models [18, 19, 29]. In the next sections of this chapter, the different microscopic models of the cortical and trabecular bone regions that have been used with Monte Carlo techniques to generate dosimetric data for nuclear medicine are presented. Because a model is based upon specific assumptions, and is derived for a reference individual, correction methods for patient-specific applications are given as well. The chapter continues with a review of current models of skeletal dosimetry developed for and adopted by the International Commission on Radiological Protection (ICRP). We close with a discussion of the challenges of used reference models of skeletal dosimetry in the prediction of marrow toxicity in the RPT patient and the many more recent advances that have been made to personalized marrow dose estimates in therapeutic nuclear medicine.

As noted earlier, the skeleton fundamentally must be treated as a collection of independent dosimetric regions with each bone site having at the macroscopic level a unique external shape, cortical thickness, and spongiosa volume, and at the microscopic level a unique marrow volume fraction and marrow cellularity. As such, radiopharmaceutical dosimetry models are first constructed independently for each bone of the skeleton, and then mass-weighted to yield whole-body skeletal averaged dosimetry quantities. As described in Appendix B of Pafundi *et al* [30], skeletal-averaged values for (1) the particle/energy-dependent absorbed fraction $\phi_{\text{skel}}(r_T \leftarrow r_S; E_i)$, (2) the particle/energy-dependent specific absorbed fraction $\Phi_{skel}(r_T \leftarrow r_S; E_i)$, and (3) the radionuclide-dependent S value $S_{\text{skel}}(r_T \leftarrow r_S; E_i)$, are computed in turn as:

$$\phi_{\text{skel}}(r_T \leftarrow r_S; E_i) = \sum_x f_{x,\ S}\ \phi(r_T \leftarrow r_S; E_i) \tag{19.2}$$

$$\Phi_{\text{skel}}(r_T \leftarrow r_S; E_i) = \sum_x f_{x,\ S}\, f_{x,\ T}\ \phi(r_T \leftarrow r_S; E_i) \tag{19.3}$$

$$S_{\text{skel}}(r_T \leftarrow r_S; E_i) = \sum_x f_{x,\ S}\, f_{x,\ T}\ S(r_T \leftarrow r_S; E_i) \tag{19.4}$$

where $f_{x,\ S}$ is the relative mass fraction of the source tissue r_S in bone site x, $f_{x,\ T}$ is the corresponding relative mass fraction of the target tissue r_T in bone site x, and the summation is then over all bone sites comprising the human skeleton. For the adult and adolescent male and female, the collection of bone sites considered when the target tissue is active marrow excludes all sites in the arms and legs with the exception of the proximal ends of the humeri and femora [3].

19.2 Early models of trabecular bone

19.2.1 Research at the University of Leeds

The foundation of all bone dosimetry modelling was established through the research of F W Spiers and his students at The University of Leeds between the years 1949 to 1981 [7, 22–26, 31–33]. From this research group, three major dissertations on bone dosimetry are of particular note: one by Darley in 1972 [24], one by Whitwell in 1973 [25], and one by Beddoe in 1976 [7]. Spiers determined through this research that the microstructure of trabecular bone could not be well described using simple geometrical shapes. Instead, he used frequency distributions of linear path lengths through trabeculae and marrow cavities as a quantitative description of the three-dimensional structure of this region of the skeleton. For this purpose, Darley and Beddoe designed an optical bone scanner capable of automatically measuring bone and cavity space pathlength length distributions of thin sections of human trabecular bone (see figure 19.2) [23, 24]. Assuming a direction of alignment of the cavities, they derived corresponding omnidirectional pathlength length distributions (figure 19.3). A total of seven bone sites were measured from a 44-year male subject including the cervical vertebra, lumbar vertebra, femur head, femur neck, iliac crest, parietal bone, and rib. Images of five bone sites were

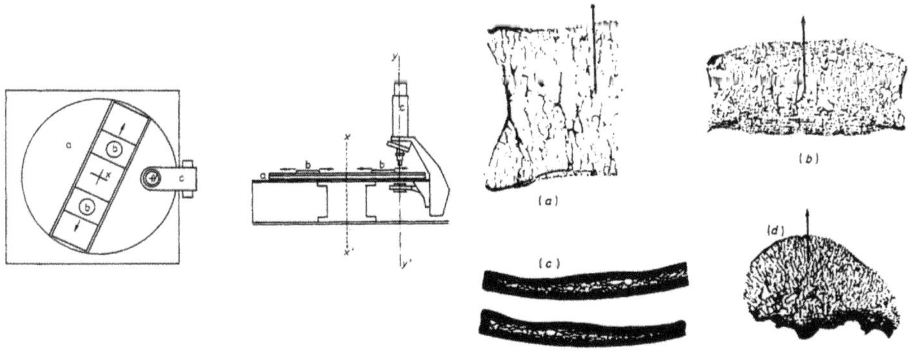

Figure 19.2. (Left) basic layout of the University of Leeds bone scanning microscope [22]. (Right) contact radiographs of thin sections of human trabecular bone: (a) left half of an adult third lumbar vertebra (mid-coronal plane) (b) third lumbar vertebrae from a 9-year child (mid-coronal plane), (c) parietal bone from a 9-year child (parasagittal plane), and (d) femoral head (proximal to epiphysis) from a 9-year child (mid-coronal plane). Reproduced from [22].

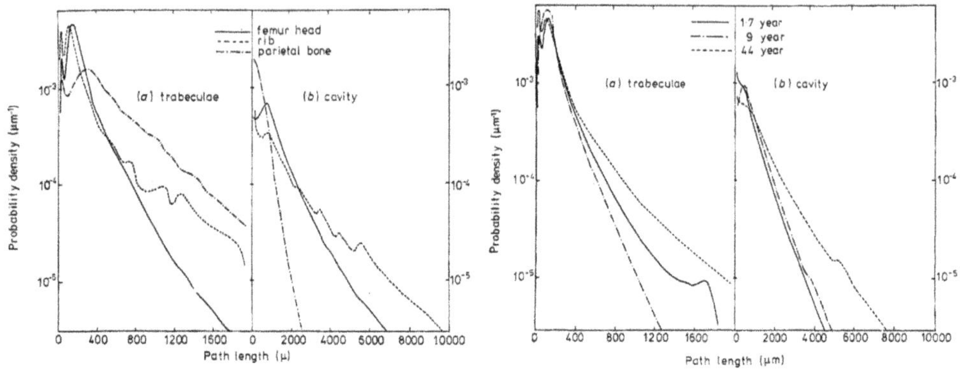

Figure 19.3. (Left) omnidirectional path-length distributions for three bones from a 44-year male. (Right) omnidirectional path-length distribution within the third lumbar vertebrae at ages of 1.7-years, 9-years, and 44-years. Reproduced from [22]. Copyright IOP Publishing Ltd. All rights reserved.

additionally taken for a 1.7- and 9-year male subject. Whitwell used the experimentally measured omnidirectional pathlength length distributions to calculate dose conversion factors (ratio of absorbed dose in the target region to the absorbed dose in hypothetical trabecular bone assuming complete absorption of the beta emissions) for seven radionuclides (^{14}C, ^{18}F, ^{22}Na, ^{32}P, ^{45}Ca, ^{90}Sr, ^{90}Y) distributed within the volume of trabeculae. A Monte Carlo process was used to sample the pathlength length distributions and select electron paths through marrow cavities and bone trabeculae. The energy deposited in those different regions was determined using range-energy relationships in both the bone trabeculae and marrow cavities (continuous slowing down approximation) assuming a straight electron path through these regions. Although these calculations were performed assuming the electron deposits all its energy in the trabecular bone, Whitwell applied correction factors to simulate the escape of electrons from the trabecular bone for high-energy electron sources. In this study, only bone- and surface-seeking radionuclides were considered, and only the marrow cavity and endosteum were chosen as targets.

19.2.2 Model of trabecular bone in MIRD Pamphlet No. 11

The research of Whitwell focused on radionuclides of interest for radiation protection and, therefore, nuclear medicine dosimetry was not of primary importance. To apply Whitwell's data to additional radionuclides, Snyder *et al* approximated the specific absorbed fractions for monoenergetic sources by expressing the skeletal-averaged dose conversion factors for the seven radionuclides as a function of their average beta energy [25, 26]. Using these specific absorbed fractions, trabecular bone *S* values were tabulated as part of MIRD Pamphlet No. 11 [17]. In these reports, four source regions were considered (trabecular bone, cortical bone, red marrow, and yellow marrow) and four target regions (skeletal bone, red marrow, yellow marrow, and skeletal endosteum). For the skeletal source–target combinations not considered by Whitwell, Snyder *et al* applied one of the following five assumptions to estimate values of absorbed fraction: (a) no energy deposition in the target region, (b) full energy deposition in the target region, (c) conservation of energy within neighbouring regions, (d) uniform energy deposited per unit mass in both the source and target regions, and/or (e) the reciprocity theorem [34, 35].

19.2.3 Model of trabecular bone in ICRP Publication 30

In 1979, the ICRP issued its Publication 30 [36], which recommended absorbed fractions of energy for beta particles for use in radiation protection. For beta particles originating in the bone volume, a single value of absorbed fraction was recommended. For beta particles originating on the bone surface, one absorbed fraction for low-energy beta particles (average beta energy less than 0.2 MeV), and one for high-energy beta particles (average beta energy greater than or equal to 0.2 MeV) were recommended. These absorbed fraction values were based on the dose conversion factors from Whitwell [25, 26], and were intended to be conservative, as their objective was to provide a system of radiation protection for workers and not for clinical radiopharmaceutical therapy. Subsequently, these relatively energy-independent absorbed fractions of energy were implemented in the MIRDOSE program [20] for use in nuclear medicine dosimetry. In this same computer program, the self-absorbed fraction to the marrow was assumed to be unity as suggested in Part 3 of the ICRP Publication 30.

The reference values of skeletal source/target absorbed fractions of the ICRP Publication 30 model are listed in table 19.2. In the dose calculation, they were associated with reference target tissue masses defined for Reference Man to be 1500 g of total active (red) bone marrow and 120 g of total bone endosteum. In Publication 30, however, the term ‘bone surfaces’ was used to refer to a potential radionuclide source region (on the surfaces of bone trabeculae in spongiosa and along the Haversian canals of cortical bone). While this nomenclature was correct, ICRP Publication 30 incorrectly reused that same term—bone surfaces—to also represent the endosteum target region. Thus, in the ICRP literature you will still see

Table 19.2. Summary of absorbed fractions applied to skeletal tissues in the ICRP 30 model. Energy ranges for β-emitters on the bone surfaces refer to the mean beta energy.

Source	Target	α-Emitter uniform in volume	α-Emitter on bone surfaces	β-Emitter uniform in volume	β-Emitter on bone surfaces (⩾200 keV)	β-Emitter on bone surfaces (<200 keV)
Trabecular bone	Endosteum	0.025	0.25	0.025	0.025	0.25
Cortical bone	Endosteum	0.01	0.25	0.015	0.015	0.25
Trabecular bone	Red marrow	0.05	0.5	0.35	0.5	0.5
Cortical bone	Red marrow	0	0	0	0	0

the acronym 'BS' to refer to bone surfaces as a target for dose estimation. Clearly, the term 'endosteum' should have been used instead.

Another important aspect of the ICRP 30 bone models was the assignment of time-integrated activity $\widetilde{A}(r_S)$. The ICRP 30 model simplistically assumes only two localizations for any bone-seeking radionuclide: (1) uniformly across all bone surfaces (initial deposition) or (2) uniformly distributed throughout the bone volumes (equilibrium deposition). In reality, there is an active and dynamic process of bone remodeling which takes initial surface deposition and 'buries' that activity over time throughout the bone volume due to osteoblastic and osteoclastic action. The 'rule of thumb' then for health physics applications of the ICRP 30 bone model was to assign as a 'surface-seeker' any radionuclide that had a physical half-life of less than 15 days. Otherwise, for longer-lived radionuclides, they were treated as 'volume-seekers' and different absorbed fractions were applied as given in table 19.2. Once that decision is made—surface versus volume activity—the division between trabecular and cortical bone must be made. For surface-seekers in the ICRP 30 bone model, the partition of $\widetilde{A}(r_S)$ is made equally (50%/50% between trabecular and cortical bone) as bone forms of mineral bone were assumed in ICRP Publication 30 to have equal total surface areas. If, however, the radionuclide was assigned as a volume-seeker, the model partitions $\widetilde{A}(r_S)$ as 20% to trabecular bone and 80% to cortical bone, in accordance with their total mass allotment in Reference Man (1000 g trabecular bone and 4000 g cortical bone). Note that this model does not provide guidance between the skeletal features of adult females, nor does it allow for any age-dependence in the assignment of skeletal tissue dosimetry to children. In the assignment of their fractional weight in the effective dose (called the effective dose equivalent in that time period), a tissue weighting factor of 0.12 was assigned to active marrow equivalent dose and a tissue weighting factor of 0.03 was assigned to the endosteum equivalent dose.

19.2.4 Spherical models of trabecular bone

Chen and Poston [37] proposed a model in 1982 that assumed the marrow spaces in trabecular bone were spherical in shape. The marrow space dimensions and the trabecular thicknesses were chosen randomly from the pathlength-length distributions and other data of Spiers and his colleagues [7, 22–26, 31–33]. A computer code, DAB-BE, was written which could be used to calculate the absorbed energies deposited by β particles or electrons inside human trabecular bone. The radiation source geometry could be either uniformly distributed in the whole bone or located at a fixed point inside the bone. Bremsstrahlung production was not considered in these calculations. Results were presented for six monoenergetic electron sources (0.1, 0.5. 0.8, 1.0, 1.5, and 2.0 MeV) distributed uniformly in a mathematical representation of the arm [16]. In addition, a fixed-point source of 1.0 MeV monoenergetic electrons was studied.

This spherical model was improved by Zuzarte de Mendoca by consideration of the endosteal layer as well as the marrow space and trabecular dimensions [38]. She used the Monte Carlo code EGS4 [39] and considered both surface and volume sources. These calculations produced absorbed fractions and specific absorbed fractions of energy for six monoenergetic electron sources (0.1, 0.2, 0.5, 1.0, 2.0 and 4.0 MeV) for source–target combinations in the spine, the cranium, and the pelvis. She found that the absorbed fractions in bone marrow were strongly dependent on the ratio of trabecular bone to bone marrow volume. The higher the ratio, the lower the absorbed fraction in bone marrow.

Parry [40] continued this research and devised a modified model of the trabecular bone for use in the EGS4 code. He calculated electron absorbed fractions for red marrow and the endosteal tissue on trabecular bone. The model was based on measured omnidirectional pathlength length distributions through trabeculae and marrow cavities in nine skeletal locations. Absorbed fractions were calculated for only two target regions and all other skeletal regions were estimated based on these results. In addition, the absorbed fractions were used to calculate *S*-values for a variety of beta emitting radionuclides at each of 15 skeletal regions containing red bone marrow. Absorbed fractions and *S*-values were calculated for sources in the marrow, on the bone surface and in the bone volume. Comparisons were made between these results and those published in ICRP Publication 30 [36] as well as those determined using MIRDOSE software [20].

19.2.5 The Eckerman trabecular bone model

Eckerman [18] presented new electron absorbed fractions of energy in the trabecular bone region based on the pathlength length distributions measured by Darley [24]. Using Whitwell's approach [25, 26] of combining Monte Carlo sampling of pathlength distributions in bone cavities and trabeculae with electron range-energy relationships, Eckerman derived absorbed fractions of energy for monoenergetic electrons for the seven adult male trabecular bone sites measured by Darley [24]. Source regions explicitly considered in this model were the marrow cavity, bone trabeculae, and bone surfaces, with target regions being the marrow cavity, endosteum, and bone trabeculae (see figure 19.4). To simulate a bone surface source, the Eckerman model assumed that electrons were emitted at the surface between the trabeculae and marrow cavities. For the endosteal target, a 10-μm layer of soft tissue was assumed on the bone surface as recommended in ICRP Publication 30 [36]. Because the radiosensitive active marrow target was part of the simulated marrow cavity, Eckerman derived the absorbed fraction to this sensitive structure by partitioning the energy deposited in the full marrow cavity by the cellularity factor (CF—percentage of total marrow mass that is assumed to be hematopoietically active). This model simulated no escape of the electron from trabecular spongiosa, and assumed that all energy emitted was deposited in sampled spongiosa. In 1994, results from this model were implemented in the MIRDOSE program [20]. Reference masses for the endosteum and active marrow in trabecular bone of the 15 skeletal regions defined in the anthropomorphic phantoms [14] were derived as described in [19] and used to convert Eckerman's absorbed fractions of energy to radionuclide *S* values. This implementation allowed for explicit calculation of regional absorbed dose in the 15 different bone sites as well as for the whole

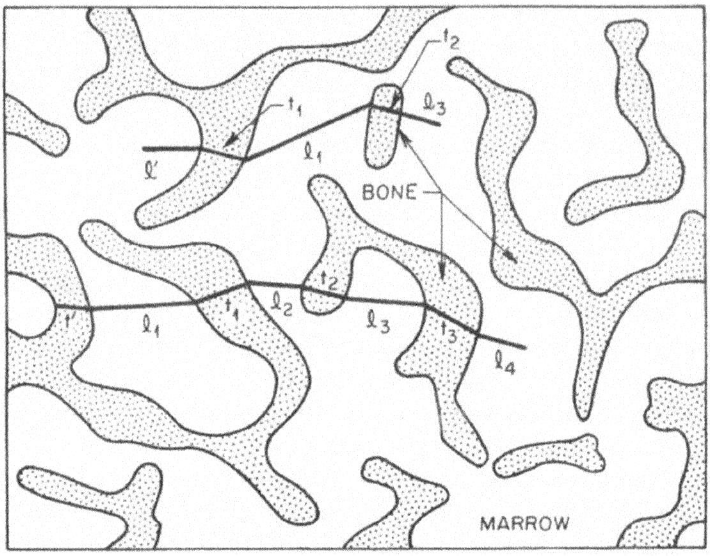

Figure 19.4. Schematic of the electron transport model of Eckerman [18]. Particle energy tracking is accomplished by alternate sampling of bone trabeculae and marrow cavity path lengths as given by the University of Leeds distributions of figure 19.3. For the lower electron path in this figure, an electron is initially emitted on the trabecular surface. With 50% probability, this electron is emitted into the bone trabecula and crosses it with a randomly sampled pathlength t'. With its residual energy, it next emerges on the far side of this trabecula and crosses the adjacent marrow cavity with randomly selected marrow pathlength of ℓ_1. With sufficient residual energy, it next enters the adjacent bone trabecula with sampled pathlength t_1. This process is continued until the original surface-emitted electron has crossed four bone trabeculae, three marrow cavities, and finally loses all its kinetic energy within the fourth marrow cavity (residual pathlength is less than the sample marrow cavity pathlength). The total energy lost to bone marrow (both active and inactive) is thus given by range-energy relationships across a total marrow pathlength of $\ell_1 + \ell_2 + \ell_3 + \ell_4$. A similar pathlength sampling process is applied to the upper electron which is shown to be emitted within the marrow space reaching the first bone trabeculae after traversing pathlength ℓ' which is computed as a random fraction (0 to 1) of the initially sampled marrow pathlength.

skeleton, for two trabecular bone target regions (red marrow and the endosteum) and for three trabecular bone sources (trabecular bone volume, trabecular bone surface, and active (red) bone marrow).

19.2.6 The Bouchet *et al* trabecular bone models

In 1999, Bouchet *et al* extended the one-dimensional transport model for electrons in trabecular bone to a three-dimensional model [41]. This model was based on the path-length distributions of the group of Spiers. A series of 3D spherical half-spaces of bone or marrow, with radii consistent with the measured pathlength distributions, were presented to the particle during Monte Carlo radiation transport. In this manner, allowances were made for potential differences between the particle pathlength and its linear displacement through the media, for backscattering electrons, as well as for the production of both delta-rays and bremsstrahlung x-rays. Electron transport in the endosteum was performed by creating a 3D spherical half-space with radii selected randomly from a random entry angle in a 10-μm thick layer. The EGS4-PRESTA Monte Carlo transport code [39] was used to simulate the electron interaction within this 3D model, assuming no escape of energy from the trabecular bone. Similar to Eckerman's simulations, Bouchet *et al* calculated absorbed fractions of energy for monoenergetic electrons for the seven trabecular bone sites for which Spiers' group tabulated the pathlength length distributions. Source and target regions considered were the trabecular marrow space, the trabecular bone volume, and the trabecular bone endosteum. Results from these simulations were combined with reference masses derived from ICRP Publication 70 for 22 skeletal sites and nine source–target tissue combinations within the trabecular bone [42].

19.2.7 Differences between the 3D model of Bouchet *et al* and the 1D model of Eckerman

The benefits of a three-dimensional transport model over a one-dimensional model were investigated by Bouchet *et al* [41]. They constructed a one-dimensional model for electron transport in trabecular bone following the pathlength length sampling methodology employed in their three-dimensional model. The energy deposited in a given region was scored assuming a straight electron trajectory through each tissue region and using the range-energy relationships adopted by Whitwell and Eckerman. They reported improvements over one-dimensional transport at low electron energies (less than ~200 keV). Large differences were observed when the source and the target were not adjacent regions (30%–50% and upward for energies less than 100 keV) due to the possible variations in the range of the electrons modelled by the EGS4 Monte Carlo transport code (i.e., range straggling).

Three additional differences in the methodologies used by Eckerman and Bouchet *et al* are associated with the selection of source and target for the active marrow and the bone surfaces. The first difference is associated with the derivation of absorbed fractions for self-irradiation of red marrow. In the Eckerman model, absorbed fractions of energy in the red marrow were calculated by multiplying the absorbed fractions in the marrow cavity by reference cellularity values for each bone region. In the Bouchet *et al* model, it was assumed that the absorbed fractions of energy to the red marrow and marrow cavity were numerically equal. As a result of this difference in methodology, skeletal average S values for an active marrow source and target calculated by Eckerman were determined to be ~50% lower than those calculated by Bouchet *et al*. Subsequent research at the University of Florida explored these differences using 3D radiation transport techniques in trabecular bone based upon images acquired through NMR microscopy [43], and based upon calculations for the femur and humerus, have concluded that: (1) the Eckerman model provides acceptable estimates of AF (red marrow) at energies above 200 keV; (2) the Bouchet *et al* model provides acceptable estimates of these same values at energies less than 20 keV; and (3) neither model accurately predicts the absorbed fraction for self-irradiation of the active marrow in the energy range of 20–200 keV. Results from these two models were combined using the University of Florida findings to generate more accurate absorbed fractions of energy for the active marrow [44].

The second difference between these involves bone surface sources. For a bone surface source, the Eckerman model uses a 2D planar surface source, selected at the interface between the trabeculae and marrow cavities as recommended in ICRP Publication 30 [36]. The Bouchet *et al* model assumes that the source is distributed throughout the 10-μm layer of soft tissue (endosteum) on the bone interface. Consequently, the Eckerman skeletal average *S* values for a 'bone surface' source and a marrow target are ~15% lower than those of Bouchet *et al* for low-energy sources.

The third difference is associated with an endosteal target. Both models assume the same 10-μm layer of soft tissue on the bone surface. However, because the Eckerman model is a 1D model, electrons are assumed to enter this target with a uniform distribution of angles as compared to the 3D-transport of Bouchet *et al* that assumes a uniform distribution of the cosine of the angle. The 1D assumption leads to a decrease in the pathlength through the endosteum. As a result of the differences associated with the modelling of the surface source and endosteal target, calculated *S* values for a surface source and an endosteal target are 60% lower in the MIRDOSE 3/Eckerman model than in the Bouchet *et al* model for low-energy sources. These discrepancies, however, became mute as described later in section 19.4 with the ICRP's redefinition of the thickness and skeletal location of the bone endosteum.

19.2.8 Model of a trabecular bone lesion

A different type of trabecular bone model was developed in 1995 by Samaratunga *et al* [45]. This group developed an anatomical model of trabecular bone designed specifically to calculate the dose to skeletal metastases from the surface seeker ^{186}Re. The model was based on anatomical measurements of pathlength length distributions of cavities and bone trabeculae on 25 samples from skeletal metastases. Bone trabeculae were represented by ellipsoids located in an infinite marrow-tissue medium. Finally, the transport of electrons was simulated in a three-dimensional geometry using the Monte Carlo electron transport code EGS4. Even though the procedure was very sophisticated, the model developed by Samaratunga was intended for transport of electrons in trabecular bone metastases, and therefore was not applicable for ICRP Reference Man tabulations.

19.2.9 The Watchman *et al* trabecular model for alpha particles

In 2005, Watchman *et al* presented new data on absorbed fractions of energy for alpha particles in the skeletal tissues [46]. Previous alpha particle models did not account for variations in either marrow cellularity or particle energy. These existing models assumed a fixed reference marrow cellularity (defined as the volume fraction of total marrow that is hematopoietically active) and assigned absorbed fractions only relevant energies in the range of 4.5 to 5.0 MeV as needed for dosimetry of occupationally important alpha-emitting radionuclides. Alpha emitters of interest in

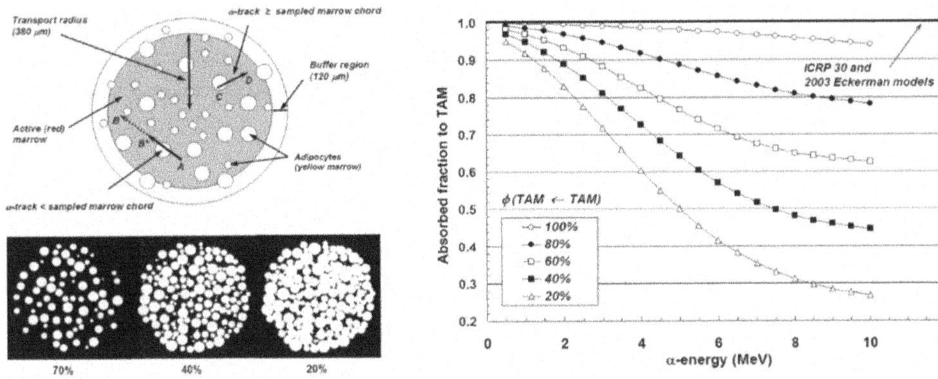

Figure 19.5. (Left) geometric model used in Watchman *et al* [46] to partition sampled marrow cavity chords into sub-trajectories of particle through active (red) marrow and inactive (yellow) marrow, the latter represented by individual adipocytes (white spheres). (Right) dependence of active marrow self-dose absorbed fraction with changes in marrow cellularity within the adult lumbar vertebrae. This research was originally published in JNM. [46]. Copyright SNMMI.

radiopharmaceutical therapy, however, are characterized by shorter physical half-lives and higher emission energies extending upwards to ~9 MeV. In this study, the authors presented chord-based radiation models (3D-CBIST) that combined (1) pathlength-based techniques for tracking alpha-particles across bone trabeculae, endosteum, and marrow cavities, and (2) a spatial model of the marrow tissues that explicitly considered the presence of marrow adipocytes. The model and absorbed fraction results for alpha-particle active marrow self-irradiation are shown in figure 19.5.

19.2.10 University of Florida models based upon NMR microscopy

An alternative to pathlength-based transport models for trabecular bone dosimetry is the use of 3D digital images of these skeletal sites coupled directly to Monte Carlo radiation transport codes. This technique offers several distinct advantages over pathlength-based models. First, absorbed fractions of energy are acquired within a skeletal site for which mass estimates are also known. In pathlength-based models, only the absorbed fraction is obtained from the transport calculations; independent assignments of target tissue mass are required to report absorbed dose. Second, the full three-dimensional structure of the anatomical site is available for particle tracking, enabling one to include energy loss to the cortical bone cortex of the skeletal site for high-energy sources. Most current pathlength-based transport models continue to utilize an infinite spongiosa transport (IST) geometry. Third, explicit partitioning of the marrow source and target tissues into active (red marrow) and inactive (yellow marrow) is accommodated within 3D models of trabecular bone, avoiding scaling approaches necessary within pathlength-based models to account for varying marrow cellularity.

Three imaging techniques have been applied to the general study of trabecular bone microarchitecture: nuclear magnetic resonance (NMR) microscopy (resolution: ~50–100 μm), micro-computed tomography (resolution: ~20–50 μm), and synchrotron microtomography (resolution: <20 μm). Jokisch *et al* investigated the use of NMR microscopy to establish 3D electron transport models of trabecular bone [47, 48]. In these studies, cadaveric sources of thoracic vertebral bodies were subjected to marrow digestion, immersion in Gd-doped water, and imaging at 14.1 T using a 3D spin-echo pulse sequence. Electron transport with the EGS4 code was conducted under a dual transport geometry: one utilizing a segmented, voxel model of the bone trabeculae and marrow cavities, and one utilizing a stylized representation of the trabecular spongiosa and cortex of cortical bone. These studies were extended by Bolch *et al* to include explicit consideration of active marrow as both a source and target region [49]. This study showed that, for non-specific uptake of radioactivity in the active marrow, marrow cellularity is an important parameter to consider in patient-specific assessments of active marrow self-irradiation for high-energy therapy radiopharmaceuticals. For low-energy emitters, reference values of cellularity are sufficient for dose estimation.

In studies by Patton *et al*, energy loss to cortical bone was explored and compared to results from infinite spongiosa transport [50]. This study indicated that S value corrections on the order of 5%–8% for ^{32}P and 8%–11% for ^{90}Y were required to properly account for the finite size of skeletal site. In a subsequent study, Patton *et al* established that the overall methodology employed in NMR microscopy and radiation transport was reproducible to ~1%–2% at energies below 1 MeV, and no more than 4% at 4 MeV [51]. Furthermore, the study established that nearly identical absorbed fraction energy profiles could be obtained via either marrow-intact or marrow-free NMR imaging, thus suggesting the potential for *in vivo* patient-specific dosimetric models of the skeleton. This conclusion was supported by the theoretical study by Rajon *et al* [52] noting that image resolutions no finer than ~300 μm were sufficient for the development of 3D dosimetry models of trabecular bone for high-energy electron emitters employed in radiopharmaceutical therapies.

19.3 Early models of cortical bone

19.3.1 Research at the University of Leeds

Similar to trabecular bone dosimetry, the foundation of cortical bone dosimetry was established by F W Spiers at the University of Leeds. In 1976, A H Beddoe used the bone scanning microscope to measure transverse chord lengths through Haversian cavities and the intervening cortical bone matrix [32]. Initially developed for path-length measurements in trabecular bone, this bone-scanning microscope was refined to measure the small cavities in cortical bone. Stained sections of transverse slices of cortical bone (20–30 μm thick) were scanned to obtain transverse pathlength distributions of the Haversian cavity sizes and cortical bone matrix; the latter defined as the distances between Haversian cavities in the plane perpendicular to the bone axis. With this optical scanning system, an effective resolution of about 8 μm was achieved. Beddoe made measurements of the cortex in three cortical bones: the humerus, tibia, and femur. In his measurements, Beddoe assumed that the Haversian cavities ran parallel to the long axis of the bone, which is generally the case for the long bones. He also assumed that the number of transverse canals

(Volkmann canals) appearing in the transverse planes was negligibly small. This last assumption was verified visually on his stained sections of cortical bones. Beddoe included measurements of the iliac crest, but his results showed irregular fluctuations due to scanning only one small section.

For dosimetry calculations, Beddoe followed the technique employed by Whitwell [25, 26] in the dosimetry of trabecular bone. Assuming a straight electron path through Haversian cavities and cortical bone matrix, a Monte Carlo sampling technique was used to randomly select electron paths through the Haversian canal and cortical bone matrix. The energy deposited in these regions was determined using range-energy relationships in bone and soft tissues, and was used to calculate dose conversion factors for several radionuclides. These dose conversion factors were only published in the dissertation of Beddoe [7], and only for a limited number of radionuclides (^{14}C, ^{18}F, ^{22}Na, ^{32}P, ^{45}Ca, ^{90}Sr, ^{90}Y). Moreover, the only source considered in his study was the bone volume, where both the endosteum and Haversian canals were of interest for nuclear medicine and health physics dosimetric applications.

19.3.2 Model of cortical bone in ICRP Publication 30

In ICRP Publication 30, recommended absorbed fractions for cortical bone are only given for the endosteum as target region and for the cortical surface and bone volume as source regions [36]. For beta emitters originating on the cortical surface, the Commission recommended an absorbed fraction to the endosteum of 0.25 for mean beta energies below 200 keV, and an absorbed fraction of 0.015 for mean beta energies greater than or equal to 200 keV. For beta emitters originating in the cortical bone volume, an energy-independent absorbed fraction of energy to the endosteum of 0.015 was recommended. These absorbed fractions values—shown in table 19.2—were based on the dose conversion factors of Beddoe, and were intended to be conservative, as their objective was to provide a system of radiation protection for occupational workers.

19.3.3 Research of Akabani

In 1993, Akabani developed an anatomical model for the dosimetry of a single Haversian canal, and derived dose factors for several beta-emitting radionuclides (^{32}P, ^{45}Ca, ^{89}Sr, ^{90}Sr, ^{90}Y, ^{131}I, ^{153}Sm) [53]. This single Haversian canal was modelled as a cylinder surrounded by an infinite cortical bone matrix. Six Haversian canal sizes were individually considered with radii of 5, 10, 20, 30, 40, and 50 μm. The Monte Carlo transport code EGS4-PRESTA was used to derive electron absorbed fractions of energy for two sources (Haversian canal and bone surface) and three targets (Haversian canal, endosteum, and the bone matrix). From these calculations, dose conversion factors were tabulated for these seven radionuclides as a function of the Haversian canal radius. Although elegant in its design, this anatomical model only considered a single size for the Haversian canal, and it did not consider the cross-osteon irradiation.

19.3.4 The Bouchet *et al* cortical bone model

In 1999, Bouchet *et al* developed a three-dimensional transport model for electrons in cortical bone [29] using the same methodology they used for trabecular bone dosimetry [41]. They modelled the transport geometry experienced by an electron travelling in the cortical bone using the transverse chord length distributions through Haversian cavities and cortical bone matrix measured by Beddoe [32]. Because the pathlength distributions were measured within the plane perpendicular to the bone axis, each region was represented by parallel portions of cylinders with radii sampled from the chord length distributions. By sampling many different chords, for many electrons, the average transport of an electron in cortical bone was simulated. However, the escape of the particle from the cortical bone was not simulated, and therefore, these cylinders were not limited by any planes along the direction of their axes. Electron transport in the endosteum was performed by creating a 3D cylindrical half-space with radii selected randomly from an entry angle in a 10 μm thick layer. The EGS4-PRESTA Monte Carlo transport code was used to simulate the electron interaction within this 3D model, thus allowing for modelling of (a) the non-linear electron trajectories, (b) full transport of the electrons with delta-rays and bremsstrahlung, (c) the distribution of sizes of Haversian canals, and (d) cross-osteon irradiation. Absorbed fraction results for source and target regions being the cortical endosteum, the Haversian canal, or the cortical bone matrix were tabulated for the three cortical bone sites for which Beddoe tabulated the chord length distributions. Results from these simulations were combined with reference masses derived from ICRP Publication 70 for 22 skeletal sites and nine source–target tissue combinations within the trabecular bone [42].

19.3.5 The Eckerman cortical bone model

In 2000, Eckerman and Stabin [19] presented absorbed fraction for monoenergetic electron sources distributed in the volume and on the surface of cortical bone matrix. Details regarding these calculations were, however, not provided in [19]. The absorbed fractions were based on a cylindrical model of the Haversian canal located within an infinite cortical bone matrix. The diameter of the Haversian canal was taken to be 40 μm and the energy deposition within the 10 μm layer lining the canal was evaluated based on the range-energy relationship (CSDA approximation).

19.4 Current ICRP reference models of skeletal dosimetry

19.4.1 Changes in the definition of skeletal sources and targets by the ICRP

Significant changes to the source and target tissue masses of the skeleton were made by the ICRP in the release of its Publication 89 in 2002 [54], thus replacing prior values given in the ICRP 30 bone model. These revised tissue masses are summarized below in table 19.3.

As seen above, some features were new and some remained constant from the prior reference skeletal model. First, the masses of the skeletal tissues were now defined across the full series of ICRP reference individuals to include both the male

Table 19.3. Summary of reference masses (grams) in the skeleton of the ICRP reference individuals given in Publication 89 [54].

Organ/tissue	0-year	1-year	5-year	10-year	15-year male	15-year female	Adult male	Adult female
Mineral bone	170	590	1260	2300	4050	3700	5500	4000
Trabecular bone	35	120	250	460	810	740	1100	800
Cortical bone	135	470	1010	1840	3240	2960	4400	3200
Bone marrow	50	170	500	1260	2560	2380	3650	2700
Active (red)	50	150	340	630	1080	1000	1170	900
Inactive (yellow)	0	20	160	630	1480	1380	2480	1800
Cartilage	130	360	600	820	1140	920	1100	900
Teeth	0.7	5	15	30	45	35	50	40
Miscellaneous tissues	20	45	55	90	155	145	200	160
Total skeleton	370	1170	2430	4500	7950	7180	10,500	7800
Average CF	1.0	0.88	0.68	0.50	0.42	0.42	0.32	0.33

and female at each of the six reference ages. Second, the fraction of total mineral bone assigned to trabecular and cortical forms was kept constant as a function of age and consistent with the assumption made in ICRP Publication 30—that is, mineral bone is found across the entire skeleton as 80% in its cortical form and 20% in its trabecular form. Third, and not shown in table 19.3, were changes in the distribution bone surfaces between trabecular and cortical bone. In the new ICRP 89 model, the total bone surface area in cortical and trabecular bone was assigned as 6.5 and 10.5 m^2, respectively in the adult male. No values were given for the reference adult female nor for any of the reference pediatric individuals. Thus, for bone surface seeking radiopharmaceuticals, total bone activity would be partitioned as 38% cortical activity and 62% trabecular activity—across the whole skeleton with no variation by age or sex. Third, the newer ICRP 89 model for reference marrow masses included a more refined consideration of age-dependent changes in marrow cellularity. Starting at unity for the newborn (all marrow in hematopoietically active across all marrow cavities), the skeletal-averaged CF is shown to decline with increasing age—being 0.88 at 1-year and only 0.32–0.33 in the adult. Accordingly, for the reference adult male, the total active (or red) bone marrow dropped from 1500 g in the ICRP 30 skeletal model to 1170 g in the ICRP 89 skeletal model. The previously higher estimate was in fact total bone marrow in those axial regions of the adult skeleton which contained some active (or red) marrow. The lower value of 1170 g thus excludes the mass of inactive marrow that is co-located in axial regions of the skeleton with some inactive bone marrow.

What is obviously missing in table 19.3 (and in ICRP Publication 89) are any references to mass of bone endosteum in either trabecular or cortical regions of the skeleton by age and sex. This curious fact can be traced to an ongoing debate about this target tissue within the ICRP at the time of the release of Publication 89. As described in Bolch *et al* [55], studies were reported by Gossner *et al* [56] and Gossner

[57] which proposed significant changes to the dosimetric target region for radiogenic bone cancer. These studies indicated that the cells at risk for bone cancer induction are (1) localized out to 50 μm from both the trabecular surfaces of spongiosa and the interior cortical surfaces of the long bone medullary cavities, and (2) not localized within the Haversian canals of cortical bone. As a result, cortical bone no longer was considered to encompass any dosimetric target tissue of interest—both active marrow and the vast majority of bone endosteum were contained in the spongiosa regions of the skeleton. Cortical bone, however, would remain a source region for radionuclide or radiopharmaceutical uptake, but any dosimetric contribution to active marrow or bone endosteum would have to come from cortical bone-to-spongiosa radiation cross-fire. Using bone-specific measurements and/or estimates of bone-specific surface-to-volume ratios, bone trabeculae volumes, and the revised 50-mm endosteal thickness, one could then provide some estimates of total endosteal tissue masses (see below).

19.4.2 PIRT—paired image radiation transport models of the University of Florida

Under support from the U.S. National Cancer Institute, the research laboratory of Dr Wesley Bolch at the University of Florida undertook new methods of skeletal dosimetry for internal beta-particle emitters. The methodology was termed PIRT for Paired-Image Radiation Transport [58, 59] in which electrons emitted within various source regions of trabecular spongiosa were tracked simultaneously in two coordinate systems—one at a microscopic level of the marrow cavities and bone trabeculae as given by segmented microCT images of cored spongiosa and one at the macroscopic level of the entire bone partitioned into regions of exterior cortical bone and interior spongiosa, as given by *ex vivo* CT images of harvested bone samples. Graphical schematics of the PIRT method are shown in left-hand panels of figures 19.6 and 19.7 for the pelvis and ribs, respectively (cranium values were also explored—not shown).

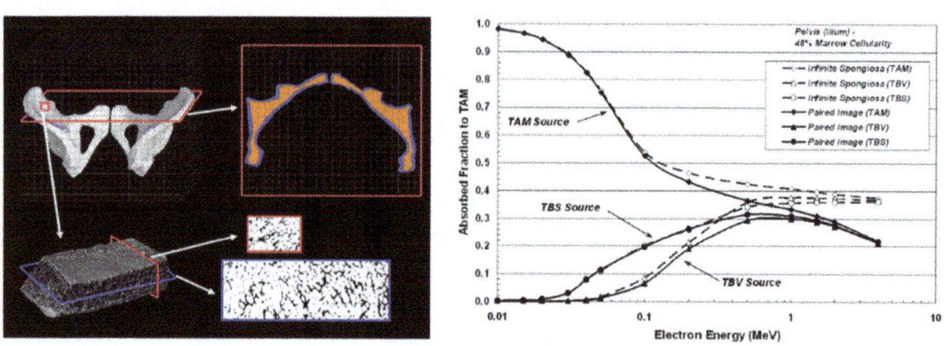

Figure 19.6. (Left) schematic of the PIRT model constructed for the pelvis. Regions of spongiosa (interior) and cortical bone (exterior) are shown in the CT macroimage (upper right). (Right) electron absorbed fractions to active bone marrow within the pelvis for three source tissues—TAM (trabecular active marrow), TBV (trabecular bone volume), and TBS (trabecular bone surfaces). Data shown by solid lines are from the PIRT model, while those given by dashed lines are from IST simulations. Reproduced from [58] John Wiley & Sons. Copyright 2005 American Association of Physicists in Medicine.

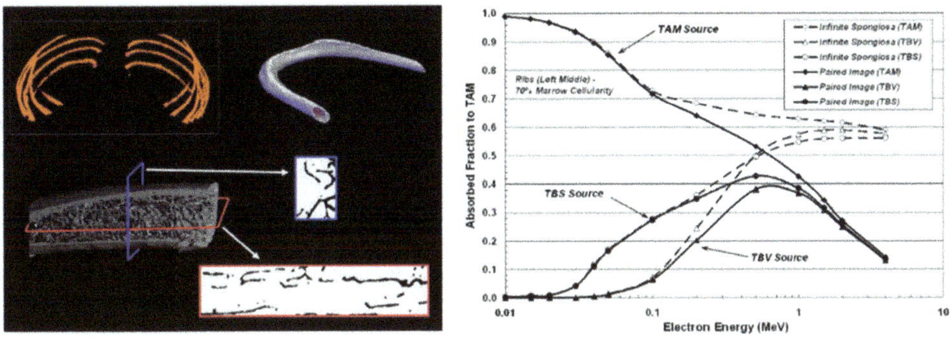

Figure 19.7. (Left) schematic of the PIRT model constructed for the ribs. (Right) electron absorbed fractions to active bone marrow within the ribs for three source tissues—TAM, TBV, and TBS. Data shown by solid lines are from the PIRT model, while those given by dashed lines are from IST simulations. Reproduced from [58] John Wiley & Sons. Copyright 2005 American Association of Physicists in Medicine.

Bone samples were taken from whole-body skeletal harvest of a 66-year male (BME of 22.7 kg m^{-2}). Results from Shah *et al* [58] were that at initial electron energies greater than 50–200 keV, a divergence in absorbed fractions to active marrow are noted between PIRT model simulations and those estimated under infinite spongiosa transport (IST) techniques. As shown in the right-hand panels of these figures, electron escape from spongiosa is greater in the ribs than in the pelvis. Calculations of radionuclide *S* values under both methodologies imply that current chord-based models used in clinical skeletal dosimetry can over-estimate dose to active bone marrow in these skeletal sites by ~2% to 28% for low-energy beta emitters (^{33}P, ^{169}Er, and ^{177}Lu), by ~4% to 48% for intermediate-energy beta emitters (^{153}Sm, ^{186}Re, and ^{89}Sr), and by ~14% to 76% for high-energy beta emitters (^{32}P, ^{188}Re, and ^{90}Y). The authors concluded that the PIRT methodology allowed for detailed modeling of the 3D macrostructure of individual marrow-containing bones within the skeleton thus permitting improved estimates of absorbed fractions and radionuclide *S* values for intermediate-to-high energy beta emitters.

19.4.3 Image-based reference models of the ICRP adults and children

Following the 2007 Recommendations of the ICRP [60], a broad program of work had been initiated by ICRP Committee 2 in the mid-2000s to update the series of internal dose coefficients for inhaled or ingested radionuclides [61]. This included the use of the ICRP Publication 110 voxel-based reference computational phantoms of the ICRP adult male and female [62] (see chapter 5) with a major revision to the specific absorbed fractions (SAFs) for internally emitted alpha particles, beta particles and electrons, and photons [63]. To support the development of new SAFs for internally emitted electrons, the University of Florida identified a 40-year male cadaver of approximately 82 kg in total mass and 170 cm in stature at the time of death (BMI of 28.4 kg m^{-2}). The subject died of sudden complications associated with cardiopulmonary arrest following myocardial infarction. As reported in Hough *et al* [64], following whole-body CT imaging, a total of 38 major skeletal sites were harvested, imaged *ex vivo* by CT scanning (macroscale models), after which microCT imaging (30 μm isotropic resolution) was performed of cored samples of spongiosa from each site (microscopic models). Voxels of the marrow cavities were then retagged as AM (active marrow) or IM (inactive marrow) in proportion to their bone-specific reference marrow cellularity (see ICRP Publication 70 [3]). PIRT models of internally emitted electrons were constructed from which energy-dependent absorbed fractions and specific absorbed fractions were computed.

Values of $\phi(r_T \leftarrow r_S)$ are shown in figure 19.8 for active marrow as the dosimetric target and either active marrow (AM) or the bone trabeculae volume (TBV) as the

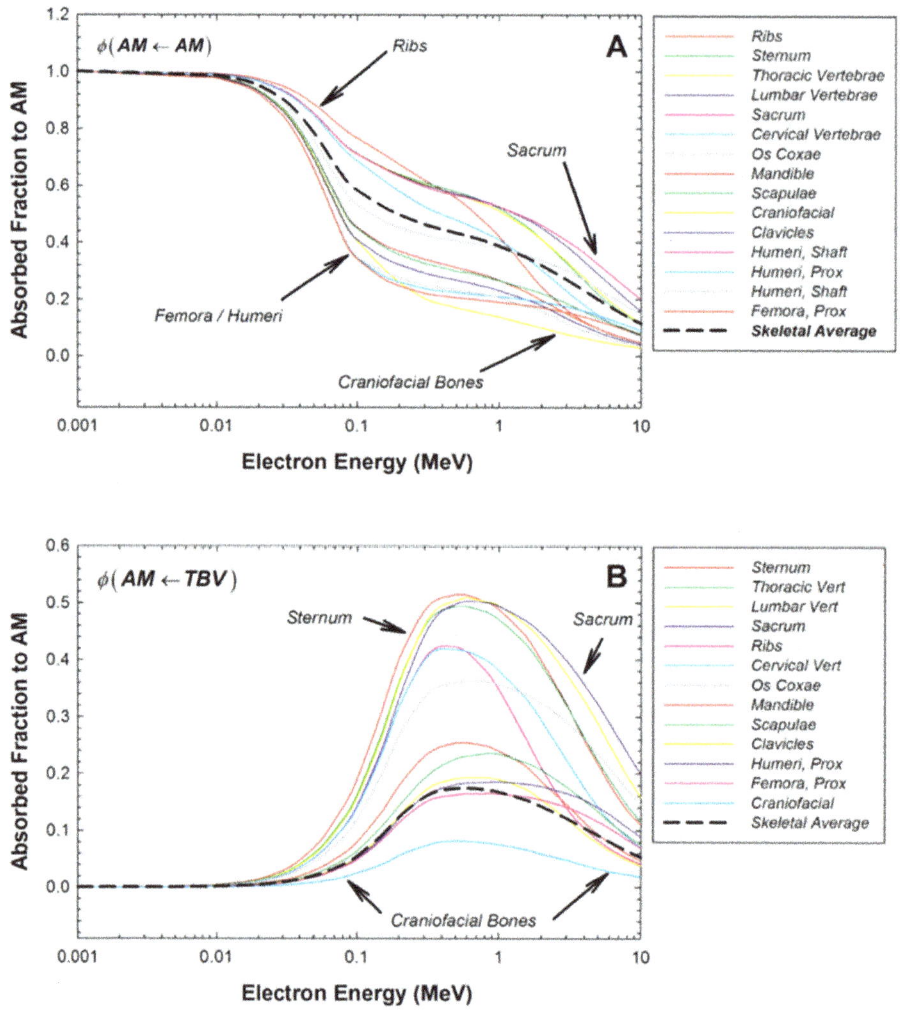

Figure 19.8. Transport results from the Hough reference adult male model showing variations in bone-specific electron absorbed fractions for (a) active marrow (AM) self-irradiation and for (b) active marrow irradiation from a trabecular bone volume (TBV) source. Dashed lines indicate the skeletal-averaged absorbed fractions for these source–target combinations. Reproduced from [64].

source region. For an AM source and at very low electron energies, the absorbed fraction is shown to be ~1.0 (source = target). As the electron energy increases further, an increasing fraction of the electron energy is lost to marrow adipocytes (defining inactive marrow) and/or the bone trabeculae. At higher energies—and depending the 3D size and shape of the spongiosa region—the absorbed fraction for AM self-irradiation further declines as the electrons reach the outer perimeter of spongiosa, enter the cortex of cortical bone, and eventually even escape the bone site entirely. Similar trends are noted for TBV sources, although absorbed fraction to AM begins at zero for very low energy electrons (source ≠ target), peaks to a maximal value that is energy and bone-site specific, and then also declines further at much higher energies. From both figures 19.8(a) and (b), it is clear that marrow dosimetry varies significantly across the skeleton. For AM self-irradiation, maximal absorbed fractions are seen in the ribs and sacrum, with minimal absorbed fractions seen in the femora, humeri, and craniofacial bones. The solid dashed lines are computed as the skeletal-averaged absorbed fraction taken as a target mass weighted average of the bone-specific absorbed fraction energy profiles. For $\phi(AM \leftarrow AM)$, the skeletal-averaged values fall intermediate to the bone-specific values, while for $\phi(AM \leftarrow TBV)$, the skeletal-averaged values are shown to be within the lower tier of bone-specific values. This latter observation is attributed to the inclusions of many null values included in the averaging process from bone sites in the arms and legs of the adult where TBV sources exist, but the adjacent marrow cavities are filled only with inactive marrow (no AM). This is the case, for example, in the distal humeri, proximal/distal ulnae and radii, wrists/hands, distal femora, and proximal/distal tibiae and fibulae and ankles/feet. The only active marrow present in the ICRP reference adults is found in the proximal humeral and femoral heads at a lower marrow cellularity of 25%. In 2016, O'Reilly published a similar and comprehensive image-based skeletal dosimetry model of the ICRP reference adult female based upon full array of bone CT and microCT images of a 45-year female cadaver [65]. Electron dosimetry of the ICRP pediatric series was the focus of UF dissertation studies by Pafundi [66] and by Wayson [67].

As one element of the studies by Hough *et al* [64] (reference adult male skeletal model) and by O'Reilly *et al* [65] (reference adult female skeletal model), the issue of the 'missing' bone endosteum mass of ICRP Publication 89 was revisited. In their analyses, these authors worked with the new definition of the bone endosteum regarding its anatomic skeletal location and revised 50-μm thickness. Additionally, they applied model-specific values of bone site spongiosa volumes, ICRP 89 reference surface-to-volume ratios in trabecular bone, and the UF reference phantom defined sizes and shapes of the long bone medullary cavities. Final estimates of total 50-μm bone endosteal masses were reported to be 447 g in the reference adult male [64] and 456 g in the reference adult [65] female (both inclusive of a portion of miscellaneous skeletal tissues). Earlier estimates made by the University of Florida and which were reported in ICRP Publication 110 were 544 and 407 g, respectively. All of these estimates can be considered provisional given the complexity of their geometric derivation, but the latter two values in ICRP

Publication 110 are now considered to be the official bone endosteum reference values adopted and used in calculations of ICRP dose coefficients [61, 68, 69]. In some literature from the University of Florida, the term 'shallow marrow' with the acronym TM_{50} can be seen. This term was applied as a better descriptor of the newer definition of bone endosteum as it (1) explicitly denotes a 50-μm cell layer thickness, and (2) implies that the target region includes both active and inactive (total) marrow within that 50-μm marrow region covering the bone surfaces in spongiosa and along the inner shafts of the long bones. It represents a dosimetric region for the true cellular target for radiogenic bone cancer risk—the osteoprogenitor cells. It is noted, however, that to date there are no definitive immunohistology data published which support experimentally the 50-μm target thickness.

19.4.4 Methods of photon dosimetry of the skeletal tissues

Given the unique microarchitecture of skeletal spongiosa in which irregular cavities of bone marrow are defined in 3D by arrays of plates-like and rod-like bone trabeculae, the assessment of photon dose to active bone marrow and endosteum is complex. During skeletal irradiation photon interaction sites act as electron sources with energy distributions governed by the type of interaction and the energy of the incident photon. Once a secondary electron is liberated with a known kinetic energy, the situation becomes analogous to that encountered in internal dosimetry and can be treated as such by using pre-calculated electron absorbed fractions (AFs) for skeletal sites of photon interaction. Since the bone trabeculae contain the higher-Z element Ca which is absent in the marrow tissues, photoelectric absorption events become significantly higher in the trabeculae whereby the resulting photoelectrons then escape these mineral bone regions and enter the adjacent marrow cavities. Thus, a 'dose enhancement' of bone marrow (including active bone marrow) is seen at photon energies where photoelectric absorption is the dominant photon interaction.

While the concept of a fluence-to-absorbed dose response function was first introduced by K F Eckerman in 1985 [18] and later applied to the tabulations of photon SAFs across the ORNL series of stylized phantoms [14] (see chapter 5), a revised tabulation of these functions was introduced by Johnson *et al* in 2011 [70]. The mathematical formulation of the photon skeletal dose-response function (or DRF) is given by equation (19.2) below:

$$\frac{D(r_T)}{\Phi(E)} = \sum_S \frac{m(r_S)}{m(r_T)} \sum_i \int_0^{E_{\max}} \phi(r_T \leftarrow r_S, E_i)\,(i/\rho)_{r_S}\, n_{r_S}(E_i)\, E_i\, \mathrm{d}E_i \qquad (19.5)$$

In this equation, $m(r_S)$ and $m(r_T)$ are the masses of the source and target regions, $\phi(r_T \leftarrow r_S, E_i)$ is the absorbed fraction from source region r_S to target region r_T for electrons of energy E_i, $(i/\rho)_{r_S}$ is the mass attenuation coefficient of region r_S, with $i = \tau, \sigma, \kappa$, and ν indicating one of four photon interactions including photoelectric absorption, Compton scattering, pair-production in the nuclear field, or pair-production in the electric field. Finally, $n_{r_S}(E_i)\, E_i\, \mathrm{d}E_i$ denotes the number of electrons between energies E and $E + dE_i$ released by process i within region r_S. In this study, an in-house MATLAB™ code was written to evaluate this equation, in which integration was handled numerically when the energy distribution of secondary electrons was continuous. This was the case for Compton scattering, pair and triplet production, where the Klein–Nishina relationship was used to describe the energy spectrum of the former and spectra for pair and triplet production were based upon a uniform distribution from zero to their maximal energies. Mass attenuation coefficients for mineral bone, AM, and IM were evaluated using the NIST Physical Reference Data Library [71]. Elemental compositions of each bone component were taken from ICRP Publication 89. Finally, source/target tissue masses and electron absorbed fractions were taken from the data published in Hough *et al* for multiple bone sites harvested from a 40-year male cadaver.

Selected results from this study are shown in figure 19.9. In panel A, a graphic display is shown for both DRF_{AM} and DRF_{TM50} as a function of photon energy entering the cranium of the reference adult male. These functions are differentiated from four different kerma coefficients—AM (active marrow), SP (spongiosa), TB (trabecular/mineral bone), and TM (total marrow = AM + IM). At roughly 200 keV, these various functions converge such that secondary electronic equilibrium is established across the marrow cavities. In panel B, percent excess dose (defined relative to that given by a kerma approximation) is shown for multiple skeletal sites across the adult skeleton. Dose enhancement is shown to be highest in the cranium (peak of 25% at 50 keV) and the cervical vertebrae (peak of 13% at 50 keV), and lowest for the proximal humerus and mandible (peak of 2% at 80 keV). During photon transport in a whole-body computational human phantom, one then computed the energy-dependent photon fluence within spongiosa regions and medullary cavities of the skeletal system, after which these fluence tallies are convolved with the DRFs to report either the active marrow or bone endosteum absorbed dose from that simulated photon exposure.

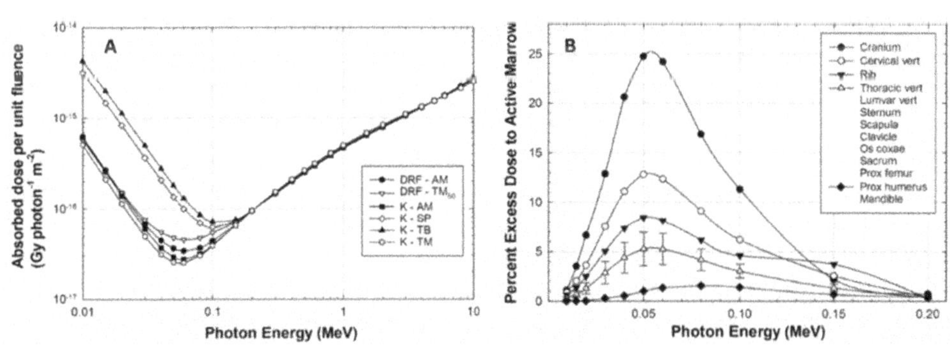

Figure 19.9. (a) Dose response functions and kerma coefficients in the cranium of the adult male. (b) Dose enhancement factors in active marrow calculated as the ratio of the DRF_{AM} to K_{AM} and formulated as percent excess dose $(DRF_{AM} - K_{AM})/K_{AM} \times 100\%$. The error bars indicate the standard deviation of the grouping of bones to include the thoracic vertebrae, lumbar vertebrae, sternum clavicles, pelvis, sacrum, and proximal femora.

19.5 Predicting patient marrow toxicity using reference models

19.5.1 The clinical experience in predicting marrow toxicity with reference skeletal models

Radiopharmaceutical therapy is used increasingly in the treatment of cancer (thyroid, leukemias/lymphoma with radioimmunotherapy, primary and secondary bone malignancies, and neuroblastomas). Marrow toxicity generally limits the amount of treatment that can be administered safely. The most difficult problem in calculating bone marrow dose arises from the complex distribution of bone marrow in the skeleton. The microanatomy of the marrow and its diffuse intra-skeletal distribution make it difficult to calculate red marrow dose accurately using only classical physical approaches, as discussed previously. Although stem cells and their primordial derivatives are presumed to be the primary targets, the stroma that supports the red and yellow marrow also plays an important role in facilitating marrow function, and is an important secondary target [72, 73]. Dose to the bone marrow comes mainly from activity within the marrow itself (specific binding to a cellular component of either the marrow or blood elements), with the next highest contribution coming from tracer accumulation, when it occurs, in the surrounding bone trabeculae, with lesser contributions from more remote sources (cortical bone and/or surrounding muscle). Due to limited spatial resolution, nuclear medicine imaging devices are unable to resolve the source distribution in bone at its microscopic scale, and other methods are needed to calculate the contributions from these sources. More recent advances in bone activity images are reviewed in section 2.6.1.

Bone marrow toxicity is the major dose-limiting factor in radioimmunotherapy (RIT), but there is little consensus on how to calculate that dose accurately, or of individual patients' ability to tolerate the planned therapy [9, 74–82]. Experience with external beam therapy has shown that there is a narrow margin between the delivered dose that kills a tumour and that which causes serious injury to the patient. Given that with modern imaging methods the dose from external beam treatments can be accurately calculated, the remaining major issues are tumour sensitivity and patient tolerance. In nuclear medicine, in contrast to external beam radiotherapy, marrow absorbed dose may not be accurately calculated, even when a treatment planning dosimetry study is performed [83]. Nonetheless, both external beam and internal emitter therapy share the same uncertainty regarding predicted patient tolerance. Nuclear medicine physicians in general have less experience in the therapy regime than those who deal with external treatment approaches, and typically undertreat patients in order to minimize the risk of even low-grade toxicity. If treatment of cancer with radionuclides and RIT are to become a primary means of treating cancer patients, the activity to be administered to individual patients should be based on accurate evaluations of the dose to be delivered to the tumour and of patient tolerance [84, 85].

Radiopharmaceutical therapy has been most effective in the treatment of thyroid cancer because of the high specific uptake of radioactive iodine by thyroid tissue, but there is still no agreement on the means for choosing an optimal therapeutic radiation dose to be delivered. The hope for RIT derives from the notion that it can achieve selective high tumour uptake based on improved targeting of tumour specific/associated antigens. Intravascular tumours, including leukaemia and lymphoma, have the highest likelihood of successful treatment with RIT, partly because tumour targeting efficiency is high, and partly because of the high radiation sensitivity of lymphocytes, and to a lesser extent of granulocytes. A major problem with RIT treatment of most solid tumours has been low uptake per unit tumour mass, so that the delivered dose is often too low for successful therapy. If one gives very large doses intending to ablate marrow and tumour followed by marrow transplantation, lungs, liver, or kidneys will become the dose limiting organs. Bone marrow transplantation (BMT) is being employed routinely in the treatment of breast cancer with high doses of chemotherapy and or of some solid tumours with RIT. A reasonable goal would be to develop an improved means of defining an appropriate intermediate dose that provides high 'cure'/complete remission rates, without the need for BMT.

Progress has been made in the development of improved calculation methods and tools for patient-specific treatment planning focusing primarily on tumour dose optimisation [86–88]. Still, problems remain in the accurate assessment of bone marrow dose and in the estimation of marrow reserve. Better patient-specific knowledge of both dose and marrow reserve is needed if one is to develop a good predictor of tolerance dose. Two analytical approaches have been taken. The first is the physical method, in which one measures the biokinetics and computes dose based on knowledge of the amount injected, the radionuclide emission characteristics, and the measured residence times in the different organs using accepted dose calculation methods [12]. A complementary approach attempts to use patient-specific characteristics to estimate residual damage from prior therapeutic treatments to estimate the amount of added dose that would be tolerated by normal organs, including bone marrow.

The importance of including pretherapy patient status in toxicity predictions has been studied by several groups [82, 89, 90], and it is clear that prior patient therapy and disease status must be factored into treatment decisions. Nonetheless, to date poor correlations have been observed between administered dose and marrow toxicity, which are somewhat improved by more carefully calculated marrow doses [91]. This suggests the need for more patient-specific information, including better dose estimates and better information on other patient characteristics, including perhaps patient-specific indicators of response (i.e., biomarkers).

Marrow toxicity from RIT is manifest by hematological changes in circulating platelets, lymphocytes, granulocytes, reticulocytes, and later changes are seen in red blood cells. Due to the high sensitivity of the marrow, second organ toxicity is not a problem unless doses, which are known to be lethal to the marrow, are intentionally given. Attempts to correlate hematological toxicity with marrow dose, when marrow cells are specifically targeted, have not been particularly successful in the past, in part due to uncertainties in the actual absorbed dose, but also due to the difficulty in assessing marrow functional status prior to therapy [9, 74–82]. The best correlations obtained to date have had a correlation coefficient r around 0.8 [76, 80, 92, 93].

Whole-body absorbed dose and red marrow absorbed dose are usually the best indicators of hematological toxicity, as measured by platelet toxicity grade, with red marrow dose being slightly better. However, the poor correlation coefficients obtained from these analyses emphasize the need to better account for prior patient therapy and bone marrow reserve. That red marrow absorbed dose is the best (albeit not appropriate on a patient-by-patient basis) predictor of hematological toxicity has been confirmed using data from other antibody trials [94]. A survey of the literature using data from different antibodies, different patient populations and different disease sites has confirmed this observation over a much larger scale [95]. Blumenthal *et al* showed that plasma FLT3-L (a stromal-cell produced positive stimulatory cytokine) levels predicted excess platelet toxicity in patients receiving RIT after chemotherapy [96]. In a 2011, Vanderhoek *et al* [97] reported on a pilot study suggesting that FLT PET imaging during induction of chemotherapy may serve as an early biomarker of treatment response in acute myeloid leukaemia. In bone marrow mean and maximum standardized uptake values were 0.8 and 3.6 for complete remission and 1.6 and 11.4 for resistant disease. With its success in monitoring bone marrow response to chemotherapy, FLT PET imaging can readily be applied to the monitoring of marrow response to treatment in RPT.

Many other studies have found little or no correlation between calculated marrow dose for large numbers of patients and observed toxicity, as determined by various hematological parameters (platelet nadir, % platelet drop, % lymphocyte drop, and % granulocyte drop). The primary reason for this is probably that all available marrow dose models are patterned after a normal healthy young adult with good marrow reserve, while the treated population is generally older (with inherently different trabecular bone structures and less marrow reserve), often with compromised marrow due to disease, prior chemotherapy, or prior radiotherapy.

19.5.2 Correcting the models using patient-specific data

It is unreasonable to expect that doses calculated from a single model representing an average healthy adult will yield good correlations of radiation dose with observed effects in a heterogeneous patient population. Many patients are older, with presumably less active marrow mass than younger people, more osteoporotic trabecular bones, and many have disease or have had prior therapies which may affect marrow status and trabecular bone microanatomy. Simple adjustments to standard model S values may be made, assuming that patient marrow mass changes in some predictable way with lean body mass, body surface area, or other easily measured anatomic features. Such approximations will of course be imperfect, but they are better than simply applying the standard model to all patients equally. Some improvements in correlations between marrow dose and observed marrow toxicity can be realized by calculating patient-specific contributions to marrow dose and by making patient-specific adjustments to relevant parameters such as marrow mass and cellularity (as implied by patient age, disease, and prior therapeutic treatment) [77, 96].

More patient-specific evaluations can be envisioned, which could be realized with some additional investigational effort. Functioning marrow could be evaluated through the use of marrow scanning agents, such as ^{99m}Tc-sulfur colloid, ^{18}F-fluorothymidine [98], or other marrow avid agents. If some quantitative measure of uptake in different regions can be obtained, model-specific trabecular bone dose conversion factors [19, 64, 65] may be modified to be more specific to the patient. For example, skeletal averaged *S* values may be calculated from skeletal averaged absorbed fractions as shown previously in section 19.1.2. The values of bone volume fractions and marrow cellularity factors may be assigned based on information gained in individual radionuclide marrow scans; that is, the fraction of marrow calculated to be in each of the model regions and the cellularities may be calculated on a patient-specific basis. In the absence of patient-specific values of cellularity derived from marrow biopsies, values from age-dependent models (e.g., [99, 100]) may be applied, or as will be discussed in the next section, magnetic resonance imaging methods as well could be used to determine patient-specific active marrow mass and location [74–77]. Bone marrow uptake of RIT agents in expected marrow sites which have diminished or absent marrow function on radionuclide scans may be assumed to be tumour, and the computed dose calculated using a more specific model (e.g., [45]). Time-integrated activities may be calculated and doses estimated for the different bone marrow regions in which marrow localization has been scored, assigning the total marrow residence time to each region using the values of the presumed skeletal-site distribution of active marrow [3, 101]. Whole marrow weighted average dose and differential and integral dose-volume histograms then may be calculated.

Determining marrow time-integrated activities for internal emitters is an area of difficulty, as marrow activity is heterogeneous even in normal subjects, and more so in patients with cancer, with or without a history of previous treatment. Different imaging methods rely on either: (1) the quantitation of activity in a certain segment of the marrow (e.g. the lumbar spine or sacrum) and extrapolation of total marrow activity assuming the total amount of active marrow in the body and the fraction of the total that resides in individual bone regions [74]; or (2) the assumption that the dose is derived only from blood contributions, with the assumption of normal marrow size [10, 11]; and/or (3) combined blood and marrow-fixed activity methods. The correlation between dose calculated in these different manners, biomarker results, and clinical predictors with observed toxicity and patient cure rates should be used to assess the potential utility of the different dose calculation and marrow indices for improving treatment outcome. This is not currently in standard practice, but should be considered if we are to seek more meaningful dose and risk assessment in individual patients, as is routine in external beam therapy [102]. Recent advances in patient-specific adjustments to reference person bone marrow dosimetry models are summarized in our next section of this chapter.

19.5.3 Potential dosimetric errors in skeletal dosimetry via use of reference skeletal models

In a patient-specific application of the MIRD schema for bone marrow dosimetry, one must address both the assessment of the time-integrated activity in the skeletal

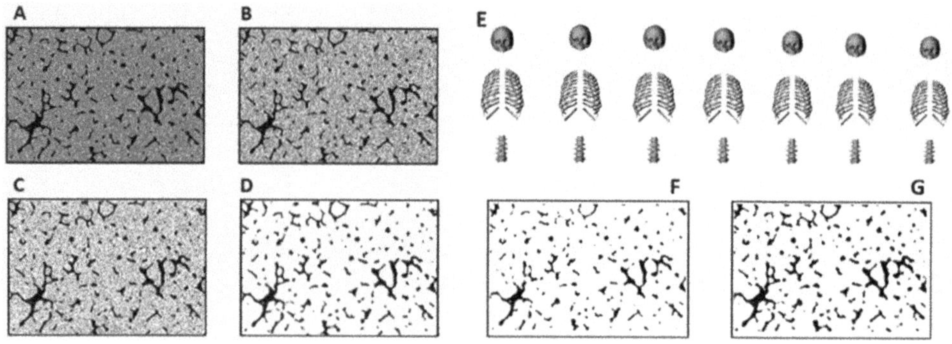

Figure 19.10. Example graphical displays of the parametric sensitivity study of Geyer *et al* [103]. Cellularity variations are shown for the L3 vertebrae in the adult male for (a) 10%, (b) 40%, (c) and (d) 100% marrow cellularity. Panel (e) shows scaled macrostructure skeletal sites of the UF reference adult female phantom are shown includes (left to right) +15%, +10%, +5%, 0% (reference size), −5%, −10%, and −15% changes in total bone volume. Finally, L3 vertebrae are shown with (f) −15% volume change and (g) +15% volume change in the trabecular bone volume fractions (both at 100% marrow cellularity for clarity). Reproduced from [103] John Wiley & Sons. Copyright 2016 American Association of Physicists in Medicine

tissues of the patient, as well as any adjustments to the radionuclide S values applicable to skeletal source and target regions. In 2017, the latter issue was address in the study by Geyer *et al* [103] in which three physiological and morphometry parameters of the ICRP reference model were varied across a range of values potentially encompassing their values in a given patient population. As shown graphically in figure 19.10, these three parameters were (1) marrow cellularity, (2) skeletal size, and (3) bone mineral density as realized by the bone volume fraction in spongiosa. These three parameters were varied in three different skeletal regions—the cranium, the ribs, and thc lumbar vertebrae.

The key conclusions of this study, looking at the potential for dosimetric errors in patient-specific dosimetry via use of a reference skeletal dosimetry model, were the following. First variations in overall bone size play a very minor role in active marrow dose variability. Second, marrow cellularity is a significant factor in dose variability for active marrow self-irradiation, but it plays virtually no role for radiopharmaceuticals localized to the trabecular bone matrix (either on their surfaces or within their volumes). Third, variations in trabecular bone volume fraction impact the active marrow dose variability for short-ranged particle emitters such as ^{45}Ca, ^{223}Ra, ^{219}Rn, and ^{215}Po in the vertebrae and ribs—both skeletal sites with small spongiosa proportions of trabecular bone. In the cranium, with its relatively high proportion of trabecular bone, significant differences in marrow dosimetry from the reference model were noted for all radionuclides included in this investigation [103].

19.6 Current advances: patient-specific bone marrow dosimetry

19.6.1 Assessment of patient-specific time-integrated activity in the skeleton

As summarized in Gear *et al* [104], one of the larger sources of uncertainty in internal dosimetry via the MIRD schema is the assignment of the time-integrated activity $\widetilde{A}(r_S)$ in the various source regions of the radiopharmaceutical therapy patient. The most recent guidance on quantitation of radiopharmaceutical skeletal activity was published by Hindorf *et al* in 2010 by the Dosimetry committee of the European Association of Nuclear Medicine (EANM) [102]. According to this EANM guidance, the first step of activity quantification should be assessment of the radiopharmaceutical activity in the whole body as a function of time post-administration. This can be done from serial whole-body scans or from serial acquisitions by external calibrated probe. The first acquisition should be performed before any excretion of activity so as to properly normalize the number of detected counts to the injected activity. Both anterior and posterior acquisitions should be performed at each time point.

When it is determined that there is no specific uptake of the radiopharmaceutical in the bone marrow cells, the activity in bone marrow cavities can be determined from the activity concentration in blood or plasma. Ideally, the activity concentration in the blood and plasma, as well as the patient's hematocrit (HCT—proportion of total blood volume occupied by blood cells) should be determined to certify the absence of specific activity uptake in any blood component. This 'blood method'—originally established by Sgouros [10]—is based on the assumption that the activity concentration in the extracellular fluid (ECF) spaces of bone marrow is equivalent to the activity concentration in the sampled plasma of the patient. The activity concentration in the bone marrow can therefore be calculated via the activity concentration in plasma combined with the red marrow extracellular fluid fraction (RMECFF) which was originally proposed to equal 0.19 [10]. The mean dose to active bone marrow from activity in the ECF is thus given as:

$$D(AM \leftarrow ECF) = \left[\widetilde{A}_{\text{plasma}}\right] \bullet \text{RMECFF} \bullet m_{AM} \bullet S(AM \leftarrow AM) \tag{19.6}$$

where $[\widetilde{A}_{\text{plasma}}]$ is the measured time-integrated activity concentration in the patient's blood plasma (as measured serially over time and then time-integrated), m_{AM} is the total body active marrow mass, and $S(AM \leftarrow AM)$ is the radionuclide for active marrow self-irradiation. If the S value is taken from that computed for anatomical reference phantom, then the AM mass term should also be that in the same reference phantom. Assuming that the patient's total body active marrow mass scales proportionally to his/her total body mass, then a patient-specific estimate of m_{AM} can be substituted but only if that same mass adjustment is made to the radionuclide S value. As such, both the latter two terms may be taken from a reference model with no impact.

Another scenario would be if the radiopharmaceutical were taken up by blood cells (erythrocytes, lymphocytes, neutrophils, etc) in addition to the free-activity within the blood plasma. This additional dose contribution to active marrow is given as:

$$D(AM \leftarrow BL_{\text{cells}}) = [\widetilde{A}_{\text{blood}}] \bullet RMBLR \bullet m_{AM} \bullet S(AM \leftarrow AM) \tag{19.7}$$

where $[\widetilde{A}_{\text{blood}}]$ is the measured time-integrated activity concentration in the blood of the patient, and RMBLR is the red marrow-to-blood activity concentration ratio which is computed as RMECFF/(1 − HCT). The RMBLR has been found to be different for different radiopharmaceuticals. Originally, the activity concentration in bone marrow was equated to the concentration in blood for RIT agents, although this approach was considered as conservative. In Sgouros [10], a value of RMBLR = 0.36 was recommended based upon theoretical investigations of radiolabeled antibodies and this value was in line with that recommended by the American Association of Physicists in Medicine (AAPM)—with values ranging from 0.2 to 0.4 [9]. A value of RMBLR of unity has been shown to be appropriate for bone marrow dosimetry of ^{131}I-NaI [105] and for ^{177}Lu-peptides [106]. Hindorf *et al*, however, showed that values of RMBLR might not be constant over time for all radiopharmaceuticals [107].

The above applications of the MIRD schema to marrow dosimetry assume non-specific uptake in the skeletal tissues, and thus activity observed within the skeleton is that of the radiopharmaceutical passing through that distributed organ via blood cells and/or blood plasma. If, however, the radiopharmaceutical has specific uptake in the skeletal tissues—either incidentally or by design—then the 'blood method' is not appropriate and quantitative nuclear medicine imaging must be performed. The two most common methods for imaging skeletal uptake of radiopharmaceuticals are (1) 2D planar imaging via gamma-cameras, or (2) 3D imaging via SPECT. PET can be used but only for those radiopharmaceuticals utilizing radionuclide decaying via positron emission. ROIs (planar imaging) and VOIs (SPECT imaging) are drawn over suitable locations where the most common are the sacrum [74] or lumbar vertebrae [108]. Some investigators, however, have quantified patient-specific skeletal uptake in other bone regions such as the upper humeri or upper femora [92]. Total time-integrated activity through the entire skeleton of the patient is then estimated by dividing the regionally measured time-integrated activity by proportion of total body skeletal tissue associated with the ROI or VOI. These scaling factors are typically taken from ICRP reference literature [3, 54]. Importantly, however, these bone site-specific measurements of time-dependent radiopharmaceutical activity cannot be resolved at the microscopic tissue level. That is, a value of $\widetilde{A}(r_S)$ assessed in the L3 vertebrae, for example, cannot be resolved as to its fractional components of cortical bone, bone trabeculae, and bone marrow. Further assumptions must be made as to bone-specific uptake such as (1) the measured activity is in the bone volumes (cortical and trabeculae) only, (2) the measured activity is on the bone surfaces (cortical and trabecular), or (3) the measured activity is in the marrow cavity tissues. Furthermore, estimating these factors would require patient-specific information on both fraction of spongiosa that is occupied by marrow or bone trabeculae, and the fraction of total marrow that is hematopoietically active. Techniques that might be applied to these patient-specific histological parameters are reviewed below.

19.6.2 Assessment of patient-specific total skeletal spongiosa volume (TSSV)

An approximate but widely adopted adjustment of reference phantom S values for patient-specific dosimetry involved simply mass scaling of reference phantom tissue masses. For example, two means of assigning total body active marrow mass in the RPT patient are:

$$(m_{AM})_{\text{patient}} \approx (m_{AM})_{\text{patient}}\left(\frac{\text{TBM}_{\text{patient}}}{\text{TBM}_{\text{phantom}}}\right) or\ (m_{AM})_{\text{patient}}\left(\frac{\text{LBM}_{\text{patient}}}{\text{LBM}_{\text{phantom}}}\right) \quad (19.8)$$

where TBM and LBM are the total or lean body mass of either the patient or reference phantom. Brindle *et al* suggested a more anatomically refined method which would take into account the summation of active marrow masses across all bone sites x in the patient's skeleton. As such,

$$(m_{AM})_{\text{patient}} \approx \sum_{x} SV^{x}\bullet MVF^{x}\bullet CF^{x}\bullet \rho_{AM} = (TSSV)\sum_{x} f_{SV}^{x}\bullet MVF^{x}\bullet CF^{x}\bullet \rho_{AM} \quad (19.9)$$

where TSSV is the total skeletal spongiosa volume, SV^x is the volume of spongiosa within bone site x, f_{SV}^x is the fraction of TSSV within bone site x, MVF^x is the marrow volume fraction for bone site x (fraction of spongiosa volume containing bone marrow), CF^x is the cellularity factor for bone site x (fraction of marrow volume that is active marrow ~ 1—marrow fat fraction), and ρ_{AM} is the mass density of active marrow. Under the assumption that all parameters of equation (19.9)—with the exception of TSSV—could be taken from reference literature (e.g., ICRP Publication 70), then one could focus could determining patient-specific values of TSSV. In Brindle *et al* [109], 10 male and 10 female cadavers were subjected to whole-body CT scans within which manual segmentation was used to estimate TSSV in all 13 active marrow-containing bone sites of the adult skeleton. The age, total body height, and 14 CT-based skeletal size measurements were obtained for each cadaver. Multiple regression was used with the dependent variables to report a predictive model of patient TSSV. The pelvis height and width were the two skeletal measurements that proved to be the most important parameters predicting TSSV. The multiple R^2 value for the statistical model with these two parameters was 0.87. The analysis revealed that these parameters predicted the patient-specific TSSV to within ±10% for 15 of the 20 cadavers and two within ±20% for all study subjects.

To expand upon this work, and to derive predictive models of TSSV for males and females independently, Pichardo *et al* later reported on a study with similar methods but with a larger sample size of 20 males and 20 females. They reported both a pooled model ($R^2 = 0.76$) and a sex-specific model ($R^2 = 0.79$). A leave-one-out analysis revealed that these models predict TSSV with less than 10% error for 50%–70% of the subjects and with less than 20% error for 70–90% of the subjects. Tables were constructed that provide the percent distribution of SV in active marrow-containing bone sites for both adult males and females.

19.6.3 Assessment of patient-specific trabecular bone volume fractions (TBVF) within spongiosa

The second term of the summation in equation (19.9) is the marrow volume fraction (MVF) which is for a given region of spongiosa equal to 1—TBVF, with TBVF defining the trabecular bone fraction of spongiosa. In 2019, Salas-Ramirez *et al* reported on the use of dual-energy quantitative computed tomography (DEQCT) for the quantification of the TBVF for bone marrow dosimetry in radiopharmaceutical therapy [110]. DEQCT analysis of material mass fractions (MF) involves solving a series of coupled integral equations as outlined by Liu *et al* [111] in which a lookup table for each x-ray tube spectrum and detector sensitivity arrangement is prepared. These provide values of effective density ρ_{eff} and effective atomic number Z_{eff} of the material to be quantified. In Salas-Ramirez *et al* [110], two numerical methods were proposed for determining MF values. The first was to use a reference ratio of mineral bone mass fraction to total bone mass fraction ($R_{HA/B}$). The second involved changing the hydroxyapatite chemical description to that for a cortical bone chemical description as given in ICRU Report 44 [112] in the material definition used in the mass fraction method.

For the first method, the authors considered bone to be composed of minerals and an organic matrix, and thus the mass fraction method of DEQCT provides directly the hydroxyapatite mass fraction (MF_{HA}). This parameter is related to the total bone mass fraction by the ratio of mineral bone mass fraction to the total bone mass fraction ($R_{HA/B} = MF_{HA}/MF_B$). The trabecular bone volume fraction (TBVF or VF_B) can be defined as:

$$VF_B = \frac{\rho_{\text{eff}} \bullet MF_B}{\rho_B} = \frac{\rho_{\text{eff}}}{\rho_B} \bullet \frac{MF_{HA}}{R_{HA/B}} \tag{19.10}$$

where ρ_B is taken to be the reference density of adult cortical bone (1.92 g cm^{-3}) [113].

For the second method, the authors noted that $R_{HA/B}$ can be quantified by measuring the relative amount of mineral and organic material in bone samples. In Quelch *et al* [114], the percentages by mass of minerals and organ material in dry, fat-free human vertebra were measured, and a percentage of mineral bone at 66.3% was documented for the adult vertebrae. This value was thus used as a reference value for $R_{HA/B} = 0.663$ in the study by Salas-Ramirez *et al*. Accordingly, equation (19.10) can be re-expressed as:

$$VF_B = \frac{\rho_{\text{eff}}}{\rho_B} \bullet \frac{MF_{HA}}{0.663} \tag{19.11}$$

Following a series of physical phantom measurements, Salas-Ramirez *et al* [110] analyzed a patient image set from a 49.5 year-old female patient with a diagnosis of osteopenia at the border to osteoporosis (DEXA T-score of −3.6 and 3.0 in the lumbar vertebrae 1 and 2, respectively). The bone volume fraction obtained by using the first calculation method are presented in table 19.4 together with the trabecular bone volume fraction obtained by the second method. Additionally, shown are the mass fractions of hydroxyapatite, the bone volume fractions of hydroxyapatite and bone, and ratio $R_{HA/B}$. The authors conclude that DEQCT using the phantom-less mass fraction method can be used to quantify TBVF in spongiosa directly with a SPECT/CT in a nuclear medicine department. This parameter provides a valuable input to patient-specific marrow dose assessment.

Table 19.4. Values of mass and volume fraction of hydroxyapatite and trabecular bone in LV_1 and LV_2 from a patient image set [110].

Parameter							
Region		MF_{HA}	VF_{HA}	MF_B	VF_B[b]	VF_B[c]	$R_{Ha/B}$
Whole Vertebra[a]	Lumbar Vertebra 1	0.138	0.052	0.124	0.249	0.148	0.554
	Lumbar Vertebra 2	0.145	0.056	0.132	0.261	0.157	0.556
Spongisa	Lumbar Vertebra 1	0.113	0.039	0.093	0.203	0.071	0.557
	Lumbar Vertebra 2	0.116	0.040	0.095	0.208	0.072	0.558

MFHA: hydroxyapatite mass fraction, VFHA: hydroxyapatite volume fraction, MFB: bone mass function, VF_B: bone volume function
[a]Execluding transverse processes.
[b]First volume fraction calculation method
[c]Second volume fraction calculation method

19.6.4 Assessment of patient-specific bone marrow cellularity factors (CF)

The third term of the summation in equation (19.9) is the bone marrow cellularity factor (CF) representing the fraction of total marrow volume in spongiosa that comprises hematopoietically active (or red) marrow. For marrow dosimetry applications in the clinic, Bolch *et al* [49] had previously proposed a working definition of marrow cellularity as:

$$\text{Marrow Cellularity Factor (CF)} = 1 - \text{Marrow Fat Fraction} \qquad (19.12)$$

Under this definition, the radiation target of active marrow would encompass all soft tissues of spongiosa with the exclusion of the marrow adipocytes. This would include not only the hematopoietic stem and progenitor cells, but also marrow support cells vascular structures, marrow stromal cells, and other marrow-resident tissues. Accordingly, if the marrow fat fraction could be measured non-invasively in the patient, the patient's unique value of CF could be computed.

In 2011, Pichardo *et al* reported on MR-based measurements of bone marrow cellularity for use in radiation dosimetry applications. Spoiled gradient-echo *in vivo* images of the femur, humerus, upper spine, and lower spine acquired from two dogs were taken using a clinical 3T MR scanner. Single-peak iterative decomposition of water and fat with echo asymmetry and least squares analysis (SP-IDEAL) was used to derive the bone marrow fat fractions in these skeletal regions. *In vivo* accuracy was validated by comparison to 'gold-standard' measurements in which marrow adipocytes were segmented in prepared *ex vivo* bone marrow histology slides of these same skeletal sites. Bland–Altman plots demonstrated excellent agreement between SP-IDEAL and histology, with a mean difference of −0.52% cellularity and most differences were found to be within ±2% cellularity (see figure 19.11). Agreement between MP-IDEAL (multi-peak) and histology was not as good with mean

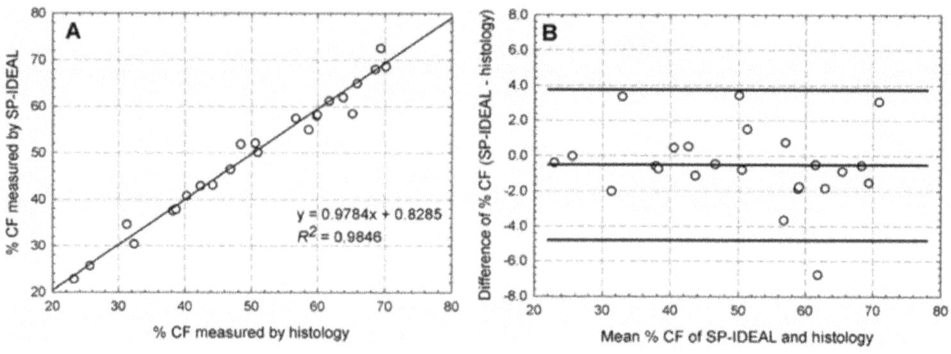

Figure 19.11. Agreement of CF measured by SP-IDEAL and histology at the same location on bone in two dogs. (a) Linear regression fit. (b) Bland–Altman plot. Middle horizontal line corresponds to mean difference, equal to 20.5% cellularity, and top and bottom horizontal lines correspond to 95% confidence interval, equal to 4% and 25% cellularity, respectively. Plot identifies an outlier, corresponding to measurement in L4 in one of the dogs. This research was originally published in JNM [115]. Copyright SNMMI.

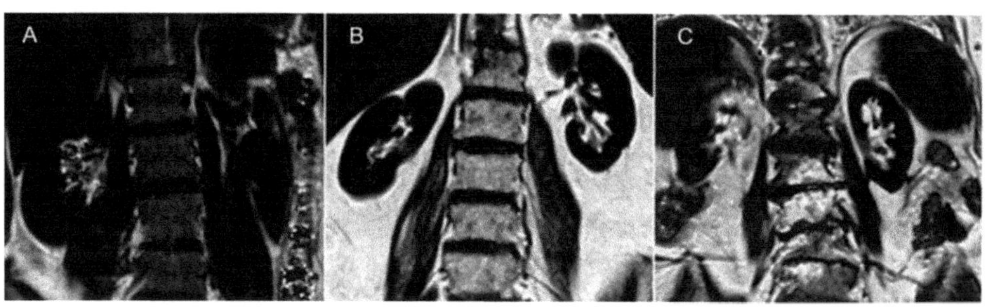

Figure 19.12. Fat fraction images of the lumbar anatomical region in three female patients: (a) 20 years, (b) 55 years, and (c) 87 years of age. Reproduced from [116]. Copyright IOP Publishing. All rights reserved.

differences of −7% and a range of from +5% to −20%. The authors concluded that SP-IDEAL measures the relative abundance of methylene lipids and this measure is shown to be equivalent to the adipocyte volume fraction.

The work of Pichardo *et al* was further explored in 2018 in a study by Salas-Ramirez *et al* in which both physical phantom and patient volunteer MR imaging were performed using an accelerated two-point Dixon MRI sequence for fat–water image separation (see figure 19.12). The authors concluded that the two-point Dixon method enables non-invasive and spatially resolved quantification of the fat fraction in bone marrow. The study provided important evidence on the differences in marrow age-dependence between both females and males. Differences were observed in marrow cellularity from that recommended by the ICRP for the reference adult. The observations lead to the conclusion that the fat fraction in bone marrow should be considered as a patient-specific variable in clinical dosimetry procedures, particularly in cases where one must compute marrow dose from self-irradiation as noted by Geyer *et al* [103].

19.6.5 Applications of dual-energy CT imaging for simultaneous assessment of CF and TBVF

While MR imaging has been shown to be a very accurate and clinically efficient measure for patient-specific and bone site-specific assessment of the marrow CF, one must also then quantify the TBVF via methods such as DEQCT. In 2014, Goodsitt *et al* [117] demonstrated that the latter modality could be use directly to fully quantify the material and elemental composition of skeletal spongiosa. A three-equation three-unknown method for determining the composition of spongiosa using DEQCT was employed in which spongiosa is considered to include only three distinct components—fatty marrow (marrow adipocytes), fat-free marrow (active or red marrow), and bone (the bone trabeculae). The outer cortex of cortical bone is thus excluded from the image analysis. The free equations to be solved are then:

$$\overline{CT\#_{\text{mixture}}}(E_1) = B\bullet\, \overline{CT\#_B(E_1)} + R\bullet\, \overline{CT\#_R(E_1)} + F\bullet\, \overline{CT\#_F(E_1)} \tag{19.13}$$

$$\overline{CT\#_{\text{mixture}}}(E_2) = B\bullet\, \overline{CT\#_B(E_2)} + R\bullet\, \overline{CT\#_R(E_2)} + F\bullet\, \overline{CT\#_F(E_2)} \tag{19.14}$$

$$1 = B + R + F \tag{19.15}$$

where B, R and F are, respectively, the computed volume fractions of bone, red marrow and fat in the imaged spongiosa region. $\overline{CT\#_{\text{mixture}}}$ is the measured mean CT number within the ROI of the spongiosa of a vertebra or other bone site of interest in the CT images. $\overline{CT\#_B}$, $\overline{CT\#_R}$ and $\overline{CT\#_F}$ are the measured meant CT numbers in pure bone pure red marrow, and pure fat calibration standards, respectively. These are also determined from ROI placed within the CT images of calibration materials of each type (see figure 19.13). E_1 and E_2 are ideally two monochromatic x-ray energies. For conventional CT scanners with polyenergetic x-ray spectra, E_1 and E_2 are the corresponding peak kilovoltages (kVp's) of the polyenergetic beams (such as 80 and 140 kVp).

Phantom-based studies conducted by Goodsitt *et al* study concluded that the use of DEQCT is a promising technique for assessing patient-specific spatial distributions of active marrow and bone trabeculae in individual skeletal regions of the radiopharmaceutical patient. The technique allows one to move away from the assumption of reference values for TBVF and marrow CF toward patient-specific

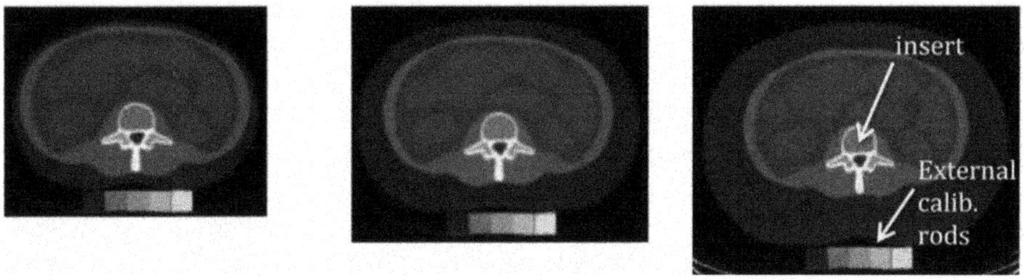

Figure 19.13. Images of the small (left), medium (center), and large (right) sized phantoms. Internal calibration (or test) inserts and external calibration rods are indicated [117].

image-based values. Correspondingly, one would then want to assemble a library of radionuclide *S* values for various skeletal source and target tissues that vary with both TBVF and marrow CF permitting a lookup-table approach to patient marrow dosimetry in opposition to the use of standardized reference phantom skeletal *S* values. This customized approach to patient marrow dosimetry had previously been proposed by Wilderman *et al* [118] in which synthetic models of skeletal spongiosa were fashioned and thus customized to the unique patient-specific spongiosa material composition. This approach is also in line with newer stylized but morphometrically variable models of skeleton macrostructure and microstructure. These Stochastic Parametric Skeletal Dosimetry (SPSD) models [119] have been developed for assessing active marrow dose uncertainties to the members of the Techa River Cohort in the USSR who were exposed to radiostrontium releases from the Mayak Plutonium Production Facility in the early-to-mid 1950s [120, 121].

19.6.6 Microscopic refinement of the target tissues for active marrow dosimetry

As noted above, a variety of imaging techniques can be applied to both (1) quantify the time-integrated activity of the radiopharmaceutical in the skeletal tissues, and (2) quantify key parameters such as TSSV, TBVF, and marrow CF in the derivation of patient-specific estimates of active marrow dose. However, all these approaches presume that the relevant dosimetry quantity is the mean absorbed dose to all soft tissues of the spongiosa marrow cavities exclusive of the marrow adipocytes (the latter important for active marrow self-dose). In reality, one would like to assess the dose to active marrow on a cellular level, with particular focus on the hematopoietically active stem and progenitor cells. By reporting the mean dose to active marrow, the presumption is that these target cells are uniformly dispersed across the marrow cavities in human spongiosa. However, previous studies in the mouse femoral shaft had demonstrated a clear spatial gradient in the distribution of different cell lines responding to marrow growth factors as a function of distance into the femoral medullary cavity [122–124].

In 2007, the study by Watchman *et al* reported the first quantitative data on spatial gradients of CD34+ stem and progenitor cells in human bone. Core biopsy samples of normal bone marrow from the iliac crest were obtained from clinical cases at Shands Hospital at the University of Florida Department of Pathology. The specimens were sectioned and immunohistochemically stained for CD34 (red) and CD31 (brown) antigens. These two stains were used simultaneously to differentiate between hematopoietic stem and progenitor cells (CD34+/CD31−) and vascular endothelium (CD34+/CD31+). Distances from the hematopoietic CD34+ cells and blood vessels to the nearest bone trabecula surface were measured digitally and then binned in 50-μm increments with results normalized per unit area of marrow tissue (see figure 19.14(a)). These distances separating CD34+ cells from blood vessels were also tallied. The study confirmed that the spatial gradient of hematopoietic stem and progenitor cells previously measured in the mouse femur is also present in human cancellous bone. Dosimetric implications of these results were potentially significant, especially for marrow exposure scenarios that deliver a nonuniform dose gradient across the marrow cavities (as would be the case for radiopharmaceutical surface-seeking short-ranged alpha-particle emitters).

In 2009, a further study by Bourke *et al* expanded this prior work at the University of Florida by (1) immunohistochemical stratification of the additional population of CD117+ stem and progenitor cells, (2) consideration of a possible bone-specific dependence of the spatial gradients, (3) use of larger field-of-view marrow specimens through autopsy harvest, and (4) explicit consideration of active marrow versus total marrow area (through adipocyte segmentation). Results from two of the three bone sites of the study (ribs and L1 vertebrae) for the CD34+ cells are shown in figures 19.15(a) and (b). A clear linear spatial gradient of the marrow target cells are shown in the ribs and iliac crest (not shown) with a slightly delayed

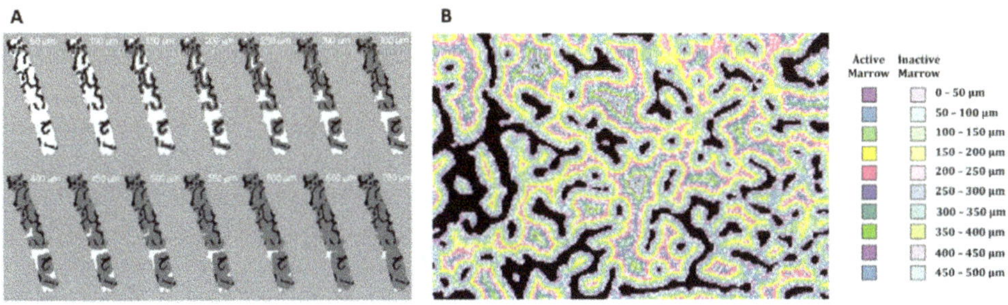

Figure 19.14. (a) Image processing of bone marrow cores to determine 50-μm areal contours from the bone trabeculae surfaces. This research was originally published in JNM [125], copyright SNMMI. (b) Spatial binning of marrow regions in a voxelized model of the L5 vertebra of the UF reference female phantom [65, 126] John Wiley & Sons. Copyright 2016 American Association of Physicists in Medicine.

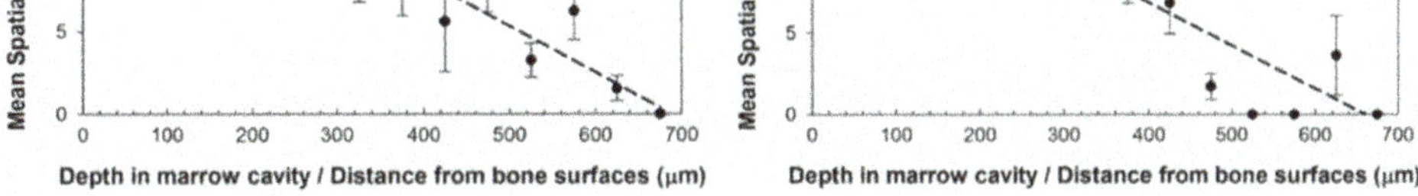

Figure 19.15. Spatial gradients of CD34+ hematopoietic stem and progenitor cells in bone marrow. Mean areal concentrations of hematopoietic cells in active bone marrow as a function of distance into the marrow cavities of (a) ribs (CD34+) and (b) L_1 vertebrae (CD34+). Data points indicate specimen-averaged mean values binned every 50 μm, whereas the dashed line represents a linear fit to the data with R^2 values shown. Also shown is the population mean average concentration (horizontal solid) ± 1σ (horizontal dashed). Not shown are results for the iliac crest which gave a linear depth profile of CD34+ cells out to ~500 μm with an R^2 value of 0.909. Reprinted from [127], copyright (2009), with permission from Elsevier.

linear fall off of stem cell concentration with depth seen in the L_1 vertebrae. Solid horizontal lines in both figures indicate the mean average concentration (total cell count divided by total marrow area seen in the histology slides).

In 2017, Geyer *et al* [126] used the spatial gradients of target marrow cells from the study of Bourke *et al* [127] to construct a library of depth-dependent radionuclide S values for a variety of radiopharmaceuticals of interest in therapy nuclear medicine (see figure 19.14(B)). Their study concluded that for both beta-emitters and alpha-emitters localized in active and inactive bone marrow depth-weighted S values were shown to differ from volume-averaged S values by only a few percent, as dose gradients across the marrow tissues are virtually nonexistent. For bone volume and bone surface sources of alpha-emitters and lower energy beta-emitters, when marrow dose gradients are expected, explicit consideration of target cell spatial concentration gradients are shown to significantly impact marrow dosimetry. For ^{153}Sm and ^{223}Ra, accounting for stem and progenitor cell concentration gradients resulted in maximum percent differences to reference skeletal S-values of ~21% and 55%, respectively. The work of Geyer *et al* [126], however, did not consider the dynamic nature of cell trafficking—that is, the movement of hematopoietic stem or progenitor cells through the marrow spaces to repopulate regions where beta or alpha cell kill had depleted that region of the irradiate marrow cavity. The impact of cell trafficking on overall marrow cell kill for ^{223}Ra had previously been explored by Hobbs *et al* [128] in a Monte Carlo-based spherical model of the marrow cavities which also applied the spatial stem and progenitor cell gradients reported in Bourke *et al* [127].

19.7 Conclusions

Skeletal dosimetry is complicated by the intricate nature of its microstructure and by the irregular distribution of the radiosensitive marrow tissues, both of which can vary greatly between patients and throughout various bone regions. Accurate patient-specific skeletal dose calculations are a primary concern in radiopharmaceutical therapy as the active marrow is usually the dose-limiting tissue. Significant improvements have been made in the use of skeletal dosimetric models for bone marrow dose calculations, mainly from the development of better models (more appropriate for therapeutic nuclear medicine dosimetry) and a better understanding of the models and how to use them. Correlations of radiation toxicity to marrow dose, however, remain weak, and will only improve as more patient-specific dosimetric models are developed and used. The increase in the availability of high-performance computing has recently permitted the coupling of high-resolution, three-dimensional digital images of the skeletal microstructures acquired *in vitro* to a Monte Carlo transport code. New imaging techniques also allow the *in vivo* distribution of marrow cellularity and trabecular bone volume fractions to be determined. Application of these and other techniques *in vivo* will allow improvement in the quality of patient-specific marrow dose calculations for a better prediction of bone marrow toxicity and an optimal selection of the amount of therapeutic radiation dose to be delivered.

References

[1] Fliedner T M, Graessle D, Paulsen C and Reimers K 2002 Structure and function of bone marrow hemopoiesis: mechanisms of response to ionizing radiation exposure *Cancer Biother. Radiopharm.* **17** 405–26

[2] Gray H 1974 *Gray's Anatomy.* (Philadelphia, PA: Running Press)

[3] ICRP 1995 ICRP publication 70: basic anatomical and physiological data for use in radiological protection: the skeleton *Ann. ICRP.* **25** 1–180

[4] Rowland R E, Stehney A F and Lucas H F Jr. 1978 Dose-response relationships for female radium dial workers *Radiat. Res.* **76** 368–83

[5] Littman M S, Kirsh I E and Keane A T 1978 Radium-induced malignant tumors of the mastoid and paranasal sinuses *AJR Am. J. Roentgenol.* **131** 773–85

[6] Rowland R E, Stehney A F and Lucas H F 1983 Dose-response relationships for radium-induced bone sarcomas *Health Phys.* **44** 15–31

[7] Beddoe A H 1976 *The Microstructure of Mammalian Bone in Relation to the Dosimetry of Bone-Seeking Radionuclides* (Leeds: University of Leeds) p 165

[8] Lewington V J 1996 Cancer therapy using bone-seeking isotopes *Phys. Med. Biol.* **41** 2027–42

[9] Siegel J A, Wessels B W and Watson E E 1990 Bone marrow dosimetry and toxicity for radioimmunotherapy *Antib. Immunoconjug. Radiopharm.* **3** 213–33

[10] Sgouros G 1993 Bone marrow dosimetry for radioimmunotherapy: theoretical considerations *J. Nucl. Med.* **34** 689–94

[11] Sgouros G, Stabin M, Erdi Y, Akabani G, Kwok C, Brill A B and Wessels B 2000 Red marrow dosimetry for radiolabeled antibodies that bind to marrow, bone, or blood components *Med. Phys.* **27** 2150–164

[12] Bolch W E, Eckerman K F, Sgouros G and Thomas S R 2009 MIRD Pamphlet No. 21: a generalized schema for radiopharmaceutical dosimetry: standardization of nomenclature *J. Nucl. Med.* **50** 477–84

[13] Snyder W S, Ford M R and Warner G G 1978 *MIRD Pamphlet No. 5, Revised: Estimates of Specific Absorbed Fractions for Photon Sources Uniformly Distributed in Various Organs of a Heterogeneous Phantom* (New York: Society of Nuclear Medicine)

[14] Cristy M and Eckerman K F 1987 *Specific Absorbed Fractions of Energy at Various Ages from Internal Photon Sources* (Oak Ridge, TN: Oak Ridge National Laboratory)

[15] Clariand I, Bouchet L G and B.W. E 2000 A new macroscopic model of the long bones for skeletal dosimetry—abstract *J. Nucl. Med.* **41** 1062

[16] Snyder W S, Ford M R, Warner G G and Watson S B 1974 *A Tabulation of Dose Equivalent per Microcurie-Day for Source and Target Organs of an Adult for Various Radionuclides* (Oak Ridge, TN: Oak Ridge National Laboratory)

[17] Snyder W S, Ford M R, Warner G G and Watson S B 1975 *MIRD Pamphlet No. 11: S, Absorbed Dose Per Unit Cumulated Activity for Selected Radionuclides and Organs* (New York: Society of Nuclear Medicine)

[18] Eckerman K F 1985 Aspects of the dosimetry of radionuclides within the skeleton with particular emphasis on the active marrow *Proc. of the Fourth Int. Radiopharmaceutical Dosimetry Symp.* (Oak Ridge, TN: Oak Ridge Associated Universities)

[19] Eckerman K F and Stabin M G 2000 Electron absorbed fractions and dose conversion factors for marrow and bone by skeletal regions *Health Phys.* **78** 199–214

[20] Stabin M G 1996 MIRDOSE: personal computer software for internal dose assessment in nuclear medicine *J. Nucl. Med.* **37** 538–46

[21] Bouchet L G, Clairand I and Bolch W E 2000 Improvement of skeletal dosimetry associated with photon sources *J. Nucl. Med.* **41** 1049

[22] Beddoe A H, Darley P J and Spiers F W 1976 Measurements of trabecular bone structure in man *Phys. Med. Biol.* **21** 589–607

[23] Darley P J 1968 Measurement of linear path length distributions in bone and bone marrow using a scanning technique *Symp. on Microdosimetry* (Ispra, Italy: EAEC Report EUR) p 3747

[24] Darley P J 1972 *An Investigation of the Structure of Trabecular Bone in Relation to the Radiation Dosimetry of Bone-Seeking Radionuclides* (Leeds: University of Leeds) p 231

[25] Whitwell J R 1973 *Theoretical Investigations of Energy Loss by Ionizing Particles in Bone* (Leeds: University of Leeds) p 268

[26] Whitwell J R and Spiers F W 1976 Calculated beta-ray dose factors for trabecular bone *Phys. Med. Biol.* **21** 16–38

[27] Spiers F W 1949 The influence of energy absorption and electron range on dosage in irradiated bone *Br. J. Radiol.* **22** 521–33

[28] Spiers F W 1966 Dose to trabecular bone from internal beta-emitters *First Int. Congress of Radiation Protection (Rome)* (Oxford: Pergamon)

[29] Bouchet L G and Bolch W E 1999 A three-dimensional transport model for determining absorbed fractions of energy for electrons in cortical bone *J. Nucl. Med.* **40** 2115–24

[30] Pafundi D, Rajon D, Jokisch D, Lee C and Bolch W 2010 An image-based skeletal dosimetry model for the ICRP reference newborn—internal electron sources *Phys. Med. Biol.* **55** 1785–814

[31] Beddoe A H 1976 A quantitative study of the structure of mammalian bone *J. Anat.* **122** 190

[32] Beddoe A H 1977 Measurements of the microscopic structure of cortical bone *Phys. Med. Biol.* **22** 298–308

[33] Beddoe A H 1978 A quantitative study of the structure of trabecular bone in man, rhesus monkey, beagle and miniature pig *Calcif. Tissue Res.* **25** 273 81

[34] Loevinger R, Budinger T F and Watson E E 1991 *MIRD Primer for Absorbed Dose Calculations* 2nd edn (New York: The Society of Nuclear Medicine) p 128

[35] Howell R W, Wessels B W and Loevinger R 1999 The MIRD perspective 1999 *J. Nucl. Med.* **40** 3S–10S

[36] ICRP 1979 ICRP publication 30, *part 1: limits for intakes of radionuclides by workers Ann. ICRP.* **3** 1–116

[37] Chen H-T and Poston J W 1982 Estimates of absorbed energy in trabecular bone due to beta particles or electrons *Health Phys.* **43** 647–53

[38] Zuzarte de Mendonca A 1993 *Trabecular Bone Dosimetry using a Monte Carlo Code* (College Station, TX: Texas A&M University)

[39] Nelson W R, Hirayama R H and Roger D W O 1985 *The EGS4 Code System.* (Stanford, CA: Stanford Linear Accelerator Center)

[40] Parry R A 1995 *Revised estimates of electron absorbed fractions and radionuclide S-values in trabecular bone, Nuclear Engineering.* (College Station: Texas A&M University) p 169

[41] Bouchet L G, Jokisch D W and Bolch W E 1999 A three-dimensional transport model for determining absorbed fractions of energy for electrons in trabecular bone *J. Nucl. Med.* **40** 1947–66

[42] Bouchet L G, Bolch W E, Howell R W and Rao D V 2000 S values for radionuclides localized within the skeleton *J. Nucl. Med.* **41** 189–212

[43] Patton P W 2000 NMR microscopy for skeletal dosimetry: an investigation of marrow cellularity on dose estimates *Nuclear and Radiological Engineering* (Gainesville, FL: University of Florida) p 426

[44] Stabin M G, Eckerman K F, Bolch W E, Bouchet L G and Patton P W 2002 Evolution and status of bone and marrow dose models *Cancer Biother. Radiopharm.* **17** 427–45

[45] Samaratunga R C, Thomas S R, Hinnefeld J D, Kuster L C V, Hyams D M, Moulton J S, Sperling M I and Maxon H R 1995 A Monte Carlo simulation model for radiation dose to metastatic skeletal tumor from rhenium-186(Sn)-HEDP *J. Nucl. Med.* **36** 336–50

[46] Watchman C, Jokisch D W, Patton P, Rajon D, Sgouros G and Bolch W 2005 Absorbed fractions for alpha-particles in tissues of trabecular bone: considerations of marrow cellularity within the ICRP reference male *J. Nucl. Med.* **46** 1171–85

[47] Jokisch D W, Patton P W, Inglis B A, Bouchet L G, Rajon D A, Rifkin J and Bolch W E 1998 NMR microscopy of trabecular bone and its role in skeletal dosimetry *Health Phys.* **75** 584–96

[48] Jokisch D W, Bouchet L G, Patton P W, Rajon D A and Bolch W E 2001 Beta-particle dosimetry of the trabecular skeleton using Monte Carlo transport within 3D digital images *Med. Phys.* **28** 1505–18

[49] Bolch W E, Patton P W, Rajon D A, Shah A P, Jokisch D W and Inglis B 2002 Considerations of marrow cellularity in 3D dosimetric models of the trabecular skeleton *J. Nucl. Med.* **43** 97–108

[50] Patton P W, Rajon D A, Shah A P, Jokisch D W, Inglis B and Bolch W E 2002 Site-specific variability in trabecular bone dosimetry: considerations of energy loss to cortical bone *Med. Phys.* **29** 6–14

[51] Patton P W, Jokisch D W, Rajon D A, Shah A P, Myers S L and Bolch W E 2002 Skeletal dosimetry via NMR microscopy: investigations of sample reproducibility and signal source *Health Phys.* **82** 316–26

[52] Rajon D A, Jokisch D W, Patton P W, Shah A P and Bolch W E 2000 Voxel size effects in three-dimensional nuclear magnetic resonance microscopy performed for trabecular bone dosimetry *Med. Phys.* **27** 2624–35

[53] Akabani G 1993 Absorbed dose calculations in Haversian canals for several beta-emitting radionuclides *J. Nucl. Med.* **34** 1361–66

[54] ICRP 2002 ICRP publication 89: *basic anatomical and physiological data for use in radiological protection—reference values Ann. ICRP.* **32** 1–277

[55] Bolch W E, Shah A P, Watchman C J, Jokisch D W, Patton P W, Rajon D A, Zankl M, Petoussi-Henss N and Eckerman K F 2007 Skeletal absorbed fractions for electrons in the adult male: considerations of a revised 50-microm definition of the bone endosteum *Radiat. Prot. Dosimetry* **127** 169–73

[56] Gossner W, Masse R and Stather J W 2000 Cells at risk for dosimetric modeling relevant to bone tumour induction *Radiat. Prot. Dosimetry* **92** 209–13

[57] Gossner W 2003 Target cells in internal dosimetry *Radiat. Prot. Dosimetry* **105** 39–42

[58] Shah A, Rajon D, Patton P, Jokisch D and Bolch W 2005 Accounting for beta-particle energy loss to cortical bone via Paired-Image Radiation Transport (PIRT) *Med. Phys.* **32** 1354–66

[59] Shah A P, Bolch W E, Rajon D A, Patton P W and Jokisch D W 2005 A paired-image radiation transport model for skeletal dosimetry *J. Nucl. Med.* **46** 344–53

[60] ICRP 2007 ICRP publication 103: *recommendations of the Int. Commission on Radiological Protection Ann. ICRP.* **37** 1–332

[61] ICRP 2015 ICRP publication 130: *occupational intakes of radionuclides, part 1 Ann. ICRP.* **44** 1–188

[62] ICRP 2009 ICRP publication 110: *adult reference computational phantoms Ann. ICRP.* **39** 1–165

[63] ICRP 2016 *ICRP publication 133: the ICRP computational framework for internal dose assessment for reference adults: specific absorbed fractions Ann. ICRP.* **45** 1–74

[64] Hough M, Johnson P, Rajon D, Jokisch D, Lee C and Bolch W 2011 An image-based skeletal dosimetry model for the ICRP reference adult male-internal electron sources *Phys. Med. Biol.* **56** 2309–46

[65] O'Reilly S E, DeWeese L S, Maynard M R, Rajon D A, Wayson M B, Marshall E L and Bolch W E 2016 An image-based skeletal dosimetry model for the ICRP reference adult female-internal electron sources *Phys. Med. Biol.* **61** 8794–824

[66] Pafundi D H 2009 *Image-Based Skeletal Tissue and Electron Dosimetry Models for the ICRP Reference Pediatric Age Series* (Gainsville, FL: Department of Biomedical Engineering, University of Florida)

[67] Wayson M B 2012 *Computational Internal Dosimetry Methods as Applied to the University of Florida Series of Hybrid Phantoms* (Gainesville, FL: Department of Biomedical Engineering, University of Florida)

[68] ICRP 2010 ICRP publication 116: *conversion coefficients for radiological protection quantities for external radiation exposures Ann. ICRP.* **40** 1–257

[69] ICRP 2020 ICRP publication 144: *age-dependent dose coefficients for external exposures to environmental sources Ann. ICRP.* **49** 1–147

[70] Johnson P B, Bahadori A A, Eckerman K F, Lee C and Bolch W E 2011 Response functions for computing absorbed dose to skeletal tissues from photon irradiation—an update *Phys. Med. Biol.* **56** 2347 65

[71] Hubbell J and Seltzer S 2004 *Tables of x-ray mass attenuation coefficients and mass energy-absorption coefficients* (Version 1.4). September 13, 2005 http://physics.nist.gov/zaamdi

[72] Breems D A, Blokland E A, Neben S and Ploemacher R E 1994 Frequency analysis of human primitive haematopoietic stem cell subsets using a cobblestone area forming cell assay *Leukemia* **8** 1095–104

[73] Neben S, Anklesaria P, Greenberger J and Mauch P 1993 Quantitation of murine hematopoietic stem cells *in vitro* by limiting dilution analysis of cobblestone area formation on a clonal stromal cell line *Exp. Hematol.* **21** 438–43

[74] Siegel J A, Lee R E, Pawlyk D A, Horowitz J A, Sharkey R M and Goldenberg D M 1989 Sacral scintigraphy for bone marrow dosimetry in radioimmunotherapy *Int. J. Rad. Appl. Instrum. B.* **16** 553–59

[75] Bolch W E, Bouchet L G, Robertson J S, Wessels B W, Siegel J A, Howell R W, Erdi A K, Aydogan B, Costes S and Watson E E 1999 MIRD Pamphlet No. 17: the dosimetry of nonuniform activity distributions—Radionuclide S Values at the voxel level *J. Nucl. Med.* **40** 11S–36S

[76] DeNardo D A, DeNardo G L, O'Donnell R T, Lim S M, Shen S, Yuan A and DeNardo S J 1997 Imaging for improved prediction of myelotoxicity after radioimmunotherapy *Cancer.* **80** 2558–66

[77] Lim S, DeNardo G L, DeNardo D A, O'Donnell R T, Yuan A and DeNardo S J 1997 Predicition of myelotoxicity using semi-quantitative marrow image scores *J. Nucl. Med.* **38** 1749–53
[78] Eary J F, Krohn K A, Press O W, Durack L and Bernstein I D 1997 Importance of pre-treatment radiation absorbed dose estimation for radioimmunotherapy of non-Hodgkin's lymphoma *Nucl. Med. Biol.* **24** 635–8
[79] Breitz H B, Fisher D R and Wessels B W 1998 Marrow toxicity and radiation absorbed dose estimates from rhenium-186-labeled monoclonal antibody *J. Nucl. Med.* **39** 1746–51
[80] Behr T M, Sharkey R M, Juweid M E, Dunn R M, Vgg R C, Siegel J A and Goldberg G R 1996 Hematological toxicity in the radioimmunotherapy of solid cancers with 131-I-labeled anit-CEA NP-4 IgG1: dependence of red marrow dosimetry and pretreatment *Proc. of the Sixth Int. Radiopharmaceutical Dosimetry Symp.* (Gatlinburg, TN: Oak Ridge Associated Universities) 113–25
[81] Juweid M E, Zhang C H, Blumenthal R D, Hajjar G, Sharkey R M and Goldenberg D M 1999 Prediction of hematologic toxicity after radioimmunotherapy with ^{131}I-labeled anticarcinoembryonic antigen monoclonal antibodies *J. Nucl. Med.* **40** 1609–16
[82] Emami B, Lyman J, Brown A, Coia L, Goitein M, Munzenrider J E, Shank B, Solin L J and Wesson M 1991 Tolerance of normal tissue to therapeutic irradiation *Int. J. Radiat. Oncol. Biol. Phys.* **21** 109–22
[83] Erdi A K, Erdi Y E, Yorke E D and Wessels B W 1996 Treatment planning for radioimmunotherapy *Phys. Med. Biol.* **41** 2009–26
[84] DeNardo G L, DeNardo S J, Macey D J, Shen S and Kroger L A 1994 Overview of radiation myelotoxicity secondary to radioimmunotherapy using ^{131}I-Lym-1 as a model *Cancer.* **73** 1038–48
[85] Sgouros G 2020 Dosimetry, radiobiology and synthetic lethality: radiopharmaceutical therapy (RPT) with alpha-particle-emitters *Semin. Nucl. Med.* **50** 124–32
[86] Sgouros G, Barest G, Thekkumthala J, Chui C, Mohan R, Bigler R E and Zanzonico P 1990 Treatment planning for internal radionuclide therapy: three-dimensional dosimetry for nonuniformly distributed radionuclides *J. Nucl. Med.* **31** 1884–91
[87] Akabani G, Hawkins W G, Eckblade M B and Leichner P K 1997 Patient-specific dosimetry using quantitative SPECT imaging and three-dimensional discrete fourier transform convolution *J. Nucl. Med.* **38** 308–14
[88] Furhang E E, Chui C S and Sgouros G 1996 A Monte Carlo approach to patient-specific dosimetry *Med. Phys.* **23** 1523–29
[89] Bentzen *et al* 2010 Quantitative Analyses of Normal Tissue Effects in the Clinic (QUANTEC): an introduction to the scientific issues *Int. J. Radiat. Oncol. Biol. Phys.* **76** S3–9
[90] Jackson A, Marks L B, Bentzen S M, Eisbruch A, Yorke E D, Ten Haken R K, Constine L S and Deasy J O 2010 The lessons of QUANTEC: recommendations for reporting and gathering data on dose-volume dependencies of treatment outcome *Int. J. Radiat. Oncol. Biol. Phys.* **76** S155–60
[91] Lim S M, DeNardo G L, DeNardo D A, Shen S, Yuan A, O'Donnell R T and DeNardo S J 1997 Prediction of myelotoxicity using radiation doses to marrow from body, blood and marrow sources *J. Nucl. Med.* **38** 1374–8
[92] Sgouros G, Jureidini I M, Scott A M, Graham M C, Larson S M and Scheinberg D A 1996 Bone marrow dosimetry: regional variability of marrow-localizing antibody *J. Nucl. Med.* **37** 695–98

[93] Wessels B, Breitz H B, Meredith R F, Juweid M, Pastor M, Colburn M and DeNardo G L 2000 Bone marrow dosimetry model adjustments based on age, gender, and prior therapy [Abstract] *J. Nucl. Med.* **41** 83P

[94] Sgouros G, Deland D, Loh A C, Divgi C R and Larson S M 1997 Marrow and whole-body absorbed dose versus marrow toxicity folling 131-I-G250 antibody therapy in patients with renal-cell carcinoma *J. Nucl. Med.* **38** 252P

[95] O'Donoghue J A, Baidoo N and Sgouros G 1998 Calculated red marrow absorbed dose predicts hematological toxicity in radioimmunotherapy *J. Nucl. Med.* **39** 186P

[96] Blumenthal R D, Lew W, Juweid M, Alisauskas R, Ying Z and Goldenberg D M 2000 Plasma FLT3-L levels predict bone marrow recovery from myelosuppressive therapy *Cancer.* **88** 333–43

[97] Vanderhoek M, Juckett M B, Perlman S B, Nickles R J and Jeraj R 2011 Early assessment of treatment response in patients with AML using [(18)F]FLT PET imaging *Leuk. Res.* **35** 310–6

[98] Shields A F, Grierson J R, Dohmen B M, Machulla H J, Stayanoff J C, Lawhorn-Crews J M, Obradovich J E, Muzik O and Mangner T J 1998 Imaging proliferation *in vivo* with [F-18] FLT and positron emission tomography *Nat. Med.* **4** 1334–6

[99] Mauch P, Constine L, Greenberger J, Knospe W, Sullivan J, Liesveld J L and Deeg H J 1995 Hematopoietic stem cell compartment: acute and late effects of radiation therapy and chemotherapy *Int. J. Radiat. Oncol. Biol. Phys.* **31** 1319–39

[100] Koo K H, Dussault R, Kaplan P, Kim R, Ahn I O, Christopher J, Song H R and Wang G J 1998 Age-related marrow conversion in the proximal metaphysis of the femur: evaluation with T1-weighted MR imaging *Radiology.* **206** 745–8

[101] Cristy M 1981 Active bone marrow distribution as a function of age in humans *Phys. Med. Biol.* **26** 389–400

[102] Hindorf C, Glatting G, Chiesa C, Linden O and Flux G 2010 EANM Dosimetry Committee guidelines for bone marrow and whole-body dosimetry *Eur. J. Nucl. Med. Mol. Imaging.* **37** 1238–50

[103] Geyer A M, Schwarz B C, Hobbs R F, Sgouros G and Bolch W E 2017 Quantitative impact of changes in marrow cellularity, skeletal size, and bone mineral density on active marrow dosimetry based upon a reference model *Med. Phys.* **44** 272–83

[104] Gear J I, Cox M G, Gustafsson J, Gleisner K S, Murray I, Glatting G, Konijnenberg M and Flux G D 2018 EANM practical guidance on uncertainty analysis for molecular radiotherapy absorbed dose calculations *Eur. J. Nucl. Med. Mol. Imaging.* **45** 2456–74

[105] Sgouros G 2005 Blood and bone marrow dosimetry in radioiodine therapy of thyroid cancer *J. Nucl. Med.* **46** 899–900 (author reply 901)

[106] Forrer F *et al* 2009 Bone marrow dosimetry in peptide receptor radionuclide therapy with [^{177}Lu-DOTA(0),Tyr(3)]octreotate. *Eur. J. Nucl. Med. Mol. Imaging.* **36** 1138–46

[107] Hindorf C, Linden O, Tennvall J, Wingardh K and Strand S E 2002 Time dependence of the activity concentration ratio of red marrow to blood and implications for red marrow dosimetry *Cancer.* **94** 1235–9

[108] Shen S, Meredith R F, Duan J, Macey D J, Khazaeli M B, Robert F and LoBuglio A F 2002 Improved prediction of myelotoxicity using a patient-specific imaging dose estimate for non-marrow-targeting ^{90}Y-antibody therapy *J. Nucl. Med.* **43** 1245–53

[109] Brindle J M, Trindade A A, Shah A P, Jokisch D W, Patton P W and Bolch W E 2006 Linear regression model for predicting patient-specific total skeletal spongiosa volume for use in molecular radiotherapy dosimetry *J. Nucl. Med.* **47** 1875–83

[110] Salas-Ramirez M, Tran-Gia J, Gbureck U, Kosmala A and Lassmann M 2019 Quantification of the trabecular bone volume fraction for bone marrow dosimetry in molecular radiotherapy by using a dual-energy (SPECT/)CT *Phys. Med. Biol.* **64** 205014

[111] Liu X, Yu L, Primak A N and McCollough C H 2009 Quantitative imaging of element composition and mass fraction using dual-energy CT: three-material decomposition *Med. Phys.* **36** 1602–9

[112] ICRU 1989 *ICRU Report 44: Tissue Substitutes in Radiation Dosimetry and Measurement* (Bethesda, MD: Int. Commission on Radiation Units and Measurements)

[113] ICRU 1992 *ICRU Report 46: Photon, Electron, Proton and Neutron Interaction Data for Body Tissues* (Bethesda, MD: Int. Commission on Radiation Units and Measurements)

[114] Quelch K J, Melick R A, Bingham P J and Mercuri S M 1983 Chemical composition of human bone *Arch. Oral Biol.* **28** 665–74

[115] Pichardo J C, Milner R J and Bolch W E 2011 MRI measurement of bone marrow cellularity for radiation dosimetry *J. Nucl. Med.* **52** 1482–9

[116] Salas-Ramirez M, Tran-Gia J, Kesenheimer C, Weng A M, Kosmala A, Heidemeier A, Kostler H and Lassmann M 2018 Quantification of fat fraction in lumbar vertebrae: correlation with age and implications for bone marrow dosimetry in molecular radiotherapy *Phys. Med. Biol.* **63** 025029

[117] Goodsitt M M, Shenoy A, Shen J, Howard D, Schipper M J, Wilderman S, Christodoulou E, Chun S Y and Dewaraja Y K 2014 Evaluation of dual energy quantitative CT for determining the spatial distributions of red marrow and bone for dosimetry in internal emitter radiation therapy *Med. Phys.* **41** 051901

[118] Wilderman S J, Roberson P L, Bolch W E and Dewaraja Y K 2013 Investigation of effect of variations in bone fraction and red marrow cellularity on bone marrow dosimetry in radio-immunotherapy *Phys. Med. Biol.* **58** 4717–31

[119] Shishkina E A, Zalyapin V I, Timofeev Y S, Degteva M O, Smith M and Napier B 2018 Parametric stochastic model of bone structures to be used in computational dosimetric phantoms of human skeleton *Radiat. Appl.* **3** 133–37

[120] Akleyev A, Deltour I, Krestinina L, Sokolnikov M, Tsareva Y, Tolstykh E and Schuz J 2016 Incidence and mortality of solid cancers in people exposed in utero to ionizing radiation: pooled analyses of two cohorts from the Southern Urals, Russia *PLoS One.* **11** e0160372

[121] Schuz J, Deltour I, Krestinina L Y, Tsareva Y V, Tolstykh E I, Sokolnikov M E and Akleyev A V 2017 In utero exposure to radiation and haematological malignancies: pooled analysis of Southern Urals cohorts *Br. J. Cancer.* **116** 126–33

[122] Frassoni F, Testa N G and Lord B I 1982 The relative spatial distribution of erythroid progenitor cells (BFUe and CFUe) in the normal mouse femur *Cell Tissue Kinet.* **15** 447–55

[123] Lord B I 1990 The architecture of bone marrow cell populations *Int. J. Cell Cloning.* **8** 317–31

[124] Cui Y F, Lord B I, Woolford L B and Testa N G 1996 The relative spatial distribution of *in vitro*-CFCs in the bone marrow, responding to specific growth factors *Cell Prolif.* **29** 243–57

[125] Watchman C J, Bourke V A, Lyon J R, Knowlton A E, Butler S L, Grier D D, Wingard J R, Braylan R C and Bolch W E 2007 Spatial distribution of blood vessels and CD34+ hematopoietic stem and progenitor cells within the marrow cavities of human cancellous bone *J. Nucl. Med.* **48** 645–54

[126] Geyer A M, Schwarz B C, O'Reilly S E, Hobbs R F, Sgouros G and Bolch W E 2017 Depth-dependent concentrations of hematopoietic stem cells in the adult skeleton: implications for active marrow dosimetry *Med. Phys.* **44** 747–61

[127] Bourke V A, Watchman C J, Reith J D, Jorgensen M L, Dieudonne A and Bolch W E , 2009 Spatial gradients of blood vessels and hematopoietic stem and progenitor cells within the marrow cavities of the human skeleton. *Blood.* **114** 4077–80

[128] Hobbs R F, Song H, Watchman C J, Bolch W E, Aksnes A K, Ramdahl T, Flux G D and Sgouros G 2012 A bone marrow toxicity model for Ra-223 alpha-emitter radiopharmaceutical therapy *Phys. Med. Biol.* **57** 3207–22

IOP Publishing

Monte Carlo Calculations in Nuclear Medicine (Second Edition)
Therapeutic applications
Habib Zaidi

Chapter 20

The use of Monte Carlo simulations to evaluate image-based absorbed dose calculation procedures

Michael Ljungberg and Yuni K Dewaraja

20.1 Introduction

Calculation of the internal absorbed dose is generally based on the MIRD formalism, which is a framework for dosimetry that starts with the decay in the source volume and includes the transport of charged particles and photons across tissue and the energy deposition in the target volume. This formalism is applicable at the whole-body level, organ/lesion-level, voxel-level, and all the way down to cellular levels. The accuracy primarily reflects the accuracy of the source definitions and how well the activity distribution has been quantified, the anatomical definition, and the accuracy of how the radiation transport is modelled (i.e., type of dosimetry method or program that has been used). The main application of the MIRD formalism has been for risk estimation in diagnostic nuclear medicine applications where organ-based absorbed dose fractions, the so-called *S*-values, which have been pre-calculated from a generalised mathematical geometry representing a reference human, are utilised. The most common dosimetry approach is to calculate the average deposited energy over the entire organ volume. Recently, there has been much interest in dosimetry for radionuclide therapy applications with 3D (voxel-level) dose assessment.

Estimates of the radioactivity in the source volume for dosimetry can be obtained from either 2D planar scintillation (gamma) camera imaging or, preferably, from 3D imaging with SPECT or PET systems. Both gamma camera and PET systems have physical effects that hamper quantitative imaging accuracy; therefore, proper corrections for various physical and camera-specific effects, such as non-uniform photon attenuation, contributions from scatter in the object, and degradation in spatial resolution, are needed. SPECT and PET technologies have an advantage over 2-D planar imaging in providing 3D information regarding the activity

doi:10.1088/978-0-7503-2694-0ch20

distribution. If a 3D attenuation distribution of the patient that is aligned to the SPECT/PET data is available, which is more or less standard in modern camera systems that are integrated with CT, these two types of information sets (physiology and morphology) can be combined and used in a calculation scheme to produce patient-specific 3D absorbed dose distributions down to the voxel-level. The main computational approaches for voxel-level dosimetry are (1) dose-point kernel convolution, (2) voxel-level *S*-values, and (3) direct Monte Carlo radiation transport, which is generally considered the most accurate method. Furthermore, when the photon contribution to the absorbed dose is small, the simple assumption of local energy deposition within the voxel where the decay originates can provide accurate estimates of the energy deposited by the charged particles, depending on the energy of the particles, voxel size, system resolution, and noise level. With advances in computational power, voxel-level non-uniform dosimetry should be the main future direction for dosimetry development in radionuclide therapy. In addition to the absorbed dose rate and/or dose maps, the absorbed dose distribution within a volume of interest can also be presented as absorbed dose-volume histograms (DVHs), often displayed as the frequency of voxels having a particular absorbed dose or higher. These histograms provide information on how the absorbed dose is locally distributed and can, in principle, provide useful information for understanding dose–effect relationships.

Because tomographic emission imaging provides the underlying information for 3D activity quantification and biokinetics, the evaluation and validation of the dosimetry should also consider methods for image reconstruction and their parameters, as well as registration methods for anatomical images and functional images (for example, when combining data from multiple SPECT/CT studies acquired at different time points). Furthermore, the conversion of CT Hounsfield units (HUs) to mass–density and attenuation coefficients for the principal photon energy, scatter and attenuation compensation methods, and depth-dependent detector response compensation should be considered. In addition, partial volume effects and inherent camera limitations, such as count-rate limitations and related pulse pile-up effects, may need to be evaluated, depending on the study. The latter is important for post-therapy imaging of patients undergoing treatment with high activities. The evaluation of 3D activity quantification accuracy can be performed by physical measurements/imaging using physical phantoms, from computer-generated virtual phantoms in combination with Monte Carlo simulations of the imaging process, or as a combination of both.

20.2 Monte Carlo simulation of imaging systems

Several Monte Carlo programs, designed to simulate scintillation camera imaging for both SPECT and PET, have been described in the literature. These programs have been used to investigate and understand the radiation transport from a decay in the source region of the object to the generation of an event/coincidence in the projection space, as well as to evaluate methods that compensate for image-degrading physical factors. A major advantage of simulation is the possibility of obtaining 'ideal images' that are infeasible to measure, by turning off effects such as photon attenuation and scatter in the object (i.e., mimicking a perfect scatter and attenuation compensation). The design of the programs varies according to the complexity and the intended application area. In this chapter, we will briefly describe three commonly used programs and show examples from one of these.

The SIMIND program [1, 2] has been developed over three decades at Lund University, Lund, Sweden, and is a Fortran code that runs on the major platforms. The program includes the capability to simulate gamma-camera projections for planar or SPECT imaging using arbitrary voxel-based phantoms. The code can be coupled with a collimator routine where photon transport is modelled realistically, including collimator scatter and penetration events. The program has been designed to simulate standard gamma cameras with different types of collimators, but recently, CZT detectors that offer improved energy resolution have been implemented, as these systems are now commercially available and used for clinical imaging. The program is driven by tools to perform post-processing, such as image display and region-of-interest (ROI) manipulations, made with external programs. It is also possible to post-process the projection data from SIMIND to a DICOM file format and import the result into a clinical workstation for further analysis. For research purposes, an off-line alternative, such as the Castor reconstruction program [3], could be useful. Recently, a conversion program to convert the SIMIND output to be compatible with the Castor SPECT reconstruction program was included in the distribution. In this chapter, some of the reconstructed images were created using SIMIND simulations coupled with Castor reconstruction software.

The SIMSET code (Simulation System for Emission Tomography) was developed by a group at the University of Washington, Seattle, USA [4]. It is a program to simulate photon transport and imaging physics for applications in both SPECT and PET. The program is written in a modular form and includes a photon history generator (PHG), a module to define the object and the source distribution, a collimator module, a detector module, and additional modules that are necessary for PET simulations, including a random coincidence generation module and a rebinning module.

The GATE program [5] is an open-source code, based on the Geant4 C++ library code designed for medical physics applications, released in 2004. GATE was initially dedicated to performing highly realistic simulation of PET and SPECT imaging systems, but it has been extended for dosimetry and radiotherapy experiments and x-ray tomography by taking advantage of the particle physics modules in the underlying Geant4 library. The user creates a file that contains macro commands defining the geometry of the simulation (phantoms, collimator, detector, etc). These objects can be described by voxels, meshes, or analytical equations. Included here are also the definitions of the composition of the materials and their densities. A part of the macro includes the physics (types of particle, energies, and abundances) and related cross-section. The description of the source includes sampling of the location and emission type. The camera can be described in terms of energy resolution and digitisation parameters (matrix size and pixel size). A powerful option in GATE is the geometry package, which allows the user to build complex geometries starting with simple modules, such as a block-based ring PET system, with a few lines of macro-code.

20.3 Monte Carlo programs for dosimetry

While Monte Carlo codes such as SIMIND and SIMSET have been developed solely for modelling nuclear medicine imaging systems, the GATE code is capable of calculating absorbed dose distribution because the program allows for the simulation of charged particles. Other general-purpose Monte Carlo programs that can be used for image-based dosimetry include the **MNCP** code [6, 7], **EGSnrc** code [8], **Penelope** code [9], **Fluka** code [10], **Peregrine** [11] and **DPM** code [12]. A description of the pros and cons of using these codes for image-based dosimetry is beyond the scope of this chapter, but studies using these codes for internal dosimetry with patient-specific activity/density maps as input have been published [13–19]. One of these publications compares voxel-level dosimetry performed with MCNP4C, EGSnrc, and GEANT4 for three different radionuclides (^{90}Y, ^{131}I, and ^{188}Re) [20]. Monte Carlo based voxel-dosimetry codes are presented in earlier chapters. In the examples provided later in this chapter, we have used an in-house-written EGSnrc user code to generate absorbed dose images from aligned voxel-based source and density images.

20.4 Image-based phantoms for use in Monte Carlo evaluations

Anthropomorphic software phantoms with internal organs facilitate clinically realistic Monte Carlo simulations. These computer phantoms are generated using either mathematically defined functions or are based on segmented digital images. The analytic phantoms consist of regularly shaped continuous objects defined by combinations of simple geometrical shapes or by the use of surface functions (NURBS or mesh). Image-based voxel phantoms, typically derived from segmented tomographic images of the human anatomy obtained by either x-ray CT or MRI, are highly clinically realistic. However, an advantage of surface-based phantoms over voxel phantoms is that changes in the geometry can more easily be implemented to allow for anatomical variability.

20.5 Digital representation of physical phantoms

Some clinics have access to standard physical phantoms for quality assurance and calibration. These phantoms are often cylindrical or elliptical, with the possibility of inserting spheres of various diameters. Digital versions of such phantoms, typically generated by segmenting high-resolution CT images of the phantom, can be useful for the validation of Monte Carlo simulation codes and for evaluation of imaging methods, including reconstruction and compensation for image-degrading factors. Thresholding or manual delineation can be used to segment different compartments such as spheres positioned at the desired position, background volumes, and the phantom walls. To make the phantom more flexible, the different segmented volumes can then be assigned a unique value, and from this 'coded' phantom, both density images and activity distribution images can be generated for input to an MC code that can accommodate voxelised phantoms.

A commonly used phantom in the clinic is the National Electrical Manufacturers Association (NEMA) phantom, where spheres of various diameters can be inserted in a ring around a centre cylinder that can be filled with Styrofoam beads and water to mimic the density of lung tissue. The digital version of the NEMA phantom was created as a digital reference object (DRO) by the Quantitative Imaging Biomarker Alliance (QIBA) and has been used in a study for harmonising reconstruction parameters across different centres [21]. This and other DROs developed by QIBA are available for download at [22]. Although the NEMA phantom is quite useful for basic comparison, it is at the same time limited because of the lack of realistic organ shapes and heterogeneities. Some more clinically realistic commercially available physical phantoms that can be digitised in a similar manner include: (a) the Data Spectrum anthropomorphic phantom, with liver, lungs, cardiac, and spine inserts; (b) the RSD torso phantom, with liver, heart, lungs, and a realistic skeleton; and (c) the Kyoto phantom, with kidneys, lungs, and liver. Figure 20.1 shows photographs of the physical NEMA phantom and the RSD Torso phantom, together with images from measurements (planar projections and SPECT) and matching Monte Carlo simulations using digitised versions of the phantoms.

There have been considerable advances in 3D printing technology in the past few years that have been exploited to generate phantoms for evaluating nuclear medicine imaging and dosimetry methods. The advantage is, of course, the ability to generate arbitrary geometries of organs and tumours to make the experimental phantom measurement more clinically relevant. Furthermore, 3D printed phantoms, often generated using in-house facilities, can be substantially more economical than purchasing a phantom from the manufacturer.

The volume shape to be printed is initially constructed from a computer-aided design (CAD) program or 3D model that then determines a process where materials are added together, layer by layer, to form the volume to be created. The CAD

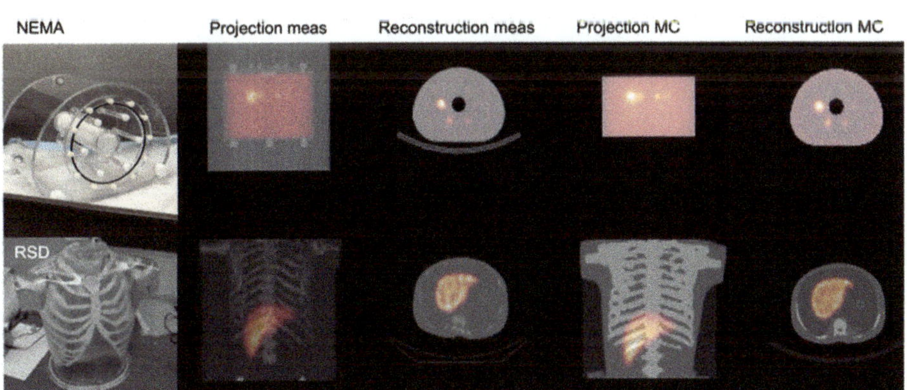

Figure 20.1. Two examples of commercially available physical phantoms that have been digitised into image-based versions to allow for comparison between measured and simulated results. Note that the bone material in the RSD digital phantom is represented by a single attenuation coefficient, in contrast to the physical phantom.

model is often stored as stereolithography (STL) files—a file format that defines a large number of triangulations that define a surface. From this, a large number of slices are created that directly determine how the material should be fused. Common materials for printing include polylactic acid (PLA) and acrylonitrile butadiene styrene (ABS), as they are similar to tissue in terms of photon interaction properties, but metals and ceramics can also be used.

As an example of the use of 3D printing of phantoms for dosimetry applications, although not involving Monte Carlo methods, is the work presented by Robinson *et al* [23]. The aim of this research work was to investigate whether or not a camera calibration procedure that is based on the measurements of simple spheres is sufficiently accurate to be used in a dosimetry procedure where *S*-values have been calculated for irregular organ-shaped sources and targets, as is the case for the data in the OLINDA/EXM software. Four organs (kidneys, pancreas, liver, and spleen) were printed, according to the equations that define the shapes for the OLINDA/EXM dose factors, using a fused filament with ABS plastic. The organs are shown in figure 20.2. These phantoms were then filled with ^{99m}Tc and ^{177}Lu, placed in the centre of a cylinder, and imaged with an Infinia^{TM} Hawkeye SPECT/CT camera (GE Healthcare). From the OSEM reconstructed images, a specific calibration factor (cps MBq^{-1}) for each of these organs was obtained and used in the absorbed dose calculation. Figure 20.2 shows the four different phantoms and a reconstructed SPECT image of a measurement with ^{99m}Tc. The results were compared with the absorbed doses, based on the use of calibration factors obtained from spherical shapes. The results from the study show that using these

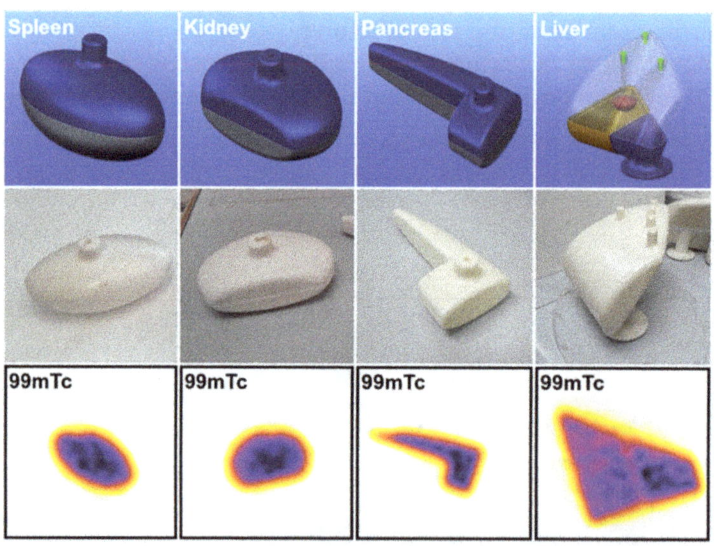

Figure 20.2. Models of the adult male mathematical phantom with, from left, spleen, kidney, pancreas, and liver. The middle row shows photos of the actual printed phantom, and the bottom row shows reconstructed slices through the centre of each phantom. The original images are included in reference [23] and reprinted under the condition of the Common Creative License.

organ-specific calibration factors, a reduction of 24%, 11%, and 8% was found regarding the calculated absorbed dose for the liver, spleen, and liver, respectively.

Recently, 3D printed kidney inserts have been used to evaluate and optimise quantitative ^{177}Lu SPECT/CT because of the interest in performing accurate imaging-based renal dosimetry in patients undergoing ^{177}Lu-DOTATATE peptide receptor radionuclide therapy for neuroendocrine tumours [24]. The two-compartment kidney phantoms were designed with separate fillable cortex and medulla regions to replicate the non-uniform distribution of ^{177}Lu in the sub-organ, as observed in patient imaging studies.

20.6 Software for generation of image-based human-like phantoms

One of the first available human-like phantoms for simulations was developed by Zubal *et al* [25]. This phantom was created using manual segmentation of x-ray CT slices from a male human. Fifty-seven major internal structures were segmented, and voxels inside each segmented structure were given a unique code. These codes were then used to assign an activity concentration value or mass–density values to each voxel within a particular structure. A further development of the phantom was then made by adding arms and legs from the Visible Human (VH) and attaching them to the original torso phantom. Zubal and colleagues also developed a high-resolution brain phantom (256 × 256 pixels) based on an MRI scan that included 62 segmented volumes.

A computer phantom constructed from a CT study of a person is very realistic, but only for that individual. The ability to change, move, or add organs or structures in an already segmented CT-based computer phantom will, however, be very complicated. To overcome this limitation of static voxel-based phantoms, mathematical technologies based on non-uniform rational B-splines (NURBS) or polygon mesh surfaces have been used. NURBS defines continuous surfaces and, by this, allows the phantom and its internal compartments to be smooth, anatomically realistic, and relatively flexible to modify. In the simulation examples shown later in this chapter, we have used the XCAT phantom, developed by Paul Segars at Johns Hopkins and Duke University. From an initial segmented template, obtained from the VH image data set [26], a software tool was constructed that allows the user to generate voxel-based versions of the VH model and to modify the model by altering an extensive number of anatomical and physiological parameters in an input file. The XCAT phantom includes temporal functions, such as modelling of a realistic heart and respiratory motion. All of the organs and skeletal models (except for the heart model) originate from segmented CT information from the VH male data set. The heart and its movements originate from the tagged MRI scans of a human, and the respiratory motion comes from a 4D respiratory gated CT. Several patient models representing various patient geometries, ages, and body-mass index (BMI) are available from the author of the XCAT program (figure 20.3).

A comprehensive topical review has recently been published by Kainz and colleagues [27] where different modelling techniques are described to produce human-like phantoms for use in various applications of medical imaging and biomedical engineering.

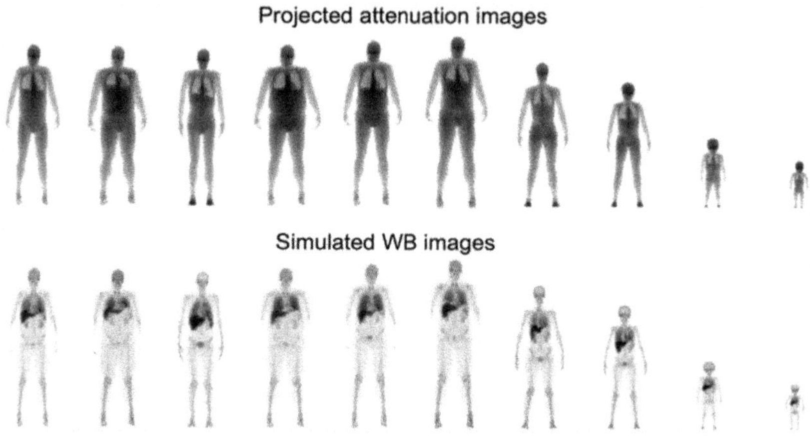

Figure 20.3. Examples of ten XCAT phantoms ranging from adults of different body-mass index down to infants. The upper row shows projected attenuation images and the lower row show simulated whole-body images.

20.7 Validation of Monte Carlo simulation codes with experimental measurement

Monte Carlo simulation combined with virtual phantoms is much more convenient to use and can be better suited for the evaluation of imaging systems/protocols compared with experimental measurements. However, careful experimental validation of the Monte Carlo code needs to be performed. Validation of the Monte Carlo-generated point-spread function and sensitivity (e.g., counts per second per MBq) with a simple measurement of a point-like source in air is a first step. Further validation for more clinically relevant phantom geometries should be undertaken to ensure that photon transport physics is well-modelled. Physical phantoms such as the RSD phantom (figure 20.1) or even simple ‘hot’ inserts in a uniform phantom can be used for these validation studies, and the corresponding image quality and quantification measures such as contrast and activity recovery can be compared. In the case of SPECT, the measured and simulated energy spectra should also be compared. Ideally, the code should be validated for each radionuclide and, in the case of SPECT, for each collimator used.

Here, we will focus on validation studies for the following radionuclides that are widely used in therapy.

^{177}Lutetium: ^{177}Lu is a β^- emitter with a maximum kinetic energy of 500 keV that also emits two principal gamma-ray photons that can be used for imaging. The energies of these two photons are 112.9 keV (6.2%) and 208.4 keV (10.4%), respectively. The therapeutic application of ^{177}Lu includes peptide receptor radionuclide therapy of neuroendocrine tumours, radioimmunotherapy of non-Hodgkin’s lymphomas, and ^{177}Lu-PSMA for metastatic prostate cancer. ^{177}Lu image quality and activity quantification is degraded by the relatively low intensity of the main gamma ray and the down-scatter associated with the presence of multiple higher-energy gamma rays.

^{131}Iodine: The therapeutic use of ^{131}I includes radioiodine for the treatment of thyroid cancer and hyperthyroidism, ^{131}I-MIBG for treatment of neuroblastoma, and ^{131}I radioimmunotherapy for the treatment of non-Hodgkin's lymphoma. ^{131}I is efficient for therapy because it emits β^- particles with a maximum energy of 608 keV. However, it is not ideally suited for gamma camera imaging because of the relatively high energy (364 keV (81.2%) of the predominant gamma-ray and the prevalence of higher-energy emissions at 637 keV (7.1%) and 723 keV (1.8%)), which are low in intensity but have a higher probability of penetrating the collimator septa, thereby contributing disproportionally to the image counts. Therefore, ^{131}I image quality is degraded by the collimator septal penetration, poor spatial resolution associated with high-energy collimators, scatter of the 364-keV photons, and down-scatter of higher-energy gamma rays into the 364-keV energy window.

^{90}Yttrium: Therapeutic applications with ^{90}Y-loaded microspheres and ^{90}Y-labelled antibodies and peptides have attracted a growing interest in quantitative imaging of ^{90}Y, an almost pure high-energy β^- emitter. The relatively high energy and range in tissue of ^{90}Y β^--particles results in absorbed dose deposition to surrounding tumour cells. Additionally, the half-life (64 h) is well-suited for a variety of radionuclide therapy applications. The β^- particles emitted by ^{90}Y lose kinetic energy as they interact with atoms, and some of these interactions result in the emission of bremsstrahlung photons, which can be used for scintillation camera imaging [28]. The energy spectrum of the bremsstrahlung photons is continuous, ranging from the maximum released energy of 2.3 MeV to nearly zero. However, bremsstrahlung photons only account for a small fraction of β^- particle interaction, as most are inelastic interactions between the β^--particles and atomic electrons. As a consequence, the ^{90}Y photon yield is very low. In the case of ^{90}Y, less than 3% of β^- interactions in water result in photons of energy > 50 keV [29] hence, bremsstrahlung imaging with a gamma camera is an inefficient process.

Validation of Monte Carlo codes requires modelling the object and the camera used for the measurement as closely as possible in the simulation. Modelling the object geometry is a relatively easy task, as an accurate digital representation of the physical phantom can be achieved using high-resolution x-ray CT from a SPECT/CT or PET/CT system. The calibration procedure used to convert the CT HUs to density must be carefully verified to ensure that the density map that is used as an input to the Monte Carlo code is accurate. When filling the physical phantom for the validation experiment, the activity measurement must be performed with a well-validated activity meter (dose-calibrator) that is traceable to a National Standards laboratory. These values are then used to generate an activity map for the simulation. The camera specifications available from the manufacturer can be used to model the imaging system; however, this is a more complex task than modelling the object. Certain parts of the system are too complex to model in detail, or information is unavailable; hence, some approximations are always required. For example, the photomultiplier tubes and electronics that are behind the scintillation crystal in a conventional gamma camera have been approximated by a layer of glass in the Monte Carlo model. The thickness of the glass layer was determined empirically based on agreement between the measured and simulated energy spectra [30, 31]. The impact of including this 'backscatter layer' in the

SPECT Monte Carlo model is especially significant in the case of ^{131}I gamma-ray and ^{90}Y bremsstrahlung imaging because of the presence of high-energy photons. Other sources of backscatter, such as scattering off of surrounding structures, including walls and floor, are more difficult to include in the camera model and could lead to some discrepancy between measurement and simulation.

Despite the complexities of carefully validating Monte Carlo models by comparison with experiments, several groups have reported on this. The original publication [5] describing the development of GATE-Geant4 for SPECT and PET includes a list of validation studies that were performed comparing spatial resolution and sensitivity for ^{18}F PET and ^{99m}Tc SPECT, scatter fractions for PET, and energy resolution and energy spectra for SPECT. They reported an agreement of within 5% between measurement and simulation for most of the metrics that were evaluated for the different systems. Validations of GATE with experimental measurements have also been published for ^{131}I [32], ^{177}Lu [33] and ^{90}Y [34]. There are early validations [35] of the SimSET code for PET and a more recent validation that included comparison to GATE simulations, showing agreement within 4%–8% for a count-rate test and 3%–11% for a spatial resolution test [36]. The SIMIND code has been validated for ^{131}I [37–39], ^{177}Lu [38, 40] and ^{90}Y [41–44]. The validation of ^{90}Y SPECT highlights the complexities of accurately modelling the bremsstrahlung production process. In the work by Rong *et al* [41] an underestimation of the abundance of bremsstrahlung photons greater than 500 keV and the impact of incorporating the distance between the ^{90}Y decay location and the bremsstrahlung generation location was identified. The refined ^{90}Y SIMIND simulation model led to better agreement with the experimental measurement for a line source. Walrand *et al* identified the importance of accounting for the process of internal bremsstrahlung generation, which contributes significantly to the total photon yield in ^{90}Y SPECT imaging [45]. Internal bremsstrahlung arises during the β^- decay process itself, compared with external bremsstrahlung, which is generated as β^- particles traverse tissue. For multiple phantom geometries, Dewaraja *et al* [46] showed that simulated SPECT energy spectra corresponding to SIMIND simulations that included the ^{90}Y internal bremsstrahlung component better agreed with the measured energy spectra (figure 20.4(c)).

Figure 20.4 compares the measured SPECT energy spectra with SIMIND simulated energy pulse-height distributions for ^{177}Lu, ^{131}I, and ^{90}Y phantoms. The modelled phantom geometries are indicated in the figures. The SPECT system was modelled based on manufacturer specifications and included a glass backscatter layer to model structures behind the crystal, as described above. However, other sources of backscatter, such as the structure surrounding the camera, are not included in the model and may explain some of the differences in the spectra, especially at lower energies. The spectra for ^{90}Y show the simulation results with and without the internal bremsstrahlung component discussed above. For the ^{90}Y study, a line profile across the centre of the sphere is shown in figure 20.5 demonstrating excellent agreement between the measured and simulated line profiles (FWHM agreed to within 5% and FWTM agreed to within 1%). Overall, these results for ^{90}Y show good agreement between measurement and simulation, especially in the energy range typically used for bremsstrahlung imaging.

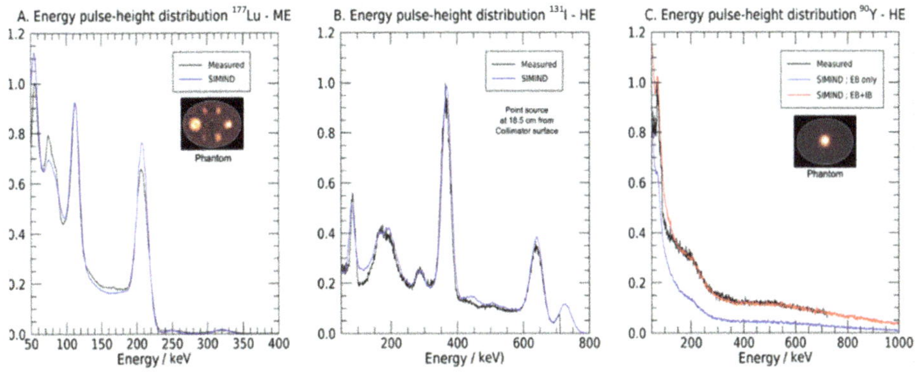

Figure 20.4. Comparison between simulated and measured energy pulse-height distributions for three different radionuclides. For ^{90}Y, distributions are shown for two Monte Carlo simulations, with and without including internal bremsstrahlung production. ME = medium energy collimator, HE = high energy collimator, EB = external bremsstrahlung photons, IB = internal bremsstrahlung photons.

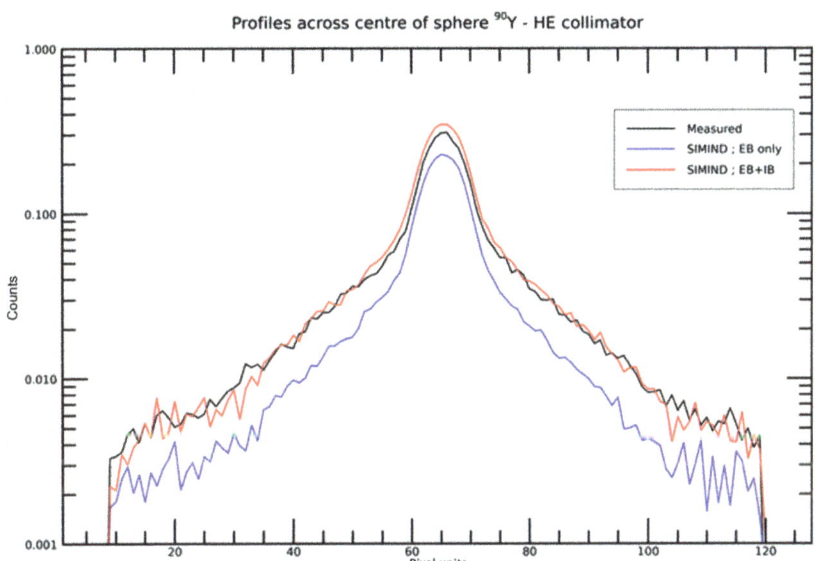

Figure 20.5. Pixel profiles extracted from the projection in a location across the centre of the ^{90}Y comparison shown in figure 20.4(c). The improvement with the inclusion of internal bremsstrahlung is clearly seen.

20.8 Evaluation of SPECT image degradation by Monte Carlo simulations

The accuracy of the activity quantification is an essential part of absorbed dose calculation. SPECT image quality and quantification accuracy are degraded by scattering, collimator penetration, photon attenuation in the object, and finite spatial resolution, dominated by the collimator resolution. Methods that compensate for these effects have mostly focused on ^{99m}Tc (140 keV), as this SPECT radionuclide is widely used in diagnostic nuclear medicine. However, quantitative SPECT imaging of therapy radionuclides with correction for image-degrading physical factors is more challenging [47]. Radionuclides that are widely used in beta-emitter therapy include ^{177}Lu, ^{131}I, and ^{90}Y, all three of which are associated with multiple gamma-rays and continuous photons in the case of ^{90}Y (see spectra of figure 20.4) that degrade SPECT activity quantification accuracy. However, the possibility of combining therapy with a β^{-}-emitter and imaging with the same radionuclide is highly desirable, as it makes it possible to monitor on-going therapy without the need for surrogate imaging agents. The 'easiest' to quantify is ^{177}Lu when imaging using only the 208-keV γ-ray because in this case there is only a minor down-scatter contribution from the higher-energy 249-keV and 321-keV gamma rays. The scatter contribution is moderate, and even using the ME collimator, as recommended in the MIRD guidelines [48], good image quality is achieved. If an ME collimator is not available, one can image using the 113-keV photopeak with a LE collimator, but then down-scatter from the 208-keV γ-ray becomes a problem. For therapy with ^{131}I, the image degradation resulting from the 637-keV and 723-keV photons is significant, despite the low abundance. This is due to septal penetration and backscatter from the interaction behind the crystal [30]. The imaging of ^{90}Y is complicated by the fact that there is no discrete photon energy for imaging. The events of primaries and non-primary events are distributed over the entire energy spectrum. Therefore, the positioning of an energy window and the use of energy-window-based scatter corrections is not trivial. Figure 20.6 shows an example of a Monte Carlo simulation of the NEMA phantom with ^{177}Lu, ^{131}I, and ^{90}Y 90.

To exemplify a more clinically realistic geometry than the NEMA phantom, a simulation study with the XCAT phantom was conducted, and the whole-body projections are shown in figure 20.7. In this case, we used SIMIND to simulate both with and without considering scatter and attenuation events in the phantom for the three radionuclides described above. No correction for the degradation caused by the collimator was made. The activity distribution used in these simulations is fictitious, with the purpose of visualising major organs to demonstrate how the images change with the properties of the radiation emission from the different radionuclides. The images shown in panel B, though not possible to be obtained with a real camera measurement, can serve as reference images when evaluating scatter- and attenuation-correction methods, applied to the type of images shown in panel A. Note the reduced visibility of the kidneys for the cases in panel A as compared with those in Panel B. The effects of septal penetration for ^{131}I and especially for ^{90}Y are also evident, where a large number of counts are registered outside the patient geometry.

The possibility of Monte Carlo simulations to separate the origin and properties of photons that contribute to the image is a great advantage in understanding how a gamma camera image has been created. Figure 20.8 shows energy pulse-height distributions separated by events coming from primary events (in this case, defined as photons that have escaped the phantom without scattering) and events that have

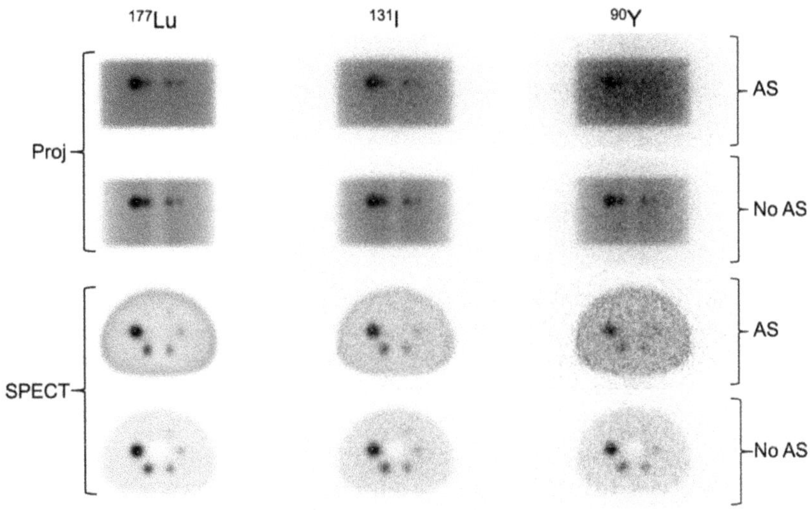

Figure 20.6. Simulation with the SIMIND code using a modified version of the digital NEMA phantom, where the sphere sizes have been increased to be more appropriate for SPECT. The relative activity concentration was set to 12 for the six spheres and 1 for the background. 120 projections where simulated for a 360° rotation and reconstructed with the Castor reconstruction software using OSEM with six iteration and eight subsets. The images show all events for the anterior projections and reconstruction (upper two rows) and for the case including scatter and attenuation (AS) and for the case with no Compton scatterings and photon attenuation (No AS) occurring in the phantom.

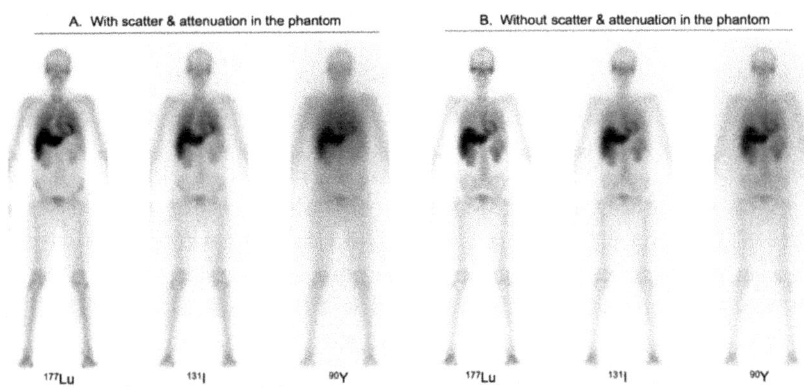

Figure 20.7. Visualisation of image quality degradation by comparing an XCAT WB simulation with scatter and attenuation in the phantom (left three images) and without scatter and attenuation in the phantom (right three images). All images include the effects of the collimator (spatial resolution and effects from septal penetration).

scattered in the phantom. The sum (representing the distribution that is obtained in a real measurement) is also shown. These curves also quantify important information, such as the scatter-to-total fraction in different parts of the energy spectrum. From these examples, it is evident that Monte Carlo simulation is a valuable tool for the evaluation of scatter correction methods.

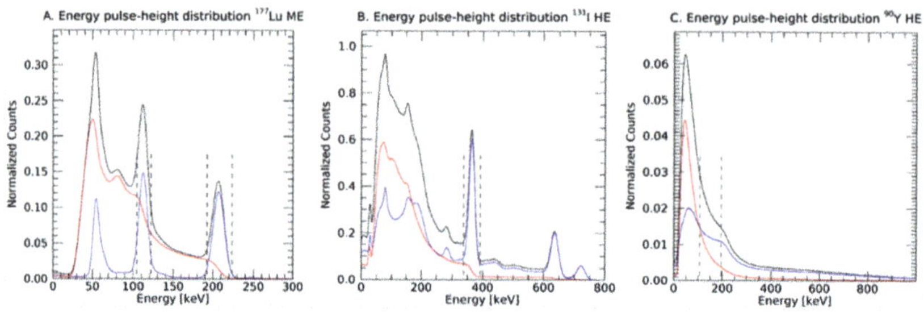

Figure 20.8. Simulated energy spectra for (a) ^{177}Lu, (b) ^{131}I and (c) ^{90}Y from the XCAT simulations shown in figure 20.7. Black lines show the total detected spectra. Red lines show the component in the spectra caused by photons that have been scattered in the phantom, and blue lines show the events from non-scattered photons that have passed through the phantom without interactions. The energy windows are defined as 20% for ^{177}Lu and ^{131}I, but for ^{90}Y, a 60% energy window centred at 150 keV was used because there is no γ emission.

20.9 Evaluation of image-based absorbed dose calculation procedures using Monte Carlo methods

The Monte Carlo method has been shown to be very efficient for both calculating the absorbed dose and also as a powerful tool for evaluating various parts of the calculation scheme. The key point is that if an accurate simulation of SPECT imaging (including all inherent limitations, such as photon attenuation, scatter contribution, and collimator and septum penetration effects) can be made, then the output from the proposed reconstruction/quantification procedure can be compared against the absorbed dose calculated directly using the known activity and attenuation distribution used in the simulation. Figure 20.9 shows a flow chart of the essential steps in such an evaluation. The scope of such an evaluation could be the whole quantification chain or individual components in the chain that affect the results (number of iterations, post-filtering, scatter correction method, etc).

To exemplify the flow chart above, a simulation of a cylindrical phantom with different spherical sources was defined. The radii of the sources were (a) 3.3 cm, (b) 1.65 cm, (c) 1.045 cm, and (d) 0.22 cm. The smallest sources (d) were seven in number and grouped symmetrically, which made the cluster of them have a radii approximately the same as (c). Uniform decays within sources of ^{99m}Tc, ^{177}Lu, ^{131}I, and ^{90}Y were simulated. Three programs were used in this example: (1) the SIMIND program, which generated SPECT projections from the photons emitted from the sources; (2) the Castor program, which was used to reconstruct the simulated SPECT projections to tomographic images; and (3) an EGSnrc-based Monte Carlo program, which used the images to sample the photons and electrons, perform radiation transport, and generate the 3D absorbed dose maps. In this example, no scatter and attenuation were simulated in the phantom for the SPECT SIMIND simulation, but collimator resolution and penetration were included. The initial source distribution, denoted below as reference images, was defined as 512^3 matrices with a voxel size of 0.11 cm, and the SPECT projections and the related reconstructed images (64 projection angles for 360°) included 128^2 images with a pixel size of 0.44 cm. Reconstruction was performed with six iterations and eight

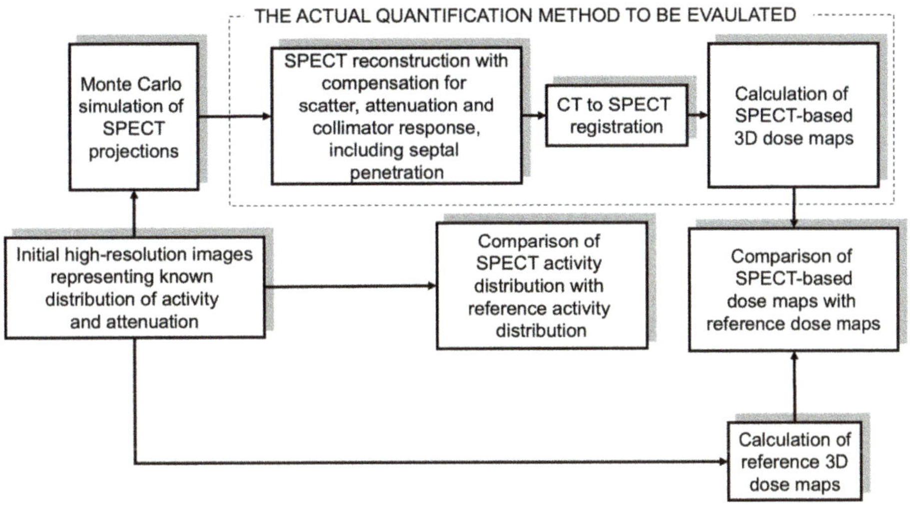

Figure 20.9. Flow chart showing steps for evaluating quantitative SPECT images in dosimetry procedures. The reference dose maps are created from an initial image-based activity distribution of high spatial resolution that represents the activity uptake. These images are then used for (1) a direct absorbed dose calculation by some dosimetry method (e.g., full Monte Carlo calculation or point-dose convolution). The high-resolution activity distribution is also used to model realistic SPECT projections used in subsequent reconstruction and compensation used to calculate absorbed dose, either with the same procedure as the reference dose images or some other method.

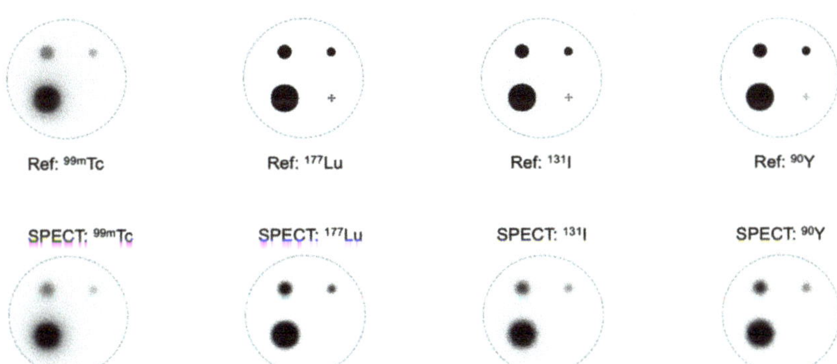

Figure 20.10. Images of absorbed doses on a voxel level are shown in the upper row for the reference case (high-resolution source images as input to the EGSnrc-based dosimetry program) and for the reconstructed SPECT images as input to the EGSnrc-based dosimetry program.

subsets. Both the initial high-resolution source distribution and the SPECT images were used to calculate the absorbed dose images presented in figure 20.10 as the distribution of the absorbed dose in a voxel per unit of activity. The aim of this example is to display the distribution of energy resulting from the characteristics of the four different radionuclides and also how SPECT imaging with its limited spatial resolution affects the distribution of the absorbed dose on a voxel level.

In addition to the three therapy radionuclides, the ^{99m}Tc radionuclide was also included to exemplify a case with only photon emission. The upper row in figure 20.10 shows the absorbed dose images when using the reference image, and the bottom row shows the absorbed dose distribution when using the corresponding SPECT reconstruction as the input to the Monte Carlo-based dosimetry calculation.

For ^{99m}Tc, both the reference and SPECT-based absorbed dose distributions are substantially spread-out outside the spheres because most energy is released by Compton interaction of the 140-keV photons, and these have a relatively long mean-free path length. The clusters of small spheres are only slightly visible, and the contrast between spheres (b) and (c) is different, even if the concentration is the same. The ^{177}Lu decay emits β particles with the lowest kinetic energy out of the three radionuclides compared, so the absorbed dose images for this source display very sharp images, with most of the energy deposited inside the source. For the reconstructed images, the images are also relatively sharp, although the cluster of small spheres is difficult to resolve, which is not the case for the reference images. The reference images for ^{131}I are also sharp, but the β particles have a higher energy than the ^{177}Lu β particles; therefore, the edges of the spheres are more blurred out because of escaping electrons. The clusters in the SPECT images are also visible, and the contrast difference between spheres (b) and (c) is large because of the spread of energy. Of the three radionuclides evaluated, ^{90}Y emits β particles with the highest kinetic energy, and the effect of this is clearly seen in the blurred reference image. Here, the sphere cluster is difficult to detect, and it is not possible to identify the uptake as a separate source.

This example was expanded by calculating the dose-volume histograms (DVHs), which are commonly used in external radiotherapy (figure 20.11). The volume-of-interest DVHs corresponding to the SPECT-based and reference dose maps can be compared. DVH has become increasingly used to describe the heterogeneity of absorbed doses within a volume of interest. These histograms can, in principle, be connected to the calculation of the biological effective dose (BED) and effective uniform dose (EUD).

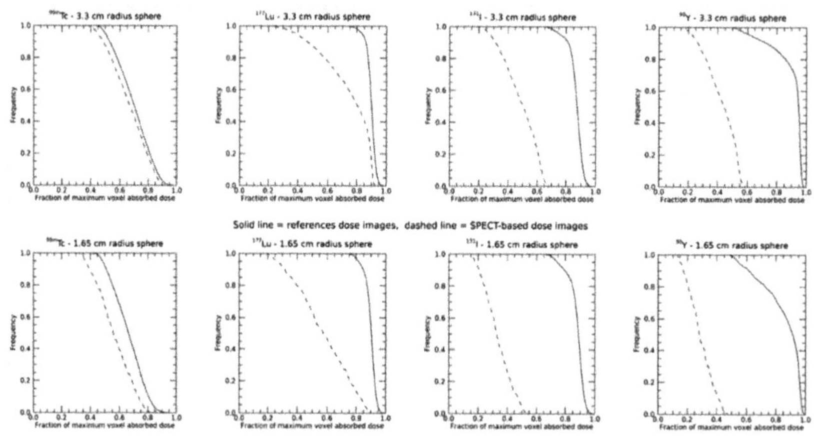

Figure 20.11. Dose-volume histogram obtained from the volume of the two largest spheres, shown in figure 20.10. The curves have been normalised to the largest calculated absorbed dose found in a voxel.

For the largest sphere, the SPECT resolution has little impact on the DVH for ^{99m}Tc because the absorbed dose originates from photons with a relatively long path length. However, for the other three that include short-range charged particle emission, it is clear that the limited spatial resolution generally reduces the absorbed dose on a voxel level for this particular case. The resolution effect is larger for smaller sources (lower row). It should be stressed that this is only an example of the principle of comparing reference absorbed dose images with reconstructed SPECT images. For example, if a type of voxel-level-based partial volume correction is applied, the shape of the DVH will change. In addition, in the example above, events in the images can also come from septal penetration through the collimator walls that may affect the shape of the DVH.

Thus, care must be taken to identify those cases when the heterogeneity in absorbed dose images on a voxel-level, obtained from an image-based dosimetry procedure, as a true effect, originating from non-uniform targeting of the radionuclide, or if the variation is due to other effects associated with imaging and radiation transport physics. Examples of imaging associated effects that can affect the DVH are the poor spatial resolution of the imaging system, effects from septal penetration, the reconstruction method and reconstruction parameters such as iteration number and subsets, artifacts resulting from collimator compensations (Gibbs artifacts) or limited angular sampling, post-filtering because of image noise, and variation related to organ motion. The relative impact of each of the above-mentioned effects is also dependent on the volume of interest in which the DVH is calculated.

20.10 An example of a Monte-Carlo-based evaluation of a dosimetry procedure

A realistic and comprehensive study that builds on these principles has been published by Gustafsson *et al* [49]. The scope of their work was to determine the uncertainty in the measurement of kidney absorbed doses for an image-based SPECT/CT dosimetry calculation for treatment with ^{177}Lu-DOTATATE. Their group was involved in a clinical trial (ILUMINET) where image-based dosimetry was performed for a number of treatment cycles, given to the patient, until the total absorbed dose and BED to the kidney reached a predefined maximum tolerated dose (MTD) limit, based on external beam radiotherapy limits. The dosimetry procedure in the trial was based on a hybrid SPECT/CT and WB/Scout method where quantitative planar imaging (conjugate-view method with correction for scatter, and scout-based attenuation and over/underlapping activity) combined with one SPECT at 24 h. The purpose of the WB images was to estimate the shape of the time-activity curve (TAC) while SPECT/CT provided the quantitative activity values and the base for dosimetry. More details about the clinical trial and the outcomes can be found in the following two references [50, 51].

In the evaluation study by Gustafsson *et al*, the SIMIND code was used to simulate SPECT-based dosimetry calculations on three XCAT patient models. It was assumed that EGSnrc Monte Carlo dosimetry calculations directly for the activity distribution in the three XCAT phantoms were the reference for absorbed doses. No experimental validation of the absorbed dose calculation was performed in the study.

The following steps, related to the different parts of the dosimetry procedure, were investigated:

(1) The camera calibration. This procedure converts measured or reconstructed counts to activity (cps MBq^{-1}). The calibration was based on the simulation of a 10-cm-diameter circular source where Poisson-distributed noise was added.
(2) CT-derived density maps. Because the original attenuation images were obtained from the XCAT software and therefore 'perfect', the images were modified to mimic the uncertainty in CT HU-based attenuation images.
(3) SPECT imaging was used to obtain quantitative images. Here, each organ in the XCAT was simulated separately and then scaled and combined according to the desired activity distribution. The time points that where investigated were 0.4 h, 21.3, 93.3 and 165.8h p.i. SPECT projections were simulated and reconstructed according to the protocol for ILUMINET (60 projections, 360 degrees and 45 s/projection).
(4) Manual delineation of volume of interest. The analysis of the uncertainty related to the delineation was based on a separate study [52].
(5) Calculation of the absorbed dose rate from the SPECT images. Here, the uncertainty in the decay data and the recovery coefficient were included.
(6) The fitting of the time-dose-rate data.
(7) The calculation of the absorbed dose and the BED.

Figure 20.12 is reprinted from the original paper [49] and summarises the different steps in the evaluation procedure.

An important part of this work was to create a realistic activity distribution over time to simulate the images for the relevant time points. Brolin *et al* [53] developed a first-order linear compartment-based pharmaco-kinetic (PK) model for simulating

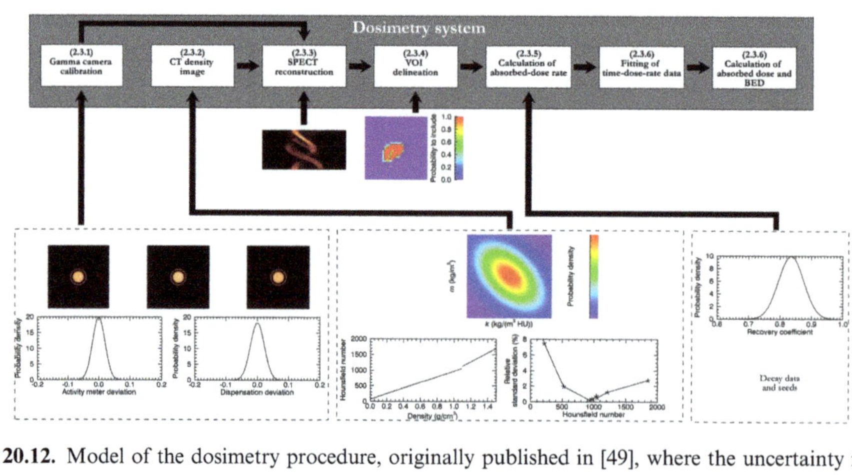

Figure 20.12. Model of the dosimetry procedure, originally published in [49], where the uncertainty in seven separate modules and its propagation were investigated. The figure has been printed in this chapter under the terms of the Common Creative License. Copyright IOP Publishing Ltd. All rights reserved.

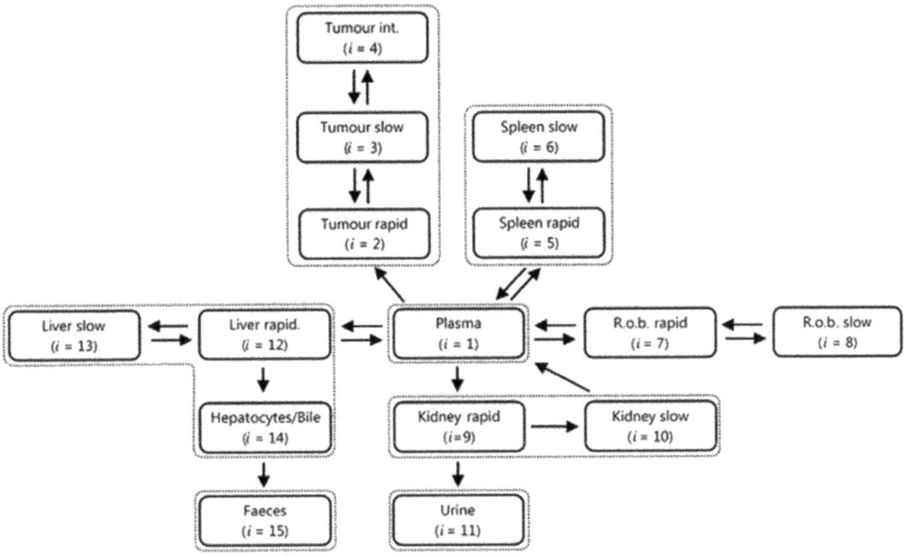

Figure 20.13. PK model for simulation of ^{177}Lu-DOTATATE pharmacokinetics. The solid frames represent individual compartments, while the dashed frames indicate groups of compartments associated with a specific set of measured data. 'R.o.b.' is an abbreviation for 'remainder-of-body', and 'int.' stands for 'internalised'. The figure was originally published in [53] and has been printed in this chapter under the terms of the Common Creative License. Copyright IOP Publishing Ltd. CC BY 3.0

the typical distribution of ^{177}Lu-DOTATATE patients. This model was used in the study by Gustafsson to define time-varying activity distributions in the patient models for Monte Carlo simulation of realistic SPECT and planar scintillation camera images. The PK model was developed as a closed system, which ensures consistency between time-activity curves for different organs and the PK curve of the total-body activity. Figure 20.13 shows the different parts of the PK model.

The radiopharmaceutical distributions, defined from the compartment model, are shown in figure 20.14 for day one for the three different patient models created by the XCAT program, together with the corresponding total intensity projection image of the density distribution.

The main conclusion from the study by Gustafsson *et al* was that standard deviations in the estimated absorbed doses and in the BED values were on the order of 6% and that compensation for the partial volume effect using a recovery coefficient and the calibration factor for converting counts to activity were the two sources of error that contributed most to the overall uncertainty.

20.11 Role of Monte Carlo in developing and testing deep learning methods in nuclear medicine imaging and dosimetry

In the past few years, deep learning methods have found widespread use in medical imaging. These methods, which typically involve the use of artificial neural networks that map inputs to desired outputs, have recently entered the nuclear medicine

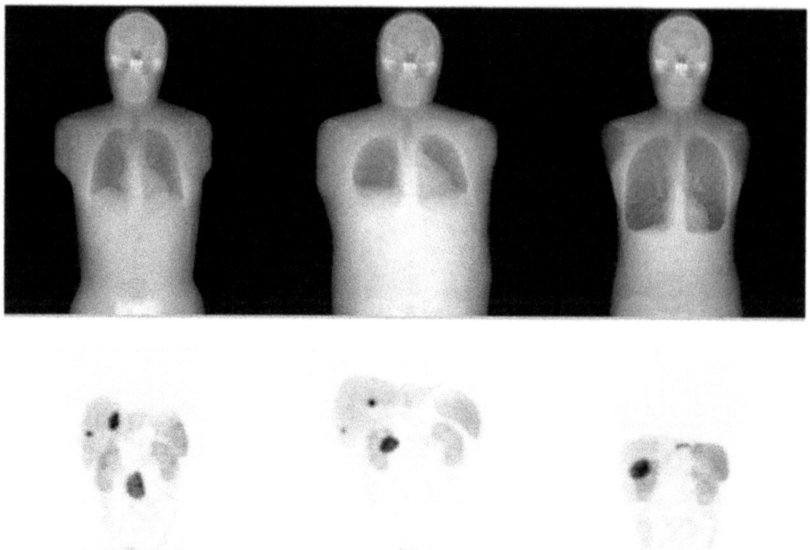

Figure 20.14. The upper row shows the total intensity projection of the mass–density distribution of three different XCAT phantom models used in the study. The lower row shows frontal SPECT images of the simulated ^{177}Lu-DOTATATE distribution one day post-injection. The images are originally published in reference [49] and printed in this chapter under the terms of the Common Creative License. Copyright IOP Publishing Ltd. CC BY 3.0

imaging arena [54–56]. Deep learning methods have been proposed for PET and SPECT image denoising, reconstruction, and compensation for image-degrading physical factors such as scatter and attenuation, as well as for absorbed dose estimation. Data used in the development of deep learning methods are divided into three sub-types: training data used to 'learn' or 'teach' the model, validation data that is used to fine-tune the model, and test data used to evaluate the performance of the trained network. In nuclear-medicine-related applications, Monte Carlo simulations can be used to generate one or more of these three data sub-types. Thus, the Monte Carlo method plays an important role not only in the evaluation step, but also in the development of the methods.

In supervised learning, the training data set used to train the network consists of pairs of input and corresponding desired output data. Examples of these pairs of low- and high-quality images include noisy (low-count) PET or SPECT images and the corresponding low-noise (high-count) images, and reconstructions without attenuation correction and corresponding attenuation-corrected reconstructions. In some applications, the data pairs for this training process can be derived from clinical patient measurements. However, in some cases, it is not feasible or practical to measure the desired output for the training process; hence, Monte Carlo simulation is essential to generate these data pairs. For whole-body ^{99m}Tc-hydroxydiphosphonate bone scans, Minarik *et al* [57] used a convolutional neural network (CNN), trained with sets of noisy and noiseless images generated by the SIMIND code to demonstrate denoising with little or no resolution loss. Training data sets consisted of whole-body bone scan images simulated using XCAT with bone metastases as well as simulations of simple cylindrical phantoms. The CNNs were evaluated on both Monte-Carlo-simulated and clinical bone scan images. This study demonstrates that it is possible to train a CNN entirely based on Monte Carlo simulated data, provided the data are generated carefully. The authors point out that the alternative of using clinical bone scan measurements with very high counting statistics as a substitute for the Monte-Carlo-generated noiseless images would require an imaging time of >1 h, which is prohibitive for most patients and clinics. This situation is also encountered in the case of ^{90}Y PET imaging, where because of the very low positron yield, prohibitive acquisition times are needed to generate low-noise patient images for network training. Hence, in the study by Lim *et al* [58] where an iterative neural network was developed to improve low-count PET reconstruction, both ^{90}Y phantom measurements and simulations relevant to the microsphere radioembolisation application were used to generate high-count training data. The performance was evaluated in phantom measurements, simulations, and clinical data over a range of count levels. The simulations in this study were generated by coupling XCAT with an analytical PET projector. In some cases, where clinical data are available for use in network training and testing, Monte Carlo simulated data is used to expand the clinically derived data, as more training data will lead to enhanced performance and generalisability. In a study by Dietze *et al* [59], a CNN was implemented to enhance ^{99m}Tc MAA filtered back-projection SPECT reconstruction; both measured clinical data and data generated using the Utrecht Monte Carlo System (UMCS) were used as training data.

While trained CNNs have been used post-reconstruction to map uncorrected PET images to attenuation and scatter-corrected images [60] the SPECT scatter correction approach of Xiang *et al* [61] relies on a CNN for estimating scatter projections, which are then used in the conventional physics-based reconstruction model. Here, SIMIND Monte Carlo simulation is used to generate the training pairs, consisting of the total emission projections (scatter + primary) and the ground truth (scatter only) projections (figure 20.15). The CNN is trained to estimate each scatter projection from the corresponding measured total SPECT projection view and the projected CT-based linear attenuation coefficient map. The attenuation projection is included as an additional input because scatter within a patient depends on both the activity distribution and the attenuation map. This approach exploits the potential of Monte Carlo to separate the scatter events, which is not feasible with phantom or patient measurements. Furthermore, the ability to use simulated data as training labels circumvents the difficulties associated with obtaining sufficient real patient data sets for training, which has been a challenge in some other applications of deep learning, such as lesion segmentation. Although generating the training data by Monte Carlo simulation with full radiation transport physics and the training process itself is computationally expensive, these steps are only performed once. In the study by Xiang *et al* [61] the trained CNN was tested for the challenging case of ^{90}Y bremsstrahlung SPECT imaging, where there is substantial down-scatter from the continuous bremsstrahlung energy spectrum to the lower-energy acquisition windows. The performance of the CNN-based method was compared with a previous

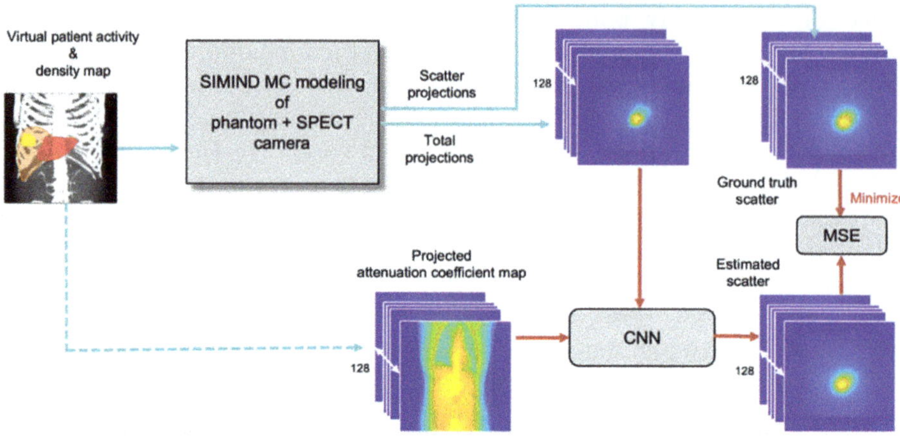

Figure 20.15. Use of Monte Carlo simulation to generate the training pairs (total and scatter projections) for deep learning based estimation of SPECT scatter projections in the work by Xiang *et al* [61].

Monte-Carlo-based approach for estimating scatter projections in simulation studies, phantom measurements, and patient studies associated with ^{90}Y microsphere radioembolisation. In all test cases, the CNN-based scatter estimation demonstrated accuracy similar to that of the Monte-Carlo-based scatter modelling, but it was generated in a fraction of the time.

In addition to deep learning methods for improving SPECT and PET image quality and quantification, CNNs have been proposed for SPECT- and PET-based patient-specific dosimetry [62–65]. Monte Carlo simulation plays an important role in these studies, as full Monte Carlo radiation-transport-based dosimetry algorithms, generally considered the most accurate for patient-specific voxel-level dosimetry, are used to generate the ground truth training labels for the CNNs and/or to generate the gold standard dose maps for testing the performance of the CNNs. In a recent publication, Akhavanallaf and colleagues [65] proposed a method to perform personalised whole-body dosimetry on an organ level that considers heterogeneity in activity and anatomy using deep neural network algorithms to predict the distribution of deposited energy and compared the results with reference images generated by Monte Carlo simulations and two methods based on pre-calculated *S*-values. The results showed that the absorbed doses obtained from the deep-neural-network-based dosimetry method were comparable with the Monte Carlo approach and reduced problems associated with conventional methods for dosimetry, based on planar whole-body images. Two other studies [62, 63] investigating CNNs for dosimetry relied on the Geant4 Application for Emission Tomography (GATE) toolkit, while a third study [64] used the Dose Planning Method (DPM)-based Monte Carlo dosimetry software. In applications ranging from absorbed dose estimates for a diagnostic PET imaging agent to therapeutic agents (^{177}Lu-PSMA and ^{177}Lu-DOTATATE), these studies demonstrate that CNN-based voxel-level dosimetry is feasible, with accuracies comparable to direct Monte Carlo radiation transport but achieved in a fraction of the time.

Thus far, testing of most deep learning methods in nuclear medicine dosimetry and imaging applications has been performed on limited data sets, and evaluation using larger data sets is needed to establish generalisability prior to the clinical adoption of these promising methods.

References

[1] Ljungberg M 2012 The SIMIND Monte Carlo code *Monte Carlo Calculation in Nuclear Medicine: Applications in Diagnostic Imaging* ed M Ljungberg *et al* 2nd edn (Boca Raton, FL: Taylor and Francis) pp 315–21

[2] Ljungberg M and Strand S E 1989 A Monte Carlo program for the simulation of scintillation camera characteristics *Comput. Methods Programs Biomed.* **29** 257–72

[3] Merlin T, Stute S, Benoit D, Bert J, Carlier T, Comtat C, Filipovic M, Lamare F and Visvikis D 2018 CASToR: a generic data organization and processing code framework for multi-modal and multi-dimensional tomographic reconstruction *Phys. Med. Biol.* **63** 185005

[4] Harrison R L, Vannoy S D, Haynor D R, Gillispie S B, Kaplan M S and Lewellen T K 1993 Preliminary experience with the photon history generator module of a public-domain simulation system for emission tomography *1993 IEEE Conf. Record Nuclear Science Symp. and Medical Imaging Conf.* 1154–8

[5] Jan S *et al* 2004 GATE: a simulation toolkit for PET and SPECT *Phys. Med. Biol.* **49** 4543–61

[6] Briesmeister J F 1970 MCNP—A General Monte Carlo N-Particle Transport Code

[7] 1979 MCNP—a general Monte Carlo code for neutron and photon transport *Los Alamos Technical Report LA-7396-M* (Los Alamos: Los Alamos National Laboratory) 44–9

[8] Kawrakow I, Mainegra-Hing E, Rogers D, Tessier F and Walter B 2015 The EGSnrc Code System: Monte Carlo simulation of electron and photon transport *Technical Report PIRS-701* (Ottawa, ON: National Research Council Canada))

[9] Salvat F, Fernández-Varea J and Sempau J 2007 Penelope. A code system for Monte Carlo simulation of electron and photon transport *NEA Data Bank, Workshop Proceeding, Barcelona* 4–7

[10] Battistoni G *et al* 2016 The FLUKA code: an accurate simulation tool for particle therapy *Front. Oncol.* **6** 116

[11] Hartmann Siantar C L *et al* 2001 Description and dosimetric verification of the PEREGRINE Monte Carlo dose calculation system for photon beams incident on a water phantom *Med. Phys.* **28** 1322–37

[12] Sempau J, Wilderman S J and Bielajew A F 2000 DPM, a fast, accurate Monte Carlo code optimized for photon and electron radiotherapy treatment planning dose calculations *Phys. Med. Biol.* **45** 2263–91

[13] Villoing D, Marcatili S, Garcia M P and Bardies M 2017 Internal dosimetry with the Monte Carlo code GATE: validation using the ICRP/ICRU female reference computational model *Phys. Med. Biol.* **62** 1885–904

[14] Yoriyaz H, Stabin M G and Santos A 2001 Monte Carlo MCNP4B-based absorbed dose distribution estimates for patient-specific dosimetry *J. Nucl. Med.* **42** 662–9

[15] Lehmann J *et al* 2005 Monte Carlo treatment planning for molecular targeted radiotherapy within the MINERVA system *Phys. Med. Biol.* **50** 947–58

[16] Dewaraja Y K, Wilderman S J, Ljungberg M, Koral K F, Zasadny K and Kaminiski M S 2005 Accurate dosimetry in ^{131}I radionuclide therapy using patient-specific, 3-dimensional methods for SPECT reconstruction and absorbed dose calculation *J. Nucl. Med.* **46** 840–9

[17] Prideaux A R, Song H, Hobbs R F, He B, Frey E C, Ladenson P W, Wahl R L and Sgouros G 2007 Three-dimensional radiobiologic dosimetry: application of radiobiologic modeling to patient-specific 3-dimensional imaging-based internal dosimetry *J. Nucl. Med.* **48** 1008–16
[18] Botta F *et al* 2013 Use of the FLUKA Monte Carlo code for 3D patient-specific dosimetry on PET-CT and SPECT-CT images *Phys. Med. Biol.* **58** 8099–120
[19] Hippelainen E, Tenhunen M and Sohlberg A 2015 Fast voxel-level dosimetry for ^{177}Lu labelled peptide treatments *Phys. Med. Biol.* **60** 6685–700
[20] Pacilio M, Lanconelli N, Lo M S, Betti M, Montani L, Torres A L and Coca P M 2009 Differences among Monte Carlo codes in the calculations of voxel S values for radionuclide targeted therapy and analysis of their impact on absorbed dose evaluations *Med. Phys.* **36** 1543–52
[21] Nakahara T *et al* 2017 Use of a digital phantom developed by QIBA for harmonizing SUVs obtained from the state-of-the-art SPECT/CT systems: a multicenter study *EJNMMI Res.* **7** 53
[22] *2016 Digital Reference Object* (Seattle, WA: University of Washington Imaging Research Laboratory)
[23] Robinson A P *et al* 2016 Organ-specific SPECT activity calibration using 3D printed phantoms for molecular radiotherapy dosimetry *EJNMMI Phys.* **3** 12
[24] Tran-Gia J and Lassmann M 2018 Optimizing image quantification for ^{177}Lu SPECT/CT based on a 3D printed 2-compartment kidney phantom *J. Nucl. Med.* **59** 616–24
[25] Zubal I G and Harrell C R 1994 Computerized three-dimensional segmented human anatomy *Med. Phys.* **21** 299–302
[26] Slavin K V 1997 The visible human project *Surg. Neurol.* **48** 638–9
[27] Kainz W *et al* 2019 Advances in computational human phantoms and their applications in biomedical engineering—a topical review *IEEE Trans. Radiat. Plasma Med. Sci.* **3** 1–23
[28] Cherry S R, Sorenson J A and Phelps M E 2012 *Physics in Nuclear Medicine* (Philadelphia, PA: Elsevier)
[29] Uribe C F, Esquinas P L, Gonzalez M and Celler A 2016 Characteristics of Bremsstrahlung emissions of ^{177}Lu, ^{188}Re, and ^{90}Y for SPECT/CT quantification in radionuclide therapy *Phys. Med.* **32** 691–700
[30] Dewaraja Y K, Ljungberg M and Koral K F 2000 Characterization of scatter and penetration using Monte Carlo simulation in ^{131}I imaging *J. Nucl. Med.* **41** 123–30
[31] DeVries D J, Moore S C, Zimmerman R E, Friedland B and Lanza R C 1990 Development and validation of a Monte Carlo simulation of photon transport in an anger camera *IEEE Trans. Med. Imag.* **4** 430–8
[32] Autret D, Bitar A, Ferrer L, Lisbona A and Bardies M 2005 Monte Carlo modeling of gamma cameras for I-131 imaging in targeted radiotherapy *Cancer Biother. Radiopharm.* **20** 77–84
[33] Zhao W, Esquinas P L, Hou X, Uribe C F, Gonzalez M, Beauregard J M, Dewaraja Y K and Celler A 2018 Determination of gamma camera calibration factors for quantitation of therapeutic radioisotopes *EJNMMI Phys.* **5** 8
[34] Kim M, Bae J K, Hong B H, Kim K M and Lee W 2019 Effects of collimator on imaging performance of Yttrium-90 bremsstrahlung photons: Monte Carlo simulation *Nucl. Eng. Technol.* **51** 539–45
[35] Badawi R D, Kohlmyer S G, Harrison R L, Vannoy S D and Lewellen T K 2000 The effect of camera geometry on singles flux, scatter fraction and trues and randoms sensitivity for cylindrical 3D PET-a simulation study *IEEE T. Nucl. Sci.* **47** 1228–32

[36] Poon J K, Dahlbom M L, Casey M E, Qi J, Cherry S R and Badawi R D 2015 Validation of the SimSET simulation package for modeling the siemens biograph mCT PET scanner *Phys. Med. Biol.* **60** N35–45

[37] Dewaraja Y K, Ljungberg M and Koral K F 2000 Accuracy of ^{131}I tumor quantification in radioimmunotherapy using SPECT imaging with an ultra-high-energy collimator: Monte Carlo study *J. Nucl. Med.* **41** 1760–7

[38] Gustafsson J, Brolin G and Ljungberg M 2018 Monte Carlo-based SPECT reconstruction within the SIMIND framework *Phys. Med. Biol.* **63** 245012

[39] Ejeh J E, van Staden J A and du Raan H 2019 Validation of SIMIND monte carlo simulation software for modelling a siemens symbia T SPECT scintillation camera *World Congress on Medical Physics and Biomedical Engineering 2018* ed L Lhotska *et al* (Singapore: Springer) pp 573–6

[40] Ramonaheng K, van Staden J A and du Raan H 2020 Validation of a Monte Carlo modelled gamma camera for Lutetium-177 imaging *Appl. Radiat. Isot.* **163** 109200

[41] Rong X, Du Y, Ljungberg M, Rault E, Vandenberghe S and Frey E C 2012 Development and evaluation of an improved quantitative ^{90}Y bremsstrahlung SPECT method *Med. Phys.* **39** 2346–58

[42] Dewaraja Y K, Chun S Y, Srinivasa R N, Kaza R K, Cuneo K C, Majdalany B S, Novelli P M, Ljungberg M and Fessler J A 2017 Improved quantitative ^{90}Y bremsstrahlung SPECT/CT reconstruction with Monte Carlo scatter modeling *Med. Phys.* **44** 6364–76

[43] Barquero R, Garcia H P, Incio M G, Minguez P, Cardenas A, Martinez D and Lassmann M 2017 ^{131}I activity quantification of gamma camera planar images *Phys. Med. Biol.* **62** 909–26

[44] Minarik D, Sjogreen Gleisner K and Ljungberg M 2008 Evaluation of quantitative ^{90}Y SPECT based on experimental phantom studies *Phys. Med. Biol.* **53** 5689–703

[45] Walrand S, Hesse M, Lhommel R and Francois J 2017 Internal bremsstrahlung: a forgotten but significant effect in ^{90}Y SPECT and Y90 imaging *Eur. J. Nucl. Med. Mol. Imaging* **44** S177

[46] Dewaraja Y, Fleming R, Simpson P, Walrand S, Ljungberg M and Wilderman S 2018 Impact of internal bremsstrahlung on Y-90 SPECT imaging *J. Nucl. Med.* **59** 577

[47] Dewaraja Y K, Frey E C, Sgouros G, Brill A B, Roberson P, Zanzonico P B and Ljungberg M 2012 MIRD pamphlet No. 23: quantitative SPECT for patient-specific 3-dimensional dosimetry in internal radionuclide therapy *J. Nucl. Med.* **53** 1310–25

[48] Ljungberg M, Celler A, Konijnenberg M W, Eckerman K F, Dewaraja Y K and Sjogreen-Gleisner K 2016 MIRD pamphlet No. 26: joint EANM/MIRD guidelines for quantitative ^{177}Lu SPECT applied for dosimetry of radiopharmaceutical therapy *J. Nucl. Med.* **57** 151–62

[49] Gustafsson J, Brolin G, Cox M, Ljungberg M, Johansson L and Gleisner K S 2015 Uncertainty propagation for SPECT/CT-based renal dosimetry in ^{177}Lu peptide receptor radionuclide therapy *Phys. Med. Biol.* **60** 8329–46

[50] Sundlov A, Sjogreen-Gleisner K, Svensson J, Ljungberg M, Olsson T, Bernhardt P and Tennvall J 2017 Individualised ^{177}Lu-DOTATATE treatment of neuroendocrine tumours based on kidney dosimetry *Eur. J. Nucl. Med. Mol. Imaging* **44** 1480–9

[51] Sundlöv A, Gustafsson J, Brolin G, Mortensson N, Hermann R, Bernhardt P, Svensson J, Ljungberg M, Tennvall J and Sjogreen Gleisner K 2018 Feasibility of simplifying renal dosimetry in ^{177}Lu peptide receptor radionuclide therapy *EJNMMI Res* **5** 12

[52] Mortensen N, Morin K, Hermann R, Brolin G, Gustafsson J, Sundlöv A, Svensson J, Bernhardt P and Sjögreen Gleisner K 2014 Inter-operator variability in manual image

segmentation and its impact on the renal absorbed dose and biologically effective dose in 177Lu-Dotatate therapy *Eur. J. Nucl. Med. Mol. Imaging* **41** S259

[53] Brolin G, Gustafsson J, Ljungberg M and Gleisner K S 2015 Pharmacokinetic digital phantoms for accuracy assessment of image-based dosimetry in ^{177}Lu-DOTATATE peptide receptor radionuclide therapy *Phys. Med. Biol.* **60** 6131–49

[54] Gong K, Berg E, Cherry S R and Qi J 2020 Machine learning in PET: from photon detection to quantitative image reconstruction *Proc. IEEE* **108** 51–68

[55] Uribe C F, Mathotaarachchi S, Gaudet V, Smith K C, Rosa-Neto P, Bénard F, Black S E and Zukotynski K 2019 Machine learning in nuclear medicine: part 1-introduction *J. Nucl. Med.* **60** 451–8

[56] Veit-Haibach P, Buvat I and Herrmann K 2019 EJNMMI supplement: bringing AI and radiomics to nuclear medicine *Eur. J. Nucl. Med. Mol. Imaging* **46** 2627–9

[57] Minarik D, Enqvist O and Trägårdh E 2020 Denoising of scintillation camera images using a deep convolutional neural network: a Monte Carlo simulation approach *J. Nucl. Med.* **61** 298–303

[58] Lim H, Chun I Y, Dewaraja Y and Fessler J A 2020 Improved low-count quantitative PET reconstruction with an iterative neural network *IEEE Trans. Med. Imaging* **39** 3512–22

[59] Dietze M M A, Branderhorst W, Kunnen B, Viergever M A and de Jong H W A M 2019 Accelerated SPECT image reconstruction with FBP and an image enhancement convolutional neural network *EJNMMI Phys.* **6** 14

[60] Shiri I, Arabi H, Geramifar P, Hajianfar G, Ghafarian P, Rahmim A, Ay M R and Zaidi H 2020 Deep-JASC: joint attenuation and scatter correction in whole-body 18F-FDG PET using a deep residual network *Eur. J. Nucl. Med. Mol. Imaging* **47** 2533–48

[61] Xiang H, Lim H, Fessler J A and Dewaraja Y K 2020 A deep neural network for fast and accurate scatter estimation in quantitative SPECT/CT under challenging scatter conditions *Eur. J. Nucl. Med. Mol. Imaging* **47** 2956–67

[62] Lee M S, Hwang D, Kim J H and Lee J S 2019 Deep-dose: a voxel dose estimation method using deep convolutional neural network for personalized internal dosimetry *Sci. Rep.* **9** 10308

[63] Götz T I, Schmidkonz C, Chen S, Al-Baddai S, Kuwert T and Lang E W 2020 A deep learning approach to radiation dose estimation *Phys. Med. Biol.* **65** 035007

[64] Li Z, Fessler J A, Mikell J, Wilderman S J and Dewaraja Y 2020 A deep residual learning network for practical voxel dosimetry in radionuclide therapy *IEEE Conf. Record Nuclear Science Symp. and Medical Imaging Conf. (Boston, MA)*

[65] Akhavanallaf A, Shiri I, Arabi H and Zaidi H 2020 Whole-body voxel-based internal dosimetry using deep learning *Eur. J. Nucl. Med. Mol. Imaging* **48** 670–82

IOP Publishing

Monte Carlo Calculations in Nuclear Medicine (Second Edition)
Therapeutic applications
Habib Zaidi

Chapter 21

Summary

Habib Zaidi

Monte Carlo calculations in nuclear medicine have been used for several decades in clinical and research settings. Over this time, their use has evolved from a tool available to only a few expert physicists to one that may reside on a desktop computer and that is now potentially available for routine use. This evolution has been driven by four fundamental and critical developments:

1. The ever-increasing sophistication of multimodality imaging instrumentation and associated advanced image analysis and processing methodologies; an area in which simulation studies have become more than essential;
2. The emergence of theranostics and targeted radionuclides as a therapeutic modality, requiring improved and patient-specific dosimetry at the macro and micro level;
3. The availability of three- and four-dimensional realistic computerized anthropomorphic models providing flexibility and accurate modelling of populations of patient anatomies, and chemical properties as well as biodistributions of radiotracers in the patients;
4. The near exponential increase in the speed and proportional reduction in size and cost of computing power, including high performance computing, grid and cloud computing. These four interrelated developments have fostered the work reviewed in this volume.

The recent development of combined modality systems (e.g., SPECT/PET/CT/MRI) has created new opportunities for Monte Carlo simulation in multimodality quantitative imaging. These new imaging modalities, in turn make it possible to meet the new requirements on dosimetry that have been driven by therapeutic nuclear medicine. The increasing availability of end-user image formation/analysis and dosimetry software packages that incorporate Monte Carlo calculations reflects all four of the developments listed above.

It is worth emphasizing that although Monte Carlo techniques have existed for decades, and despite the progress achieved during the last decade, their use in therapeutic nuclear medicine, and in particular, in imaging-based patient-specific dosimetry is still in its infancy. This volume may, therefore, be viewed as an updated album containing several of the many and varied snapshots of this rapidly growing area.

In summary, the use of Monte Carlo calculations in therapeutic nuclear medicine remains an area of considerable research interest in academic and corporate settings, leading the nuclear medicine community to predict promising progress during the next few years.

www.ingramcontent.com/pod-product-compliance
Ingram Content Group UK Ltd.
Pitfield, Milton Keynes, MK11 3LW, UK
UKHW051910100926
8556IPUK00005B/837

9 780750 326926